PRACTICAL PROCESS INSTRUMENTATION AND CONTROL

PRACTICAL PROCESS INSTRUMENTATION AND CONTROL

Edited by

The Staff of Chemical Engineering

André Briand
02/21/83

McGraw-Hill Publications Co., New York, N.Y.

Printed in the United States of America

Library of Congress Cataloging in Publication Data

Main entry under title:

Practical process instrumentation and control.

1. Chemical process control. 2. Engineering instruments. I. Chemical engineering.
TP155.75.P73 660.2'83 79-27225
ISBN 0-07-606651-7 (paper)
0-07-010712-2 (case)

CONTENTS

FOREWORD

Control systems in process plants are changing so dynamically—growing ever more versatile and complex—that even trained engineers in these plants are having difficulty keeping abreast of the field.

Furthermore, these systems are becoming more critical to the profitable operation of process plants. So much so, that no engineer in the process industries can afford not to be knowledgeable about them. Because of rising energy costs, limited availability of raw materials, and tighter safety and environmental regulations, the process industries now function under operating constraints that are far more stringent than ever before. These factors have placed unprecedented demands for greater proficiency upon engineers concerned with designing, selecting, operating, maintaining and improving process control systems.

"Practical Process Instrumentation and Control" offers to these engineers the means for advancing their knowledge about process instrumentation and control. Written by practical engineers thoroughly experienced in the field, it encompasses the whole of instruments and control systems—from pneumatic to electronic, and beyond to control via computers (analog, digital, hybrid and networked), and even to distributed digital control (the latest breakthrough in control technology). It does this in addition to covering the wide range of instrument and controller functions, including the measurement and control of flow, level, pressure and temperature, as well as the specialized applications of "online" analyzers.

Besides explaining the fundamentals of control theory, this book analyzes the economics of advanced instrumentation and computer control. Indeed, the cost and profit aspects of instruments and control system receive constant and paramount attention throughout this book. It also provides insights into instruments and control systems that will enhance trouble-shooting skills and communication with control specialists. It embraces the control of batch operations, as well as of continuous processes. It also deals with the safety aspects of instrumentation and control systems—certainly one of the more important functions—and with environmental monitoring and control.

Engineers in the process industries—whether they be in process design, research and development, technical service, operations or maintenance—will find in this book a vast amount of information that will widen and deepen their knowledge of process instrumentation and control.

Section I
INTRODUCTION

Trends in process control

Trends in process control

Aiming for greater efficiency in process operations, the new technology in sensors, controllers, computers, distributed systems and data paths will cause major changes in process control schemes.

***Cecil L. Smith**, Louisiana State University and **Marcel T. Brodmann**, Economation AG*

□ In the chemical process industries, the keyword today is productivity, and the future should see an even greater emphasis on it.

One reason for the interest in process control is because plant managers have seen improved control as one route to greater productivity. Tighter environmental regulations, more-expensive raw materials, processes designed with less excess capacity, and other such trends will cause process control to continue to receive as much, if not more, attention.

In what follows, we give our ideas on where the present trends will lead us in process control. We will try to avoid broad generalities and instead be specific. We also give the rationale for some of our thinking, and in places discuss the alternatives we will face and the factors we will be considering.

We also want to avoid leaving the impression that problems will gradually disappear and everything will be great. Of course, we may turn out to be wrong on some points, and we may have omitted some significant developments.

Forces shaping process control systems

Many interesting forces are affecting the development of process control. In addition to the pressures of the market, the availability of new technology is likely to determine in which direction and how far we will move.

We have already mentioned environmental regulations and rising raw-material prices as two of the basic market factors at work. Together with the need to conserve energy, they limit the degrees of freedom available to operate a plant. Increased consumer awareness is another factor that may lead to higher-quality standards. All this adds up to considerable direct pressure for tighter process control.

Economic considerations have caused, and will continue to cause, changes in plant design and production technology. Among these are more tightly designed and highly heat-integrated plants. In some industry sectors, batch production is finding renewed interest, since it provides the flexibility to change production rates and product grades easily and without losses in off-specification material. With rising costs of transportation and other logistics, decentralized batch plants, capable of producing a limited range of products, may become more attractive than large continuous monoproduct plants. As a result, more-complex tasks will be expected from process control.

Competent personnel have been in short supply traditionally—even in periods of recession—and there is no reason why this should change. With manpower availability a critical factor, there will be continuing demand for simple systems that are easy to understand, operate and maintain.

These basic market forces are complemented by the results of technological advances available to the designers of future control systems. Although it is commonplace, we should just mention the overwhelming technological factor—i.e., LSI (large-scale integration) technology, with its resultant cheap computing and memory devices.

If we look at the interdependence of all these factors, we can readily recognize that economics is the common driving force. We shall proceed on the assumption that if in the future there is any change from process control as we know it today, the basis for this change will be economic.

Economic aspects of process control

There are two possibilities for improved economics in process control:

1. Future hardware (having the same capabilities as today's hardware) will be less expensive.

2. Processes will be operated in such a manner as to produce a larger economic return through improved applications of process control technology.

Of course, some combinations of these, such as new

Originally published June 21, 1976.

hardware developments (for example, analyzers), will permit control of process variables that are currently impractical to control.

Of the two possibilities, advances in hardware usually prove to be the easiest to put into practice because their economic advantages can be readily identified. However, the largest returns are likely to be from control improvements that lead to more-efficient process operation.

The objective of improved control is normally to reduce the variance in the control error. Said another way, we want to keep the controlled variable closer to the setpoint a greater percentage of the time. The difficulty is in attaching an economic value to this.

For a given process, the economics are established by the base operating conditions of the plant. That is, the setpoints for the major process variables determine the economic return. Small control errors that result in variance of the controlled variables around these setpoints do not usually contribute significantly to the economics.

In most plants, the optimum operating condition requires some process variables to be kept at a limit—for example, at a purity constraint. We are not able, however, to put the setpoints exactly at the limits, since normal variance in the controlled variable would then result in an unsalable product or other such loss. We must allow a safety margin, which is usually set by operating experience, so that the control errors resulting from upsets normally encountered do not violate the constraint. The poorer the control, the wider the safety margin. Therefore, the incentive for improved control is to permit the plant to be operated closer to its limiting conditions. This leads to greater production, less raw-material consumption, lower production costs, and so on.

Plants have usually operated in one of two modes:

1. Market limited, where sales determine the production data.

2. Production limited, where the company can sell all that is produced.

The last few years have seen the rise of yet a third mode:

3. Resource limited, where the availability of a raw material or energy limits production.

Of these, the production-limited mode is the easiest in which to attempt optimization strategies.

In the past, the plant switched from a market-limited to a production-limited situation rather infrequently. Therefore, the control strategies designed initially could be maintained over years. With the resource-limited situation a possibility at almost any time, more-frequent switches can be expected. To maintain the highest possible return from plant operation, the control strategy must be adjusted accordingly.

An important economic consideration is the reliability or vulnerability of a control system or proposed control strategy. In plant operation, shift foremen intuitively prefer a "safe" operating mode over an apparent "optimum" mode. For example, in an exothermic catalytic reaction, they would keep temperature and yield lower rather than risk a temperature runaway that could wipe out the incremental return accumulated over months or even years of "optimum" operation.

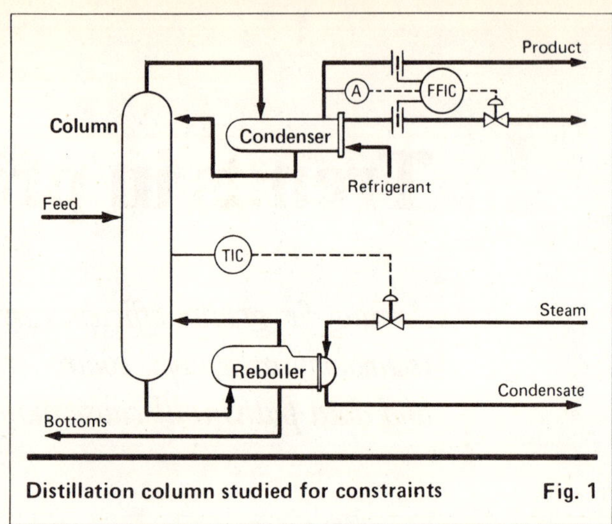

Distillation column studied for constraints **Fig. 1**

Since no control system can be expected to function perfectly all the time, there is a similar tradeoff between its economic advantages and the risks and consequences of a malfunction. Stated bluntly, a control system capable of running a plant optimally is not worth much if half the time it does not work at all. The real value of a control system is determined by the amount of improvement it can achieve over some other system (its quality), the percent of time it can be expected to work at optimum (its reliability), and the amount of degradation it suffers during nonoptimal operation (its vulnerability).

Economic pressures of the future will force careful analysis of various operating modes, and design of control schemes that lead to the most efficient process operation, yet that are inherently simple, robust and reliable.

Process analysis and control schemes

In the future, the objective of process analysis will be to identify how varying market conditions will affect optimum plant operation, and to invent control strategies that maximize economic return.

To operate efficiently under various modes, the plant itself must be operable over wider ranges, and the control systems must remain functional over the entire operating range. This will increase the need to integrate control-system design and dynamic considerations into the plant-design stage. More attention must be paid to the degree of freedom left to the operations personnel.

In recent years, optimization schemes have been implemented by using a computer to adjust the setpoints of conventional controllers. Under certain situations, the basic control system can be modified to do the same. In many cases, this approach results in a simpler, less vulnerable, or cheaper overall system.

Let us illustrate with an example. Fig. 1 presents a simplified schematic of a column studied by Baxley [*1*]. Under the production-limited situation, the objective is to produce as much overhead product as possible with the existing process equipment. Baxley considered three equipment constraints:

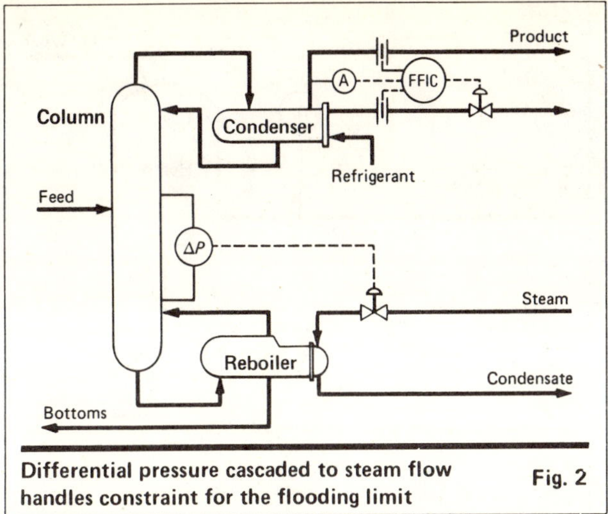

Differential pressure cascaded to steam flow handles constraint for the flooding limit **Fig. 2**

1. The condenser (refrigerant consumption is maximum allowable).

2. The reboiler (steam valve is fully open).

3. The tray capacity (column pressure drop, ΔP, is at the maximum allowable, i.e., flooding is approaching).

Baxley presented a procedure whereby the setpoint of the side temperature controller is adjusted to maintain the column operating at one of these constraints.

If the column always operates against the same constraint, then simple procedures can be used. For example, if the constraint is always at the reboiler, then why not simply do away with the side-temperature controller and run with the steam valve wide open? If steam pressure varies widely, or other upsets enter in the bottom of the tower, this may not be satisfactory; otherwise, it achieves the objective simply.

If the flooding limit is the constraint, then a ΔP controller cascaded to the steam-flow controller is appropriate. As shown in Fig. 2, the flooding condition is approached in the stripping section of the column.

A similar approach occurs in what is often called floating-pressure control in a column. A common approach to pressure control is to manipulate the condenser cooling water to maintain the desired column pressure. In some columns, such as depropanizers, the separation is easier (relative volatility is greater) at lower pressures. In such columns, why not operate the condenser at full condenser cooling-water flow so as to maintain the lowest possible column pressure? Such approaches do not produce straight lines on stripcharts, but they make money. Of course, when the constraints change, the control scheme itself may need to be changed.

The paper industry now widely uses multivariable control, and the chemical and petrochemical industries apply it whenever necessary. The approach normally taken is via the decoupler. Although successful, this approach has certain drawbacks, and may never be applied on a routine basis.

Frequently, a careful process analysis makes it possible to redefine the controlled and/or manipulated variables in order to achieve effective control and avoid the need for a decoupler. For example, the level-control loop and the concentration-control loop on the evaporator in Fig 3a may interact, depending upon the size of the vessel. Instead of designing a decoupler, a simpler approach is the scheme in Fig. 3b, whereby the product flow is ratioed to the feed flow, with the ratio being reset by the composition controller. Interaction is no longer a problem. This approach is often equally as effective on more complex units, and will lead to simple, reliable systems.

The feedback control loop

The proportional-integral-derivative (PID) control algorithm will continue to be the workhorse of the industry. As microprocessor technology continues to develop, we can expect the internals of even a single loop controller to be digital in nature. However, this will not affect the way in which the device is applied from a process point of view.

In recent years, we have seen an increase in the popularity of control techniques such as cascade, ratio and feedforward. However, it is now recognized that all such "advanced" control techniques impose a greater dependence upon the primary sensors. To illustrate, the outlet temperature of the hot-water heater in Fig 4a is controlled by a simple feedback controller that manipulates the steam valve. If the demand for hot water varies frequently, superior performance will be realized from the system in Fig. 4b, whereby the steam flow is ratioed to the cold-water flow, with the ratio constant being adjusted by the temperature controller. Notice that the two flowmeters and the temperature sensor must operate for the system in Fig. 4b to function.

Improved sensor reliability would immediately lead to a wider acceptance of advanced control concepts. As is now the case with computers, our situation will not be "If the sensor fails" but "When the sensor fails." In future control systems, we will see an extension of the rudimentary sensor-checking facilities now incorporated in digital computer-control systems. Not only will we be able to detect the sensor malfunction, but with flexible control hardware we will be able to switch to an alternate control configuration that does not depend upon the sensor. This is known as "soft degradation."

The past few years have witnessed a baffling array of composition sensors and analyzers. Currently, a small percentage of these instruments provide the feedback signal from which automatic control action is taken. Again, the primary concern has been reliability. Some analyzers such as chromatographs are now quite reliable, but these devices require a sample system, whose record of performance is not nearly as good.

Whenever available, process engineers strongly prefer an onstream analyzer that does not require a sampling system. We will see more developments in this area, but an onstream replacement for the chromatograph may be a long time coming. In any case, product specifications will be extended to encompass many more components than are covered by present specifications, and the limits will be tighter.

An overriding consideration regarding control systems in the chemical process industries has been simplicity. This basic philosophy will not change. However, the flexibility of future hardware will enable us to

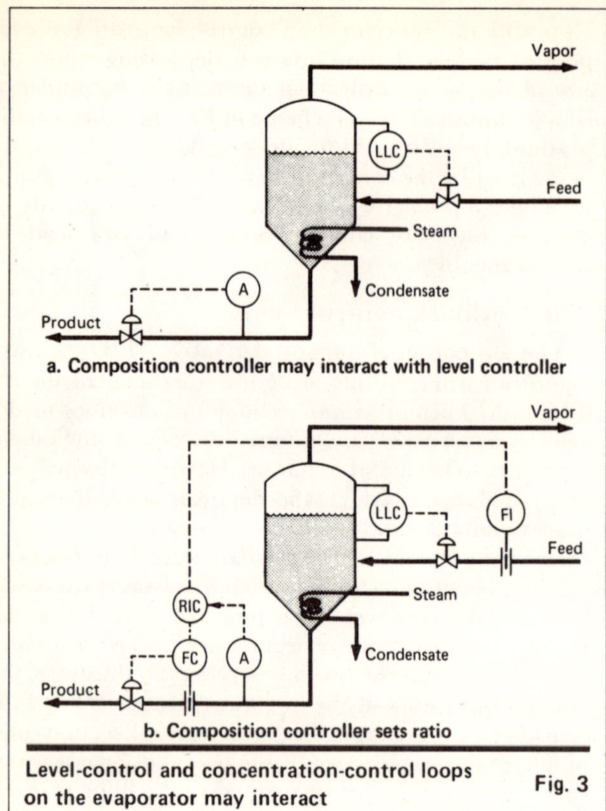

Level-control and concentration-control loops on the evaporator may interact

Fig. 3

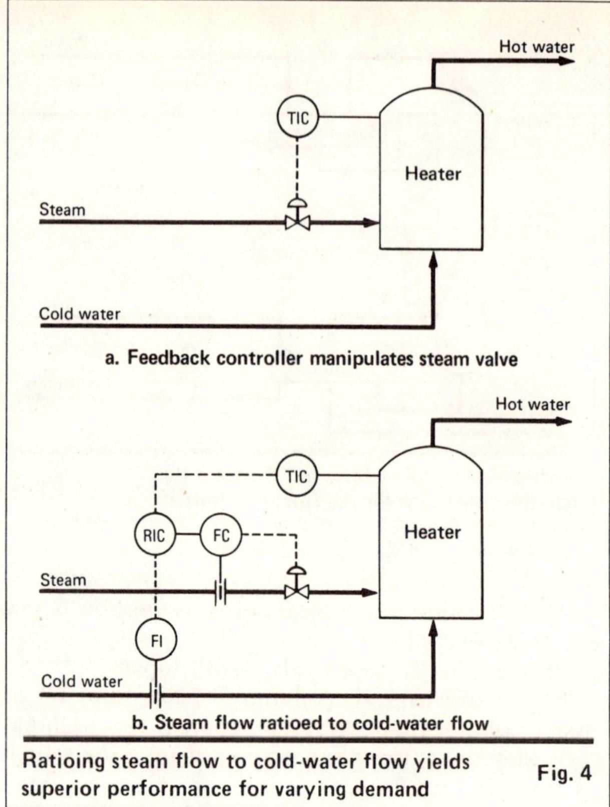

Ratioing steam flow to cold-water flow yields superior performance for varying demand

Fig. 4

readily implement almost any control strategy. For example, the limiting relays and special computing elements required to implement a fuel-oil viscosity-control system [*2*] in conventional control hardware makes the system appear more complex than it really is.

In the future, we will be able to implement such a control strategy in a single general-purpose controller. The hardware will have the inherent flexibility to support various modes of operation. We can forget the "how do we implement it?" question, and devote our attention to "does it lead to greater productivity?"

Finally, where are we going in regard to modern or optimal control? The future is not clear at all. A few industrial applications have been reported over the past few years, but it is still not clear how the use of these techniques leads to greater productivity.

Supervising the entire process

It is now generally recognized that data logging is of no value in the day-to-day operation of a plant. Sometimes management can use such data in allocating a scarce raw material to various plants in some optimal way. In most cases, it is not at all clear that the data are used effectively.

Even so, data logging is experiencing a resurgence. One reason for this is the influence exerted by regulatory agencies. Often, a computer may convince these agencies that the data they request will be obtained. And where accurate data have to be kept for legal reasons, the computer may be the cheapest method.

Alarms constitute a facet of control systems that have not received due consideration. If anything, too many alarms have been included in most systems, often to the detriment of the system. Interest is now being shown in trying to determine which alarms can be removed, in establishing more-appropriate alarm limits, and in analyzing alarm occurrence-patterns. In many plants, combined alarm/operator action logs have proven very useful for continued operator education.

In the past, most plants have been overdesigned, a few to such a degree as to present serious problems. But in most, the overdesign has made the overall plant less sensitive to upsets. If problems are encountered in one section of the plant, a holding tank or other capacitance is usually present between this section and the next. This has smoothed the effect of the upset, and, coupled with the overdesign in a downstream unit, has made the overall plant insensitive to the upset.

In future plants, holdup tanks or other capacitances will not be included. Each section will be designed close to specifications. Heat integration to save energy will be used more frequently, adding interaction between plant sections that used to be independent. An upset in one section will quickly be felt by the downstream (or upstream) section that is not designed for such conditions. This requires close control and supervision in each and every section of the process, and may necessitate a special control mode for automated startup and shutdown.

Distributed systems for process control

We have already mentioned the anticipated impact of the microprocessor on the future of the feedback controller. Even should the system design not change functionally, the location of the hardware and the human interface to it will change drastically.

Let us review how we got to where we are. Even in the late 1950s (possibly even today), plants were being constructed with a distributed-control system in the sense that the individual controllers were located on or near the piece of process equipment with which they interacted. To make adjustments in the operating conditions, the process operator had to visit each area and prescribe the adjustments. His ability to make an overall assessment of the process operation was impaired, especially in bad weather.

To enhance the operator's performance, the concept of the centralized control room was developed. In essence, the controllers were all mounted in a single room so that the operator could quickly assess the status of the process and prescribe the adjustments directly. With the control hardware available in the 1950s and 1960s, the signal processing or computing elements were also located in the control room. Using one controller per loop, such a system was functionally distributed, but geographically centralized.

While this approach has proved successful, there are some problems. One is that the installation of the signal paths is complex and costly. As the number of control loops becomes large, the number of displays in the control room also becomes large. The result not only is costly, but it tends to overwhelm the operator with information. In the normal operation of the process, the operator needs to see only a small percentage of the displays.

Several concepts are now being explored to provide a more effective and less costly centralized control room. It has been recognized that the display function is separate from the signal processing or computing function. In fact, in Foxboro's SPEC-200, Fischer & Porter's Series 4000, Honeywell's TDC-2000 and Taylor's MOD III, the display and signal processing functions are separated.

Another approach being explored, especially in computer systems, is the use of a remote multiplexer. This unit consists of a minicomputer or microprocessor, an A/D (analog to digital) converter, and a multiplexer. The process signals are scanned, converted to digital, and transmitted to the central computer, normally located in or adjacent to the central control room. Thus, a single signal path is shared for all of these process signals, as opposed to an individual signal path for each. The result is a device hanging on an "umbilical cord" to the computer. This begs the question, "What happens when the umbilical cord is severed?"

To date, most of these devices have been installed in supervisory computer-control systems. Thus, a conventional control system exists, which means that the control signals have already been installed to the central control room. Many of these signals are routed to the computer via a local multiplexer located with the central computer system. Therefore, the signals routed to the remote multiplexer are those needed only for logging, for display, or for supervisory control calculations.

Since none of these functions are critical to continued process operation, several remote multiplexers have been installed and operated successfully. The present devices are "dumb," but many (such as General Automation's Remote Intelligent Multiplexer) can be programmed. In fact, a minicomputer can be installed for about the same cost as the remote multiplexer. Therefore, the incentive to install the remote multiplexer is not to reduce costs by using the computational power of the central control computer, but instead to save wiring costs. The remote multiplexer can be justified only when the data are needed in the central computer.

A logical next step is to place some feedback-control logic in the remote multiplexer. With this approach, some (if not all) of the conventional controllers can be removed from the central control room. The control calculations would be done in the remote unit, but setpoints, tuning parameters, auto/manual switches, and other monitoring commands from the process operator would be received through the central computer in the control room. One would obtain a geographically distributed system, with display/access capability in the control room, and control capability in the field. Thus, the advantages of the centralized control room are not lost.

Reliability of remote controllers

At this point, the remote multiplexer, or perhaps we should say "remote controller," becomes a more critical element. There are actually two facets of the reliability of the system—i.e., the reliability of the:

1. Remote controller itself.
2. Data path from the remote controller to the central control room.

As for the reliability of the remote controller, current practice is to put only about eight loops in a single remote unit. While the loss of such a device would be a serious problem, it would be manageable and should not lead to a plant shutdown. As costs of microprocessors, converters and associated electronics continue to decrease, the number of loops in a single remote controller will decrease—eventually reaching the single-loop level. The reliability will be equivalent to, if not better than, that of present controllers. We will have a system that is not only geographically but also functionally (in a way) distributed and, therefore, much less vulnerable.

The integrity of the data path to the remote controller will be of considerable concern. Presumably, count fields and other check information will accommodate all transmissions, so it is extremely unlikely that a transmission error or system fault could result in unwanted commands to the remote controller. Hence, if the data path becomes inoperable, the remote controller should continue to operate. The problem is that the operator cannot make adjustments from the control room, but presumably such adjustments could be made at the remote controller itself.

Data paths and carriers

As long as eight or so control loops are implemented in a single remote controller, the reduction in signal wiring will be appreciable. However, when we reach the one-for-one level, the practice of installing a dedicated path from each remote controller to the central system will prove to be not much better, if any, than the situation today. It will be necessary to devise some means to communicate to several remote controllers over a single transmission path that they all share.

This raises the question of a communication protocol and a carrier. Several protocols have appeared and more will appear. For the carrier, cable TV is frequently suggested, and is capable of doing the job. But it probably has much more capability than is needed, and a less expensive means may appear.

If cable TV is adopted, the data path will be used for purposes other than process control and supervision. For example, badge readers are now being installed in some plants to keep track of who is in the process area, for how long, and to record the concentrations of air pollutants while the workers are there.

Digital data paths are so flexible that a wide variety of protocols could be implemented to do the job. However, standardization is needed. With pneumatic instrumentation, the standard 3- to 15-psi signal level permits products from several suppliers to be integrated into the overall control system. Many companies regularly purchase a sensor from Company X, the controller from Company Y, and the valve from Company Z. With electronic instrumentation, compatibility problems arose because signal levels of 1 to 5 ma, 4 to 20 ma and 10 to 50 ma were offered. (Fortunately, the 4- to 20-ma signal was finally adopted as the standard.)

With digital systems, the number of possible protocols is astounding. Unless one of these is adopted as the industry standard, the resulting state of confusion will retard acceptance of such systems. Whatever protocol is adopted, it must be flexible enough to permit virtually any data to be transmitted, even to the point of reprogramming the remote controllers from the central system. This is within the capabilities of some protocols available today; for example, the synchronous data-link control (SDLC) has these capabilities [*3*].

The use of remote controllers will also necessitate changes in control room displays. While dedicated recorders and indicators could be interfaced to the system, less-expensive and more-effective approaches will be sought. Cathode-ray-tube (CRT) display devices will play an important role. Through these devices, the operator can: interrogate values of process variables, obtain status summaries on the plant as a whole or on an individual unit, display a trend history of one or more process variables, specify changes in setpoints or other operating parameters, and interact with the system in other ways. However, the replacement of the display panels, as now used in control rooms, with a CRT console will probably meet considerable opposition.

Analog vs. digital

When distributed architecture becomes accepted, the entire control system will be digital. Does this mean that the analog vs. digital argument will disappear? Probably not.

First of all, pneumatic systems will still be produced and used in some plants. Second, only the name of the argument may change. Although it is digital, the remote controller as described above is not programmable, but instead offers a fixed set of capabilities from which the user can select. In this respect, it is more similar to the conventional PID controller than it is to the process control computers. While the remote controller offers a wider spectrum of capabilities (various versions of the PID algorithm, ratio control, feedforward capabilities, cascade, dead-time compensation, etc.), the user will not expect to change the progam itself, but will instead select from this spectrum of capabilities when applying the device to the process.

Programmable control computers will continue to be available and will be used, especially in a supervisory role. The computers will be more powerful than those available today, will have better applications-software support, and will differ architecturally from those now available. For example, the disk that is an integral part of today's systems will be replaced by a large directly addressable memory.

Implementation of these systems will also become much easier. Very little, if any, programming will be done in languages such as FORTRAN. Instead, software systems will appear that are even easier to use than the fill-in-the-forms systems now available [*4*].

Thus, the potential will exist to argue for the use of the fixed-function devices as opposed to the programmable computers. Hopefully, this argument will not be pursued with as much vigor as was the analog vs. digital argument.

References

1. Baxley, R. A., Local Optimizing Control for Distillation, *Instrum. Technol.*, Oct. 1969, p. 75.
2. Kostiw, W. B., How To Instrument Fuel Oil Systems, *Instrum. Technol.*, Nov. 1974, p. 31.
3. Gray, J. P. and Blair, C. R., IBM's Systems Network Architecture, *Datamation*, Apr. 1975, p. 51.
4. Smith, C. L., Fill-in-the-Forms Computer Languages for Process Control, *Chem. Eng.*, Mar. 3, 1975, p. 151.

The authors

Cecil L. Smith is Professor and Chairman, Dept. of Computer Science, and Professor, Dept. of Chemical Engineering, Louisiana State University, Baton Rouge, LA 70803. His interests lie mainly in the areas of digital control and simulation of industrial processes. He holds B.S., M.S. and Ph.D. degrees in chemical engineering from Louisiana State and is the recipient of the 1972 Donald P. Eckman Award for outstanding contributions to the field of automatic control. He is a member of AIChE, Instrument Soc. of America, and Assn. for Computing Machinery, and is a registered professional engineer in Louisiana.

Marcel T. Brodmann is President of the firm Economation AG, Postfach 246, CH-7002 Chur, Switzerland, a company specializing in industrial computer applications. A chemical engineering graduate of the Swiss Federal Institute of Technology, he gained his first industrial experience with Emser Werke AG. Later, he worked for Exxon in various managerial positions related to computing and process computer control. Dr. Brodmann is a member of AIChE, International Management Assn., and the Schweizerische Ingenieur- und Architektenverein.

Section II
ECONOMICS OF PROCESS INSTRUMENTATION

Higher Profits Via Advanced Instrumentation

Better instrumentation can increase the capacity of existing plants, and decrease production costs, all for a relatively minor capital investment.

BÉLA G. LIPTÁK, Lipták Associates

In most chemical process plants, production can be increased by about 20% solely through the updating of instrumentation. This increase needs merely a small investment and can take place in a short time because only the instrumentation has to be updated—all other equipment remains unaltered.

During times of economic slowdown, one can consider this approach as an alternative to building new plants.

In this article, only one type of chemical process plant will be discussed, but the approach used for systems evaluation and cost-benefit analysis can be applied to any other type of plant.

What we will consider here is a wastewater-treatment plant. First, the main unit-operations of such a plant will be described, together with the alternative control strategies available for each. Later, an example will show how a cost-benefit analysis is performed by quantitatively identifying both the expenses and the economic benefits associated with, a traditional and, an advanced method of controlling the same unit operation.

DISCUSSION OF UNIT OPERATIONS

Fig. 1 depicts the main unit-operations found in a typical industrial (or municipal) wastewater-treatment plant. Each of these unit operations will be briefly reviewed from the point of view of instrumentation. First, the purpose of the control system will be described. Second, the traditional methods of instrumentation will be compared with the capabilities of more-advanced control systems.

Prechlorination for Odor Control

When wastewater enters the treatment plant, it is likely to be putrefactive. The purpose of prechlorination is to kill the odor-causing bacteria and to destroy odor-causing compounds such as hydrogen sulfide. This can be done either by adding chlorine compounds that will selectively remove sulfur (ferrous chloride, for example), or by breakpoint chlorination. In breakpoint chlorination,

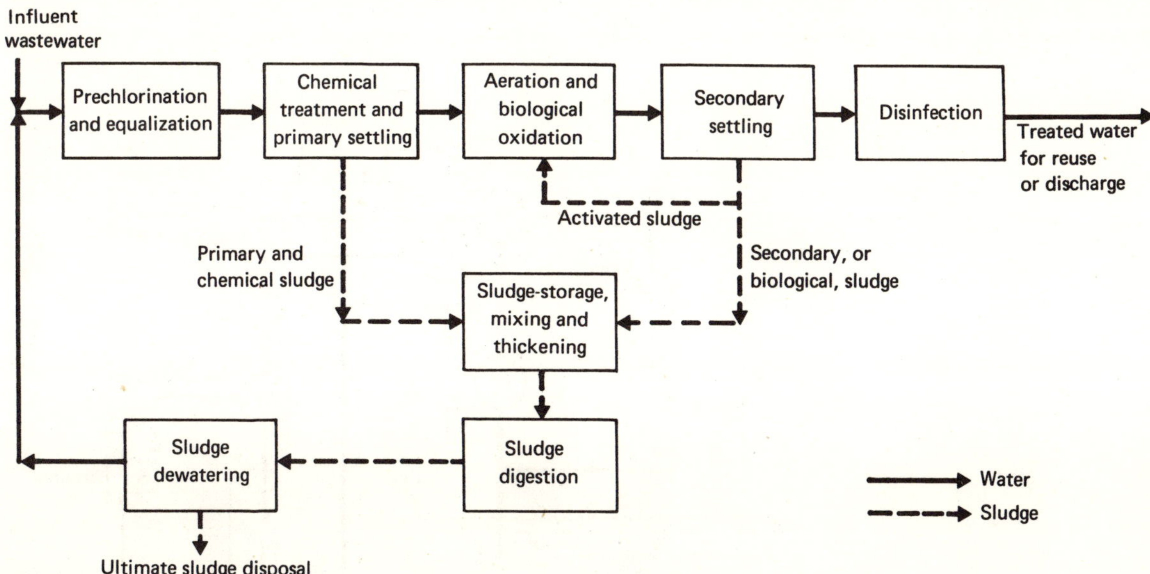

MAJOR unit operations that comprise a typical wastewater-treatment plant—Fig. 1

Originally published June 23, 1975.

enough chlorine is added to indiscriminately oxidize all substances that produce a chlorine demand. In installations where the total chlorine demand is high (or is unpredictably variable), $FeCl_2$ can be a better choice. This is both because less chemical is required for the selective removal of sulfur and because overchlorination—which can slow biological activity—is less likely to occur.

For breakpoint chlorination (the most frequently used method of odor control) the control system must set the rate at which chlorine is added. If the chlorine dose is too low, the odor problems will persist. If too much chlorine is added, this will not only increase the plant's chlorine consumption but will also interfere with the biological activity in the aeration tank.

Traditional prechlorination control-systems charge chlorine either at a fixed rate or in proportion to the volumetric flowrate of the incoming wastewater. Unfortunately, neither the flowrate nor the pollutant concentration of the incoming wastewater is constant. Hence, at times of low influent concentrations, the traditional control systems will charge more chlorine than is needed.

With a 5-ppm Cl_2 dose [*1,2*], the yearly cost is:

Plant Size, 10^6 gal/d	Chlorine Cost, $
1	1,825
5	9,125
10	18,250
50	91,250
100	182,500

The potential yearly savings, resulting from the use of an improved control system, are directly proportional to these figures.

Improved Control of Prechlorination

To add the proper amount of chlorine, one must vary the dosage as a function of total chlorine demand. The dosage adjustment cannot be based on residual chlorine measurement, because after deodorization there should be no residual chlorine.

A suitable control parameter is the oxidation-reduction potential (ORP) of the sewage. Odor-causing, anaerobic, bacteria thrive at negative ORP values; chlorine in sub-residual quantities can be used as a chemical oxidizing agent to change the ORP of the sewage. Thus, continuous odor control can be achieved on a load-following basis by maintaining the ORP of the sewage at a value that is unsuited to the life processes of the odor-causing bacteria.

This improved control system is shown in Fig. 2, where the chlorinator is paced by the raw-sewage flow signal [*3*] to give an approximate Cl_2 feedrate, which is then feedback-trimmed by ORP measurements. In a 100-Mgd (million gal/d) plant, the annualized total cost of this improved control system is $20,232, while the yearly savings of chlorine amount to $64,000 [*4*].

Chemical Treatment—Phosphate Precipitation

Chemical treatment of wastewater can serve many purposes, including neutralization, precipitation, coagulation, flocculation and colloid destabilization. Because phosphate removal is one of the most common aims of such treatment, this unit operation will be used to illustrate the potential of advanced instrumentation. We will consider phosphate removal by means of alum treatment. (Toward the end of this article, phosphate removal by lime treatment will be used in describing the step-by-step preparation of a cost-benefit analysis.)

Aluminum, calcium and iron ions all remove phosphorus from solution by forming insoluble precipitates with it. With lime treatment, the phosphate and carbonate anions compete for the calcium ions. This results in more lime being consumed (and more chemical sludge being produced) than would be needed for phosphate removal

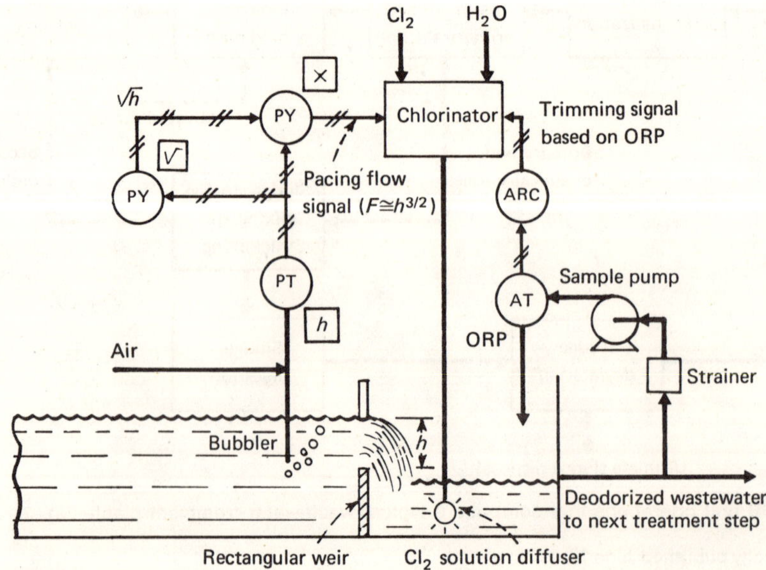

INSTRUMENTS for improved prechlorination control—Fig. 2

only. (In addition to phosphate removal, the water is simultaneously softened.)

When aluminum or iron ions are used for phosphate removal, there is also a competition for them, but this time it is between the phosphate and the hydroxide anions. To minimize this competition (and maximize phosphate-removal efficiency), the hydroxide ion concentration should be low (that is, the pH should be low). For phosphate precipitation by these metallic ions, the desired pH ranges are: aluminum—4 to 6; ferric—4.5 to 5; and ferrous—7 to 8.

Phosphate removal by aluminum is efficient because (under favorable pH conditions) the molal ratio of Al to PO_4 approximates 1:1. The most available chemical source of aluminum ion is alum, which is hydrated aluminum sulfate, and has the formula: $Al_2SO_4 \cdot 14H_2O$. For 95% phosphorus reduction, the weight of alum required is 22 lb/lb phosphorus. This corresponds to a mole ratio of aluminum to phosphorus of 2.3:1.

If alum is charged at a constant fixed rate, its yearly cost as a function of plant size is given below. These figures are based on an alum dose of 200 ppm and on a delivered cost of $80/ton alum [*5,6*].

Plant Size, 10^6 gal/d	Alum Cost, $
1	24,500
5	122,500
10	245,000
50	1,225,000
100	2,450,000

When alum is charged either at a fixed maximum rate, or based on the wastewater volumetric flow (hydraulic load-following), more chemical will be used than necessary. An advanced control system can operate in a total load-following mode, where a drop in the phosphate concentration of the influent will result in a lowering of the alum dose and will therefore conserve this valuable additive.

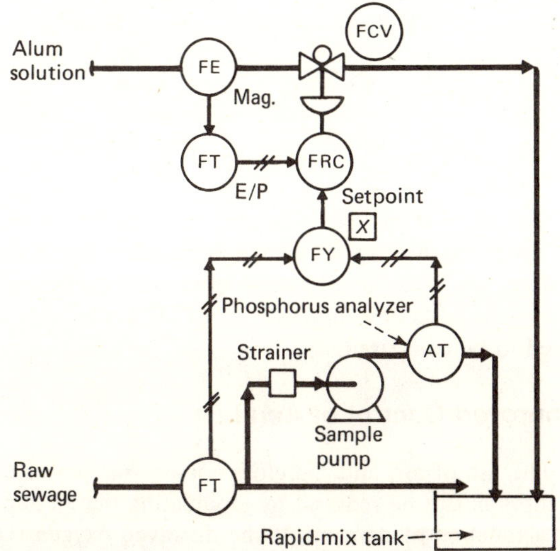

CONTROL of phosphorus removal by alum addition—Fig. 3

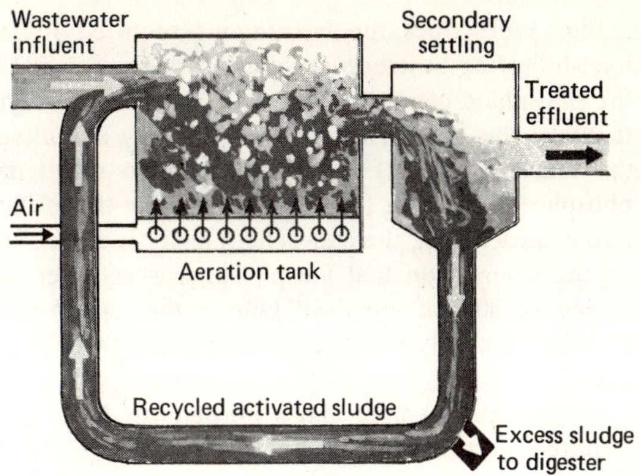

ACTIVATED-SLUDGE process works this way—Fig. 4

Terminology of Wastewater Treatment

Advanced Wastewater Treatment—Usually refers to processes that are used to remove nutrients, trace organics and dissolved minerals.

Biological Wastewater Treatment—Usually means that the organic pollutants are biologically oxidized by bacteria.

BOD (Biochemical Oxygen Demand)—The amount of oxygen used by bacteria to oxidize the organic material in a liter of wastewater.

Conventional Wastewater Treatment—Usually means that the treatment processes are directed toward the reduction of suspended-matter concentration, biochemical oxygen demand and bacterial content.

Physical-Chemical Wastewater Treatment—Usually refers to unit operations, including: chemical (such as coagulation), physical (such as activated-carbon adsorption), or thermal (such as heat treatment). Depending on its purpose, the treatment can be conventional or advanced, and can be applied either in addition to or in place of biological treatment.

Primary Treatment—Usually means the removal of suspended solids only.

Secondary Treatment—Treatment directed toward the reduction of organic pollutant concentration (BOD) in the wastewater.

Tertiary Treatment—Usually means the almost complete removal of all solids and organic matter from the wastewater.

These terms have frequently been misused or misunderstood. For example, "physical-chemical" treatment has been used interchangeably with "advanced" treatment. This is not correct. Physical-chemical treatment means only that the particular unit operation is nonbiological, but its purpose can be either conventional or advanced treatment.

Similarly, "biological" treatment should not be used interchangeably with either "conventional" or "secondary" treatment, because its purpose can also be advanced treatment. On the other hand, the terms "advanced" and "tertiary" mean practically the same thing, and the term "conventional" can usually be taken to mean "primary plus secondary" treatment.

Precipitation by Alum—Improved Control

Fig. 3 describes the instruments required for total load-following based on detecting both the flowrate and the phosphate concentration in the raw sewage. Unfortunately, this feedforward control strategy is limited by the unavailability of a reliable and low-maintenance phosphorus analyzer [7]. However, taking this difficulty into consideration, the cost-benefit analysis can be based on the assumption that the phosphorus analyzer is out of service 20% of the time. During these periods, it is assumed that only hydraulic load-following will be practiced.

We can also assume that a phosphorus analyzer will require 150 h of maintenance per year [7,8]. (To date, only automated wet-chemistry procedures are available for onstream phosphate analysis. This makes for low reliability and high maintenance.)

In a 100-Mgd wastewater-treatment plant, the annualized total cost of this improved control system is $53,480, while the resulting yearly savings in alum come to $790,000 [4].

In this case, the benefits are so much greater than the costs that one can consider making further capital investments by installing redundant (or even two-out-of-three voting-type) sensors to increase measurement reliability and operator confidence in this advanced control system. Further possible improvements over the system shown in Fig. 3 include the use of self-calibrating and self-diagnosing features in the analyzer, and the addition of another phosphorus analyzer downstream of the rapid-mix tank to provide feedback trimming based on residual-phosphate measurement. The need for such a residual phosphate analyzer is more pronounced in installations where the pH cannot be kept at around 5 during the precipitation reaction.

Nomenclature

ARC	Analysis recorder-controller
AT	Analysis transmitter
BOD	Biochemical oxygen demand
C_a	Average concentration
C_p	Peak concentration
DO	Dissolved oxygen
E/P	Voltage-to-pressure transducer
F_a	Average flow, Mgd
F_p	Peak flow, Mgd
F/M	Food-to-microorganism ratio = organic loading in lb/d of BOD divided by lb/d of MLSS
FAL	Flow alarm, low
FCV	Flow-control valve
FE	Flow element
FRC	Flow recorder-controller
FSL	Flow switch, low
FT	Flow transmitter
FY	Flow relay
HIC	Hand (manually initiated) indicator-controller
K	F_p/F_a
LIC	Level indicator-controller
Mgd	Million gal/d
MLSS	Mixed-liquor suspended solids
ORP	Oxidation-reduction potential
pHAL	pH alarm, low
pHE	pH element
pHRC	pH recorder-controller
pHSL	pH switch, low
pHT	pH transmitter
PY	Pressure relay
Q	Wastewater flowrate, Mgd
V	Aeration tank (contactor) volume, 10^6 gal

Aeration and Biological Oxidation

The activated-sludge process illustrated in Fig. 4 requires three things for its operation: microorganisms, food and oxygen. The microorganisms use the organic substances (pollutants) in the wastewater as their food. The end-products of this natural process are carbon dioxide, water and new microorganisms (excess sludge). In the aeration tank, the wastewater (food) is mixed with both oxygen and a high concentration of microorganisms to achieve oxidation of the organic pollutants without nuisance and within a very short time. In addition to good mixing, it is necessary to maintain the organic loading, or "food-to-microorganism ratio" (F/M) in the aeration tank between 0.1 and 0.5 for 90%-or-better BOD reduction.

$$F/M = \frac{\text{(Organic concentration)(wastewater flow)}}{\text{(Microorganism concentration)(tank volume)}} = \frac{Q(\text{BOD})}{V(\text{MLSS})}$$

The F/M ratio can be maintained by adjusting the rate at which sludge is recycled; adequate oxygen can be guaranteed by modulating the rate at which air is introduced.

When aeration airflow is set at a fixed rate corresponding to both the maximum wastewater flowrate and maximum BOD concentration during the daily cycle, the result is minimum equipment life and maximum power consumption.

The yearly operating cost can be calculated based on continuous power consumption for each Mgd unit being 20 kW [9], and a cost of electricity ranging from 3 to 5¢/kWh (as a function of quantity used).

The annualized total cost [4] of fixed-rate aeration as a function of plant size is:

Plant Size, 10^6 gal/d	Aeration Cost, $
1	10,400
5	45,340
10	80,250
50	343,800
100	591,300

Some 25% of the above costs can be saved by supplying air in proportion to the actual demand rather than at a fixed rate. This savings is the potential benefit against which the costs of a more-advanced instrument package need to be compared.

Improved Control of Aeration

The use of air, and therefore compressor power consumption, can be reduced by modulating the air supply as a function of demand. If the dissolved oxygen (DO) in the aeration tank is maintained at a 2-mg/l level, this will guarantee an adequate supply of oxygen to the microorganisms in the contactor. Fig. 5 illustrates an advanced control system that regulates the air supply in proportion to the process demand.

The control loop consists of a feedforward and feedback section. The feedforward portion ratios the air-supply-flow to the wastewater-flowrate into the aerator (hydraulic loading). The feedback signal adjusts the ratio setting so as to maintain the dissolved-oxygen concentration in the aerator at 2 ppm. In other words, the hydraulic load is followed in a feedforward, and the organic load in a feedback, manner. If the dynamic tuning is correct, this approach results in total load-following, which is the most economical mode of operation. A minimum airflow is always provided (even at zero load) to guarantee good mixing at all times.

In a 100-Mgd wastewater-treatment plant, the annualized total cost of this improved control system is $24,880, while the resulting yearly savings in electric power represent a benefit of $43,000 [*4*].

Postchlorination for Disinfection

The purpose of disinfection is to prevent the spread of water-borne diseases by eliminating pathogenic organisms. This is accomplished by adding enough chlorine to provide a residual chlorine concentration of under 1 ppm. Because the chlorine itself is toxic (and some of its organic compounds are carcinogenic), it must be neutralized into HCl by the addition of SO_2 before the effluent leaves the treatment plant, in order to protect the receiving waters.

Chlorine is the most widely used chemical disinfectant and, therefore, this control-system analysis is based on its use. An alternative disinfectant is ozone. It takes about 12 kWh to generate a pound of O_3.

The potential advantages of ozonation include:

■ Destruction of viruses, not only bacteria.

■ The leaving of no residual ozone in the water, and the formation of no carcinogenic compounds.

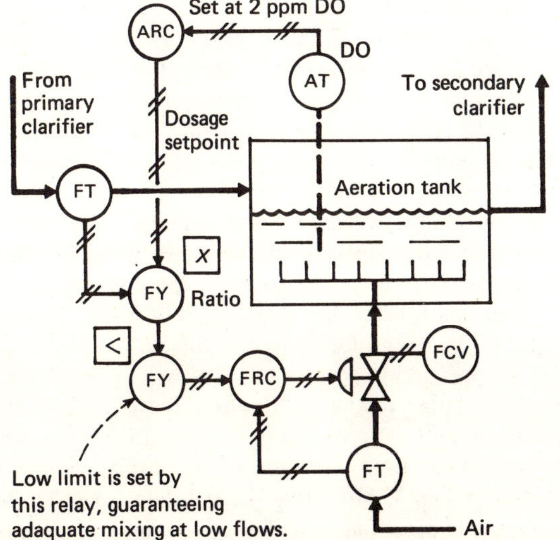

IMPROVED control in aeration tank—Fig. 5

■ Lack of need for the dechlorination and aeration steps that are used to remove excess Cl_2 and restore dissolved oxygen when chlorination is practiced.

■ Payment of 60 to 75% of O_3-generator equipment costs by the federal government, if a demonstration grant is provided.

The highest consumption of Cl_2 results when the chlorinator is manually set for charging chlorine at a fixed rate that corresponds to the product of maximum wastewater flowrate and maximum chlorine demand (assumed to be a dose of 15 ppm) in the daily cycle. The yearly total costs for various plant sizes are:

Plant Size, 10^6 gal/d	Chlorination Cost, $
1	5,480
5	27,400
10	54,800
50	274,000
100	548,000

It is assumed that plants practicing fixed-rate chlorination will also practice fixed-rate dechlorination, using liquid SO_2 at $75/ton. The combined economic benefits of load-following controls for both Cl_2 and SO_2 additions can be assumed to be 10% higher than for Cl_2 alone.

Improved Control of Disinfection

A control system that charges only as much chlorine as is needed for disinfection is beneficial for several reasons. First, it conserves chlorine, which is an expensive chemical. Second, it requires less chemical additive in the dechlorination step. Finally, it protects against the discharging of residual chlorine into the receiving waters.

Total load-following is accomplished by modulating both the Cl_2 and SO_2 feedrates in accordance with both wastewater flow and concentration. Fig. 6 depicts a control system in which both the chlorinator and sulfonator are paced by the effluent-flowrate signal to yield an approximate Cl_2 feedrate, and are then trimmed in accordance with the residual chlorine concentration. The pre-contact residual chlorine analyzer (ARC-1) adjusts the chlorinator dosage, and the post-contact chlorine analyzer (ARC-2) trims the setpoint of ARC-1 in a cascade manner. When, due to a sudden upset or to misoperation, ARC-2 cannot hold its setpoint below some preset limit, its output signal will be sent to the sulfonator, activating the normally inactive SO_2 charging loop.

In a 100-Mgd wastewater-treatment plant, the annualized total cost of this improved control system is $59,728, while the resulting yearly savings in chemicals represent a benefit of $209,000 [*4*]. The sampling system and chlorine analyzers are fairly reliable but do require daily checks and 140-h/yr maintenance. This amount of attention is not required if self-diagnosing and self-calibrating features are provided.

Sludge Digestion

Anaerobic digestion is a complex biological process that operates in the absence of molecular oxygen to convert organic matter to methane and carbon dioxide.

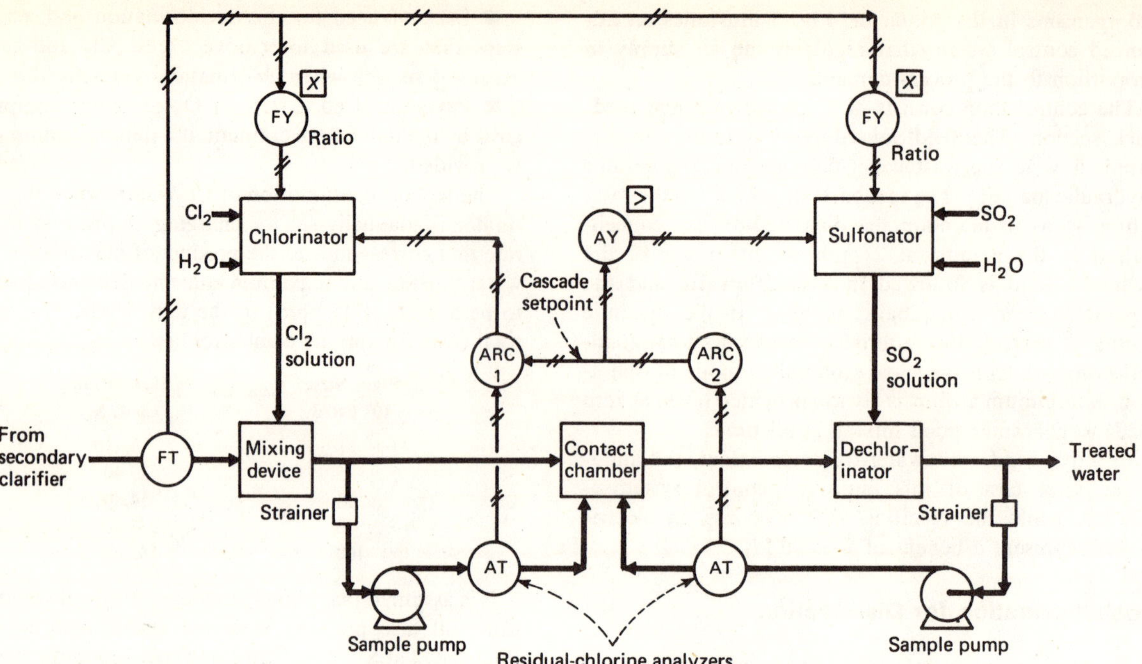

POSTCHLORINATION for disinfection requires modulation of both Cl_2 and SO_2 feedrates—Fig. 6.

The process is widely used in the stabilization of domestic and industrial wastewater sludges. The goal of digestion is to produce an easily dewaterable sludge that can be safely disposed of without environmental nuisances or hazards.

The digestion operation tends to be unstable, unreliable and troublesome. However, this is primarily due to improper operation and control, rather than to any inherent instabilities of the digestion process.

The efficiency and stability of digesters is influenced by the food-to-microorganism ratio, volatile-acid concentration, alkalinity, retention time, biomass concentration, loading rates, temperature and pH. (The three common causes of digester failures—hydraulic, organic and toxic overloadings—all result in pH changes.) If the foregoing variables are manually controlled, or are allowed to drift, the digesters must be oversized by as much as 25 to 30% to compensate for the resulting low efficiency of operation. Advanced control offers a method of improving the reliability of digesters by continuously and automatically regulating some of these variables.

Digester systems represent 30 to 40% of the total capital investment in a wastewater-treatment plant [*10*]. Therefore, if the digester size can be reduced because of improved efficiency, this represents an economic motivation to install improved instrumentation. A 30% reduction in digester size can correspond to a 10% reduction in the total capital expenditure for the wastewater-treatment plant.

Improved Control of Sludge Digestion

In a traditional treatment plant, digestion equipment is sized so that for each 1 Mgd of capacity there are 5,000 ft^3 in gas-holder capacity, 10,000 ft^3 in primary-digester capacity, and 15,000 ft^3 in secondary-digester capacity [*11*]. Advanced controls can substantially reduce these sizes.

Fig. 7 illustrates an improved digester-control system. The controlled variables include:

- Operating temperature.
- Operating pH.
- Rate of methane generation.

The manipulated variables are:

- Heat input.
- Caustic-reagent addition rate.
- Rate of sludge recycling.

This control system assumes that the sludge feed is to be accepted as a continuous (or intermittent) uncontrolled stream. Of course, if there is sufficient storage capacity upstream of the primary digester, the sludge feed can be used as another manipulated variable.

The rate of biological digestion increases when the operating temperature is increased. Fig. 7 shows a continuous, external sludge-circulation system, controlled by a temperature cascade loop. The continuous operation is advantageous because it keeps the digester temperature at a fixed value (instead of cycling between limits) and because it contributes to good mixing and agitation.

One can further improve the automatic control of digesters by continuously monitoring and controlling the pH of the circulated sludge. This will prevent the pH from dropping to the point where the growth of methane-forming microorganisms is inhibited by excessive acidity. The control system to accomplish this is also shown in Fig. 7. Mounting the pH detector in an easily isolated bypass line of well-mixed and continuously flowing sludge sample, and providing it with automatic

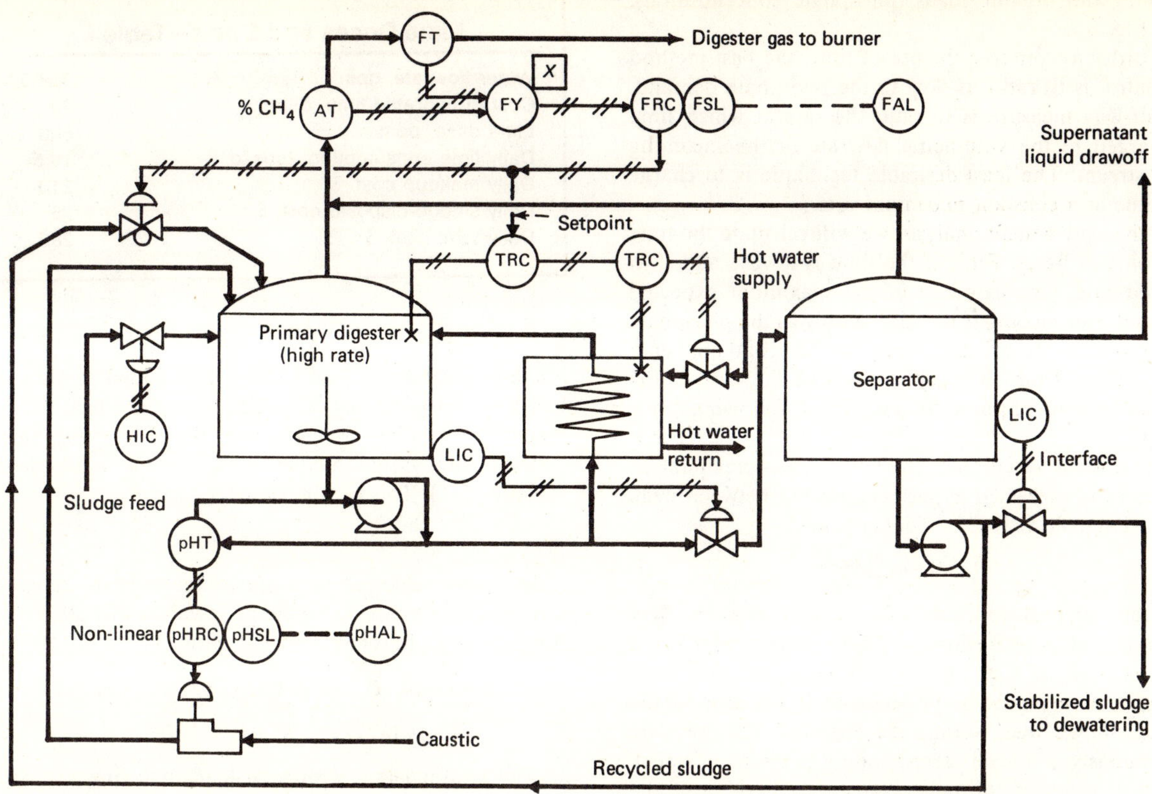

ADVANCED control of two-stage digester involves measuring and controlling total methane production—Fig. 7

cleaning, will increase the mean time between the required preventive-maintenance checks. The nonlinear controller [*12*] should also contribute to stable control performance.

In addition to pH and temperature, the quantity of methane gas generated is another important indicator of digester performance. If, at constant feedrate (fixed hydraulic and organic loading), the methane production drops off, this can be taken as an early signal of a toxic overloading episode. Automatic controls can respond to this by increasing the operating temperature to increase the growth rate of methane formers, or by recycling more methane bacteria from the second-stage separator.

If the dropoff in methane production occurs as a result of increased hydraulic feedrate (resulting in bacterial washout), the automatic instrumentation can reduce or stop the feed (in addition to the corrective actions listed above for toxic overloading).

The instantaneous response to organic overloading is an increase in methane production (which lasts until the resulting low pH starts to inhibit the growth of the methane formers). Therefore, organic overload is best controlled in a feedforward manner by measuring the incoming flow and concentration and keeping their product nearly constant. In the absence of such measurements, pH can be used as a fairly sensitive feedback control to slow or stop the digester feed.

In the system of Fig. 7, the total methane production is measured and controlled. If the rate is dropping, the operating temperature can first be increased, and then additional methane-forming microorganisms can be recycled. Alarms are actuated at low values of pH or of methane production. They will alert the operator to slow or terminate feeding the digester (through the manual loading station, HIC) until the abnormal condition is corrected.

In a 100-Mgd wastewater-treatment plant, the annualized total cost of this improved control system is \$58,320, while the corresponding annualized savings in capital expenditure is \$540,000 [*4*].

COST-BENEFIT ANALYSIS

The cost-benefit analysis described below is an actual one, which involves a precipitation unit-operation.

Precipitation through chemical treatment requires the addition of a chemical reagent that lowers the solubility of the substance being precipitated. This previously soluble material is made insoluble, and is removed as a solid precipitate, owing to the addition of a chemical reagent.

In this example, the precipitation of phosphate by lime addition will be considered. The maximum water flowrate is assumed to be 3,500 gpm. The costs will be associated with the cost of new instruments, and the benefits will be related to the amount of lime saved.

Load-Following

All precipitation-type unit operations of this kind experience cyclic variations in their hydraulic loads (water volume) and organic loads (phosphate concentration). (See Fig. 8.)

In order to conserve the use of lime, the best method of control is to ratio its flow to the phosphate demand. A half-way measure is to ratio the rate at which lime is charged to the volumetric flowrate of the incoming total stream. The least-desirable technique is to charge the lime at a constant maximum rate.

In this cost-benefit analysis, we will calculate the lime cost on two bases. First, if the lime is always added at a fixed rate corresponding to the maximum expected demand, and second, if it is added only as the phosphate load requires. We will then compare the economic benefits resulting from lime savings against the cost of the instrumentation required for a total load-following mode of operation.

The normal variation of the flowrates in precipitation processes is expressed as the relationship between peak and average flows [13] by Eq. (1) below:

$$F_p = 1.84(F_a)^{0.92} \qquad (1)$$

The ratio of peak-to-average flow decreases as the flow increases; at a peak flow of 3,500 gpm (5 Mgd), the ratio is $F_p/F_a = K = 1.62$. If the daily hydraulic load variation is assumed to be sinusoidal [14] around the average flows, then, setting the chemical addition rate continuously at a level corresponding to the plant's peak flow will result in the unnecessary overcharging of the chemical additive by $(K - 1)/K = 38\%$.

The phosphate concentration of a wastewater stream varies according to a diurnal schedule [14]. This variation can also be considered sinusoidal in character, and in phase with the hydraulic-load variation. In other words, at low hydraulic loads the concentration is also likely to be low, and at peak volumetric flows the phosphate concentration of the wastewater stream is also likely to be high.

Therefore, savings in the consumption of chemical additives will be even greater when the feature of phosphate-concentration load-following is added to that of hydraulic flow ratioing. This amplification effect can be made quantitative by assuming that the phosphate-concentration variation equals 75% of the hydraulic-load variation [15]. This can be expressed as:

$$C_p = (0.75(K - 1) + 1)C_a \qquad (2)$$

If lime were charged at a maximum constant rate, its flow would correspond to the peak values of both water flowrate and phosphate concentration:

$$\text{Maximum constant rate-setting} \simeq (F_p)(C_p) = (KF_a)(0.75(K - 1) + 1)C_a \qquad (3)$$

If it is assumed that with total load-following, the rate of chemical additive usage corresponds to F_aC_a, then the peak-to-average chemical use rate is:

$$F_pC_p/F_aC_a = 0.75K^2 + 0.25K = A = 2.38 \qquad (4)$$

The percent savings accomplished by total load-following can be calculated as $(A - 1)/A = 58\%$.

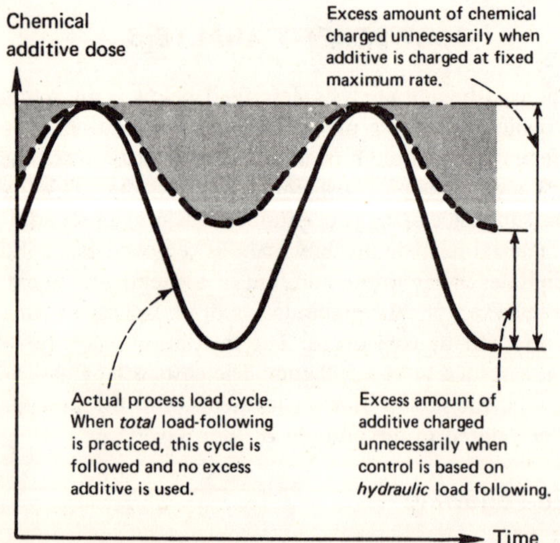

CHANGING additives to match total load of process—Fig. 8

Lime Doses and Costs—Table I

Water flowrate, gpm	3,500
Cost of hydrated lime, $/ton	20
Lime dose, ppm	500
Daily lime consumption, tons/d	10.5
Daily makeup cost, $	210
Daily sludge-disposal cost, $	75
Daily total cost, $	285

Quantity and Value of Lime Saved

Having approximated the lime-saving potential of the load-following strategy as 58%, we can now proceed to convert that into cost units. Table I gives some of the quantitative data for phosphate precipitation with lime, assuming its maximum fixed-rate charging.

The daily cost of $285, given in Table I, includes not only the cost of lime at $20/ton but also the cost of disposing of the sludge that is generated. The daily cost of chemical sludge disposal has been estimated to be $5/wet ton. It was further assumed that the treatment of each 1 million gal of water with lime will produce 3 tons of sludge [2].

Having determined the daily cost of lime at $285 and the savings potential of total load-following as 58%, we have quantitatively identified the potential economic benefits of better control instrumentation. The yearly benefit in this case is:

$$(0.58)(285)(365) = \$60{,}400/\text{yr}$$

Cost of the Required Instrumentation

Fig. 9 describes the instruments required for total load-following, based on pH trimback of a hydraulic-flow ratio. The corresponding capital and operating costs [7,8,16] are given in Table II.

The annualized cost of this control system, based on 10% interest, is:

$$\text{Annual cost} = 1{,}510\frac{(0.1)(1.1)^5}{(1.1)^5 - 1} + 4{,}000\frac{(0.1)(1.1)^{10}}{(1.1)^{10} - 1} +$$

$$7{,}900\frac{(0.1)(1.1)^{15}}{(1.1)^{15} - 1} + 3{,}700 + 3{,}700 = 1{,}150\ (0.264) +$$

$$4{,}000\ (0.163) + 7{,}900\ (0.131) + 3{,}700 = \$5{,}800/\text{yr}$$

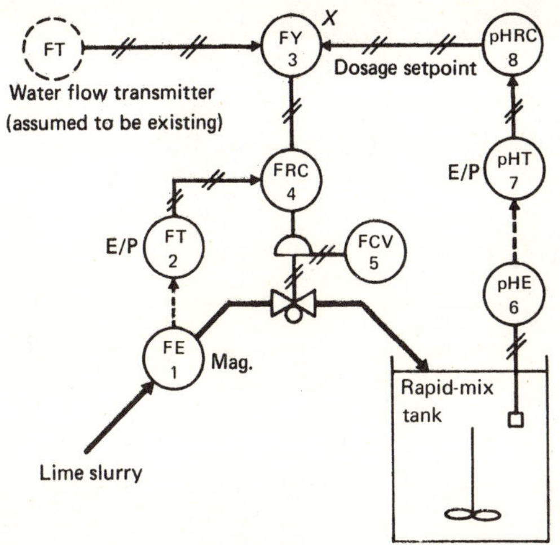

PHOSPHATE precipitation control (lime addition)—Fig. 9

Costs of Total Load Following (Phosphate Precipitation)—Table II

Cost Components	Capital Cost, $	Reqd. Maintenance, (h/yr)	$
5-Year depreciation period			
pHE-6	510	50	1,000
pHT-7	1,000	10	200
	1,510		
10-Year depreciation period			
FT-2	1,200	10	200
FY-3	500	10	200
FRC-4	760	16	320
CV-5	720	12	240
pHRC-8	820	16	320
	4,000		
15-Year depreciation period			
FE-1	1,400	12	240
Installation materials	500		
Control-panel section	600	30	600
Design and Engineering	2,400	166 h/yr	$3,320
Installation and startup labor	3,000		
	7,900	Yearly cost of replacement parts is assumed to be $380.	
Total installed cost per loop	13,410		
Total annual operating cost per loop			$3,700

Ref. 7, 8, 16

Cost-Benefit-Analysis Results

The yearly saving resulting from the introduction of this control system is $60,400. The annualized total instrument cost is $5,800. Therefore the yearly profit resulting from better control is $54,600. The total capital investment for the instruments needed is $13,410. This investment will be paid for by the profits in 13,410/54,600 = 0.24 yr.

Conclusions

Instrumentation is a powerful tool in increasing the profitability of processing plants. Before proceeding to evaluate the potentials of advanced controls—such as feedforward, computer, multivariable, adaptive or optimizing—one should make sure that the conventional loops are fully utilized.

Whenever a new control strategy is suggested, a cost-benefit analysis should be included in the proposal. This analysis should identify the economic and other benefits to be gained from the new control concept, and it should give the associated capital and operating costs.

References

1. Liptak, B. G., "Environmental Engineers' Handbook," vol. I, Sect. 5.23, Chilton, Radnor, Pa., 1974.
2. Ibid, Sect. 5.11.
3. Shinskey, F. G., "pH and pIon Control in Process and Waste Streams," Wiley, New York, 1973.
4. Liptak, B. G., "Cost-Benefit Analysis, Instrumentation of Wastewater Treatment Facilities," U.S. Environmental Protection Agency, *Technology Transfer* (in print).
5. Roesler, J. F., "Factors in the Selection of a Control Strategy," Office of Research and Monitoring, Environmental Protection Agency, Cincinnati, Ohio.
6. Liptak, B. G., (1974), op. cit., Sect. 6.22.
7. Molvar, A. E., "Survey of Instrumentation and Automation Experiences in Wastewater Treatment Facilities," EPA Program Element No. 1BB043, Project Officer: Joseph Roesler.
8. Upfold, A. T., "Manhour Ratings Standardized for Instrument Maintenance," *Instrum. Technol.*, Feb. 1971.
9. Molvar, A. E., "Instrumentation and Control of Biological Wastewater Treatment Plants," Raytheon, Feb. 4, 1972.
10. U.S. Environmental Protection Agency, "Capital and Operating Costs of Pollution Control Equipment Modules," EPA-R5-73-023, July 1973.
11. Liptak, B. G. (1974), op. cit., Sect. 5.17.
12. Liptak, B. G., "Instrument Engineers' Handbook, vol. II," Chilton, Radnor, Pa., 1970, p. 1552.
13. Boyle Engineering and Lowry and Associates, "Master Plan Trunk Sewer Facilities" for District No. 3 of Orange County, Calif., June 1968.
14. Smith, R., Eilers, R. G., "Simulation of the Time Dependent Performance of the Activated Sludge Process Using the Digital Computer," U.S. Dept. of the Interior Cincinnati, Ohio, Oct. 1970.
15. Liptak, B. G., (1974), op. cit., Sect. 2.12.
16. Liptak, B. G., Costs of Process Instruments, *Chemical Engineering*, New York, N.Y., Reprint 101 (originally published in *Chem. Eng.*, Sept. 7, 1970, p. 60; Sept. 21, 1970, p. 175; Oct. 5, 1970, p. 83; Nov. 2, 1970, p. 94).

Meet the Author

Bela Liptak is a principal of Liptak Associates, 84 Old Stamford Rd., Stamford CT 06905. The firm provides consulting and engineering services in the fields of instrumentation and environmental engineering. He attended the Technical University of Budapest and has an M.E. from Stevens Institute of Technology and an M.M.E. from City University of New York. He has published seven books and a number of technical articles. He is also an outside consultant to Crawford & Russell, Inc.

Tighter process operation via computer control

J. PATRICK KENNEDY, Taylor Instrument Process Control Division

Soaring costs of energy, fluctuating prices, environmental regulations, and shortages of materials and manpower, all put a premium on process and equipment design. Here's what this means to control and operation.

Originally published March 17, 1975.

The chemical and petroleum industries are markedly reacting to the real and imagined effects of the oil embargo.

New plants must use more-expensive feedstocks, capital equipment and energy. This is bringing about major design and operating changes, which often put more-stringent demands on control systems. These demands, plus the remarkable flexibility of the process computer, have caused significant changes. At least five basic trends, caused by the higher cost of energy, have resulted in new control systems. These trends:

- Larger sizes for process units.
- Less overdesign.
- A higher degree of automation.
- Variable feedstocks and products.
- Integration of heat- and material-flow systems.

Although construction is at an all-time high, by far the largest capacity and investment remain in existing plants. Advanced control systems are finding extensive but varied use in these plants for five reasons:

1. Much higher operating costs.
2. Different operating philosophies (e.g., maximum throughput versus minimum raw-material consumption).
3. Variable product and feedstock prices.
4. Competition from new processes.
5. Environmental regulations.

The net result has been a rapid rise in control projects. A recent survey* on process computer installations showed an 8% increase in refineries, 13% increase in pipelining, 44% in petrochem installations, and 52% in oil and gas production. The 1974 totals were 481 installations for refineries, 1,029 for petrochem, 353 for pipelines and 216 for production. The same type of growth is being experienced in food processing, mining, pulp and paper, and many other industries.

In order to design such new control systems, it is important to understand how basic trends influence controllability, and how a process computer can be used to counteract negative effects. Such understanding may be helped through examples of the trends and computer justifications. Also, specific examples of both a batch process (PVC resin plant) and a continuous process (recovery boiler) may serve to show the inherent simplicity of the control solutions, once the logic abilities of the computer are made available.

Trends in New Plants

New ways to reroute process streams for energy and raw-material savings, in addition to changes in equipment, are now being studied by many design groups. This has resulted in a high degree of *heat and material integration* via new heat exchangers and recycle streams. Although integration can save money, it can lead to a major coupling of disturbances. Consider the case where hot product streams from a fractionating column are used to preheat the feedstream. Disturbances in the product temperatures are reflected in the inlet temperature of the feed, and unless good regulation can be maintained, the operation can degrade to the point where any savings are lost. The resulting cycles or oscillations can be on the order of many hours duration, so that their identification may be difficult, further complicating the control problem. Several computational techniques are available for either avoiding this type of problem or solving it once it occurs.

Existing Plants

Improved control systems on existing plants must work around present equipment limitations. This has not slowed down implementation of projects however. The average cost of energy when most of the chemical-process-industries plants were constructed was 10-25¢/million Btu. Today $1.50/million Btu is not uncommon (equivalent to $9/bbl for oil), and future prices for good fuels, such as natural gas, will probably be around $2.50/million Btu. Clearly, payouts of advanced control will become easier to justify.

The relationship of capital to operating costs has been changed dramatically by the energy crisis. To minimize the effect of this change, different control strategies must be employed. Not much can be done about the capital investment with existing equipment, but a modern control system can reduce operating costs. This is often the only economical way to make an old plant more competitive.

Energy costs are the most probable source of savings, but often there are other areas, such as number of personnel and automated maintenance. The latter is often overlooked, but over the course of a year a plant may be shut down or cut back needlessly. Vibration monitoring, wall-thickness gauging, motor-load monitoring and many other measurements are available as maintenance checks. Tracing the trends of heat and material balances is also important to show up problems. Leaks and partially open valves are the type of problem often uncovered by such trending. The first detection of a hydrocarbon leak in a furnace-pass can pay for several computer systems.

The market today moves rapidly because of the unsettled economy caused by the oil shortage. If prices force a change in operating philosophy, this can require a significantly different control system. Distillation columns, for example, are operated quite differently for maximum throughput, or minimum energy, or minimum raw-material loss. Some advanced systems go one step further and consider the current prices, so as to run at maximum profit. Very seldom can this be done without considering the synergistic effects of many units.

In addition to all the other problems a plant might have, two considerations will prevent business-as-usual. These are competition from modern plants, and environmental regulations. The PVC plant is a prime example of both of these. This last six months, four plants capable of producing 25% of the total U.S. capacity of PVC have come online. Three of these plants are completely automated and require no manual operations during the run of a batch. These plants use reactors much larger than

*Oil and Gas J., Dec. 9, 1974, p. 53.

those currently in service (as much as ten times larger) and can produce a high-quality PVC resin at a much lower cost. In addition, remote operation allows these reactors to meet the EPA regulations on VCM (vinyl chloride monomer) much more easily. The smaller plants will have to revamp their control systems to be competitive in a soft market.

Computer Control Cuts Costs

The process computer is by far the most important aspect of the modern control system. With the reduction in cost of computers and the increase in cost of energy and raw materials, the process computer can be the best industrial purchase of all. For around $100,000 one can buy an installed process-computer system that compares to a $500,000 system of ten years ago. To put this in perspective, a $100,000 computer system will lease for around $3.50 an hour. There are very few processes on which a computer will not save many times that much money. Assuming $1.50/million Btu, $3.50/h represents 2.3 million Btu/h, a very small amount of heat in the chemical process industries. If furnaces of 60% efficiency are used, this means that a savings of 1.4 million BTU/hr of process heat will pay for a computer system. On an average depropanizer with 80°F reflux, this means cutting the reflux by 200 bbl/d, less than 5%.

Many companies have shied away from process computers because of their purported complexity. Computer control is actually much simpler than normal methods, because modern software allows a process engineer to do the implementation.

Take, for example, the role of the orifice plate in process control. The precise metering of flows is essential in many processes. The relationship of pressure-drop to flow is derived from an energy balance, and the equations show clearly that the orifice plate is a weight-flow device, i.e., the pressure drop is correlated with the velocity and density of the fluid. In most cases, the major error in flow measurement is the assumed density. This parameter is generally measured once in the lab or computed from correlations, and then rarely changed. Using a computer, it is easy to compute a gas density from the pressure and temperature, while liquid density is well correlated with temperature. There are many applications where better flow measurement alone can justify a process computer.

The greatest detriment to using computers in the field has been their previous misapplication. Two axioms for applying these systems are:

- Digital loops are more expensive than analog loops. A computer is usually not justified on the basis of simply replacing analog equipment. The calculational ability must be used to do the job better.
- Plantwide optimizations are rarely worthwhile until the basic process is under control.

There are certainly exceptions to these rules, but many expensive installations have been unsuccessful because they failed to observe them.

Batch-Process Example

The control problems of batch processes differ greatly from those of continuous processes. Batch processes tend to be less well instrumented, and existing batch plants are very expensive to adapt to an integrated control system. Thus, the outstanding applications are on new plants designed for computer control. Functions handled by the basic computer system are better defined than in continuous processes because continuous plants have always had essentially digital control, even when run by an operator. The replacement of manual operations, sequencers, and programmable controllers by a fully automatic computer system has been a natural evolution.

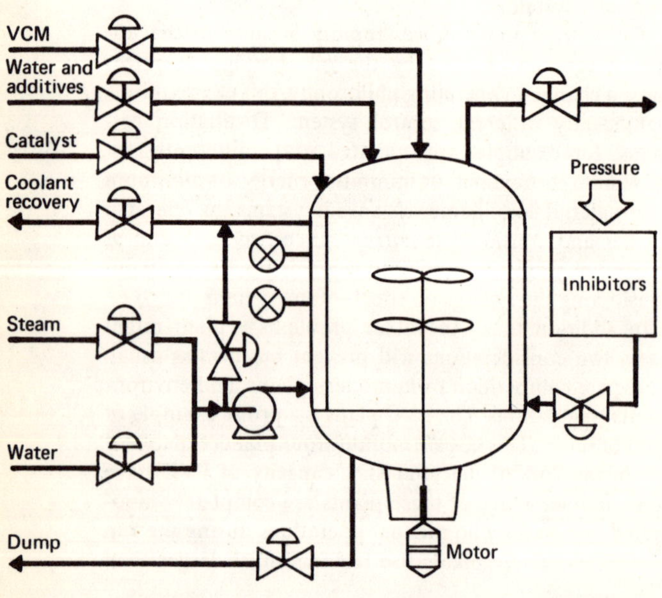

SUSPENSION PVC reactor input/output system—Fig. 1

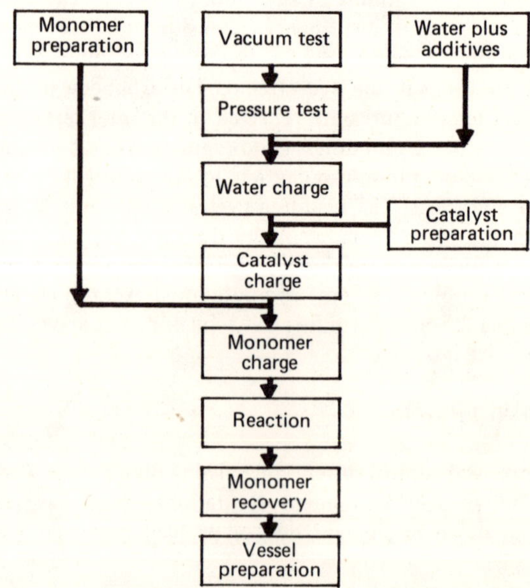

BASIC sequencing control for PVC reactor—Fig. 2

The reactors used to manufacture PVC are an example of a batch process. Fig. 1 is a schematic of a reactor for a suspension PVC process, the most common process for larger plants coming online this winter. New reactors of this type can be as large as 50,000 gal, whereas older plants used much smaller reactors, 2,000 to 4,000 gal. The cost of steel, utilities and process equipment has forced the vessels to be larger and larger. By designing these plants for complete computer control, however, throughput has been increased 30-40%, based on decreases in cycle time. This is a clear savings of capital, which brings simultaneous savings in energy, due to more-efficient use of resources.

To understand how process computers and instrument systems were used to improve efficiency, we must look at how PVC is made. Fig. 2 shows the normal operation of a PVC reactor. During the preparation and charging of a reactor, a number of procedures run in parallel. The metering of components, testing of the vessel, filling of the vessel—all are going on at the same time. In addition, the computer continually runs checks to make sure valves and switches are always in the correct pattern. Numerous other safety checks are run continuously. Since there are from 5 to 20 vessels that could be starting up or in various stages of operation, this sequencing will result in as many as 100 parallel operations.

Procedures running plantwide also govern the shared equipment in the plant. In addition to the sequencing control, reactor temperature is controlled continuously during the cycle. Fig. 3 shows the reactor temperature control. The loop on each portion is different.

During warmup, a proportional-plus-integral-with-output-limiters algorithm is used. This will start to throttle the steam, before the setpoint is reached, by an amount of time determined by the reset rate. After the setpoint is attained, a more-amenable reset rate is used for the reaction run, and a derivative mode is added to take care of the load changes caused by the nature of the PVC polymerization toward the end of the cycle. The cooldown portion is started at a specific % conversion, calculated by the integral of the total heat removal. At this point, the setpoint is ramped down and the vessel is dumped.

This is clearly a loop done best by a process computer. The total automation of the process results in a reduction in cycle time from 14 h to a minimum of around 8 h. If monomer supplies and sales prove no problem, this is a 30-40% reduction in capital costs per pound of PVC resin produced.

The control equipment required for the total automation is shown in Table I. To obtain the best advantage from these control systems, the plant should be designed from the beginning for a process computer. Table I shows the inputs brought into the computer, and underneath that, the outputs. In addition to sequencing control and temperature control, these inputs are used to heat-balance the reactor and to estimate the end point as described above.

To put this in perspective, assume stabilized values as: 400,000 tons/yr PVC, 14-h cycle, 12¢/lb monomer, 26¢/lb product.

On this basis, every minute trimmed from the cycle is

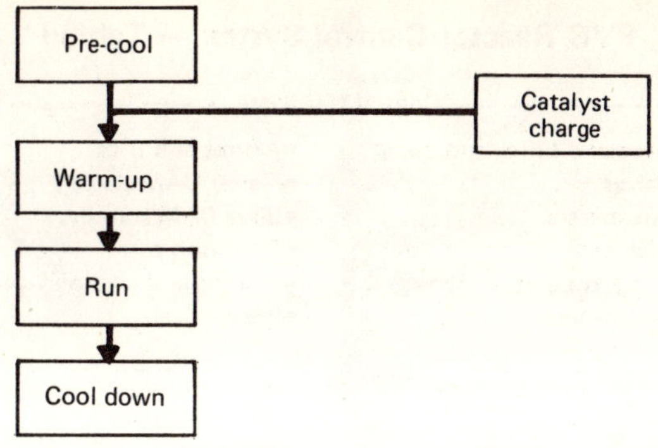

TEMPERATURE control for a PVC reactor—Fig. 3

worth about $125,000/yr from increased throughput, with no increase in capital equipment. The advanced control of the warmup is thus worth $1.5 million/yr, and the total automation of the sequencing is worth $30 million/yr. Certainly, no new large PVC plants will be built without process computers, and when the price of PVC softens, these more-efficient plants will cause the smaller plants to close.

In this case, the implementation of computerization was simple and provided an assured payout. For example, how does an analog sequencer close a valve? It puts out a *close* signal, assumes the valve closes and proceeds to the next step after a fixed amount of time. A computer-operated valve will probably have four signals associated with it; two outputs from the computer (open, close) and two inputs from the limit switches on the valve (open, close). The computer sends out a *close* signal, starts a timer when the top switch operates, goes to the next step as soon as the bottom switch operates. If the valve does not close within a specified time, the appropriate message and emergency procedure are initiated. This is not only a safer way to operate the valve, it is faster; the operation is advanced as soon as the valve closes.

In charging a PVC reactor, as many as 50 devices must operate. Just a few seconds per device will quickly show large payouts. Since one system may handle the charging and running of as many as 20 reactors, the logic ability and the speed of a computer soon pay off. This is only one of the many places in the sequence where time is saved.

Many other processes use batch operations that will be all computerized for reasons of efficiency. Probably the best examples are coal gasification or liquefaction plants. In theory, they are continuous, but the gasifiers tend to coke and must come down every two weeks or so, and it takes as much as two days to get them back on line. With these plants producing 200 million ft^3/d of gas, increased capacity is very valuable. Material-handling equipment will also be controlled by computers. Because of the danger of equipment failure, extensive monitoring will also be done.

Other examples are: dye becks for textiles, smelters, steel mills, blast furnaces, and the list could go on. In each case, however, the computerized plant will produce

PVC Reactor Control System — Table I

Control Hardware	
•Automatic filling and dump valves	•Alarms and trips
•Flow meters	•Master CAM station
•Weigh cells	•Slave CAM station
•RTD Temperature Sensors	•Deviotion trip
•Transmitters	•Heating and cooling valves
	•Pressure cell
Computer Input	
•Flow rates	•Alarm contacts
•Valve limit switches	•Setpoints
•Weights	•Pressure
•Temperatures	•Acknowledge switches
•C-A switches	•Agitator speed/load
Computer Output	
•Temperature controller output	•Annunciators
•Valve open/close	•Sequence lights
•Trends	•Inhibitor system

Control Problems Caused by Energy Shortage

1. *Larger Units*—This results in large lags, less excess design in cooling systems, slower-operating valves. Less responsive than smaller units.
2. *Integrated Units*—Integrated heat exchangers, cross-coupled disturbances, intermediate-product recycles. Slow (8-24 h) oscillations likely.
3. *Less Surge Capacity*—This means more flow variations, faster transmission of disturbances. Could combine with integrated units to cause trouble.
4. *Widely Varying Feedstocks*—Units must be flexible enough to run whatever feeds are available.
5. *Lower-Quality Feedstocks*—High-sulfur oil, pyrolysis oils from shale and coal must be used. There will also be looser specs on impurities, as competition for intermediate feedstocks warms up.
6. *New Measurements for Process Control*—Oxygen, combustibles, viscosity, density, analysis measurements will be used for more-efficient operation.
7. *Reduced Energy Consumption*—Minimum energy consumption often results in a control problem. For example, lowering hot-oil temperature on distillation processes and raising chilled-water temperature on reactors.
8. *Rapidly Varying Relative Cost of Utilities*—Electricity, steam, gas and oil will all vary with respect to each other in their price per Btu. To take advantage of rates, units must be under control while moving from one operating-point to another.
9. *Fluctuating Market Conditions*—The best example of this is the gasoline blender. Only under computer control is it feasible to change final properties to take advantage of a special market.
10. *Varying Heating Value of the Fuel*—By venting gases of largely varying heating values, the plant fuel may fluctuate rapidly. For example, producer gases run about 100 Btu/scf compared to 1,000 Btu/scf for methane and 1,800 Btu/scf for ethane.

more and better product that conserves raw material and capital.

Continuous Processes

Continuous processes present a completely different problem for computerization. The payouts are both increased throughput and reduced waste, as in the case of a batch process, but they are attained in completely different ways. For example, the computer system that runs a $250-million vinyl chloride monomer plant is less than one-sixth as complex as the system that runs a $25-million PVC plant. The continuous plant is scanned at the rate of 5 points/s while the batch unit must process loops as fast as 400/s. The major new incentives in continuous processes are conservation of fuel and elimination of losses of product or raw material. New processes are becoming more interactive and difficult to control because of heat-recovery programs. Some operating conditions that have made control more difficult: raised chilled-water temperatures, lowered heat-medium temperature, lowered refluxes, closed bypasses.

The concept of entropy is very useful in analyzing where a move toward a more difficult control situation is worth money. For example, one Btu in 500°C oil costs more than a Btu in 400°C oil.

A good example of a continuous process is the recovery boiler in a kraft paper mill. This unit contributes heavily to the economics of a paper mill. Not only does it convert a waste stream to large amounts of energy, it also converts expended chemicals back to a useful form.

Black liquor from the pulp digesting process is burned in the presence of air, yielding sodium sulfide and sodium carbonate in smelt-form at the base of the furnace. The gaseous combustion products pass across steam tubes, through dust collectors, through additional heat exchangers and then out to the stack. The dust collectors are electrostatic precipitators; the heat exchangers, if used, cause additional concentration of the black liquor prior to its being burned.

The recovery boiler is highly interactive and careful balances must be maintained for it to function in a satisfactory manner. Off-performance causes a generous production of mercaptans and sulfides, both of which carry highly penetrating obnoxious odors into the atmosphere. Dust burden is quickly added, and the smelt reaction can generate black spots and come to a stop. On the other hand, maintenance of a black-liquor solids-to-liquid ratio, good temperature control of the liquor prior to its entrance into the spray nozzles, flow control of both air and black liquor, and careful splitting of primary and secondary air, will facilitate good production of both smelt-product and steam, and at the same time keep odor-producing compounds to an acceptable minimum.

Fig. 4 shows the instrumentation and controls associated with combustion and smelt formation. Boiler level controls and steam pressure control both have been omitted since they are not part of the combustion control system. It is assumed that all of the recovery-furnace production is to be used, and hence there is no throttling of black-liquor flow as a function of steam pressure.

Many analog-controlled recovery boilers are in oper-

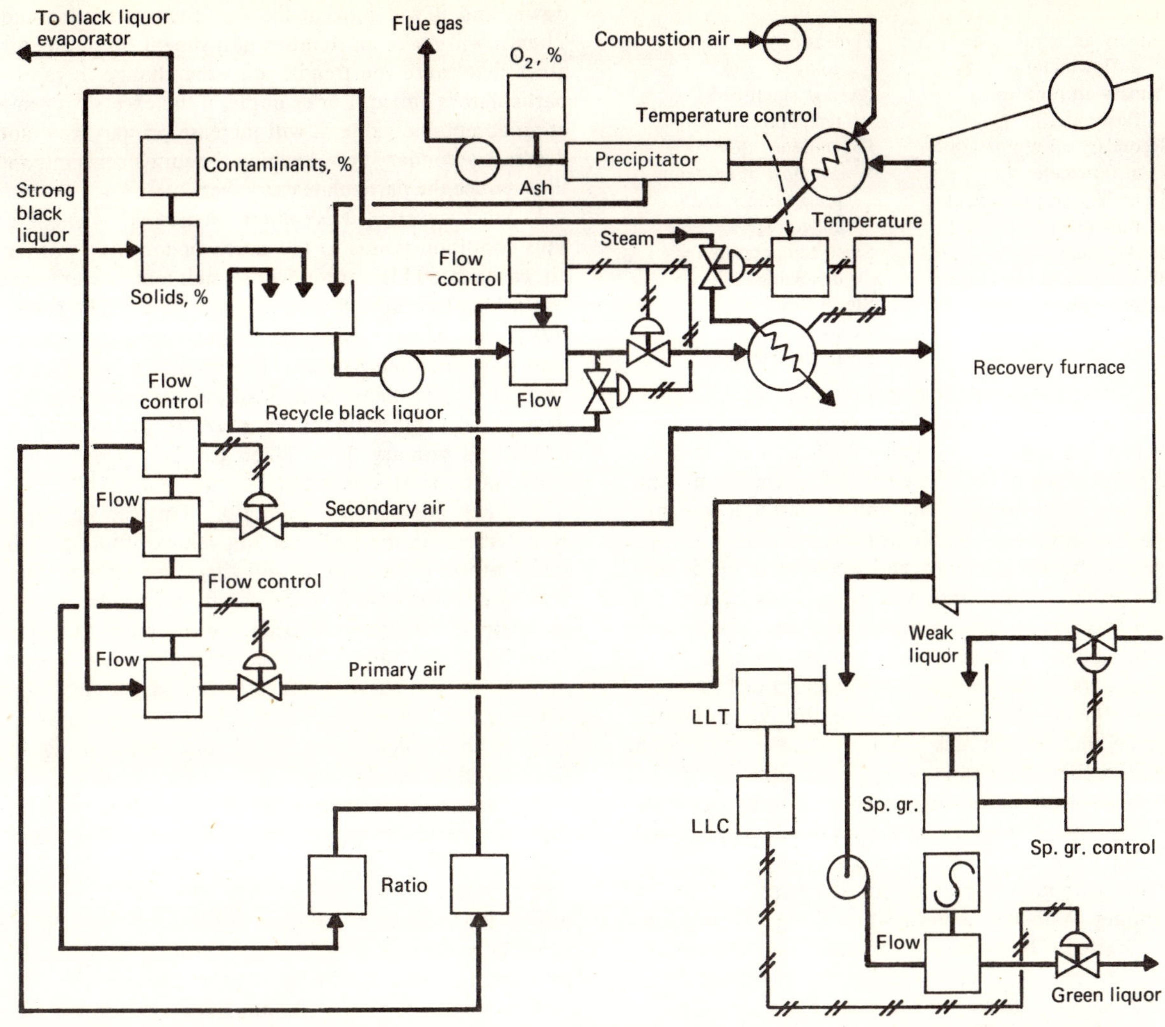

FLOW and instrumentation diagram for a kraft pulping process recovery furnace—Fig. 4

ation today, as attested to by reduced odors and ash. As mentioned earlier, a broad interaction between variables must be recognized if efficient operation is to be achieved. For example, the size of the black-liquor droplets leaving the spray nozzle is of vital importance. If they are large, they will fall to the bottom of the furnace before being vaporized, and give up their moisture to the smelt bed. If they are too small, the updraft will carry them away, and their ash will go upward to ultimately deposit on the steam coils or to burden the dust precipitator, with the attendant loss of available thermal energy.

The balance of primary air and secondary air is another example. The sum of the two is directly related to black-liquor flow, while the primary air controls smelt temperature and the secondary air the combustion of the liquor droplets. This duality of effects causes analog control to fall short and requires the flexibility of digital control.

Digital control does not reduce the number of measurements needed. As a matter of fact, it generally improves with an increase in measurements. Various capabilities, such as arithmetic, storage, sensing, and alarm analysis, provide the basis for interrelating variables to achieve optimum combustion with stable smelt production, minimum excess oxygen, minimum particulate carryover, maximum steam production, as well as generation of hard copy using line printers that accurately spell out pounds of steam per pound of liquor, stack-gas oxygen content, pounds of particulate per pound of liquor, smelt production, etc.

The function of the optimization program is to perform an operation over and above the regular control program. The concept here is to have the entire recovery boiler on digital control, which can be either supervisory or total. To achieve a satisfactory order of success, the following loops are controlled:

Primary-air flow
Seconadary-air flow
Black-liquor flow to
spray nozzles
Black-liquor temperature
at spray nozzles
Black-liquor temperature
at concentrator
Smelt density
Smelt level
Furnace draft pressure
Salt-cake mix-tank level
Boiler-drum liquid level

Measurements required for calculation purposes are these:

Primary-air temperature a at flowmeter	Percent oxygen in gas going to stack
Primary-air pressure at flowmeter	Average smelt-bed temperature
Secondary-air temperature at flowmeter	Green-liquor flow
Secondary-air pressure at flowmeter	Weight flow of ash from precipitators and collectors
Black-liquor percent solids	Smelt temperature in dissolving tank
Black-liquor percent water	Steam flow
Black-liquor density	Feedwater flow
	Steam temperature

The strategy is to put the recovery boiler on total automatic control. This implies, for example, that the computer receives data to calculate weight flow of black liquor, weight flow of both primary and secondary air, and that, based on a fuel-air ratio entered through the console, it supplies these ingredients to be burned.

The split between primary and secondary air is dependent on smelt temperature and excess oxygen, and the control here can be quite complex. The computer is also controlling concentrated black-liquor temperature, holding it at a value that is specified through the data-entry console. Many additional statements on control status could be included at this point, but these have been omitted since the emphasis is on optimization.

As mentioned earlier, optimization programs operate over and above the currently controlled systems. All loops are stable and linearized, so setpoint changes can be made without the problem of sustained oscillations developing. With this condition prevailing, the program can be started. This is done at the data-entry console, either by depressing a dedicated button, or calling out a specific display page on the cathode-ray display tube and typing OPT on a specific line identified with the quantity to be optimized.

Here are four specific programs important to the efficient operation of the recovery boiler. Assume that:

1. *Excess oxygen in the gas tends to increase*: This would appear to be the result of a falling fuel value in the black liquor. However, before taking any action, the computer will check the particulate trend and note whether it is increasing or decreasing. Particulate accumulation rate is indicative of carryover due to the updraft in the boiler's combustion section. If the rate is increasing, a secondary-air flow should be decreased, which in turn reduces excess oxygen. If the rate is decreasing, the smelt temperature is checked and if this is decreasing, the air taken away from the secondary system should be supplied to the primary system, keeping the total a constant. If the smelt temperature is not changing, the primary-air flow is correct, and the secondary air should be used to adjust the stack excess oxygen. Steam production is also noted and if tending to decrease, a loss in fuel value is substantiated.

2. *Particulate burden is increasing*: This is traceable to excessive smelt temperature, a decrease in black-liquor surface tension, or excessive secondary air. The computer will look at excess oxygen; it will look at the black-liquor temperature and at nozzle backpressure histories, to determine whether the trends in these values were up or down, and it will look at the smelt-temperature trend. Then it will make an iterative adjustment on those variables that cause the trends, until the change in rate of particulate is halted. For example, if the excess oxygen is at an acceptable value, it will increase primary flow and decrease secondary flow, keeping the sum a constant, and thus reduce the particulate carryover.

3. *Stack emission is becoming increasingly obnoxious*: This condition points to too much or too little primary air, causing too high or too low a smelt temperature; or it could be a too-rapid drying of black liquor. The corrective procedure here would be to note trends in excess oxygen and smelt temperature. If both appear satisfactory, or if the oxygen shows some increase, the procedure is to decrease the black-liquor temperature and slowly increase the primary flow. When excess oxygen falls to 2-3%, the odors will decrease.

4. *Steam production is falling off*. This can be caused by a decrease in the fuel's heating value or fouling of the boiler tubes. The computer will check the ratio of steam flow to total air flow. If within limits, no corrections will be made. If the ratio is reduced, the computer will check excessive oxygen and adjust secondary-air flow if necessary. It also will check the furnace pressure to determine whether any induced-fan difficulties are present; and if there is no problem there, it will start soot blowing and note the effectiveness.

The above optimizations are carried out in small discrete steps every 30 min unless otherwise set. The alarm-detection portion of the program is continuously functioning with appropriate message printout in the event any of the measured variables falls out of range. If the sum-alarm state reaches a value indicating danger to life or property, a fast shutdown program is begun.

A vast amount of housekeeping is needed to keep the recovery boiler in a safe production state; this calls for a digital computer with a flexible online programming capability.

Conclusion

In this time of scarce and expensive raw materials, high energy costs, and high construction costs, the companies that completely automate will have the competitive edge. A larger margin is obtained by capital credits from increased production. Many improvements will also come as a result of the careful study that must precede any computer installation.

Meet the Author

J. Patrick Kennedy is field systems specialist located at Taylor Instrument Process Control Div., Sybron Corp., 1661 Timothy Dr., San Leandro, CA 94577, where he has just moved from a position as systems engineering supervisor for Taylor at Rochester, N.Y. Before joining Taylor, he worked for Shell Development Co., as research engineer in advanced control applications. An author of several papers on computer control, he is a member of the AIChE, ISA and AAAS. He holds a PhD (1970) and BS (1964) in chemical engineering from the University of Kansas.

Cost comparisons of analog-control and computer-control systems

Here is an analysis for advanced control schemes in which economic justification includes factors such as fail-safe logic, onstream reliability, and supervisory control.

*A. Eli Nisenfeld, Olin Corp.**

☐ The question of backup frequently occurs in cost comparisons between digital-computer and analog-control systems for processes. Including backup will increase the cost of the digital system and therefore place it at a disadvantage, because no backup is normally suggested for the analog-control scheme.

Backup considerations for digital computer systems have continued unabated for more than 12 years. The environment in which these concepts were originally developed was one of low reliability of the computer hardware. As hardware reliability grew, software complexities became the limiting factor. Today, with excellent hardware reliability and straightforward software systems having been developed for process control, the question of backup must be reexamined.

This discussion can have several orientations: direct-digital control (DDC), supervisory computer control, data-logging and alarming systems. In this article, we will discuss primarily the criteria for backing up a computer control system for supervisory control. The DDC system requires special considerations and is a subject unto itself. The question of backing up a data-logging system is so seldom raised that discussing it here is not justified.

Hundreds of advanced control systems based on analog hardware are now in operation. These systems perform functions that a supervisory computer might do. The thought of backing up these analog systems is seldom considered. Even though the risks are now no greater in a digital system than in an analog system, a proposal for a digital system generally includes funds for a measure of backup, either analog or digital. This extra cost, and it can be large even for minimum backup, can make justification of the digital system marginal. Furthermore, the increased complexity of the system, especially in the area

* The author, who is no longer with Olin Corp., now works as an independent consultant.

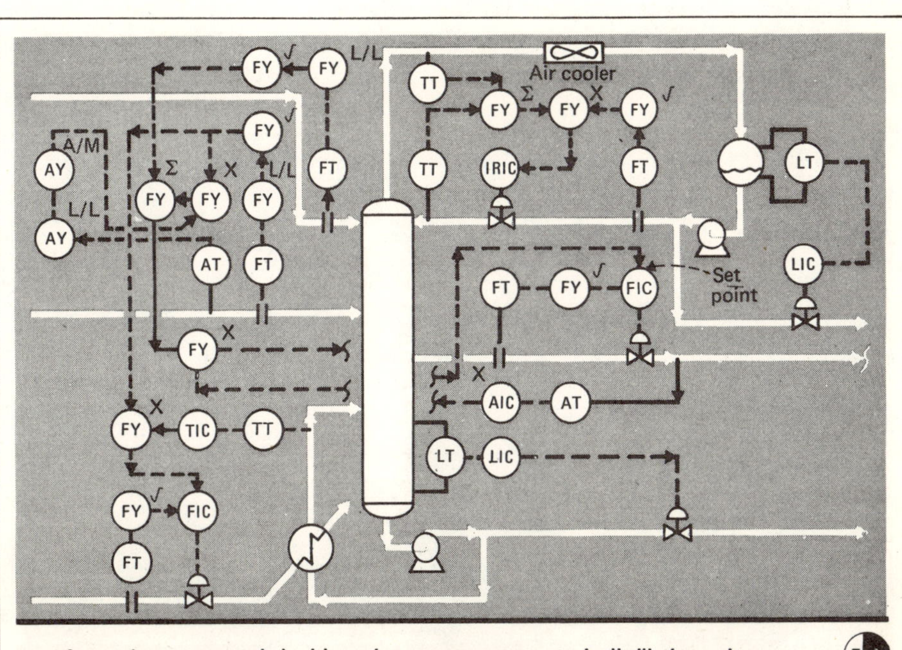

Feed-forward system coupled with analog components controls distillation column

Originally published August 18, 1975.

of programming, may actually decrease the overall system reliability.

Control schemes: analog vs. digital

We can see what happens when the same standards of reliability and safety are applied, by examining an advanced control system for a train of distillation columns. In Fig. 1 (F/1), we find a distillation column with an analog feedforward system; and in F/2, the same column with the same feedforward system implemented on a digital computer. The column shown is a deisobutanizer (one of six columns in the distillation train).

Justification for the advanced control system is dramatic. The original backup philosophy for digital-computer control (in F/2) was a fail-safe system. If the computer were to fail, its outputs would be frozen, and the column transferred to manual control. Similar backup was not provided for the analog system.

On this basis, it is less expensive to implement the advanced control system by using analog hardware. In fact, it seems less expensive to implement advanced control for all six columns with analog hardware, as shown by the cost data in Table I (T/I). The backup philosophy, shown in F/2, is the one commonly used, and is probably an expensive method.

Reliability of analog-control systems

To make a fair comparison, we must question the reliability of the analog systems and the results of failure. It is apparent from inspection of F/1 that the analog system does not have the protective features of the digital system should any one of the computing components fail. There are 12 computing relays in this control system; the possibility of failure is real. Therefore, fail-safe intelligence is required for the analog system to provide the same protection as the digital ones.

Comparative costs of basic analog- vs. digital-control systems for a six-column distillation train (T/I)

Distillation column	Analog Hardware	Analog Other*	Digital Hardware	Digital Other*
1	$9,300	$3,500	$51,000	$8,000
2	3,950	1,750		1,800
3	4,300	2,200	3,100	1,250
4	4,300	2,200		300
5	3,950	1,750		300
6	4,300	2,200	2,300	450
Subtotal	30,100	13,600	56,300	12,100
Total	$43,700		$68,400	

*Other costs for analog systems result from special calibration and documentation, and for digital systems, from programming and documentation.

Note: Installation costs between the two types of systems are about equal and are not included.

F/3 shows part of one of the analog loops with fail-safe logic added. This system is much more complex and expensive than the original analog system of F/1. The cost of adding this fail-safe function to the advanced control loops on all six columns is $26,504, including engineering, and is based on the hardware additions shown in F/3. The total cost of the analog system is now $70,204, or $1,804 more than digital. Besides the basic computing relays, the analog system with backup components now has 52 separate logic elements and 8 special controllers.

Control schemes

The safety of the analog and digital control schemes of F/2 and F/3 are now almost equivalent if the logic elements are considered perfect. These equivalent systems would be a reasonable basis for a cost comparison. However, the digital system still performs functions that the analog system cannot. The digital system increases the overall reliability of the control scheme by checking the validity of analyzer signals, detects impending failures, and actually initiates calibrations.

When these features are included in an evaluation, a properly designed digital system is competitive with an analog system—maybe even for the control of a single distillation column. T/II shows a comparison of the costs and justifications for the two systems on the single column. The basic justification uses the same operating improvements, that is, increased recovery of

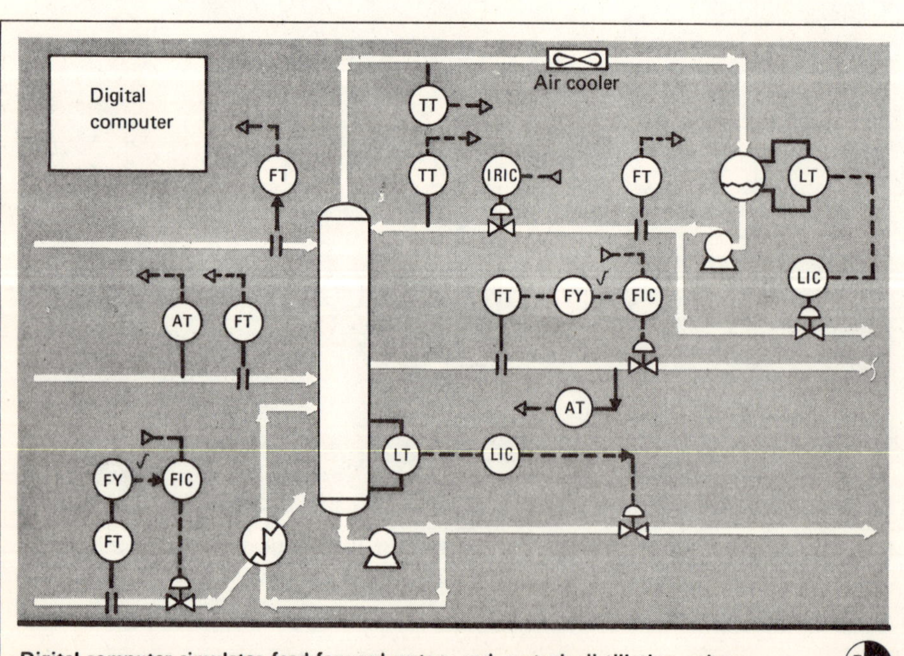

Digital-computer simulates feed-forward system and controls distillation column (F/2)

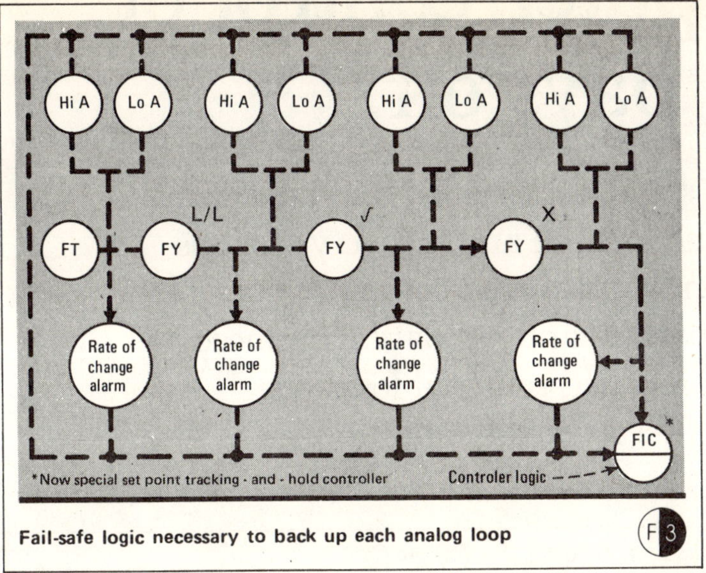

Fail-safe logic necessary to back up each analog loop — F3

Nomenclature

AIC	Analysis, indicator/controller
AT	Analysis, transmitter
AY	Analysis, relay or computer
IRIC	Internal reflux, indicator/controller
FIC	Flowrate, indicator/controller
FT	Flowrate, transmitter
FY	Flowrate, relay or computer
HiA	High alarm
LoA	Low alarm
LIC	Level, indicator/controller
LT	Level, transmitter
TIC	Temperature, indicator/controller
TT	Temperature, transmitter

Function designations

A/M	Automatic/manual
L/L	Lead/lag
X	Multiply
Σ	Sum
√	Square root

more-valuable products and energy savings.

Justification for the analyzer validating and calibrating functions of the computer can be taken in either of two ways: manpower savings, or improvement in the onstream factor of the control system. It is difficult to estimate manpower savings, so an estimate of the increase in onstream time will be used. By monitoring analyzer performance, the computer provides for scheduling of preventive maintenance and for online calibration to improve the analyzers' onstream factor.

The onstream factor of the analog system with fail-safe logic is 0.68 (analyzer downtime is the major problem). The onstream factor for the digital system is 0.92. Here, too, analyzer downtime is the major problem—even with the improvements effected by the digital computer. As T/II shows, the return on investment of the digital system is about the same as the analog system for control of a single distillation column. If a multiple-train distillation system were being evaluated, the justification for the digital system over the analog system would be even greater, as shown in T/III.

In this analysis, justification of the computer system was based on its control functions only. Although the computer will be doing a reasonable amount of monitoring of process variables, no attempt was made to include the value of this monitoring in the cost evaluations. By removing the intangible value for data monitoring and logging from the justification, the control engineer is able to guarantee the integrity of the control functions against the computer's time-and-memory demands for logging and monitoring.

The logic presented in this article and the cost comparisons of T/II and T/III show that the cost of digital computer control is competitive with an equivalent-quality analog control.

Performance comparison of digital vs. analog control on one distillation column — T/II

	Digital	Analog
Hardware	$51,000	$16,005
Programming and documentation	8,000	–
Special engineering, calibration and documentation	–	5,176
Total cost	**$59,000**	**$21,181**
Potential increase in annual profit	$150,000	$150,000
Onstream factor	0.92	0.68
Net improvement in profit (before taxes)	$138,000	$102,000

Performance comparison of digital vs. analog control on six-column distillation train — T/III

	Digital	Analog
Hardware	$56,300	$51,332
Programming and documentation	12,100	–
Special engineering, calibration and documentation	–	18,908
Total cost	**$68,400**	**$70,240**
Potential increase in annual profit	$475,000	$475,000
Onstream factor	0.89	0.63
Net profit before taxes	$422,750	$299,250
Return on investment: $\frac{\text{(50\% of net return)}}{\text{(Total cost)}} \times 100$	309%	213%

The author

A. Eli Nisenfeld is an engineering consultant in process control, 355 Purdy Hill Road, Monroe, CT 06468. Previously, he was manager of the control and statistics group at Olin Corp. He also worked for a large engineering contractor for five years, and spent an additional three years with a major instrumentation firm. Mr. Nisenfeld has a B.A. and the B.S. in chemical engineering from Drexel University, and is a member of AIChE, Instrument Soc. of America, and the International Federation of Automatic Control.

Advanced computer control of ethylene plants pays off

In naphtha-fed plants, the authors see a 100% return on investment (discounted cashflow) for placing the furnace and recovery sections and the overall material balance under advanced computer control. Benefits include improved yield reporting and regulation, reduced reprocessing and overfractionation, lessened material losses via vent gases, and more-efficient use of feeds, energy and utilities.

J. L. Hammett, Jr., and L. A. Lindsay, The Foxboro Co.

☐ Over the past 10 years, nearly 40 computer-control systems have been installed in ethylene process units [*1*]. Large plant size, interactive plant operation, and high product value have provided the chief incentives for bringing ethylene processing under the close control offered by computer schemes.

On a worldwide basis, the throughput equivalent of 52 one-billion-lb/yr ethylene plants has been forecasted for construction by 1980 [*2*]. This article is intended to guide management in evaluating the potential cost-benefits that could accrue from the inclusion of computer controls in the designs for these anticipated installations.

The process

The process flowscheme and material balance used for this analysis are summarized in Fig. 1 and Table I. Recovery-section product purities and cost-benefit data have been derived from in-plant studies, whereas typical furnace yields and the rest of the ethylene plant flowscheme have been developed from literature data [*3, 4*].

The example chosen was a naphtha-fed plant producing 1 billion lb/yr of ethylene. Obviously, not all plants use this configuration, but for similar arrangements and capacities, cost benefits can be adjusted linearly according to size. For example, in a 1.2-billion-lb/yr plant the benefits should be adjusted upward by 20%.

The incentives

Benefits evaluated here include only those that offer improvements to continuous control. For this study, the comparison was made between a conventional analog-controlled plant and a process computer-controlled plant. Although some of the improvements can be achieved by using advanced analog-control schemes [*5*], a digital computer is required to achieve all the benefits.

The specific plant operations that were evaluated are as follows:

- Regulatory control of the recovery section.
- Regulatory control of the furnace section.
- Optimization of the overall material-balance.

Table II summarizes the economic values of feedstock, products and utilities used in the benefit analysis. Although these prices may vary, the ones listed are typical of those that have been used for project evaluation in the industry.

In addition to the process benefits summarized in the analysis, there are other benefits for which no credit was taken. Management should evaluate these additional benefits in light of its operating objectives and philosophies. Depending upon the installation, computer control may be advantageous for improving management reporting, yield accounting, automatic furnace decoking, control enforcement, and operator enforcement.

Recovery-section benefits

For this example, feedforward control strategies are used in the following five areas of the plant's recovery section:

- Demethanizer feed chilling-train.
- Demethanizer.
- Deethanizer.
- Acetylene converter.
- C_2 splitter.

The expected increased plant revenue from implementation of these strategies is $809,000, as shown in Table III.

By manipulating the heat and material balances, the goal of these control strategies is to hold the "hard spec" stream at its maximum allowable impurity con-

Originally published November 8, 1976.

Plant material balance for a typical naphtha-fed ethylene plant **Table I**

	Furnace effluent		Hydrogen offgas		Methane offgas		Ethylene product		Ethane recycle	
	wt%	million lb/yr	wt%	million lb/yr	wt%	million lb/yr	wt%	million lb/yr	wt%	million lb/yr
Hydrogen	1.0	37	35.5	37	–	–	–	–	–	–
Methane	16.1	592	57.0	59	94.0	533	0.02	–	–	–
Acetylene	0.6	22	–	–	–	–	–	–	–	–
Ethylene	28.3	1041	7.5	8	6.0	34	99.95	1000	1.6	3
Ethane	4.1	151	–	–	–	–	0.03	–	91.8	167
Propylene	12.3	452	–	–	–	–	–	–	6.6	12
Propane	1.1	40	–	–	–	–	–	–	–	–
Butadiene	4.3	158	–	–	–	–	–	–	–	–
Other C_4's	3.7	136	–	–	–	–	–	–	–	–
Gasoline	23.9	879	–	–	–	–	–	–	–	–
Fuel oil	4.5	165	–	–	–	–	–	–	–	–
CO & CO_2	0.1	4	–	–	–	–	–	–	–	–
Total	**100.0**	**3677**	**100.0**	**104**	**100.0**	**567**	**100.00**	**1,000**	**100.0**	**182**
Stream		1		3		4		7		5

	Propylene product		Propane recycle		C_4 product		Pyrolysis gasoline		Pyrolysis fuel oil	
	wt%	million lb/yr	wt%	million lb/yr	wt%	million lb/yr	wt%	million lb/yr	wt%	million lb/yr
Hydrogen	–	–	–	–	–	–	–	–	–	–
Methane	–	–	–	–	–	–	–	–	–	–
Acetylene	–	–	–	–	–	–	–	–	–	–
Ethylene	–	–	–	–	–	–	–	–	–	–
Ethane	0.05	2	–	–	–	–	–	–	–	–
Propylene	99.90	439	3.0	1	–	–	–	–	–	–
Propane	0.05	2	94.0	38	–	–	–	–	–	–
Butadiene	–	–	3.0	1	53.6	157	–	–	–	–
Other C_4's	–	–	–	–	46.4	136	–	–	–	–
Gasoline	–	–	–	–	–	–	100.0	879	–	–
Fuel oil	–	–	–	–	–	–	–	–	100.0	165
CO & CO_2	–	–	–	–	–	–	–	–	–	–
Total	**100.00**	**443**	**100.0**	**40**	**100.0**	**293**	**100.0**	**879**	**100.0**	**165**
Stream		8		6		11		9&10		2

tent, and the "soft spec" stream at its target value as specified by higher-level control strategies. For example, the hard spec for the demethanizer is the methane content of the bottoms stream, whereas the soft spec is the ethylene content of the methane offgas. The hard-spec stream is held at its maximum impurity so that the unit will not waste capacity or utilities in overfractionation. The soft-spec stream is held at a target value selected so that the individual process units are operated in accordance with a local optimization plan [*6*], or with the overall strategy for maximum plant profit.

The revenue increases for items 1 and 2 in Table III are due to reduced ethylene losses to offgas. Benefits from items 3, 4 and 5 derive from the reduced reprocessing costs in recovering propylene, ethane and ethylene, respectively.

In the demethanizer feed-chilling train, the setpoints of the five refrigerant liquid-level controllers are adjusted by the process computer so as to hold the ethylene losses in the hydrogen offgas stream at the value specified by a higher-level control strategy or by the plant operators. As shown in Table III, the computer should reduce the ethylene losses from 7.5 to 5.5 wt%, yielding a cost benefit of $126,000.

For the case of the methane offgas, the computer adjusts the distillate takeoff and boilup rates in the demethanizer so as to hold the methane-to-ethylene ratio in

Typical ethylene-plant price data **Table II**

Stream	Value, ¢/lb
Naphtha feed	5
Hydrogen offgas	4
Methane offgas	4
Ethylene	10
Propylene	8
C_4's	4
Pyrolysis gasoline	6
Pyrolysis fuel oil	2
Operating cost	1*

*Per pound of ethylene

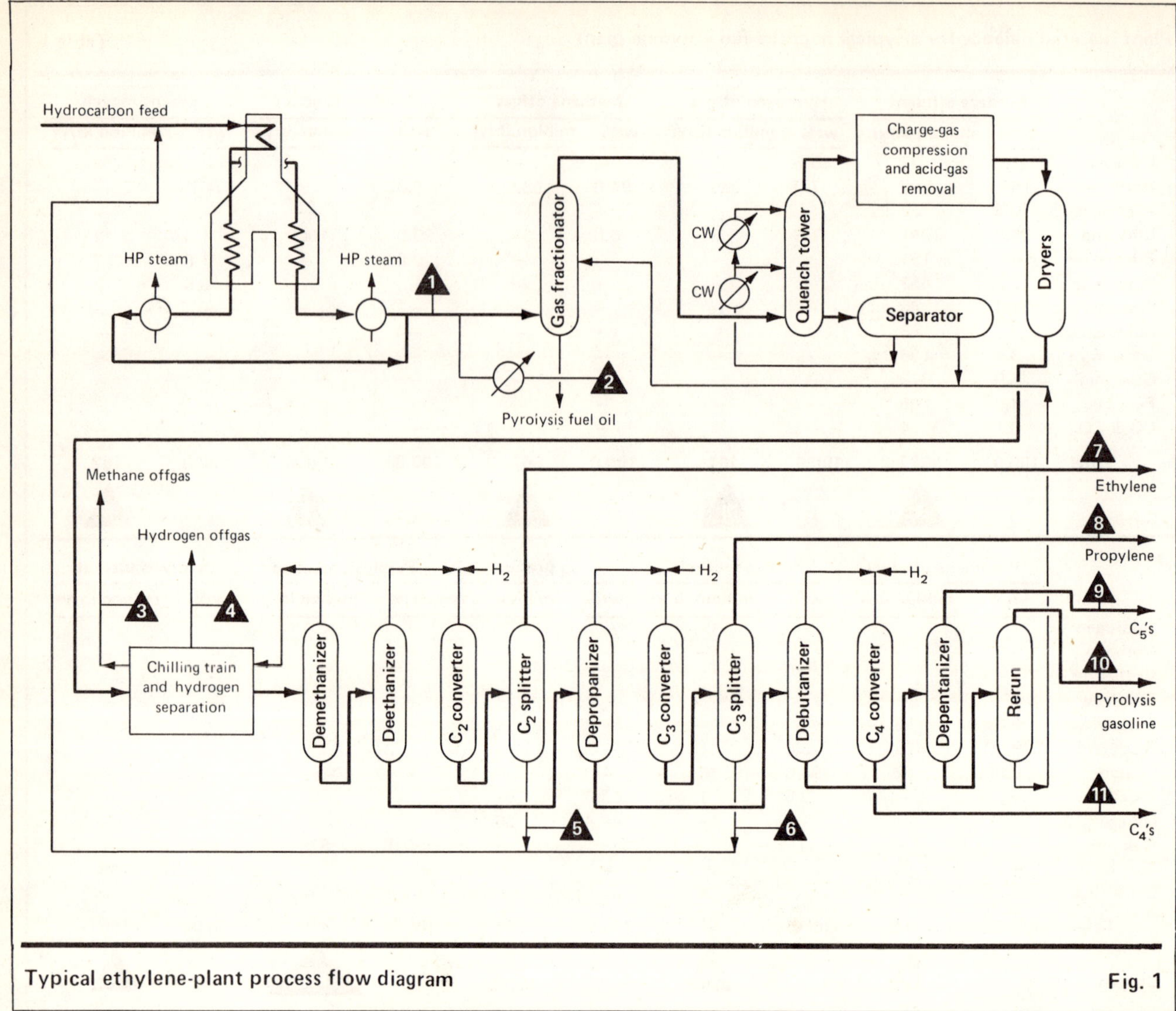

Typical ethylene-plant process flow diagram **Fig. 1**

the bottoms stream just below its maximum allowable value, while keeping the ethylene loss in the methane offgas at the value specified by a higher-level control strategy or by the plant operators. As shown in Table III, the computer should reduce the ethylene loss from 6.0 to 5.0 wt%, with a corresponding benefit of $342,000.

Similarly, in the deethanizer, the computer adjusts the setpoints of the bottoms takeoff rate and reflux rate to maintain the ethane-to-propylene ratio in the bot-

Increased ethylene-plant revenue from regulatory computer control of the recovery section **Table III**

Item	Product recovered	Normal loss without computer, wt %	Expected loss with computer, wt %	Increased product, million lb/yr	Incentive value, ¢/lb	Increased revenue, $1,000/yr.
1. Hydrogen offgas	Ethylene	7.5	5.5	2.1	6.0	126
2. Methane offgas	Ethylene	6.0	5.0	5.7	6.0	342
3. Deethanizer overhead	Propylene	1.0	0.6	4.7	1.0	47
4. Acetylene converter	Ethylene	—	—	7.0	1.0	70
5. C_2 splitter bottoms	Ethylene	1.6	0.3	2.4	1.0	24
					Subtotal	609
6. Reduced operating cost (2% reduction)						200
		Total benefit — recovery section				809

Total revenue from overall plant throughput — **Table IV**

Product	Rate, million lb/yr	Value ¢/lb	Annual revenue, $1,000/yr
Hydrogen offgas	104	4	4,000
Methane offgas	567	4	23,000
Ethylene	1,000	10	100,000
Propylene	443	8	35,000
C_4's	293	4	12,000
Pyrolysis gasoline	879	6	53,000
Pyrolysis fuel oil	165	2	3,000
Total product	3,451		230,000
Feed	3,451	5	(173,000)
Operating costs		(1¢/lb C_2H_4)	(10,000)
Total revenue			47,000

toms stream just below its maximum allowable value, while keeping propylene losses in the overhead stream at the target value—again specified by a higher-level control strategy or by the plant operators. The computer should reduce the propylene loss from 1.0 to 0.6 wt%—worth $47,000 as illustrated in Table III.

For the acetylene converter, the setpoint of the flow-rate controller for the hydrogen and combined ethane-ethylene streams is adjusted in a feedforward manner to maintain the desired target value of the hydrogen-to-acetylene ratio (moles) in the converter feed. The target value is periodically recalculated according to the acetylene content of the converter exit stream. In addition, the feed inlet temperature is adjusted as a function of the calculated hydrogen-to-acetylene mole ratio in order to compensate for changes in the catalyst activity.

These control actions keep the acetylene content of the C_2 stream entering the fractionator just below its maximum allowable value, while minimizing undesired hydrogenation of ethylene to ethane in the converters, and reducing the amount of methane impurity introduced into the system at this point via the hydrogen stream. With this strategy, it is expected that the average inlet hydrogen-to-acetylene mole ratio can be reduced from 2.0 (normal plant performance) to 1.7, cutting the ethylene hydrogenation rate by 7 million lb/yr, yielding a $70,000 benefit.

Finally, for the C_2 splitter, the computer adjusts the setpoints of the production rate (side-stream), distillate rate, and boilup rate so as to maintain the impurity content of the side-stream just below its maximum allowable value, while maintaining the ethylene content of the ethylene recycle stream at the value specified by a higher-level control scheme or by plant operators. As indicated in Table III, the computer should reduce the ethylene loss from 1.6 to 0.3 wt%—worth $24,000.

Because the feedforward control strategies include logic to minimize the utilities, operating costs should also be reduced. Typically, it is found that the overall operating cost can be cut by about 2% through computer control of the recovery section. Using an operating cost of 1¢/lb of ethylene product, the increased revenue is $200,000 per year, as shown in Table III. Therefore, the total annual benefit for the recovery section comes to $809,000.

No credit has been taken for other recovery-unit operations, such as the C_3 converter, C_3 splitter, or the gasoline fractionator. These areas should be evaluated for additional benefits, depending upon which flow configuration is used.

Furnace-section benefits

For this study, three strategies were adopted for controlling the furnace section: residence-time control, conversion control, and constraint control. These regulatory strategies are run in an interactive manner, supplying setpoints for steam, fuel and hydrocarbon rates, which will maximize furnace profit.

Residence-time control maintains a constant furnace exit-velocity (operator-entered setpoint) by using the hydrocarbon rate, and the coil-outlet temperature and pressure to adjust steam rate. Control of the furnace exit-velocity at a constant, maximum-allowable rate minimizes spalling and, when combined with conversion control, tends to maintain constant yields. Conversion control calculates adjustments to the fuel rate to maintain a target conversion level.

For residence time and conversion control, the expected yearly increase in plant throughput revenue is 1.0%, and expected reductions in plant operating costs are 1.5%. Throughput revenue is calculated in Table IV, and a total operating cost is also shown. Benefits deriving from these two areas are $470,000 and $150,000, respectively.

Constraint control maximizes the furnace hydrocarbon feed, subject to certain constraints related to the furnace and the entire plant. Some typical ones:

Furnace constraints	Plant constraints
Tube-wall temperature	Feed availability
Coil-outlet temperature	Compressor capacity
TLE (transfer-line exchanger) outlet temperature	Tower flooding
Fuel supply rate	Reboiler capacity
Gross furnace heat-release	

For constraint control, the expected increase in plant throughput revenue and the reduced plant operating costs are likewise estimated to be $470,000 (1.0%) and $150,000 (1.5%), respectively. This brings the total furnace-section benefits to $1,240,000.

Plant optimization

The two most common optimization methods are furnace optimization using nonlinear techniques, and overall plant optimization using linear programming. The former incorporates a mathematical model of the furnace, an objective function, a simple recovery model, and major constraints due to equipment characteristics, management policy, ambient conditions and degradations of unit efficiency. A general-purpose nonlinear program is used for furnace optimization.

Overall plant optimization draws on linear equations to represent operating yield, recovery level and utility consumption. An objective function and the major process constraints are incorporated into the gen-

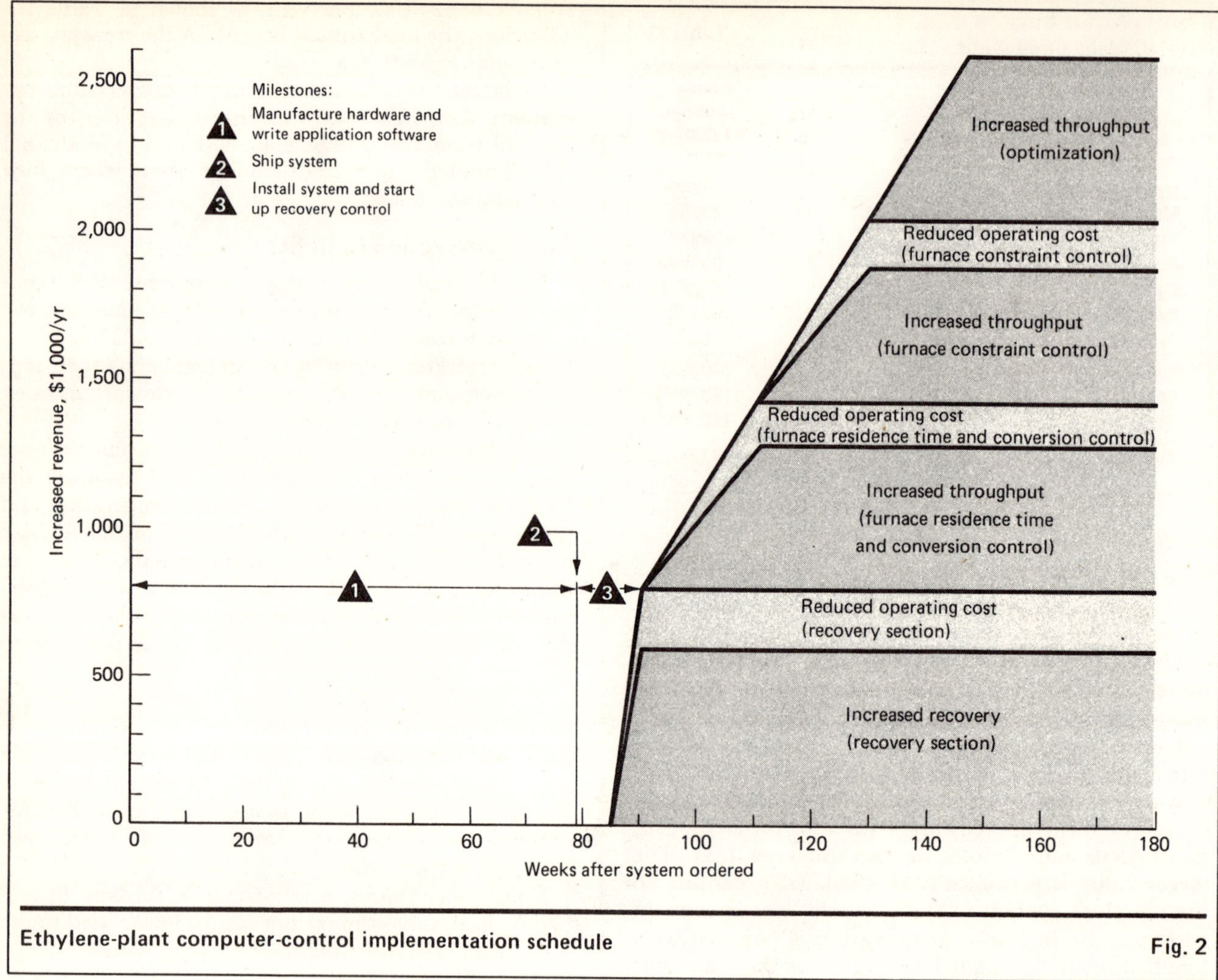

Ethylene-plant computer-control implementation schedule **Fig. 2**

eral-purpose linear programming package. Plant optimization can assist feedstock selection, evaluate operations for a desired product mix, and provide online reviews of plant performance.

Increases in total plant throughput of 1–5% are typically obtained with furnace and overall-plant optimization. A conservative value of a 1% increase in plant throughput is used for this evaluation, resulting in an annual benefit of $470,000.

When this benefit is added to those achieved by previously described improvements ($809,000 for regulatory control in the recovery section and $1,240,000 for regulatory control in the furnace section), the total annual project benefit comes to $2,519,000.

System cost

To implement the improvements defined above, the system used was a typical, large-sized process-control computer with core, drum and disk configuration, analog and digital multiplexers and hardcopy, and CRT interfaces. The total capitalized price of the system was estimated at $825,000. This includes $475,000 for standard digital hardware and standard software, plus $350,000 for application programs, data-base generation and system commissioning.

Traditionally, the application software has been implemented by the computer supplier, software houses, or the ethylene plant's own engineering organization. More recently, engineering firms who design and build ethylene plants have formed their own software development and project implementation groups in order to supply such systems with the plant [7]. These alternatives must be evaluated according to the particular circumstances involved.

Certain other costs are not included in this analysis and must be evaluated-for over the entire project payout. These include items such as personnel training, installation, added control-room space, and any additional analog equipment. Costs for these should be evaluated on a per-project basis.

Cashflow analysis

Fig. 2 shows the project implementation schedule. Delivery time for the system was estimated at 78 weeks from order entry, and cost benefits were assumed to begin about 12 weeks later. It was also assumed that the proposed applications would be commissioned starting with the recovery area, then the furnace area, and ending with the plant optimization. Other schedules could be used; however, this schedule has the advantage that the applications are placed in service starting with the least-complicated strategy. The 78-week delivery is a

Ethylene-plant computer system cashflow analysis. (All cash values are in thousands of dollars) Table V

End of year	Capital expended[a]	Gross revenue[b]	Operating cost[c]	Depreciation[d]	Pretax profit[e]	Taxes[f]	After-tax income[g]	Cashflow from operations[h]	Net annual cashflow[i]
1	495	0	0	0	0	0	0	0	(495)
2	330	315	0	52	263	131	132	184	(146)
3	0	2,024	46	103	1,875	937	938	1,041	1,041
4	0	2,519	100	103	2,316	1,158	1,158	1,261	1,261
5	0	2,519	100	103	2,316	1,158	1,158	1,261	1,261
6	0	2,519	100	103	2,316	1,158	1,158	1,261	1,261
7	0	2,519	100	103	2,316	1,158	1,158	1,261	1,261
8	0	2,519	100	103	2,316	1,158	1,158	1,261	1,261

Notes:

[a] Total system price = \$825,000 with payments of 30% at 6 mo, 30% at 12 mo and 40% at 18 mo after order entry. (System delivery = 18 mo.)

[b] From Fig. 2.

[c] 2 men at \$50,000/year each for the purpose of maintaining the system. (Years 1 and 2 no maintenance, and year 3 partial maintenance.)

[d] Eight-year, straight-line depreciation. Depreciation = 0.125 X accumulated capital expended X % of year in service. Salvage value assumed to be 0.0%.

[e] Pretax profit = gross revenue less operating cost and depreciation.

[f] Income taxes assumed to be 50%. Taxes = 0.5 X pretax profit.

[g] After-tax income = pretax profit less taxes.

[h] Cashflow from operations = after-tax income plus depreciation.

[i] Net annual cashflow = cashflow from operations less capital expended.

conservative estimate. It was assumed that, during this time, the programs for all three application areas would be developed. Another possible plan would be to develop only the recovery area before system shipment and develop the other programs at the plant. This would have the advantage of reduced delivery time and, therefore, earlier benefit return.

Table V shows the project cashflow analysis. The financial analysis was made using the internal rate-of-return, discounted cashflow method (*DCF*), with interest rates from Ref. 8. This method was employed because it provides a basis for project comparison that accounts for the time value of money.

For this project, the *DCF* (interest rate at which the net present-value—the sum of the discounted net annual cashflows—equals zero) works out to be approximately 100%. This is a very sizable return in view of the fact that many computer projects are justified with *DCF*s close to 30%.

There are other factors not included in Table V that if applied to the analysis would yield a still higher *DCF*. For example, if the plant were built in the United States, this project could qualify for investment tax credit, thus decreasing income tax and increasing net cashflow. Additionally, straight-line depreciation was assumed. An accelerated method such as sum-of-the-years digits would improve the cashflow and yield a higher *DCF*.

References

1. The International Petroleum Encyclopedia, The Petroleum Publishing Co., Tulsa, Okla., 1976, pp. 232–248.
2. 1977 HPI Market Data, Gulf Publishing Co., Houston, Tex., Aug. 1976.
3. Maddock, M. J., and Dorn, R. K., Ethylene Processes Profit From SRT Heaters, *Oil and Gas J.*, Feb. 1, 1971, p. 65.
4. *Hydrocarbon Processing*, Nov. 1975, p. 127.
5. Fauth, C. J., and Shinskey, F. G., Advanced Control of Distillation Columns, *Chem. Eng. Prog.*, June 1975, pp. 49–54.
6. *Ibid.*, p. 52.
7. Asgari, Maxwell, and Messina, John B. J., "Status Report On Computers In the Process Industries . . . From A Contracting Engineer's Point of View," presented by C. E. Lummus Co. at the VII Interamerican Congress of Chemical Engineering, Caracas, Venezuela, July 13-16, 1975.
8. "Financial Compound Interest and Annunity Tables," 5th ed., Financial Publishing Co., Boston, Mass., 1972, pp. 733–734.

The authors

James L. Hammett, Jr., is Senior Systems Sales Application Engineer (Chemical Industry) at The Foxboro Co., Foxboro, MA 02035, and is responsible for control-system market development. He has prior experience with Exxon Research and Engineering Co., where he developed process control systems for ethylene and other petroleum and petrochemical processes. An M.B.A. candidate at Babson College in Wellesley, Mass., he received his M. S. in mechanical engineering and control from Cornell University and a B. S. in mechanical engineering from Worcester Polytechnic Institute. A registered P. E., he belongs to the Instrument Soc. of America.

Lisle A. Lindsay is Principal Product Development Engineer with The Foxboro Co., managing the development of a new large computer system to be used for industrial process control. He has also managed the installation of several computer systems used to control ethylene plants. While employed at C. F. Braun & Co., he designed ethylene plants and was involved in developing pyrolysis technology. A graduate of the University of California, he holds a B. S. in chemical engineering and has taken graduate courses in business management for technical personnel through that school's extension divisions.

Justifying a Minicomputer For Process Control

A process computer easily duplicates—even easily exceeds—conventional instrument control. Even so, an exact financial justification remains elusive. But the minicomputer makes a qualitative justification more acceptable.

JOHN R. HILEMAN, Hercules Inc.

How can a manager know whether or not his engineer is being overly enthusiastic when he urges investing money in a minicomputer for process control?* Unfortunately for the manager, he cannot become an instant computer expert. But, if he is to keep his plant modern and his costs down, he must face the problem.

Exploiting Computer's Full Potential

It was in the face of such a problem that managers of Hercules' Central Engineering Dept. early in 1970 authorized a trial installation of a minicomputer for the primary purpose of gaining the experience that would serve as the basis for evaluating any such proposal.

The installation, which now directly controls a distillation column, proved that computer control can be reliable, maintained at reasonable cost, superior to conventional instrumentation, and acceptable to operations personnel. It also provided information on investment costs for both software and hardware, and programming experience useful for other applications.

The complex control techniques that the computer made possible opened the door to better product quality, a higher production rate and lower operating costs.

Whether or not a computer's maximum potential is approached depends on the programmer's ability and how well the process is understood. But even without accurate process models, a clever programmer can, by trying different control techniques, bootstrap his way to better product quality, higher production and lower costs.

Because the computer's maximum benefits must be won through the fullest exploitation of its power, it is important to have a feeling for its capability. The example that follows will show how easily a computer can duplicate conventional analog control. A subsequent example will go further and illustrate how easily computer control can exceed the practical limits of conventional analog control.

After the computer's power has been demonstrated, economic considerations will again be taken up; that is, the justification for installing a process-control computer.

Duplicating Analog Control

If a digital computer only duplicates a control scheme that could be set up with analog controllers, the programmer has overlooked the computer's power, and less than maximum benefits can be expected.

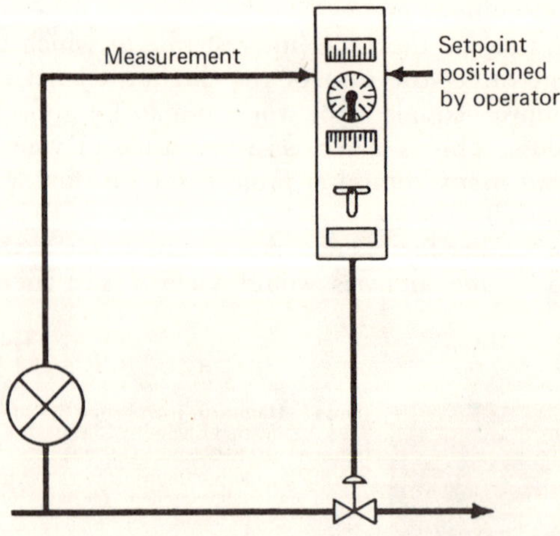

CONTROLLER is arranged for feedback control—Fig. 1

*A minicomputer is defined as a computer that costs from $5,000 to $20,000, not including support equipment. The trend in process control is to mini- and small-scale computers, with the cost of the total system (computer with interface and support equipment) ranging from $40,000 to $140,000.

Originally published May 29, 1972.

Conventional analog control consists of such instruments as the on-off controller (to cut off the filling of a vessel at a high level), and the usual proportional controllers: (1) straight-proportional (such as gas-pressure regulators); (2) proportional-plus-integral (such as steam-flow controllers); and (3) proportional-plus-integral-plus-derivative (such as reactor-temperature controllers).

Generally, these controllers are arranged for feedback control, as in Fig. 1. If the measurement does not match the setpoint, the error is used to modify the output signal to the control valve. The valve's movement changes the process, and the effect is fed back through the process to correct the measurement. The controller works continuously in this way to make the measurement equal to the setpoint.

Three-Mode Analog

The example will show how simply the computer can duplicate the most complex of the aforementioned controllers: the three-mode, or porportional-plus-integral-plus-derivative, controller.

The equation for an ideal three-mode controller is:*

$$P = K_c\left(e + \frac{1}{T_R}\int edt + T_D\frac{de}{dt}\right) \qquad (1)$$

In this equation, P = controller output; K_c = controller gain ($100/K_c$ = proportional band); e = error (measurement minus setpoint); T_R = integral (or reset) time, min./repeat; T_D = derivative (or rate) time, min./repeat; and t = time, min.

A glance reveals that the equation breaks up into three parts: a proportional, an integral, and a derivative term. Fig. 2 shows graphically the effect of these three terms (or modes) on the controller output. Most analog controllers are not ideal; there is interaction from one mode to another (e.g., changing integral time affects derivative time). However, the computer can match the ideal analog-controller equation, Eq. (1).

Three-Mode Digital

The computer system, using direct digital control, can take the place of the controller in Fig. 1. A set of instructions in the computer program tells the computer to periodically perform the following computation:

$$\Delta P - K_c\left[(e_0 - e_1) + \frac{1}{T_R}(e_0)(\Delta t) + T_D\frac{(m_0 - 2m_1 + m_2)}{\Delta t}\right] \qquad (2)$$

In this equation, ΔP = incremental change in controller output; e_0 = error at present time; e_1 = error at previous check (10 sec. ago); m_0 = measurement at present time; m_1 = measurement at previous check (10 sec. ago); m_2 = measurement at second-previous check (20 sec. ago); and Δt = time interval between checks (10 sec.).

To make the computation, the computer gathers the following data from its memory: (1) the value of the measurement 20 sec. ago, (2) the value of the measurement and the error 10 sec. ago, and (3) the value of the measurement and the error at the present moment.

*Harmott, P., "Process Control," McGraw-Hill, 1964, p. 12.

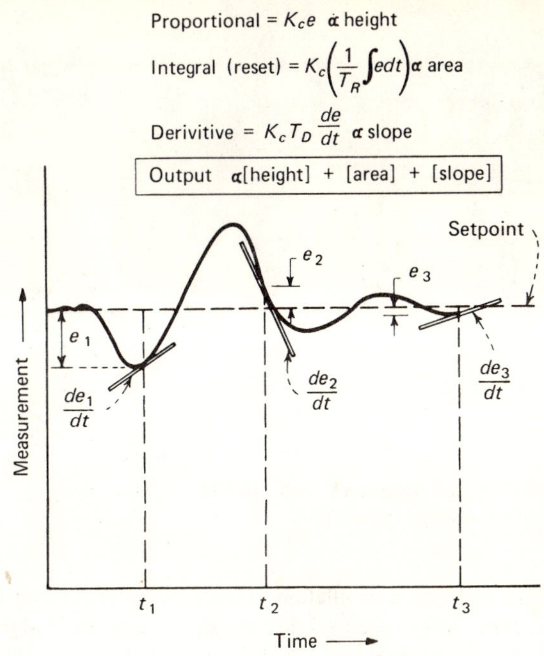

OUTPUT shows effect of three-mode control—Fig. 2

It plugs the data into the equation and computes the change in the valve output. (Note that both the controller and the computer equation contain three terms: a proportional, an integral, and a derivative term.)

By means of simple mathematics, the equation used by the computer can be derived from the ideal equation; the equivalent terms are shown below:

	Proportional	Integral	Derivative
Ideal equation	$K_c e$	$\frac{K_c}{T_R}\int edt$	$K_c T_D\left(\frac{de}{dt}\right)$
Computer equation	$K_c(e_0 - e_1)$	$\frac{K_c}{T_R}(e_0)(\Delta t)$	$\frac{(K_c T_D)(m_0 - 2m_1 + m_2)}{\Delta t}$

An examination of the equivalent terms will reveal that the term used by the computer is the first derivative of the term used by the analog controller; for example, to get the computer's integral term from the controller's integral term, just drop the integral sign from the controller's term.

Because the computer's control equation can be derived from the controller equation, the two control methods are mathematically equivalent. So, the first objective has been accomplished. It has been shown that the computer can easily duplicate the most ideal form of the three-mode analog controller.

Exceeding Analog Control

Now it will be shown how easily a computer can go beyond conventional analog-control techniques. The example will be kept simple (the point is not difficult to make).

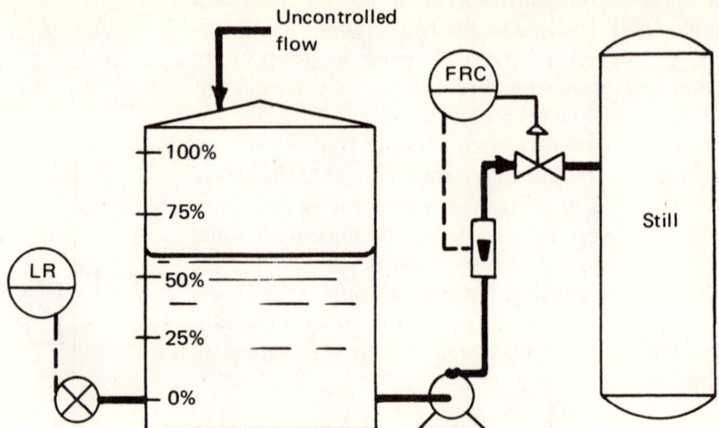

CONTROLLER regulates feed flow to still from storage tank—Fig. 3

Fig. 3 shows a distillation column being fed from a feed tank. A feed pump, a flow meter and a control valve are in the line between the discharge of the feed tank and the inlet to the feed tray. The feed tank, in turn, is fed by a line from a process unit. Because the feed to the tank can vary considerably, it is advantageous to have a control system that will minimize the changes in the feedrate to the column by allowing the level in the tank to vary within limits. Because a step change in feedrate can upset a distillation column, the control system will ramp the feedrate to the new setting whenever a change is made; however, even these ramped changes are to be kept to a minimum.

The scheme will control as follows:

1. As long as the level in the feed tank is above the 25% mark and below the 75% mark, there will be no change in the feedrate to the still.

2. If the level is below the 25% mark and the tank's net flow* is negative (tending to make the level drop even further), the feedrate to the still will be slowly decreased. However, if the net flow is positive (tending to let the level rise), the feedrate to the still will not be changed.

3. If the level is above the 75% mark and the net flow is positive (tending to make the level rise further), the feedrate to the still will be slowly raised. However, if the net flow is negative (tending to let the level drop), the feedrate to the still will not be changed.

4. The feedrate should have a high limit (a maximum rate of flow) and a low limit (a minimum rate of flow).

Computer Flowchart

The next step is to translate this description into a flowchart (Fig. 4). Such flowcharts are essential to process management because they force the computer programmer to document what is in the computer in a fashion that can be easily understood by managers. With such flowcharts, they can dictate changes, additions or deletions to the program to suit themselves.

*Net flow = constant (level now − level at last check); or, NETFLO = C (LVLNOW − LVLLAST). For a given time interval, net flow is proportional to the change in level.

A quick run through several paths of the flowchart will illustrate its use.

To start, the programmer decides to have the computer do its job every 30 sec. In the first diamond-shaped "decision box," the question is asked, "Is the last change in the feedrate complete?" (If the computer had previously decided to change the feedrate to the still, such a change was to be made slowly, i.e. ramped from the old value to the new.)

If the ramping is still going on, the answer will be "no," so exit is from the righthand side of the box. This path leads to two rectangular-shaped "operation boxes," which update the computer's memory before proceeding to the next job.

Specifically, the first box replaces the last level-measurement data (LVLLAST) with the level at the moment, so that, on the next check, the computer will have valid data. Also, a counter is reset to zero (the purpose of the counter will be explained in the discussion of the next path).

If the answer to the first decision is "yes," the lefthand path leads to a decision box that asks, "Is the counter reading 10?" It permits a change in flowrate only once every 300 sec. (30 sec. times 10). If 300 sec. have not elapsed, the "no" path to the left is taken (and the counter incremented) to the next task.

If, on the other hand, the counter reads 10 (300 sec. have passed), the "yes" path to the right is taken.

The program is now into the heart of the control scheme. After the counter is reset to zero (for the next time around), NETFLO is computed and a battery of questions are asked:

- Is the level above the 25% mark?
- Is the level below the 75% mark?
- Is NETFLO positive?
- Is NETFLO negative?

If everything is normal (the level between the 25% and 75% marks), the program keeps passing through the decision boxes to the left side of the flowchart (i.e., on the "yes" lines) and proceeds to the next task without making changes.

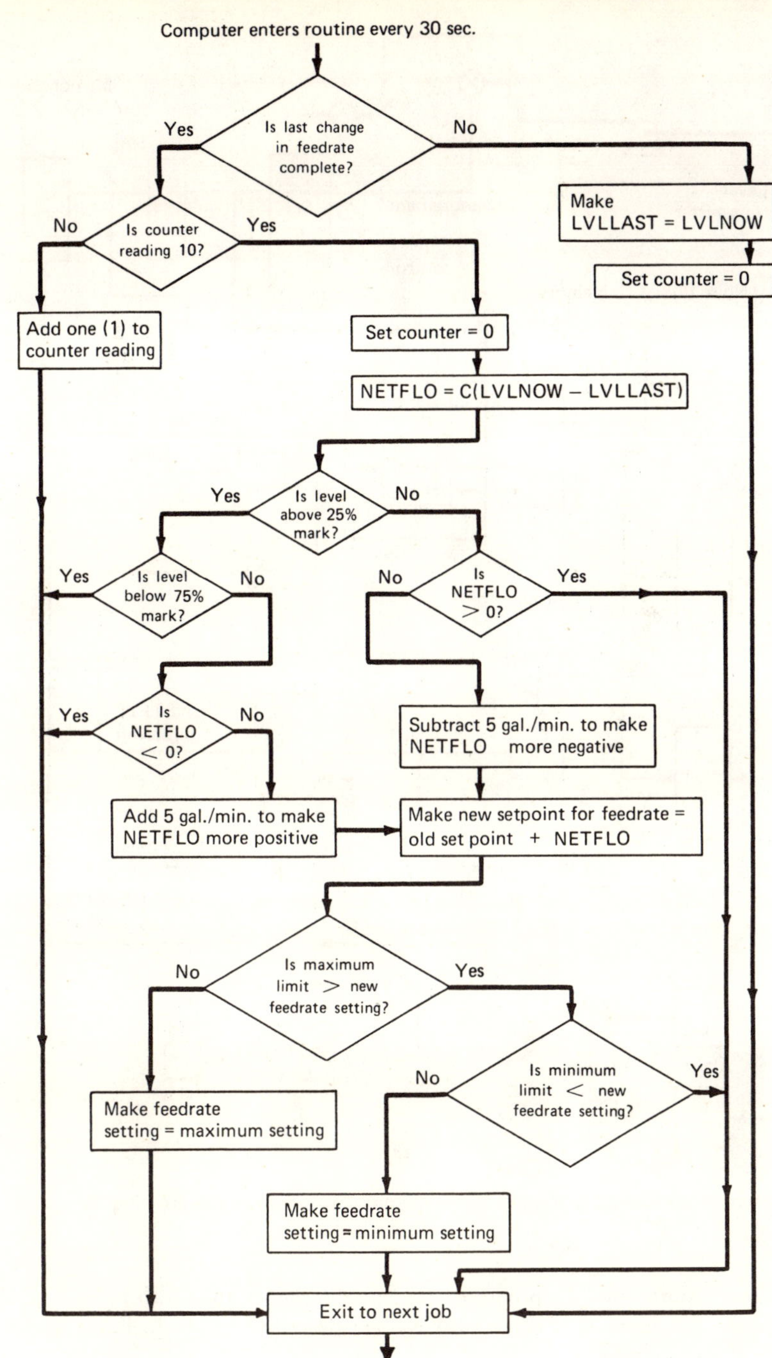

FLOWCHART depicts the computer control scheme of the process described in a manner that can be easily understood by personnel who are not expert in computer programming, enabling them to make changes in the program—Fig. 4

However, if the level is not normal, a "no" line from one of the boxes is taken to boxes that will check NETFLO and reset the feedrate (within limits) if necessary.

After the programmer has completed the flowchart and it has been approved, he translates the flowchart into a computer program.

The Analog Equivalent

The block diagram (Fig. 5) of the analog equivalent of the computer control scheme shows that the hardware to do this fairly simple job requires 29 devices! Without even considering the cost of mounting, wiring, calibrating and checking out these devices, it would cost roughly $6,750 just to purchase the five analog summing units, six alarm units, two multipliers, two memory units, input filter, lag unit, low-limit unit, high-limit unit, two timers and eight relays. This $6,750 for hardware is the equivalent to the 39 computer instructions.

It is unlikely that such a system would ever be used. It is too complicated. The cost of keeping such a system calibrated and in service would be prohibitive. A comparison of the 39 instructions to a computer against the 29

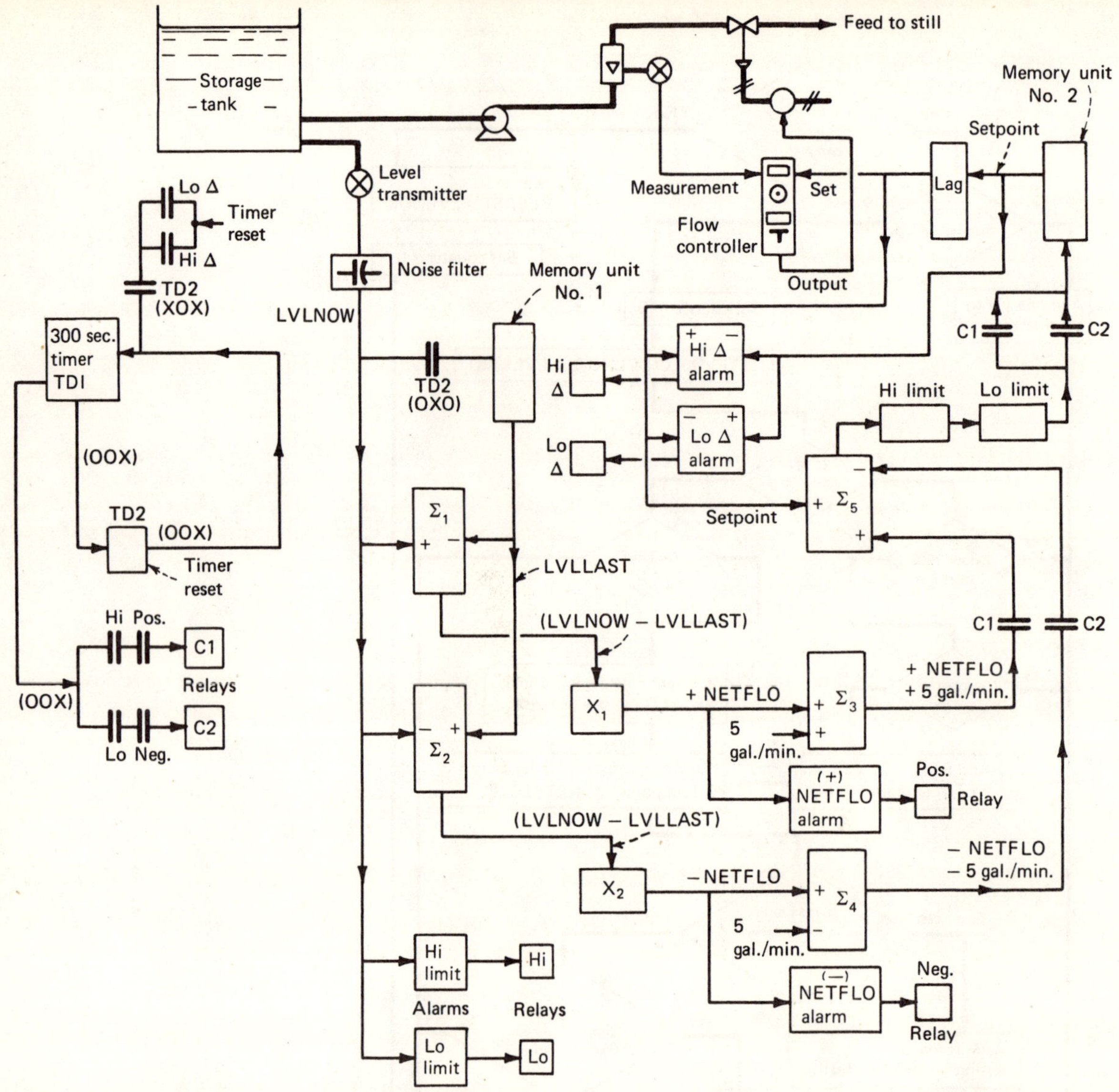

EQUIVALENT of computer control scheme would require 29 pieces of conventional instrument hardware—Fig. 5

analog devices makes clear the power of the computer. In fact, if a higher-level programming language were used, only 24 (or even fewer) instructions would be needed.

What gives the computer such an advantage over normal analog devices? The answer lies in its arithmetic, logic and memory capabilities.

Its logic capability is involved in such decisions as: Is the level less than 75%? Is the newly computed feedrate setting greater than the maximum limit?

Its arithmetic capability is used to compute NETFLO (subtraction and multiplication).

Its memory capability is relied on to accurately store such data as: the counter reading (how many seconds have elapsed); the present level reading; and the level reading 300 sec. ago.

In the beginning, an envisioned control scheme may not work too well in practice. The control scheme described here was modified by a change in the computer program after installation. It ended up being similar to the control system described by A. L. Pekar and J. H. Sayles.*

What is important, however, is not the details but rather that a change was made. Such a change in a hard-wired analog system would have required ordering new devices, making field revisions and updating construction drawings. With the computer, however, the change was quickly and easily made without any hardware alterations.

Return on Investment

It would be helpful to have an accurate appraisal of each of the benefits possible from a computer installation

*Pekar, A. L. and Sayles, J. H., Computerized Material Balance Achieves Level Control, *Instrumentation Technology,* December 1971.

before it is approved. For example: How much will product quality be improved, yield increased, production boosted, maintenance costs raised or lowered, safety and process stability enhanced, and money saved on fuel, power and raw materials?

Of course, to determine the return, the cost of the computer installation must also be known, including expenditures for purchasing and installing the hardware and developing the software.

Generally, hardware costs are easy, software costs difficult, and the value of benefits next to impossible, to estimate.

The Benefits

Benefits are so difficult to estimate because simply applying the computer to the process will uncover half-forgotten data, prompt the collection of additional data, and give new insight into the process. This better understanding can save money.

The benefits to be gained from the preliminary design work on an application vary, of course, with the degree of sophistication about process control. If good analyses already have been performed, a study will not likely point the way to significant savings. In such cases, a process computer usually can be justified only in terms of its online regulation or optimization functions.

Some processes present difficult control problems because of frequent disturbances, complex interactions between process variables, and continually changing plant characteristics. The computer's ability to deal rapidly with such variables, and the ease with which its operations can be modified, often permit it to be the source of significant benefits in the form of increased production, reduced quality variations and lower operating costs.

Data usually will be better because of more-careful instrument maintenance. Process control will be tighter. Averaging, and other data-enhancing, techniques can be used. Also, experiments can be run more easily and less expensively.

More On Flexibility

A discussion of benefits would be incomplete without further commenting on the computer's flexibility. Changing a computer control scheme is not difficult. On the other hand, changing an analog control scheme requires the addition, deletion and change of equipment, wiring and tubing—which must be ordered and delivered, and installed by mechanics. If the new scheme does not work, it will have been an expensive, time-wasting exercise.

A computer scheme can be tried, and it is does not work, it can be easily replaced with a new scheme.

Still another benefit is "control enforcement." A computer can be an around-the-clock extension of a plant's best supervision. The process can be made to operate the way the plant manager, production superintendent, plant engineer, process engineer or instrument engineer wants it to. It will lessen the amount of operator attention required and will not allow operators to arbitrarily make changes. In short, a plant's supervisors can enforce proper control.

Costs of Developing the Computer System

In addition to the difficulty of accurately estimating the benefits of a process computer, there is also the problem of accurately estimating the costs of developing the computer system. It is, of course, the software that causes the difficulty.

The more intricate and innovative the application, and the more limited the user's experience, the more likely it will be that the precise requirements and detailed operating characteristics called for will take shape only after some trial. Indeed, it is only infrequently that the final system does not differ significantly from what was originally contemplated.

Furthermore, changes will continue to be made, because these operations are usually dynamic. As problems are encountered, changes must be made to rectify flaws. Certain of the original operating characteristics may later prove unnecessary or uneconomical. These must be eliminated. Also, new features will be added.

It would be pointless to go to inordinate lengths seeking explicitness and painstaking accuracy in installation plans or analyses.

The two problems (of not being able to predict all benefits and software costs) appear to force the conclusion that an accurate return on the investment generally cannot be made! If this is so, what should management's approach be?

Rather than spending a lot of time and money trying to pinpoint savings, gains and costs, management should gamble if rather broad studies show the odds to be favorable. Rather than trying to economically justify the computer by estimates made from costly preliminary studies, its installation should be approved if the application is likely to succeed. It would then be the job of the engineers to make it pay off.

The minicomputer makes the decision easier. Although large sections of a plant may be adequately controlled with analog instruments, there may be a process unit that, for example, is extremely critical or complicated, that demands too much operator attention, or that often turns out a poor product. If broad studies by process and control engineers indicate that the unit could benefit from computer control, it is easier to gamble $40,000 to $80,000 for a minicomputer system than a much larger expenditure for a computer installation for the entire plant.

Meet the Author

John R. Hileman, a senior engineer with Hercules Inc. (Wilmington, DE 19899), handles project instrumentation from design and procurement through installation and startup. He is now assigned to the world's largest dimethyl terephthalate plant.

He previously worked for Hercules as a maintenance supervisor, for Foxboro as a sales engineer, and for Du Pont as a production group supervisor. He holds a B.S. in chemical engineering from Cornell University, and is a senior member of the Instrument Soc. of America.

Section III
THEORY OF INSTRUMENTATION AND CONTROL

Fundamentals of control theory

Understanding the process dynamics is essential for applying automatic control to process plants. Here is detailed information on how to select the operating mode for controllers in actual plant environments, and how to tune these controllers after they have been installed.

Cecil L. Smith, Cecil L. Smith, Inc.

☐ The process control field today is more dynamic than ever before. Production plants are facing even more stringent operating constraints, precipitated by rising energy costs, limited raw-material availability, pollution regulations and personnel attitudes. When energy was cheap, distillation columns with simple controls could be conservatively operated to give gross over-separations. Energy conservation means reducing the over-separation, which reduces the margin of allowable operational errors. At the same time, sensitivity of the process to upsets increases [*1*]. The net result is that the control system that operated the column under the conservative mode proves inadequate for the new mode of operation.

Manufacturers of process-control equipment have responded by upgrading their offerings, both in their analog and digital lines. To meet the new requirements, the systems must be highly flexible, yet easy to maintain. However, the real problems have not been with the hardware, but with the process analysis required for a successful application.

Many engineers have traditionally approached process control using the "cookbook" method, which ideally would have the control configuration for distillation columns on one page, for heat exchangers on another page, and so forth. This never really worked well, and is less acceptable today. In this respect, control is for the thinker and the innovator, not one seeking comfort in the status quo.

The needs of many engineers would be satisfied by enough insight to troubleshoot the routine problems that arise with their controls and to "communicate" with the control experts within the company. As many have found, process control is not an easy subject to present in "condensed form," and the connection between what is covered in many textbooks and what is practiced in the plants is obscure.

To some extent, the steady-state or equilibrium point of view (i.e., input equals output) presented in courses such as unit operations and plant designs becomes an obstacle. Process control is inherently concerned with process behavior under dynamic or unsteady-state conditions, where the accumulation terms in mass and heat balances cannot be ignored. The added dimension has many side-effects, such as new terminology (time constant, gain, etc.), more complex mathematics (differential equations and Laplace transforms), and others.

The focus of this article will be on the basic principles of process control, which will enhance our understanding of the subject and provide a basis for topics such as mode selection and tuning. To be useful, our discussion must encompass several supporting topics—an important one being process dynamics. The level of mathematics will be kept as low as possible, and potentially unfamiliar topics such as Laplace transforms are included as necessary. Although the article will end with a treatment of controller tuning, the selection of topics is such that the reader should be able to better understand other material on process control.

Example of a controlled system

Fig. 1 illustrates about as simple a process system as we can imagine—a hot-water heater that mixes cold water with steam to produce hot water. Also illustrated is a temperature controller that accepts a signal from the hot-water temperature transmitter and produces a

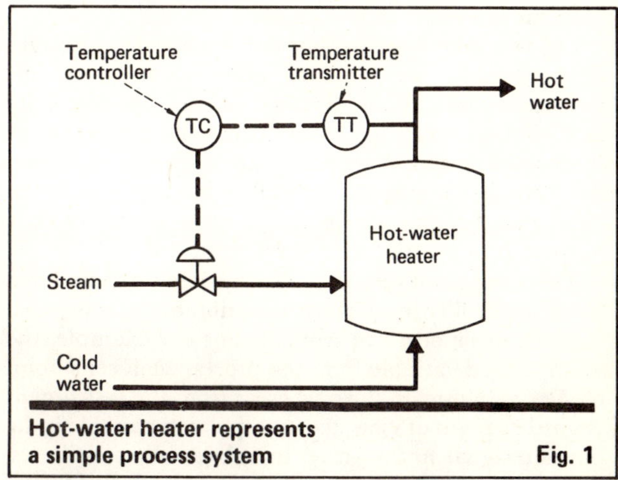

Hot-water heater represents a simple process system **Fig. 1**

Originally published October 15, 1979.

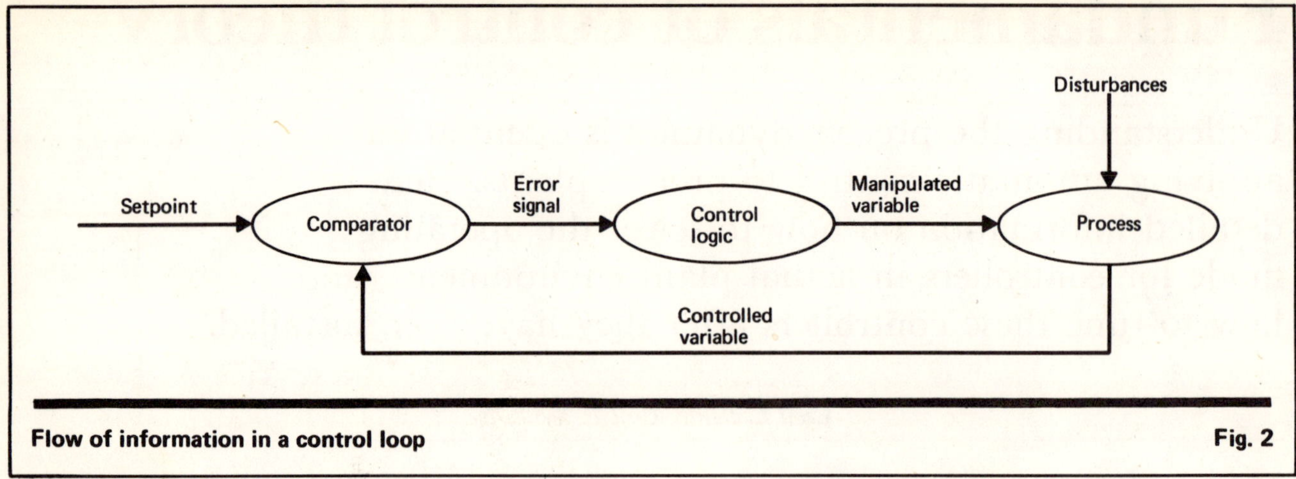

Flow of information in a control loop **Fig. 2**

signal to position the steam valve. This control configuration is referred to as simple feedback control. Fig. 1 will be used to introduce the terms controlled variable, manipulated variable, and disturbances that are prominent in discussions of process control.

From a process engineer's point of view, the *controlled variable* is the process variable that we would like to maintain at or near some desired value (usually called the setpoint). For the hot-water heater, the controlled variable is the hot-water temperature. Many instrument engineers would take issue with designating the hot-water temperature as the controlled variable. In reality, the temperature controller will maintain (or at least try to maintain) the signal it receives from the temperature transmitter at or near the setpoint for the hot-water temperature. Thus, it could be argued that the "real" controlled variable is the measured value of the hot-water temperature.

This distinction can be quite important where the measured signal is not linearly related to the process variable. Two examples are:

1. Control is based directly on the millivolt output from a thermocouple.

2. Control is based directly on the pressure-drop (ΔP) measurement across an orifice meter.

Let us remember that special nonlinear scales and chart paper are available for indicators and recorders for such signals. With electronic technology (especially in digital systems), linearization circuits are being incorporated into most systems, so the distinction between the "process engineer's controlled variable" and the "instrument engineer's controlled variable" serves little purpose beyond confusing the uninitiated.

The *manipulated variable* is that variable manipulated by the controller in its efforts to maintain the controlled variable at or near the setpoint. For our example, the manipulated variable from the process engineer's point of view is the steam flow; whereas from the instrument engineer's point of view, the manipulated variable is the position signal to the valve. In valves with positioners, the valve position will be close to the value called for by the position signal, but the relationship between the valve position and flow is frequently complex and nonlinear. This can have a significant influence on the performance of the control system.

Disturbances are those variables other than the manipulated variable that affect the controlled variable. For the hot-water heater, disturbances could include the cold-water flow, cold-water temperature, steam pressure, steam enthalpy, heat losses from the tank, and others. In most industrial control systems, there exist numerous sources of disturbances, some of which may be poorly understood even if recognized at all. Thus, the usual practice is to focus attention on the principal or "important" disturbances, which usually reduces the number considerably.

Load changes affect a control system in the same manner as disturbances, and in fact the two terms are practically interchangeable in control circles. For the water heater, the load can be thought of as the demand for hot water. Changes in the hot-water demand (or load) will result in changes in cold-water flow, which were previously listed as a disturbance.

Fig. 2 illustrates the flow of information in the temperature-control loop in Fig. 1. The relationships between the manipulated variable and the controlled variable and between the disturbances and the controlled variable are embodied in the process. In most industrial control systems, the setpoint is compared to the controlled variable to produce the error signal. Control logic is applied to the error signal to produce the manipulated variable.

Fig. 2 shows that the relationships of the various components of the control system produce a loop in the flow of information. Fig. 2 also illustrates that there are two external inputs to the loop, namely the setpoint and the disturbances. Most industrial control systems will experience inputs from both sources, i.e., the operator will change the setpoint, and external factors will produce changes in the disturbances. However, in most applications one of these will be more important to the operation of the plant than will the other.

The classical *servomechanisms* control problem considers only changes in the setpoint. This point of view is most appropriate for the batch type of process plant. A typical requirement in a batch unit is to vary the temperature in a batch reactor according to a pre-established profile with time. This is conventionally implemented by using a cam programmer to generate the setpoint for the temperature controller. Thus, the pri-

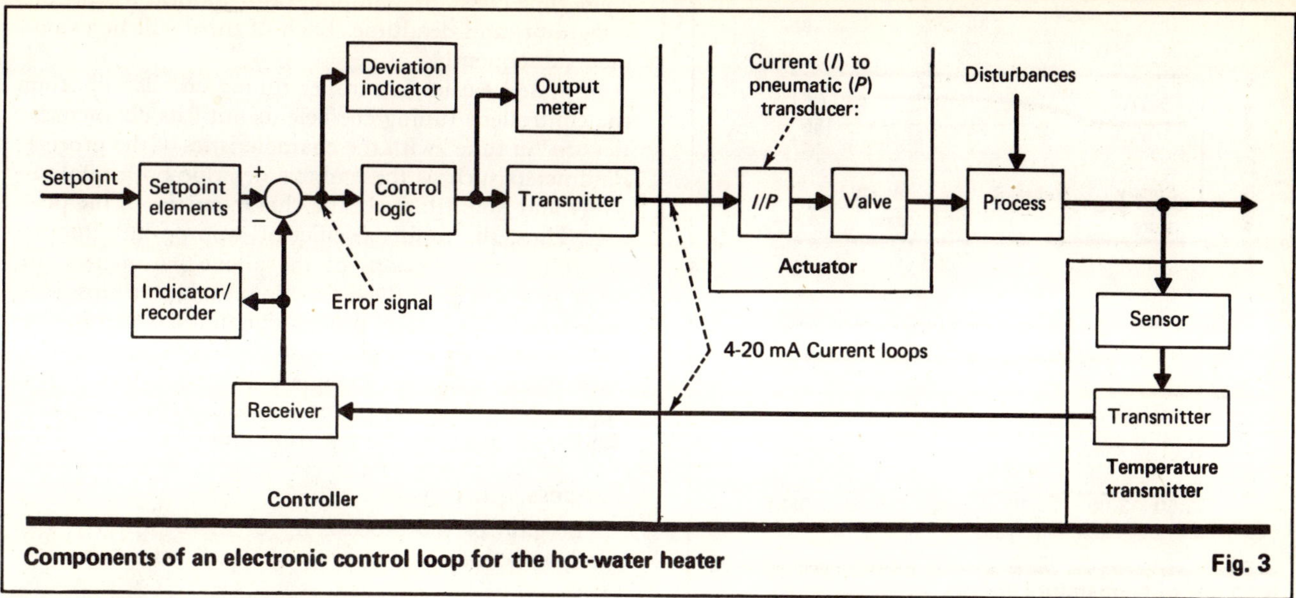

Components of an electronic control loop for the hot-water heater **Fig. 3**

mary consideration for the control system is its response to setpoint changes.

The classical *regulator* control problem considers only changes in the disturbances. Most hot-water heaters operate in this manner. The setpoint for the hot-water temperature is changed infrequently if at all, but the hot-water demand (or cold-water flow) changes every time someone opens or closes a valve. Thus, the primary consideration is the control system's response to disturbances. Control systems for most continuous processes fall into this category.

Although various approaches are possible for the control logic, its objective in Fig. 2 is very simple, namely, to drive the error signal to zero. An error signal of zero means that the controlled variable equals the setpoint, which is the desired objective. An advantage of using the approach in Fig. 2 for the control logic is that it will function for both the servomechanisms and regulator control problems, i.e., it will work for both setpoint and disturbance changes.

Fig. 3 presents a block diagram that shows in more detail the various components of the control loop for the hot-water heater of Fig. 1. The components in Fig. 3 reflect the use of an electronic controller, although the configuration would be similar for a pneumatic system. In addition to the process, the three other major components of the control loop are the temperature transmitter, controller, and actuator.

A typical temperature transmitter might use a thermocouple as the sensor or transducer to generate a millivolt signal that is related to the process temperature. The transmitter would include circuits to linearize the thermocouple signal and convert it to a 4–20 mA current signal that can be transmitted over a considerable distance.

In many controllers, the "receiver" is simply a resistor (called the range resistor) that converts the current signal into a voltage. The indicator or recorder for the process temperature would be driven by this voltage signal. The setpoint is normally specified by positioning some mechanical knob or dial. The purpose of the "setpoint elements" is to convert the position of the mechanical element to a voltage signal that can be compared to the voltage signal from the receiver. The output of the comparator is the error signal, for which many controllers provide an indicator. The control logic operates on the error signal to produce a voltage signal that represents the desired valve position. Most controllers provide an indicator, called the output meter, to display this signal. In order that the controller can be located a substantial distance from the valve, the output is converted to a 4–20 mA current signal.

As most control valves are pneumatic, the current signal is converted to a pneumatic signal by the *I/P* transducer. In many installations, the output of the *I/P* transducer is not applied directly to the diaphragm of the pneumatic valve, but instead is the input or "setpoint" to the valve positioner (not shown in Fig. 3). The valve positioner is in effect a control-loop internal to the final actuator that accurately positions the valve to the value specified by the input.

Process dynamics

If the only requirement were to purchase appropriate hardware for each component in Fig. 3 and install it properly, the instrument engineer's job would be relatively easy. Several factors complicate the job—one of which is that the control logic may require that some parameters, called tuning coefficients, be specified. The tuning coefficients essentially specify performance parameters, such as sensitivity and response speed, for the control logic. Someone has to assume the responsibility for adjusting these coefficients so that the controller's behavior (sensitivity, response speed, etc.) is "in tune" with that of the process.

The analysis of process dynamics can proceed via two approaches:

- Experimental, i.e., test the operating process in some appropriate manner.
- Theoretical, i.e., write the equations that describe the process and analyze their characteristics.

As the latter can be applied only to simple (typically

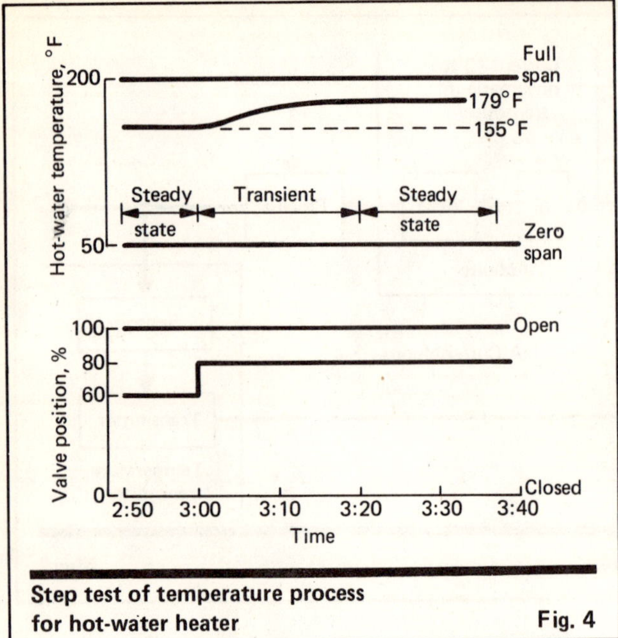

Step test of temperature process for hot-water heater **Fig. 4**

textbook) problems, the former is used in practice. A large number of options are available for conducting process tests. One of the simpler to execute proceeds as follows:

1. Place the controller on manual and hold all inputs at fixed values until the process "lines out".
2. Change the output of the controller to a new value, holding all other inputs fixed.
3. Record the response of the process.

Fig. 4 illustrates the results from such a test. An input generated by changing as quickly as possible from one value to another is called a step input. Thus, the test illustrated in Fig. 4 is frequently called a step test.

In Fig. 4, the valve position was initially at 60%, and was changed to 80%. The hot-water temperature was initially at 155°F, and eventually increased to about 179°F. Although the valve was changed from 60% to 80% in a matter of seconds, the hot-water temperature required about 15 min to change from 155°F to 179°F.

This region of the response is called the transient-response period. Prior to introducing the step change in the valve position, the process was "lined out" or at steady state with a 60% valve position giving a process temperature of 155°F. Eventually, the process reached a new steady state, with an 80% valve position, giving a process temperature of 179°F.

The conditions at the respective steady states are dictated by the steady-state equations that describe the process. For the hot-water heater, the only steady-state relationship required would be an enthalpy balance. The response during the transient region is described by the unsteady-state equations. For the hot-water heater, an unsteady-state enthalpy balance would be required.

For control purposes, certain parameters have been developed to describe behavior of processes under both steady-state and unsteady-state conditions. For the steady state, the only parameter of interest is the gain. For the unsteady state, several parameters may appear, depending upon the process. Possible parameters include time constant, damping ratio/natural frequency, integrator, and deadtime. Each of these will be examined subsequently.

As stated earlier, controller tuning entails adjusting the controller's tuning coefficients until its characteristics are "in tune" with the characteristics of the process. Parameters such as those above describe both qualitatively and quantitatively the characteristics of the process. Thus, the tuning coefficients appropriate for the controller are functions of the above parameters. In order to appreciate these functional relationships, it is necessary to appreciate process dynamics, which is why most control texts begin with process dynamics. An understanding of this topic is useful regardless of the approach used to tune controllers, i.e., seat-of-the-pants (knob-twiddling) or tuning techniques.

Process gain

The gain of the process, K, (or more properly, the steady-state gain of the process) is defined as:

$$K = \frac{\text{Change in output}}{\text{Change in input}}$$

where "change" is understood to mean the change from one steady-state to another.

For the hot-water-heater test in Fig. 4, the output is the hot-water temperature and the input is the steam-valve position. The process gain is computed as:

$$\begin{aligned} \text{Change in output} &= 179°\text{F} - 155°\text{F} = 24°\text{F} \\ \text{Change in input} &= 80\% - 60\% = 20\% \\ \text{Process gain, } K &= \frac{24°\text{F}}{20\%} = 1.2°\text{F}/\% \end{aligned}$$

This value indicates that for every 1% increase in valve position, the hot-water temperature should increase 1.2°F. If the change in input had been 30%, this number indicates that the change in output would have been 36°F.

A prudent observer would note that the value determined above is not the true process gain. This is probably best seen from Fig. 5, in which the detailed control loop in Fig. 3 has been re-drawn showing only the controller, valve, process, and transmitter. The valve position signal in the test in Fig. 4 is actually the input to the valve. Thus, the gain as computed above is that of the valve-process combination.

Using the definition of the process as per Fig. 5, the process output is the hot-water temperature, but the process input should be the steam flow. This gives the following expression for the process gain K:

$$K = \frac{\text{Change in hot-water temperature}}{\text{Change in steam flow}}$$

If the units for the steam flow were lb/h, then the units for K would be °F/(lb/h). Unfortunately, the data in Fig. 4 do not permit this gain to be computed. To obtain the data would require a measurement for the steam flow. A possible alternative would be to make some gross approximations about the valve characteristics in order to estimate the steam flow from the valve position, but this usually does not prove to be useful.

Although the above limitation normally does not permit the "true" process gain to be calculated, this proves not to be a problem. As stated earlier, the main objective in determining the process gain is for controller-tuning purposes. Again referring to Fig. 5, the "process" gain as seen by the controller is that of the valve-process-transmitter combination. In fact, it could be noted that the test in Fig. 4 is exactly for this combination, since the "hot-water temperature" is in reality the output of the transmitter. Instead of using °F for the units of the output, % of span should really be used, with 155°F corresponding to 70% of span and 179°F corresponding to 86% of span. The gain is now:

$$K = \frac{\text{Change in output}}{\text{Change in input}} = \frac{86\% - 70\%}{20\%} = 0.8\%/\%$$

which includes the valve, process, and transmitter. Although the units are usually written as %/%, this gain is actually dimensionless, which should be expected since the input and output from the controller are both 4–20 mA signals. In fact, the gain could be written as 1.6 mA/mA, or even 0.8 psi/psi for a pneumatic system.

It is possible to attach some significance to the overall gain of the valve-process-transmitter. Ideally, the lower span of the transmitter should be set at the value of the controlled variable when the valve is closed, and the upper span of the transmitter should be set at the value of the controlled variable when the valve is fully open. With these settings, a valve change of 0 to 100% would give a transmitter output change of 0 to 100%, and the "process" gain would be 1.0. Several factors make it impractical to set the span of the transmitter exactly in this way, so the actual gain will usually differ from 1.0. Some variation from 1.0 is certainly tolerable, but extreme variations may cause problems. If the process gain is in the range $0.5 \leq K \leq 2$, no difficulties will likely be experienced. In fact, many systems are operating successfully with gains outside this range, so the above limits are by no means hard and fast rules. However, if the gain is outside these limits, we should not be surprised if some difficulties are encountered.

Large and small process gains

Fig. 6 illustrates a system in which a large process gain created problems. The process was a reactor producing a product for the manufacture of explosives. The system illustrated was installed on several reactors, and was found to work fine on reactors with large heat duties but very poorly on those with small heat duties. Those with unacceptable performance exhibited a temperature cycle of approximately 10°C on a 15-min cycle. A closer examination revealed that the cooling-water valve was operating very close to its seat, and was not behaving linearly in this range. At the bottom of the temperature cycle, the valve would be closed. The exothermic reaction would slowly increase the process temperature. When the reactor temperature exceeded the setpoint, the controller would slowly increase the signal to the valve. Eventually, the signal would become sufficient to open the valve, but instead of moving off its seat smoothly, it would "jump" to a small opening. This was aggravated by the fact that the valve was not sized for the normal cooling load in the reactor, but instead was designed for emergency cooling in case of a runaway reaction. Thus, the small opening produced a large cooling-water flow relative to the heat-transfer requirements, which in turn caused the temperature to drop rather quickly and the valve to close.

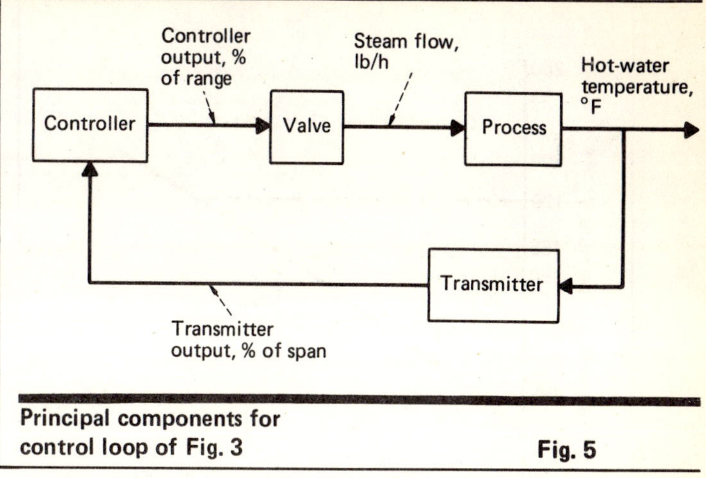

Principal components for control loop of Fig. 3 **Fig. 5**

For valves to behave in a nonideal manner when nearly closed is to be expected. The real problem in this example was not the nonideal behavior of the valve near its seat, but that the valve was grossly oversized with respect to the normal cooling duty. In fact, the gain was so large that it could not be measured (the minimum flow was sufficient to reduce the reactor temperature to below the freezing point of the product). As the large valve was required for safety considerations, it should remain, but a small bypass valve provided for the temperature controller to manipulate. This would reduce the gain to an acceptable value.

An example of a small process gain occurs in temperature controllers for a zone in an extruder. Suppose the temperature in an extruder zone normally operates in the vicinity of 500°F. Once the extruder is operating, most of the heat comes from the input of mechanical energy from the screw, and very little from the heater bands. If the heater band were left off, the temperature in the zone would perhaps fall to only 475°F. If left on, the temperature might rise to 525°F. Thus, the process gain would be 50°F/100% or 0.5°F/%.

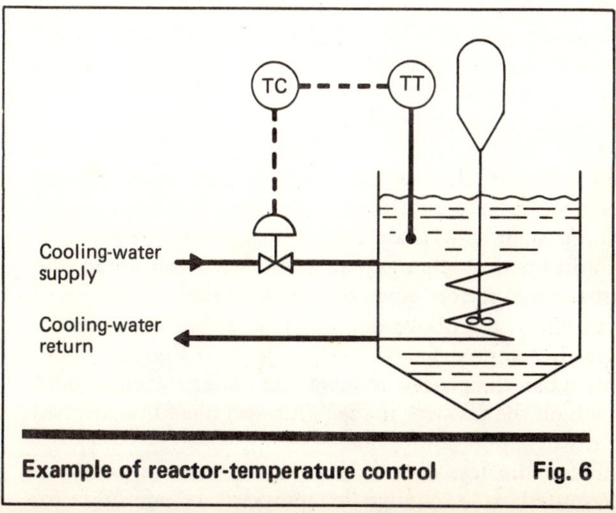

Example of reactor-temperature control **Fig. 6**

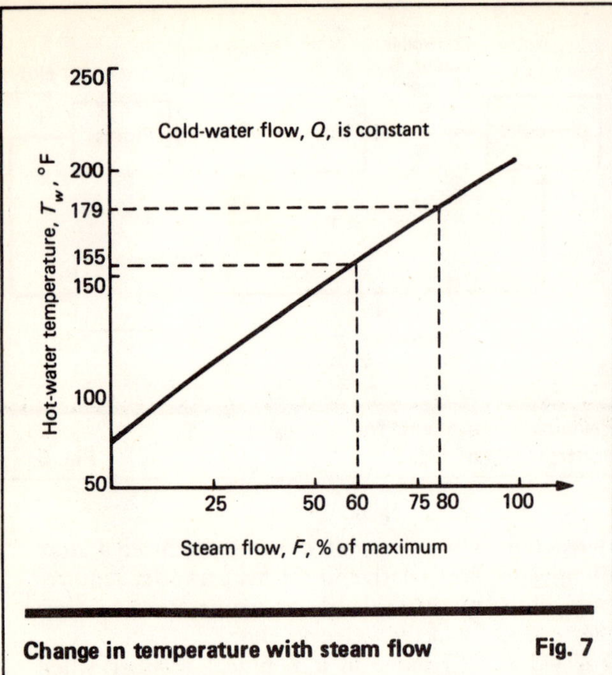

Change in temperature with steam flow **Fig. 7**

For startup purposes, the span on the controller would extend down to ambient temperatures. Therefore, a typical span would be 0°F to 600°F. Thus, a change of 50°F would be 8.33% of span, which gives a process gain of 8.33%/100% = 0.0833%/%. As the process has a very low gain, it would be rather insensitive to changes in its inputs. This must be countered with a controller with a very high gain. Most controllers have a gain adjustment of approximately 0.1 to 10, which proves inadequate for this application. Manufacturers who cater to this market typically produce controllers with a gain adjustment of approximately 1 to 100. Since the process gain is about ten times lower than normal, the controller gain is about ten times higher than normal.

Test limitations

In practice, we have to appreciate the limitations of gains determined in this manner and interpret them accordingly. In effect, the gain is a measure of the "sensitivity" of the process. In a test such as in Fig. 4, the value obtained for the gain is an average of the sensitivity of the process over the operating conditions experienced. That is, if another test were made with the steam-valve position changed from 60% to 65%, a slightly different value for the gain should be expected. Theoretically, the value could be grossly different, and even could have a different sign. Although a few applications such as pH control are highly nonlinear, most exhibit modest changes in gain, but even these can impact the performance of the controller.

In effect, the above discussion says that the value of gain determined from a test such as in Fig. 4 is valid only when the process is operating under the conditions at which the test was made. This also has to be applied to disturbances and loads.

From the test data in Fig. 4, the process gain was computed as a change in temperature of 24°F (or change in span of 16%) divided by the change in valve position of 20%. At no point was any consideration given as to why a valve position of 60% gave a temperature of 155°F. In fact, if the change in valve position had been from 10% to 20%, the value calculated for the gain would have been exactly the same.

Effect of operating parameters

In the discussion that follows, we shall make the assumption that the steam flow varies linearly with valve position. This will permit the use of the steady-state enthalpy balance to investigate the effect of certain operating parameters on the gain.

Fig. 7 illustrates the variation of the hot-water temperature with steam flow at a constant cold-water flow. Also noted are the two steady-state points determined in the test in Fig. 4. The process gain, K, (i.e., change in output/change in input) is the slope of this line:

$$K = \frac{\Delta T_w}{\Delta F} \approx \frac{\partial T_w}{\partial F}$$

Since the plot in Fig. 7 is very nearly a straight line, the gain K is practically constant for changes in the operating temperature.

Fig. 8 repeats the plot of Fig. 7, along with the plot of T_w vs. F for another cold-water flowrate, namely half of the value for Fig. 7. Again, the line has only a very slight curvature, indicating a constant gain with operating temperature. However, the slopes of the two lines in Fig. 8 are significantly different, indicating that the gain is a strong function of cold-water flow.

These conclusions can also be drawn from the steady-state enthalpy balance:

$$FH_F + QC_p(T_i - T_R) = (Q + F)C_p(T_w - T_R) \quad (1)$$

where F = steam flow, lb/h; H_F = steam enthalpy, Btu/lb; Q = cold-water flow, lb/h; C_p = heat capacity of water, Btu/(lb)(°F); T_i = cold-water temperature, °F; T_w = hot-water temperature, °F; T_R = reference temperature, °F. Heat loss to surroundings and other small effects have been neglected.

As noted above, the gain K is $\partial T_w/\partial F$. This term can be obtained by taking the partial differential of Eq. (1) with respect to F:

$$\frac{\partial}{\partial F}[FH_F + QC_p(T_i - T_R)] = \frac{\partial}{\partial F}[(Q + F)C_p(T_w - T_R)]$$

$$H_F = C_p(T_w - T_R) + (Q + F)C_p\left[\frac{\partial T_w}{\partial F}\right]$$

$$\frac{\partial T_w}{\partial F} = \frac{H_F - C_p(T_w - T_R)}{(Q + F)C_p} = K$$

Since H_F will be about 1,000 Btu/lb, the term $C_p(T_w - T_R)$ will be small in comparison, and changes in T_w will have a small effect on K. However, Q will be large with respect to F, so changes in Q will have a very large effect on the gain K. In fact, the gain K is essentially inversely proportional to Q, as evidenced by the relationship in Fig. 9.

With a little reflection on the hot-water heater, the relationship in Fig. 9 should seem plausible. Suppose a 1-in. steam valve is opened an additional 10% when the flow through the system is that of a garden hose. The

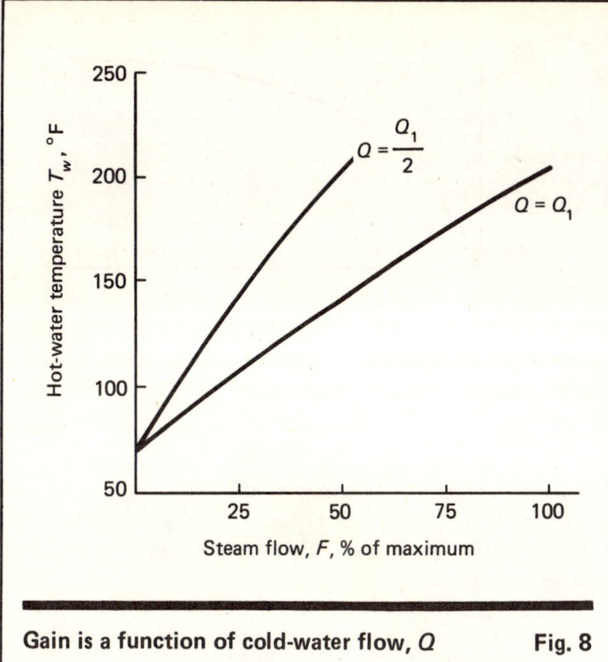

Gain is a function of cold-water flow, Q **Fig. 8**

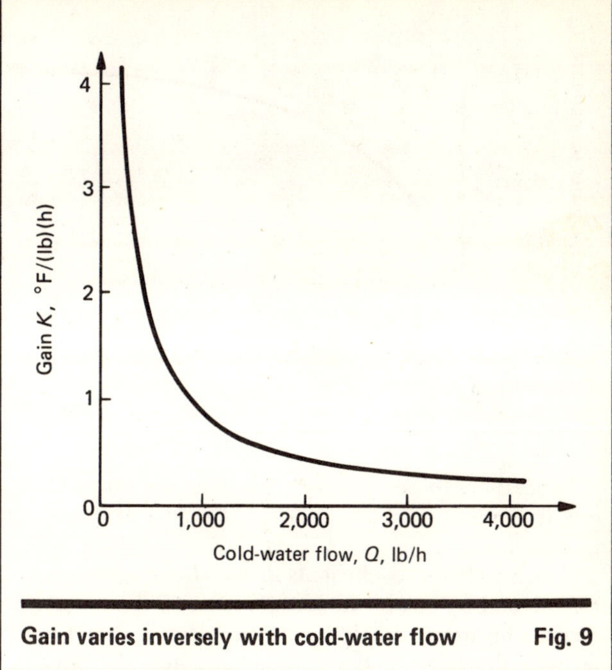

Gain varies inversely with cold-water flow **Fig. 9**

change in hot-water temperature should be large and noticeable. But if the same change is made with the Mississippi River at New Orleans flowing through the system, the increase in hot-water temperature would be nil.

Factors such as illustrated in Fig. 9 are not always detected from an examination of the test data alone. Even in cases where the material and energy balances are not readily written, a good insight into the nature of the process and an appreciation of the cause and effect relationships is the best way to detect such characteristics.

Time constant

The analysis of the transient portion of the response in Fig. 4 entails an examination of the unsteady-state enthalpy balance for the hot-water heater. The equation is:

$$MC_p\left[\frac{dT_w}{dt}\right] = FH_F + QC_p(T_i - T_R) - (Q + F)C_p(T_w - T_R) \quad (2)$$

where M = mass of water in the heater, lb.

The analysis of Eq. (2) begins by noting which parameters vary with time and which are constants. It will be assumed that the mass of water, M, in the tank is constant, that C_p is constant, and that the steam enthalpy, H_F and cold-water temperature, T_i, vary very slowly and by such small amounts that they can be assumed constant. This means that T_w, Q, and F are functions of time. Thus, the term $(Q + F)C_p(T_w - T_R)$ contains the products QT_w and FT_w. This term is therefore nonlinear, which complicates the analysis of the unsteady-state enthalpy balance.

The above form of the equation relates the actual values of the variables Q, F, and T_w. An alternative form of this equation will relate small changes in Q, F, and T_w. The term "change" as used here means a change in the sense of the differential, and not a rate of change in the sense of the derivative. It turns out that this alternative form of the equation will always be linear. For the analysis of the behavior of a control system within a relatively narrow operating region, this alternative form of the equation will be quite satisfactory.

A small change in a variable can be considered a differential, or vice versa, a differential can be approximated by a small change. To obtain the enthalpy balance written in terms of small changes, the total differential of the original enthalpy balance must be determined:

$$d\left[MC_p\left(\frac{dT_w}{dt}\right)\right] = d[FH_F + QC_p(T_i - T_R) - (Q + F)C_p(T_w - T_R)]$$

Using the properties of the differential gives:

$$MC_p\left[\frac{d(dT_w)}{dt}\right] = H_F dF + C_p(T_i - T_R)dQ - C_p(T_w - T_R)(dQ + dF) - (Q + F)C_p dT_w$$

In this article, lower case letters will designate the small changes; upper case letters, the actual values. Thus, $dT_w = t_w$, $dF = f$, $dQ = q$; and the preceding equation becomes:

$$MC_p\left(\frac{dt_w}{dt}\right) = H_F f + C_p(T_i - T_R)q - C_p(T_w - T_R)(q + f) - (Q + F)C_p t_w$$

Collecting terms gives the following:

$$MC_p\left(\frac{dt_w}{dt}\right) + WC_p t_w = [H_F - C_p(T_w - T_R)]f - C_p(T_w - T_i)q$$

where $W = Q + F$.

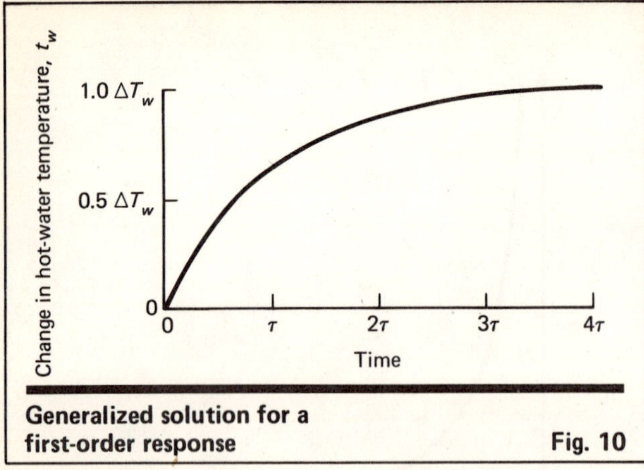

Generalized solution for a first-order response Fig. 10

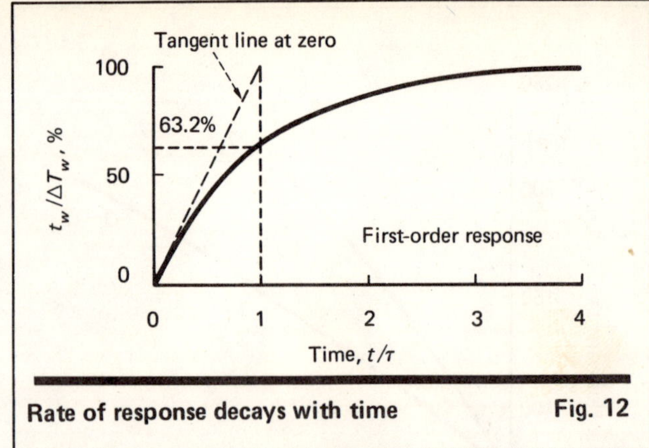

Rate of response decays with time Fig. 12

Models composed of such equations are called differential models, perturbation models, or various other terms.

The number of coefficients in the above equation can be reduced by one by dividing by the coefficient of any term. The most judicious one to select is that for the term of the controlled variable, t_w. Thus, dividing the above equation by WC_p gives:

$$\frac{M}{W}\left(\frac{dt_w}{dt}\right) + t_w = \left[\frac{H_F - C_p(T_w - T_R)}{WC_p}\right]f - \left(\frac{(T_w - T_i)}{W}\right)q \quad (3)$$

In this form, observe that the coefficient f is the process gain K as determined earlier, namely the gain with respect to the manipulated variable F. The coefficient of q, namely $-[(T_w - T_i)/W]$, is the gain with respect to the disturbance Q. Observe that this gain is negative, i.e., an increase in the cold water flow Q will cause a decrease in the hot-water temperature.

However, the coefficient of interest is the coefficient of the derivative term, namely, M/W. The units of M/W are (lb)/(lb/h), or h, i.e., time. This is no accident, for in order to have dimensional consistency in equations, such as Eq. (3), the units on the coefficient of the derivative term must always be time.

Next, it should be observed that this coefficient, M/W, will have no effect on the steady-state solution. At steady state, dt_w/dt must equal zero, and thus the term $(M/W)(dt_w/dt)$ will equal zero regardless of the value of M/W. However, this coefficient does affect the transient response of the process.

The coefficient M/W is called the *time constant* of the process. The time constant is customarily represented by τ, which permits Eq. (3) to be written as:

$$\tau\frac{dt_w}{dt} + t_w = K_F f + K_Q q \quad (4)$$

where K_F is the gain with respect to changes in F, and K_Q is the gain with respect to changes in Q.

To illustrate the characteristics of the time constant, the solution to Eq. (4) will be determined for the test in Fig. 4. The process was initially lined out with $F = 60\%$, $T_w = 155°F$, and Q equal to some unknown but constant value. These conditions will be used as the zero points for f, t_w, and q. At 3:00, the flow F was changed to 80%, which means that f becomes 20% and remains constant. The value of Q is unchanged, so $q = 0$. Eq. (4) becomes:

$$\tau\frac{dt_w}{dt} + t_w = K_F\Delta F \quad (5)$$

where ΔF is the change made in F (for Fig. 4, $\Delta F = 20\%$ and $K_F = 1.2°F/\%$).

Since K_F and ΔF are both constant, the solution of Eq. (5) is:

$$t_w = (1 - e^{-t/\tau})(K_F\Delta F) \quad (6)$$

Noting that $K_F \cdot \Delta F = \Delta T_w$ will be the value that t_w approaches as t becomes large, the solution to Eq. (6) can be presented as in Fig. 10. The vertical axis is calibrated in a fraction change in ΔT_w, and the time axis is calibrated in units of the time constant τ. Since Eq. (5) is a first-order differential equation, its solution is often called a first-order response. Systems described by Eq. (5) are often called first-order lags.

Figure 11 presents the first-order response for various values of τ. In effect, the time constant determines the speed of response of the process. For small τ, the system responds quickly; for large τ, the system responds slowly. For the hot-water heater, this could be reasoned by noting that $\tau = M/W$, or the time constant is the ratio of holdup to throughput. A large tank with a small flow would be expected to respond slowly. Its time

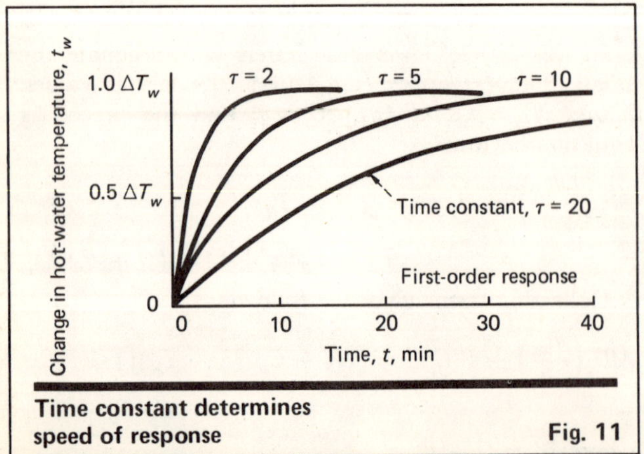

Time constant determines speed of response Fig. 11

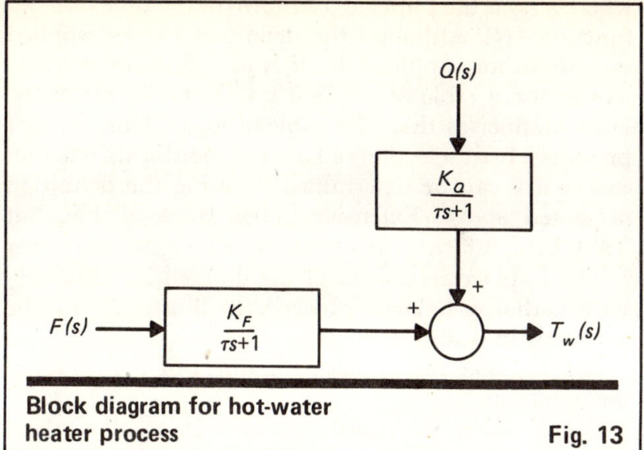

Block diagram for hot-water heater process Fig. 13

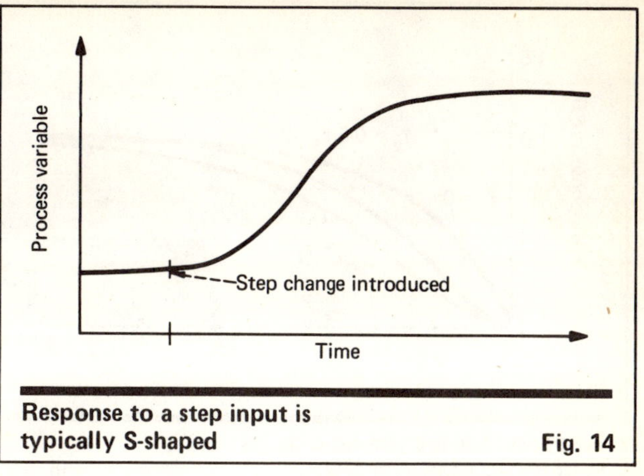

Response to a step input is typically S-shaped Fig. 14

constant, M/W would be large, and Fig. 11 confirms that it would respond slowly. A small tank with a large flow would be expected to respond quickly. Its time constant, M/W would be small, and Fig. 11 confirms that it would respond quickly.

As shown in Figs. 10 and 11, the first-order response assumes a nonzero rate of response immediately after the input changes. What if the system had continued to respond at this rate? Fig. 12 shows the first-order response with a line drawn tangent to the response curve at time $t = 0^+$. This line intersects the final value ($t_w/\Delta T_w = 100\%$) at $t = \tau$. That is, if the response had continued at its initial rate, it would have reached the final value in one time constant.

A feature of the first-order response is that the rate slows (or decays) as the response gets closer to its final value. Because of this, only 63.2% of the total response will be achieved in one time constant. The response values achieved after 1, 2, 3, 4 and 5 time constants are:

Time	Response
τ	63.2%
2τ	86.5%
3τ	95.0%
4τ	98.2%
5τ	99.3%

Hence, to achieve 99% of the total response requires 5 time constants.

In the previous section, we learned that the gain is not constant but changes with process conditions. The same is generally true of the time constant. For the hot water, the time constant τ equals M/W, and thus varies with the demand for hot water.

Transfer functions

The block diagram, presented in Fig. 3, for the hot-water-temperature control loop was largely qualitative, illustrating how the components related to one another but not showing the quantitative relationships. To be quantitative requires the incorporation of relationships such as the differential form of the enthalpy balance, previously presented as Eq. (4):

$$\tau\left(\frac{dt_w}{dt}\right) + t_w = K_F f + K_Q q$$

Although various representations are suitable for this purpose, the customary approach in process control is to use the Laplace transform. Unfortunately, this topic has proven to be a stumbling block for many who wanted to gain some appreciation of the subject but did not wish to devote too much time to the topic. For this individual, most control texts devote far too much attention to Laplace transforms. At the simplest level, Laplace transforms can be used as little more than a notational convenience. In this article, the Laplace transform will be used essentially in this context. Hence, its coverage will be brief.

The Laplace transform is to functions what the loga-

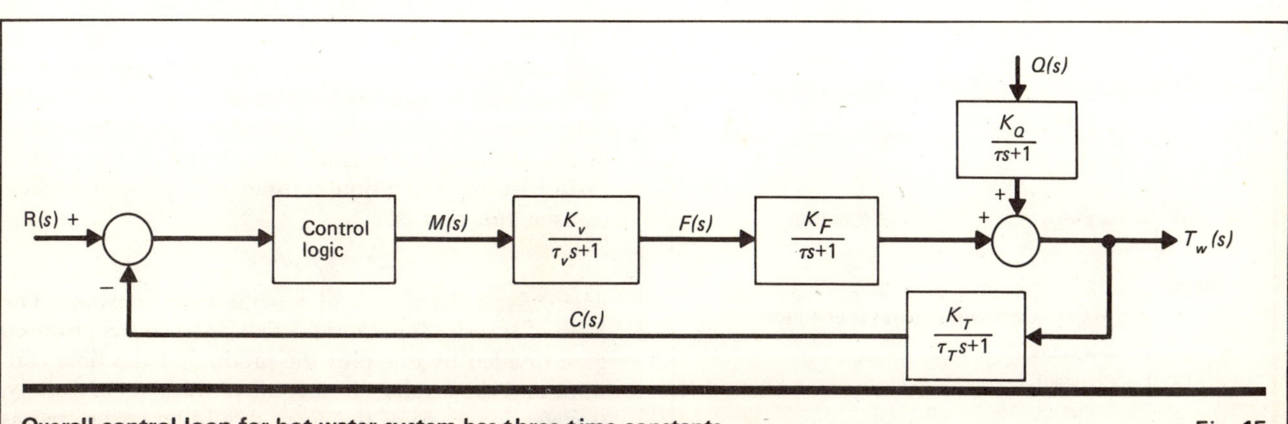

Overall control loop for hot-water system has three time constants Fig. 15

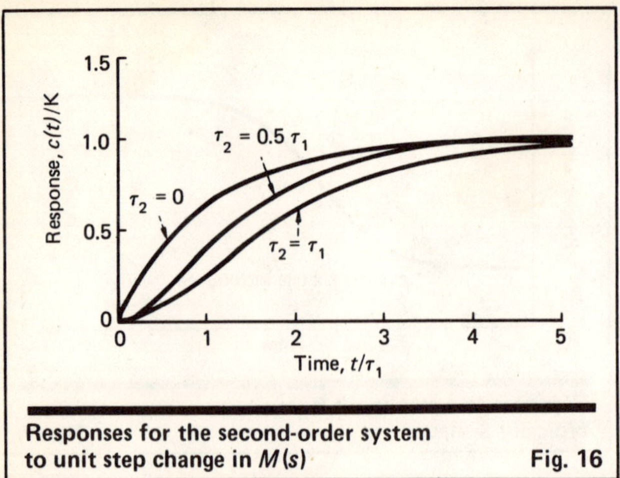

Responses for the second-order system to unit step change in $M(s)$ **Fig. 16**

rithm is to numbers, namely an alternative expression with more convenient properties. The Láplace transform expresses a differential equation with t as the independent variable as an algebraic equation with s as the independent variable.

The Laplace transform is an integral transform defined as:

$$F(s) = \mathcal{L}[f(t)] = \int_0^\infty f(t)e^{-st}dt$$

Selected Laplace transforms **Table I**

$f(t)$		$F(s)$
$u(t)$	[1]	$1/s$
e^{-at}		$1/(s+a)$
$\omega_n/(s^2+\omega_n^2)$		$\sin \omega_n t$
$\dfrac{s}{s^2+\omega_n^2}$		$\cos \omega_n t$
$u(t-\theta)$	[2]	$e^{-\theta s}$
$f(t)+g(t)$		$F(s)+G(s)$
$kf(t)$	[3]	$kF(s)$
$df/dt = f'(t)$		$sF(s) - f(0)$
$d^2f/dt^2 = f''(t)$		$s^2F(s) - sf'(0) - f(0)$

Notes

[1] $u(t)$ = Unit step function (change of 1 unit)

[2] θ = Deadtime

[3] k = Constant

where $F(s)$ is the Laplace transform of the time domain function $f(t)$. Although the definition can be applied directly in any application, it is usually more convenient to use a table of Laplace transforms in much the same manner as that of a table of logarithms. Table I presents a brief selection of Laplace transforms, wherein each entry can be determined by using the definition presented above. Extensive tables are available, but Table I will suffice for all except more-complex problems.

The Laplace transform of the differential unsteady-state enthalpy balance proceeds as illustrated by the five steps in Table II.

Step 1—Take the Laplace transform of both sides of the equation.

Step 2—Write the transforms in order to allow them to be determined term by term.

Step 3—Determine the transform for each term.

Step 4—Replace each term in the equation by its transform.

Step 5—Solve for the transform of the controlled variable.

The fifth step is nothing more than an algebraic manipulation of the equation, which is perfectly legal since s is a variable in the same sense that t is a variable.

The block diagram of the hot-water-heater process in Fig. 13 is based on the Laplace transform equation derived in Table II. The block diagram consists of a summation element and two blocks, one containing $K_F/(\tau s + 1)$ and the other containing $K_Q/(\tau s + 1)$.

The block containing $K_F/(\tau s + 1)$ relates the hot-water temperature $T_w(s)$ to the steam flow $F(s)$, which may be written as:

$$\frac{T_w(s)}{F(s)} = \frac{K_F}{\tau s + 1} = G_F(s)$$

Note that $T_w(s)/F(s)$ is in output/input ratio, with both output and input expressed as Laplace transforms. Such expressions are called *transfer functions,* and we say that the transfer function relating the hot water temperature to steam flow is $G_F(s)$, which is $K_F/(\tau s + 1)$.

Similarly, the transfer function relating hot-water temperature to cold-water flow is:

$$\frac{T_w(s)}{Q(s)} = \frac{K_Q}{\tau s + 1} = G_Q(s)$$

The overall relationship for Fig. 13 is:

$$T_w(s) = G_F(s)\,F(s) + G_Q(s)Q(s)$$

The output, $T_w(s)$, equals the sum of the product of each input and its respective transfer function. This form applies to systems in general.

In the previous section, we noted that the relationship of the hot-water temperature to the steam flow was a first-order lag (i.e., a time constant). The corresponding transfer function is:

$$G_F(s) = K_F/(\tau s + 1)$$

where K_F is the gain and τ is the time constant. The form of transfer functions for first-order lags is always a gain divided by one plus the product of the time constant and the Laplace transform variable s. Conversely, transfer functions of this form should be recognized as first-order lags.

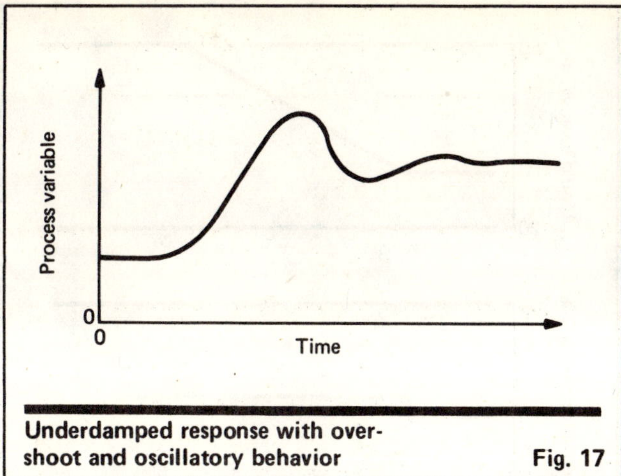

Underdamped response with overshoot and oscillatory behavior **Fig. 17**

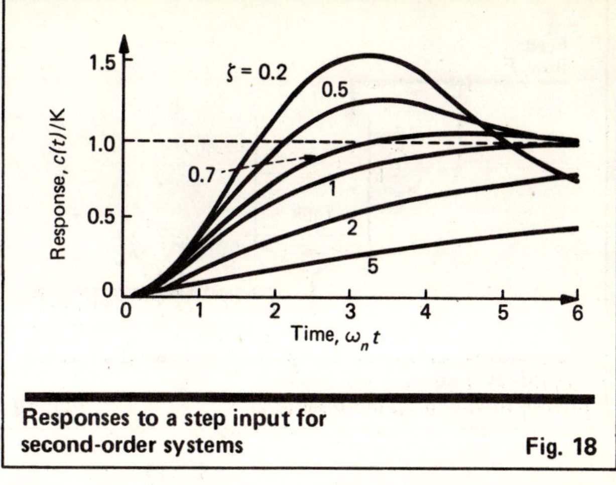

Responses to a step input for second-order systems **Fig. 18**

Although the Laplace transform is far more powerful than what has been presented here, this limited description is sufficient for a general appreciation of process control.

Higher-order systems

Often, the responses in Fig. 10, 11, and 12 for first-order lag are not quite what is observed for industrial processes. Instead, an S-shaped curve, such as in Fig. 14 is common. In fact, the response (Fig. 4) for the hot-water heater also exhibited such a curve—even though the mathematical model was first order.

Several factors could account for the discrepancy between the model response and the observed response for the hot-water heater. As indicated earlier, the characteristics of the valve and sensor also contributed to the test data in Fig. 4. Many temperature sensors and valves will exhibit first-order characteristics. Assuming each is first order, the overall control loop would be represented by the block diagram in Fig. 15. The control loop now contains three time constants.

Another possible explanation for the non-first-order behavior of the hot-water heater is that the tank contents are not perfectly mixed. Imperfect mixing can exhibit a variety of characteristics, one of which is that of multiple time constants.

For the system in Fig. 15, the transfer function relating the transmitter output, $C(s)$, to the valve position, $M(s)$, is:

$$\frac{C(s)}{M(s)} = \frac{K_V}{\tau_V s + 1} \cdot \frac{K_F}{\tau s + 1} \cdot \frac{K_T}{\tau_T s + 1}$$

$$\frac{C(s)}{M(s)} = \frac{K_V K_F K_T}{(\tau_V s + 1)(\tau s + 1)(\tau_T s + 1)}$$

Applying the reverse of the Laplace-transform procedure (Table II) to this expression would reveal that the above transfer function represents a third-order differential equation.

Any combination of two or more time constants in series will produce the S-shaped response curve. Fig. 16 illustrates the response of the second-order system:

$$\frac{C(s)}{M(s)} = \frac{K}{(\tau_1 s + 1)(\tau_2 s + 1)}$$

By arbitrarily defining τ_1 as the larger of the two time constants, the response can be presented as a family of curves for various values of τ_2, where $0 \leq \tau_2 \leq \tau_1$. The time constant τ_1 is the scale factor for the time axis.

Although rarely encountered for the process itself, Fig. 17 illustrates a response that is frequently observed for process-control systems when the controller is in automatic. The response exhibits overshoot and oscillatory behavior, which should be contrasted with the first-order responses (Fig. 9–11), and the second-order responses (Fig. 16). Neither of these exhibit overshoot or oscillatory behavior, and are referred to as overdamped responses. The response in Fig. 17 is said to be underdamped.

Laplace transform of the enthalpy balance **Table II**

Step	Procedure
1.	$\mathcal{L}\left[\tau \frac{dt_w}{dt} + t_w\right] = \mathcal{L}\left[K_F f + K_Q q\right], \quad t_w(0) = 0$
2.	$\mathcal{L}\left[\tau \frac{dt_w}{dt}\right] + \mathcal{L}\left[t_w\right] = \mathcal{L}\left[K_F f\right] + \mathcal{L}\left[K_Q q\right]$
3a.	$\mathcal{L}\left[\tau \frac{dt_w}{dt}\right] = \tau \mathcal{L}\left[\frac{dt_w}{dt}\right] = \tau\left[sT_w(s) - t_w(0)\right] = \tau s T_w(s)$
3b.	$\mathcal{L}\left[t_w\right] = T_w(s)$
3c.	$\mathcal{L}\left[K_F(f)\right] = K_F \mathcal{L}[f] = K_F F(s)$
3d.	$\mathcal{L}\left[K_Q(q)\right] = K_Q \mathcal{L}[q] = K_Q Q(s)$
4.	$\tau s T_w(s) + T_w(s) = K_F F(s) + K_Q Q(s)$
5.	$T_w(s) = \frac{K_F}{\tau s + 1} F(s) + \frac{K_Q}{\tau s + 1} Q(s)$

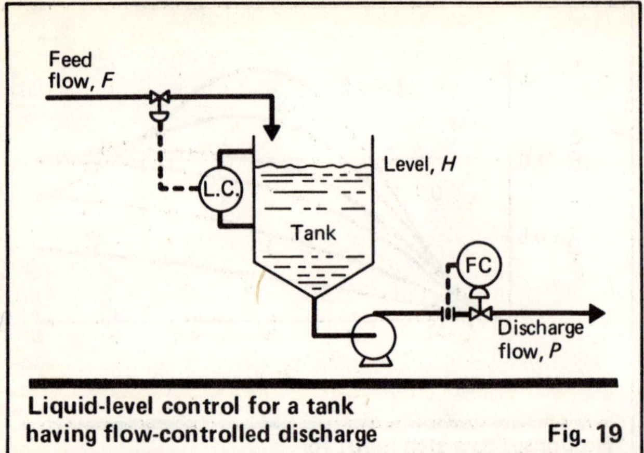

Liquid-level control for a tank having flow-controlled discharge **Fig. 19**

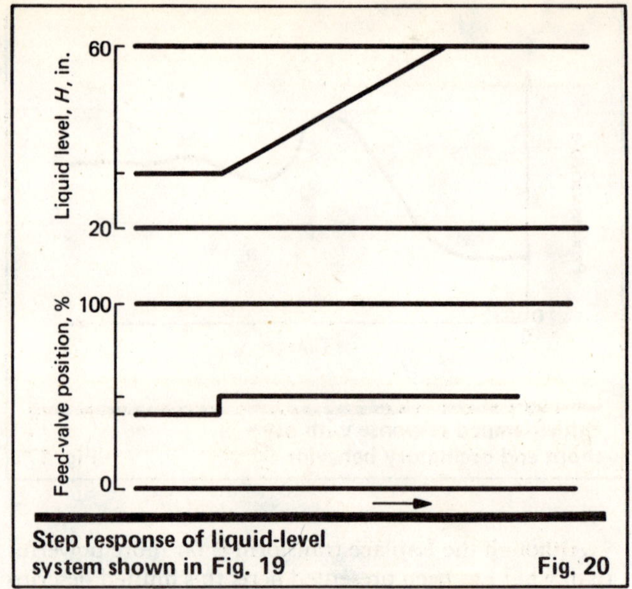

Step response of liquid-level system shown in Fig. 19 **Fig. 20**

The responses in Fig. 16 comprise only a subset of the possible second-order responses. That is, the transfer function for these responses was always of the form:

$$\frac{C(s)}{M(s)} = \frac{K}{(\tau_1 s + 1)(\tau_2 s + 1)}$$

$$= \frac{K}{\tau_1 \tau_2 s^2 + (\tau_1 + \tau_2)s + 1}$$

where τ_1 and τ_2 were restricted to real numbers. An alternative form for representing a second-order response involves the damping ratio ζ and the natural frequency ω_n:

$$\frac{C(s)}{M(s)} = \frac{K}{\dfrac{s^2}{\omega_n^2} + \dfrac{2\zeta s}{\omega_n} + 1}$$

The relationship between τ_1, τ_2 and ζ, ω_n can be established by equating the coefficients in the above two transfer functions:

$$\tau_1 \tau_2 = \frac{1}{\omega_n^2}$$

$$\tau_1 + \tau_2 = \frac{2\zeta}{\omega_n}$$

Relationships for the expressions for ζ and ω_n in terms of τ_1 and τ_2, and vice versa, are:

$$\omega_n = \frac{1}{\sqrt{\tau_1 \tau_2}} \qquad \tau_1 = \frac{\zeta + \sqrt{\zeta^2 - 1}}{\omega_n}$$

$$\zeta = \frac{\tau_1 + \tau_2}{2\sqrt{\tau_1 \tau_2}} \qquad \tau_2 = \frac{\zeta - \sqrt{\zeta^2 - 1}}{\omega_n}$$

The expressions for τ_1 and τ_2 reveal that τ_1 and τ_2 will be real values only when $\zeta \geq 1$. When $\zeta > 1$, τ_1 and τ_2 will be real and unequal; when $\zeta = 1$, τ_1 and τ_2 will be real and equal; when $\zeta < 1$, τ_1 and τ_2 will be complex conjugates. While there is mathematically no problem with τ_1 and τ_2 being complex conjugates, the physical interpretation is lost. For the hot-water heater in Fig. 1, the time constant was derived to be the capacity M divided by the throughput W. The result of such an analysis will always be a real time constant.

Instead of attempting an interpretation of complex-conjugate time constants, a more convenient approach is to provide an interpretation for the damping ratio and natural frequency. When the damping ratio ζ is zero, the second-order transfer function would reduce to:

$$\frac{C(s)}{M(s)} = \frac{K}{\dfrac{s^2}{\omega_n^2} + 1} = \frac{K\omega_n^2}{s^2 + \omega_n^2}$$

Table I indicates that a Laplace transform expression with $s^2 + \omega_n^2$ in the denominator corresponds to either $\sin \omega_n t$ or $\cos \omega_n t$, depending on the numerator of the transfer function. In either case, the response would be sinusoidal with a constant amplitude and frequency ω_n. Since the amplitude is constant, the response is said to have no damping, i.e., the damping ratio ζ equals zero.

For values of ζ between zero and 1, the response would have a component of the form $e^{-\zeta\omega_n t} \sin \omega_n(1 - \zeta^2)^{1/2}t$. As illustrated in Fig. 18, the response is sinusoidal, but is bounded by the decaying envelop provided by $e^{-\zeta\omega_n t}$ and $-e^{-\zeta\omega_n t}$. Thus, the damping ratio ζ is a measure of how rapidly the sinusoid is "damped out", with $\zeta \to 0$ meaning no damping and $\zeta \to 1$ meaning heavy damping. For $\zeta \geq 1$, the damping is sufficient to completely eliminate the sinusoidal component.

The natural frequency ω_n is the frequency that the sinusoidal component would exhibit if there were no damping, i.e., $\zeta = 0$. For $\zeta \neq 0$, the frequency of the sinusoid is the damped frequency $\omega_n \sqrt{1 - \zeta^2}$.

Fig. 18 presents the family of responses to a step input for a second-order system. The time axis is normalized by using the product $\omega_n t$. Observe that as ζ decreases, the oscillations increase. Also note that the speed of response increases with increasing ω_n. These observations will be used subsequently in our discussion of controller performance.

Integrators

A transfer function of the form $K/(\tau s + 1)$ was observed to be that of the first-order lag, with τ being the

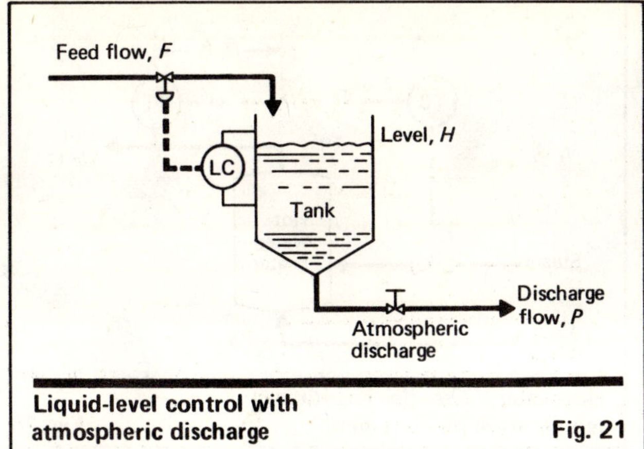

Liquid-level control with atmospheric discharge **Fig. 21**

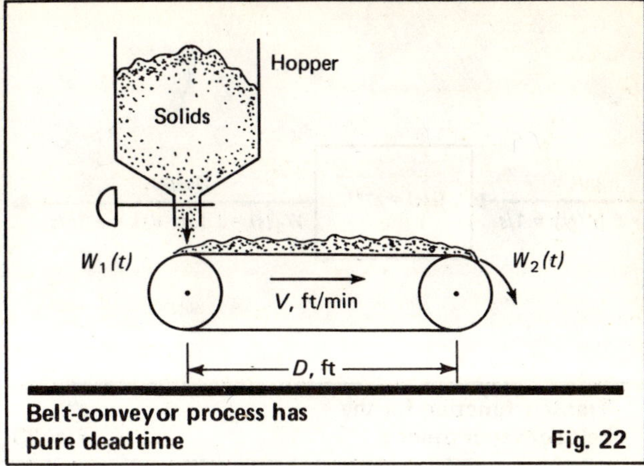

Belt-conveyor process has pure deadtime **Fig. 22**

time constant. Another form that may be encountered for the transfer function is that of K/s. Systems of this type do not exhibit the time constant type-of-behavior, but instead act as integrators.

Systems of this type are frequently encountered in liquid-level control problems. Fig. 19 illustrates such a liquid-level control system. The level in the tank is controlled by manipulating the feed flow (or technically, the feed-valve position). The discharge from the tank is pumped out through a flow controller. We shall assume that the flow-control loop responds very fast and that its setpoint is never changed, which permits P to be treated as a constant.

Suppose a test such as that used for the hot-water tank is applied to the system in Fig. 19. The system will only "line out" if the feed flow F equals the discharge flow P. From these conditions and with the level controller on manual, the feed-valve position is changed very quickly.

Fig. 20 illustrates the response to an increase in this valve position. Soon after the valve position is changed, the level begins to rise. Unlike the time constant, the rate of increase in the level remains essentially constant with time. This continues in Fig. 20 until the upper span of the level transmitter is reached, and unless corrective action is taken, the tank will eventually overflow.

The mathematical model of the level process in Fig. 19 is obtained by writing the unsteady-state material balance:

$$F - P = \frac{d}{dt}(\rho AH)$$

where ρ = liquid density and A = cross-sectional area of the tank. In the constant cross-sectional area of the tank, this equation is linear, but it is still useful to obtain the differential form of the model:

$$dF - dP = \rho A[d(dH)/dt]$$

Since P is constant, dP is zero, and the above equation becomes:

$$f = \rho A\, dh/dt \qquad (8)$$

Alternatively, Eq. (8) can be integrated to obtain:

$$h = (1/\rho A) \int f dt$$

which indicates that the process is acting like an integrator.

The Laplace transform of Eq. (8) can be derived by following procedures similar to those in Table II. The transform is:

$$F(s) = \rho AsH(s)$$

which can be rearranged to give the transfer function:

$$\frac{H(s)}{F(s)} = \frac{1}{\rho As}$$

This transfer function has the form K/s, which is that of an integrator.

Systems such as the level-control system (Fig. 19) that give responses such as in Fig. 20 are often called non-self-regulated systems. If the input to the system, namely F, is set at some arbitrary value, the system will not "line out", unless this value is judiciously chosen as being equal to P.

On the other hand, the hot-water heater discussed earlier behaves quite differently, and is called a self-regulated system. For any combination of the cold-water flow Q and steam flow W that is maintained for a sufficiently long period of time, the hot-water temperature will "line out." That is, the self-regulated system will seek an equilibrium (or steady-state) point of its own accord. The non-self-regulated system requires control action to seek the equilibrium point. The integrator-type of process behaves in the non-self-regulated manner, whereas the time-constant-type of process behaves in the self-regulated manner.

If the level control system in Fig. 19 is modified slightly, its characteristics can be changed from non-self-regulated to self-regulated. As presented in Fig. 21, the discharge arrangement is changed to an atmospheric discharge through a manual valve. Instead of behaving like an integrator, the system will behave like a time constant.

In the flow-controlled discharge system in Fig. 19, the discharge flowrate is independent of the liquid level in the tank. For the atmospheric-discharge arrangement (Fig. 21), the discharge flow will increase as the liquid level increases. Thus, if the inlet flow F is increased, the liquid level will increase until the discharge flow equals the new feed flow.

In either case, the increase in the feed flow to the

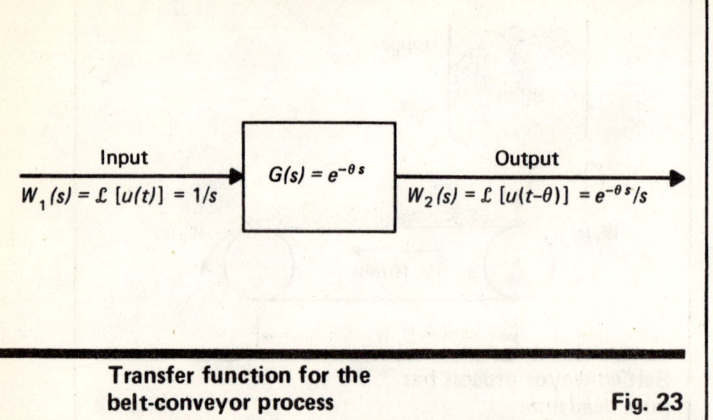

Transfer function for the belt-conveyor process Fig. 23

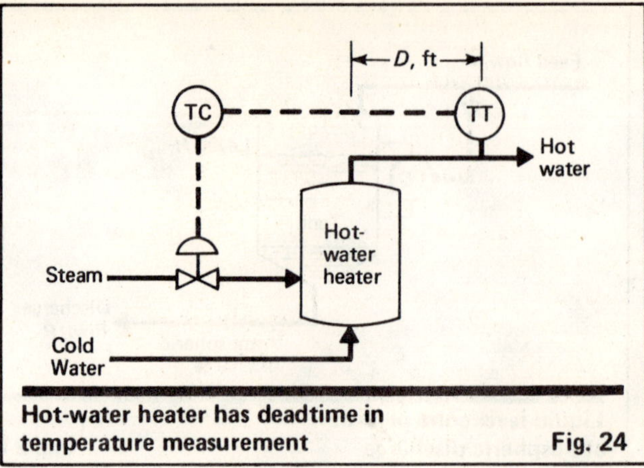

Hot-water heater has deadtime in temperature measurement Fig. 24

system causes an imbalance in the material balance. For the non-self-regulated system in Fig. 19, the response of the system (i.e., the increase in level H) has no effect on the imbalance. For the self-regulated system in Fig. 21, the response of the system (the increase in level H) produces other changes (namely, the increase in discharge flow P) that reduce the imbalance in the material balance. From one point of view, the time-constant type of system can be said to have internal feedback, whereas the integrator has none.

Deadtime

Fig. 22 illustrates a system whose dynamics are characterized by pure deadtime. A solids-hopper discharges onto a moving conveyor belt at a rate $W_1(t)$, lb/min. Solids fall off the other end of the conveyor belt at a rate $W_2(t)$, lb/min. The conveyor of length D, ft moves at velocity V, ft/min, so the transportation time is: $\theta = D/V$.

Suppose a unit step-change is made in the flow $W_1(t)$ at time zero. This change in the flow will not be seen in $W_2(t)$ until the material has been transported the entire length of the conveyor. Thus, the same step change will be seen in $W_2(t)$, but it will occur at time θ instead of at time zero.

The unit step change $u(t)$ is defined as:

$$u(t) = \begin{cases} 0 \text{ for } t < \theta \\ 1 \text{ for } t \geq \theta \end{cases}$$

The change is made when the argument t of the unit step function $u(t)$ is zero. Thus, a unit step change occurring at time θ would be defined as:

$$u(t-\theta) = \begin{cases} 0 \text{ for } t < \theta \\ 1 \text{ for } t \geq \theta \end{cases}$$

Again, the change occurs when the argument of the step function is zero.

Fig. 23 illustrates a block diagram of the belt conveyor. The input is shown as $u(t)$, whose Laplace transform is $1/s$. The output is shown as $u(t-\theta)$, whose Laplace transform (see Table I) is $e^{-\theta s}/s$. The transfer function $G(s)$ is the Laplace transform of the output divided by the Laplace transform of the input, or:

$$G(s) = \frac{\mathcal{L}[u(t-\theta)]}{\mathcal{L}[u(t)]} = \frac{e^{-\theta s}/s}{1/s} = e^{-\theta s}$$

Thus, the transfer function of a pure deadtime process is $e^{-\theta s}$.

Deadtimes are commonly encountered in sheet-processing industries, such as paper making, plastic film, textiles, etc. However, deadtimes can also be encountered in fluid-flow systems. Fig. 24 illustrates the hot-water heater of Fig. 1 with the temperature transmitter located some distance, D, down the pipeline. The deadtime, θ, for this system is:

$$\theta = \frac{D}{W/\rho A}$$

where ρ = density and A = cross-sectional area of the pipe. Observe that the deadtime for this system varies with the water flow, i.e., the deadtime is longer for low flows.

The transfer function of the hot-water tank was previously derived as: $K_F/(\tau s + 1)$. By locating the sensor downstream (Fig. 24), a deadtime term $e^{-\theta s}$ is added to the transfer function, giving:

$$G(s) = \frac{K_F e^{-\theta s}}{\tau s + 1}$$

This transfer function is commonly called a first-order-lag-plus-deadtime, and is often used to approximate the dynamic characteristics of complex processes.

Deadtime is referred to by a variety of terms, including time delay, delay time, and transportation lag. The term transportation lag is not equivalent to first-order lag, the latter meaning a time-constant type system and not deadtime. Unfortunately, the term "lag" is also used at times. The meaning could be either transportation lag or first-order lag, so its meaning can only be inferred from the context in which it was used.

Deadtime is relatively common in process systems. In fact, the existence of significant deadtimes is one factor that tends to distinguish process control from the practice of control in other disciplines. As will be discussed in a subsequent section, deadtime proves to be a very difficult element to control.

Control of processes

The selection and adjustment of process controllers rely very heavily on the characteristics of the process itself. The discussion that follows will consider both on-off and proportioning controllers, and will attempt to relate their characteristics to those of the process.

Adjustment of any controller requires that a compromise be made between desirable and undesirable effects. The discussion is focused on an understanding of the various effects, and how they relate to the available adjustments. The intent is to provide us with the insight to make our own compromises. No hard and fast rules will suffice because the compromises depend upon a variety of factors such as the nature of the process, operator attitudes and personal preference.

Stability of control

One of the first topics usually considered in a discussion of control is that of stability. Fig. 25 illustrates three responses, one of which is indicated as stable, one as conditionally stable, and one as unstable. In practice, the conditionally stable response could be called stable or unstable, depending upon the definition of stability.

One concept of stability is that of asymptotic stability, which is based on the following:

$$\lim_{t \to \infty} C(t) = M$$

where M is a finite number. This condition says that as time increases, the response of the system must approach some finite value. Only the upper response in Fig. 25 meets this condition.

Another concept of stability is that of a bounded response, which is based on the following:

$$\lim_{t \to \infty} |C(t)| < M$$

where M is again a finite value. Any response that meets the conditions for asymptotic stability will also be stable according to the above condition. However, the conditionally stable response (Fig. 25) would be stable by the above definition.

For industrial processes, the acceptability of the conditionally stable response (Fig. 25) would depend upon the application. As time becomes large, the response assumes a low-amplitude, sinusoid-like response referred to as a limit cycle. Responses of this type occur when simple on-off controllers are used. These are often quite acceptable for the loops of lesser importance in a process, but would not be acceptable for the critical process variables.

The lower response in Fig. 25 meets neither of these conditions for stability. The response is characterized by a sinusoidal response of growing amplitude that will usually cause the process operator to switch the controller from automatic to manual.

On-off controller

The on-off controller may be acceptable for the loops of lesser importance in the process. The control logic in an on-off controller is represented as:

$$m = \begin{cases} 100\% \text{ for } e \geq 0 \\ 0\% \text{ for } e < 0 \end{cases}$$

This relationship indicates that the output is switched from on to off or vice versa when the error signal passes through zero.

Most industrial controllers of this type provide for a

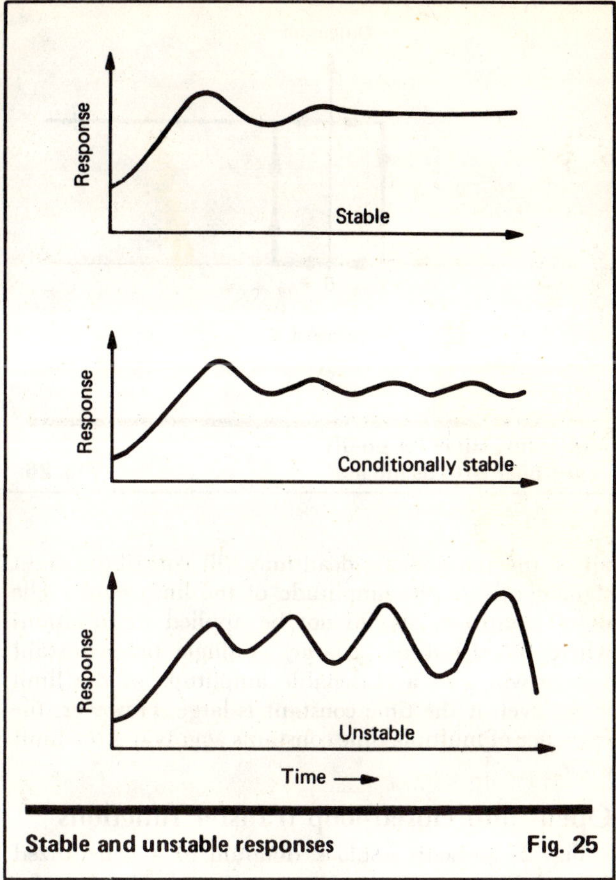

Stable and unstable responses **Fig. 25**

deadband adjustment that will cause the output to be switched at errors slightly above or below zero. The characteristics of this controller are described by the switching curve in Fig. 26. If $e \geq e_{DB}$, then $m = 100\%$. If $e \leq -e_{DB}$, then $m = 0\%$. If $-e_{DB} < e < e_{DB}$, the output remains at whatever value it is presently in. That is, if the error is decreasing from a large positive value, then $m = 100\%$ and will not be switched to 0% until $e \leq -e_{DB}$.

The deadband adjustment produces two effects:

1. Time between switches increases as e_{DB} is increased.

2. Amplitude of the limit cycle increases as e_{DB} is increased.

The former is desirable as it reduces the wear on the final actuator; the latter is normally undesirable. Thus, the adjustment is a compromise between these two.

Advantages of the on-off controller include:

- Controller is relatively inexpensive.
- Solenoid valve is less expensive than a positioning valve.
- System is reliable.
- System is easy to install and adjust.

Where the limit cycle can be tolerated, an on-off controller should be seriously considered. Unfortunately, the amplitude of the limit cycle is difficult to predict. Its amplitude is clearly related to the process gain, time constants, and deadtime. While relationships can be developed, the importance of the loops where it is applied is so low that the effort required to determine the process characteristics cannot be justified. Of

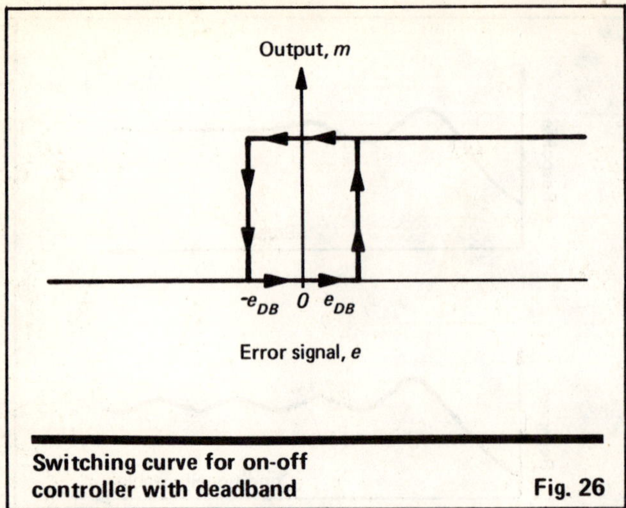

Switching curve for on-off controller with deadband Fig. 26

all of the parameters, deadtime will contribute most significantly to the amplitude of the limit cycle. The on-off controller should not be applied in situations where the deadtime is large. A single time-constant system will give a reasonable amplitude of the limit cycle, even if the time constant is large. However, the existence of multiple time constants aggravates the limit cycle.

Open- and closed-loop transfer functions

Fig. 27 presents a block diagram of a generalized control loop containing a process with the transfer function $G(s)$ and a controller with the transfer function $G_c(s)$.

The transfer function $G(s)$ relates the controlled variable, $C(s)$, to the manipulated variable, $M(s)$, and is properly called the open-loop transfer function. When the controller is in manual, the manipulated variable $M(s)$ can be adjusted by the operator, and the response $C(s)$ depends only upon $G(s)$. The transfer function, $G_c(s)$, of the controller is not involved in any way.

When the controller is in automatic, the operator adjusts the setpoint $R(s)$. The response $C(s)$ now depends upon both $G(s)$ and $G_c(s)$. The closed-loop transfer function, $C(s)/R(s)$, can be determined by first writing the following relationship for the block diagram in Fig. 27:

$$C(s) = G(s)M(s)$$

$$C(s) = G(s)G_c(s)E(s)$$

$$C(s) = G(s)G_c(s)[R(s) - C(s)]$$

Solving for $C(s)/R(s)$ gives:

$$\frac{C(s)}{R(s)} = \frac{G(s)G_c(s)}{1 + G(s)G_c(s)} = G_{CL}(s)$$

It is this transfer function that determines the characteristics of the control loop when the controller is in the automatic mode.

The terms open-loop time constant and closed-loop time constant are also used. The open-loop time constant is the time constant of the open-loop transfer function $G(s)$. The closed-loop time constant is that for the closed-loop transfer function $G_{CL}(s)$.

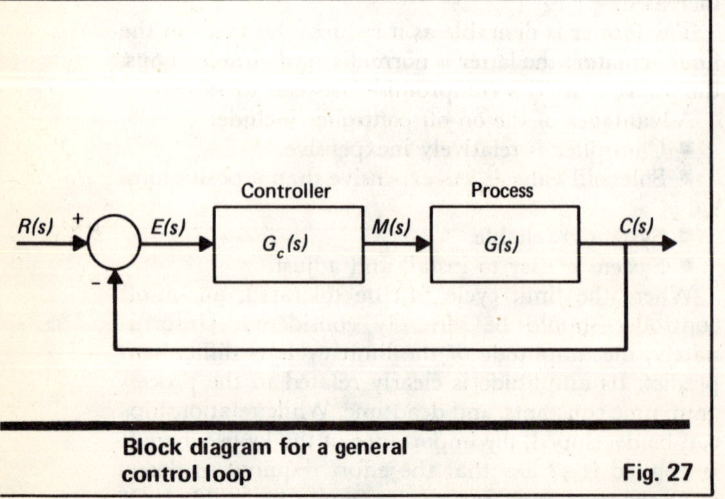

Block diagram for a general control loop Fig. 27

Proportional control

The simplest form of a proportionating controller is the proportional-only controller, which is described by:

$$M(s) = K_c E(s)$$

or,
$$m(t) = K_c e(t)$$

The output of this controller is proportional to the error $e(t)$.

The term $m(t)$ in the above equation represents the change in the output from some initial or reference value. That is, the actual output $M(t)$ is given by:

$$M(t) - M_R = K_c e(t)$$

where M_R is the initial output or bias term, i.e., M_R will be the output when the error is zero.

The value of M_R is initialized in this and other controllers of this general type during the transition from manual to automatic. The older controllers had to be "balanced" manually when switching from manual to automatic. The procedure was as follows:

1. Make the setpoint equal to the process variable, i.e., set the error to zero.
2. Switch the controller to automatic.

The value of the output at the time the switch is made to automatic became the value for M_R. Thus, at time 0^+, $e(0^+) = 0$, and the proportional-control equation reduces to:

$$M(0^+) - M_R = 0$$

or,
$$M(0^+) = M_R$$

Since M_R equals the controller output just prior to switching to automatic, the transfer is "bumpless."

Most newer controllers provide bumpless transfer by internal initialization procedures that do not require any action by the operator. Approaches other than the above can be used for balancing, but in any case the bias term M_R will be initialized.

The coefficient K_c in the proportional controller is called the proportional gain or proportional sensitivity. Like the process gain K, its units are normally in %/%. Most controllers provide an adjustment scale of approximately 0.1 to 10, although some provide even wider ranges.

In some controllers, the proportional adjustment is the proportional band in %. The concept of the proportional band is that the proportional mode will be active over only some percentage of the chart span. For example, suppose the proportional gain is 5. A change in error of 20% would produce a change in output of 100%, i.e., the valve would move from fully closed to fully open when the error changed by 20% of the chart span. Thus, the controller effectively proportionates the controller output over only 20% of the chart span, so the proportional band is 20%.

If the proportional gain is 1, a change in the error of 100% would be required to move the output from one extreme to the other. In this case, the proportional band is 100%. Thus, the proportional band (*PB*) and K_c are inversely related:

$$PB \text{ (in \%)} = 100/K_c$$

Both proportional gain and proportional band are currently in common use, but the trend appears to be toward the proportional gain. In this article we will consider the proportional setting to be the proportional gain.

We will now relate the loop-response characteristics to the proportional gain, by beginning with the first-order system, illustrated in Fig. 28. The closed-loop transfer function can be determined as:

$$G_{CL}(s) = \frac{G_c(s)G(s)}{1 + G_c(s)G(s)}$$

$$G_{CL}(s) = \frac{K_c\left(\dfrac{K}{\tau s + 1}\right)}{1 + K_c\left(\dfrac{K}{\tau s + 1}\right)}$$

$$G_{CL}(s) = \frac{KK_c}{\tau s + 1 + KK_c}$$

$$G_{CL}(s) = \frac{KK_c/(1 + KK_c)}{\left(\dfrac{\tau}{1 + KK_c}\right) s + 1}$$

Let us observe that the order of the closed-loop transfer function is the same as that of the open-loop transfer function, which is always true for proportional control.

Gain and time constant for the closed loop system are:

$$\text{Closed-loop time constant} = \tau/(1 + KK_c)$$

$$\text{Closed-loop gain} = KK_c/(1 + KK_c)$$

Since KK_c is a positive number, the closed-loop time constant, $\tau/(1 + KK_c)$, must be less than the open-loop time constant, τ. Thus, the closed-loop system will respond faster than the open-loop system. Furthermore, the closed-loop time constant decreases as the controller gain K_c is increased, so larger controller gains lead to faster responses. For large, slow processes, this can be an important benefit from closed-loop control.

The closed-loop gain, $KK_c/(1 + KK_c)$, will be a positive number less than one. For example, if the product $KK_c = 4$, then the closed-loop gain would be 0.8. Note that the closed-loop gain relates changes in the controlled variable $C(s)$ to changes in the setpoint $R(s)$. Using the hot-water heater as an example, if the setpoint were increased by 10°F, the hot-water temperature would only increase 8°F when the closed-loop gain is 0.8.

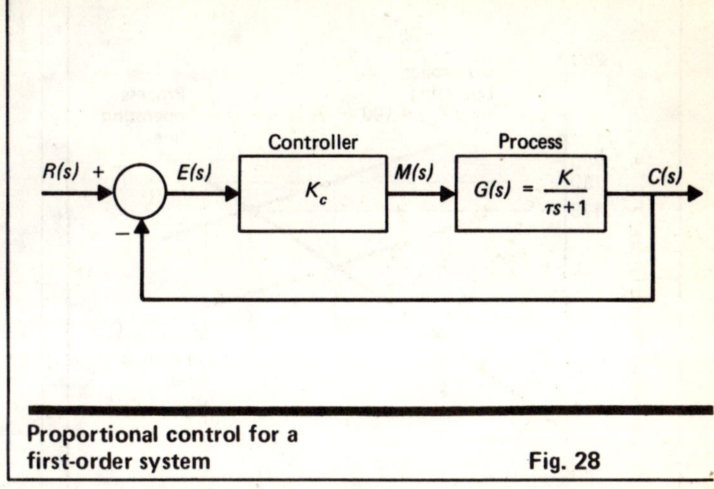

Proportional control for a first-order system Fig. 28

Fig. 29 illustrates the response of the hot-water heater (first-order behavior assumed) for various values of KK_c. It is assumed that the proportional controller is balanced at 150°F. Although the setpoint and controlled variable are equal at the balance point, they are not equal when controlling away from the balance point.

The difference between the setpoint and the controlled variable at the line-out point is referred to as offset. When the setpoint of the proportional controller is increased from the balance point of 150°F to 190°F (Fig. 29), the controller must respond by opening the valve. However, in order for the proportional controller to open the valve above M_R, the error must be positive. The system will line out at a temperature at which the difference between the setpoint and that temperature (i.e., the error) will produce the change in valve position required to give the new temperature.

In effect, the line-out point will be where the operating line of the process intersects the operating line of the controller. Fig. 30 reproduces the operating line for the hot-water heater originally presented as Fig. 7. The operating line for the proportional controller is:

$$F - F_B = K_c(T_{SP} - T_w)$$

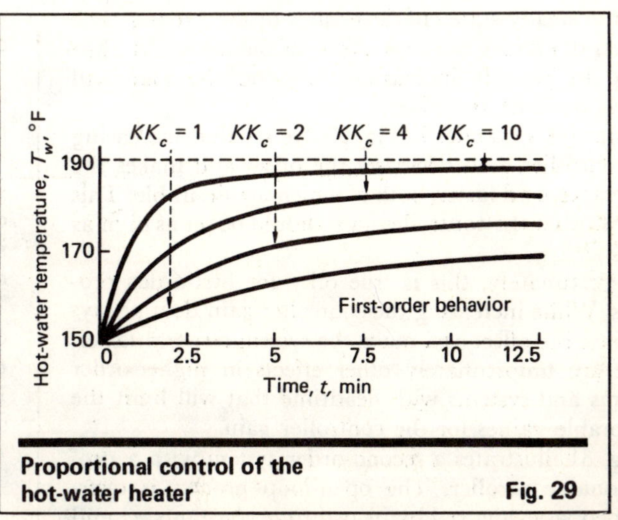

Proportional control of the hot-water heater Fig. 29

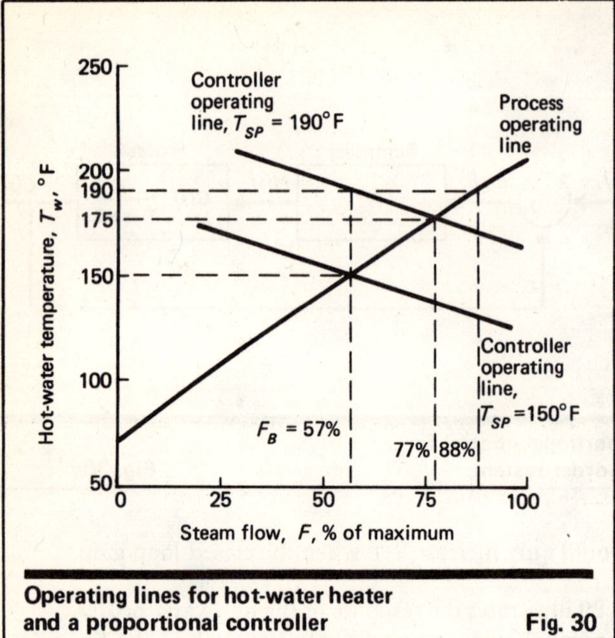

Operating lines for hot-water heater and a proportional controller **Fig. 30**

If the controller is balanced at $T_{SP} = 150°F$, then F_B equals 57%. The operating line for the controller is a straight line with slope $-1/K_c$, and passing through the point (F_B, 150°F). For Fig. 30, the controller gain is 2%/%, or 1.33%/°F. As illustrated in Fig. 30, the operating lines of the process and the controller both pass through the balance point.

When the setpoint is moved to 190°F, the operating line shifts to pass through the point (F_B, 190°F), with the same slope as before. As shown in Fig. 30, the operating lines for the process and the controller no longer intersect at the setpoint of 190°F. Instead, they intersect at a temperature of 175°F and a valve position of 77%. The offset is 190°F − 175°F = 15°F, and this error times the controller gain K_c produces the change in valve output from the bias of 57% to the new value of 77%. That is, the equality resulting from the proportional control equation holds:

$$77\% - 57\% = (1.33\%/°F)\,(190°F - 175°F)$$

The offset is solely determined by the controller gain and the steady-state characteristics of the process. The process dynamics have no effect on the offset. As illustrated in Fig. 29, increasing the controller gain will always decrease the offset.

From the responses in Fig. 29, we see that increasing the controller gain decreases the offset and makes the process respond faster, both of which are desirable. This suggests that the controller gain should be set as high as possible.

Unfortunately, this is true only for first-order processes. While increasing the controller gain does always decrease the offset and make the system respond faster, there are unfortunately other effects in higher-order systems and systems with deadtime that will limit the acceptable values for the controller gain.

Fig. 31 illustrates a second-order system with a proportional controller. The open-loop process is overdamped, as evidenced by its two time constants, τ_1 and

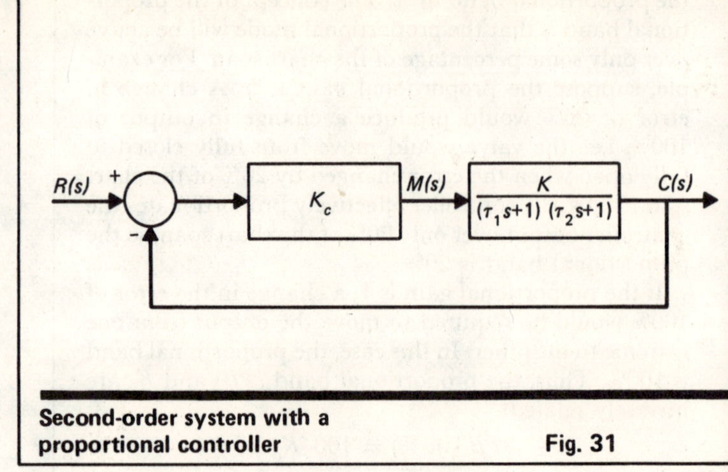

Second-order system with a proportional controller **Fig. 31**

τ_2. The closed-loop transfer function is obtained as follows:

$$G_{CL}(s) = \frac{\dfrac{KK_c}{(\tau_1 s + 1)(\tau_2 s + 1)}}{1 + \dfrac{KK_c}{(\tau_1 s + 1)(\tau_2 s + 1)}}$$

$$G_{CL}(s) = \frac{KK_c/(1 + KK_c)}{\left(\dfrac{\tau_1\tau_2}{1 + KK_c}\right)s^2 + \left(\dfrac{\tau_1 + \tau_2}{1 + KK_c}\right)s + 1}$$

The closed-loop gain is the same as before, which confirms that the offset is unaffected by the presence of the second time constant. For the closed-loop transfer function, the natural frequency, ω_n, and damping ratio, ζ, are computed as:

$$\omega_n = \sqrt{\frac{1 + KK_c}{\tau_1\tau_2}} = \omega_{n(OL)}\sqrt{1 + KK_c}$$

$$\zeta = \frac{\tau_1 + \tau_2}{(1 + KK_c)} \cdot \frac{\omega_n}{2} = \frac{\tau_1 + \tau_2}{2\sqrt{\tau_1\tau_2}} \cdot \frac{1}{\sqrt{1 + KK_c}} = \frac{\zeta_{OL}}{\sqrt{1 + KK_c}}$$

where $\omega_{n(OL)}$ and ζ_{OL} are the natural frequency and damping ratio of the open-loop transfer function.

The effect of K_c on the speed of response can be deduced from the effect of K_c on the closed-loop natural frequency. The above equation shows that ω_n is essentially proportional to $(K_c)^{1/2}$ for $KK_c >> 1$, i.e., large gains. Thus, the natural frequency increases with the controller gain, which means that the response speed increases with controller gain. This is consistent with the observation for first-order systems.

The second equation shows that the damping ratio decreases with increasing gain. That is, for $KK_c >> 1$, the damping ratio ζ is essentially proportional to $1/(K_c)^{1/2}$. The closed-loop transfer function becomes underdamped when $\zeta < 1$, or when:

$$KK_c > \frac{(\tau_1 + \tau_2)^2}{4\tau_1\tau_2} - 1 = \zeta_{OL}^2 - 1$$

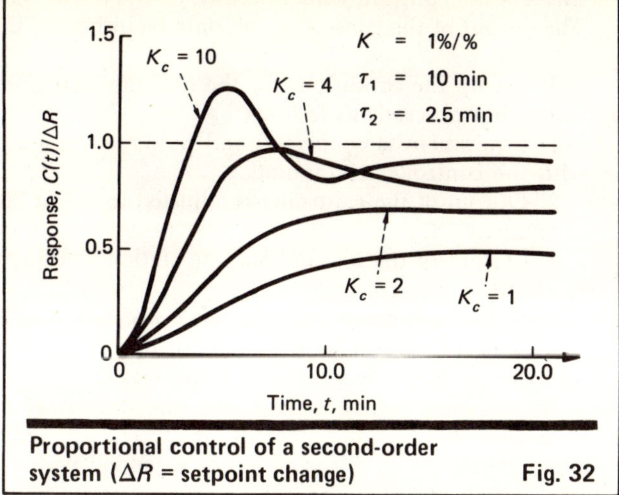

Proportional control of a second-order system (ΔR = setpoint change) **Fig. 32**

As the gain is increased beyond this value, the damping ratio decreases further, which means that the oscillations and overshoot will increase.

Fig. 32 illustrates the response of the system for various values of K_c. This confirms the previous observations that:

- Offset decreases with increasing gain.
- System responds faster with increasing gain.
- Overshoot and oscillations increase with increasing gain.

While the first two are desirable effects, the third is not. The usual practice is to increase the gain until the amount of overshoot and oscillations reach the limit of acceptability.

Although the behavior of the second-order system is more akin to that of industrial processes, even it differs in one respect. No matter how much the gain is increased, the system will never go unstable (a negative damping ratio would be unstable). For most industrial processes, the gain can be increased enough to produce an unstable response. A third-order system or a system with deadtime will behave in this manner. Fig. 33 illustrates an unstable proportional controller for a process with deadtime. But in other respects, the behavior is similar to that of a second-order system. In practice, the limit on the acceptable level of oscillations will be reached before the loop becomes unstable.

In most industrial applications, the gain cannot be set high enough to reduce the offset to a negligible amount. The operator could compensate for the offset by raising or lowering the setpoint, but many controllers provide an adjustment labeled "bias" or "manual reset" which permits the operator to adjust M_R to eliminate or at least reduce the offset.

Reset action

For most systems, proportional control would be quite acceptable except for the existence of offset. Although the process operator can eliminate the offset by adjusting the bias or manual reset, this duty becomes a nuisance, especially in larger plants. Thus, most process controllers contain an additional mode, namely the reset mode, that will automatically eliminate the offset. Whereas the output of the proportional mode is based

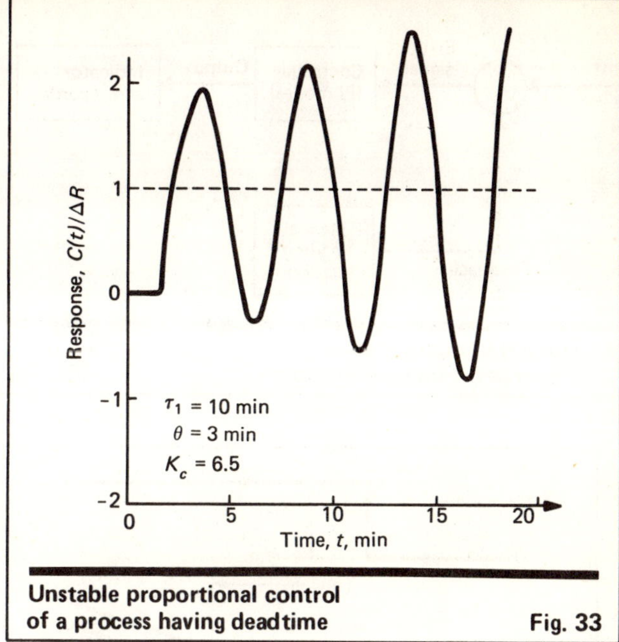

Unstable proportional control of a process having deadtime **Fig. 33**

on the error, the output of the reset mode is based on the integral of the error.

Although reset or integral action used alone will provide control action, the response speed is usually quite slow. Thus, reset action is normally provided in addition to proportional action, the result being the proportional-integral (PI) controller.

The proportional-integral controller is described by the following:

$$m(t) = M(t) - M_R = K_c\,[e + (1/T_i) \int edt]$$
$$M(s) = K_c\,[1 + (1/T_i s)]\,E(s)$$

The tuning coefficient for the reset mode is the reset time, normally in minutes. Most process controllers provide for reset times varying from less than a minute to a maximum of about 30 min. Although analog controllers can be obtained with narrow spans for the reset time (e.g., 0.1 to 3 min), digital technology is normally required if one is to obtain reset times greater than one hour.

Although reset time will be used as the reset adjustment throughout this article, the reset coefficient in many controllers is the reset rate, which is simply $1/T_i$. The coefficient on the integral mode decreases as the reset time increases, which results in an inverse relationship between reset time and reset action ($T_i = \infty$ means no reset). When reset rate is used, the relationship is direct ($1/T_i = 0$ means no reset).

To give some insight into the behavior of reset action, a simple experiment on a PI controller will be discussed. Perhaps the best way to envision this experiment is to consider that the instrument technician has completely removed the controller from the control-room panelboard, and will conduct this test at the workbench. As the "process" is now missing from the control loop, changes in the output of the controller have no effect on the feedback.

As illustrated in Fig. 34, the feedback for this test will be provided by a manual loading station (the output of

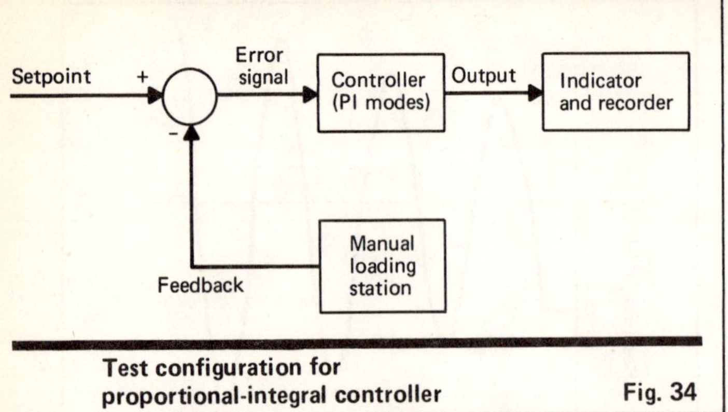

Test configuration for proportional-integral controller **Fig. 34**

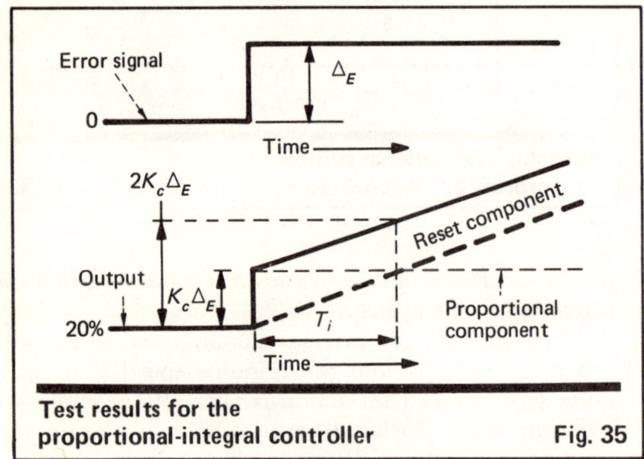

Test results for the proportional-integral controller **Fig. 35**

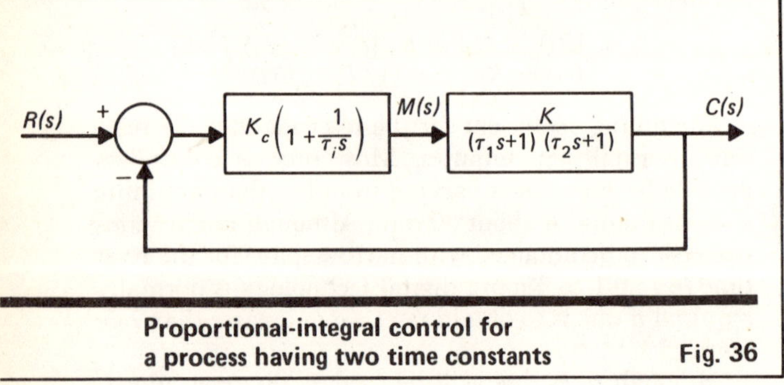

Proportional-integral control for a process having two time constants **Fig. 36**

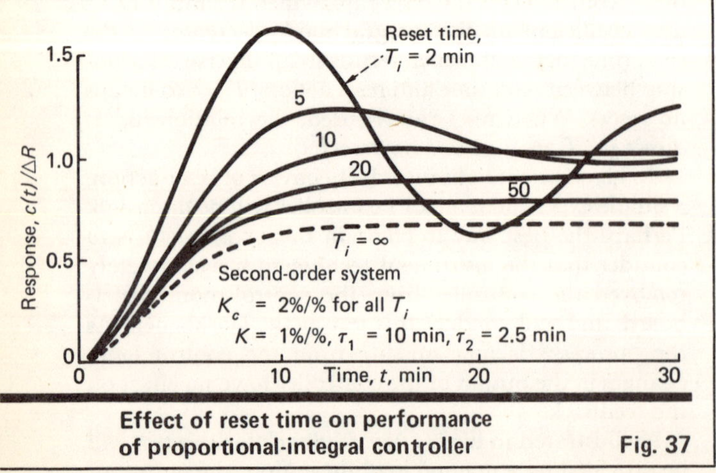

Effect of reset time on performance of proportional-integral controller **Fig. 37**

this unit is a constant value specified by the technician). The output of the controller will only be indicated and recorded.

To set up the conditions for the test, the instrument technician proceeds as follows:

1. Equipment is connected and power is turned on with the controller in manual.
2. Output of the controller is adjusted to about 20% of range.
3. Output of the manual loading station is adjusted to about midrange.
4. Setpoint is made equal to the "process variable" (which is the output of the manual loading station), and the controller is switched to automatic.

As the error signal is now zero, the output of the controller should remain steady at 20% (if an analog controller is left in this state for an extended period of time, small errors in the balancing may cause a slow drift in the output).

The next procedure in the test is to introduce a step change in the error signal. This may be accomplished by adjusting either the setpoint or the output of the manual loading station. Choosing the former, the setpoint will be increased by some amount, say Δ_E. As shown in Fig. 35, a step change in the error signal results.

Fig. 35 also presents the response of the controller for this test. When the error changes from 0 to Δ_E, the proportional mode immediately changes the output by $K_c\Delta_E$. Thereafter, the output from the proportional mode is constant. The output from the reset mode does not change instantly with the change in error, but the nonzero error causes the output of the integrator to change with time. Since the error is constant at Δ_E, the output will be the ramp $\Delta_E t$.

Fig. 35 illustrates the proportional and reset components of the output. The proportional component is constant at $K_c\Delta_E$, and the reset component is the ramp $K_c\Delta_E t/T_i$.

An important determinant of the effectiveness of any controller is its initial reaction to a change. For the test in Fig. 35, the output of the proportional mode changes immediately as the error changes. However, the output of the reset mode does not change immediately, but changes with time at a rate proportional to the error. Thus, the proportional action is said to lead the reset action, which means that the proportional mode will be more effective than reset in responding quickly to process upsets.

For the test in Fig. 35, we note that the proportional action immediately changes the output by a certain amount, namely $K_c\Delta_E$. The time required for the reset action to change the output by this amount is the reset time, T_i. The reset action is said to "repeat" the proportional action, and the units on T_i are frequently written min/repeat. Similarly, the units for reset rate would be min/repeat.

It is probably not evident from Fig. 35 that reset action eliminates offset. However, this can be deduced from the PI equation:

$$m(t) = M(t) - M_R = K_c\,[e + (1/T_i) \int e\,dt]$$

At steady state, the process must be "lined out," i.e., no

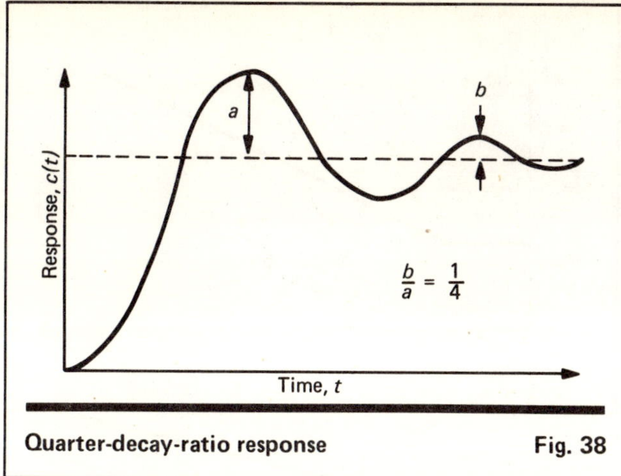

Quarter-decay-ratio response **Fig. 38**

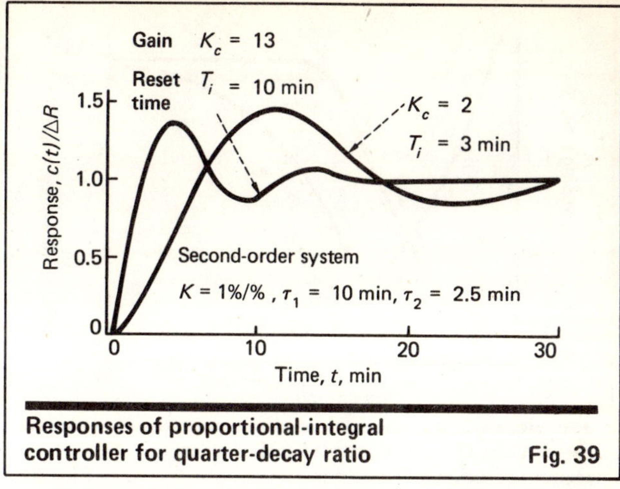

Responses of proportional-integral controller for quarter-decay ratio **Fig. 39**

variables are changing. For $m(t)$ to be constant, then $K_c[e + (1/T_i)\int edt]$ must also be constant. If the setpoint and process variable are constant, then the term $K_c e$ will be constant. However, the integral term $(K_c/T_i)\int edt$ will be constant only if e is zero, i.e., if there is no offset.

This argument is correctly stated as follows: if a control loop "lines out", the error (and the offset) must be zero when the controller contains integral action. The existence of reset action does not guarantee line-out (and thus a stable loop), but only guarantees that there will be no steady-state offset if the loop does attain steady-state conditions.

Fig. 36 illustrates a PI controller being used to control a second-order system. The presentation of the performance of a PI controller is complicated by the fact that there are two tuning parameters. One approach to tuning a PI controller is to first tune the proportional mode, and then tune the reset. We shall follow this procedure.

Fig. 32 presented the performance of a proportional-only controller on the second-order system of Fig. 36. The selection of the appropriate controller gain depends upon the desired performance of the loop. For purposes of illustration, suppose we desire a response with minimal overshoot. From Fig. 32, a proportional gain of 2 %/% gives the fastest possible response with no overshoot (for proportional-controller responses such as in Fig. 32, the overshoot should be evaluated relative to the line-out value of the process variable, and not relative to the setpoint).

Fig. 37 presents the response of the control loop in Fig. 36 for a proportional gain of 2 %/%, and various values of the reset time. Although it is not entirely evident from the responses in Fig. 37, all would eventually attain the setpoint, but quite a long time would be required when $T_i = 20$ min, and even longer when $T_i = 50$ min. For long reset times, the response has a "tail", and approaches the setpoint slowly. For short reset times, the response overshoots and oscillates.

If the primary purpose of reset action is to eliminate offset and if any reset setting other than $T_i = \infty$ would eliminate offset, then what determines the proper reset time? In Fig. 37, the responses for $T_i = 10$ through $T_i = 50$ have basically the same "shape" as the proportional-only response, i.e., their damping ratio is essentially that of the proportional-only response. For reset times less than 10 min, the responses exhibit more overshoot and oscillations, and in fact their damping ratio decreases as the reset time increases. The proper value of the reset time is the lowest value (or "shortest time") that does not significantly affect the damping ratio. Using this approach, the "shape" of the response is basically determined by the proportional setting, and the reset time is adjusted so as to remove the offset as quickly as possible without affecting the damping ratio.

Another performance criterion for control loops is the quarter-decay-ratio criterion, illustrated in Fig. 38, for which the ratio of the second overshoot to the first is 1/4. The underdamped component of the response has a damping ratio of approximately 0.215.

For the PI control loop of Fig. 36, let us consider the following two approaches for attaining a 1/4-decay-ratio response:

1. From Fig. 37, a reset time of somewhere between 2 and 5 min should give a 1/4-decay-ratio response for a controller gain of 2 %/%.

2. From the proportional-only responses in Fig. 32, the gain of 10 %/% gives a response somewhat more damped than 1/4 decay (again, the line-out value and not the setpoint is used to determine the overshoot). A gain of approximately 13 would give a response with a 1/4 decay ratio, and reset action could be added to eliminate the offset.

Fig. 39 illustrates the results using both approaches. The response with the larger gain is clearly the fastest.

In this example, for any gain less than 13 %/%, the reset time can be adjusted to give a response with a 1/4-decay ratio. For any process, there will be a gain for which a proportional-only controller will give a response with a 1/4-decay ratio. For this or any smaller gain, the reset time can be adjusted to give a 1/4-decay-ratio response. However, when speed of response is an important consideration, then the largest gain consistent with the response objectives should be used.

For any response criterion (minimal overshoot, 1/4-decay-ratio, or other), the following tuning procedure will give the fastest response from a PI controller:

1. Remove any reset action, and adjust the proportional mode to give the desired response, ignoring the offset (Fig. 32).

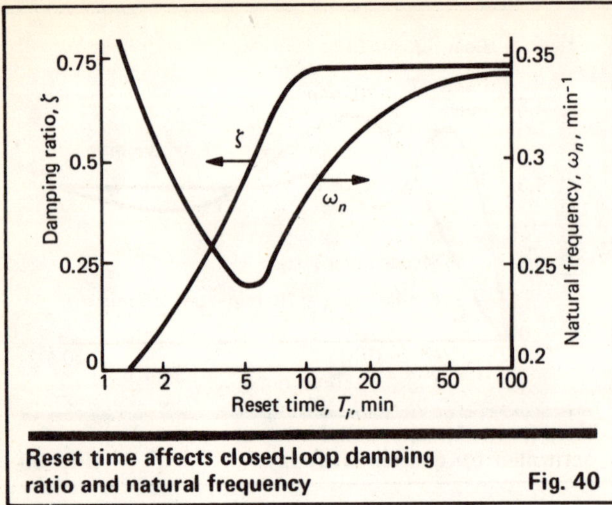

Reset time affects closed-loop damping ratio and natural frequency **Fig. 40**

2. Adjust the reset time to eliminate the offset (Fig. 37). Thus, the adjustment of the PI controller is a two-step procedure.

Poles and zeros of transfer functions

For the PI control loop in Fig. 36, the closed-loop transfer function can be written as:

$$\frac{C(s)}{R(s)} = \frac{K_c\left(1 + \frac{1}{T_i s}\right) \cdot \frac{K}{(\tau_1 s + 1)(\tau_2 s + 1)}}{1 + K_c\left(1 + \frac{1}{T_i s}\right) \cdot \frac{K}{(\tau_1 s + 1)(\tau_2 s + 1)}}$$

$$\frac{C(s)}{R(s)} = \frac{KK_c\left(s + \frac{1}{T_i}\right)}{\tau_1\tau_2 s^3 + (\tau_1 + \tau_2)\, s^2 + (KK_c + 1)\, s + (KK_c/T_i)}$$

$$\frac{C(s)}{R(s)} = \frac{A(s)}{B(s)}$$

The closed-loop transfer function is thus the ratio of a numerator polynomial, $A(s)$, to a denominator polynomial, $B(s)$. Let us assume that no root of $A(s)$ equals a root of $B(s)$, i.e., no cancellation of factors is possible.

The polynomials $A(s)$ and $B(s)$ have roots, i.e., values of s for which the polynomials are zero. For $A(s)$, the root is $s = -1/T_i$. For $B(s)$, the roots are more difficult to determine since $B(s)$ is cubic, but even trial-and-error procedures work rather quickly on a programmable calculator.

The *zeros* of a transfer function are those values of s for which the value of the transfer function is zero. The zeros are the roots of the numerator polynomial $A(s)$.

The *poles* of a transfer function are those values of s for which the value of the transfer function is undefined. When s equals one of the roots of $B(s)$, a division by zero occurs, and the result is undefined. Thus, the poles are the roots of the denominator polynomial $B(s)$.

Let us suppose that the response to a step change in the setpoint is to be computed. Then:

$$C(s) = \frac{1}{s} \cdot \frac{A(s)}{B(s)}$$

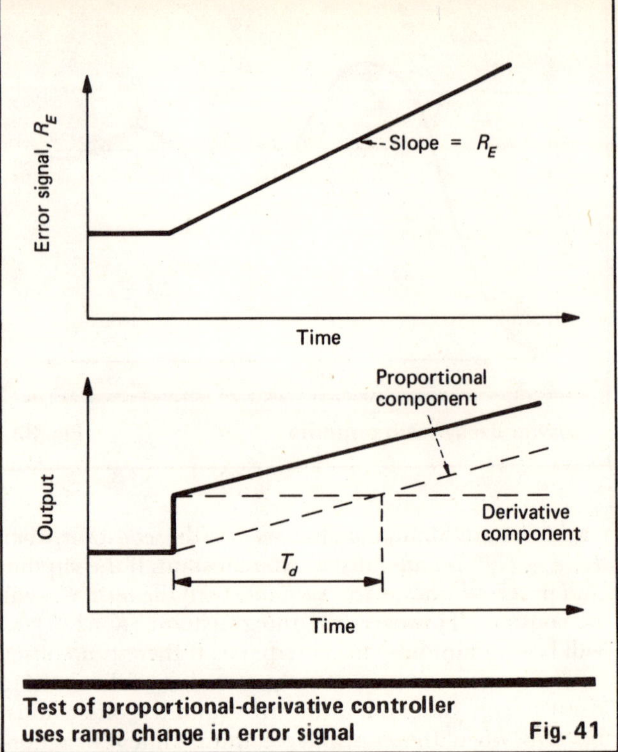

Test of proportional-derivative controller uses ramp change in error signal **Fig. 41**

For the PI control loop, $B(s)$ is cubic, and its three roots, say r_1, r_2 and r_3, are the poles of the closed-loop transfer function.

Then $C(s)$ can be written as:

$$C(s) = \frac{A(s)}{s(s - r_1)(s - r_2)(s - r_3)}$$

For a cubic polynomial, one of the roots must be real; but two may be complex conjugates. For simplicity, we shall assume that no roots of $B(s)$ are repeated, i.e., have the same value. (The following development can be applied to the more general case to obtain the same conclusion, but the mathematics is more complicated.)

Using the technique of expansion by partial fractions, $C(s)$ can be written as:

$$C(s) = \frac{a_0}{s} + \frac{a_1}{s - r_1} + \frac{a_2}{s - r_2} + \frac{a_3}{s - r_3}$$

Although the details are not important for our consideration, the coefficients, a_0, a_1, a_2 and a_3 can be readily calculated from $A(s)$.

In Table I of Laplace transforms, the term $1/s$ in the s-domain corresponds to the unit step function $u(t)$ in the time domain, and the term $1/(s - r_j)$ corresponds to $e^{r_j t}$. Thus, the response $c(t)$ would be:

$$c(t) = a_0 u(t) + a_1 e^{r_1 t} + a_2 e^{r_2 t} + a_3 e^{r_3 t}$$

The coefficients, r_1, r_2 and r_3, in the exponentials are the roots of $B(s)$ or the poles of the closed-loop transfer function. The zeros of the closed-loop transfer function, i.e., the polynomial $A(s)$, affect only the coefficients a_0, a_1, a_2 and a_3.

If the real part of any pole is positive, the corresponding term $e^{r_j t}$ in the expression for $c(t)$ will become larger as t increases, which means the loop is unstable. Thus, for a system to be stable, all of its poles must have negative real parts.

When two of the poles are complex conjugates, the exponential form of the expression is still valid. The complex exponentials can be expressed in terms of sines and cosines, and vice versa. Thus, when two roots are complex conjugates, the response will have an underdamped component.

For the PI control loop, it is convenient to express the two, complex-conjugate roots in terms of the damping ratio and natural frequency. Thus, the denominator $B(s)$ could be factored as:

$$\frac{C(s)}{R(s)} = \frac{A(s)}{(s + p_1)\,(s^2 + 2\zeta\omega_n s + \omega_n^2)}$$

The response will exhibit two components, namely a component $e^{-p_1 t}$ corresponding to the term $(s + p_1)$, and a damped sinusoid corresponding to the term $(s^2 + 2\zeta\omega_n s + \omega_n^2)$.

For the PI control example in Fig. 37, the damping ratio and natural frequency can be determined as functions of T_i. Fig. 40 presents this plot using a logarithmic scale for T_i. Note that values of T_i above 10 min have essentially no effect on the damping ratio. As T_i is decreased from large values, the natural frequency first decreases (slower response), passes through a minimum, and then increases for very small values of T_i. However, the increases are only in the region where the damping ratio is low.

Although our discussion focused on poles and zeros of closed-loop transfer functions, it applies equally to open-loop transfer functions such as controllers and processes. The PI controller can be written as:

$$G_c(s) = K_c\left(1 + \frac{1}{T_i s}\right) = \frac{K_c\left(s + \dfrac{1}{T_i}\right)}{s}$$

This controller has a pole at $s = 0$, and a zero at $s = -1/T_i$.

The relationship between poles/zeros and control-system performance can be analyzed by several techniques—one of which is frequency response. Although this technique formed the basis of classical control theory and is very useful in analyzing the effects of deadtime, chemical engineers have traditionally preferred time-domain analysis over frequency-domain analysis.

As s is a true variable, algebraic manipulations such as cancellation of poles and zeros is permissible. For example, let us look at a controller-process combination, i.e., in $G_c(s)G(s)$. For the control loop for Fig. 37, $G_c(s)$ and $G(s)$ are:

$$G_c(s) = 2(1 + \frac{1}{T_i s}) = 2\left[\frac{(T_i s + 1)}{T_i s}\right]$$

$$G(s) = \frac{1}{(10s + 1)\,(2.5s + 1)}$$

$$G_c(s)G(s) = 2\left[\frac{T_i s + 1}{T_i s}\right] \cdot \frac{1}{(10s + 1)\,(2.5s + 1)}$$

Note that a pole-zero cancellation would occur for $T_i = 10$ or $T_i = 2.5$.

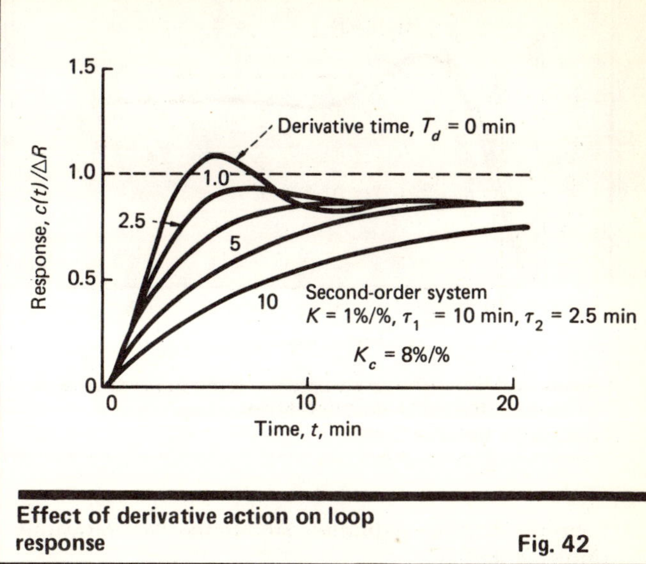

Effect of derivative action on loop response **Fig. 42**

In Fig. 40, it is no coincidence that the damping ratio begins to decrease for reset times less than 10 min. The point at which $T_i = 10$ min corresponds to the larger of the two process time constants. From a pole-zero cancellation point of view, the reset time should be set equal to the largest of the process time constants. This is a reasonable starting value for reset time for most processes, although some may require shorter values.

Derivative mode

The third mode of control available for industrial controllers is the derivative mode. For reasons that will become evident, this mode is by far the most difficult to adjust.

Unlike the reset mode, the derivative mode cannot be used alone. In practice, it is invariably coupled, at least, with the proportional mode, giving as a minimum the proportional-derivative (PD) controller. In almost all cases, reset is also used, giving the proportional-integral-derivative (PID) controller. As the tuning of the PD controller is a subset of the tuning of the PID controller, the simpler PD controller will be considered first.

In the test of the PI controller (Fig. 35), a step change in the error signal was used as the input. A similar test can be done on the PD controller, except that a ramp in the error signal must be generated instead of the step. Fig. 41 illustrates the error signal and the response of the PD controller. The derivative component is constant at $K_C T_D R_E$, and the proportional component is the ramp $K_C R_E t$. The derivative time is the time required for the proportional output to "repeat" the derivative output.

Since the output of the derivative mode is based on the derivative of the error, the ideal equation for a PD controller is:

$$M(t) - M_R = m(t) = K_C\left[e + T_d\left(\frac{de}{dt}\right)\right]$$

or

$$M(s) = K_C(1 + T_d s)E(s)$$

In practice, problems are encountered when imple-

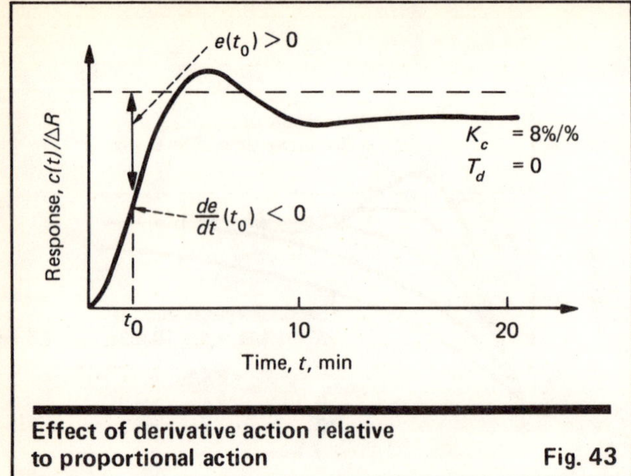

Effect of derivative action relative to proportional action **Fig. 43**

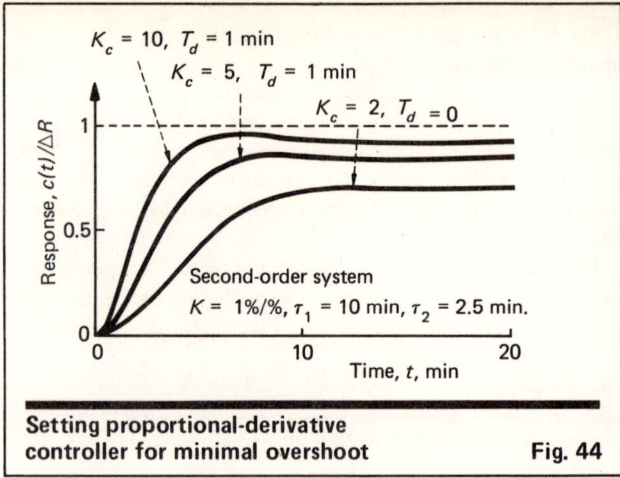

Setting proportional-derivative controller for minimal overshoot **Fig. 44**

menting the above equation. An alternative that can be implemented is:

$$\frac{M(s)}{E(s)} = K_C\left[\frac{1 + T_d s}{1 + \alpha T_d s}\right]$$

which is a lead-lag circuit.

In digital systems, the problems encountered are somewhat different, and the following modification has been used:

$$m(t) = K_C\left[e - T_d\left(\frac{dc}{dt}\right)\right]$$

Here, the derivative is based on the feedback variable instead of the error. Some analog systems have also incorporated this modification.

The test in Fig. 41 also reveals that the derivative output leads the proportional output. If reset had been included, its component would have been 0.5 $K_C R_E t^2/T_i$, which is a parabola. The reset output starts at zero with a zero slope, which makes its initial response slowest of all. Thus, the initial response of the various modes has derivative leading proportional, which in turn leads reset.

Fig. 42 illustrates the effect of derivative for a PD controller on the second-order system previously considered for proportional-only and PI modes. The proportional gain has been arbitrarily set at 8%/%. For this example, the effect of derivative can be seen to be beneficial to stability, but generally at the expense of response speed.

Although the test in Fig. 41 is the usual approach to introducing derivative time, it does not reveal why derivative action should enhance stability. To examine this, the response for $T_d = 0$ in Fig. 42 has been redrawn in Fig. 43. Let us examine the conditions at time t_o when the response is about half-way to the setpoint. At t_o, the error is positive, but decreasing. Thus, the output from the proportional mode, $K_C e$, is positive. Although no derivative action was included for the response in Fig. 43, what would its output have been if it were present? Since the error is decreasing, its derivative is negative, so the derivative component of the output would be negative. As this is opposite from the sign of the proportional component, the output of a PD controller at time t_o would be less than the output of a pure proportional controller. With a smaller output, the PD controller should overshoot less than the pure proportional controller, and as Fig. 42 illustrated, it is possible to adjust the derivative so as to eliminate all overshoot for this example.

Is the derivative mode beneficial in Fig. 42? The answer depends on the response criterion selected for this loop. If the response criterion is 1/4-decay ratio, the answer is no, because the derivative action is increasing the damping. If the response criterion is minimal overshoot, the answer is yes, as the response with $T_d = 1$ min comes close to meeting this criterion.

Suppose the desired performance is minimal overshoot. For the second-order process ($\tau_1 = 10$ min, $\tau_2 = 2.5$ min) in Fig. 31, Fig. 32 showed that a pure proportional controller with a gain of 2%/% would meet this criterion. In Fig. 42, a PD controller with a gain of 8%/% and a derivative time of 1 min would also meet this criterion. Since the proportional gain is higher for the PD controller, two benefits accrue: (1) the response is faster, and (2) the offset is less.

Fig. 44 illustrates minimal overshoot responses for various settings of the PD controller.

Tuning a PD controller is more difficult than tuning a PI controller. A PI controller can be tuned by first adjusting the gain and then adjusting the reset—a simple two-step procedure. For a PD controller, the proportional and derivative settings must be adjusted simultaneously, which is far more difficult.

Using Fig. 44 as the basis for discussing the tuning of a PD controller, the first step would be to set $T_d = 0$ and tune the proportional mode, which gives a gain of 2%/%. The next step would be to increase the gain, which would increase the overshoot beyond what is desired. However, the derivative time could then be adjusted in hopes of reducing the overshoot to an acceptable amount. If successful, the gain could again be increased, and the derivative re-tuned.

The procedure for tuning a PD controller is summarized as:

1. Remove all derivative action, and tune the proportional mode.

2. Increase the proportional gain, and attempt to restore the desired response characteristics by adjusting

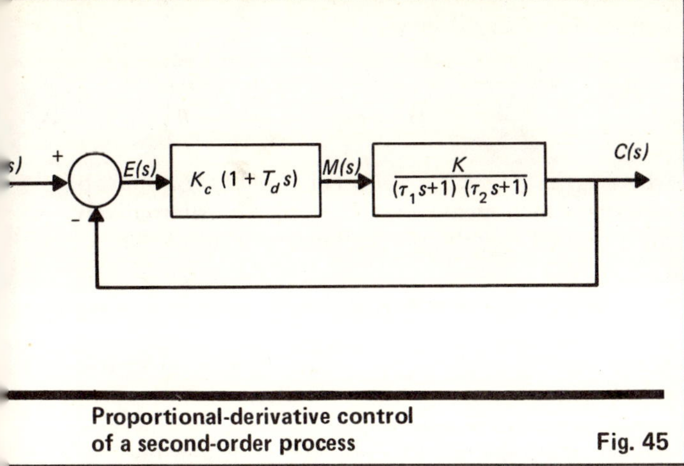

Proportional-derivative control of a second-order process **Fig. 45**

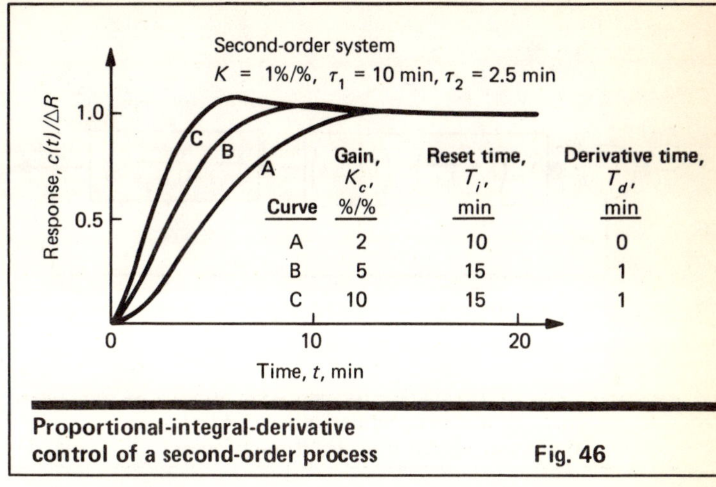

Proportional-integral-derivative control of a second-order process **Fig. 46**

the derivative time. Repeat until the proportional gain is as large as possible.

For most processes, a 25% to 50% increase in the gain would be a reasonable first attempt for Step 2. The three responses in Fig. 44 would result from such a procedure.

Fig. 45 illustrates a loop with a PD controller and a second-order process. In order to have a pole-zero cancellation, the derivative time should be set equal to either τ_1 or τ_2.

In either case, the net result would be proportional control of a first-order system (the derivative term in the controller cancels one pole of the second-order process). It is best to set the derivative time equal to the smaller of the two time constants. In practice, the derivative time is usually set somewhat less than the smaller of the process time constants. (In Fig. 44, the smaller time constant was 2.5 min, but the derivative time was set at 1.0 min.)

Although we shall omit the details, derivative action will not be beneficial in any of the following cases: (1) where the process is predominantly first order, (2) where the process deadtime is large, (3) where the measurement of the controlled variable is noisy. As will be discussed in the next section, temperature loops rarely exhibit any of the preceding properties.

PID controller

The PID controller combines the better stability feature of the PD controller and the no-offset feature of the PI controller into a single controller. There are two formulations for this controller:

1. *Time domain*—The terms for the proportional mode ($K_C e$), reset mode [$(K_c/T_i)\int e dt$], and derivative mode [$K_C T_d(de/dt)$] in the PID controller are the same as the corresponding terms in the PI and PD controllers.

2. *Frequency domain*—The poles and zeros for the PID controller are the same as those for the PI (pole at $s = 0$; zero at $s = -1/T_i$) and PD (zero at $s = -1/T_d$) controllers.

As shown in Table III, the two formulations are very similar but are not quite the same. The zeros of the time-domain formulation are not simple functions of T_i and T_d, whereas the proportional term of the frequency-domain formulation depends upon T_i and T_d.

In practice, most analog controllers use the frequency-domain formulation, which is much easier to implement in analog circuitry. Most computer systems have implemented the time-domain formulation, although the frequency-domain formulation is equally easy to implement. In reality, both exhibit PID characteristics, and will do the job.

As discussed in the previous section, the approach to tuning a PID controller is to first tune PD and then add the reset action. Fig. 44 presented PD responses for various values of the proportional gain, and derivative time.

The addition of reset action to the PD controller is essentially the same as adding reset action to a proportional-only controller. Fig. 46 illustrates the PID responses obtained by adding reset action to the PD responses in Fig. 44.

To summarize, the tuning of a PID controller should proceed as follows:

1. Remove all reset and derivative action, and tune the proportional mode to give the desired response characteristics, ignoring any offset.

2. Increase the proportional gain, and attempt to restore the response characteristics by adjusting the derivative time. Repeat until the proportional gain is as large as possible.

3. Adjust the reset time to remove the offset.

Of these, Step 2 is by far the most difficult, primarily because the two mode adjustments have to be made simultaneously.

Since the derivative mode significantly complicates the tuning procedure, its use should be limited to those loops where improved loop performance justifies the extra effort. The rule-of-thumb generally followed is to use PID control only in temperature loops, using PI elsewhere.

This rule-of-thumb can be rationalized on two factors. First, temperature loops are usually relatively slow, so any improvement would be very beneficial to process operation. Second, the previous section noted that the derivative mode was needed only when the process had a major time constant and a minor one. Temperature loops are most likely to be of this type. Using a reactor as the example, the reacting mass usually produces the major time constant. For temperature control, the reac-

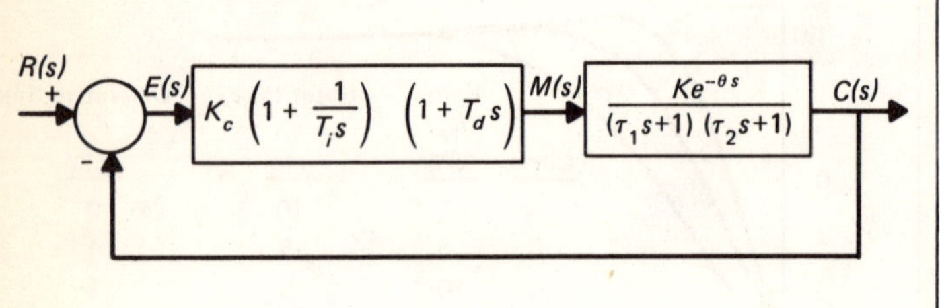

Proportional-integral-derivative control of a second-order plus deadtime process **Fig. 47**

tor must also have a jacket, as internal coil, or other arrangement for heat removal. The minor time constant associated with this equipment is usually significant.

Fig. 47 illustrates PID control of a two time-constant plus deadtime process. Using the pole-zero cancellation concepts and assuming that $\tau_1 \geq \tau_2$, the following settings result for T_d and T_i:

$$T_i = \tau_1$$

$$T_d = \tau_2$$

After these pole-zero cancellations, the closed-loop transfer function is (with $T_i = \tau_1$):

$$G_{CL}(s) = \frac{\dfrac{KK_C e^{-\theta s}}{\tau_1 s}}{\dfrac{KK_C e^{-\theta s}}{\tau_1 s} + 1} = \frac{e^{-\theta s}}{\dfrac{\tau_1}{KK_C}s + e^{-\theta s}}$$

When the deadtime, θ, is small, the approximation $e^{-\theta s} = 1 - \theta s$ is acceptable. Replacing $e^{-\theta s}$ in the denominator of $G_{CL}(s)$ gives:

$$G_{CL}(s) = \frac{e^{-\theta s}}{\dfrac{\tau_1}{KK_C}s + (1 - \theta s)} = \frac{e^{-\theta s}}{\left(\dfrac{\tau_1}{KK_C} - \theta\right)s + 1}$$

where $\tau_{CL} = [(\tau_1/KK_C) - \theta]$. For stability, τ_{CL} must be positive, which means that $K_C < \tau_1/K\theta$. Alternatively, for a desired τ_{CL}, the proportional gain would be:

$$K_C = \frac{\tau_1}{K(\theta + \tau_{CL})}$$

Using the equations $T_i = \tau_1$ and $T_d = \tau_2$ gives relationships for tuning a PID controller.

Although the details will be omitted, this development can be extended to provide guidelines on when derivative should be used. It turns out that the derivative mode will be effective only when the minor time constant τ_2 exceeds the deadtime θ, i.e., use $T_d = \tau_2$ provided $\tau_2 > \theta$; otherwise, omit the derivative mode. This will be the case for most temperature loops, which justifies the use of the derivative mode in those loops.

Controller tuning

The final step in the commissioning of a control loop is usually the adjustment of the tuning parameters. In reality, this constitutes a test on the design of the control loop. If the controller can be tuned to give satisfactory performance, it is customarily assumed that all elements of the loop are suitable to the task and in good working order.

Where the controller cannot be satisfactorily tuned, the selection of the remaining elements of the loop must be reviewed, and the overall configuration (selection of controlled and manipulated variables) examined. The "untunable controller" is a side effect of a flaw in the design of the loop. Loops that are properly designed

Formulations for the proportional-integral-derivative controller **Table III**

	Time Domain	Frequency Domain
Differential equation	$K_C\left[e + \frac{1}{T_i}\int e\,dt + T_d\left(\frac{de}{dt}\right)\right]$	$K_C\left[\left(1 + \frac{T_d}{T_i}\right)e + \frac{1}{T_i}\int e\,dt + T_d\left(\frac{de}{dt}\right)\right]$
Transfer function	$K_C\left(1 + \frac{1}{T_i s} + T_d s\right)$	$K_C\left(1 + \frac{1}{T_i s}\right)\left(1 + T_d s\right)$
Proportional term	$K_C e$	$K_C\left(1 + \frac{T_d}{T_i}\right)e$
Zeros	$\dfrac{-1 \pm \sqrt{1 - 4\left(\frac{T_d}{T_i}\right)}}{2T_d}$	$-\frac{1}{T_i};\ -\frac{1}{T_d}$

and applied will be tunable; those that are misapplied or have a design mistake will not be tunable until the problem is corrected.

Tuning a controller is complex and involves:

1. Process characteristics (gain, time constants, deadtime, etc.).
2. Controller capabilities (PI or PID).
3. Desired response (minimal overshoot, quarter decay, or other).

The most common approach to tuning controllers is popularly referred to as the "seat-of-the-pants" method, knob-twiddling, or other. Previous sections have suggested approaches for the various types of controllers. With some experience, one can become rather proficient with this approach. However, many pitfalls await the inexperienced.

An alternative approach is to rely on a tuning technique. These methods embody:

1. A test of some type to obtain some numbers that are indicative of the process characteristics.
2. A set of equations that relate the controller settings to the numbers determined above.

In Step 2, the equations are based on an objective of providing a response with certain characteristics.

In practice, tuning techniques usually do not give exactly the type of response desired. Thus, some final adjustments to the tuning parameters must be made manually. However, we should at least obtain a reasonable starting point from which the desired response characteristics can be obtained.

Tuning techniques can be broadly classified as closed-loop and open-loop methods. Closed-loop methods check the process with the controller in automatic. Open-loop methods check the process with the controller in manual.

Closed-loop tuning methods

The Ziegler-Nichols closed loop method [2] requires finding the gain of a proportional-only controller that will cause the loop to cycle indefinitely with a constant amplitude. This gain is the upper limit of the gains for which the loop is stable, and thus is called the ultimate gain. The method proceeds as follows:

1. Tune out all reset and derivative action ($T_i = \infty$, $T_d = 0$), and place the controller on automatic.
2. Introduce an upset into the loop, and adjust the proportional gain, K_C, until the loop cycles continuously.
3. Record the value of K_C from Step 2 as the ultimate sensitivity, S_u, and record the period as the ultimate period, P_u.
4. Determine the settings from the equations in Table IV.

The objective of this method is to give a response with a quarter-decay ratio.

The upset in Step 2 can be introduced in a variety of ways. One approach is to simply move the setpoint up or down by some arbitrary amount, wait until the process begins to respond, and then return the setpoint to its previous value.

Fig. 48 illustrates determining the ultimate gain for a temperature loop. With the gain set at 2%/%, the set point is increased from 174°F to 180°F. When the

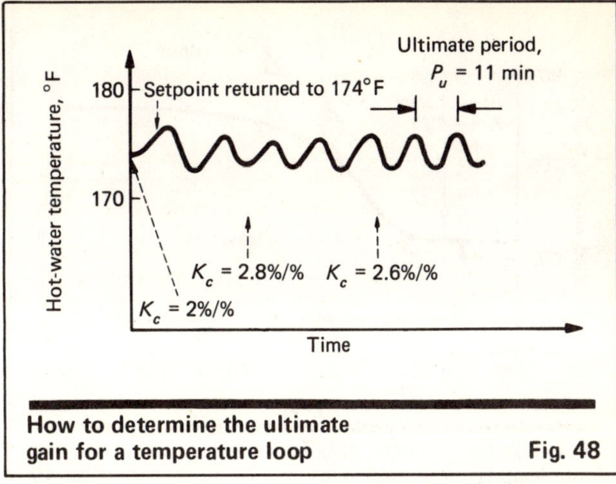

How to determine the ultimate gain for a temperature loop **Fig. 48**

process begins to respond, the setpoint is returned to 174°F. The initial gain is observed to give a response with a decaying amplitude, so the gain is increased to 2.8%/%. This gives a response with a slightly increasing amplitude, so the gain is decreased to 2.6%/%. The amplitude is essentially constant, so the ultimate gain S_u is 2.6%/% and the ultimate period P_u is 11 min. The amplitude of the cycle is immaterial, but all elements of the loop must be within their operating range (i.e., the valve must not go full open or full closed).

The equations in Table IV can be used to compute controller settings. For a PI controller, the settings would be $K_C = 1.2\%/\%$ and $T_i = 9.2$ min.

As for the effectiveness of this method, the results are normally quite good for a PI controller. For a PID controller, the ratio of the derivative-time to the reset-time settings, resulting from the tuning relations in Table IV is:

$$\frac{T_d}{T_i} = \frac{P_u/8}{0.5P_u} = \frac{1}{4}$$

This ratio is not always appropriate.

In practice, the major obstacle to applying this method is the nature of the test itself. For most major loops in a production plant, an oscillatory behavior such as that in Fig. 48 cannot be tolerated for very long. For a slow process, the test could last for quite a while. The test in Fig. 48 requires about 1.5 h for a process with a period of oscillations of just over 10 min. Furthermore, the test in Fig. 48 went very smoothly, requiring only three trials to obtain the proper gain, and at most three cycles at each setting. One will rarely be more fortunate.

The severity of the test can be reduced somewhat by proceeding as follows:

1. Start with a proportional-only controller, adjust the gain to give a response with a quarter-decay ratio. Call this gain $K_{1/4}$, and the period $P_{1/4}$.
2. Make the following approximations to give:

$$S_u = 2K_{1/4}$$
$$P_u = P_{1/4}$$

The first equation follows directly from the tuning equation for a proportional-only controller. The latter

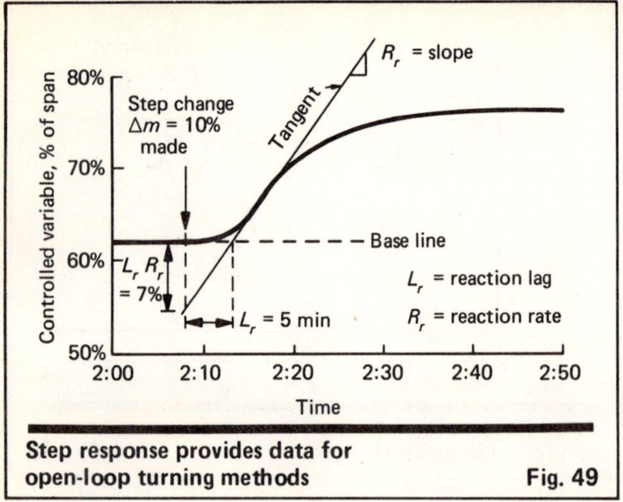

Step response provides data for open-loop turning methods **Fig. 49**

equation is an approximation, since P_u should be less than $P_{1/4}$.

3. Use the equations in Table IV with these values of S_u and P_u. For a PI controller, the net result is: $K_C = 0.9K_{1/4}$ and $T_i = P/1.2$.

In practice, this conforms very closely to the previously recommended procedure for tuning a PI controller. The proportional mode is first tuned, the gain is reduced by only 10% (which is marginally significant), and the reset time is adjusted beginning with a value about 20% less than the period of the proportional response.

Open-loop tuning methods

Open-loop tuning methods use a step-response test of the process, and proceed as follows:

1. Place the controller on manual and allow the process to line out.

2. Make a step change in the controller output (valve position).

3. Record the response.

This test is exactly the same as that shown in Fig. 4 for the hot-water heater.

While this test is quite simple, there are some practical problems. First, extreme care must be taken not to allow the test data to be influenced by disturbances, one source of which is corrective actions made by the operator in neighboring process equipment. Second, the customary strip-chart recorder is at best marginally adequate for recording the response. A faster chart drive is only a partial solution. To be really effective, a strip-chart recorder with adjustable speed, adjustable sensitivity, and adjustable zero is required. Of course, these are much easier to connect to electronic controllers than to pneumatic ones. Some process-computer systems provide special software to collect such data and plot it with whatever resolution is needed.

Fig. 49 illustrates a typical step response obtained from such a test. The various methods use different parameters as determined from the test. However, all begin by constructing a line tangent (Fig. 49) to the response at the point of steepest ascent (the inflection point).

The Ziegler-Nichols open-loop tuning method [*2*] uses the parameters L_r and R_r to characterize the process. The reaction rate, R_r, is the slope of the tangent line in Fig. 49. The reaction lag, L_r, is the time that the tangent line intersects the original base line, where time is measured from the introduction of the step input. It turns out that the tuning equations never use R_r alone, but always use the product L_rR_r. This product can be read directly in Fig. 49.

Table IV also presents the equations that relate the tuning coefficients to the parameters L_r and L_rR_r, which were determined in Fig. 49 to be 5 min and 7%, respec-

Tuning equations to compute controller settings **Table IV**

Controller	Setting	Ziegler-Nichols Closed Loop	Ziegler-Nichols Open Loop	ITAE Open Loop*
Proportional, P	K_C	$0.5S_u$	$\Delta m/L_rR_r$	$\frac{0.490}{K}\left(\frac{\theta}{\tau}\right)^{-1.084}$
Proportional + integral, PI	K_C	$0.45S_u$	$0.9\Delta m/L_rR_r$	$\frac{0.586}{K}\left(\frac{\theta}{\tau}\right)^{-0.916}$
	T_i	$P_u/1.2$	$3.33L_r$	$\frac{\tau}{1.03 - 0.165\,(\theta/\tau)}$
Proportional + integral + derivative, PID	K_C	$0.6S_u$	$1.2\Delta m/L_rR_r$	$\frac{0.965}{K}\left(\frac{\theta}{\tau}\right)^{-0.855}$
	T_i	$0.5P_u$	$2.0L_r$	$\frac{\tau}{0.796 - 0.147\,(\theta/\tau)}$
	T_d	$P_u/8$	$0.5L_r$	$0.308\tau\left(\frac{\theta}{\tau}\right)^{0.929}$

*ITAE = integral of time and absolute error

tively. For a PID controller, the settings would be:

$$K_C = 1.2\Delta m/L_r R_r = 1.7\%/\%$$
$$T_i = 2L_r = 10 \text{ min}$$
$$T_d = 0.5L_r = 2.5 \text{ min}$$

These settings are intended to give a 1/4-decay ratio.

In practice, this method is usually quite effective in tuning a PI controller. For the PID, this method also fixes the ratio T_d/T_i at 1/4. Furthermore, the reaction lag, L_r, is directly related to the process dead time. Thus, the equation $T_d = 0.5L_r$ for the derivative mode in effect makes the derivative setting proportional to the process deadtime, which conflicts with the previous observation that the derivative mode is ineffective for processes with large deadtimes.

For those cases where a 1/4-decay-ratio response is too oscillatory, an alternative tuning method [*3*] is one based on minimizing the integral of time and the absolute error (*ITAE*) defined as:

$$(ITAE) = \int_0^\infty e(t) \cdot tdt$$

This criterion penalizes very heavily for small errors, occurring late in time. Thus, the controller settings that minimize this criterion usually produce responses close to critically damped.

The relationships in Table IV for this method require the parameters K, τ, and θ, i.e., the process is approximated by a first-order lag plus deadtime model. The gain K is determined as $\Delta c/\Delta m$ in the usual manner. As illustrated in Fig. 50, the dead time θ is the same as the reaction lag L_r, and thus is determined by constructing the tangent line. The time constant τ is determined from the time required for the response to attain 63.2% of the change Δc. That is, the response should attain 63.2% of the total change Δc in time equal to $\theta + \tau$.

From the graph in Fig. 50, the process parameters are $K = 1.4\%/\%$, $\tau = 8$ min, and $\theta = 5$ min. For a PID controller, the settings for an (*ITAE*) test would be:

$$K_C = (0.965/K)\ (\theta/\tau)^{-0.855} = 1.03\%/\%$$
$$T_i = \tau/[0.796 - 0.147(\theta/\tau)] = 11.4 \text{ min}$$
$$T_d = 0.308\tau(\theta/\tau)^{0.929} = 1.6 \text{ min}$$

The Ziegler-Nichols method gave the settings 1.7%/%, 10 min, and 2.5 min, respectively.

In practice, this method generally works well for PI, but not for PID. If the exponent in the equation for T_d were 1.0 instead of 0.929, the equation would reduce to $T_d = 0.308\theta$. Thus, the derivative time is related to the process deadtime, which is not fundamentally sound.

The basic problem with both of these methods in the tuning of the PID controller is with the parameters that characterize the process. Basically, both utilize a first-order (single time constant) approximation to the process. Our previous discussion, relative to the PID controller, indicated that the derivative time should be related to the second or minor process-time constant. Neither method evaluates this parameter, so their effectiveness in tuning the derivative mode suffers.

While graphical methods do exist for approximating responses such as in Fig. 49 and 50 with a second-order plus deadtime model, their detail is beyond what can reasonably be used on realistic process responses. An alternative is to use a nonlinear regression program on a computer system. If the present trend toward microprocessor-based controls continues, this could become a realistic technique.

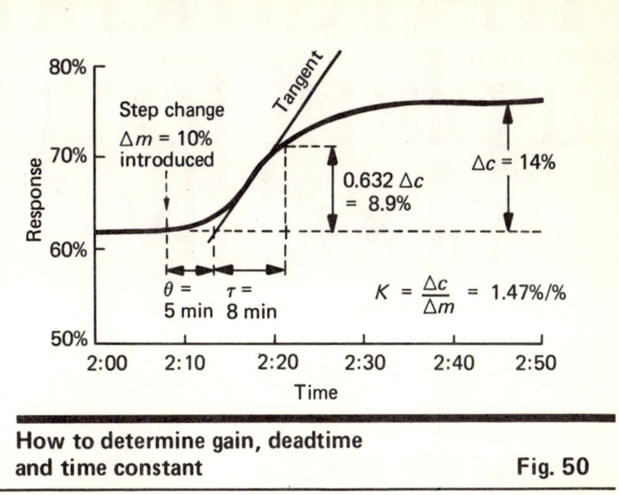

How to determine gain, deadtime and time constant **Fig. 50**

Summary

The subject of process control encompasses many topics not included in this article. Control equations other than the PID are more appropriate for some applications such as pH control. Concepts such as cascade control, feedforward control, deadtime compensation, noninteracting control, and others are finding wider applications as the demands on control-system performance increase. Thus, this article is not a comprehensive treatment of process control, but an understanding of the basics is essential in order to appreciate the other topics.

References

1. Smith, D. E., Stewart, W. S. and Griffin, D. E., Distill With Composition Control, *Hydrocarbon Process.*, Feb. 1978, p. 99.
2. Ziegler, J. G. and Nichols, N. B., Optimum Settings for Automatic Controllers, *Trans. ASME*, Nov. 1942, p. 759.
3. Rovira, A. A. and others, Tuning Controllers for Set Point Changes, *Instrum. Control System.* Dec. 1969.

Acknowledgement: This article is based on the author's presentations in the short course, "Automatic Control of Processes," offered by the AIChE Educational Services Dept.

The author

Cecil L. Smith is a consulting engineer (4620 Bluebell St., Baton Rouge, LA 70808, phone 504/344-2321) specializing in process automation technology, and provides both process automation services and software packages for process control systems. He has authored nine books and over 100 articles. A former professor at Louisiana State University, he is a member of AIChE and the Instrument Soc. of America, and is registered as a chemical engineer in the State of Louisiana and as a control systems engineer in the State of California.

Instrumenting a plant to run smoothly

The instrumentation that controls a process plant must be designed with the problems of the operator in mind. Also, every control scheme has advantages and disadvantages. Here are ways to strike a balance that will help your plant to run smoothly.

Norman Lieberman, *Adrian R. Davis Associates*

☐ The most important aspect of any process design, as far as operating personnel are concerned, is the process control scheme.

The operator should feel confident that the designer is using the best practical technology in designing the "brain" of the unit that the operator is to run. To do this, the process designer must be familiar with the day-to-day problems (such as locating the liquid level in a drum) that the operating superintendent must live with.

It is the process designer's job to make sure that a unit can be started up, shut down, and run during upset conditions, by the average hourly operating person. To do this, a process unit must be designed with adequate instruments and controls. Remember four things:

1. Controls must deal with the unexpected, not just day-to-day trimming of operating parameters.
2. Analysis of an operating problem in the field is much harder than analysis in the office.
3. The hourly operating people are not engineers. Control schemes must be kept simple. The cleverest control scheme will not work if the average operator cannot understand it.
4. Operating personnel will have to live with the designer's mistakes for many years.

A process design is not complete until the control scheme has been developed. Heat integration, equipment sizing, pressure and temperature levels are all functions of the control scheme chosen. Critical controls must be specified, and reasons for the controls clearly explained. The practicality of any control scheme should be reviewed with an instrument engineer before detailed heat and material calculations are made for the final processs design. The final control scheme will often be a compromise between the process designer, the instrument engineer and the process operating man—each of whom may view the project from a slightly different perspective.

After operability, the second objective of process controls and instruments should be to optimize energy utilization. Unless the operators actually know what is going on and have a good "handle" to control the process, the best-thought-out heat integration system will not work.

Originally published September 12, 1977.

Parameter-measuring equipment

To properly control a process facility, five parameters normally are followed. These are temperature, pressure, liquid levels, flows, and composition-related factors. These parameters are measured both for local display and for transmission to a remote location (usually in the unit control room).

Local display of parameters

Temperature—A dial thermometer in a thermowell is used to read local temperatures. This type of installation is inexpensive and does not detract from the mechanical reliability of a unit, since thermowells are not prone to develop leaks.

Pressure—A pressure gage coming off a small connection is used to read local pressures. Pressure gages are not reliable in vacuum service, so mercury manometers should be employed. A pressure-gage installation in liquid hydrocarbon service is always a potential hazard, because it requires a small connection to process piping or vessels.

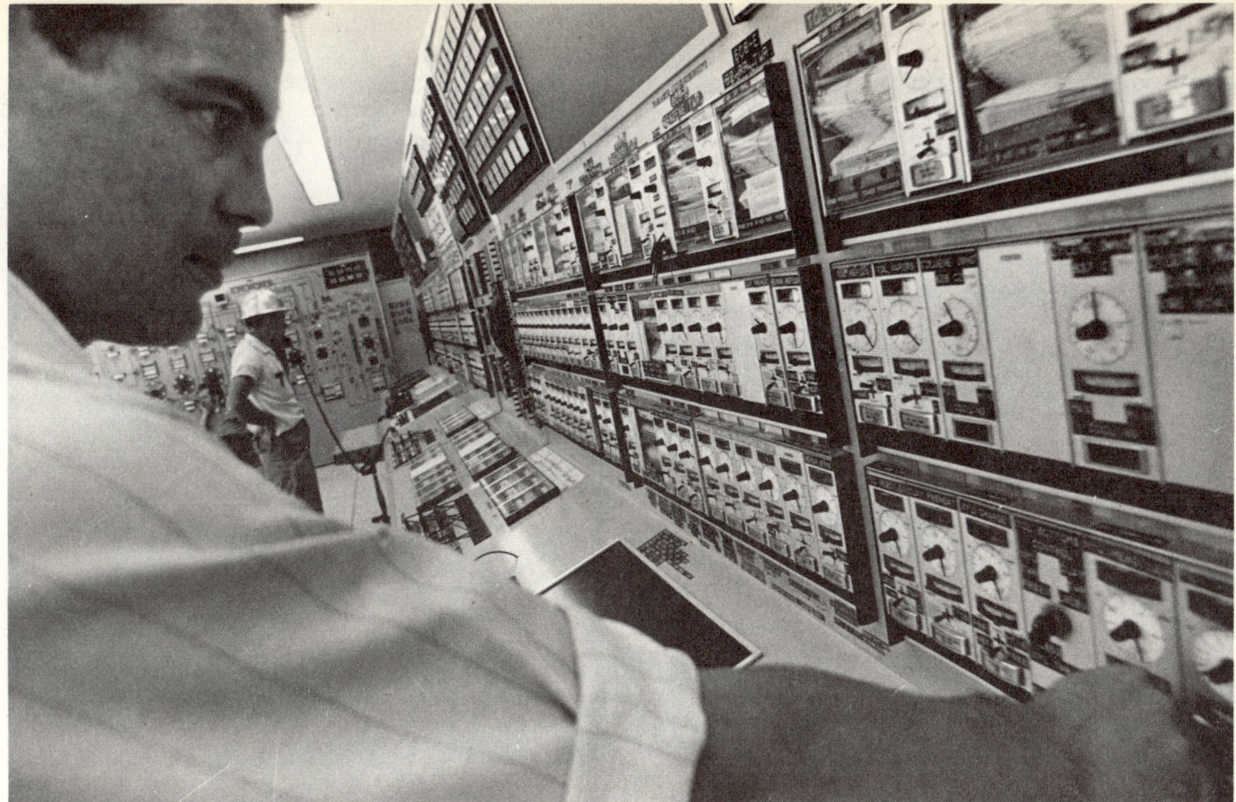

All small connections may eventually start leaking.

Liquid levels—A gage glass is used to observe the liquid levels in vessels. Many upsets are caused on process units because operators cannot visually observe the liquid levels. Fig. 1 shows a properly installed gage-glass assembly.

Liquid levels are normally *very* hard to see in gage glasses. Facilities must always be provided for an operator to drain down a gage-glass assembly so that he can pick up the level by movement. Also, liquid-level taps tend to plug up. Therefore, facilities must be provided to block in and either blow out or blow back all liquid-level taps.

Flows—Flows are generally measured by means of an orifice plate. Other types of local flow indicators are rotameters, and sonic meters that measure the velocity of fluids in a pipe without any internal interference with the flowing fluid.

Remember that flowmeters can use up a substantial amount of energy, when used in low-pressure-vapor service. Many flowmeters installed on process units are of very limited value to operating personnel. Flowmeters should be provided only where they will really be needed.

Composition-related factors—Some of the more common instruments available for hydrocarbon service are listed below:

- pH analyzer—usually reliable.
- Gas chromatograph—usually reliable.
- Colorimeter—moderately reliable.
- Viscometer—moderately reliable.
- Sulfur analyzer—poor reliability.
- Flash-point analyzer—moderately reliable.
- Oxygen analyzer—poor reliability (sample-taps frequently plugs up).
- Conductivity (to measure water in hydrocarbons)—usually reliable.
- Misc.—pour point, cloud point, octane, ASTM endpoint.

Remote display of parameters

Temperature—Except in furnaces, thermocouples are extremely reliable. For accurate measurements of furnace firebox temperatures, a "velocity" thermocouple is used. This sucks hot flue gas over a thermocouple to reduce the error caused by reradiation of heat from an ordinary thermocouple.

Pressure—The pressure on a process vessel or line is sensed directly through a small connection. The pressure can be converted into a pneumatic or electronic signal and transmitted to the remote location.

Liquid levels—A typical installation for sensing liquid levels in a vessel is shown in Fig. 2.

The liquid level pushes up the float, which changes the tension of the spring, which causes a change in the pneumatic or electronic signal from the transmitter, which sends the signal to a remote location.

Flows—For an orifice or sonic-type flowmeter, the local display is operated by a form of signal that can also be transmitted to the control room. Flows measured by rotameter are normally only for local display.

Control valves

Any type of valve (with a suitable operator) can be made to open or close by a pneumatic signal transmitted from a remote location. However, the most common type of control valve resembles a plug valve. The plug of the control valve is moved up or down by means of a rod

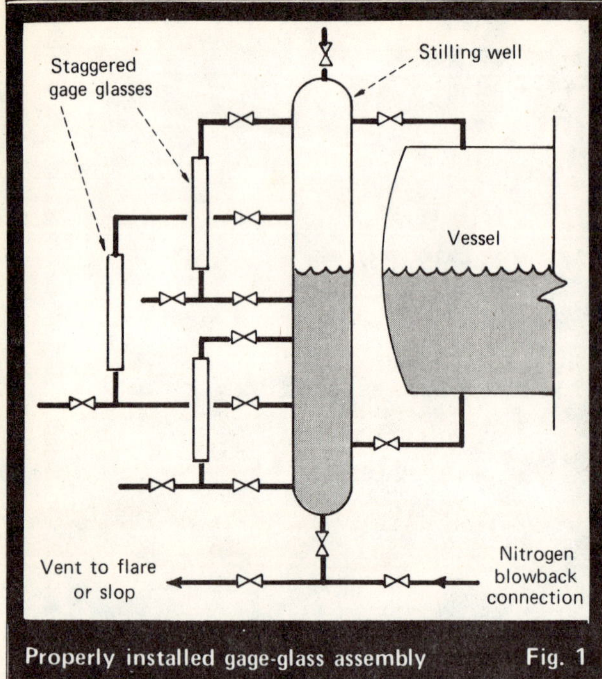

Properly installed gage-glass assembly Fig. 1

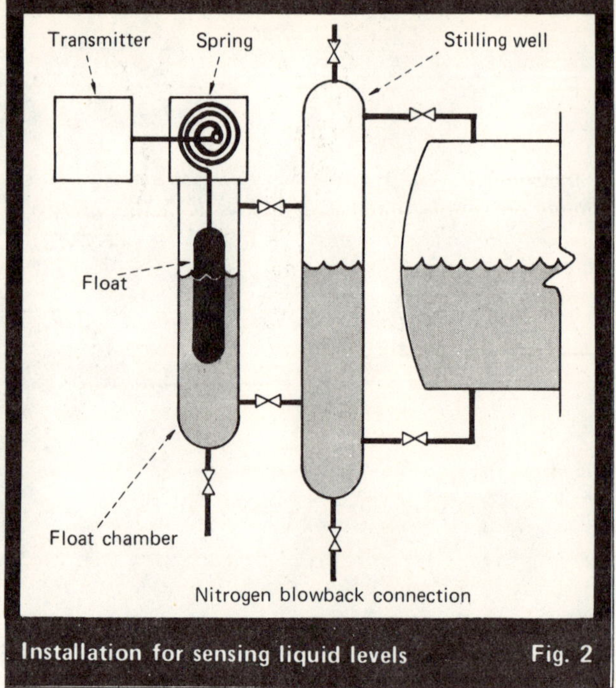

Installation for sensing liquid levels Fig. 2

extending through the valve body. This rod is moved up or down by air pressure. The air pressure to the rod is controlled by a remote signal. Note that the air signal from the remote sensing device is *not* directly moving the rod.

The control-valve I.D. is usually much smaller than that of the line that the control valve is in. Normally, control valves are sized for 50% of the design-pressure-drop of the system they are supposed to be controlling. For example, if a control valve is designed to control the flow in a line that has a 10-psi pressure drop, the control valve will be sized to have a 10-psi pressure drop when halfway open.

When the control-rod and plug move up, some control valves *open* whereas others close. Air pressure can either open or close a control valve. It is vital that the process designer decide whether he wants a particular control valve to fail open or closed in case the motive air pressure to the valve is lost. This is an important safety consideration; for instance, a control valve feeding fuel to a furnace should close if it loses air pressure.

The designer should always insist that all control valves come equipped with a hand-jack. A hand-jack will allow an operator to open or close a control valve in the field without the aid of air pressure.

Omission of block valves and a full-size bypass around a control valve is a sure sign that the designer has no intention of ever visiting the unit he designed. Control valves are prone to sticking shut, sticking open, and developing bad packing leaks around their rods. Unless the facility the control valve is part of can easily be taken out of service, blocks and a bypass are necessary.

Most control valves have a limited range. A control valve that can pass 10,000 bbl/d when fully open has difficulty controllng flows of less than 2,000 bbl/d. Control valves can be obtained that have wider operating ranges, but these are speciality items that present special maintenance problems.

If low pressure-drop is required for a system, a butterfly control valve should be used. This will not provide as fine a control as a plug-type control valve, but is suitable for many applications. Butterfly valves should not be used if tight shutoff is ever needed, as they usually leak.

If a valve is to be fully opened or fully closed by a signal from a remote location, a "motor operated valve" (MOV) should be used. MOVs are quite expensive, but do give a tight shutoff quickly, with minimum operator effort.

For local hand-control of a stream, a ball valve should be specified. Never plan to use an ordinary gate-type block valve to routinely throttle the flow of a stream. The gate and seat will be eroded away.

It is not necessary to have any particular parameter

Glossary

APC	Air pressure closes (valve)
APO	Air pressure opens (valve)
ARC	Analysis recorder/controller
C.W.	Cooling water
FC	Flow controller
FIC	Flow indicator/controller
FR	Flow recorder
FRC	Flow recorder/controller
HCV	Hand-controlled valve
HLA	High-level alarm
LG	Level glass
LI	Level indicator
LIC	Level indicator/controller
LLA	Low-level alarm
LRC	Level recorder/controller
PRC	Pressure recorder/controller
TRC	Temperature recorder/controller

operate a valve. Often, an operator working from the control room can directly manipulate the motive air pressure to a control valve by using a manual loading station (MLS).

Symbols for process flowsheets

Instruments and controls used on process flowsheets should be simple and clear. Leave as many details as possible for the instrument engineer to specify. For instance, transmitters are not normally shown on a process flowsheet.

Alarms and trips

The purpose of a trip is to automatically shut down part or all of a process when an unsafe condition has developed. For example, when the charge pump to a furnace is lost, the fuel to the furnace should be automatically shut off to prevent burning out the furnace tubes.

Except in the case of measuring flows, the sensing point for an alarm or trip should *not* be shared with the sensing point used to indicate or record the parameter in the control room. For example, suppose the pressure controller on a vessel fails to control the vessel pressure, due to plugging of the pressure-sensing tap. If the high-pressure alarm on the vessel sensed pressure from the same tap, there would be no way for the operators to know that the vessel was being overpressured. For really critical services, four separate sensing points should be used for:

1. Local indicator
2. Control-room recorder
3. Alarm
4. Trip

When an alarm is set off, a buzzer will sound in the control room and a light will come on at the alarm panel indicating which alarm has sounded. The alarm can be silenced by pushing a button, but the alarm light will stay on until the trouble that activated the alarm is corrected.

When specifying alarms, the designer should keep in mind that operators do not always respond quickly to trouble. It is not uncommon for a recorded parameter to be outside its permissible value for many hours before it is noticed, even with an alarm light on.

When specifying trips, the philosophy should be that, if it is physically possible for an unsafe condition to develop through a combination of equipment failure and operator error, the unsafe condition will eventually come about. Specifiying trips is an important part of a final process design. However, remember that, on occasion, inadvertent tripping can be more dangerous than the actual malfunction. Design philosophy should also assume that an improper trip will eventually occur.

Controlling levels in vessels

Level control is the simplest but most important function of process control. Fig. 3 illustrates how a parameter-sensing point works to operate a control valve.

In the figure, the vessel level is being recorded in the control room. An air signal from the level-sensing device is transmitted to a control valve. The control valve opens when the level in the vessel rises above an adjustable setpoint. The flow from the discharge of the pump is recorded in the control room.

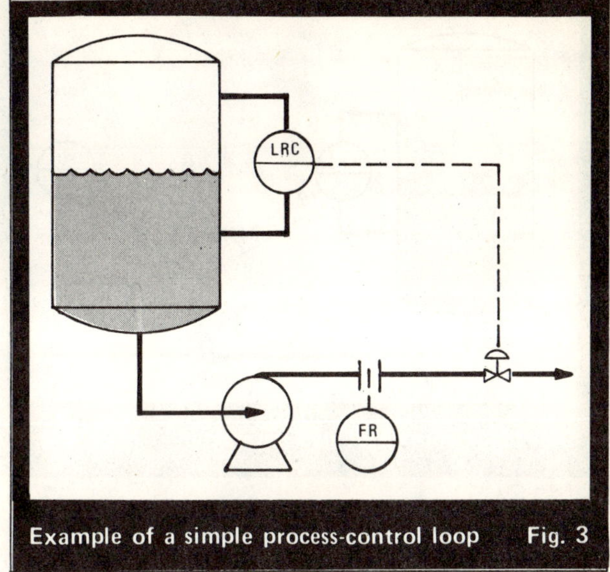

Example of a simple process-control loop Fig. 3

Split-range control

Split-range control is a simple and common method of increasing the flexibility of process equipment. There are two types of such control:

a. A parameter signal is used to control the operation of two control valves, as in Fig. 4. The pressure recorder controller (PRC) is attempting to hold a constant pressure in the vessel by admitting or venting gas when the pressure changes. Both control valves are prevented from opening at the same time because one control valve is *opened* by air pressure (APO) and the other is *closed* by air pressure (APC). The numbers in parenthesis (3-8) and (8-15) are air-signal pressures. When the air signal pressure is between 3 and 8 psi, the gas inlet control valve is partly open and the gas outlet control valve starts to open up when the air-signal pressure is 8 psi, at which point the gas-inlet control valve is completely closed.

b. One control valve is operated by two different parameter signals. For example, see Fig. 5. The control valve will be half-open with 8-psi air pressure. When the air signal from the level controller exceeds 8 psi, the level

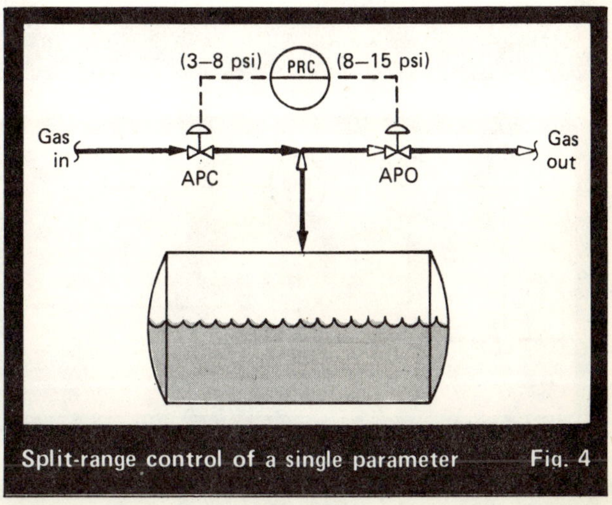

Split-range control of a single parameter Fig. 4

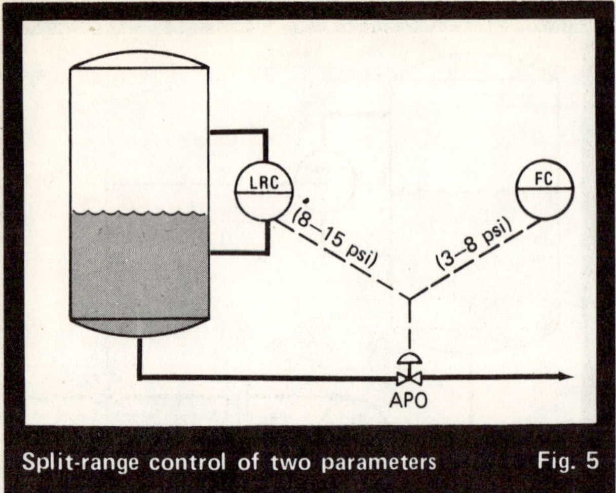

Split-range control of two parameters Fig. 5

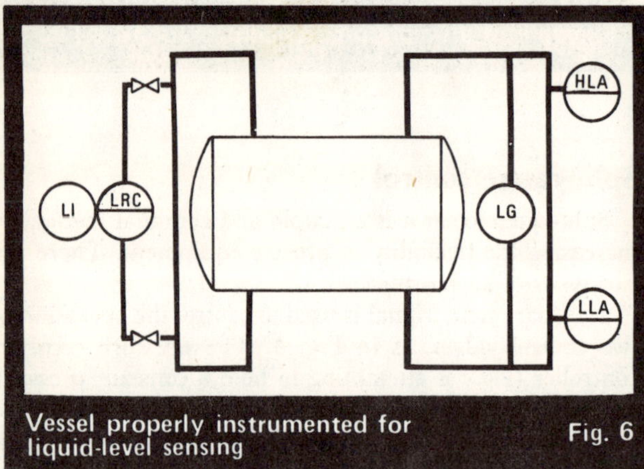

Vessel properly instrumented for liquid-level sensing Fig. 6

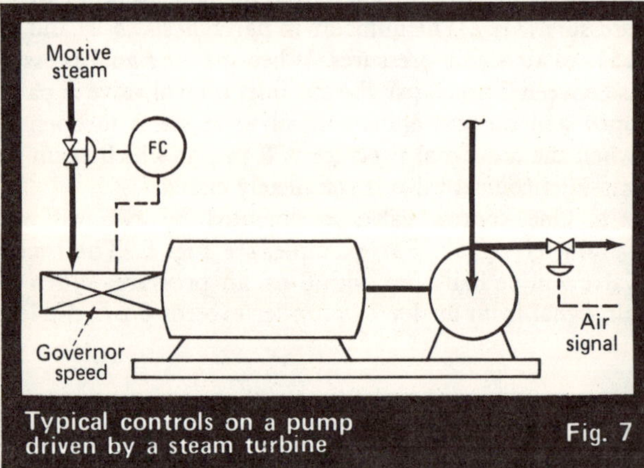

Typical controls on a pump driven by a steam turbine Fig. 7

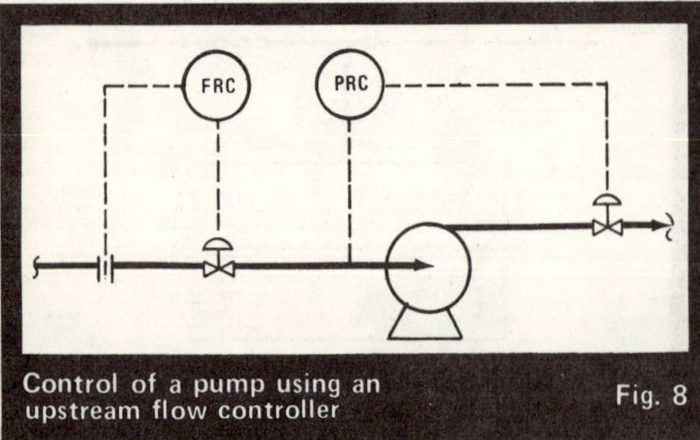

Control of a pump using an upstream flow controller Fig. 8

controller takes over operation of the control valve from the flow controller.

Liquid-liquid and three-phase systems

Instruments for controlling the level between hydrocarbon and aqueous phases are identical to those for liquid-gas phases, with one exception. The process designer must take greater care in specifying the density of each phase. The instrument engineer will use these densities to select the density of the level-chamber float.

Controlling two interfacial levels in a vessel in no way changes the control concepts outlined in the preceding sections. However, a separate float chamber is required for each interfacial level.

Instruments for liquid-level sensing

Many more dangerous situations result from operators losing track of the true liquid level in a vessel than from any other single cause. A properly instrumented vessel starts with properly designed stilling wells (sometimes called overall bridles). (See Fig. 1 and 2.) Two stilling wells are required for a vessel. One will be used to sense the liquid level for transmission of a signal to a control valve and to the level recorder. The readout from the float chamber on this stilling well should also be displayed on a dial located at the vessel. A second well will have staggered gage glasses connected to it. Also connected to this stilling well are the two small float chambers for the high- and low-level alarms.

These facilities will enable the outside operator to cross-check the visible vessel level (which may not be correct) with the level controlling the process (and being indicated in the control room), without having to contact the control room operator. The latter will have two completely independent checks of the vessel level also—the level being recorded from the first stilling well, and from the level alarms connected to the second stilling well.

Fig. 6 summarizes these requirements.

Liquid-level control—loop-seal method

Whenever liquid from one vessel is to be transferred to another vessel, by means of gravity and/or a small pressure difference, a level-control valve should not be used on the liquid outlet-line from the first vessel. Indeed, no attempt should be made at all to hold a liquid level in the first vessel. A loop seal, in the liquid outlet-line, from the first vessel to the second vessel is all that is required. The loop seal will prevent gas flow through the liquid line between the two vessels.

Pumps—process control and instruments

There are two methods for controlling the volume of liquid delivered by a pump—throttling the pump discharge or varying the pump speed. Small positive-displacement or diaphragm-type pumps usually have their speeds manually set and are adjusted as required by operators. Steam-driven reciprocating pumps are usually controlled by varying the steam rate to the driving cylinders; this in turn changes the speed of the pump.

Steam-driven centrifugal pumps are normally controlled as shown in Fig. 7. The governor maintains a constant turbine speed, while the discharge of the pump is throttled to obtain the desired flow on the process side.

Motor-driven centrifugal pumps operate at constant speed, but are also normally controlled by throttling on the discharge. Appreciable electrical savings can be obtained by use of a "variable-speed fluid coupling," which translates the rotational rate of a constant-speed motor to whatever speed is required by a pump to deliver the desired flow of liquid.

On occasion, it is necessary to control the flow upstream of a centrifugal pump. This is an undesirable situation, as it makes it difficult to guarantee that the pump has sufficient NPSH (net positive suction head) to prevent cavitation. If this situation is unavoidable, the control scheme of Fig. 8 should be used. In this scheme, the operator has to set the minimum suction-pressure to the pump to prevent cavitation. Again, it should be emphasized that this arrangement does not represent good design practice, as it is difficult for an operator to know the proper minimum suction-pressure required.

Compressors

Compressor controls usually must maintain a steady, or minimum, upstream pressure. For a turbine-driven centrifugal compressor, the speed of the compressor is varied to maintain a constant suction pressure, as shown in Fig. 9.

For motor-driven centrifugal compressors, the constant-speed characteristic of the driver complicates the problem. Fig. 10 and 11 illustrate two usual control schemes. (Note that it is extremely wasteful of energy to locate the flowmeter orifice plate upstream of the compressor.)

Reciprocating compressors are controlled in much the same way as are centrifugal compressors. However, motor-driven reciprocating compressors do have the advantage over centrifugal compressors, in that "pocket" clearances can be adjusted. This allows reduced motor amperage, at reduced compression requirements.

Instrumenting furnaces

Furnace instrumentation and controls should be designed with three objectives in mind:

1. Avoid overheating furnace tubes and coking internal material.
2. Obtain desired furnace duty.
3. Accomplish the first two objectives with minimum excess air.

The most direct way to avoid overheating furnace tubes is to employ "tube-skin thermocouples." These are thermocouples connected directly to the outside of a furnace tube at the points where the highest tube-temperatures are anticipated.

Such thermocouples have two serious drawbacks. First, they tend to pull away from the tube, and thereafter sense the firebox temperature instead of the tube temperature. Second, a section of tubing three feet away from the skin thermocouple may be burning up, without the skin thermocouple giving any indication of trouble.

Firebox thermocouples are the most common and reliable measure of fireside furnace conditions. A firebox thermocouple is shielded in a ceramic sheath and inserted about a foot in front of the radiant-section tubes. These thermocouples will always read lower than actual firebox temperatures due to reradiation of heat. For technical work, a "velocity thermocouple" should be used. The

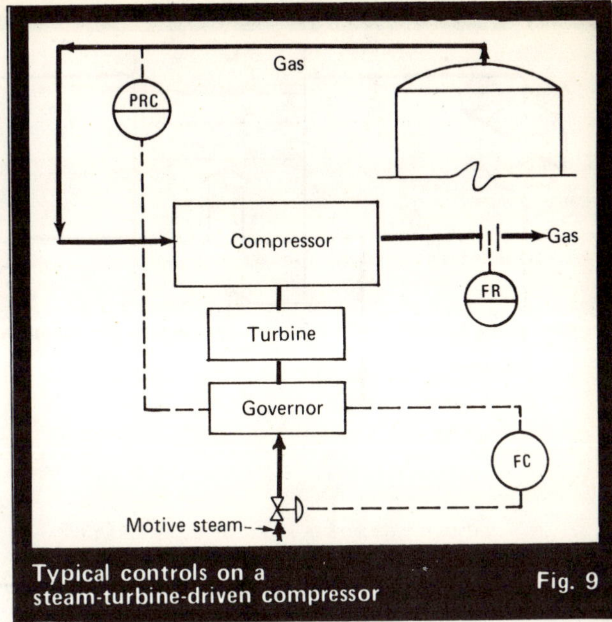

Typical controls on a steam-turbine-driven compressor Fig. 9

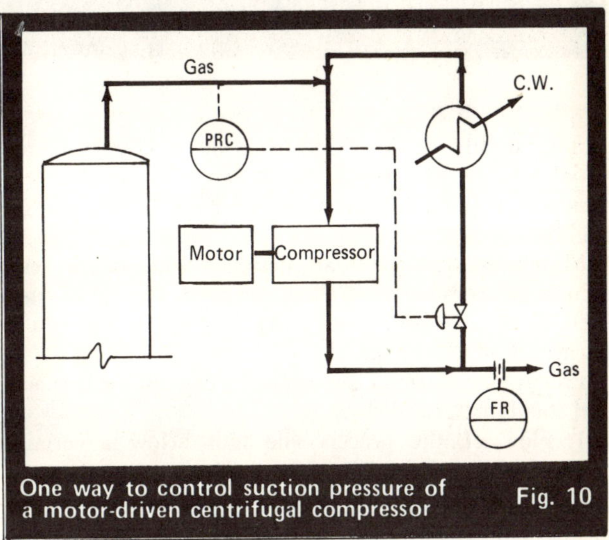

One way to control suction pressure of a motor-driven centrifugal compressor Fig. 10

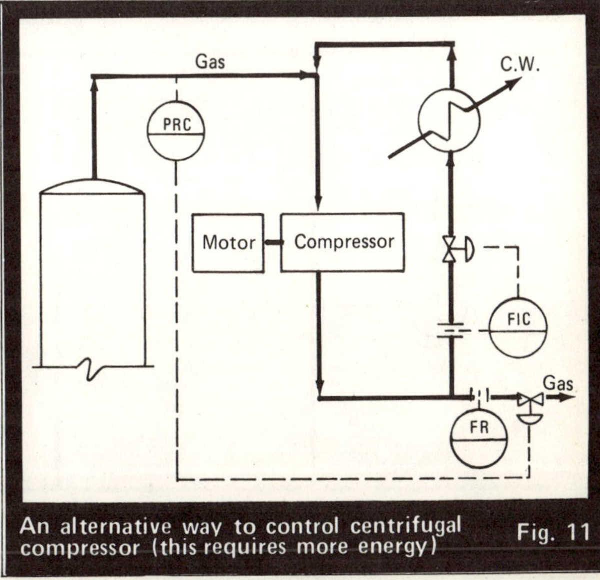

An alternative way to control centrifugal compressor (this requires more energy) Fig. 11

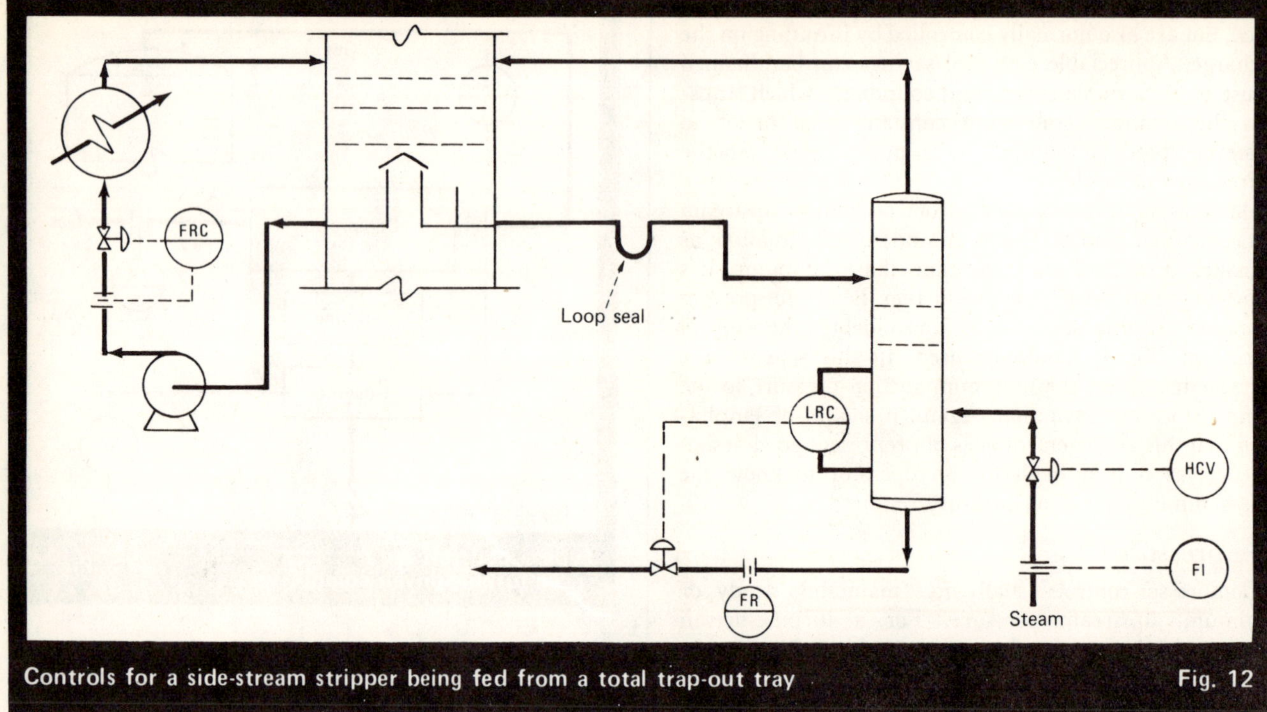

Controls for a side-stream stripper being fed from a total trap-out tray Fig. 12

ceramic sheaths on sheathed firebox thermocouples occasionally break and should be designed to be changed onstream, without an instrument mechanic having to risk his life.

Most process furnaces are designed with the process effluent temperature controlling the gross amount of fuel fired by the furnace (using a TRC). Fuel to individual burners is adjusted manually.

Provision to alarm or automatically trip out the furnace fuel should be provided when:

1. Flow on the process side falls below a certain point.

2. The furnace "flames out," and the source for fuel ignition is gone.

To minimize fuel consumption in furnaces, a stack-oxygen analyzer should be installed immediately above the radiant tubes. Automated controls of dampers and of secondary air registers should also be considered for large installations. Keep in mind that closed-loop control of secondary air registers by oxygen analyzers is not a well-established practice.

Fractionators

The term fractionators, as used here, applies to the main towers in crude units, fluidized catalytic cracking units (FCCUs), and cokers. These towers all have side-stream draws with strippers, pump-around streams, and no bottom reboilers.

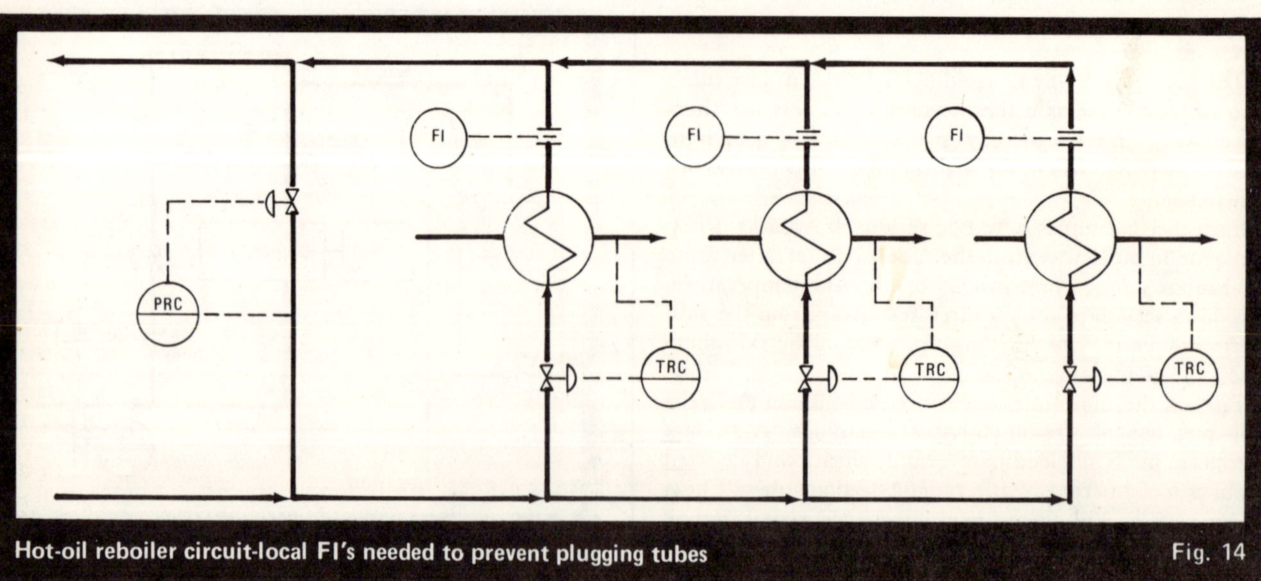

Hot-oil reboiler circuit-local FI's needed to prevent plugging tubes Fig. 14

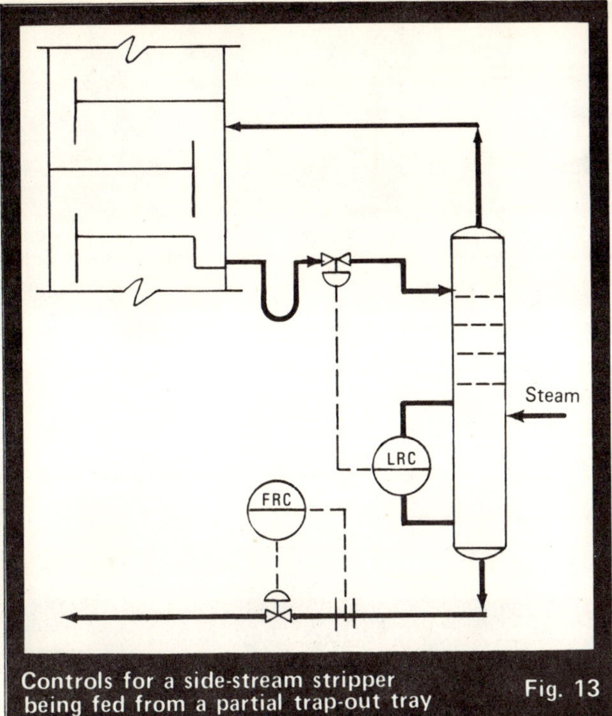

Controls for a side-stream stripper being fed from a partial trap-out tray Fig. 13

Control of trim gas oil—(also called wash oil and reflux below the pan). The purpose of this stream is to clean up the heaviest distillate cut made in the fractionator. A continuous colorimeter, which indicates to the operator whether he should increase or decrease the trim gas oil production, has worked well in this service.

Pump-around streams—Pump-arounds are invariably flow controlled. Manual adjustments are made by operators to change the heat duty as required.

Side-stream strippers—If the feed to the stripper is withdrawn from a total trap-out in the fractionator, Fig. 12 illustrates the method of control. If a pump-around stream also originates from the total trap-out pan, the basic control scheme is unaffected. Note that this method of control has proven in the field to require a higher-than-normal liquid hold-time in the stripper bottom for smooth operation.

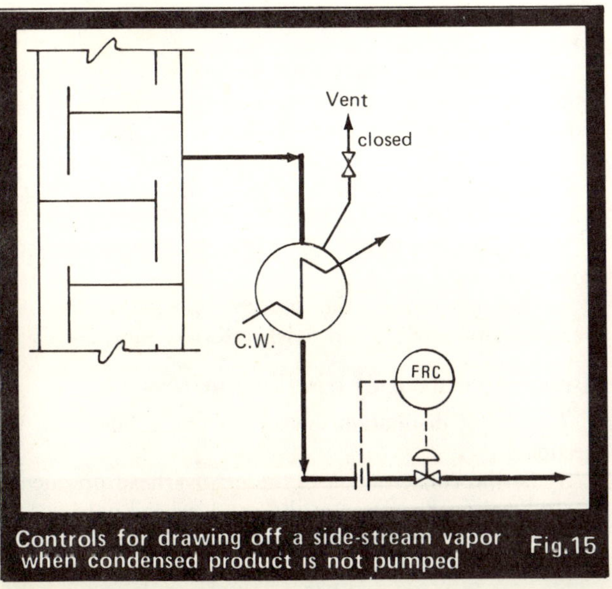

Controls for drawing off a side-stream vapor when condensed product is not pumped Fig. 15

When only a portion of the available liquid on a tray is to be withdrawn as product, the design in Fig. 13 will give the smoothest control. (Note that operator adjustment is relied on to prevent withdrawing excess distillate off the fractionator.)

Overhead system of fractionators

The overhead-system controls consist of having the reflux on flow or temperature control and the net liquid overhead-product on reflux-drum-level control.

Pressure control of fractionators

If compression of the wet gas is not required, pressure on the fractionator is simply held by a pressure-control valve that throttles the wet-gas flow. However, a compressor is normally required for this service. The method of control then becomes a function of the type of compressor purchased. See the section on compressors for further discussion.

Hot-oil-system controls

Many vapor-recovery units have several tower reboilers that receive their heat from circulating hot oil (or tower pump-around) systems. These reboilers are typically arranged in parallel. Fig. 14 illustrates the preferred control scheme.

Note that the process designer *must* allow for some flow through the pressure control line (10% of total flow), so that a constant pressure upstream of the TRCs may be maintained. Local FIs on each reboiler branch are necessary to ensure that a minimum velocity of several feet per second is maintained in the reboiler tubes. In many services, severe fouling and plugging can be encountered at low tube-velocities.

Side-stream vapor draw-offs

Vapor may be drawn off from an intermediate tray in a distillation column, using flow control. If the condensed vapor does not have to be pumped to the next process vessel, the scheme of Fig. 15 should be employed.

In Fig. 15, as the FRC closes the control valve, condensate will back up over the condenser tubes, reduce the condensation surface area and, therefore, restrict the flow of vapor drawn off the tower. This scheme requires that the condenser be oversized to ensure that total condensation can *always* be obtained.

If the condensed vapor must be pumped to the next process vessel, the scheme shown in Fig. 16 should be used.

The pumping drum may normally run 100% full of liquid. However, the pump must be protected against cavitation due to loss of liquid level in the drum. In the Fig. 16 scheme, the condenser may be sized somewhat tightly.

Refrigeration loop

Fig. 17 illustrates the controls for a typical refrigeration loop.

The reader is asked to explain the need for Drum No. 1. Why doesn't the pump have a flow controller on the

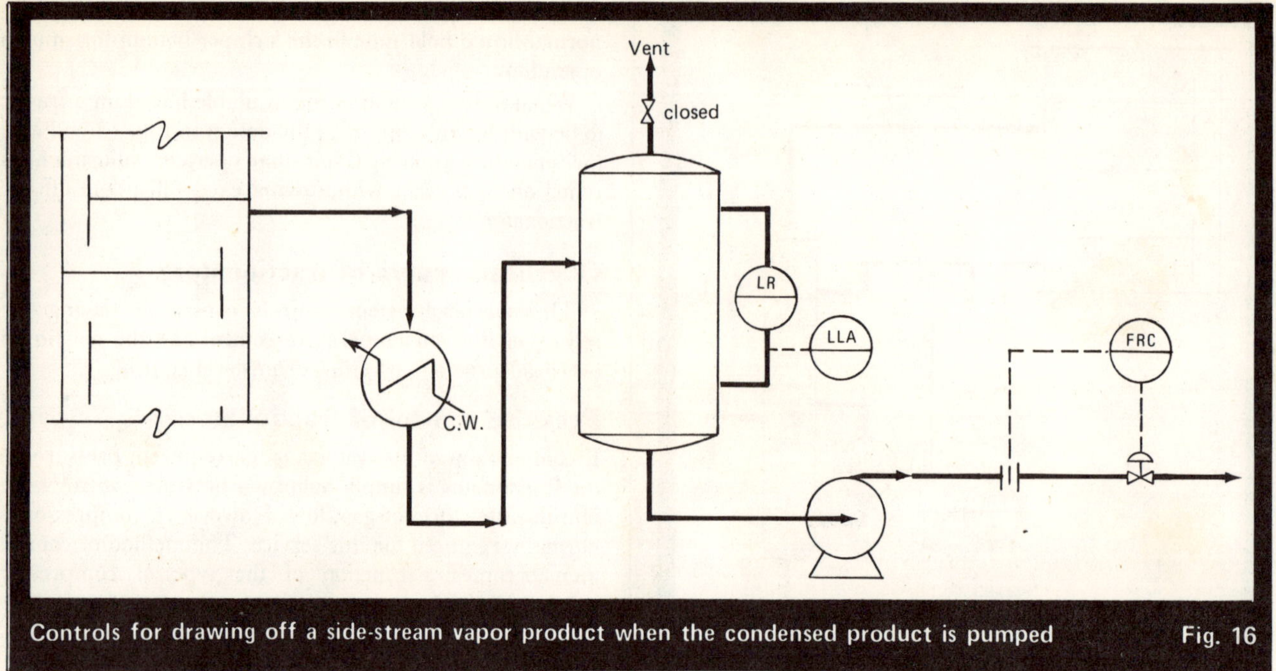

Controls for drawing off a side-stream vapor product when the condensed product is pumped Fig. 16

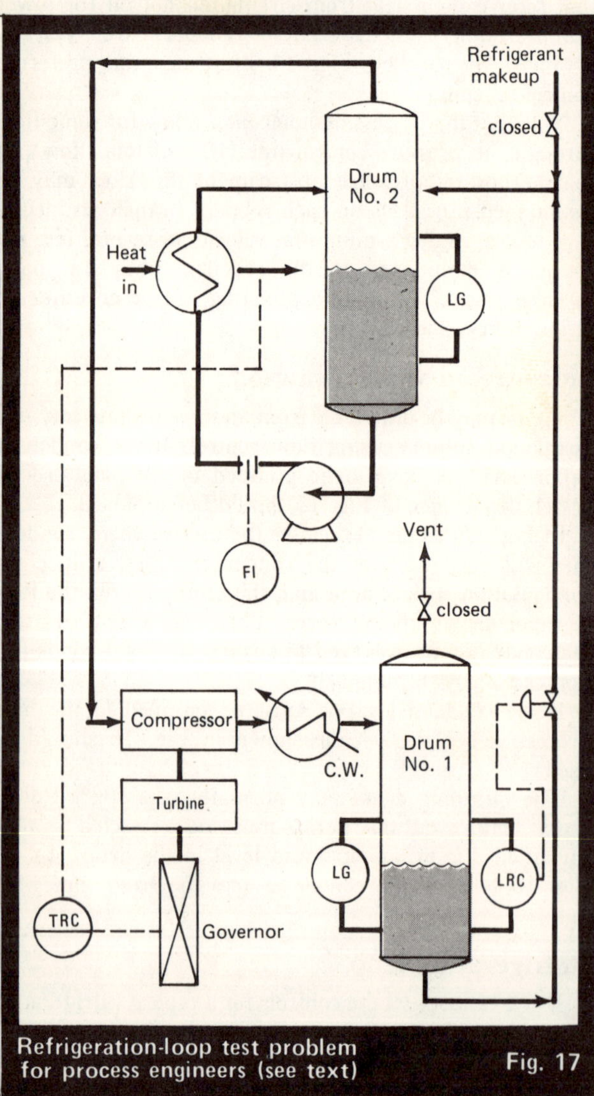

Refrigeration-loop test problem for process engineers (see text) Fig. 17

discharge? What is maintaining the level in Drum No. 2? If you know the answer to these questions, you are a real process engineer.*

Control between process units

No area of control is more important to the overall smooth operations of a refinery than control of run-down streams between units. A problem arises not because of any inherent difficulty in control but because of poor communications between operators working in different control rooms.

The purposes of the controls on a stream running down directly to another unit should be to maintain constant temperature, flowrate, pressure and composition.

Of these four factors, maintaining constant flowrate of the rundown stream is certainly the most important. It is also relatively easy to achieve if the following guidelines are followed:

1. See that run-down streams are on FRC whenever practical.
2. Vessels from which products are to be run down to other units should be 50% larger than vessels that will run down to facilities inside battery limits.
3. Make sure the pressure drop through the product run-down control valve is large compared to upstream pressure fluctuations.
4. Display and record all product rundown rates in the control room.
5. Do not design the process in such a way that variations in product run-down rates are inevitable.

Process control of distillation towers

Control of distillation towers may be divided into two sections:

1. Partial condensation of the net overhead product.
2. Total condensation of the net overhead product.

Let us first consider partial condensation:

*If not, the answer will be found in the box at the end of this article.—Ed.

Partial condensation

When a significant volume of net overhead product is certain not to condense, the pressure on the tower must be held by throttling the gas from the reflux drum. Despite any possible theoretical benefits, most distillation towers have been put on manual reflux-control by the operators. Therefore, the designer might as well assume that tower reflux will usually be on flow control. The net overhead liquid product will then of necessity be on reflux-drum level-control. Heat flow to the tower reboiler will control the split between the key components.

For light hydrocarbon service, having a gas chromatograph *directly* controlling steam-flow to a reboiler has worked well at some locations. The chromatograph may measure a key component in either the top or bottom tower products, and control the reboiler with either measurement.

Many reboilers are controlled by an internal tower temperature. This method of control will *not* work well if the quantity of a *non-key component* in the tower feed is too variable. For example, a tower fractionating between isobutane and normal butane had its split controlled by an internal tower temperature. Whenever the propane content of the tower feed rose, the normal-butane content of the overhead product increased.

The bottom level of a tower is normally on level control. If it is required to use flow control on the stream leaving a tower, due to downstream considerations, the liquid residence-time in the tower bottoms should be substantially increased.

A sketch of a typical control scheme of a tower using partial overhead condensation is shown in Fig. 18.

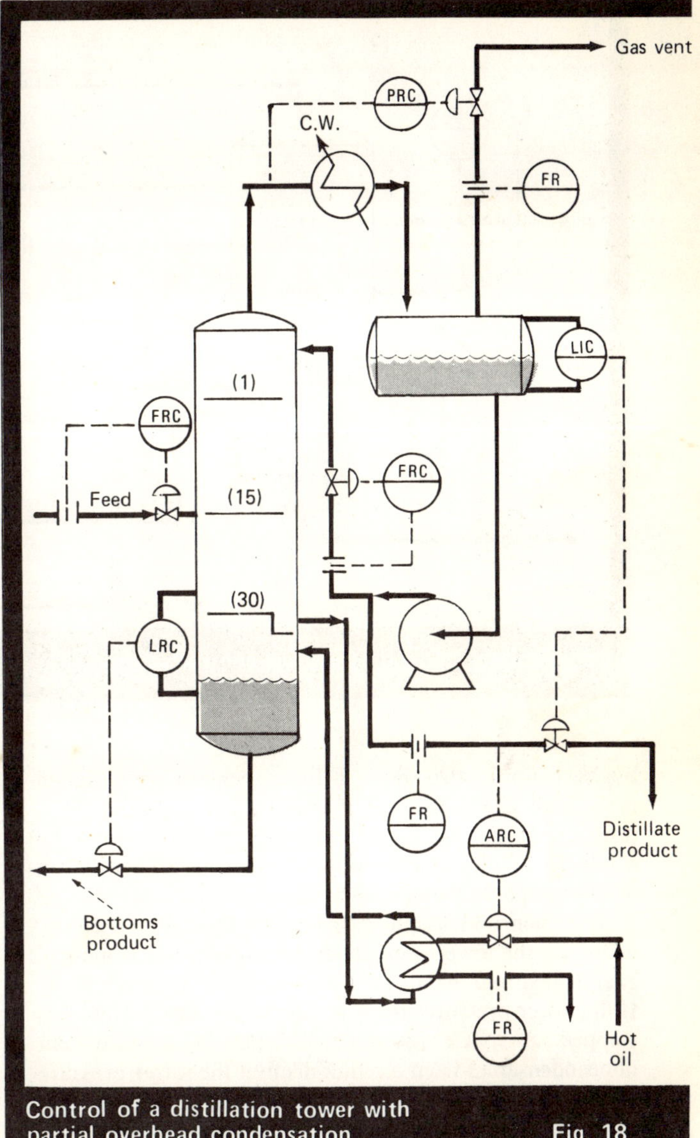

Control of a distillation tower with partial overhead condensation Fig. 18

Total condensation of net overhead product

When all, or almost all, of the overhead product of a tower is expected to condense, controlling the tower pressure becomes an especially interesting problem. Control of both the net overhead-product rate and the reboiler duty is influenced by how the tower pressure is to be held. A partial tabulation of the possible schemes follows:

1. The old-time method was to throttle the cooling water to the condenser. When the tower pressure dropped, cooling water flow to the condenser would be reduced. This would warm up the contents of the reflux drum and increase the bubble-point pressure of the liquid in it. Net, liquid, overhead product would then be controlled by the reflux-drum level, and the reboiler would be controlled to maintain the desired split. This method has a serious drawback in that reducing cooling-water tubeside velocity too much will greatly increase rates of tubeside fouling.

2. A bypass may be built around the condensers, allowing uncondensed vapors to avoid them entirely. A control valve is installed in this bypass. As the tower-pressure drops, the control valve opens, allowing vapors to enter the reflux drum and heat its contents, which raises the bubble-point pressure of the liquid in the drum. This is the so called "hot-vapor bypass" method of tower pressure control. The method has a serious drawback; it does not always work. A substantial pressure drop is required in the condensers to force the vapor through the

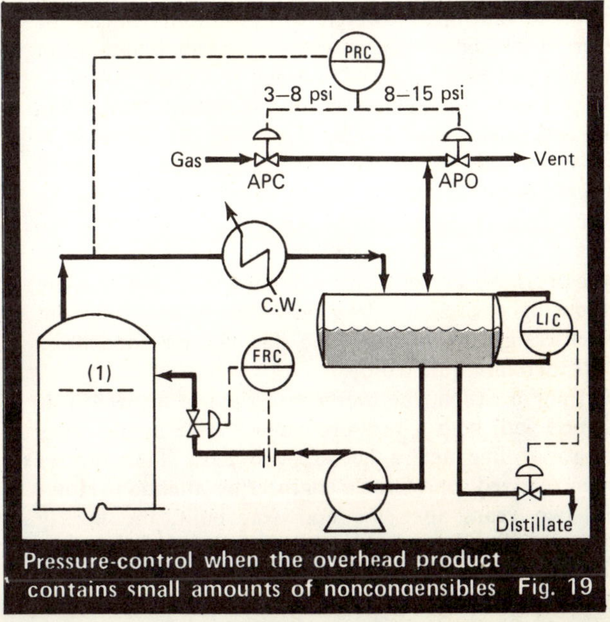

Pressure-control when the overhead product contains small amounts of noncondensibles Fig. 19

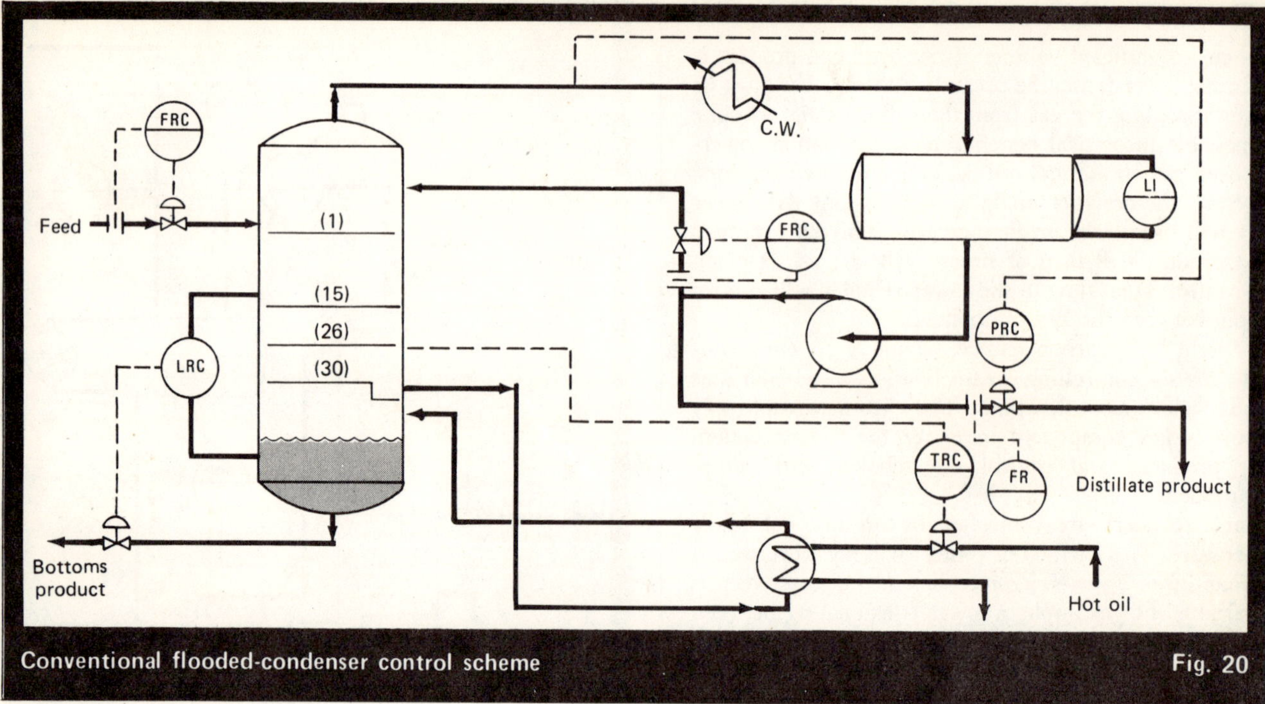

Conventional flooded-condenser control scheme Fig. 20

bypass control valve. Even if the condensers are designed for a high pressure-drop, at reduced loadings—which is just when the hot-vapor bypass is most needed—the condensers will probably not have sufficient pressure drop to force vapors through the bypass.

3. A noncondensable gas may be bled into a reflux drum. As the tower pressure drops, a gas (such as natural gas) is bled into the tower to maintain the tower pressure. If the tower pressure rises, the flow of gas into the tower is stopped. A split-range controller may be used to vent noncondensables from a reflux drum if the tower pressure continues to rise. Fig. 19 illustrates this scheme.

This technique has one drawback. The so called "noncondensable" gas is bound to be somewhat soluble in the net overhead liquid, and may flash out in lower-pressure downstream equipment. Here, gas accumulations may be difficult to vent off. However, this technique can be the preferred method of control when a *very* small amount of noncondensables is expected to occur in a tower's feed.

4. A rarely used method of tower-pressure control is to place a pressure-control valve directly in the tower overhead-vapor line. As the tower pressure drops, the control valve closes to restrict the flow of vapor exiting from the tower. This is the preferred method of control, if the reflux drum pressure is to be kept substantially below the tower pressure. For ordinary circumstances, intentionally introducing a pressure drop upstream of the overhead condensers greatly increases the size of these condensers and is therefore quite costly.

On one occasion, the overhead system of a tower was designed with both a pressure-control valve in the overhead-vapor line and a hot-vapor bypass. This control system received considerable engineering attention. However, two years after the unit was built, the hourly operators still did not understand it. They finally blocked in the hot-vapor bypass and made manual adjustments to the overhead system to maintain reflux-drum pressure. The designer is urged to remember that hourly workers, not engineers, have to operate process units.

5. The most common method of controlling a tower in recent years has been the "flooded condenser" scheme. In contrast to the preceding four methods of tower control, no liquid level is maintained in the reflux drum of a flooded-condenser controlled tower. The reflux drum is kept full of liquid at all times (note Fig. 20).

As the tower pressure falls, the flow of distillate product is reduced. This forces liquid to back up into the condenser. As it backs up, some of the tubes in the condenser are submerged and, consequently, cannot condense any of the overhead vapors. As the rate of condensation falls, the tower pressure rises. This is a smooth and simple method of controlling tower overhead pressure. The major disadvantage is that the overhead condensers must be sized quite generously to make sure that the entire tower overhead can *always* be condensed.

If the overhead from a tower on flooded-condenser control cannot be 100% condensed, the tower will overpressure almost immediately and the relief valves will open up.

6. A variation of flooded-condenser control is the "pressure control boilup" method. This somewhat complex, but actually rather smooth, method of control is shown in the Fig. 21.

Note that the reboiler heat-input is controlling the tower pressure. Therefore, the rate of vapor generation, rather than the rate of vapor condensation, controls the tower pressure. The distillate product-rate controls the split between the key components. The mechanism for controlling the split is as follows:

- The gas chromatograph on the overhead product senses that the percent of the heavy key-component in the distillate is too low.
- The distillate product control-valve is signaled to open farther.

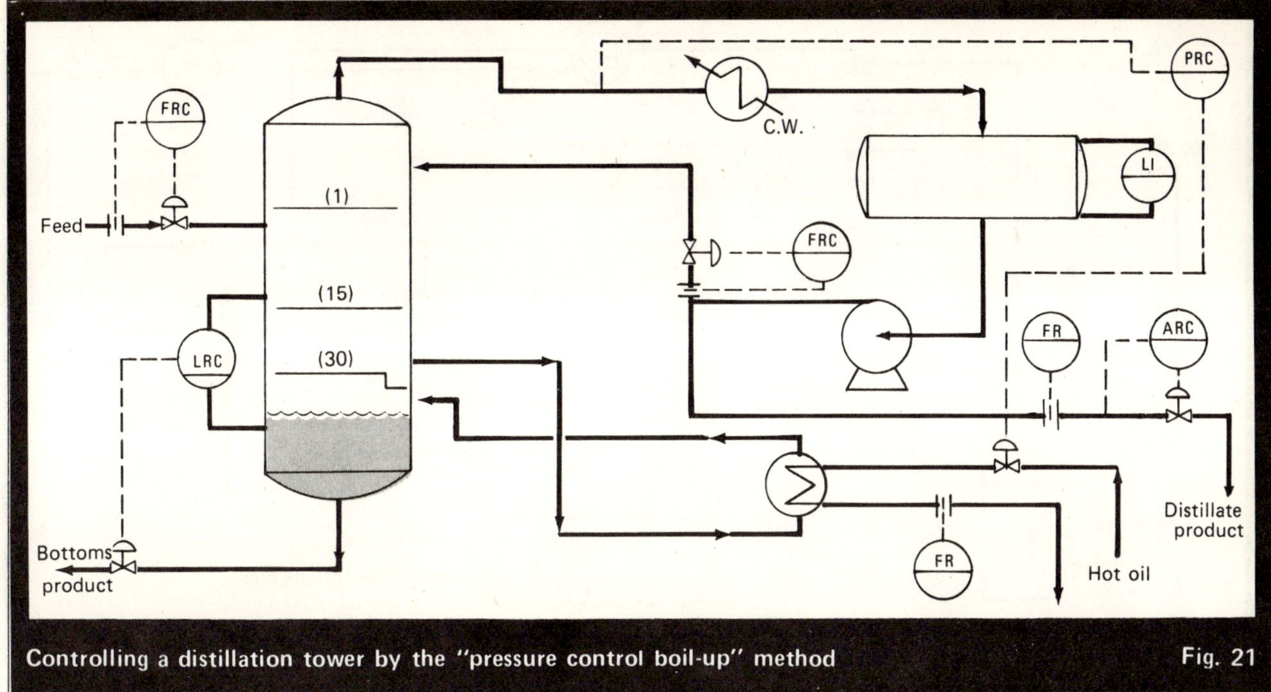

Controlling a distillation tower by the "pressure control boil-up" method Fig. 21

■ The increased rate of distillate production uncovers tubes in the condenser that had previously been submerged.

■ Tower pressure starts to drop because the rate of vapor condensation has been increased.

■ Heat flow to the reboiler increases to maintain the tower pressure.

■ More vapors are driven up the tower. This increases the percent of the heavy key-component in the tower overhead.

This method of control has one drawback. Under some circumstances, the distillate product control-valve is prone to dumping liquid out of the reflux drum faster than the reboiler can boil vapors up to replace the liquid. The reflux drum may be drained dry before the operators can manually reduce the reflux rate to correct the temporary problem.

A combination of the better features of some of the preceding schemes is discussed below. This method was developed and applied on a large alkylation unit.

Control with total overhead condensation

Many distillation columns are designed to totally condense the overhead product. The most common method of controlling the tower pressure when the overhead product is totally condensed is to partially flood the condensers. Most flooded condensers normally work fine when they are generously sized. However, flooded condensers have two serious drawbacks:

1. Even a properly sized flooded condenser will occasionally become undersurfaced owing to: a feed containing too many light ends, fouling of condenser tubes, or unusually hot weather. When a flooded condenser cannot provide at least two or three degrees of subcooling of the overhead product, the reflux drum will quickly empty, and further operation of the tower becomes extremely difficult.

2. Many flooded condensers are grossly oversurfaced much of the time for their required duty. Theoretically, this excess surface-area should be used up by throttling back on cooling water or shutting down air fans, thereby saving energy or releasing cooling water for a profitable use. But, practically speaking, it is impossible to use up the excess surface area in a flooded condenser in the above manner, because an operator cannot see the liquid level inside the condenser.

An alternative control scheme to circumvent both of the above problems is shown in Fig. 22. The main idea is to hold the liquid level in the reflux drum while the tower pressure is controlled by the steam rate to the tower's reboiler. The distillate product rate is controlled by the required product specification.

If the liquid level in the reflux drum gets too high, it means that the condensers are subcooling the overhead, so water is throttled to the condenser to hold the liquid level by reducing subcooling. The hand-jack, on the cooling-water-inlet control-valve, is set to maintain a minimum cooling-water flow. This is required to prevent fouling of the condenser tubes due to low cooling-water velocity through the tubes.

If the liquid level in the reflux drum gets too low, it means that the condensers are not subcooling the tower overhead enough, or that the total overhead may not even be fully condensed. There are two ways out of this problem:

1. Increase the tower-top temperature (this lowers the reflux-drum temperature).

2. Reduce the heat load on the condensers.

Decreasing the reflux rate will accomplish both of the above. The hand-jack on the reflux control valve is set to maintain a minimum flow of reflux to the tower, to prevent damage to the reflux pump.

An interesting question that follows from the preceding discussion is whether or not a reflux drum is really

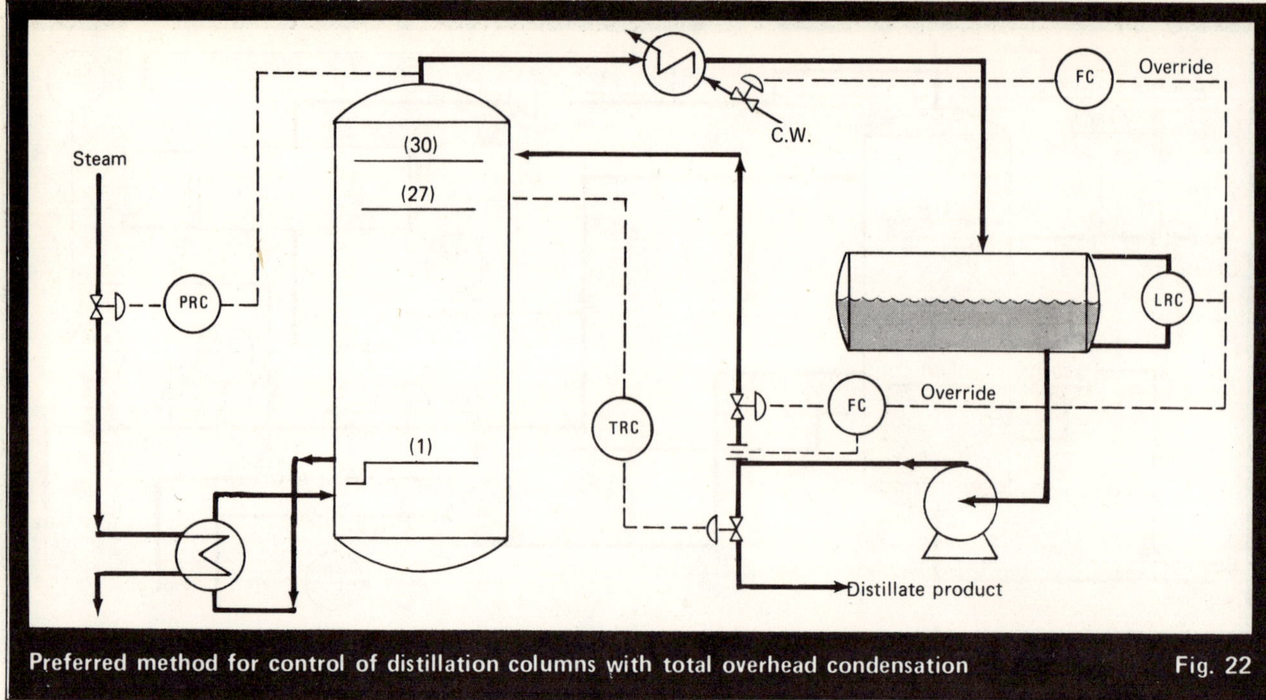

Preferred method for control of distillation columns with total overhead condensation Fig. 22

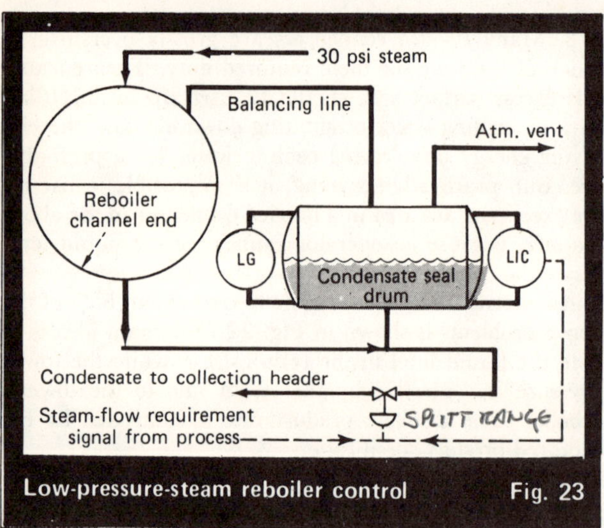

Low-pressure-steam reboiler control Fig. 23

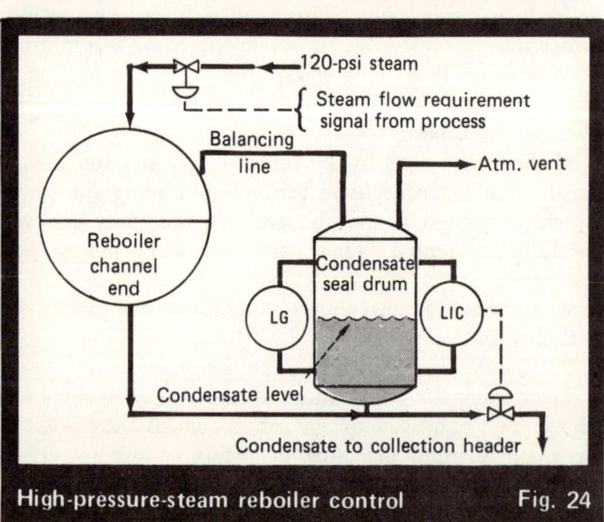

High-pressure-steam reboiler control Fig. 24

required when a flooded condenser is employed. When a flooded condenser fails to provide adequate subcooling and there is a properly instrumented reflux drum (i.e., with a level-alarm and board-mounted level indicator), there is still hope. The operators may reduce the reflux rate and vent the reflux drum to the flare to save the tower.

However, if no reflux drum is present and a condition of inadequate subcooling occurs, the reflux pump will almost instantly lose suction, the reflux flow will become quite erratic and drop precipitously, tower pressure will rise, and the relief valve will pop. This entire sequence will probably occur within a few minutes. The only indication the operators will have that the condensers are overloaded will be the erratic flow of reflux to the tower.

Control of steam reboilers on stills

It is critical that the proper control scheme be used when designing the steam-condensate system for a distillation tower's reboiler.

For low-pressure (15 to 35 psig) steam, a steam-inlet control valve should never be used. When the control valve pinches, to reduce the steam flow to the reboiler, the reboiler steam-side pressure (i.e., the channel end) naturally drops. As the steam condensate from the channel is normally drained to a condensate flash drum, along with condensate from other, higher-pressure reboilers, some minimum presssure must be maintained that will allow condensate to drain out of the channel. If this pressure is not maintained, condensate will back up in the channel, and stop steam from flowing into the reboiler. The steam-inlet control valve will then open up and the cycle will be repeated. The result will be unstable tower operation.

The proper way to control steam flow for low-pressure steam reboilers (see Fig. 23) is to control the flow of steam by controlling the flow of condensate out of the reboiler channel. This, in effect, will reduce reboiler surface-area by backing up condensate in the tubes when less heat-duty

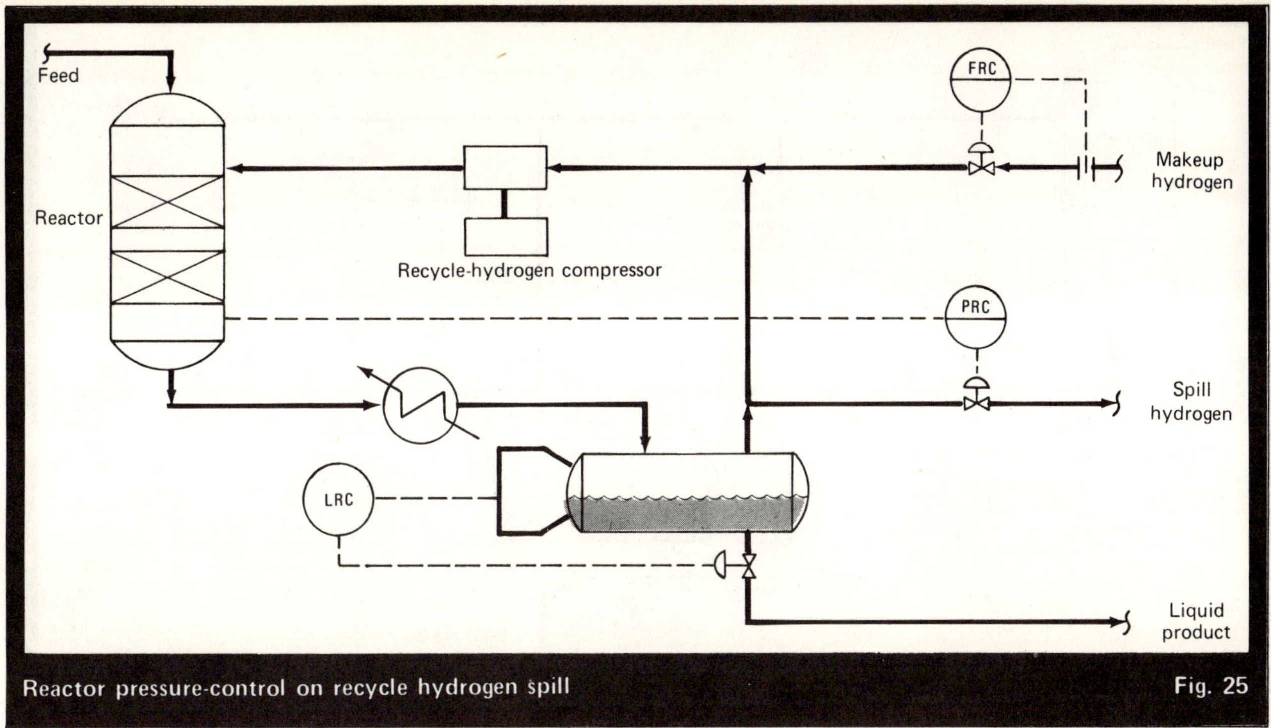

Reactor pressure-control on recycle hydrogen spill Fig. 25

is required. A large condensate-seal drum should be installed between the reboiler channel outlet and the condensate control valve. This drum will have its midpoint at the same height as the bottom of the reboiler channel. When the condensate level in the drum falls below its midpoint, a level-transmitter will take over control of the condensate-outlet control valve. This will maintain a minimum condensate-level in the drum, even though the reboiler controls are calling for more steam. Without this level-override feature, the condensate-seal-drum would go empty and the condensate seal on the reboiler channel outlet would be lost. Uncondensed steam could then pass through the condensate-outlet control valve, and the effectiveness of the reboiler in transferring heat to the process side would be greatly diminished.

For reboilers using 120-psig steam or higher, the flow of steam to the reboiler should never be controlled by backing condensate up in the channel. This will very often cause the channel-head to tubesheet gasket surface to start leaking (probably due to continual and rapid changes of local temperatures). A steam-inlet control valve should be used (see Fig. 24). Condensate drainage from the reboiler channel should be controlled by a condensate control valve; a small vertical drum should be installed. This drum should be about 4 ft high, with its midpoint at the same height as the bottom of the reboiler channel. The condensate level in the drum should always be kept below its midpoint, but still high enough to maintain a seal.

Instead of using a drum and a control valve on the reboiler channel outlet, many engineers would like to use a steam trap. This is not acceptable. It is important that the operators be able to see the actual condensate level in the reboiler channel in a gage glass. Also, steam traps are prone to plug up and stick wide-open.

To ensure that the level in the condensate drum corresponds to the level in the reboiler channel, a balancing line is required. This is simply a 1-in. line directly connecting the top of the reboiler channel with the top of the condensate drum.

A small atmospheric vent on top of the condensate drum should be provided and always left open. Noncondensables can build up in the vapor space at the top of the reboiler; they will reduce reboiler efficiency. Also, significant concentrations of CO_2 can build up in the vapor space and corrode the carbon-steel reboiler tubes.

No vapor-recovery unit operating with steam reboilers can be any better than its steam-condensate-collection system. Remember that, normally, condensate will partially flash across the condensate control valve. The steam-condensate collection system line-sizing must take this factor into account.

Pressure control of reactors

A few examples of some of the commoner type of reactors in the refining industry are discussed below:

Hydrotreating a liquid with once-through hydrogen—The question arose of how the pressure should be controlled in a fixed-bed, downflow, vapor-liquid reactor, operating at 2,000 psig. The reactor effluent was to be flashed at 50 psig, and no hydrogen from the reactor effluent was to be recycled. The obvious solution to the problem was to throttle the reactor effluent through a control valve that would hold pressure on the reactor. This meant throttling a vapor-liquid mixture through a control valve with a normal pressure drop of about 1,900 psi. Built this way, reactor pressure control has been satisfactory.

Hydrotreating a liquid with recycle hydrogen—There are two methods for controlling reactor pressure:

1. Reactor pressure control on hydrogen spill rate (Fig. 25).

2. Reactor pressure control on hydrogen makeup rate (Fig. 26).

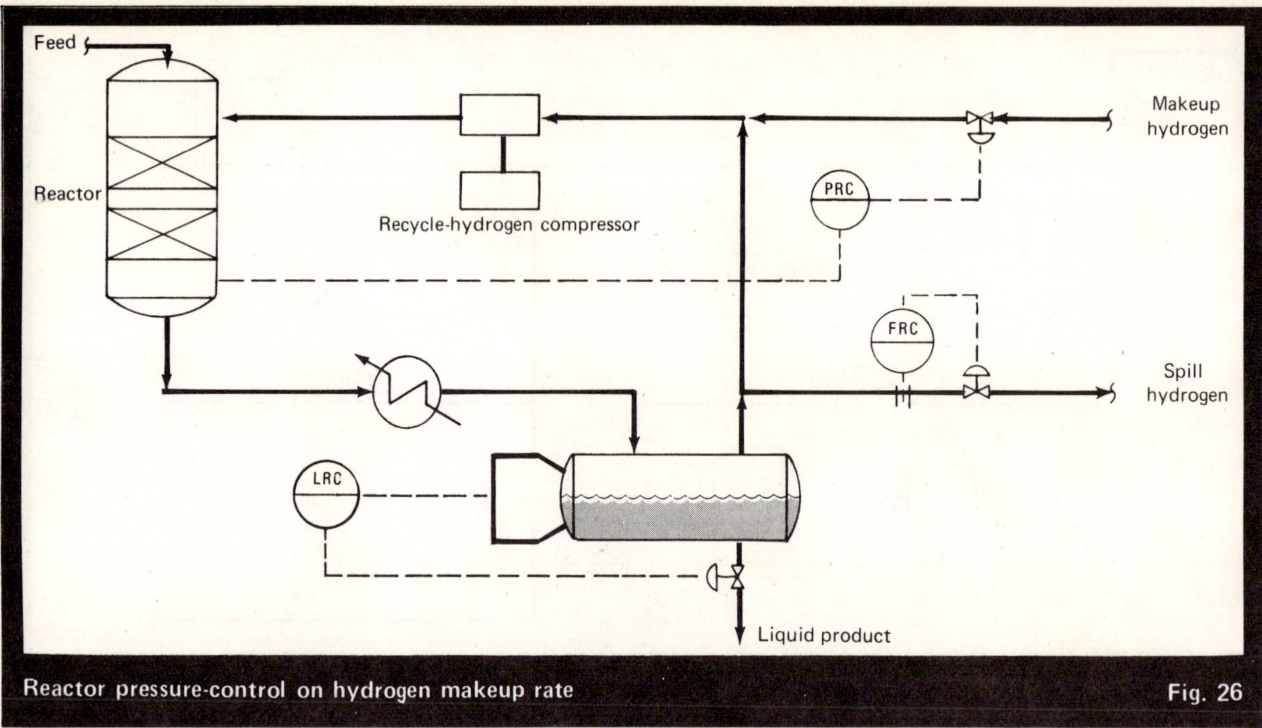

Reactor pressure-control on hydrogen makeup rate Fig. 26

Generally, the first scheme is preferable in terms of smoothness of operation, while the latter is better from the point of view of minimizing hydrogen consumption.

Liquid-filled reactors (Stratco alkylation units, treaters)— Controlling the pressure in a liquid-filled reactor is straightforward, if the majority of the charge to the reactor is coming from centrifugal pumps. A pressure-control valve is simply installed on the effluent line from the reactor, as shown in Fig. 27.

As the reactor pressure drops, the PRC will pinch back. The FRC will open up. The total system pressure-drop will have remained the same, because the pressure drop across the PRC will have increased, and the pressure drop across the FRC will have decreased. Liquid flashing across the PRC will not interfere with this method of control.

Thermal cracking units—(Cokers, fluidized catalytic cracking units (FCCUs). Pressure of the reactors is held, quite simply, by adjustments to the suction pressure of the wet-gas compressor.

Low-pressure gas-phase reactors—(acid plants, sulfur plants). Reactor pressure normally floats on atmospheric pressure with no attempt made to stabilize reactor pressure fluctuations due to variations in throughput.

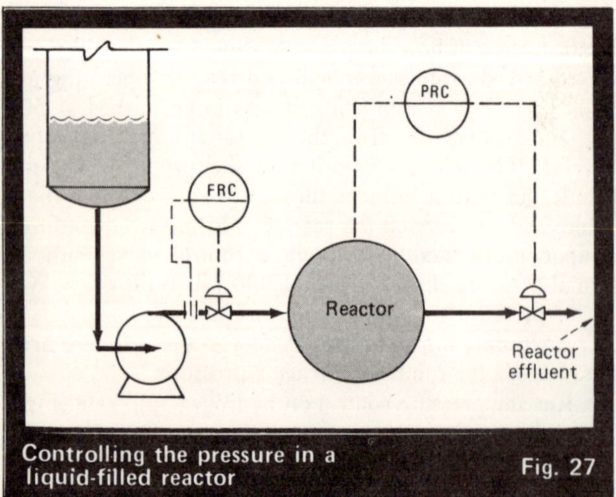

Controlling the pressure in a liquid-filled reactor Fig. 27

Answer to problem (refrigeration loop)

1. Drum No. 1 ensures that the refrigerant condenser is kept well drained of liquid, but that no uncondensed refrigerant can leave this condenser.

2. Refrigerant flow through the refrigerant evaporator should always be maximized to maximize the evaporator heat-transfer coefficient (which, in turn, will minimize compressor horsepower requirements).

3. The liquid level in Drum No. 2 is maintained by the system inventory of refrigerant.

The author

Norman P. Lieberman is Projects Manager for Adrian R. Davis Associates, P. O. Box 184, Matteson, IL 60443, a consulting firm that specializes in the process design of refinery energy-conservation projects. He was formerly a design coordinator for process energy-conservation projects for Amoco Oil's Chicago Process Engineering Div. He holds a B.S. in chemical engineering from The Cooper Union (New York City) and an M.S. in chemical engineering from Purdue University.

For Process Control . . . Select the Key Variable

The key variable that should be controlled if the entire process is to be most efficient depends on the objectives of the process, and the problems of measurement.

PAUL G. FRIEDMANN and JOHN A. MOORE, Leeds & Northup Co.

Automatic control is applied to chemical processes to help maximize profit by producing top-quality product at the most efficient level of production. The goal can only be achieved if the proper variable is selected as the basis for control.

The signal representing the selected variable may be derived from a single measurement or computed from several measurements. In either case, the selected variable must be represented in the control loop with sufficient accuracy to enable control to bring about the desired result. And it must do this despite the limitations imposed by measurement and computational error. Selection of the apparatus needed must satisfy the restraints of economics as well as those of practical instrumentation and control theory.

Criteria for Selecting the Key Variable

Reasons for applying automatic control to a chemical process are listed in Table I. Control action takes place as a function of the changes of a variable with respect to a reference value, or setpoint. The selected variable should be that process parameter which, if held at the reference value, is most likely to bring about the required objectives.

A common approach to selecting the key variable for control is to ask the two questions, "What can we measure?" and, "Which, variable, of those we can measure, will have the most influence on the specifications we must meet?" This practice may condemn the plant to less than optimum performance because the truly significant variable may not be directly measurable.[1] In fact, the key variables will be those used by the process designer in his selection and sizing of equipment, utilities and throughput. Rather than a temperature or a pressure, the design variables are likely to be such things as heat-transfer rate, or a ratio of mols of A produced per mol of B consumed. The design variables are usually calculated from heat and material balances. Since the performance of the process depends on maintaining these balances, it would seem logical to argue that the best bases for process control are those quantities used in the balances.

Objectives of Process Control—Table I

To create an atmosphere of operation where process equipment and reactions will perform at highest efficiency—closest to design level—within limits of safety.

To compensate for process disturbances, changes in operating conditions, and utility limitations.

To provide an orderly progression from one phase of operation to the next.

To maintain product purity, and agreement with product specifications.

The key variable, then, should be selected after close examination of the true objectives of the process, as well as the measurable sources of variability. It may turn out that a process state, for example the concentration of product in a batch reactor, should be controlled to achieve maximum yield.

Perhaps a variable limit sets the point of greatest efficiency of operation. An example is the pressure ratio of a centrifugal compressor; below this ratio, for a specific flowrate, safe compressor operation is not possible, while above it, efficient operation is sacrificed. The key variable may be a process load—for example, the reagent requirement in a waste neutralizer, variable not only because of unpredictably varying conditions of waste flowrate but also because of changes in pH of the incoming material.

Originally published June 12, 1972.

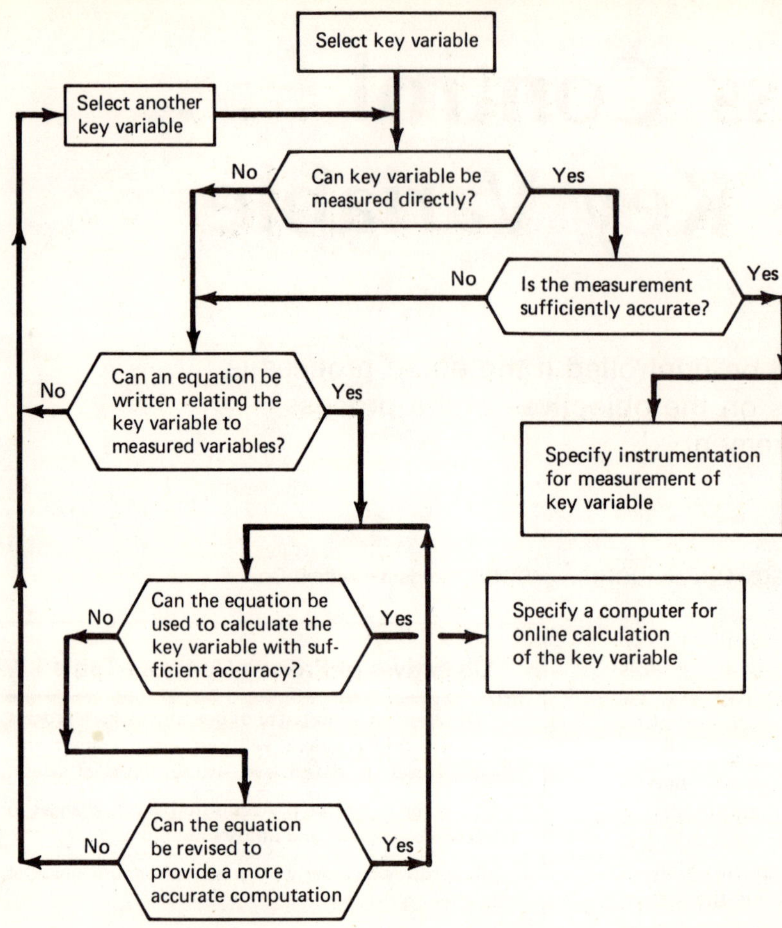

DECIDING between measurement and computation—Fig. 1

Once the variable that is the actual basis of design is chosen, a way must be discovered to represent it, either by measuring it directly or by computing it from direct measurements of other variables of which it is a function.

Selecting the Method of Representation

Fig. I is a flowchart that graphically presents a step-by-step procedure for selecting the method of representing the key variable. It emphasizes the twin problems of finding the correct variables to measure, and of measuring or calculating a true representation of the variable. To reach a yes-or-no decision for each section of the flowchart requires careful consideration of many factors. Of particular concern are:

Can the Key Variable be Measured Directly?—The signal obtained from measurement of a variable can be categorized as exact or inferential. The position of a float is an exact representation of liquid level. Alternatively, liquid level can be inferred from a measurement of hydrostatic head.

The advantages of using signals obtained by direct measurement are obvious. If the value must be inferred from the signal, more care is required to get an accurate representation. A hydrostatic head measurement will be affected not only by the level of the liquid but by its specific gravity and temperature, as well as by ambient temperature. Unfortunately, there are very few truly direct measurements that can be used in process control. The limitations of inferential values must be accepted or compensated for.

The alternative to directly measuring the selected variable is to compute it. But even if the variable is computed, measurements must be made to obtain the elements of the computation. These will, in general, produce inferential values.

Is the Signal Obtained From the Measurement Sufficiently Accurate for Adequate Control?—Factors limiting the effectiveness of inferential measurements are listed in Table II. Some of these factors can apply to exact measurement as well. For example, turbulence (noise) in a stilling well may be faithfully reproduced by float level, but the resulting signal will not be a true representation of liquid level in the tank.

Can an Equation Be Written Relating the Key Variable to Measurable Variables?—Among the principal tools of the chemical process designer are the energy balance and the material balance. Selected operating conditions will establish the amount of energy or mass of material, but the final equipment and plant facility design is based on the balances, not the operating conditions. When steam is used in a heat exchanger as a heating medium, heat-transfer design is based on the input and output enthalpy of the steam, which are computed from measured steam temperatures and pressures.

Because chemical process design makes such use of material and energy balances, and because the parameters of the balance equations are generally measurable, it is usually possible to write an equation from which the key variable for control of a specific chemical process can be computed.

A waste-treatment plant offers measurements of pH and flowrate. The plant designer uses these two quantities to write material balances that permit him to express his key variable, plant load.

For an operating regenerative furnace, one can measure fuel rate, air flowrate, and air temperature. When selecting burners, the furnace designer bases his choice on the heat available in a pound of fuel when burned with a particular amount of excess air. The relevant ratio of fuel to air is a weight ratio in this case, not a volume ratio. At the low-temperature end of the regeneration cycle, the volume airflow rate to provide the desired weight of air is very different from that at the high-temperature end. The designer must size his air ducts and fans with this consideration in mind. His key variable, Btu./hr., is expressed as a variable computed from temperature and flow rates.

Internal reflux rate,[2] the key variable of certain distillation columns using air condensers, cannot be measured directly. It can be computed, however, by an ingenious combination of energy and material balances around the top plate of the distillation column.

Design of a batch chemical reactor is directed toward maximum production rate of a product conforming to specified characteristics. In this case, there are two considerations. One is maintaining quality of product as a function of reaction rate and heat input. The other is obtaining maximum production (within the limitations of safe operation) as a function of heat generated. In each case, the measurable variables are flowrates of feed and of heating and cooling media, as well as temperatures, and pressures. These are related by the computations of the process designer to the key variable of instantaneous concentration of product (which he computes from the rate of enthalpy change expressed as Btu./hr. increase or decrease).[3]

Can the Equation Be Used to Calculate the Key Variable with Sufficient Accuracy?—In addition to the transducer characteristics that affect measurement accuracy (listed in Table II), many process characteristics influence the validity of a computed value. A computation is usually based on an idealized process model, and departure from ideality affects validity. A number of contributing influences are listed in Table III.

The mechanics of the computation must also be considered. This is particularly true if the computation involves a small difference between large numbers. If $X = Y - Z$, then the standard deviation of X is $\delta X = (\delta y^2 + \delta z^2)^{1/2}$, which seems innocent enough. But if $X \approx 0.1Y$, then 1% accuracy in the measurements of Y and Z ($\delta y = \delta z = 0.01Y$) results in $\delta X = 1.4\, \delta y =$

Conditions Limiting the Effectiveness of Measurement To Represent a Variable Table II

1. Transducer characteristics.
   a. Nonlinearity.
   b. Irreproducibility.
   c. Reference voltage (or regulated air supply) drift.
   d. Line voltage (or air supply) changes.
   e. Load-change effects.
   f. Hysteresis.
   g. Deadband.
   h. Response time.
   i. Environmental effects.
      (1) Ambient temperature.
      (2) Static pressure.
      (3) Vibration.
      (4) Humidity.
2. Installation characteristics.
   a. Dynamics of measurement—for example, the signal from a thermocouple may be a false representation of temperature because of the velocity of flow past the couple. (Heat transfer can be poor at low velocities.)
   b. Noise—for example, the signal seen by a differential-pressure transmitter may include the effect of turbulence caused by poor tap installation or inadequate approach piping.

Considerations Affecting Validity Of Computed Values—Table III

1. Design based on simplifying assumptions—the assumption of no heat loss from a jacketed vessel is a common convenience for superficial study of a physical process, but some loss does take place. It may not be possible to ignore it in a practical design. Other typical simplifying assumptions include:
   a. Use of empirical or extrapolated values.
   b. Ignoring inefficiencies of major equipment.
   c. Ignoring nonlinear effects, such as those taking place in mixing.
   d. Ignoring the interaction of process variables.
2. Design based on constants that vary during the course of a reaction or during the life of equipment. The effects include:
   a. Scale formation.
   b. Plugging.
   c. Oscillations caused by process or controller dynamics.
   d. Changing process conditions:
      (1) External—for example, the effect of barometric pressure on a mass-flow computer using a gage-pressure measurement.
      (2) Internal—load changes, changes in heating values, composition changes.
3. Incompletely representative computations. As examples:
   a. Approximation of the inferential relationship—e.g., using a square-root relationship to compute the variable when the function is really $a+bx+cx^2$.
   b. Failure to take side-reactions into account.

0.14*X*, or 14% accuracy for the computed value of *X*. Every effort should be made to select a method of computation that avoids such problems.

Values computed from heat-and-material balances are valid during steady-state operation but may contain serious temporary errors if the measurements used in computation have different response times. Flowrate of a particular component may be computed by multiplying total flow measured by a flowmeter (with response time of a few seconds) by the concentration of the component measured by a chromatograph (with several minutes' time delay). The product will be a spurious value resulting from the combination of a present flow signal with previous concentration signal.

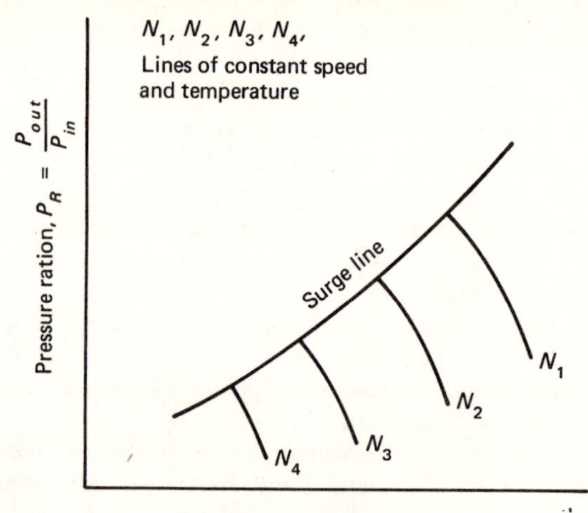

COMPRESSOR—flow and pressure-ratio relationship—Fig. 2

Computer Options

If application of the procedure shown in Fig. 1 has resulted in a decision for online calculation of the key variable, the next step is to specify a computer. Here again, the design engineer has many options. He may select either an analog or a digital computer, and can decide whether to do the programming and interfacing job himself or buy a packaged system.

The analog computer will consist of a group of modules designed to perform one particular computation, and will not be a general-purpose machine. The purchaser may elect to buy the modules and have them put together at the plant or go to a systems house for a complete, packaged product. The choice will probably be made on the basis of the electronic capability available at the plant.

The digital computer is more likely to be a general-purpose machine. If so, it probably will be a minicomputer dedicated to the particular task, although an already existing online computer with some idle time can be used. Although the purchaser will not build his own digital computer, he may construct some of the input-output hardware. He does have a choice between doing the programming in-house or contracting it out (either to the hardware supplier or a software house). Again, the choice depends on the level of in-house capability.

Recent technological developments have opened up another digital option, using LSI (large-scale integrated-circuit) chips. Each chip performs a particular computation such as addition, multiplication or integration, and chips can be combined to perform more-complex computations. Read-only memories can be used for table-lookup of functions. The design engineer again must choose between buying the chips and buying a packaged system.

Criteria for Selection Among Options

The first, and often overriding, consideration in deciding between minicomputer and hard-wired execution is cost. Schagrin[4] has estimated the installed cost of a minimal minicomputer-based data-acquisition system at $42,700. Addition of outputs to the process would raise this to at least $50,000. Unless the return from improved control can justify this expenditure, the minicomputer is excluded from consideration, and hard-wired analog or digital computation must be used (unless an existing digital computer is available).

The best prospect for justifying a digital computer is multiplexed computation. Digital costs consist, typically, of a relatively large initial expenditure and small additional costs per computation, especially if little additional programming is required, while analog costs increase almost linearly with number of computations. Calculation of the surge point for one compressor would almost certainly be executed by an analog or a hard-wired digital computer, while for a dozen compressors, a minicomputer might be called for.

Other considerations that may affect the choice of implementation include the accuracy and speed requirements, the nature of the variables, and the complexity of the computation. In general, a high-accuracy requirement favors digital execution, and may in some cases require floating-point computation. An application that requires a fast response may favor analog execution, particularly if the digital hardware is not dedicated to a single application but is multiplexed among several. A computation involving only continuous variables represented by continuous signals from transducers can usually be implemented easily by analog hardware. The presence of discrete variables, such as contact or switch closures, suggests that digital execution of some sort should be considered.

Computational complexity refers chiefly to the number of branches within the computational procedure, rather than the mathematical difficulty of the algorithm. A straight computational structure tends to favor analog execution, while a highly branched structure, indicating many decisions, favors digital implementation. If the structure is not firm and may be changed significantly, a minicomputer should definitely be considered, since only the application program need then be changed.

The tradeoffs involved in selecting a key variable, computing it, and specifying a computer for a particular

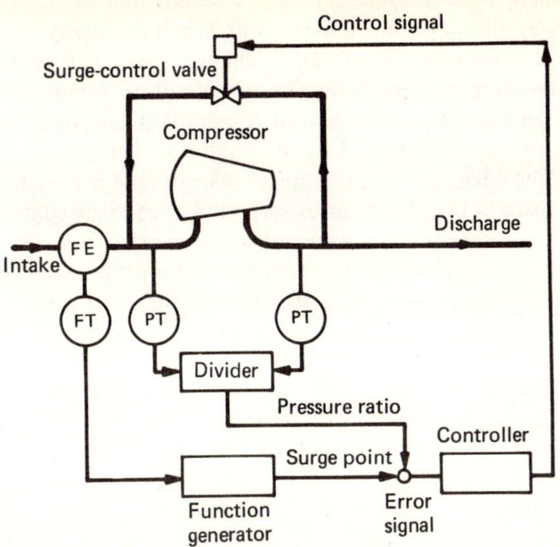

SURGE control by computation of surge line—Fig. 3

application can best be illustrated by considering specific examples. Two are discussed below.

Surge Control in Gas Compression

An inherent characteristic of centrifugal and axial compressors is that performance becomes unstable at some minimum flowpoint. This occurs when backpressure in the system cannot be overcome by the head produced, and a reversal of flow results. The process becomes cyclic with increasing frequency, and mechanical as well as thermal damage can occur to the compressor. To prevent surge, the most commonly used control action is to open a vent or bypass valve in order to maintain flow above the critical point.

Select the Key Variable—The variables of particular interest that can be measured are: inlet and discharge pressures, volumetric flowrate, and compressor speed. Their relationship is shown in Fig. 2. The surge line is usually determined by the compressor manufacturer, and is the locus of "surge points," each representing a combination of speed, pressures and flowrate at which instability occurs.

The key variable is the limit representing surge point. Safe operation is achieved by maintaining flow to the right of the surge line. Efficient operation is achieved by operating as close to the surge point as possible. Gas can be vented or bypassed to increase flowrate to a safe level for a particular pressure ratio, but economy of operation is sacrificed by doing so.

Can the Key Variable Be Measured Directly?—If the compressor has a single speed and if the temperature of the entering gas is constant, there will only be one surge point, and measurement of one variable (flow or pressure-ratio) can suffice to maintain flow above this point. In the case of an electric-motor-driven compressor, the motor current may vary sufficiently with flow, so that its measurement may be used. This is usually more economical than flow measurement because no device is required in the piping to produce a differential pressure.

Can an Equation Be Written Relating the Key Variable to Measured Variables?—Surge point cannot be measured directly if the compressor is a variable-speed machine. In this case, the surge point must be computed as a function of the other variables—inlet and discharge pressures, flowrate, and speed.

A number of equations have been derived from empirical and theoretical concepts to relate the variables.[5,6] Alternatively, the equation of the relationship of volumetric flow to pressure ratio can be developed from the compressor manufacturer's calibration data. Then, a function of pressure ratio, computed from a measurement of flow, can be compared to actual pressure ratio, to provide a continuously varying setpoint to the surge-control loop (Fig. 3).

Can the Equation Be Used To Calculate the Key Variable With Sufficient Accuracy?—How accurately the surge point must be computed will depend on the economics of operation. The more sophisticated the compressor design, and the more important its efficiency of operation, the closer to the surge point the operation must be maintained (without endangering equipment). This will determine the accuracy required of the measured variables, and the validity of the function relating the variables. An approximate function can contribute as much error as an inaccurate measurement.

For example, the relationship:

$$P_R = K(F_v)^2 + 1$$

(where P_R is the pressure ratio, and F_v is the volumetric flow) has been used to represent the surge curve. This approximation has the advantage that no square-root extraction need be performed to calculate the required pressure ratio. Only $(F_v)^2$ is needed and this is directly proportional to pressure drop across an orifice.

Unfortunately, it is rare that the surge curve actually follows a square-root relationship. Use of this function can produce safe but inefficient operation. Improved operation, at greater initial cost, is obtained by approximating the relationship with a quadratic:

$$P_R = a_0 + a_1 (F_v) + a_2 (F_v)^2$$

A function generator made by programming a multiplier and some operational amplifiers can develop an accurate setpoint from the flow measurement to match against the P_R measurement, so as to get economical and safe operation at every speed and inlet-temperature condition.

Implementation

For each method of measurement and computation mentioned, analog instrumentation is indicated for a single-compressor installation. The computations are not difficult, and the extra benefits of a digital computer would be hard to justify on an economic basis.

A digital computer should be considered for the computation if any of the following sets of circumstances occur:

1. Computed surge point is just one of many measurements made by the entire instrumentation facility of a

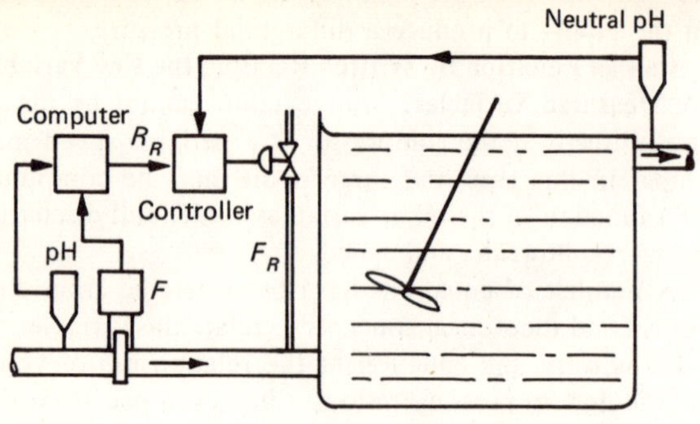

NEUTRALIZATION of waste by reagent *R*—Fig. 4

very expensive compression system, where significant payout could be realized by this computation.

2. A time-shared installation is set up for a large number of compressors. The time period allowable for identifying and correcting a dangerous situation must be a consideration in selecting analog or digital implementation. Continuous measurement of variables will be mandatory for an installation that could "run away" in a time interval shorter than the sampling period of a time-shared digital system.

3. In a compressor design- and test-facility where payout could be realized by cycling each compressor through a set of test programs including rapid, frequent changes to allow investigation of many actual and hypothetical process conditions, the digital computer becomes attractive if it can be justified economically.

Waste-Neutralizer Load

A typical waste neutralizer (Fig. 4) must accept almost random inputs from an entire plant complex, and produce an effluent that satisfies legal limits on pH. The feedback control system will do its best to control effluent pH by manipulating reagent flow, but it can only react to disturbances as they appear at the process output. Hence, rapid influent changes can produce significant deviations from the effluent pH setpoint. If the influent titration curve of pH versus reagent addition has the usual S-shape, process load and therefore reagent requirement can be calculated from influent flow and pH as:

$$R_R = KF10^{-\text{pH}}$$

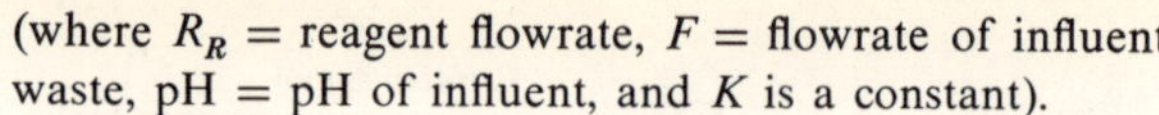

(where R_R = reagent flowrate, F = flowrate of influent waste, pH = pH of influent, and K is a constant).

As shown in Fig. 4, the calculated signal can be fed forward to reagent flow. The signal will, in general, not be an exact representation of reagent flow and therefore is trimmed or adapted by the feedback signal.

The effect of using calculated reagent load is to reduce the size of the disturbances seen by the feedback controller. Another way to reduce disturbance size is to use a larger tank, so the cost of computation must be balanced against the cost of tankage. Unless the tank must be constructed from expensive materials, it seems unlikely that minicomputer implementation can be justified.

The calculation requires only moderate accuracy since feedback adaptation will be used in any case. Rangeability may pose a problem since the signal can vary over a range of 1,000/1 for a change of 3 pH units. The range problem can be circumvented by using logarithms for the computation, but then some exponential device must be used to produce the required flow.[7] Speed is no problem. The input variables of flow and pH are both available as continuous signals from conventional transducers. Computation is a straight-through procedure unless input pH can vary from acid to basic. To neutralize a basic feed, not only must a different reagent be used, but the algorithm changes to:

$$R_R = KF10^{14-\text{pH}}$$

Since cost excludes minicomputer execution, the choice usually lies between hard-wired analog and digital systems. For a limited range of influent pH, an analog system would appear preferable. A wide range of pH, particularly if the range included both acidic and basic influent, might be better suited to a digital circuit. The exponential function could be handled by table lookup and interpolation, perhaps built into the analog-to-digital converter.

References

1. Byer, G. T., Control the Real Variable, *Chem. Eng.*, July 6, 1964, p. 125.
2. Lupfer, D. E., Berger, D. E., Computer Control of Distillation Reflux, *ISA (Instr. Soc. Am.) J.*, June 1959.
3. Tolin, E. D., Fluegel, D. A., An Analog Computer for On-line Reactor Control, *ISA (Instr. Soc. Am.) J.*, October 1958.
4. Schagrin, E. F., How Much Do Minicomputer Control Systems Cost?, *Chem. Eng.*, Mar. 22, 1971, p. 103.
5. Magliozzi, T. L., Control System Presents Surging in Centrifugal Flow Compressors, *Chem. Eng.*, May 8, 1967, p. 139.
6. Hatton, C., Controlling a Gas Compressor Station Electronically, *Instr. Control Systems*, May 1967.
7. Shinskey, F. G., "Process-Control Systems," 1967, McGraw-Hill, N.Y.

Meet the Authors

◀ **Paul G. Friedmann** is a senior scientist in the Systems Analysis Dept. of Leeds & Northrup Co., Dickerson Rd., North Wales, PA 19454. He attended the University of Michigan, and holds an M.S. in chemical engineering from the University of Pennsylvania. He is a member of Instrument Soc. of America, and has written papers on simulation and distillation control.

John A. Moore is a member of the Analog Control Div. of Leeds & Northrup Co., where he specializes in flow and pressure measurement and control. He has spent more than 25 years selecting and applying instruments in the chemical, metallurgical and power industries. He received his B.S. in chemical engineering from Drexel Institute of Technology and is a registered professional engineer in the State of Pennsylvania. ▶

Analyzing process-control loops

Even without having an exact mathematical model of a process-control loop, you can use a computer to identify critical and noncritical parameters and boundaries for tuning constants.

G. Edward Graham, Computer Sciences Corp.

☐ If you are a process-control engineer, how will you go about designing and tuning the control loops on the new digital control system your company is installing? If this system is replacing a not-very-large analog system that has been in use for several years, the design problem may be fairly straightforward. You can probably use the same control algorithms and tuning constants used in the analog system.

But what if this is a brand-new control problem? A complex one with several interacting parameters? A control loop wherein slight variations from optimum can have a very significant effect on cost or product quality? How do you know what algorithms will give you a stable process? Given some combination of algorithms, how do you select the necessary tuning parameters?

One approach is to analyze the open-loop transfer function of the process with one or more of several widely used methods—Nyquist plots, Bode plots, root-locus, analog simulation, etc. Of course, the results of any of these approaches are only as accurate as the transfer function. Perhaps you do not know the process well enough to write a model of it. No matter! You probably know enough to create a model that is similar in form to the actual process. While you may not know exact values of various coefficients in the model—thermal lag, transport deadtime, motor time-constant, etc.—you can still make useful progress if you know maximum and minimum values. If you evaluate a number of combinations of coefficients in your model, you can develop an understanding of how the loop will respond to changes in tuning parameters. This understanding should help you identify noncritical parameters and pinpoint the measurements and adjustments that will have to be made when the actual system starts operating.

The analysis can show the effects of the digitizing process and will help you select the necessary scanning frequencies for the loop. This can be particularly useful when the computer is running out of real-time for processing all the input data, and you are searching for ways to reduce the processing load. Perhaps a one-second loop will respond acceptably at a two- or five-second scan rate if you use the proper tuning parameters. You can look at the effect of various degrees of filtering, or at the results of a loop upset as compared to a setpoint change.

Historically, the approach we are recommending—parametric analysis—has been a laborious one. Bode plots were drawn on semilog paper with a ruler; one searched for a portion of the root locus with a sheet of graph paper and a spirule; the wiring and scaling of the analog computer took considerable time. The amount of work involved in analyzing a particular loop was one of the important factors behind the demand for an accurate model of the process.

Originally published August 2, 1976.

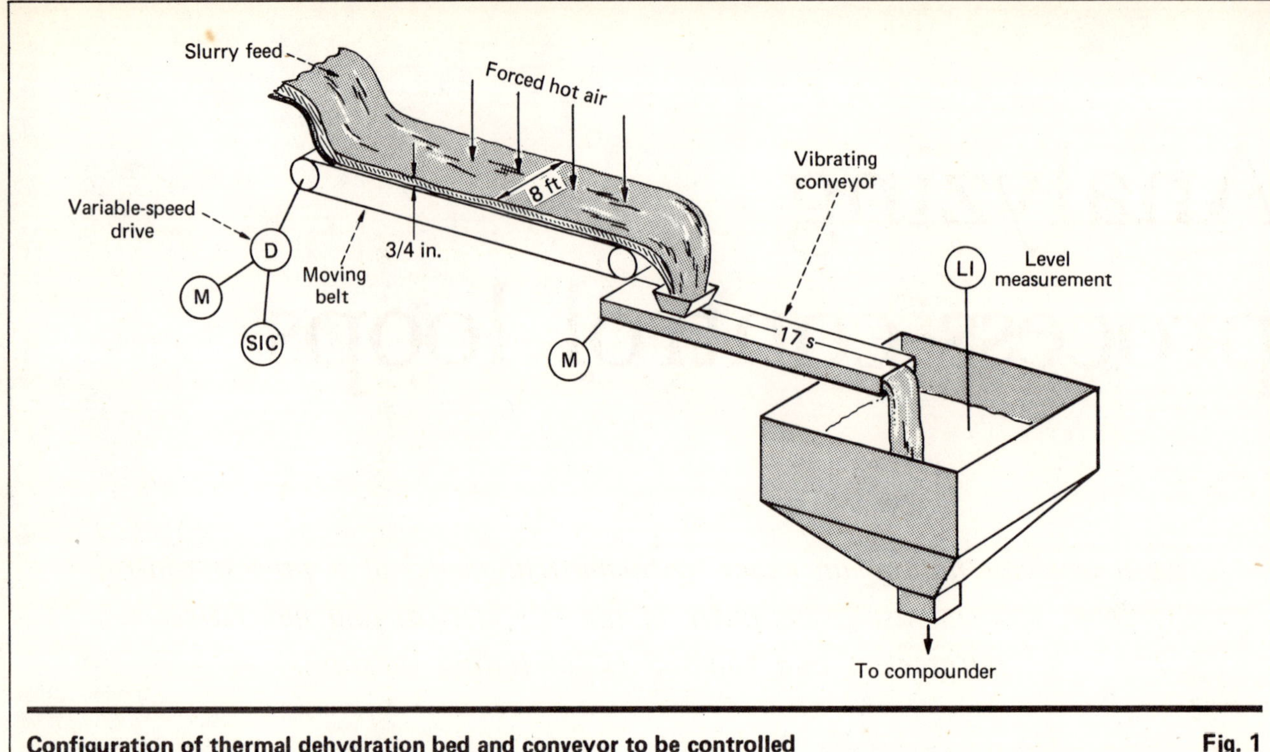

Configuration of thermal dehydration bed and conveyor to be controlled **Fig. 1**

Now, the process-control engineer should think in terms of letting the computer grind out the calculations for him. He can concentrate on improving his mathematical description of the control loop and evaluating the results of a number of potentially useful control schemes. He can calculate inverse transforms quickly to get time-response information without having to patch and scale an analog computer. While the demand for an accurate model is as great as ever, the engineer can look at a number of inaccurate models to pinpoint the boundaries of the control problem.

The following discussion demonstrates this analytical approach to loop design by using z-transform root loci and inverse z transforms. The control loop presented is in a plant the U.S. Army is building to manufacture certain nitrocellulose-based propellants. This plant will be almost completely automated and will be the first of its kind. While certain process units were evaluated in a pilot-plant operation, the full-scale units in the line will be the first to operate under closed-loop computer control.

The control problem

The basic component that is processed through the line is nitrated cellulose (NC). Since dry NC is rather flammable, it is stored in a water slurry until processing begins. The first step in the process is to remove at least 99.9% of the water in a thermal dehydration unit by using forced hot air. The slurry is spread from a weir onto a slowly moving belt, where most of the water runs off. The wet NC fibers form a cake on the belt and are dried by the circulating air as they pass under the hot-air hood. Just before falling off the belt onto a conveyor, the NC cake is sprayed with alcohol to prepare it for the next processing step.

The conveyor carries the NC to a hopper that feeds the next process unit, the compounder. The control problem is to keep this hopper half full. The hopper's function is to provide a buffer between the two process units, so that changes in feedrates of one unit will not require rapid changes in setpoints on the other unit. Consequently, the hopper level can deviate from the 50% point so long as it does not overflow or become empty.

The control scheme that is used varies the belt speed as a function of hopper level. It is assumed that the control loop that regulates the flow of slurry onto the belt will maintain a constant thickness of NC cake on the belt at all belt speeds; i.e., the dynamics of the thickness-control loop are not included in this analysis. Fig. 1 shows the essentials of the belt-speed control loop.

The mathematical model

The mathematical model for this initial evaluation is relatively simple. It includes: a proportional integral (PI) compensation algorithm; an integrator to convert the velocity-mode commands from changes in setpoint to bed-speed setpoint; a dead time for the conveyor; and an integrator and constant to convert cubic ft/min flowrate into hopper level.

The simplicity of this model was justified on the basis of the relatively long sampling period of the hopper level (30 s) and the assumption that the acceleration of the moving bed would match speed to setpoint in just a

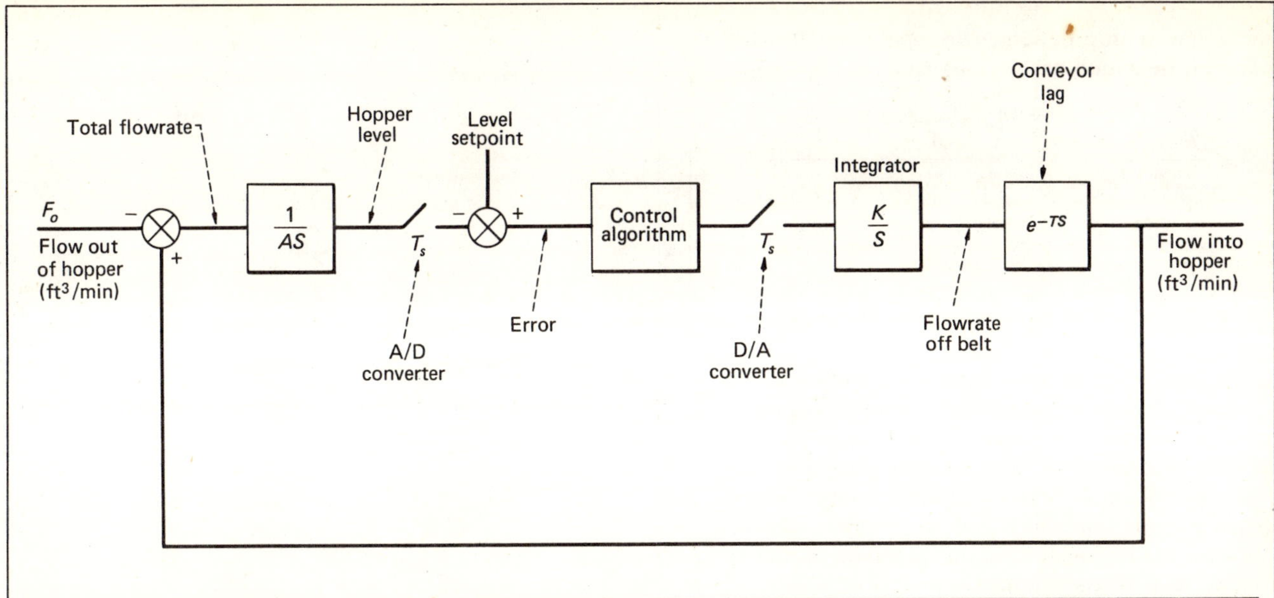

Mathematical model of system for the control of bed speed **Fig. 2**

few seconds. Fig. 2 represents the components of this relatively simple model.

The flow out of the hopper is subtracted from the flow into the hopper to obtain a net flowrate to the hopper (in ft^3/min). Hopper level is obtained by integrating this volumetric flowrate and dividing by the hopper cross-sectional area. The level is sampled once every T seconds, and the measurement is compared with the setpoint to obtain the error. A PI velocity-mode algorithm was selected to provide the necessary control compensation. The output of the algorithm then drives an integrator that causes the belt speed to change by some increment. The volumetric flowrate is delayed by $T(B)$ seconds in the constant-speed conveyor before it is added to the flowrate out of the hopper.

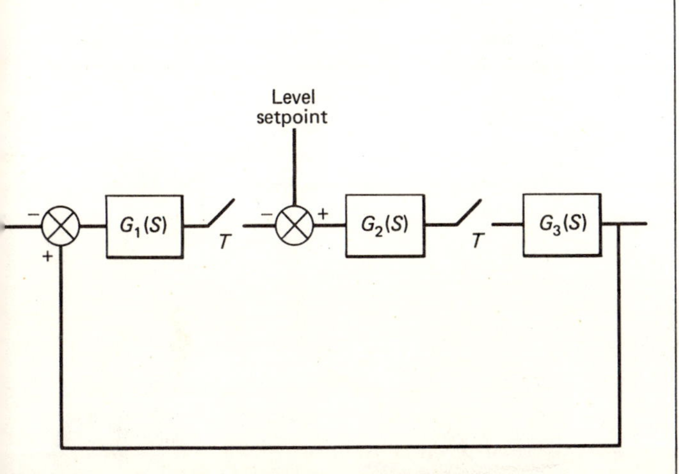

Bed-speed control Laplace transforms **Fig. 3**

Fig. 3 is a simplified schematic of the control loop, using Laplace transforms.

where:

$$G_1(s) = \frac{1}{As}$$

$$G_2(s) = \left[(1 - e^{-Ts}) + \frac{T}{T_R}\right]\frac{100}{P_B} \ldots$$ Laplace transform of the PI compensation algorithm

$$G_3(s) = \frac{Ke^{-T_B s}}{s}$$

e^{-TS} = Laplace transform of delay between sampling instants

A = Compounder hopper cross-sectional area, 4.17 ft^2

T = Sampling period, 30 s

T_R = Reset time-constant

P_B = Proportional band

K = Conversion constant between hopper level error (ft) and bed speed change (ft/min) = 3.27

T_B = Conveyor dead time, 17 s. See box on p. 000

s = Laplace complex frequency

The sampled open-loop transfer function can thus be written:

$$G^*(s) = \frac{3.27}{4.17}\frac{100}{P_B}\left[(1 - e^{-Ts}) + \frac{T}{T_R}\right]\frac{e^{-T_B s}}{s^2} \qquad (1)$$

where the G^* is used to denote the sampled-data form of the transfer function.

Eq. (1) can be converted into a z-transform format by using a modified z transform to express the transport lag introduced by the conveyor.

After a few manipulations, this open-loop transfer function can be stated as:

$$G(z) = \frac{78.5\,(1 + a)\,mT}{P_B} \cdot \frac{\left(z - \dfrac{1}{1 + a}\right)(z + b)}{z(z - 1)^2} \qquad (2)$$

where:

$a = T/T_R$

$m = 1 - (T_B/T)$

$b = (1 - m)/m$

The first zero of the transfer function is caused by the reset portion of the compensator. Changing T_R moves this zero to modify the dynamic performance. The second zero is caused by the conveyor lag. The two poles at $z = 1.0$ result from the double integration of the velocity-mode controller and the hopper volume. The pole at zero is related to the previous sampled-error-value used in the compensator.

There are two parameters that can be changed to modify the dynamic characteristics of this control loop: proportional band and reset time-constant. With a different compensation scheme, there would, of course, be other adjustable parameters. The effects of changing these parameters can easily be evaluated from the root locus of the characteristic equation.

For those who may be unfamiliar with the mathematics of digital control loops, the root locus for a sampled-data loop is directly analogous to the root locus for an analog loop. Analog loops can be analyzed using Laplace transforms and the complex variable, s. The sampling process associated with digital control introduces the term e^{-Ts} into the Laplace transfer function, where T is the period of the digitizer or sample-and-hold network. This nonlinear term prevents direct application of the theory developed for linear, continuous control loops.

The complication can be removed by making the mathematical substitution of $z = e^{Ts}$ and restating the Laplace transforms in terms of the variable z—thus the name z transform. The location of the sample-and-hold devices in the loop must be considered in grouping the Laplace transforms of the linear portions of the loop before performing this transformation. A number of textbooks on sampled-data control systems present the details of this mathematical process. For example, see Chap. 6 of Tou's book.*

The z-transform process results in the following modifications to the root locus:

1. The region of stability is transformed from the left half of the complex plane into the inside of the unit circle, i.e., the closer the locus gets to the unit circle, the more lightly damped is the closed-loop response. Outside the unit circle, the loop is unstable; on the real axis, the loop is overdamped.

2. Lines of constant damping are transformed from straight lines into log spirals inside the unit circle.

3. Lines of constant decrement factor are transformed into circles concentric with the unit circle.

Once these transformations have been made, the process of generating and interpreting the root locus for

*Tou, Julius T., "Digital and Samples-Data Control Systems," McGraw-Hill, New York, 1959.

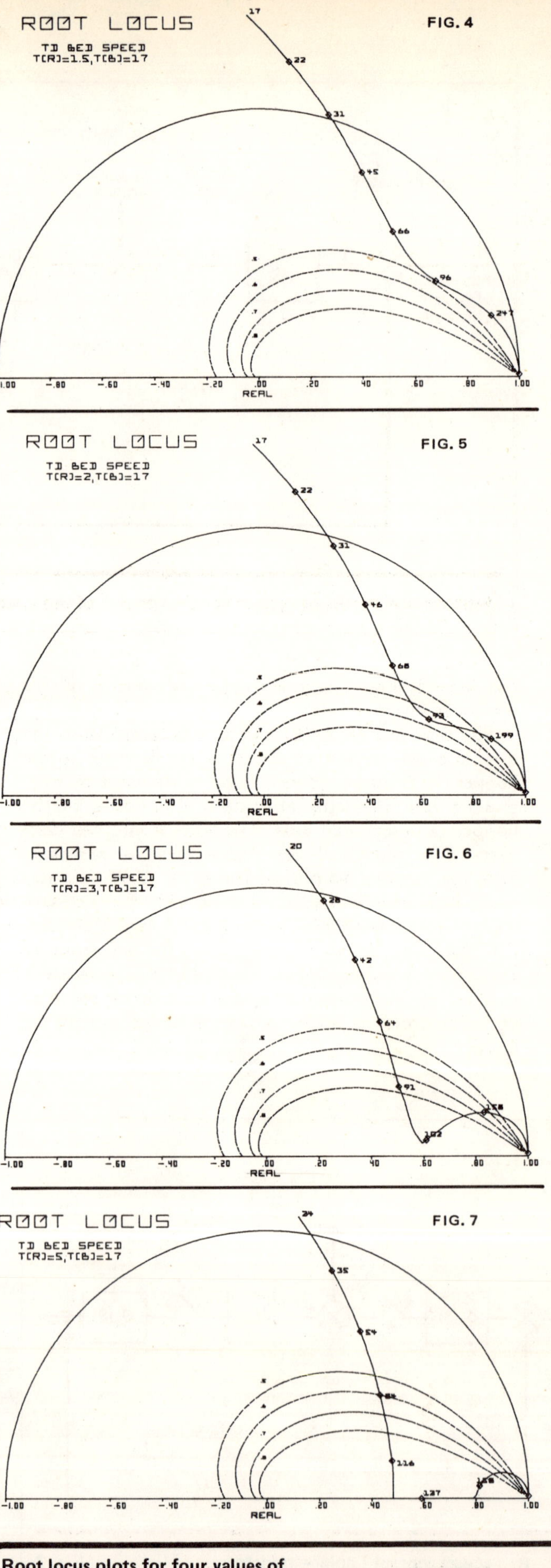

Root locus plots for four values of reset time constant

Fig. 4-7

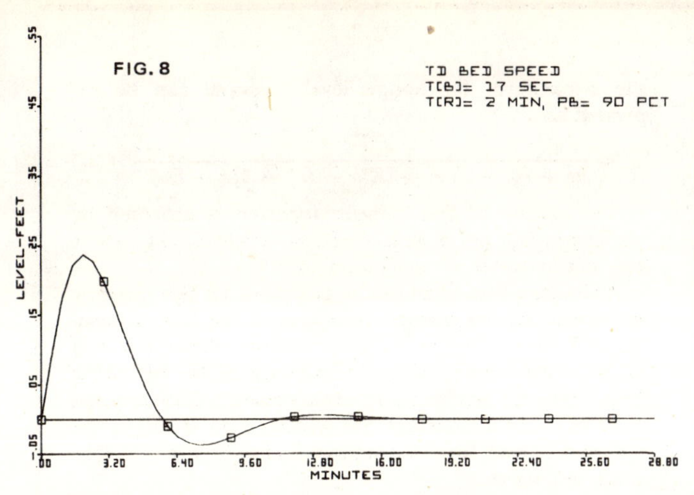

Hopper level response to change in throughput **Fig. 8**

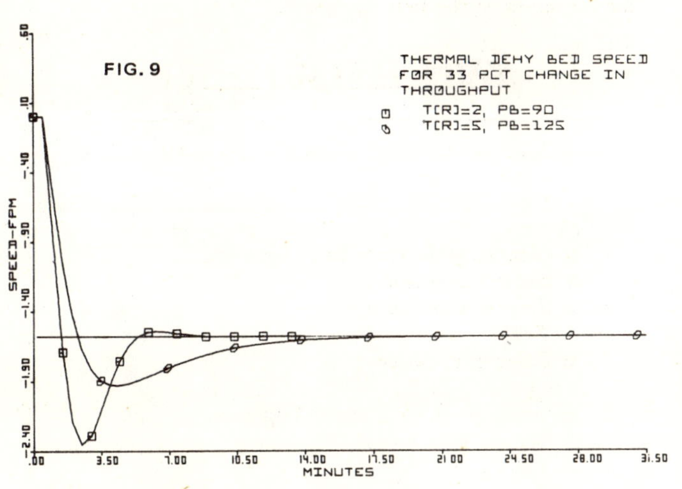

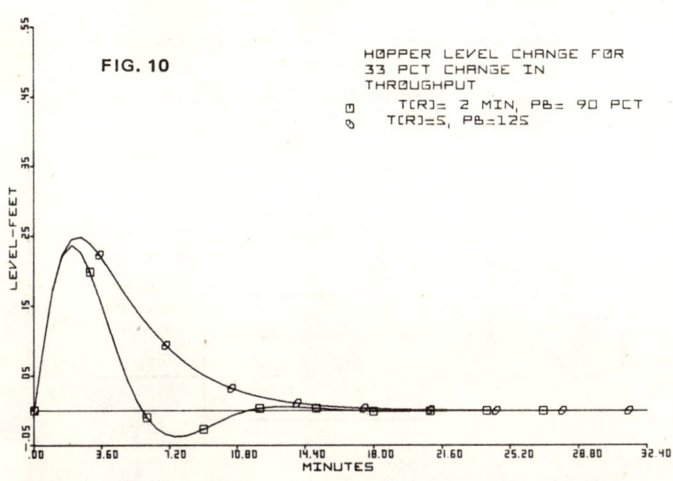

Bed speed and hopper level response to 33% change in throughput **Fig. 9-10**

sampled data loops becomes identical to that for continuous control loops.*

Results for belt-speed control model

The first step in the study was to generate the root locus for each of four values of reset time-constant: 1.5, 2, 3 and 5 min. (See Fig. 4 to 7.) From these, it appeared that a reset of 2 min. and a proportional band of 90% would yield desirable characteristics. A proportional band anywhere between 82 and 110% would result in a damping factor no less than 0.6 (a damping factor of 0.707 causes a response that settles to its final value in the minimum time). If a reset of 1.5 min. were used, see Fig. 4, the best damping factor that could be achieved would be slightly less than 0.5. This begins to be fairly lightly damped. If, on the other hand, a reset of 3 min. were used, see Fig. 6, the damping factor would become fairly sensitive to changes in loop gain.

The next step in the analysis was to evaluate the time-domain response of the loop to a step change in the flowrate out of the compounder hopper. The size of the step change was chosen to be 33% of the maximum throughput rate. This is probably the largest rate change that will occur at any time under digital control. A larger rate change can occur if the compounder stops operating; but in that event, other computer control logic will intervene to stop the thermal dehydration.

The time-domain response was determined by evaluating the inverse z transform for the closed-loop response, i.e.:

$$C(t) = z^{-1}\left\{\frac{G(z)}{1 + GH(z)}\right\}$$

The algebraic and numeric manipulations to arrive at this inversion for the transfer function of the bed-speed loop are shown in the box on p. 77.

The interesting result from the time-domain response is that the hopper level increases only about 3 in. before settling back to its original setpoint. This response takes about 11 min. to damp out (see Fig. 8).

Since the hopper-level dynamic response will apparently be fairly small, the bed speed becomes the quantity of real interest in this loop. Acceleration of the bed should be minimized so as to avoid upsets in the bed thickness. Going back to the root-locus plots, we find that a reset time-constant of 5 min. has a range of gain over which it remains on the real axis—i.e., the loop is overdamped. The maximum gain for this overdamped condition occurs with a proportional band of 125%. This is at the point where the locus leaves the real axis and heads for the unit circle. The dynamic response for this pair of parameters provides an interesting surprise. The maximum rate of change of bed speed is reduced from 1.4 fpm/min. to 0.8 fpm/min. but the maximum hopper level encountered increases less than half an inch! See Fig. 9 and 10.

*To work with the gain and phase plots of the Bode method for designing compensation schemes, there is an analogous set of plots for a sampled-data system. The bilinear transform $w = \frac{z-1}{z+1}$ will convert the coordinate system of the unit circle in the z plane into a set of rectangular coordinates in the w plane. With a little experience, it becomes a fairly straightforward task to devise a control design and then work backward through two inverse transforms—w to z and z to time—to identify the time response of the loop.

Transient response

The transient response of the control loop can be determined by calculating the inverse of its closed loop z-transform. This results in an infinite series of terms that can be represented by Eq. (A1):

$$F(t) = \sum_{n=1}^{\infty} C_n Z^{-n} \tag{A1}$$

The coefficients of this infinite series are merely the sampled magnitude of the resulting time function at each sampling instant.

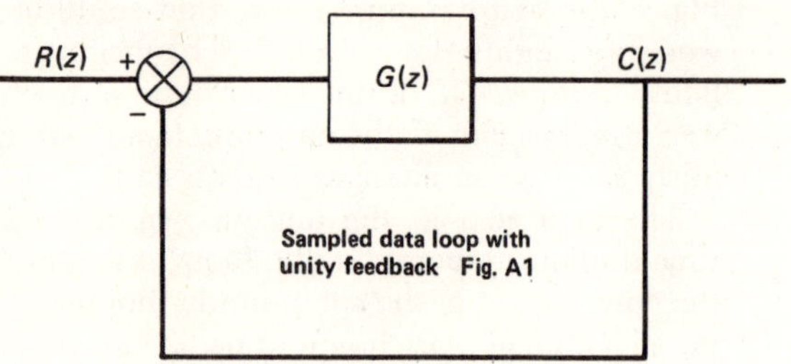

Sampled data loop with unity feedback Fig. A1

Fig. A1 represents a feedback control loop with unity feedback. From this, it is easy to show that:

$$\text{Error} = \frac{R(z)}{1 + G(z)} \tag{A2}$$

In the belt-speed control loop, a change in level is caused by a step change in feedrate out of the compounder hopper. The way the mathematical model is designed, this results in a ramp function for changes in hopper level. Expressed mathematically, this is:

$$R(t) = F_o t \tag{A3}$$

$$R(z) = \frac{F_o T z}{(z-1)^2} \tag{A4}$$

where F_o is the change in feedrate, T is the sampling period.

In the speed-control loop, the error is just the error in level, so Eq. (A2) can be used to define the behavior of level.

$$\Delta L(z) = \frac{F_o T}{(z-1)^2} \frac{1}{1 + G(z)} \tag{A5}$$

where $G(z)$ is the open-loop transfer function.

Since

$$G(z) = \frac{K(z - D)(z + b)}{z(z-1)^2} \tag{A6}$$

the z-transform for hopper-level response can be expressed as:

$$= \frac{F_o T z^2}{z^3 + (K-2)z^2 + [K(b - D) + 1]z - DbK} \tag{A7}$$

The inverse of this transfer function is obtained by performing the polynomial division indicated, to generate the infinite series of coefficients of z^{-n}.

A similar z-transform can be generated for belt speed by manipulating the transfer functions of the loop components. Referring to Fig. A2, the output of transfer function G_3 is the volumetric feedrate of material off the end of the belt. Since the analysis assumes a constant bed-thickness (0.75 in.) and a constant bed-width (8 ft), dividing the volumetric rate by cross-sectional area of the NC cake gives belt speed.

$$\text{Speed}(z) = \frac{V(z)}{\text{Cake area}} =$$

$$\frac{1}{0.5} \cdot \frac{F_o T z}{(z-1)^2} \cdot \frac{G_1(z)\, G_2(z)\, G_3(z)}{1 + G_2(z)\, G_1 G_3 G_4(z)} \tag{A8}$$

Substituting the various transfer functions and appropriate constants into Eq. (A8) results in the z-transform for the speed of the belt, Eq. (A9):

$$S(z) = \frac{\dfrac{F_o T(1 + a)\,157}{P_B}\left[z^2 - \left(\dfrac{1}{1 + a}\right)z\right]}{(z-1)[z^3 + (K-2)z^2 + (K(b - D) + 1)z - DbK]} \tag{A9}$$

where:

F_o = Change in flowrate from hopper
T = Sampling period
P_B = Proportional band
a = T/T_R
T_R = Reset gain factor
$K = \dfrac{78.5(1 + a)mT}{P_B}$ ... open-loop gain
$m = 1 - T_B/T$
T_B = Conveyor transport lag
$b = (1 - m)/m$
$D = 1/(1 + a)$

Note that because the denominator contains the term $1 + G(z)$, the denominator coefficients for the speed functions can be obtained easily from those calculated for the level function, by multiplying by $(z - 1)$.

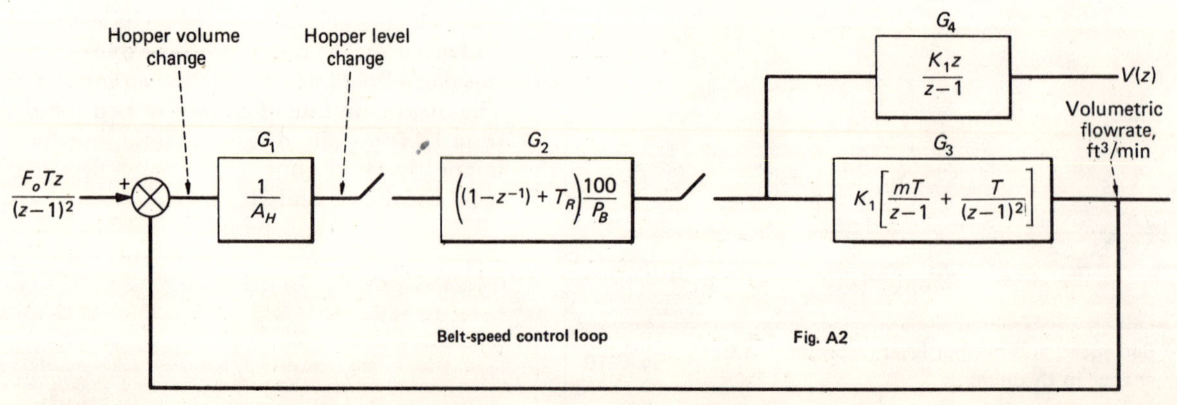

Belt-speed control loop Fig. A2

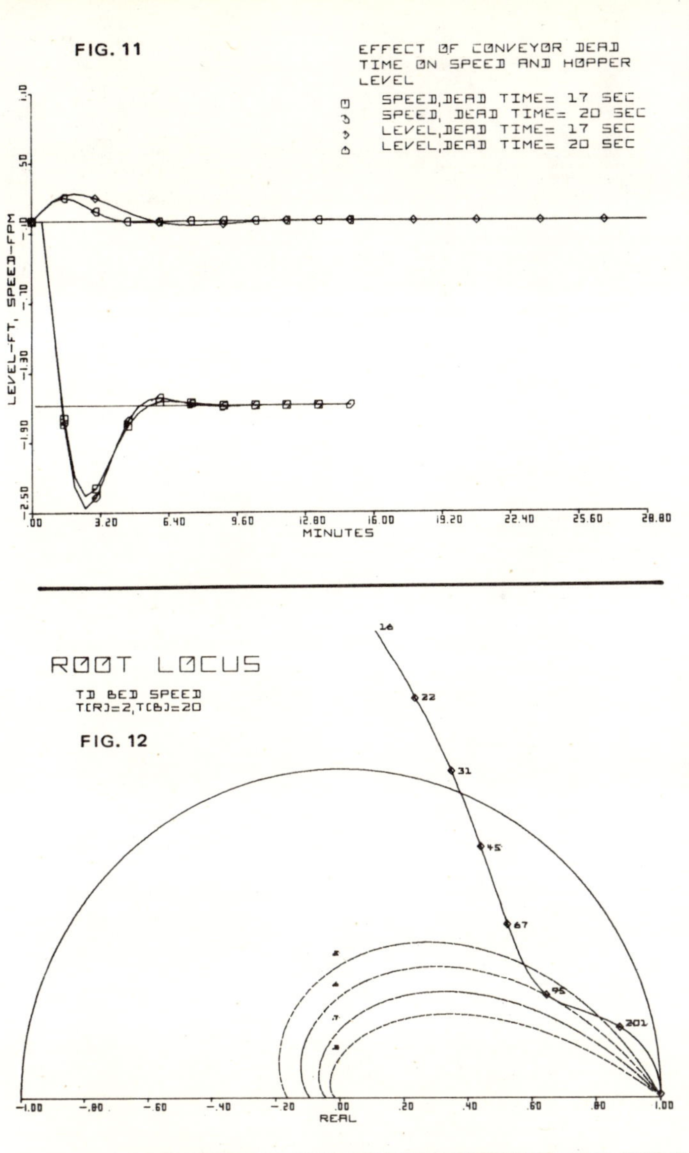

Effect of conveyor dead time on speed and hopper level, with root locus plot **Fig. 11—12**

As another experiment, the conveyor lag $T(B)$ was assumed to increase to 20 s, from its specified value of 17. If the proportional band remains set at 90%, the response is only slightly changed. The magnitude of the change in the hopper level is reduced, and the maximum acceleration of the bed is essentially the same, but the maximum change in bed speed increases. See Fig. 11. From the root locus of this case (see Fig. 12), we can deduce that a slightly larger reset time-constant, perhaps 2.2 or 2.3 min., would create a response similar to that achieved with a 17-s conveyor and a reset of 2 min.

Conclusions

1. The bed-speed control loop should be tuned to be overdamped, so as to minimize rates of change of bed speed. A reset time-constant of 5 min. with a proportional band of 125% would seem to be acceptable.

2. Hopper-level changes in response to significant changes in throughput rate are very slight. This means that NC residence time in the hopper can be fairly accurately predicted as a function of throughput. This observation has potential significance for another control loop.

3. Small changes in the lag introduced by the conveyor to the compounder hopper are apparently acceptable, with no serious consequences on loop performance.

4. The performance of the bed-thickness control loop should be analyzed to determine the effect of step changes in bed-speed commands. Remember, while the curves presented here for bed speed appear essentially continuous, the bed-speed changes result from discrete commands that occur every 30 s. Depending on the dynamics of the hardware controlling the bed speed, the speed changes can vary from pseudo step changes to reasonably linear variations.

Implications for the process engineer

As you may have recognized, the root-locus plots and the time-response functions used in this analysis were generated by a computer and an X-Y plotter. These automated design tools considerably simplify the control engineer's design task. Since they are relatively inexpensive to use,* a number of alternatives can be evaluated to find the best solution for a particular problem. So, even if you do not have an exact model for your process, you can well afford to look at a wide range of potential models to see whether or how certain parameters effect the dynamics of the control problem.

For instance, the bed-speed model discussed here does not include the dynamics of a pneumatic controller located at some distance from the motor drive. However, before adding this component to the model and repeating the analysis, it would probably be worthwhile to analyze the slurry-control response. If, indeed, the slurry-control loop can be made tight enough to maintain bed-depth constant over all belt speeds and accelerations, actual performance of the speed loop may not be critical. After all, hopper level changes only about 3 in. for both an overdamped response and for a moderately damped response.

It is observations and conclusions such as these that can help you identify the really important control problems in your process. At startup, you will be better prepared to select and adjust the various tuning parameters and get the process into economic operation.

*The root locus requires 6 to 7 s of computer time to compute and plot on a large time-sharing computer. The inverse transform takes slightly less.

The author

G. Edward Graham is a Senior Member of the Technical Staff at Computer Sciences Corp., 6565 Arlington Blvd., Falls Church, VA 22046. He works in the Industrial Process Control operation, where his recent experience included developing portions of the control software for a large propellant-manufacturing facility. He holds a B.S. in engineering science from Pennsylvania State University, an M.S. in engineering mechanics from Stanford University and an M.S. in engineering administration from George Washington University.

Section IV
A GENERAL GUIDE TO SELECTING INSTRUMENTS

Measuring Process Variables

Here is a guide to selecting and operating instruments for measuring five key process variables: temperature, pressure, flow, level, density and viscosity.

DENNIS E. ZIENTARA, Taylor Instrument Process Control Div., Sybron Corp.

TEMPERATURE

Temperature is based on inferential measure of heat. Scales such as Fahrenheit, Centrigrade and Kelvin are arbitrary and defined under standard conditions of pressure. Ways in which temperature may be inferred are:

- Change in volume.
- Change in electrical resistance.
- Voltage created at the junction of two dissimilar metals.
- Intensity of emitted radiation.
- Resonant frequency of a crystal.

Accuracy is defined as the limits within which the measured temperature value varies relative to the true value. Generally, it is expressed in percent of full scale and includes three elements: conformity, closeness to which a device will deliver a signal proportional to the actual relationship; hysteresis, the difference in the measurement signal when approaching a given value from below and above its position on the scale; and repeatability, an instrument's capability of generating the same signal under identical process conditions occurring at different times.

Characteristics of principal temperature measuring elements used in the chemical process industries are shown in Fig. 1 and Table I.

Application Tips

The temperature sensing element should be located as close as possible to the spot where the temperature reading is desired. A remote location, even if there is no error in the steady state reading, gives large errors through dynamic dead time and automatic control may be impossible. A small bulb gives faster response than a large bulb, regardless of the type. To reduce film effects, always install the sensing element where the gas or liquid is moving rapidly.

When steam and water or other liquids are mixed directly, the sensing element must be far enough away to get a true mixed temperature. At the same time, however, the transport log (distance times velocity) must be kept at a minimum.

A temperature well for the sensor does not give optimum automatic control. If one must be used because of pressure, vibration or corrosion problems, care must be given to cleaning the process side. For optimum results from a bulb in a well, use a heat transfer medium such as aluminum powder, mercury or oil. Of course, the temperature range must not exceed the boiling or freezing point of the heat transfer medium.

Thermocouple lead wires should be enclosed in a grounded, separate conduit away from AC power lines. The same is true with resistance thermometers. In addition, when there is a significant distance between the element's resistance and the resistance to the current converter, a third wire should be used for compensation.

Thermocouple extension wire must be specified by the manufacturer. Use of other types will cause unpredictable errors.

PRESSURE

Simply stated, pressure is the force acting on a surface divided by the area of the surface. Zero absolute pressure is usually defined as a perfect vacuum; barometric pressure is the force exerted by the atmosphere at a given location and time. "Gage pressure" is a scale based on local barometric pressure, which of course varies with altitude and atmospheric conditions. As an approximation, gage pressure can be converted to absolute pressure by adding the value of one standard atmosphere, 14.696 psi.

"Static pressure" in a fluid stream is the pressure acting on a sensor moving with the fluid so it is at rest or static with respect to the fluid. In practice, static pressure is usually measured by tapping into a vessel or pipe perpendicular to the flow direction. This method assumes that fluid disturbances caused by the tap are small enough to ignore. Experience shows that a burr-free tapping made with reasonable care will produce adequate results. Maximum size of static pressure taps recommended by ASME are:

Pipe Size, In.	Tap Dia., In.
2	0.25
3	0.375
4 to 8	0.50
10 and larger	0.75

Originally published September 11, 1972.

Velocity or dynamic pressure is the energy contained in a moving fluid. It can be calculated from the equation:

$$P_v = v^2W/2g$$

where:

P_v Velocity pressure
v Fluid velocity, ft./sec.
W Fluid density, lb./cu. ft.
g Gravity acceleration, ft./sec.^{2}

Total or stagnation pressure is the combination of static pressure and velocity pressure. Differential pressure, another commonly used term, is the difference between static pressure at two points.

Sensing Elements

The bourdon tube is the most common pressure sensing element. This is a flattened metal tube, sealed at one end and bent in the shape of a 'C' or spiral. Since the inner and outer surfaces of the tube have different areas, a force imbalance caused by pressure will make the tube unwind. This motion can be read directly on a gage or transduced to an electrical or pneumatic signal proportional to the pressure (Fig. 2). Care must be taken to avoid corrosion or deposits in the tube that could affect its characteristics.

Bellows are thin-walled cylindrical shells with transverse corrugations. Deflection is related linearly to the applied pressure, even over relatively large distances. They are usually specified according to their effective area and spring rate. Effective area is determined by applying a known pressure to the bellows and measuring the resulting force. Spring rate is defined as the quotient of deflection and force. As with bourdon tubes, motion may activate a gage or pneumatic or electronic transmitter.

Diaphragms are widely used as sensing elements in high-accuracy instruments. They are either flat or corrugated depending on the pressure range to be handled, and are suitable for measuring pressures from a few millimeters of water to thousands of psi. Significantly, diaphragms can be used as elastic seals or to separate two media.

Most frequently, diaphragms are designed to transmit forces or very limited motion. They are superior to bellows or bourdon tubes because they have comparatively

Characteristics of Typical Temperature Measuring Elements—Table I

	Range, F.	Accuracy, F.	Advantages
Glass stem thermometers	Practical: −200 to 600 Extreme: −321 to 1,100	0.1 to 2.0	Low cost Simplicity Long life
Bimetallic thermometers	Practical: −80 to 800 Extreme: −100 to 1,000	1.0 to 20	Less subject to breakage Dial reading Less costly than thermal or electrical
Filled thermal elements	Practical: −300 to 1,000 Extreme: −450 to 1,400	±0.5 to 2.0% of full scale	Simplicity No auxiliary power needed Sufficient response times
Resistance thermometer	−430 to 1,800	0.1 (best)	System accuracy Low spans (10 F.) available Fast response, small size
Thermocouples	−440 to 5,000	0.2 (best)	Small size, low cost Convenient mount Wide range
Radiation pyrometer	0 to 7,000	±0.5 to 1.0% of full scale	No physical contact Wide range, fast response Measure small target or average over large area
Thermistors	−150 to 600	0.1 (best)	Small size, fast response Good for narrow spans Low cost, stable No cold junction

few areas where process fluids can deposit and may be made from corrosion-resistant metals or coated with elastomers such as Teflon. On the other hand, diaphragm actuated instruments are more expensive.

The "U Tube" manometer is the simplest of the devices that rely on a change in liquid height to indicate pressure. Process instruments based on this principle use a float or various electrical techniques to detect changes in liquid level rather than visual observation. For precise readings, corrections must be made for temperature and the local value of gravity acceleration.

Installation Practice

When measuring liquid pressures the main considerations are the affect of the process fluid on the sensing element and the influence of the sensor's location above or below the measuring point. The sensing element can be isolated with a sealing liquid as shown in Fig. 3. The sealer should be immiscible in the process fluid, not react chemically with it and have a higher density. A mechanical seal is a better approach since it does not require selecting the proper sealing fluid.

On the liquid service, the pressure sensor can be located at any convenient elevation relative to the measuring point provided the effect of the location on the "zero" of the instrument is acceptable or can be adjusted out. If the sensor is mounted below the measuring point the static head in the connecting line will require zero suppression. Conversely, a sensor placed above the pressure point will require elevation.

When measuring the pressure of gases the primary consideration is to prevent accumulation of entrained and condensed liquids and solids in the sensor or connecting lines. For this reason the sensor is usually located above the process connection with lines arranged to facilitate drainage. Because of the relatively low density of gases, the degree of elevation does not have a significant influence on the zero reading.

Steam installations must be designed to keep the sensor's temperature within operating limits. Usually the sensor is mounted below the pressure tap and the connecting line filled with water. Heat losses from the line will condense enough steam to maintain the liquid level. As on other services with a sealing liquid, adjustment must be made for zero suppression.

Disadvantages	Operating Principle
Difficult to read. Only local measurement, no automatic control or recording capability.	Mercury or other liquid in glass bulb expands into capillary bore of stem as temperature increases. Stem is calibrated in degrees.
Less accurate than glass stem thermometer. Changes calibration with rough handling.	Two metals with different coefficients of expansion are bonded together. Temperature change moves strips in direction of metal with lower expansion rate. Deflection proportional to length and inversely with thickness of each metal.
Larger bulb size than electrical systems, and greater minimum spans. Bulb to readout distance is maximum of 50 to 200 ft. Factory repair only.	Sealed bulb connected to spiral-wound bourdon type by capillary tube. Expansion effect of the bulb on the bourdon is displayed with a pointer moving over a calibrated temperature scale.
Self heating may be a problem Long-term drift exceeds that of thermocouple. Some forms expensive, difficult to mount.	As temperature changes, resistance or conductance of a metal changes accordingly. With a current converter, an indicating or control signal can be obtained.
Not as simple as direct reading thermometers. Cold working of wires can affect calibration; 70 F. nominal minimum span.	When heat is applied at junction of two dissimilar metals a small electromotive force is generated. EMF is proportional to temperature.
More fragile than other electrical devices. Nonlinear scale, relatively wide span required.	Thermal radiation is a universal property of matter, absent only at absolute zero or when the material is in an inert gas. Thus it is possible to infer temperature of an object without physical contact.
Very nonlinear response. Stability above 600 F. is a problem, not suitable for wide spans. High resistance makes system prone to pickup noise from power lines.	Thermistors are semiconductors made from mixtures of pure oxides of nickel, maganese, copper, iron, magnesium, titanium and other metals sintered above 1,800 F. Detecting resistance is a function of absolute temperature. Coefficient may be positive or negative.

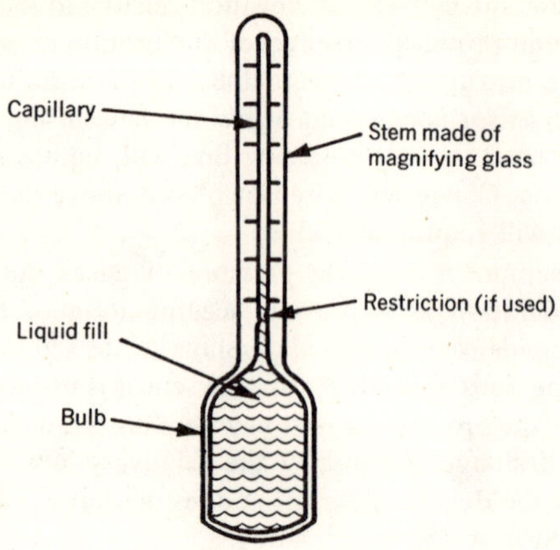

Glass Stem Thermometer

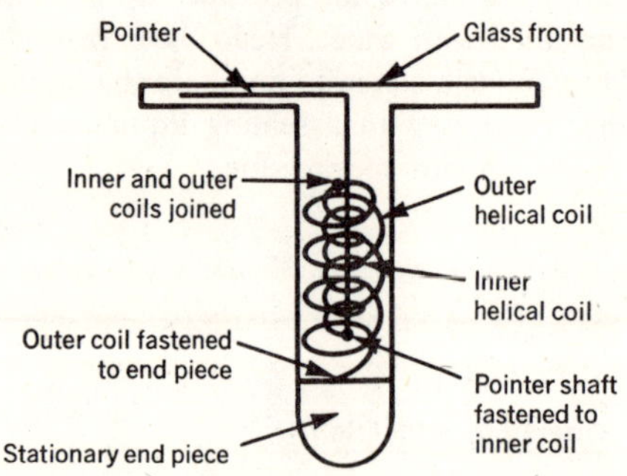

Bimetallic Thermometer

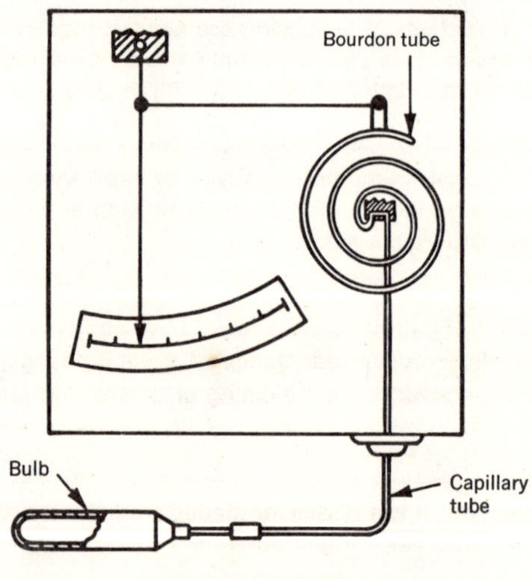

Filled Thermal Element

Thermocouple

FOUR WAYS to measure temperature—Fig. 1

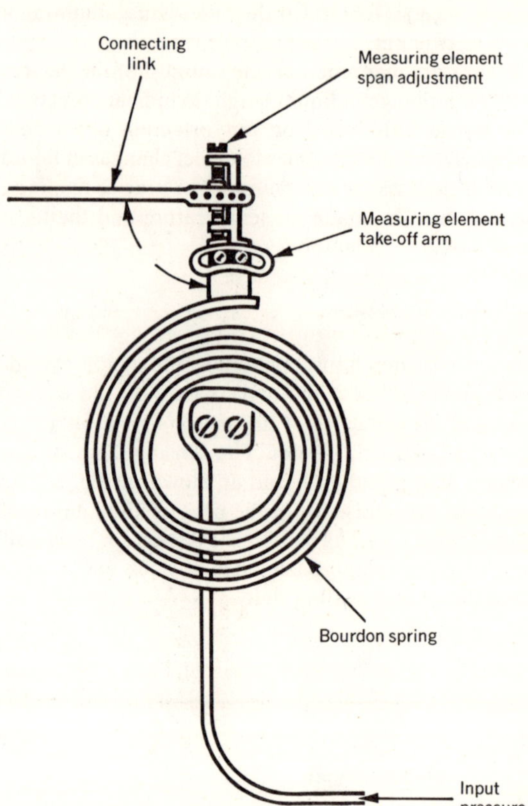

BOURDON tube is most common pressure sensing element—Fig. 2

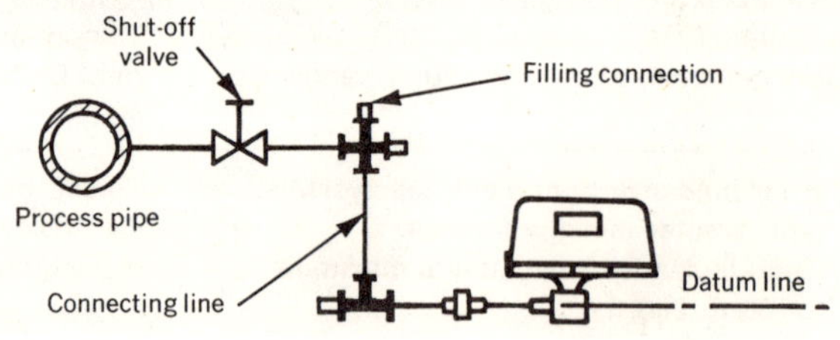

SEAL protects sensor from process fluid—Fig. 3

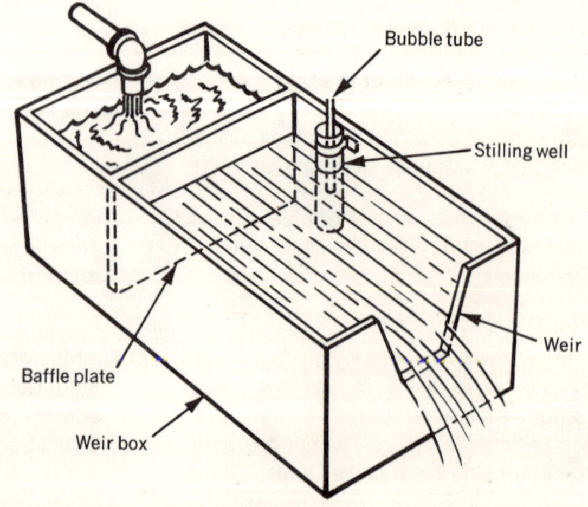

WEIR measures flow in open channel—Fig. 4

FLOW

Almost all fluid metering systems have two parts: a primary element that produces measurable phenomena such as pressure difference, turbine rotation, voltage, etc., and a secondary device to detect the phenomena and change it into force, motion or signal for remote transmission. Secondary devices vary almost without limit. Primary elements, however, depend on a few simple physical principles. Their proper installation and operation are usually quite distinct and independent from those of the secondary element.

Flow In Pipes

Differential pressure producers are the most common flow measuring devices. The orifice plate, a flat disc with a machined hole, is most popular along with the venturi and flow nozzle. For a given flow rate the differential produced is inversely proportional to the size of the restriction. For example, if the primary element has an area one half that of the pipe in which it is installed, a non-compressible fluid doubles its velocity while passing through the orifice.

Since the change in velocity through an obstruction gives an increase in kinetic energy, there will be a decrease in other forms of energy, i.e., static pressure. Flow rate is related to the pressure differential, fluid properties, size of the orifice and a discharge coefficient. Values for the discharge coefficient have been determined experimentally as a function of meter type, pressure tap location, pipe size, Reynolds number and ratio of orifice diameter to pipe diameter. The most complete tabulations are published in "Fluid Meters" by the American Society of Mechanical Engineers and "Orifice Metering of Natural Gas" by the American Gas Association.

Many flow measurements are best made without an obstruction in the stream. The magnetic flowmeter, operating on the same principle as an electric generator, is one such device. It consists of a permanent magnet or electromagnet placed so that fluid flow induces an electromotive force across the fluid. A secondary element converts the low level signal into one suitable for process control. Signal output is linearly proportional to flow rate and is independent of density and viscosity variations.

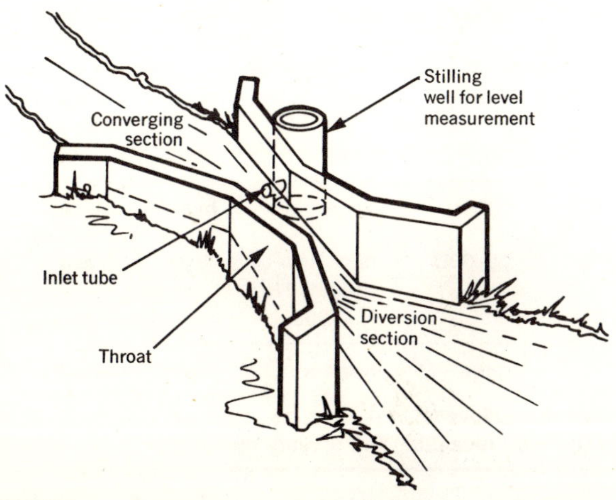

PARSHALL flume has flat bottom—Fig. 5

Magnetic flowmeters cover an extensive size range. One application is on a chemical additive feed line with a diameter of 0.1 in. At the other extreme, a 9-ft. meter operates at a sewage plant. Maximum line pressure depends on the rating of the metering tube and flanges. Safe maximum operating temperature ranges from about 170 F. to 300 F. depending on the magnetic coil insulation and temperature rating of electronic components in the primary element.

Of course, the fluid to be metered must be conductive. Gases and most petroleum derivatives are below the practical threshold conductivity, which is about 10 micro ohms/sq. cm. for most magnetic flowmeters.

Weirs and Flumes

Weirs and flumes can be used to measure flow in open channels such as plant effluents, sewage flows, cooling water discharge, irrigation systems, etc. The choice between a weir or flume depends on relative installed costs and solids content of the stream. Flumes are normally more costly but their flat bottom design allows for free flow of solids.

The most common weirs are triangular and trapezoidal (Fig. 4). Flow is determined by the relationship between flow rate and liquid height above the weir. As yet there is little standardization in flumes and many proprietary designs are marketed. The Parshall flume in Fig. 5 and the Palmer-Bowlus flume are emerging as most common because of the amount of public information available on their design and use.

Point Velocity Devices

Point velocity instruments such as the pitot tube, venturi tube and hot wire anemometer measure velocity or mass flow at a single point in a moving stream. They may be applied in both open channels and closed circuits, and are most often found in large ducts where other devices would be prohibitively expensive.

Since total average flow is derived inferrentially, a key factor is establishing by experience the relationship between the measured value and the average value. This is normally done by making a traverse—several measurements across the stream at a known average flow rate. This traverse and the initial calibration of the primary element determine the accuracy of the metering system.

Four common methods of establishing the relationship between point and average flow are the Centroid of Equal Areas, Newton-Cotes, Chebyshef and Gaussian. In general, the Gaussian method is most accurate, but velocity distribution and channel shape influence selection of the best method. Details can be found in ASME Report on Fluid Meters, 6th Ed.

Except by accident no two meters, even of the same type, are likely to give the exact same indication at equal flow rates. Tolerances assigned to the factors in the metering situation describe the unavoidable differences between ostensibly duplicate elements, not the accidental errors of observation.

Primary Element Characteristics—Table II

Primary Element	Type of Fluid	Pressure Loss	Flow Rangeability	Error	Upstream Piping	Viscosity Effect	Cost	Type of Readout
Concentric orifice	Liquid, gas Steam	50–90%	3:1	¾%	10–30D	high	low	sq. rt.
Segmental orifice	liquid slurries	60–100%	3:1	2½%	10–30D	high	low	sq. rt.
Eccentric orifice	Liquid-gas comb.	60–100%	3:1	2%	10–30D	high	low	sq. rt.
Quadrant edged orifice	Viscous liquids	45–85%	3:1	1%	20–500	low	med.	sq. rt.
Segmental wedge	slurries & viscous liquids	30–80%	3:1	1%	10–30D	low	high	sq. rt.
Venturi tube	Liquid & gas	10–20%	3:1	1%	5–10D	very high	very high	sq. rt.
Dall tube	liquids	5–10%	3:1	1%	5–10D	high	high	sq. rt.
Flow nozzle	liquid gas and steam	30–70%	3:1	1½%	10–30D	high	med.	sq. rt.
Elbow meter	liquid	none	3:1	1%	30D	negli-gible	med.	sq. rt.
Rotameter	all fluids	1–200″WG	10:1	2%	none	medium	med.	linear
V-notch weir	liquids	none	30:1	4%	none	negli-gible	med.	5/2
Trapezoidal weir	liquids	none	10:1	4%	none	negli-gible	med.	3/2
Parshall flume	liquid slurries	none	10:1	3%	none	negli-gible	high	3/2
Magnetic flow meter	liquid slurries	none	30:1	1%	none	none	high	linear
Turbine meter	clean liquids	0–7 psi	14:1	½%	5–10D	high	high	linear
Pitot tube	liquids	none	3:1	1%	20–30D	low	low	sq. rt.
Pitot venturi	liquids & gases	none	3:1	1%	20–30D	high	low	sq. rt.
Positive displacement	liquids	0–15 psi	10:1	½–2%	none	none	high	Linear totaliza-tion
Swirlmeter	Gases	0–2 psi	10:1 to 100:1	1%	10D	none	high	linear
Vortex shedding	liquids, gases	0–6 psi 0–5″WF	30:1 to 100:1	¼%	15–30D	min. Rey-nolds No. 10000	high	linear
Ultrasonic	liquids	none			none	none	high	linear

Pressure loss percentages are stated as a percent of differential pressure produced.
Upstream piping is stated in the number of straight pipe diameters required preceeding the primary element.

To provide a uniform basis for assigning numerical values to measurement uncertainty, the committee on fluid flow measurement of the International Organization for Standardization (ISO/TC-30) adopted the following procedure:

1. The numerical value of a tolerance (limit of accuracy) shall be twice the standard deviation.
2. Standard deviation is computed by summing the squares of the deviations with respect to the most probable value, and dividing by the number of observations minus one, and taking the square root of this quotient.

After setting individual tolerances by this method, the overall uncertainty is found by taking this square root of the sum of the squares of the tolerances, weighted according to the exponent of the value in the flow rate equation.

Other Flow Meters

Primary elements discussed here represent a small segment of those available. The wide variety on the market include turbine meters, vortex shedding meters, swirlmeters, ultrasonic meters, positive displacement meters, rotameters, and nuclear magnetic resonance meters. As a selection aid, Table II lists characteristics of 22 flow measuring elements.

LEVEL

For level measurement in the chemical process industries, a particularly important consideration is to have a chemically inert seal between the process fluid and the primary sensor. This has led to wide use of the diaphragm-sealed transmitter, employing a corrosion-resistant process seal. These devices transpose a liquid level to a pressure measurement, usually ranging from 20 to 800 in. H_2O.

Level measurements on tanks vented to the atmosphere are made as shown in Fig. 6. The minimum level must be at or above the datum line. When the dimension "S" has a value other than zero, the zero reading of the transmitter must be suppressed accordingly.

With closed vessels where the pressure is other than atmospheric, level measurements must compensate for the pressure above the liquid by using a differential pressure sensor, illustrated in Fig. 7. If the tank vapors are non-condensible, the connecting piping (wet leg) can be empty. Instrument span and zero suppression are calculated as for open vessel installations.

When vapors are likely to condense in the wet leg, the piping should be filled with a suitable seal liquid. As an example of instrument calibration, assume: A = 100 in., S = 10 in., E = 130 in., tank liquid with S.G. = 0.9, wet leg S.G. = 1.1.

Span = $100 \times 0.9 = 90$ in. H_2O
Wet leg zero suppression equals
$130 \times 1.1 = 143$ in. H_2O
Effect of "S" raises zero by
$10 \times 0.9 = 9$ in. H_2O
Net negative effect on the zero is
$143 - 9 = 134$ in. H_2O

Bubbler Method

When it's permissible to bubble air or another gas through the process liquid, a dip tube can be submerged in the vessel. The pressure required to produce a continuous stream of bubbles from the tube is a measure of the hydraulic head or liquid level above the dip tube (Fig. 8).

A needle valve regulates gas flow through the tube with the quantity observed by a rotometer or sight-feed bubbler. The bubble pipe should be at least 1-in. dia. to attenuate the pressure pulse. A "v" notch at the tube opening provides a location for bubble formation. The rate need not exceed 2 bubbles/sec.

Bubblers can be used in both open and closed vessels, with the latter requiring a differential pressure transmitter to detect back pressure. They are not recommended for measuring levels in highly viscous liquids because the bubbles do not readily separate from the dip tube.

Displacement Sensors

Displacement level sensors detect the change in buoyant force on a body immersed in the liquid. If the body

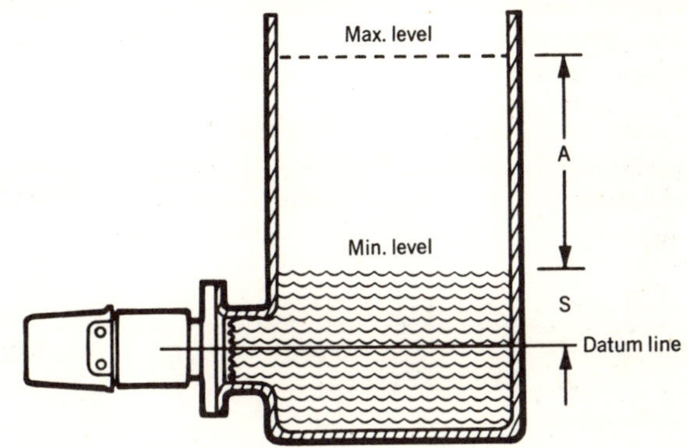

DIAPHRAGM-SEALED transmitter for level sensing—Fig. 6

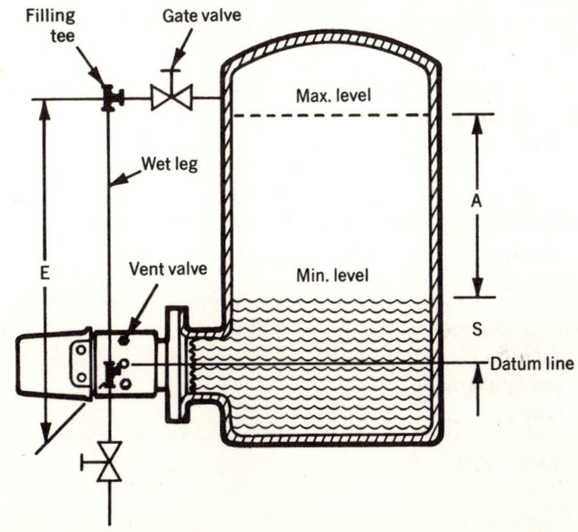

WET LEG compensates for pressure above liquid—Fig. 7

has a uniform cross section along the axis of level change, its suspended weight will decrease linearly as level increases.

A secondary element transduces the force change to a gage reading, recording or signal transmission. A typical installation is depicted in Fig. 9.

Displacement sensors are generally installed in a side arm or vessel well to avoid the effect of fluid motion on the float. Compared to other level measurement systems, displacement units are generally less expensive. Their primary disadvantages are that their calibration span cannot be changed and major zero-point adjustments require relocation in the vessel.

DENSITY

With increasingly sophisticated process control systems, online density measurement is becoming a necessary tool for the chemical engineer. Since density varies

Density Measurement Techniques—Table III

Method	Accuracy (Sp. Gr.)	Minimum Span (Sp. Gr.)	Principle of Measurement
Hydraulic head single bubble tube	0.005	0.01	Pressure = height × density
Hydraulic head dual bubble tube	0.005	0.01	Pressure = height × density
Hydraulic head direct pressure measurement type	0.005	0.01	Pressure = height × density
Fixed volume weighing	0.0005	0.05	$\frac{\text{Weight}}{\text{Volume}}$
Radiation	0.0001	0.05	Absorption of gamma radiation varies inversely with material density
Buoyancy	0.0001	0.005	Change in displacement force on a submerged body = change in submerged volume of displacer × density
Refractometer	0.0001	0.004	Refraction of light
Boiling Point Rise		—	Boiling Point of Solution compared to boiling point of water at the same pressure as the solution
Vibration amplitude of U-tube	0.001	0.05	Vibration of U-Tube is function of mass of process material within the tube
Force balance-U-tube and straight thru type	0.0005	0.05	$\frac{\text{Weight}}{\text{Volume}}$
Gravitometer			Gas density vs. standard gas

with temperature, it is specified at a base value, usually 32 F. or 60 F. (petroleum industry). Common density units are:

- Specific gravity (S.G.), the ratio of a material's weight to an equal volume of water at a specified temperature.
- Degrees API, the density unit used by the American Petroleum Institute.
- Degrees Baume, a dual scale for acid density with one for liquids lighter than water and one for those heavier than water.
- Twaddle, a scale divided into 200 equal parts over the S.G. range of 1.0 to 2.0 S.G. of 0.005 equals 1 degree Twaddle.
- Brix, percent sucrose by weight in a pure sugar solution at 17.5 C.
- Richter, Sikes and Tralles are scales reading directly in percent ethyl alcohol by weight in water.

Density Measurement Techniques—Table III

Advantages	Disadvantages	Typical Applications
Simple, inexpensive, relatively fast, may be mounted directly in the process vessel.	Cannot be used in pressurized vessel, level must be constant, air added to process, sensitive to velocity effects, subject to plugging.	Any open tank-non agitated measurement acid concentration, any solution.
Can be used where level varies, can be used in pressurized vessels, simple.	Subject to plugging, level must be above the higher tube, sensitive to velocity effects.	Coating solution makeup, lime slurry makeup, acid concentration.
Uses a diaphragm seal, eliminates plugging, no air introduced into the process.	Slow response, large time constant.	Specific gravity of fluid products, plasticizer solutions, clay, water mixtures, etc.
Can be used as direct, in line measurement. Sanitary construction not affected by fluid viscosity or suspended solids. Permits measurement of hazardous products.	Sensitive to vibration in the verticle axis. Sensitive to pressure variation where gases are present.	Tomato paste, solids control of coating solution.
Can be used for on line measurement-mounts directly on pipeline or vessel. Does not contact the process material.	Affected by entrained air or gases.	Control of centrifuge feeds, acid concentration, percent solids of waxes and oils, spent acid recovery control, lime slurries.
May be mounted directly in the process vessel.	Displacer must be kept submerged. Sensitive to velocity effects, can not be used where the process fluid coats the displacer.	Acid concentration control, Salt solution concentrations, Streptomycin, Butyl Lactone, Nylon, etc.
Insensitive to flow rates. Insensitive to undissolved solids. In line measurement. Unaffected by entrained air.	Prisms may coat from process fluid-(cleaning systems available.	Black liquor in paper industry, sugar solution concentrations, evaporator product concentrations, acid making, ethylene glycol concentration, etc.
Simple, in-line or vessel mounting.	Requires steam at the same pressure as process.	Evaporator end products.
In-line measurement, can be used under high pressure.	Mounts through pipe or vessel wall, if fluid tends to coat U-Tube cannot be cleaned while process is operating.	
In-line measurement can be used for sanitary applications. Direct measurement of density.	Accuracy affected if fluid coats or seals tube walls.	Slurrys such as cement, paper coating solutions, etc.
Fast response. Simple.	Inferential measurement.	Fuel gas density, hydrogen recycle gas, gas mixing, feed gas in ammonia synthesis.

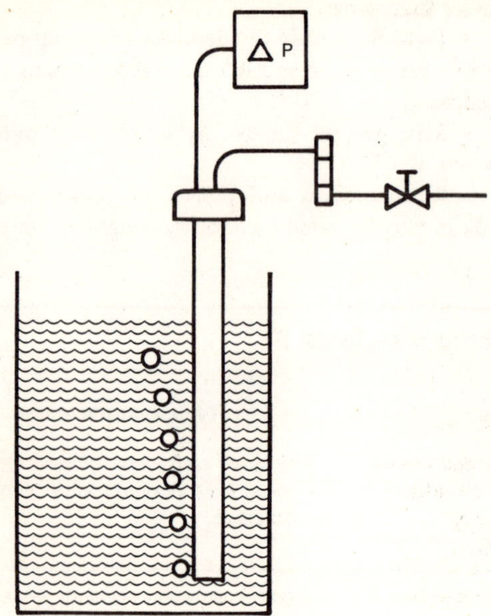

BUBBLE rate is measure of liquid level—Fig. 8

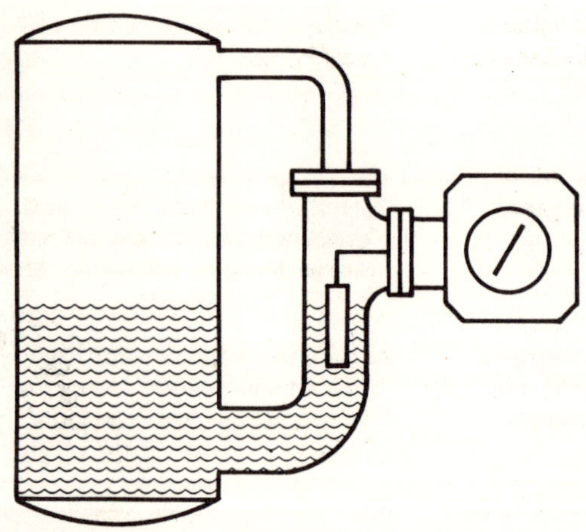

DISPLACEMENT sensor is another level technique—Fig. 9

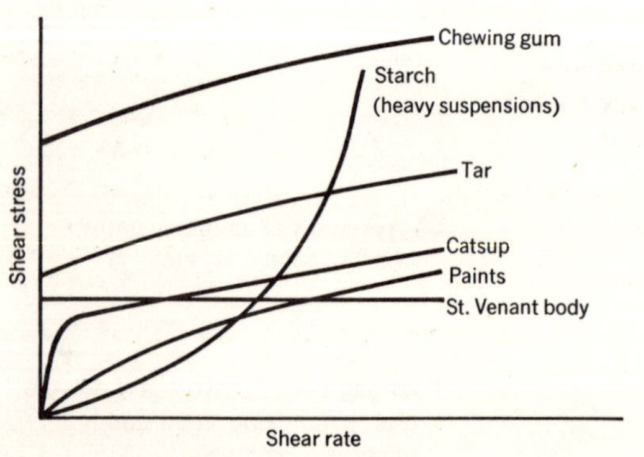

RHEOGRAMS for some non-Newtonian fluids—Fig. 10

Measurement Methods

Characteristics of density measurement methods are listed in Table III. The hydraulic head technique is based on the principle that pressure is equal to liquid height times density. At constant height, pressure varies directly with density. With fixed volume weighers, process fluid continuously passes through a container of known volume. The container's weight gives a direct density measurement.

Radiation devices pass gamma rays through the fluid to be measured. The amount of radiation picked up by the detector varies inversely with the material's density. The variable actually measured is the degree of radiation absorption.

The principle of buoyancy also can be used to determine density since the force on a submerged body is related to fluid density. Another method is based on refractometry: the velocity of light travelling through a substance is reduced in proportion to the optical density of the material.

Differential density refractometers are limited to clear fluids. Critical angle types have a tungsten light source centered on the angle at which the beam converts from relection to refraction. The collumnated light beam is focused on a rotating prism, which causes the beam to sweep the process stream. Refraction, detected by a photocell and amplified into a usable signal, occurs when the light ray sweeps down to the critical angle.

With a force balance U tube, fluid circulates through a U-shaped hollow beam attached to a null-balance servo system. A force equal to the confined fluid weight pushes downward through the tube center. Since the volume within the tube is constant, any density change causes a weight change. Tube ends are supported on a pivot. A counterbalance causes an upward force equal to the weight of the tube plus the process fluid. Any density change repositions the beam, causing the servo system to rebalance. The output signal repositioning the servo can be calibrated in terms of density.

Vibration of a U tube filled with liquid is also a function of the mass of the material within the tube. While the tube is vibrated by a pulsating current through a drive coil, the vibration is sensed and converted into a milliamp output signal. Since the vibration varies with fluid mass, the output signal can be calibrated in density units.

The gravimetric method measures torque on an impeller being driven in a fixed volume sample chamber compared with torque developed by driving the impeller in air. The torques are converted through mechanical linkages and transduced into standard instrument signals. The measuring system divides the torque developed in the gas by that developed in air, giving the same reading as dividing gas density by air density, i.e., specific gravity.

VISCOSITY

Viscosity measurements are needed in process control for many reasons, though most fall into one of two categories: to control end product quality or to maintain desirable processing conditions that depend on viscous properties of fluids.

Newtonian fluids have a linear relationship between applied shear stress and shear rate. Others have non-linear slopes with origins not necessarily starting at zero, as represented in Fig. 10. Non-Newtonian fluids are classed according to the relationship between shear stress and shear rate. For example:

• True plastic materials (Bingham body) have an initial shear stress above which flow begins. Examples are tar and chewing gum.

• Pseudoplastic materials show an initial slow shear rate with an increase in shear stress until an apparent yield stress is exceeded. Examples: catsup and printer's ink.

• Dilatant materials require ever-increasing shear stress per unit increase in shear rate. Examples: starch solutions and peanut butter.

• St. Venant Body describes substances similar to a Bingham body in that they have a yield stress, but practically no viscosity slope. Paper pulp is an example.

• Thixotropic materials are time dependent. Flow rate increases with agitation time and shear stress. Examples: glue and most paints.

• Rheopectic materials are also time dependent and exhibit a hysteresis effect. However, flow rate decreases with agitation time and increased shear stress. Fudge is an example.

Measurement Methods

Almost all viscosity measurement methods are designed to detect resistance to fluid displacement. The difference between methods lies in the techniques used to displace the fluid or to measure resistance. Methods described here are primarily intended to measure Newtonian fluid viscosity, but may be used to analyze other materials if a viscosity determination at one shear rate provides adequate information for process control.

Rotational torque viscometers apply a constant shear rate to a process fluid and then measure the resistance or "viscous drag" generated by the fluid. In practice, a disc or spindle rotating at constant speed produces torque directly proportional to fluid viscosity. Torque is measured as rotational deflection against a spring between the driving motor and the measurement spindle or disc.

Displacement time devices place a constant force on a displacing fluid and determine shear rate by measuring the time to displace a given quantity. A sample in a closed tube is expelled by the force of gravity acting on a piston as the piston falls through the tube. Time of fall is directly proportional to viscosity. The measuring orifice is the clearance between the piston and tube wall.

Vibrational amplitude instruments are based on the principle that a mass subject to viscous damping vibrates at a certain amplitude when excited at a selected frequency. For a particular frequency and mass, amplitude varies inversely with the viscous drag of the system.

Differential pressure viscometers stem from Poiseuilles law, which states that flow through a smooth, round tube is directly proportional to pressure differential and inversely proportional to viscosity. In practice, flow is fixed while the instrument measures differential pressure across a section of tubing.

Axial displacement devices are based on the fact that viscous drag creates a force on a body, and for every force there is an opposite and equal force. For example, if fluid flows upward in a tapered tube, a spindle inserted in the stream will balance at the point where upward and downward forces are equal. At constant flow, vertical displacement of the spindle is a direct measure of fluid viscosity.

Selection and Installation

Manufacturers should be consulted to obtain the best type of viscometers for a particular job, especially on non-Newtonian applications. Other considerations are materials of construction, use of purged or sealed cases with corrosive or explosive atmospheres, compensation for ambient temperature and the instrument's effect on fluid temperature, and sealed systems on differential pressure types to eliminate dead ending if polymerization or solidification occurs.

Instrument location should be readily accessible, free from vibration and sheltered from harmful environments. As with most measuring devices, the most critical installation problem is obtaining a truly representative sample. The best approach is to take measurements directly in the process vessel or line. If a sampling system must be used, keep lines short and small diameter to minimize time lags. Use a strainer where solids may damage the instrument.

A high percentage of instrument error or failure is caused by improper maintenance. Viscometers must be kept clean. Almost all are sensitive to buildup of interior coatings since this affects the mass of spindles and vibrating parts. Maintaining critical interior dimensions also depends on cleanliness, and a good flushing system goes a long way towards insuring satisfactory life and service.

Acknowledgement

Stephen Bonacci, Harold Hendler, and Michael Mihalick assisted in preparing this article.

Bibliography

1. "Fluid Meters, Their Theory and Application," 6th Ed., American Society of Mechanical Engineers, New York, 1971.
2. "Measurement of Fluid Flow by Means of Orifice Plates and Nozzles," International Organization for Standardization, 1967.
3. "Orifice Metering Of Natural Gas," American Gas Assoc., New York, 1969.
4. "Process Instruments and Controls Handbook," McGraw-Hill, New York, 1957.
5. Reamer, R., Measurement and Control and Liquid Density or Specific Gravity, Taylor Instrument Co.
6. Liptak, Bela G., "Process Measurements," Chilton Book Co., 1969.
7. Fribance, Austin E., "Industrial Instrumentation Fundamentals," McGraw-Hill, 1962.

Meet the Author

Dennis E. Zientara is a senior systems engineer for Taylor Instrument Process Control Div., Sybron Corp., 95 Ames St., Rochester, NY 14601. He currently heads a group in the systems engineering department responsible for development of advanced control concepts. Mr. Zientara has a B.S. in mechanical engineering from the Rochester Institute of Technology.

Liquid-measurement technology

Over the past decade or so, most process-industry operations, and especially the growing area of automated process control, have benefitted greatly from developments in semiconductor technology and related electronic innovation. This has produced significant advances in liquid-measurement techniques, particularly where conventional approaches do not work.

C. L. Smith, Louisiana State University

☐ Despite many recent advances in liquid-measurement technology, engineers have not actually found their jobs any easier. If anything, this work is more difficult today because we require so much more from sensing elements than we did even ten years ago. The principal reasons for these increased demands include, among others:

■ Tighter environmental controls. (Closed-vessels—used to contain pollutants—need sensors for accurate "blind" liquid-level measurements.)

■ Increased energy costs. (New process economics demand closer control of both operations and product specifications.

■ Public health and safety. (OSHA and EPA regulations require elimination of fluids leakage, to prevent both operator exposure and community pollution.)

■ Greater use of coal as an industrial feedstock. (Moving slurries instead of liquid streams mandates new technologies for both materials-handling and measurement.)

In effect, present-day experience indicates that current technology will prove less and less applicable. For example, flow measurement with the orifice meter, and level measurements with a displacer, have proven reliable and cost-effective. However, with an extremely hazardous process fluid, a clamp-on ultrasonic flowmeter and a nuclear level gage show definite advantages over traditional devices, since the newer instruments require no process connections that might leak; also, they can be serviced without the risk of employee exposure.

Process conditions, rather than instrument wiring and installation, are becoming the major problem sources in liquid-measurement systems. As a result, process engineers must get more involved in selecting measurement technologies and in analyzing instrument failures.

Factors in sensor selection

A number of technical factors can be important in proper selection of any sensor:

■ *Span* specifies engineering values that correspond to upper and lower limits of a sensor's output. In most cases, sensor accuracy is expressed as percent of the span—so choosing a sensor with too large a span will decrease the *absolute* accuracy.

■ *Accuracy* defines the limits within which the actual values of a process variable may differ from the sensor readings. With accuracy of a sensor stated as percent of span, *relative* accuracy (expressed as percent of *reading*) is greater at high readings than at low readings. Accuracy becomes especially important when the readings are used for billing purposes.

■ *Repeatability* refers to an ability to give consistent readings over a series of runs. Suppose a flowmeter with a span of 100 gpm reads 80 gpm at both time t_1 and time t_2, at identical process conditions. If repeatability of the instrument is 0.1% of span, the actual flow at time t_1 differs from that at t_2 by no more than 0.1 gpm. "Repeatability" says nothing about the accuracy: e.g., the actual flows could have been 40 gpm and 40.1 gpm respectively, or even 4.0 and 4.1.

In many applications, repeatability becomes more important than accuracy. Example: when operating a large continuous process. Here, the operating philosophy involves maintaining constant conditions (i.e., the plant should run smoothly). Having achieved this, operating personnel will make the necessary small adjustments in controller set-points in order to get maximum performance. But once these set points are established, the controller has to rely upon repeatability of the instrument to maintain constant process conditions.

■ *Environment,* as defined here, means the process conditions under which the sensor must operate. Unfortunately, decisions in this area tend to be rather fuzzy. A basic question arises: "How long will the sensor work before it needs repair?" Then comes: "Is this time interval acceptable?" Most production people will accept a shorter interval for a really critical measurement than for a less important measurement.

■ *Readout* relates to signals from instruments in re-

Originally published April 3, 1978.

sponse to process conditions. A sensor must produce signals compatible with the recorder, controller, computer, or whatever device receives the signals. Industry standards call for either a 3- to 15-psi air signal or a 4- to 20-ma electric-current signal. Some sensors, particularly those monitoring rotational speed, produce a pulse output. A computer readily accepts this signal, but conventional recorders or controllers do not. A frequency-to-current converter can work in such cases, but at additional expense and some loss in accuracy.

■ *Linearity* in a response signal refers to a straight-line function, not exponential (logarithmic). The best-known example of a nonlinear sensor is the orifice meter, where flow depends upon the square root of differential pressure across the orifice. Another nonlinear sensor is the ZrO_2 oxygen analyzer, which gives an output related to the logarithm of oxygen concentration. If this output goes only to a recorder, then a special nonlinear chart paper can be used. But, being a linear device, the conventional three-mode controller performs best with linear systems. As the sensor is actually a part of the system being controlled, it can affect overall linearity of the system.

Nonlinear signals can be linearized electronically or pneumatically, but also at a cost, and with a reduction in accuracy.

■ *Speed of response.* The importance of a sensor's speed of response is heavily dependent on what use is made of its output. When used for closed-loop control, the speed of response is always important. The slower the sensor, the poorer the performance of the control loop. The speed of response of any system is indicated by a parameter called the time constant. Normally, the desired objective is to design the installation so that the time constant of the sensor is less than one tenth of the time constant of the process.

For applications such as emergency-shut-down systems, the speed of response of the sensor is also of primary importance. But for applications in which the output of the sensor is used to generate production records, environmental emissions records, etc., response time is usually a secondary consideration.

Process flow-measurement applications

One rather obvious flowmeter application: measuring consumption rates of raw materials and production rates of products. Although other alternatives exist (such as the change in level in tanks or the weight of trucks entering and leaving the plant), flowmeters provide instantaneous readings of consumption and production rates for most modern plants. When these measurements provide the basis for billings and payments, accuracy of flow measurement becomes of paramount commercial importance.

Many companies today use material and energy balances around their plants as a check on performance; accurate flow measurement is obviously essential here. To achieve closure for material or energy balances, it is necessary to recalibrate or replace those meters that are inaccurate.

In pilot-plant work also, accuracy of flow measurement becomes critically important. A material balance, made around the unit in question, provides a check on the accuracy of information taken from a particular run.

Increased levels of instrumentation, coupled with computer-based data-collection systems, can increase productivity of pilot-plant activity. Installing more than one flowmeter on a given stream will ensure accuracy of flow measurements; in fact, the two (or more) flowmeters that are used may not necessarily be of the same type.

Metering systems for charging batch reactors represent yet another case where flowmeter accuracy is extremely important. In such reactors, the recipes must be followed as closely as possible. Since different recipes will frequently call for different amounts of various reagents, accuracy of the flowmeter that is selected may very well determine the resulting quality of the batch-unit product.

However, in many continuous-operation plants, the flowmeter accuracy takes a back seat to repeatability. Especially in control systems, where higher-level controls adjust set points of the flow controllers, accuracy is less important in flowmeters. Since the higher-level controls frequently take direct readings of product quality or composition, this feedback-control equipment will adjust the set points of flowmeters and compensate for any measurement error. However, repeatability of the flowmeters can have a very significant effect on the performance of these higher-level controls.

As an example, Fig. 1 illustrates a control system for level and composition in an evaporator. For maximum production, the steam valve is opened fully. Changing the feed and product flows controls both the level and composition.

In the control configuration shown in this figure, the level controller determines product flow and the composition controller determines the ratio of feed flow to product flow. If either (or both) of the flowmeters is inaccurate, the only result will be a commensurate difference in the ratio setting provided by the composition controller.

Control configurations can, of course, be developed for the evaporator in Fig. 1 that do not involve flowmeters. Fig. 2 illustrates one such configuration. Several

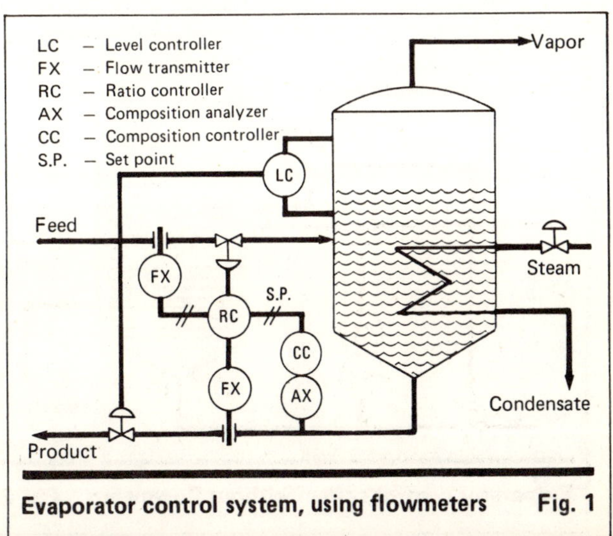

Evaporator control system, using flowmeters **Fig. 1**

problems exist with this simple configuration, three major ones being:

1. The level and composition loops interact.

2. The composition loop is very nonlinear with production rate.

3. Dead time in both process and analyzer make normal proportional/integral (PI) control difficult.

Any of these problems is likely to result in controller-tuning difficulties and poor control performance.

With industry's increasing needs for better instrumentation and tighter control, the configuration in Fig. 1 is superior to that in Fig. 2. Widely fluctuating steam input may also call for feedforward compensation to assure adequate steam availability. This necessitates a third flowmeter, to check the steam rate.

To do a complete material and energy balance around the evaporator requires a fourth flowmeter, which must measure the vapor rate. Given the problems of installing a flowmeter to measure vapor rate (especially when the evaporator is operated under vacuum), it may be much easier to use a sensor to determine condensate rate from the vapor condenser.

Orifice meter

Over the years, many techniques have come forth for measuring liquid flow.

Of all the techniques, the orifice meter has been by far the most popular, and its popularity should remain high for the next several years.

To review briefly the principle behind orifice meters, a differential pressure transmitter measures pressure drop across a restriction in the line. The restriction is usually a concentric orifice, with the orifice diameter being 10% to 75% of the inside pipe diameter. The pressure drop can be measured with flange taps, *vena contracta* taps, or other variations—flange taps being most common.

While precise flow equations are complex, most installations use the simplified equation,

$$F = C_1\sqrt{\Delta P} \tag{1}$$

where F = volumetric flow, ΔP = differential-pressure measurement, and C_1 = orifice coefficient.

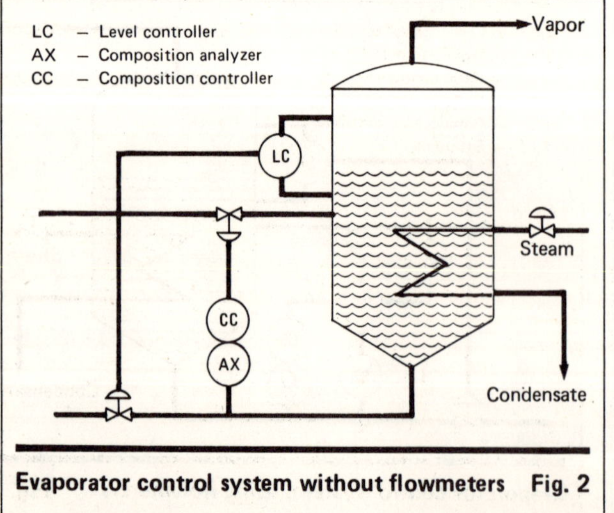

Evaporator control system without flowmeters Fig. 2

If ΔP is expressed as a percent of full scale (i.e., it varies from 0.0 to 1.0), then C_1 represents the full-scale value for flow in whatever units are used for F.

The orifice meter provides a time-proven and relatively low-cost approach for measuring most flows. Instrument shops are generally familiar with the well-developed technology of orifice meters. This fact is a significant stimulus to the use of orifice meters in those situations where they give adequate performance.

Unfortunately, the orifice meter does have its limitations, in that under certain situations, for various reasons, it will not work. For liquid service, these cases include systems where:

1. The necessary pressure drop is not available. (This eliminates it from consideration in most gravity-flow applications.)

2. The fluid will flash at the reduced pressure that occurs in the throat of the orifice meter.

3. Highly corrosive fluids may erode the orifice plate.

4. The flow involves suspended solids—e.g., a 2% slurry of paper stock would probably plug at the throat of the meter. Eccentric or segmental plates can perform satisfactorily in some applications.

5. The plant produces a number of products or multiple grades of the same product. (Certain flows may be large for some products or grades, but small for others. The "rangeability" or turndown ratio—the ratio of maximum measurable flow to minimum measurable flow—is limited to about 3 to 1 for the orifice meter. At low rates of flow, the accuracy of orifice meters decreases considerably.)

6. Flowrates are high. (Here, permanent pressure loss, due to the presence of the orifice unit itself, can be significant. Consider a venturi tube or other alternative.)

This list could be longer, but it does include the major limitations in regard to liquids. Gas flows call for additional considerations: compensating for pressure, temperature, and gravity. For best results, vapor-flow measurements must be made at relatively constant pressures and molecular weights.

As practiced today, the approach to flow measurement normally involves determining whether an orifice meter (as first choice) will be satisfactory for a particular process system. If not, then there is no choice but to seek other measurement technology. In fact, as the industrial economy makes the projected shift from oil-based feedstocks to other feedstocks such as coal, instances where the orifice meter does not apply (e.g., due to suspended solids) will probably increase and alternative measuring techniques will have to take over.

About the orifice meter: First, the orifice meter is a volumetric flowmeter. The fluid density must be known in order to convert the volumetric flow to a mass flow. If the liquid stream has a constant density, then no problem exists. However, varying density must be taken into consideration when converting from volumetric flow to mass flow. The most direct approach would be to install a density transmitter in addition to the orifice meter. However, the density transmitter costs substantially more than the orifice meter. In those situations where temperature of the fluid is the main variable affecting density, then a density/temperature correlation can be used to infer the fluid density from its temperature. In

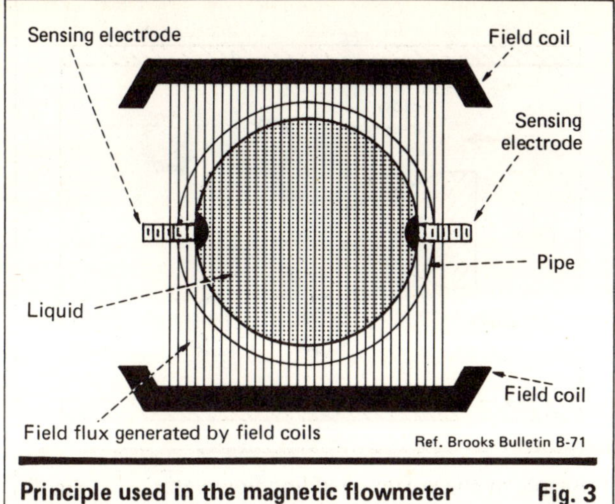

Principle used in the magnetic flowmeter **Fig. 3**

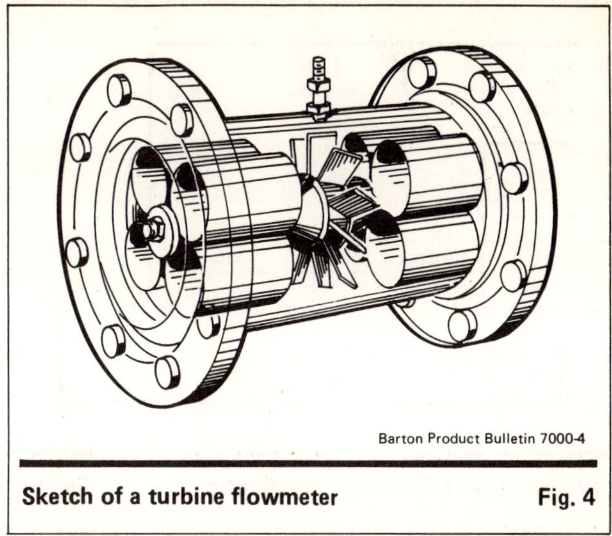

Sketch of a turbine flowmeter **Fig. 4**

simple cases, a linear relationship will be applicable. Second, since flow is related to the square root of the pressure drop, see Eq. (1), it requires a square-root extractor between the output of the differential pressure transmitter and the flow controller in order to give a linear set-point-to-flow relationship. Computers have traditionally incorporated the square-root relationship into software, and recently introduced microprocessor-based control systems also provide for square-root extraction as part of their software, with no additional hardware costs involved.

Recent technology for flow measurement

Many other flowmeters—such as rotameters, pitot tubes, weirs, etc., are in wide use and are well known to engineers. However, over the last decade or so, several techniques for measuring flow have appeared in commercial equipment, including the magnetic flowmeter, turbine flowmeter, vortex-shedding flowmeter, and ultrasonic flowmeter.

Magnetic flowmeter

As illustrated in Fig. 3, the principle of the magnetic flowmeter is basically the same as that for a generator: a conductor moving at right angles through a magnetic field induces a voltage. The magnitude of the potential is proportional to the magnetic-field intensity and the velocity of the moving conductive fluid. Maintaining the intensity of the magnetic field at a constant value permits the meter to be calibrated so that the velocity of the fluid can be determined from the induced potential. Commercial units convert the output to a 4- to 20-ma signal linearly related to the volumetric flow.

In deciding to use a magnetic flowmeter, the following points become important:

1. The fluid must have *some* minimum level of conductivity, because of a basic measurement problem associated with the potential. (Even if the fluid has zero conductivity, the electric potential is theoretically generated, but in order to measure the potential, a small current must flow. Thus, the potential generated by an element with a zero conductivity cannot be measured.) As magnetic-flowmeter technology has continued to develop, the minimum required conductivity has steadily decreased.

2. Being a zero-pressure-drop meter makes it applicable to gravity flow and to fluids close to their bubble point.

3. The flowmeter senses the average velocity through its throat.

(a) Since the meter calibration basically multiplies the average velocity by the cross-sectional area of the meter to get the volumetric flow, the assumption is made that fluid occupies the entire cross-section. In other words, the pipe must be full.

(b) Since the meter senses the average velocity, its output is therefore insensitive to the velocity profile, or in other words, it is applicable to both turbulent and to laminar flow. In particular, this makes the meter reading independent of viscosity changes.

4. Electrode coating can be a problem. In effect, coating on electrodes offers an electrical resistance, and the resulting potential drop causes erroneous meter readings. Some manufacturers provide techniques (such as ultrasonic cleaners) to aid in keeping the electrodes clean.

5. The pipe in the throat of the meter can be manufactured from a wide variety of materials, which gives this type of meter distinct advantages when it is used with corrosive liquids.

6. Having no obstruction in its throat, the meter works well with slurries. In fact, magnetic flowmeters have measured flow of paper stock in the paper industry for a number of years.

7. The basic technology of the meter permits it to handle bi-directional flow.

8. The temperature limitation on most industrial meters approaches 500°F, above which heat dissipation from the electromagnets is not adequate.

Magnetic flowmeters are at least as accurate as, and in most cases slightly better than, orifice meters as normally installed and maintained. This assumes regular calibration of the pressure transducer, as well as periodic inspection of the orifice plate itself for wear and deformation.

The cost comparison between magnetic flowmeters

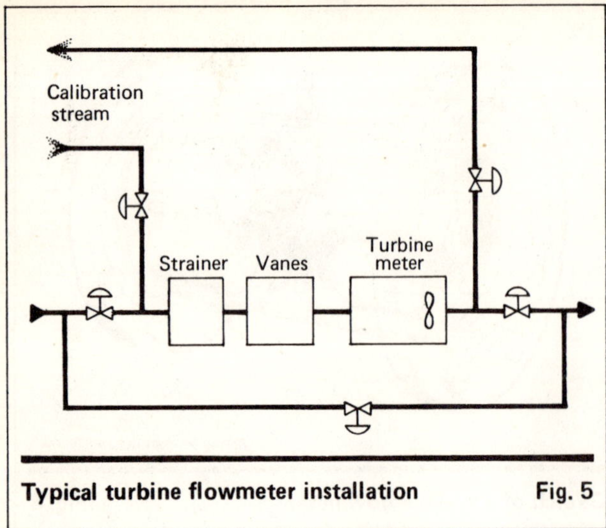

Typical turbine flowmeter installation **Fig. 5**

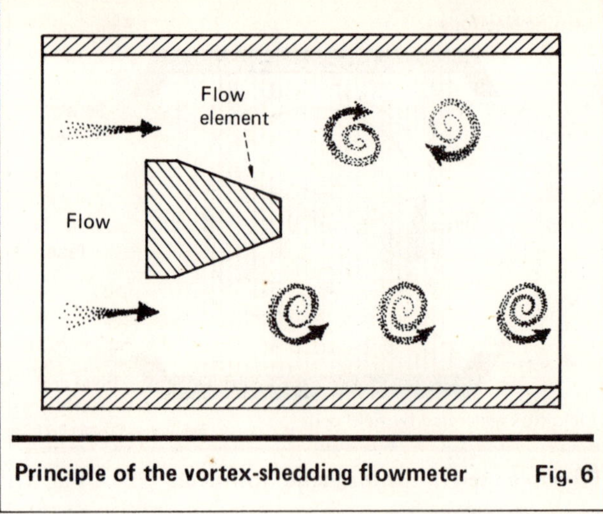

Principle of the vortex-shedding flowmeter **Fig. 6**

and orifice meters depends on the size of the pipe. For large meters, the size of the electromagnets becomes large; thus the cost of magnetic flowmeters increases with line size at a fairly rapid rate. (For a 3- or 4-inch installation using normal materials of construction, the installed cost of a magnetic flowmeter will be approximately 2.5 to 3 times that of the orifice meter. This ratio will be larger in the larger line size, but does not change much for smaller line sizes.) The minimum size of commercially available magnetic flowmeters is about 0.1 inch, which gives a maximum flow of some 0.1 gpm on the magnetic flowmeter.

Turbine meter

As illustrated in Fig. 4, the principle of the turbine meter is essentially the inverse of that of a pump. In the turbine meter, the fluid velocity causes the rotor to spin, and the rotational speed of the rotor is a measure of the volumetric flow of the fluid. Fig. 5 illustrates a typical turbine-meter installation.

Several basic considerations affect the use of turbine meters:

1. Operational problems with turbine meters are almost invariably associated with the bearings. There is always the question of whether they will hold up in service, since they are exposed to the process fluid. Depending upon the application, a manufacturer may recommend either ball bearings or sleeve bearings. In either case, the fluid must be free of particulate matter that would cause bearing wear.

2. The turbine meter is the most accurate device in available commercial metering technology, and offers the advantages of wide linear range, fast response, and excellent repeatability.

3. Large meters show a permanent pressure loss of about 5 psi. However, to prevent cavitation within the meter, the downstream pressure should be at least 10 psi above the vapor pressure. Cavitation can be destructive to the meter, even to the point of breaking rotor blades.

4. The turbine meter is rather delicate, in comparison with other flowmeters. In particular, excessive velocities may damage the bearings.

5. The turbine meter is actually a digital meter. Rotation is detected by a magnetic pick-up coil, which emits a pulse for each rotation. In practice, the calibration factor for the meter is the volume of fluid per output pulse. However, when utilizing the usual flow controller, the pulse output from the meter is first sent to a frequency-to-current converter to produce the 4- to 20-ma signal for use with conventional electronic controllers. Since the turbine meter is usually installed to obtain highly accurate flow measurement, the use of a pulse counter is preferable for computer-based control systems.

Compared with the orifice meter, the turbine meter clearly excels in accuracy. But the orifice meter is the more economical, both in initial cost and maintenance. In a 3-inch installation, the cost of a turbine meter—as of a magnetic flowmeter—is two to three times that of an orifice meter. In larger line sizes, the multiple is much higher.

For uses requiring high accuracy, the turbine meter should be considered. In many batch plants, a turbine meter forms the backbone of the reactor-charging system. In applications such as motor-gasoline blending, again the accuracy of the turbine meter makes it very attractive.

Vortex-shedding meter

As illustrated in Fig. 6, the basic principle behind the vortex-shedding flowmeter involves formation of localized regions of high velocity, called vortices (or eddies), in the vicinity of an obstruction inserted in the flow in a pipe. The rate of formation of these vortices is a function of flow velocity and the geometry of the meter itself.

Several points are important in the installation and use of vortex-shedding flowmeters:

1. In most versions, the meter has no moving parts. A heated thermistor detects the vortices, although elevated temperatures require a shuttle-ball arrangement.

2. The meter looks interesting for those cases where flows vary over wide ranges. (Turndown ratios of up to 100 to 1 are claimed, and one manufacturer claims an accuracy of 0.5% of meter reading; accuracy of other flowmeters is expressed as percent of full scale.)

3. When used as a detector, the thermistor has a temperature limit of about 400°F. However, special designs using the shuttle-ball can be used where temperatures are higher.

4. Basic measurement output from the meter is a periodic signal originating from the heated thermistor. Electronic shaping circuits easily convert this signal into a pulse output, making it a digital meter in the same sense as the turbine meter.

5. Since calibration of this design involves only geometry and flow velocity, volumetric readings from the meter do not depend upon fluid density, viscosity, and/or temperature.

6. The flow-obstructing element can be made of extremely hard material, to resist wear from abrasive material in the stream. However, anything in the process fluid that would tend to coat on the obstruction would, in effect, change the geometry of the meter and affect its calibration.

As for accuracy and cost relative to the orifice meter, the vortex-shedding flowmeter has a clear advantage in turndown ratio. For reasonably uniform flow, the accuracy of the orifice meter approximately equals that of the vortex-shedding meter.

From the economic standpoint, the installed cost for the vortex-shedding flowmeter is competitive with that of the orifice meter.

Ultrasonic flowmeter

Ultrasonic flowmeters come in two basic types—transmission and reflection. The transmission type depends on a sound-path being established through the liquid in the pipe. The reflection type depends on some particulate matter being present in the fluid at all times, and this matter must be of a kind that reflects sound to the receiver.

Fig. 7 illustrates the structure of the transmission-type meter. Two transmitters and two receivers are required, and two sound paths are established in the fluid. In one path, the sound travels with the direction of fluid flow; in the other, the sound moves against the direction of flow. In each path, there will be a frequency shift caused by the fluid motion. The frequencies at the two receivers are as follows:

$$f_A = \frac{Y + V\cos\alpha}{X}$$

$$f_B = \frac{Y - V\cos\alpha}{X}$$

$$f_A - f_B = \frac{2V\cos\alpha}{X}$$

where f_A = frequency at Receiver A, f_B = frequency at Receiver B, V = fluid velocity, X = path length, α = angle of path with direction of flow, and Y = speed of sound in the fluid.

The reflection-type meter works on a very similar principle, with the frequency shift occurring in the sound reflected from particles presumably moving at the same velocity as the fluid itself.

Several considerations help determine selection of ultrasonic flowmeters:

1. Accuracy of the ultrasonic flowmeters is currently around 1 to 2%, but accuracy figures are continuing to improve. (The Alyeska Pipe Line uses ultrasonic flowmeters to check for leaks.)

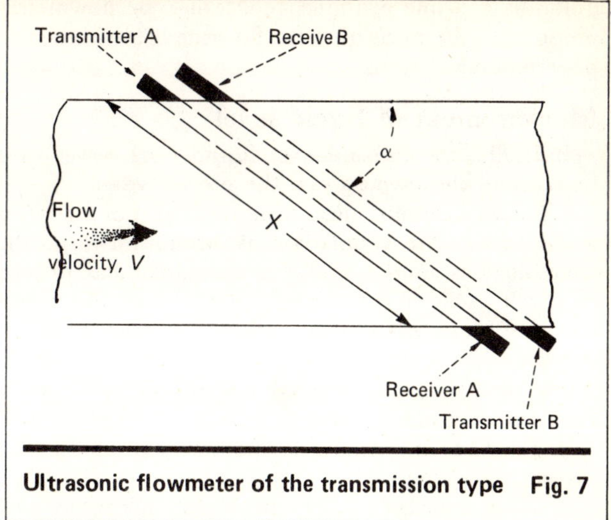

Ultrasonic flowmeter of the transmission type Fig. 7

2. This is a zero-pressure-drop meter, and has no moving parts; it comes in line sizes from about 3 in. and up.

3. The meter comes in two styles:
(a) Wetted surfaces for transmitter and receiver, and
(b) clamp-on.

In the clamp-on versions, the transmitter/receiver surfaces are bonded to the pipe with an epoxy cement; therefore, clamp-on flowmeters are not readily relocated. However, it is possible to use a multiple-head arrangement with single read-out. Although the clamp-on version generally is not as accurate as the wetted-surface version, its completely external nature makes a clamp-on extremely attractive for measuring hazardous or toxic materials.

4. The electronics of receiver and transmitter establish the temperature limit of the flowmeter (approximately 400°F).

5. In the transmission flowmeter, the liquid must not absorb sound. Furthermore, suspended solids or entrained gas bubbles may scatter the sound.

6. In the reflection models, effectiveness depends upon the presence of suspended particulates to reflect the sound. The absence of interfacial discontinuities created by suspensoids in the fluid will make the reading of a reflection-type meter unreliable.

While accuracy of the ultrasonic meters has traditionally been poorer than that of orifice meters, it has improved in recent years to a point where both types now show accuracies on the same order of magnitude.

An extremely attractive feature of the ultrasonic flowmeter is that its cost is essentially independent of line size. In the transmission type of ultrasonic flowmeters, a 30-inch meter costs less than twice that of a 3-inch meter. In the reflection variety, the unit comes with a single head which can be bonded to any size pipe. Therefore, the cost is always the same, completely independent of line size.

As for applications, the reflection type of ultrasonic flowmeter is extremely attractive in some difficult-to-measure applications involving suspended solids in the fluid. As a prime example, reflection-type flowmeters are successfully measuring the flowrate of stock used in papermaking.

Measurement of liquid level

Basically, the indication of liquid level serves as a measure of the inventory in the process vessel.

The most common measuring method uses a float or a displacer. The installation is straightforward, the technology is simple, and the equipment has proven relatively reliable. However, floats and displacers do have some operational limitations and problem potentials:

1. The seals can be troublesome at high pressures or in corrosive applications.
2. Buildups and deposits on floats and displacers can present problems.
3. Displacers have a practical span of liquid-level measurement not exceeding eight feet.

The use of differential pressure is probably equally reliable and equally simple, but the sensitivity of the measurement is not adequate for many applications. In addition, if the density of the liquid varies, then some compensation technique must be employed when the level is calculated from the differential pressure.

The *air bubbler* is really another variation of a differential-pressure measurement. Among its advantages is the ever-present flow of air, which tends to keep the unit free of foreign matter. In some applications, such as the measurement of liquid level in sugar-juice tanks, this purging nature is highly advantageous.

As with flow measurement, our approach to liquid-level measurement will emphasize recent technology typically used in some more difficult measurements of liquid level. Again, the usual procedure is to determine whether a float, a displacer, or some equivalently simple technology may be applicable. If not, then a more sophisticated approach will be required.

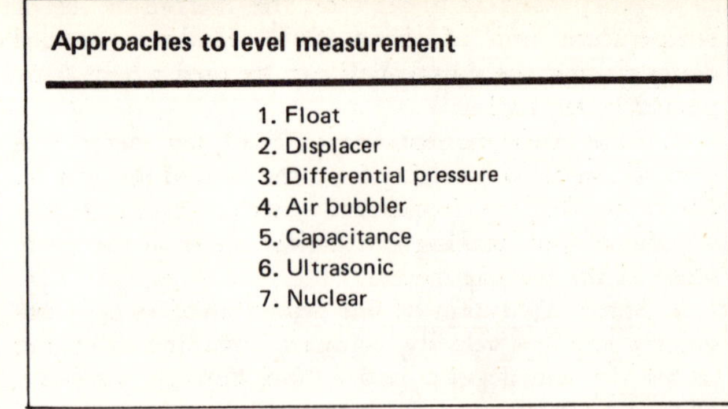
Approaches to level measurement

1. Float
2. Displacer
3. Differential pressure
4. Air bubbler
5. Capacitance
6. Ultrasonic
7. Nuclear

Capacitance level gage

This device consists essentially of an electrical probe, normally inserted down the center of the tank. This creates a capacitor, with the probe serving as one plate and the tank walls as the other. Three types of probes are used in capacitance level gages:

1. Bare probe
2. Insulated probe
3. Probe with concentric shield

Fig. 8 illustrates a typical installation of a capacitance level gage with a bare probe. The probe is insulated from the metal tank by the coupling used to mount it. The capacitance is measured using a bridge circuit excited by a high-frequency oscillator (on the order of 1 MHz). These electronics ultimately produce the 4- to 20-ma signal compatible with most controllers, recorders, etc.

When the level of a non-conductive liquid, (e.g., propane) remains below the probe, the system still has some capacitance, due to the insulator and because the air above the surface has a known dielectric constant. As the level rises, air is displaced by the liquid with its higher dielectric constant. This causes a change in the capacitance measured at the terminals. For vessels with constant diameters, capacitance varies linearly with the level.

For conductive liquids, current will flow between the two plates, destroying the linearity of the level/capacitance relationship. Covering the probe with a coating (normally Teflon) that serves as an insulator will prevent this problem.

For very large tanks, changes in the level produce relatively small changes in the capacitance. Fitting the probe with a concentric shield will increase its sensitivity. The capacitor is now formed by the probe and the shield.

A non-conductive tank cannot serve as one of the plates of the capacitor. In this case, the probe with the concentric shield will work, but the others will not.

Although most probes are rigid, some are flexible. If

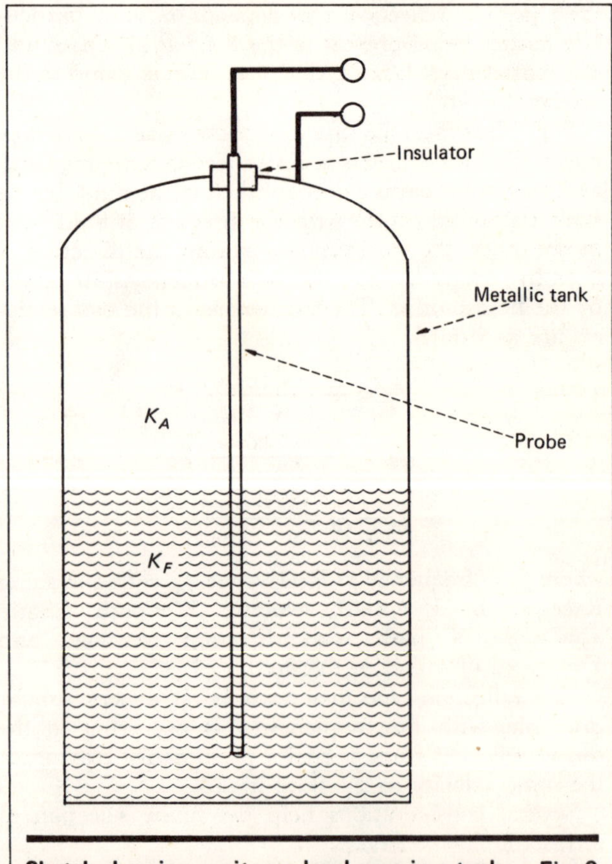

Sketch showing capitance level gage in a tank Fig. 8

the liquid is non-conductive, a probe can consist of a chain with a weight on the end. Conductive liquids require a cable with an insulated coating.

The capacitance level gage can also be used to detect interface levels. Of course, the dielectric constant of the two liquids must be different and, depending upon the respective values, the capacitance can increase or decrease as the interface level rises.

In considering the use of a capacitance level detector, note the following points:

1. Although the system is relatively simple to design, the assistance of the instrument's manufacturer can be very useful.

2. This device does not have any seals around moving parts—it has no moving parts. It does, however, require a non-conducting seal between the probe and the tank. Also, grounding problems sometimes occur in the head assembly.

3. The probe (and its coating, if necessary) can be made of a wide choice of corrosion-resistant materials of construction.

4. Any process change that affects the dielectric constant of the liquid will cause an error in the level measurement. When the culprit is a temperature variation, most commercial systems provide the capability of measuring temperature and then adjusting the measurement by using a linear relationship for the dielectric constant as a function of temperature.

5. Build-ups on the probe itself can present problems. Normally, the system continues to provide a reading for the level in the tank, but it will be in error because of the dielectric constant—and possibly the electrical resistance, as well—of the build-up.

Industry has used capacitance-type level detectors for a number of years. Experience with the devices shows them to be relatively reliable, and—assuming that no build-ups or similar problems occur on the probe—they have proven almost maintenance-free. Normally, the problem lies in justifying the additional expense of a capacitance system over the conventional float or displacer.

Ultrasonic level-measurement systems

Some applications call for measuring liquid level in a vessel without making any physical contact with the liquid. For example, consider measuring the level of molten metal. Clearly a float or a displacer would not last long in such service. The molten metal would either build up on a capacitance probe or corrode it away. A differential-pressure system would quickly plug up. As for the air bubbler, it is not clear what would happen if air (or nitrogen) were bubbled through a tank of molten metal, but certainly the splashing and other considerations make this an undesirable approach.

Fig. 9 shows an ultrasonic transmitter, mounted above the liquid, transmitting ultrasonic waves to the surface. From the speed of sound in the air or vapor space above the liquid, and the time required for the waves to reach the surface and be reflected to the transmitter, the distance of the transmitter/receiver from the surface of the liquid can be calculated.

In applying ultrasonic level-measurement technology, a number of points should be considered:

1. An ultrasonic system has no moving parts, which makes it relatively maintenance-free, provided the electronic parts of the device are not exposed to elevated temperatures or to other undesirable conditions.

2. The system is normally easy to install, if access is available above the surface of the liquid.

3. Requiring no physical contact with the liquid itself, ultrasonics are highly attractive in certain applications.

4. Also, isolation from the liquid makes measurement completely independent of the physical properties of the liquid.

5. Since sonic velocity depends on physical properties of the air or other gas between the transmitter/receiver and the liquid surface, changes in air temperature and other variables will affect the measurement reading.

6. If the liquid surface is turbulent, or covered with froth or foam, it may not adequately reflect sound to the receiver. Also, most materials reflect sound better at some wavelengths than at others; consequently, many recent commercial units provide for user-adjustment of the frequency of transmitted sound.

7. Also, since ultrasonic devices depend on the air or gas to transmit sound, presence of dust or other fine particulate matter can disperse the sound waves and disturb the transmission path.

Traditionally, the main disadvantage of ultrasonic systems has been their cost. Today, however, a relatively sophisticated ultrasonic system can be purchased for only about $4000.

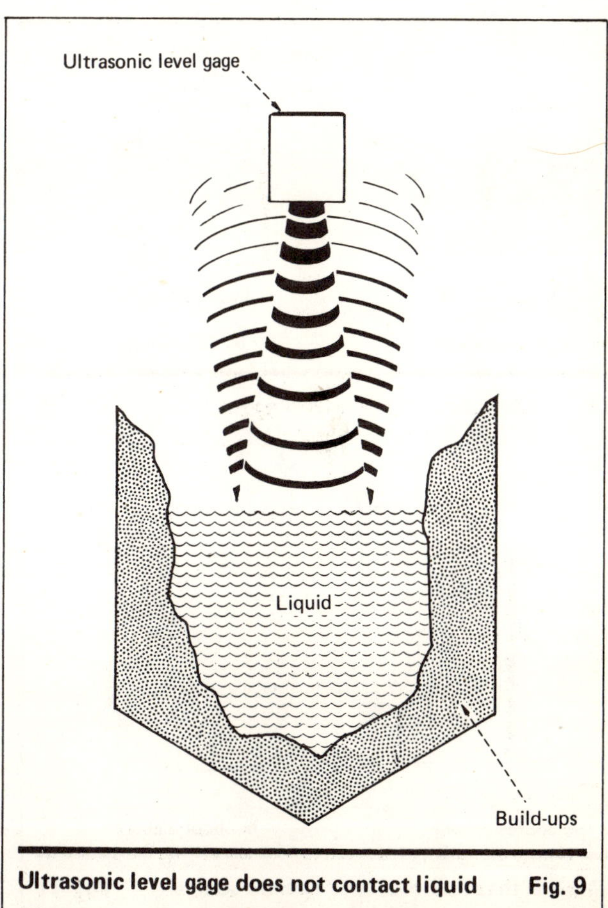

Ultrasonic level gage does not contact liquid **Fig. 9**

Nuclear level-measurement systems

Another non-contacting level-measurement system is the nuclear. The technology used in nuclear level-measurement is essentially the same as that used in nuclear density-measurement systems. In effect, both measure the amount of mass between the source and the detector. Knowing the density of the material permits calculation of its volume. Conversely, if the volume between the source and detector is known, then the density can be calculated.

Fig. 10 illustrates two arrangements for installing source and detectors for continuous measurement of level. The advantage of the point source is its easier installation and somewhat lower cost. However, on the other hand, the strip source provides a much more linear output.

The cost of a nuclear system depends largely on the size of the nuclear source required. (As the size of the vessel becomes larger, the radiation must go through more material, so the source must be stronger.) Special modifications to the installation will reduce the required size, and therefore the cost, of the source. As illustrated in Fig. 11, one approach involves inserting the source in a well inside the vessel; another approach shoots across a chord instead of across the entire diameter of the vessel.

A few basic points become significant when considering nuclear level-measurement technology:

1. This system has no contact whatever with the process fluid. This permits servicing of the instrument without breaking into the vessel itself, which is highly advantageous in situations involving corrosive or otherwise hazardous fluids.

2. Radioactivity of the source will decay with time. Most industrial systems utilize cesium-137, with half-life of 33 years (equivalent to decay rate of 2% to 3% annually). Re-calibration procedures are relatively straightforward, and automatic re-calibration is certainly within the state of the art.

3. Density variations affect the level-reading. Of course, if temperature varies, then the output can be compensated for by using a linear density/temperature relationship.

4. The nuclear approach can work even if the vessel is insulated or if it has a steam or cooling-water jacket. Hot process fluid will probably call for mounting the detector assembly outside the insulation. If necessary, the source part of the device can be mounted outside the insulation, but it can withstand higher temperatures than can the detector. The problem with this approach is cost: the thicker jacketing material requires a larger, and thus a more expensive, source.

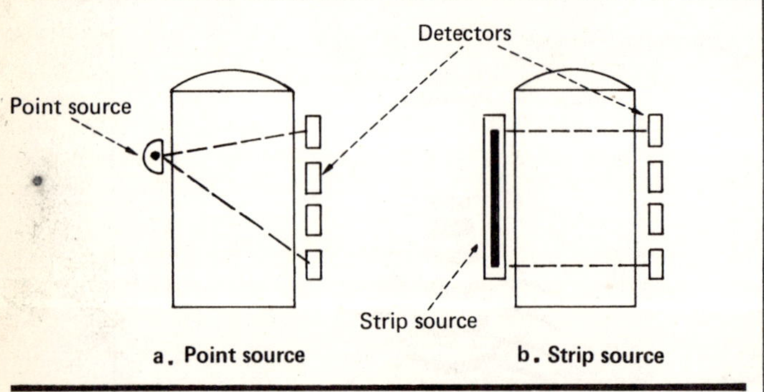

Typical installations of nuclear level gages Fig. 10

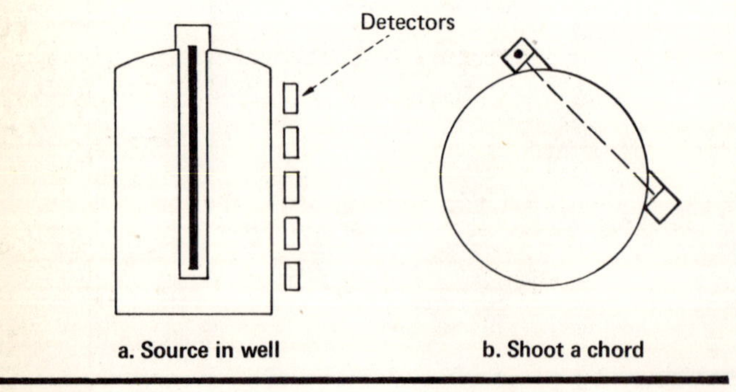

Installations that reduce size of nuclear source Fig. 11

From a technical point of view, the nuclear measurement concept can solve all of the technical problems in most applications, even in the most difficult measurement applications. However, the main problem with nuclear systems stems from their radioactive materials. The manufacturers of commercial systems have taken this into consideration, providing the source with a container that permits its withdrawal and even complete removal when personnel must enter the vessel for any reasons.

One problem with any nuclear device is the administrative overhead involved. Many of these devices require some type of license (e.g., from the Nuclear Regulatory Commission, for plants in the U.S.). Furthermore, reports or other actions may be required on an annual basis (including the "wipe test," in which the instrument is wiped with a cloth that is then submitted for analysis). This means that at least one individual must become the resident radiation expert and keep up with the various regulations. The involvement is very high for a firm's first nuclear device, but incremental work that will be required for additional nuclear devices is relatively low.

The two major sources of opposition to nuclear technology inside a plant are safety people and plant management. Some safety people will endorse nuclear systems, but only when the process fluid is extremely hazardous, to the extent that it presents a greater danger to employees than would the nuclear source under the worst possible conditions. However, plant management generally views nuclear systems as a potential source of some very sticky legal proceedings.

In the past, nuclear-measurement systems have been relatively expensive, although their cost is becoming more competitive with other technologies. It is not uncommon to find the nuclear system costing $10,000 or more, but some installations will cost only $2,000 to $3,000—it depends on the size of the source. However, in the more difficult measurement applications, cost often becomes a relatively minor consideration.

Level switches

Some applications do not need continuous measurement and control of fluid levels. Fig. 12 illustrates one such application in an oil/water separator. The entering oil/water mixture separates into two phases, with the oil on top. Manipulating the water-removal rate controls the interface level.

One way to do this is to install a continuous level-measurement device for measuring the interface level, a conventional level controller, and a proportioning valve. This will maintain the interface level at any desired position within the range of the level gage.

Fig. 12 illustrates a control arrangement, using two level switches and a solenoid valve, where an exact level is not necessary. To keep the interface level somewhere between the two level switches, the system works like this (starting with the discharge valve closed):

1. When the upper-level switch first indicates the presence of the water phase, the solenoid valve opens.
2. This valve remains open until the lower-level switch indicates the presence of the oil phase, at which time the solenoid valve closes.

If a pump is required to remove the water, the output of the level switches can also be used to turn the pump on or off.

In all respects, the level-switch system is less expensive than the continuous-control sytem. Two level switches cost less than the continuous level gage, and the solenoid valve less than the proportioning valve. Also, the discrete logic is relatively easy to implement, and maintenance of the level-switch system is simpler.

Level switches come in a number of types, the most common ones being vibrating element, capacitance, and ultrasonic. These switches find a range of applications: alarms, emergency shut-down systems, and—to a limited extent—control systems. (E.g., batch-control logic may call for a vessel to be filled until a level switch trips.) In general, their low cost and simplicity make them attractive in many cases not requiring continuous monitoring.

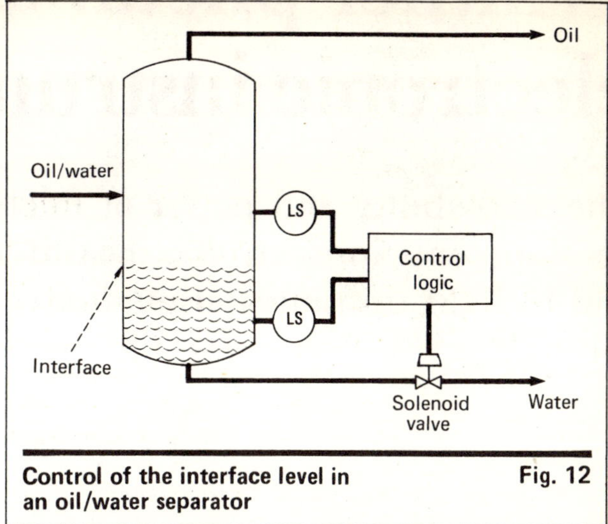

Control of the interface level in an oil/water separator **Fig. 12**

Pressure measurement

There is a wide range of choices available for measuring pressure.

Over the years, the diaphragm, bellows, or Bourdon-tube arrangements have served industry widely. A few years ago, strain-gage devices began to appear, the early ones being noted primarily for their high cost at that time. In certain applications, they did prove more reliable than their earlier competitors.

With semiconductor technology advancing at a rapid pace over the past ten years, new developments have led to much more cost-effective use of basic strain-gage technology. Essentially, these newer devices apply the process pressure to one side of a semiconductor element with the other side open to the air for gage-pressure measurement—or subject to another process-fluid pressure, if the measurement is to be a differential pressure.

These systems show the following technical characteristics:

1. Very low (3 in. H_2O) to very high (200,000 psig) spans are available.
2. Maximum permissable temperature is about 200°F. The temperature is limited by use of the semiconductor element.
3. Accuracy of the units approximates 0.1% of span, certainly comparable to the competition.
4. The solid-state system usually requires compensation for process-temperature variations, but this is done internally.
5. A bridge circuit, required for readout, can now be incorporated relatively inexpensively into the unit.

Low cost is probably the most attractive feature of these devices. Currently, the price range is about $200 to $600, and still decreasing. Some forecasts see the price decreasing to $20 or less.

This device puts out a standard 4- to 20-ma signal. And with the power supply located in the receiver instead of the transmitter, only a single pair of wires must be run to the pressure transmitter.

Acknowledgment

The material in this article is based on presentations by the author in the professional-development seminar, "On-Line Process Measurements," offered by the Continuing Education Department of the American Institute of Chemical Engineers. Appreciation is also expressed for the many helpful suggestions from Mr. R. E. Klie (Shell Oil Co., Wood River, Ill.), who also lectures in the course.

The author

Cecil L. Smith is professor of Chemical Engineering and of Computer Science, Louisiana State University, Baton Rouge, LA 70803. His interests lie mainly in the areas of control and simulation of industrial processes. He holds B.S., M.S. and Ph.D. degrees in chemical engineering from Louisiana State, and is a member of AIChE, Instrument Soc. of America, and Assn. for Computing Machinery. He is a registered professional engineer in the state of Louisiana.

Whither pneumatic and electronic instrumentation?

The availability and power of microprocessors, new developments in electronic measurement techniques and in light-transmission methods—these are shifting the balance toward electronic instrumentation.

Myles J. Marcovitch, Fischer & Porter Co.

☐ Chemical and process engineers generally choose pneumatic or electronic instrument systems by weighing factors such as:

- Environmental hazards.
- Maintainability.
- Computer interface requirements.
- Availability of compressed air.
- System interface requirements.
- Transmission distance.
- Response time.

However, three new developments, covered later in this article, are beginning to force a change in the pneumatic vs. electronic decision:

1. All of the newer measuring concepts use electronic principles.

2. The microprocessor invasion is making drastic changes, which have only just begun.

3. The use of optical transmission, although not available yet, is so close that it deserves considerable thought.

The following discussion details the current criteria for instrument selection, describes some newer instrument types (including the way they function), and suggests some future directions that the pneumatic vs. electronic decision may take.

Let us begin by examining the traditional criteria for the pneumatic/electronic instrument decision.

Pneumatic instruments

Pneumatic instruments are inherently safe. There is no chance for an explosive hazard with pneumatics unless the instrument "air" used is not air at all but part of the process itself, such as natural gas. Electronic instruments can only approach the safety of pneumatics in three ways. Each way requires some modification of the standard instrument, such as: explosion-proof housings (good only if all enclosures are properly maintained), intrinsically safe (limiting the energy in the system to below that required to ignite the explosive environment), or use of an inert gas purge of the electronic enclosure (to exclude the explosive environment).

Service personnel often consider pneumatic instruments easier to understand and therefore find them easier to maintain. Pneumatics operate without care for long periods, requiring only a clean, dry air supply. Also, these instruments can operate in some very obnoxious environments.

Pneumatic instruments interface readily with pneumatic valve-actuators. Pneumatic diaphragm-type control valves are currently the predominant type found in the chemical process industries. This type of valve is rugged and simple, and is spring-loaded for fail-safe operation.

On the other hand, to use electronic instruments with pneumatic valve actuators, a transducer is needed—another piece of equipment to buy and maintain.

Electronic instruments also exhibit a number of distinct advantages:

Electronic instruments are fast. The electrical signal travels, essentially, at the speed of light; a pneumatic signal travels at the speed of sound. This speed difference means relatively little over short distances, say,

Photo: Yokogawa Corp. of America

Vortices such as these, produced by a bluff body in the stream, are detected and used to measure flow — Fig. 1

Originally published October 15, 1979.

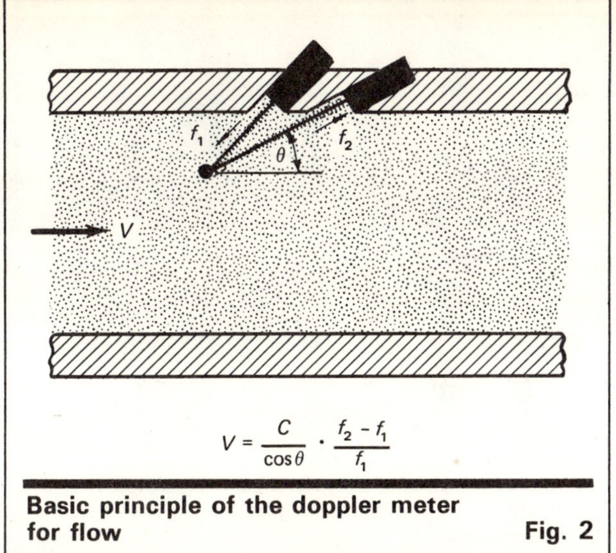

$$V = \frac{C}{\cos\theta} \cdot \frac{f_2 - f_1}{f_1}$$

Basic principle of the doppler meter for flow **Fig. 2**

under 500 feet. But beyond 1,000 feet the signal delay in a pneumatic system will be measured in seconds which, in the case of a critical reactor, could cause problems.

Electronic instrument signals can be transmitted over long distances (measured in miles) without serious degradation of signal strength. Pressure loss in a pneumatic line restricts pneumatic signal transmission distance to approximately one thousand feet. Pneumatic repeaters are available if this distance limit must be exceeded.

Electronic instruments can perform more functions than their pneumatic counterparts. You can easily get multipoint recorders and function generators as electronic instruments. They do not exist as pneumatics.

Electronic instruments interface more readily with digital computers than do pneumatics.

For many years the traditional criteria listed above held true. Three developments, however, are changing the selection process.

New measuring concepts

New electronic measuring concepts have progressed steadily over the past decade. The following primary transducers apply these new measuring concepts:

- Magnetic flowmeters.
- Ultrasonic flow and level meters.
- Vortex-shedding flowmeters.
- Turbine meters.
- Swirlmeters.®
- Nuclear density-meters.
- Temperature detector systems.

(There has not been a significant new measuring concept in pneumatics in a quarter of a century. Pneumatic instruments per se *have* been developed to a high degree but the actual measuring *fundamentals* have not changed.)

Each of the above listed instruments takes advantage of a different natural phenomenon that would go undetected if it were not for electronic augmentation. It will be helpful at this time to explain the operating concepts behind each of the meters above, and to show why each promotes the selection of an electronic system.

Magnetic flowmeters

Magnetic flowmeters measure the flowrates of some very objectionable (corrosive, non-Newtonian, abrasive, or sludgy) fluids. Such a meter has no obstruction to flow and is also immune to viscosity changes in the process liquid.

The meter applies the principle of Faraday's law, which states that a conductor moving through a magnetic field generates a voltage. Using a mildly conductive fluid as the conductor in a pipe wrapped with a substantial magnetic field coil, the meter generates a small voltage across two electrodes mounted at right angles to the field coils. This voltage is linear and proportional to the velocity of the flowing liquid.

The cross sectional area of the meter pipe is a known constant. Therefore, the electronics associated with the meter can compute the flowrate by using the velocity signal as measured across the electrodes. The output voltage of the meter is an a.c. or quasi-a.c. signal that the system must condition or convert to the conventional 4-20 mA d.c. signal required by standard process instruments. The converter can also develop a digital pulse signal for direct computer interface.

Ultrasonic metering systems

Ultrasonic flow-measuring systems provide obstructionless flow measurement for nonconductive fluids in various ways:

- Doppler. The velocity of the liquid causes a doppler shift in an ultrasonic signal propagated across the pipe through the flowing liquid. (Fig. 2)
- Time of flight. Two or more transponders (even numbers) send and receive an ultrasonic signal diagonally across the pipe. The signal is first sent by the downstream transmitter to the upstream receiver and the elapsed travel-time is measured. The upstream receiver then transmits to the downstream receiver and the elapsed travel time is again measured. The flowing liquid adds to the downstream signal velocity (reduces elapsed time) and subtracts from the upstream signal velocity (increases travel time). The difference in time is directly proportional to the flowrate. (Fig. 3)
- Echo ranging. Open channel ultrasonic flowmeters use microprocessor technology in the application of the echo ranging concept.

The microprocessors used with these echo rangers perform the following functions: control propagation of initial sound pulse; time the return of the echo from the surface of the flowing liquid; interrogate a temperature sensor to calculate the actual speed of sound at the time of echo detection; evaluate a series of echos to negate effects of floating debris; and calculate actual flowrate by using one of a number of formulas held in memory for a specific open channel configuration.

Obviously ultrasonic metering would be impossible without highly advanced electronic circuits including microprocessor technology.

Vortex-shedding flowmeters

The vortex-shedding flowmeter is gaining prominence because of its ease of application. Vortex meters can measure the flowrates of many different fluids provided they are Newtonian. (Examples of Newtonian fluids are water, alcohol, and petroleum extracts. Non-Newtonian fluids include ketchup, thin oatmeal, and cement.) These meters can replace the complex installation of differential pressure transmitters and associated paraphernalia.

Vortex shedding occurs whenever a liquid or gas travels around a bluff obstruction in the fluid stream. The fluid, passing the obstruction, creates vortexes or eddies when the fluid rejoins the main stream. The vortexes form first on one side then on the other, alternating in a regular manner proportional to the fluid's velocity. By detecting the vortex alternations, the meter's electronics can compute an accurate flowrate.

Vortex-shedding meters might use strain gages, reluctance sensors, or thermistors to detect the vortices. However, the vortex eddies at their maximum amplitude move the detection mechanism no more than 25 millionths of an inch. Only electronic methods permit such a minute motion to be detected and amplified.

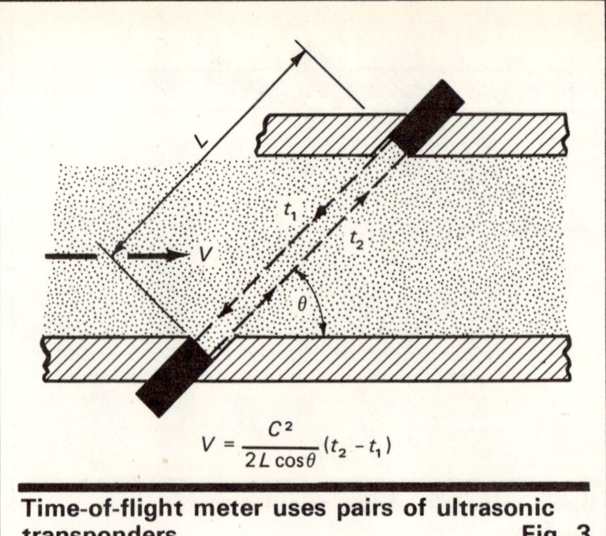

Time-of-flight meter uses pairs of ultrasonic transponders **Fig. 3**

Turbine flowmeters

Turbine meters are another example of a measuring concept impossible without electronics. The turbine meter represents the standard by which other meters are judged. Capable of extreme accuracy, turbine meters handle special tasks such as monitoring jet fuel consumption on engine test stands, as master meters in calibration rigs, and as the U.S. Government standard for alcohol taxation in the beer industry. Major limitation for turbine meters is its requirement for clean fluids, especially those with some lubricating properties. Turbine meters are also costly and, therefore, used only when the process requires their inherent accuracy. Turbines have high maintenance costs because of their mechanical nature.

The meter has a precision-machined turbine in a machined bore. The turbine turns on precision bearings—either ball, plain or ceramic, depending on service requirements. Fluid (either liquid or gas) passing through the meter spins the turbine.

The turbine's rotational speed is proportional to the fluid flowrate; the speed is detected by either a magnetic or radio-frequency pickup. (The pickup senses the disturbance caused by each turbine rotor tip.) Larger bore meters have an outer ring surrounding the rotor with drilled holes that produce a higher disturbance rate (frequency). The pickup generates pulses for each blade or hole disturbance sensed. The turbine meter, by initially generating a pulse rate proportional to flow, allows easy computer interface, as well as standard analog output.

Swirlmeters®

The Swirlmeter is to gas flow as the turbine meter is to liquid flow. It is a very accurate instrument for precise gas flow measurement. This is the way it operates: A set of stationary swirler blades, at the meter's inlet, causes the entering gas to rotate around the central axis of the instrument. Behind the swirler blade assembly, the meter body widens causing the rotating gas flow to begin to precess, or spiral around the inside wall of the meter. This behavior resembles a tornado where the air rotates wildly in the funnel cloud itself while the funnel also moves and twists as it travels across the landscape. The spiraling gas stream crosses a sensor mounted in the instrument wall at a rate proportional to the gas velocity.

The Swirlmeter uses a piezoelectric sensor, which detects the gas flow by changes in pressure on the sensor. In its basic state, the Swirlmeter outputs an analog or digital signal indicating the actual cubic feet per minute (acfm) of gas flow through the meter. With the addition of a pressure and temperature sensor, the meter's electronics compute standard cubic feet per minute (scfm), using the perfect-gas laws.

Nuclear density-meters

These meters use the ability of gamma radiation to ionize a gas, thereby creating a current flow to measure the density or specific gravity of a substance. The meter focuses a source (a radioactive isotope such as cesium-137), through the sample to be measured. The ionization cell is on the opposite side of the sample. (These two units may simply clamp on the process pipeline.) The greater the density (more molecules per unit volume), the more gamma radiation is absorbed and the less reaches the ionization cell. The meter amplifies and conditions the sensor's signal to give the standard instrument output.

Electronic temperature detectors

Thermocouples, resistance temperature detectors (RTD), thermistors, infrared pyrometers, etc., all detect temperature change by electrical/electronic means. Each finds use depending on various conditions such as range of measurement, span requirements, operating environment, and response time. Electronic temperature devices take advantage of the physical effect that temperature has on various substances including generating a minute voltage as the temperature changes (thermocouple), changing resistance with temperature change (RTD by thermistor).

Mechanical instrumentation, on the other hand, relies basically on two effects; the differential expansion of dissimilar materials (the bimetallic strip) or the change

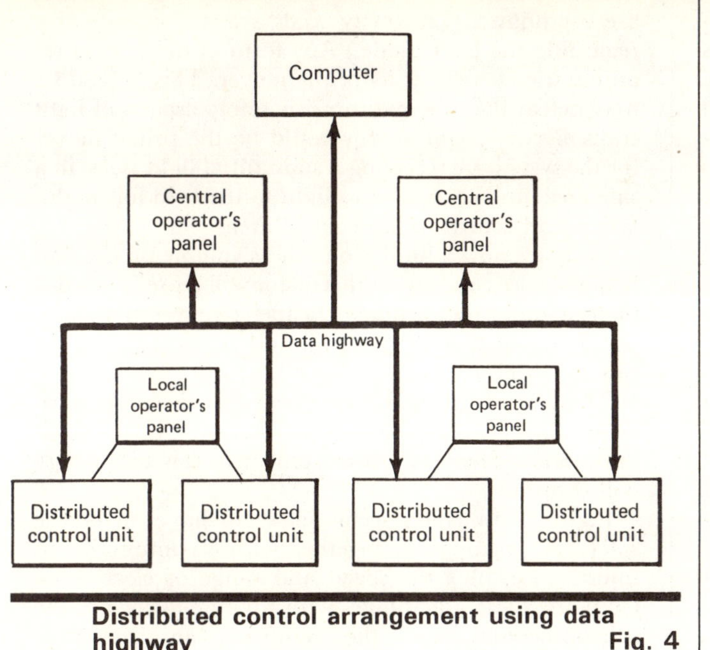

Distributed control arrangement using data highway **Fig. 4**

in volume a substance incurs when heated (filled temperature systems). Though nonelectronic temperature systems are reliable and accurate, electronic temperature allows greater flexibility and ease of application.

All of the process instruments described in this section owe their existence to electronic concepts. The moment you choose one of these metering concepts for an instrumentation task, you have started in the direction of an electronic instrument system.

Microprocessor invasion

Microprocessor/microcomputer technology, developed during the 1970's, now exerts a strong force towards electronic-instrument selection. The microprocessor permits almost every kind of device to exhibit a certain degree of decision-making capacity. Thousands of electronic components fit onto one silicon wafer allowing functions only imagined 15 years ago, when computers filled rooms.

Let us first define what a microprocessor is before delving deeper into its applications. A microprocessor represents the CPU (central processing unit) of a traditional computer system. Containing the registers, decoders, accumulator control circuitry, arithmetic logic units and internal memory, the microprocessor represents the active portion of the computer.

A microprocessor that contains both memory and functional units to connect it to the outside world (input/output, or I/O, devices) is a microcomputer. A silicon chip 1/10-in. square can hold a fully operational microcomputer.

The microprocessor performs logic and arithmetic operations. Microprocessors can store and retrieve, reshuffle, and organize data. They can compare two values and make a programmed decision based on the outcome of this comparison. Instruction sets allow almost every conceivable movement of data, limited only by the designer's creativity.

Now on to applications as they pertain to the process industry. Microprocessors make the universal instrument a reality. Unlimited flexibility permits application of the microprocessor to any kind of process measurement. Instrument manufacturers can now produce instruments capable of major changes in operating parameters.

Microprocessors perform computational manipulation such as linearization of signals, square root extraction, and integration. Operators can numerically enter numbers on a keyboard to change K-factors. The output is a digital display or a transmission of data to other intelligent devices via a "data highway."

The microprocessor combined with conventional process instrumentation results in a new form of control instrument, unavailable before, known as a "distributed control system." By extensive use of programmable read-only memory (PROM), the instrument appears (to the process engineer) like an entire analog instrument catalog. Each "analog" instrument in this distributed control system is connected into an instrument loop by a technique known as "softwiring." "Softwiring" is the functional linking up of the "analog" instruments located in memory by the entering of specific software commands. Common language programming-techniques permit simplified system configuration. In fact, the language is the same familiar instrument terminology employed every day by the process engineer or instrument technician.

This type of microprocessor-based distributed-process-control instrument allows the engineer, for the first time, to modify extensively an existing process-control system without the major rewiring required when a conventional analog system requires change. It also permits massive system changes without the need to know high-level program language such as FORTRAN or COBOL, as is required to make changes on existing minicomputer-based systems.

Process industries may now be constructed; primary sensors put into place and connected to the field input terminals of the microprocessor-based system, piped and valved; all without actually setting up the control system. Then, engineers can design the control configurations and "softwire" them right at the control console. If the control system does not perform as planned, changes are made simply.

The age of the "smart" instrument is here, and it is irreversible. Instruments that can talk to plant operators when something is wrong, and instruments that plant operators will talk to when making required adjustments are not fantasy, only the natural outgrowth of application of new memory technology, such as bubble memories, laser memories, and Josephson junctions. This new memory technology combined with microprocessors offers memory capacity of millions of bits on wafers no bigger than the microprocessor itself. The conventional analog device, whether pneumatic or electronic, compared to microprocessor-based instruments, will seem very dull indeed.

The "light" fantastic

The third force pushing new technologies further into the process instrument world is the transmission of data by modulated light through fiber-optic highways. In a matter of a few years, this technology has made immense strides. Originally suffering great signal losses over small distances (measured in feet), fibers now are available with minute losses over kilometers. Detailed studies in optics have led to the development of multiphase glass and plastic fibers. This type of fiber keeps bouncing the light beam back and forth internally, allowing only minute quantities to escape sideways, but offering no effective resistance down the length of the fiber.

What does light transmission offer the process control industry? It offers:

- Transmission of process signal at the speed of light.
- Complete lack of combustion-inducing energy in the light transmission system, i.e., intrinsic safety without additional hardware.
- Simultaneous transmission of many signals, owing to light's ultrahigh frequency and large bandwidth.
- Easy modulation of digital information.
- Absolute immunity to electromagnetic radiation (EMR).
- No radio frequency interference (RFI).
- Immunity to environmental extremes, up to the limitations of glass.
- No shock hazard.
- No short-circuit hazard.
- Use of readily available domestic materials for fiber construction.

Light transmission's adoption is restricted by certain limitations. These limitations are technological in nature, characteristic of a young technology. Let us review some of these limitations, their effects and their possible correction:

Splicing one fiber to another has previously been art, not technology. An electrical connection's only barrier to current flow is an oxide film, easily broken down by solder flux or a strong mechanical type of connection. A light fiber on the other hand, must be bonded to another at an optically flat surface to avoid reflection back into the first fiber with subsequent signal loss. This bonding process was extremely difficult. There are now, however, many companies developing connection systems similar to those found in electronics. Connectors now join light fibers with only minor effort compared to previous methods. New developments are sure to further reduce this difficulty.

There is little standardization between manufacturers concerning almost every parameter of light transmission. Standardization belongs to mature technologies. The electrical standards, which are so familiar today, took many years of trial before they were fully established. Light transmission is a very young technology, and not ready for standardization. All diverse alternatives from many sources require testing before the industry can decide how to systematize this concept. But standards will eventually come.

Modulation and demodulation devices require further refinement. The possibility exists for using light on the circuit board as well as in the fiber between units, thereby permitting the control system to be totally light operated.

The light sources for use in operational systems remain to be decided. Both light-emitting or laser diodes are candidates. (Longevity trials are still inconclusive regarding the laser diode.) Any form of diode requires an electric current as its prime mover. This, of course, may defeat the inherent intrinsic safety aspects of light transmission. Light energy could be the prime mover for the system by receiving a substantial light pulse in a safe area, and carrying this light to the field for modulation and return to the control area.

Costs are now a limitation but as volume production begins to rise concern for this factor will cease. The glass fiber actually uses a much cheaper resource than copper. It is much more plentiful and requires much less raw material.

The benefits of light transmission to the process industry far outweigh the costs necessary to overcome the limitations. Therefore, development of this technology will continue.

Light transmission seems likely to have all of the safety of pneumatics, together with its immunity to interference, plus the speed and range of electronics. Combine these advantages with the intelligence of the microprocessor, and the result is an irresistible combination.

Summary

Pneumatics will probably always have some use—whether based upon its ultimate safety (e.g., for the manufacture of explosives), or upon tradition. However, there will come a time when the selection of pneumatics based upon traditional reasoning will become a moot point as new technicians who will have been reared only on electronic concepts enter the instrumentation field.

Obviously, the microprocessor invasion is currently exerting tremendous market pressure, while light transmission is a development lurking in the future. But the ultimate impact of the use of light is likely to be as great as any preceding development.

Moving in quantum leaps, the technological progress of light transmission is behaving in the characteristic way of all modern developments, where many companies begin to see the opportunity for a breakthrough in new markets and the resultant rewards.

Although many of the traditional reasons for making electronic vs. pneumatic instrument decisions will remain valid for some years to come, the three developments detailed in this article are going to force electronics to greater industry dominance.

The author

Myles J. Marcovitch is manager of the Technical Training Center for Fischer and Porter Co., 39 E. County Line Rd., Warminster, PA 18974. He manages inplant production skill training and customer product training. He has taught a pneumatic and mechanical instrument-maintenance course, among others. He holds a B.S. in industrial education from Michigan State University and an Ed.M. in industrial education from Temple University. He is a member of the Philadelphia Chapter of Instrument Soc. of America.

Section V
FLOW MEASUREMENT WITH ORIFICES

Measuring flow in pipes with orifices and nozzles
Designing orifices for flow regulation
Estimating pipe sizes for orifice runs
Designing critical flow orifices

Measuring Flow in Pipes With Orifices and Nozzles

A properly chosen flow device must develop the maximum differential pressure and be sized for the correct Reynolds number in order to obtain an accurate reading of pipeline rate of flow.

ROBERT KERN, Hoffmann - La Roche Inc.

To measure flow accurately, the designer of flow systems must provide a pipe diameter of sufficient size and, equally important, a suitable configuration for the piping.

We will closely examine these parameters in relation to the sizing of orifices and flow nozzles in piping systems. Furthermore, we will take a look at associated requirements to ensure:

- Adequate straight-run of piping before and after the flow device.
- Economy of the piping layout.
- Provision for orifice taps, straightening vanes, and separator chambers.
- Accessibility to the flow device, and instruments connected to it, when installed in the piping system.

The most common device for measuring flow is a thin plate with a square-edged hole in the center, held between a pair of flanges. Usually, this orifice is a stainless-steel plate, ⅛-in thick (for lines 16 in, or larger, ¼-in thick). Minimum orifice bore is usually ¼ in. If required, a small vent hole and drain hole are drilled in the orifice plate—slightly overlapping the internal pipe wall at the top and bottom. A pair of jack screws, installed in the flanges, force the flange faces apart for replacing the orifice plate. Pressure taps through the flanges provide the means for connecting the orifice plate to indicating, recording or transmitting instruments. The entire assembly is shown in Fig. 1.

If an orifice is placed in a pipeline, with fluid flowing through it, the pressure will vary along the orifice pipe-run, as shown in Fig. 2. For a selected installation and fluid, the pressure difference between the inlet and outlet sides of the orifice varies in proportion to the flowrate. This pressure variation is sensed by a suitable instrument—the simplest of which is a U-tube manometer.

Piping and Orifice Sizing

We will review the fundamental relations for flow through an orifice by using Darcy's equation, $h_L = Kv^2/2g$, (*Chem. Eng.*, Jan. 6, 1975, p. 117) to express the flow velocity as:

$$v = \sqrt{1/K}\sqrt{2gh_L} \qquad (1)$$

where $\sqrt{1/K} = C$, the orifice flow coefficient.

From Part 1 of this series (*Chem. Eng.*, Dec. 23, 1974, p. 64), we find the velocity-of-flow formulas:

$$v = 0.408(Q/d_o^2) \qquad (2)$$

$$v = 0.0509\,W/(d_o^2\rho)$$

By setting Eq. (1) and (2), and Eq. (1) and (3), equal to each other, we find flowrate through the orifice proportional to:

$$Q = 19.67Cd_o^2\sqrt{h_L} \qquad (4)$$

$$W = 157.66Cd_o^2\sqrt{h_L\rho^2} \qquad (5)$$

where Q, gpm, or W, lb/h, is the flowrate capable of passing through a given orifice bore (d_o, in.) with a pressure differential (expressed as the head of flowing fluid between the inlet and outlet sides of the orifice), h_L, ft.

To change Eq. (4) and (5) into a form usable for selecting the orifice bore and metering range, or for sizing pipe, the following changes are necessary:

1. Most orifice manometers (recording and transmitting instruments) are calibrated to indicate the pressure

Originally published February 3, 1975.

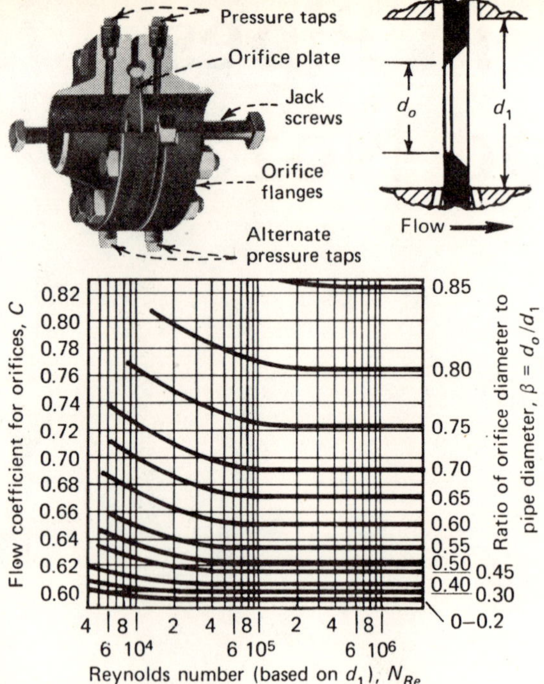

ORIFICE mounts between pair of flanges—Fig. 1

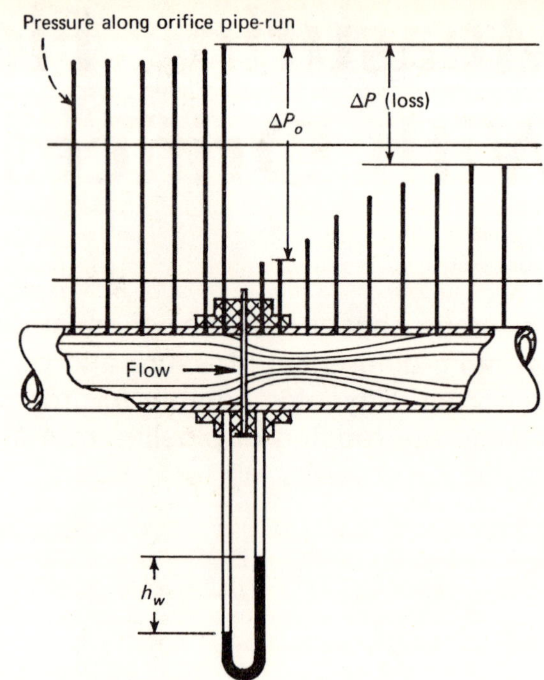

PRESSURE distribution along orifice run—Fig. 2

differential across the orifice as a head of water at 60°F, in. Thus, h_L has to be replaced by h_w in inches of water: $h_L\rho = (h_w/12)\rho_{60w}$, or:

For any fluid:

$$h_L = (h_w/12)(\rho_{60w}/\rho) \tag{6}$$

For liquids:

$$h_L = (h_w/12)(1/S) \tag{7}$$

2. For pipe sizing, the orifice diameter must be replaced by the internal diameter of the pipe. This is accomplished by using the ratio of the orifice bore (d_o) to that of the pipe inside diameter (d_1), i.e. $\beta = d_o/d_1$, or $d_o = d_1\beta$.

Inserting these values for h_L and d_o into Eq. (4) and (5) yields:

$$Q = 5.68\beta^2 C d_1^2(\sqrt{h_w}/\sqrt{S}) \tag{8}$$

$$W = 359.43\beta^2 C d_1^2\sqrt{h_w\rho} \tag{9}$$

Eq. (8) and (9) are convenient formulas for orifice pipe sizing with any chosen β ratio. A practical range is: $\beta = 0.25$ to 0.75.

(Instrument engineers multiply the righthand side of Eq. (8) by S/S_{60} if manometer indication is required at a standard 60°F liquid-flow condition. Also, they provide a more detailed evaluation of the flow coefficient. These refinements do not concern piping-design and associated fluid-flow calculations.)

When using Eq. (9) for finding the weight flow of vapor or gas, we assume that the density stays constant while the gas is flowing through a restriction. Strictly speaking, this is not true. However, the reduction in density due to a decrease in pressure can be neglected, especially if the line pressure is high compared to the pressure differential across the orifice.

Values for the orifice flow coefficient, C, are established by experiment and can be obtained from a chart such as that in Fig. 1. Up to an N_{Re} of 10,000, the flow coefficient changes greatly with varying Reynolds numbers and flow capacities. This makes for inaccuracies in flow measurements. Between an N_{Re} of 10,000 to 100,000, the flow coefficient decreases about 4% to 5% with increasing Reynolds numbers. Above an $N_{Re} = 100{,}000$, the value for C remains constant. For reasonable accuracy, pipe sizes should be selected so that $N_{Re} > 20{,}000$ for Reynolds numbers calculated with the internal diameter of the pipe.

In practical applications, β is usually 0.7 or a maximum of 0.75. The corresponding flow coefficients for $N_{Re} \geqq 100{,}000$ are $C = 0.692$ or 0.722. With these values, the capacity coefficients, $\beta^2 C$, and sizing formulas, Eq. (8) and (9), become:

For $\beta = 0.7$, $\beta^2 C = 0.339$, and:

$$Q = 1.926 d_1^2(\sqrt{h_w}/\sqrt{S}) \tag{10}$$

$$W = 121.87 d_1^2\sqrt{h_w\rho} \tag{11}$$

For $\beta = 0.75$, $\beta^2 C = 0.406$, and:

$$Q = 2.31 d_1^2(\sqrt{h_w}/\sqrt{S}) \tag{12}$$

$$W = 145.93 d_1^2\sqrt{h_w\rho} \tag{13}$$

The term h_w in Eq. (10) through (13) has two meanings. First, it is the head loss across the orifice. Expressing it as a pressure differential: $\Delta P_o = (h_w/12)(62.37/144)$, or $\Delta P_o = 0.0361 h_w$.

The permanent pressure loss of an orifice-flowmeter installation is less than the pressure differential measured across the orifice plate. As the high-velocity jet from the orifice impinges upon the slower downstream fluid, some of the jet's kinetic energy converts back to pressure. Thus, the downstream pressure becomes higher than the pressure existing at the orifice outlet. The amount of ΔP_o permanently lost is a function of the β ratio, and can be obtained from Fig. 3. For example, at $\beta = 0.7$, the permanent loss is 52% of the orifice pressure differential.

Second, h_w is the deflection of the manometer (or any other differential-pressure device) taken at the flowing condition of the fluids. For good readings or reliable operation, the selected instrument should have a scale or measuring range greater than the calculated deflection at maximum flow. At normal flow, the deflection should be roughly between one-third and two-thirds of the measuring range. Practical instrument calibrations range from 20 to 400 in; the most common is 100 in.

A liquid near its boiling point when flowing through an orifice should have a minimum of pressure drop to avoid vaporization. Sufficient positive liquid head upstream of the orifice can overcome possible vaporization. Liquid-vapor mixtures cannot be reliably measured with differential-pressure producing restrictions.

Nomenclature

C	Flow coefficient for orifice or flow nozzle
d_o	Diameter of orifice or flow nozzle, in
d_1	Inside diameter of pipe, in
g	Gravitational constant, 32.2 ft/s^2
h_L	Head loss in terms of flowing fluid, ft
h_w	Manometer deflection, or head loss across orifice [see text following Eq. (13)], in of water at 60°F
ΔP_o	Differential pressure across orifice or flow nozzle, psi
Q	Volume flowrate at flowing temperature, gpm
S	Specific gravity of liquid at flowing temperature
S_{60}	Specific gravity of liquid at 60°F
v	Mean velocity of fluid, ft/s
W	Weight flowrate, lb/h
β	Ratio of orifice (or flow nozzle) diameter to inside diameter of pipe
β^2C	Capacity coefficient for orifice or flow nozzle
μ	Viscosity, cp
ρ	Fluid density at flowing condition, lb/ft^3
ρ_{60}	Liquid density at 60°F, lb/ft^3
ρ_{60w}	Density of water at 60°F, 62.37 lb/ft^3

Flow Capacities of Orifices

Often, pipe sizes are determined before orifice sizing is done. Hence, adjustments have to be made in the design in order to get reliable flow metering. As Eq. (8) and (9) reveal, three adjustments to orifice flow capacities can be made:

1. Increase Line Size—Increasing the line size for the entire straight-run of orifice piping is the most expensive adjustment. However, this is often necessary to accommodate large flows. For piping up to 12-in dia., an increase of one pipe-size is usually made. For larger pipe diameters, an increase of two pipe-sizes is also possible. Any increase in pipe diameter should be closely followed with a check on the Reynolds number.

2. Change Manometer Range—Any change in manometer range is coupled with an altered pressure drop. The change in pressure loss must be accounted for in the overall flow-system design. If pressure differences are available in a piping system, h_w can be increased for larger flow capacities. If the available pressure difference is limited, an increase in pipe diameter and a decrease in manometer deflection might be necessary. In a pump discharge, a manometer with a high deflection might not be economical because of the high cost of utility power to overcome the permanent pressure loss across the orifice.

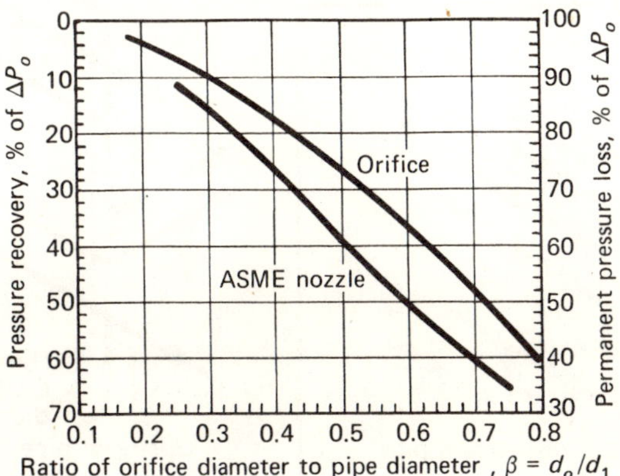

PERMANENT pressure loss through orifices—Fig. 3

The formulas for estimating orifice deflections from Eq. (8) and (9) are:

For liquid flow:

$$\sqrt{h_w} = 0.176Q\sqrt{S}/(d_1^2\beta^2C), \text{ in}^{1/2} \tag{14}$$

[If instrument deflection is calculated for a standard 60°F liquid-flow calibration, h_w is multiplied by $(S_{60}/S)^2$.]

For vapor and gas flow:

$$\sqrt{h_w} = 0.00278\,W/(d_1^2\beta^2C\sqrt{\rho}), \text{ in}^{1-2} \tag{15}$$

3. Change β Ratio—Any change in the β ratio is coupled with a corresponding value of the orifice flow coefficient. An adjustment can be made (generally to reduce capacity) by using capacity coefficients, β^2C, ranging from 0.04 to 0.4. If the ratio is less than 0.7, the orifice pressure differential and the percentage of permanent pressure loss will increase for the same flowrate.

Example Illustrates Procedures

Let us design an orifice installation for a 3-in Schedule 40 ($d_1 = 3.068$ in, $d_1^2 = 9.413$) pump-discharge line. Flow data are: $Q = 160$ gpm of kerosene, $\rho = 50$ lb/ft^3, $S = 0.8$, $\mu = 1.3$ cp, and $\beta = 0.7$. $\beta^2C = 0.339$.

First, we evaluate the Reynolds number by inserting the appropriate values into:

$$N_{Re} = 50.6(Q/d_1)(\rho/\mu)$$

$$N_{Re} = 50.6(160/3.068)(50/1.3) = 101{,}500$$

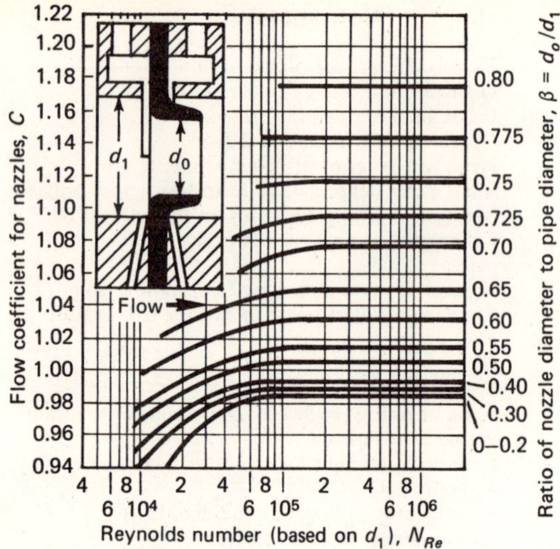

FLOW NOZZLE mounts between flanges—Fig. 4

Since the calculated Reynolds number is considerably in excess of 20,000 (the minimum value previously suggested), we find the 3-in pipe size suitable.

Next, we compute the manometer deflection by using Eq. (14) with the values for this problem:

$$\sqrt{h_w} = 0.176(160|\sqrt{0.8})/(0.339)(9.413) = 7.9 \text{ in}^{1/2}$$

$$h_w = 62.4 \text{ in.}$$

On the basis of this value for h_w, the manometer range selected is 100 in.

Since $\beta = d_o/d_1$, we can compute d_o as 0.7(3.068), or 2.15 in. The differential pressure across the orifice is calculated: $\Delta P_o = 0.0361(62.4) = 2.25$ psi. Finally, we establish the permanent pressure loss by obtaining its relation to ΔP_o from Fig. 3. For $\beta = 0.70$, 52% of actual ΔP_o is the loss. For this example, we now find ΔP for:

For a deflection of 62.4 in:

$$\Delta P = 0.52(2.25) = 1.17 \text{ psi.}$$

Deflection at assumed value of 100 in is:

$$\Delta P = (100/62)1.17 = 1.88 \text{ psi}$$

Flow Nozzles as Measuring Devices

Between line-size flanges, a flow nozzle is held in place in a manner similar to that for an orifice plate, as shown in Fig. 4. A short cylindrical section, well rounded at the inlet, provides the flow restriction. Two taps lead to the indicating, recording or transmitting instrument. The advantage of a flow nozzle is that its flow coefficient (and, consequently, flow capacity) for a given ratio is about 60% greater than that of the same size square-edged orifice. For the same flowrate and similar manometer deflection, the flow nozzle requires a smaller β ratio than does an orifice plate. Consequently, the permanent pressure loss could be roughly the same for both devices.

Flow nozzles are more expensive than orifice plates for the same pipe size. Their larger capacities, smaller lines and shorter straight length can provide cost compensation.

Because of its width, a flow nozzle is more difficult to replace than the thin, flat-plate orifice. To replace flow nozzles, the piping must be sprung well-apart where flanged and bent pipe sections are not provided.

Flow nozzles can handle liquids with high viscosities, and fluids with some entrained solids. They are suitable for high-pressure and high-temperature services, for saturated steam, and for high-velocity fluid measurements. Their application might be useful at existing installations where pipe sizes are too small for square-edged orifices.

Flow-nozzle capacity and piping are sized in the same way as the components of orifices by using Eq. (8), (9), (14) and (15), as applicable.

The Reynolds number should be equal to or greater than 50,000, a value usually attained without difficulty. Flow coefficients are readily found from the diagram in Fig. 4. The percentage of permanent pressure loss is obtained from Fig. 3. For feasibility in manufacturing, commercial sizes of flow nozzles are limited. The follow-

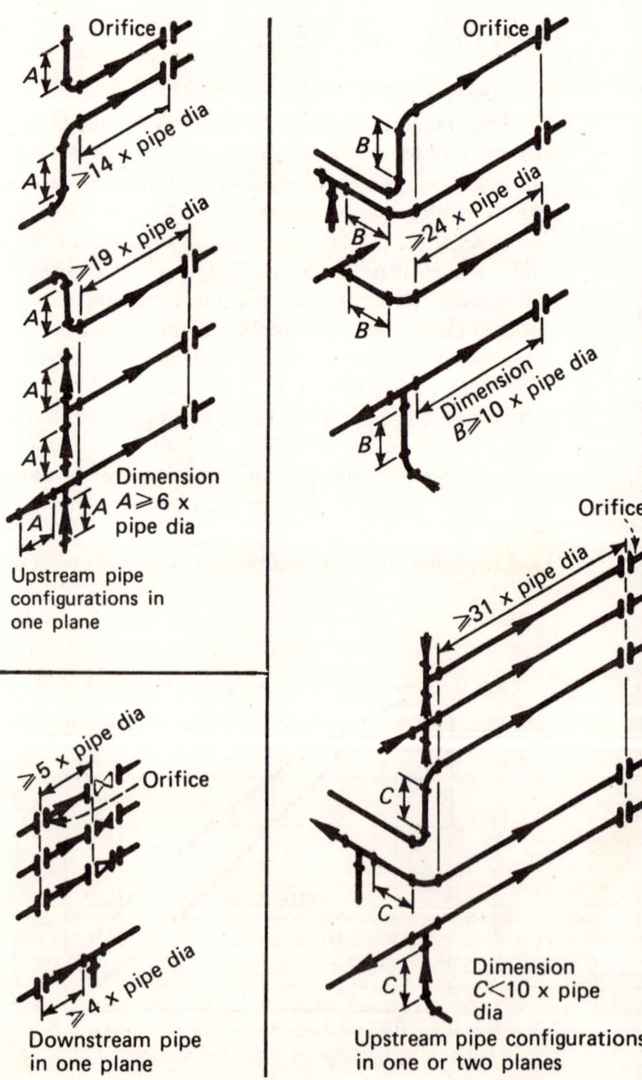

STRAIGHT-RUN needs for orifice piping—Fig. 5

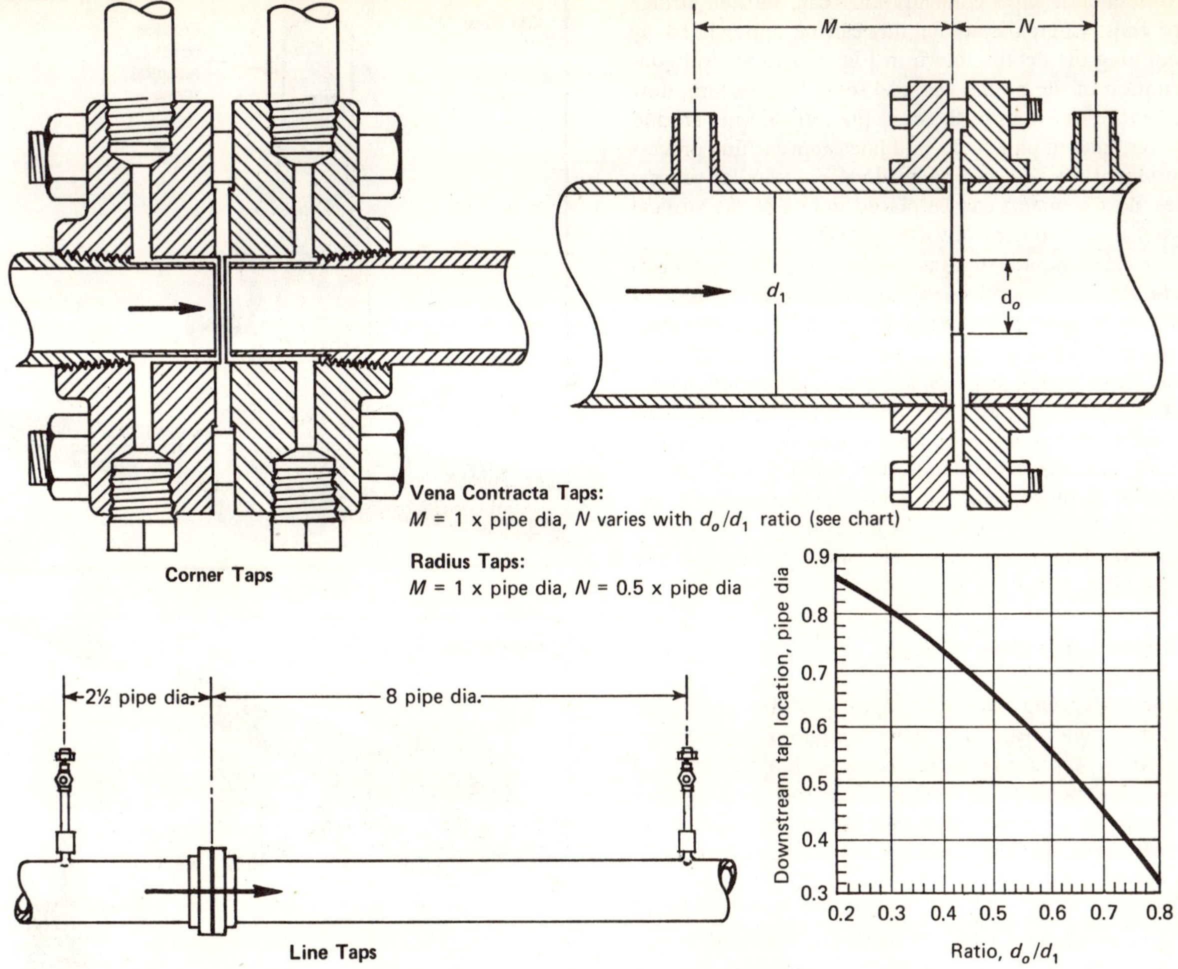

TAP TYPES that can be installed in existing pipelines without special flanges—Fig. 6

ing average β ratios and capacity coefficients β^2C can be taken for ASME nozzles:

β	0.4	0.45	0.5	0.55	0.6	0.65	0.70	0.75
β^2C	0.15	0.2	0.25	0.3	0.38	0.46	0.55	0.67

Piping Design and Pipe Configuration

For reliable, accurate and consistent flow metering, adequate straight-length of piping must be provided before and after an orifice plate, or any other differential-pressure flow-measuring element.

A straight pipe-run is more critical at the inlet side of the orifice. The straight length increases with increasing β ratio (i.e., d_o/d_1). The minimum straight length before the orifice is affected by pipe configurations and the location of valves and fittings, just before the run.

The straight-length requirements after the orifice also increase with increasing β ratio. As a conservative dimension, use five times the pipe diameter for all β ratios, as the minimum requirement. For orifices with flange taps, the minimum size for the orifice pipe diameter equals 2 in. For orifice runs smaller than 2 in, install calibrated piping.

Recommendations for straight run of piping for various piping configurations have been given by many experimenters. The American Gas Assn. (AGA)-American Soc. of Mechanical Engineers (ASME) Committee on Orifice Coefficients [*1*] has published standard arrangements for flow-meter piping. There are eight diagrams called schedules—seven are for orifices and one for venturi meters. These show piping configurations and required straight length of piping for orifices, flow nozzles and venturies.

The piping configurations of Fig. 5 are based on the AGA-ASME schedules for a d_o/d_1 ratio of 0.7. Practical orifice-piping arrangements usually fall into one of these configurations. The dimensions shown in Fig. 5 are also suitable for β ratios smaller than 0.7.

Economy of Piping Layout

Short and simple piping is desirable—expecially for large-diameter piping with heavy wall thicknesses or for expensive alloy piping. Occasionally, equipment locations, pipe connections, and predetermined distances can also influence orifice-piping dimensions. The minimum straight-length requirements are only possible if the d_o/d_1 ratio is between 0.25 to 0.40. In these cases, the piping designer should refer to the AGA-ASME schedules for minimum dimensional requirements.

Well-chosen pipe configurations can shorten orifice pipe runs. Comparisons for this can be appreciated by examining the details shown in Fig. 5. The natural configuration of the piping can also serve for inserting flow elements. For example, these are the vertical lines around a tower, lines in pipe racks, and lines approaching process equipment for yard-piping headers. For smaller process lines, flow elements can be placed in one of the vertical legs of a U-type control-valve assembly.

For clean liquid, dry gas or air, horizontal piping is preferred for the orifices or flow nozzles. For saturated steam, wet gas or air, vertical downflow is preferred; however, horizontal piping is usually also acceptable. With clean liquid, dry gas, air and superheated steam, vertical upflow can also be considered. For liquids containing suspended solids, the orifices or flow nozzles should have large diameters, and piping should be in downflow arrangement. For good pipeline drainage, eccentric orifices can be used in horizontal slurry lines—thus providing a common low point for the pipe and orifice bore.

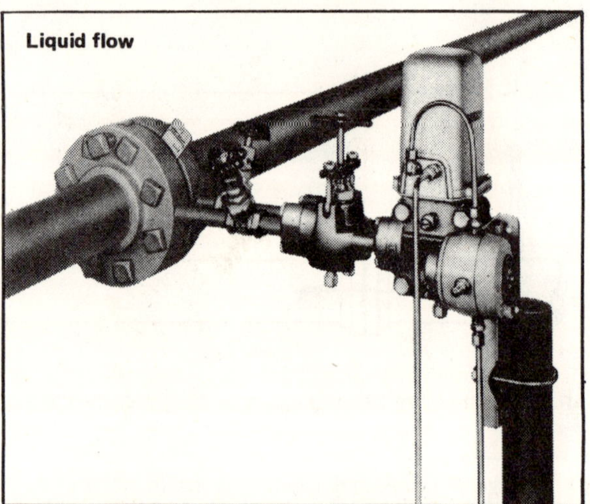

DIFFERENTIAL pressure cell mounts nearby to minimize length of interconnecting tubing—Fig. 7

Locations for Orifice Taps

For measuring the differential pressure across an orifice, a high-pressure (upflow) tap and a low-pressure (downflow) tap are provided. Tap sizes range from $\frac{1}{4}$ in to $\frac{3}{4}$ in, depending on nominal size and flange rating. Taps can be in a horizontal position for liquids; and in the case of horizontal pipe, in a vertical upward position for vapors.

In most cases, taps are located in the flow-meter flanges, as shown in Fig. 1, for lines 2 in. and larger. Flange taps are the least sensitive to viscosity changes.

Corner, vena-contracta, radius and line taps are shown in Fig. 6. Corner taps connect to the corner of the inside wall of an orifice plate. These taps are used in orifice flanges in lines smaller than 2 in. Vena-contracta taps are located in the pipeline. The high-pressure point is located one pipe diameter upstream, and the low-pressure point is at the minimum pressure point. This varies with the d_o/d_1 ratio. Dimensions can be obtained from the diagram in Fig. 6. Vena-contracta taps give the largest manometer deflection. Radius taps are a close approximation of vena-contracta taps. Line taps are 2.5 pipe diameters upstream (high-pressure point), and 8 pipe diameters downstream (low-pressure point). Line taps give the smallest deflection.

The advantage of vena contracta, radius and line taps is that they can be installed in existing pipelines, and the orifice plate can be placed between standard flanges. Straight length of piping should be measured from the taps. The usual tap size is $\frac{1}{2}$ in.

Separator Chambers

Dirty liquids, moist or condensing gases and corrosive fluids require that separation chambers or driplegs be installed between the orifice tap and the manometer or pressure transmitter. These chambers are closely mounted to both orifice taps and instruments. Sediment chambers with drain valves collect solids suspended in liquids, or moisture carried with noncondensing gases. Air chambers with vent valves, installed at high points, collect air entrained in liquids. Condensing chambers are used in steam service. Sealing chambers between orifice taps and instruments separate corrosive chemicals from contact with the instrument components.

Accessibility to Instruments

The piping designer must consider access and space requirement to orifice-tap valves and to instruments connected to orifices.

The minimum elevation of orifice runs is approximately 2 to 2.5 ft above grade. Where heavy snow fall is common at outdoor installations, a higher minimum elevation is chosen. The recommended elevation for piping with orifice flanges is 7 ft above grade or platform elevation. This is the case for orifice runs in pump-discharge lines, and in exchanger inlet and outlet lines.

In pipe racks, straight runs are easily provided. Hence,

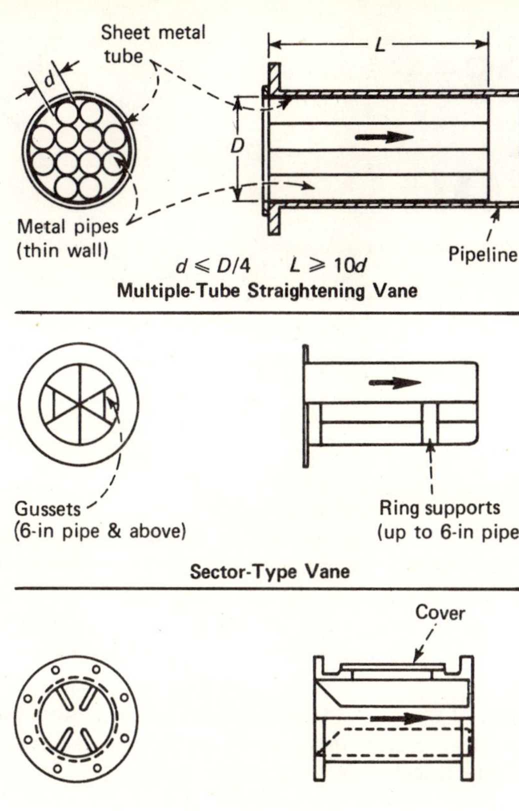

STRAIGTENING VANES shorten straight run—Fig. 8

orifice flanges in the piping are arranged at the edge of rack about 1.5 to 2 ft from supporting columns. If two or three pairs of orifice flanges are grouped side by side (or in two levels) in yard piping, the minimum horizontal or vertical distance between orifice flanges should be about 2 to 3 ft. Orifice flanges placed in the center of pipe runs between pipe supports can be a source of pipe vibration.

Where only metering flanges and taps are provided and only occasional flow indication is needed, access by portable ladder is sufficient.

Locally mounted indicating and measuring flowmeters are more frequently inspected. If necessary, permanent platform and ladder access is provided. In this category are instruments for measuring flow in process feed lines, product lines, and utility lines.

With automatic flow control, there is permanent instrument wiring and tubing between the measuring element, transmitter, recorder, controller and the control valve. These components should be closely arranged where possible. Flange or line taps are permanently open. Reasonable temporary access to these valves is sufficient.

Locally mounted indicating, recording and transmitting instruments should be visible from the operating aisle. The inclined type of U-tube manometer offers a more compact unit vertically. Dial-type indicators and recorders, calibrated in flowrates, are also available. These can be locally and remotely mounted.

For pressure transmitting, an often-used device is the differential-pressure cell. This instrument is mounted in the proximity of the orifice flanges (Fig. 7) in an accessible location.

Differential-pressure cells and manometers should be located relative to orifice flanges so that interconnecting tubing can be provided without a loop or pocket. A loop must be vented, and a pocket can collect sediment. This can affect trouble-free instrument operation.

A flow controller is also usually incorporated in an instrument loop. This controller is often mounted on a support at grade or platform elevation, and reasonably close to the orifice and control valve. Space requirement is about 2 ft square and 3.5 ft high. Transmitters and controllers should be accessible.

Straightening Vanes

The straight-length requirements enable the development of a symmetrical velocity pattern and steady flow in the moving mass of fluid. Hence, when the fluid meets the flow restriction, stable measuring conditions are present. The same flow conditions may also be developed with straightening vanes, which require a shorter straight run (Fig. 8). Straightening vanes work well while they are clean and new, and most important, if piping is suitably designed and the vanes well positioned. Even slight corrosion, erosion or deposits will hamper their function, and hence measurements will be inaccurate.

If a straightening vane is preceded by an elbow, it might defeat its own purpose. The distorted velocity distribution developed in the elbow is captured in the straightening vane and only slightly corrected in the short upstream orifice pipe-run. A distorted flow pattern will result through the orifice, and a true value of pressure difference will not likely exist. By replacing the elbow with a tee (capped at one end), a more symmetrical velocity distribution can be obtained than by a straightening vane preceded by an elbow. Manufacturers of these devices can recommend proper installation practices. Location and dimensional details are given in the AGA-ASME schedules [*1*].

References

1. Sprenkle, R. E., Piping Arrangements for Acceptable Flow Meter Accuracy, *Trans. ASME,* **67,** 345 (1945).
2. Terrell, C. E. and Bean, H. S., "AGA Gas Measurement Manual," American Gas Assn., Arlington, Va., 1963.

The author

Robert Kern is head of the plant-layout section in the corporate engineering department of Hoffmann-La Roche Inc., Nutley, NJ 07110. He is a specialist in hydraulic-systems design, plant layout, piping design and economy. He is the author of a number of articles in these fields, and has taught several courses for the design of process piping, plant layout, graphic piping and flow systems, both in the U.S. and South America. Previously, he was associated with M. M. Kellogg Co. in England and the U.S. Mr. Kern has an M.S. in mechanical engineering from the Technical University of Budapest, and is a member of AIChE.

Designing Orifices For Flow Regulation

Here is how you can calculate orifices for permanent pressure drop—by hand, with a time-shared computer program, or by making your own program.

FRANK G. McNULTY, General Electric Co.

Most chemical engineering experience with orifices is probably limited to their use as flow-measuring devices. However, orifices hold the potential for another important industrial application—flow regulation.

In this service, the sharp-edged orifice has advantages over a valve, since there are no moving parts, and the orifice is tamper- or adjustment-proof. Flow-reducing or flow-regulating orifices may be used with either gases or liquids. Generally, their only limitation is that of line size; they are not normally used in lines smaller than 1.0 in. I.D., since the plate thickness should not exceed 5% of the orifice diameter.

The distinction between an orifice as flow regulator and an orifice as flow measurer is the partial recovery of fluid pressure downstream of the orifice, as illustrated in Fig. 1. Flow-regulating orifices are sized for the permanent loss in pressure that is experienced more than 8 pipe diameters downstream, as is done for flow-measuring orifices when these are equipped with pipe taps (Fig. 1).

SHARP-EDGED ORIFICES: Relation of piping locations to static pressure in the pipe—Fig. 1

Originally published November 12, 1973.

To install a flow-regulating orifice, the engineer chooses a piping configuration that will allow at least 100 diameters of straight pipe upstream and 25 diameters downstream, so that the orifice's characteristics will not be affected. He notes the average temperature in this section of piping and, with the piping diameter, the characteristics of the fluid, the operating flowrates, and the desired pressure drop, there are enough known values for a slide-rule calculation.

Unfortunately, this calculation must be done by trial-and-error and is somewhat lengthy. The relation between

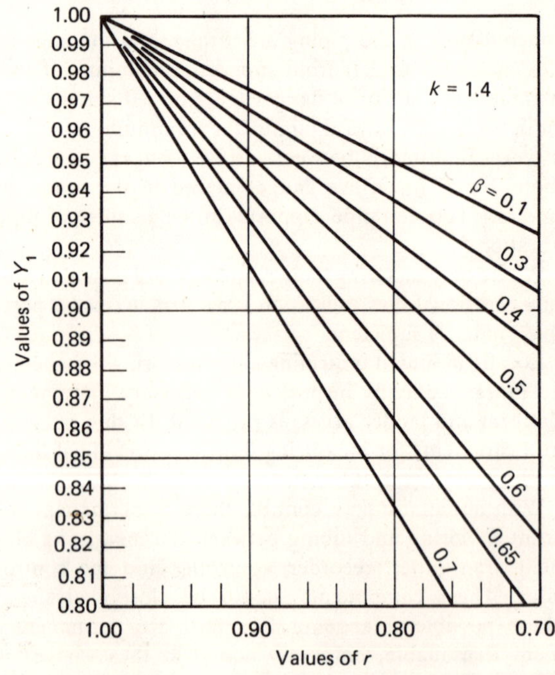

EXPANSION FACTORS for sharp-edged orifices when these are equipped with pipe taps (k = 1.4)—Fig. 2

```
00 'A FØRTRAN PRØGRAM TØ SIZE DIAMETERS ØF ØRIFICE TYPE FLØW REDUCERS
01 'BASED ØN PIPE TAP EQUATIØNS FRØM ASME PUBLICATIØN "FLUID METERS".
02  'FILL IN DATA FØR LINES 04 THRU 10.  FRANK MCNULTY AUG. 1970.
04   D=1.38              'PIPE I.D., INCHES
05   P1=25.23            'STATIC PRESSURE UPSTREAM ØF ØRIFICE, PSIA
06   P2=20.18            '   "        "   DØWNSTREAM "     "  ,  "
07   W=0.125             'FLØW RATE, PØUNDS PER SECØND
08   T1=225.             'TEMP. AT P1, DEG. F
09   XMU=0.053           'VISCØSITY AT T1, LB./HR.`FT.
10   RATIØ=1.4           'RATIØ ØF SPECIFIC HEATS, C(P)/C(V)
11   P=P2/P1
12   IF (P-.53) 200, 200, 7   'CRITICAL PRESS. RATIØ .53 IS FØR AIR
13   7  T=T1+459.67
14   RHØ=2.702*P1/T               'CALCULATE DENSITY
15   Q=W/RHØ
16   AREA=0.7854*((D/12.)**2)
17   VEL=Q/AREA                   'CALCULATE VELØCITY
18   REN=(RHØ*VEL*(D/12.)*3600.)/XMU    'CALCULATE REYNØLDS NØ.
19   HW=27.684*(P1-P2)
20  YKD2=W/(SQRT(RHØ*HW)*0.0997)
21  X=(P1-P2)/P1
22  B=0.7                         'ASSUME BETA RATIØ
23  KØUNT=0
36  IF (D-2.036) 9, 8, 8
37  8  DP=D
38  GØ TØ 10
39  9  DP=2.067
43  10  DØ=B*D
47  Y=B*B+0.7*B**5                'CALCULATE EXPANSIØN FACTØR, Y
48  Y=Y+12.*B**13
49  Y=Y*1.145+0.333
50  Y=Y*X/RATIØ
51  Y=1.-Y
57  XKE=0.5925+0.0182/DP          'CALCULATE FLØW CØEFFICIENT, K
58  XKE=XKE+(.44-(.06/DP))*B*B
59  XKE=XKE+(.935+.225/DP)*B**5
60  XKE=XKE+1.35*B**14
63  S=0.25-B
65  IF (S) 13, 13, 12
70  12 XKE=XKE+(1.43/SQRT(DP))*(S**2.5)
75  13 CØNTINUE
80  A=DØ*(905.-5.E3*B+9.E3*B*B-4200.*B**3+875./DP)
85  XKØ=XKE*((1.E6*DØ)/(1.E6*DØ+15.*A))
90  XK=XKØ*(1.+(B*A/REN))
91  V=XK*Y
92  Z=YKD2/V
93  DØ=SQRT(Z)                    'CALCULATE ØRIFICE DIAMETER
94  B1=DØ/D
109  KØUNT=KØUNT+1
111  IF (KØUNT-10) 28, 28, 165   'CHECK NUMBER ØF ITERATIØNS
112  28  CØNTINUE
115  DIFF=B-B1    'CALCULATE DIFFERENCE BETWEEN ØLD AND NEW BETA RATIØS
119  EPS=0.0025
120  IF (ABS(DIFF)-EPS) 100, 100, 50
125  50 B=B1
141  GØ TØ 10
180  100 PRINT 120, DØ
181  120 FØRMAT (20X, "ØRIFICE DIAM., INCHES=", F8.4/)
190  PRINT 121, Y, XK, B1
191  121 FØRMAT (12X, "Y=", F6.4, 8X, "K=", F6.4, 8X, "BETA=", F6.4/)
195  PRINT 130
196  130 FØRMAT (12X, "FLØW CHARACTERISTICS UPSTREAM ØF ØRIFICE ARE:")
197  PRINT 135, VEL, RHØ
198  135 FØRMAT (12X, "VEL, FT/SEC=", F5.2, 10X, "LB/CU FT=", F7.4)
199  PRINT 137, REN, Q
200  137 FØRMAT (12X, "REYNØLDS NØ.=",F10.0,6X,"CU FT/SEC=",F7.3/)
203  PRINT 140, HW
204  140 FØRMAT (9X, "PRESS DRØP ACRØSS ØRIFICE, IN. ØF WATER=", F6.2)
210  165 PRINT 170, KØUNT
211  170 FØRMAT (20X, "NØ. ØF ITERATIØNS=", I3)
212  GØ TØ 205
214  200 PRINT 201
216  201 FØRMAT ("PRESS. DRØP EXCEEDS CRITICAL RATIØ.")
219  205 CØNTINUE
220  STØP
221  END


RUN

ØRF      12:18  05TUES 09/25/73

                    ØRIFICE DIAM., INCHES=    .7081

            Y= .9054          K= .7403          BETA= .5131

            FLØW CHARACTERISTICS UPSTREAM ØF ØRIFICE ARE:
            VEL, FT/SEC= 120.87            LB/CU FT=  .0996
            REYNØLDS NØ.=     94004.       CU FT/SEC=  1.255

         PRESS DRØP ACRØSS ØRIFICE, IN. ØF WATER= 139.80
                    NØ. ØF ITERATIØNS=  4
```

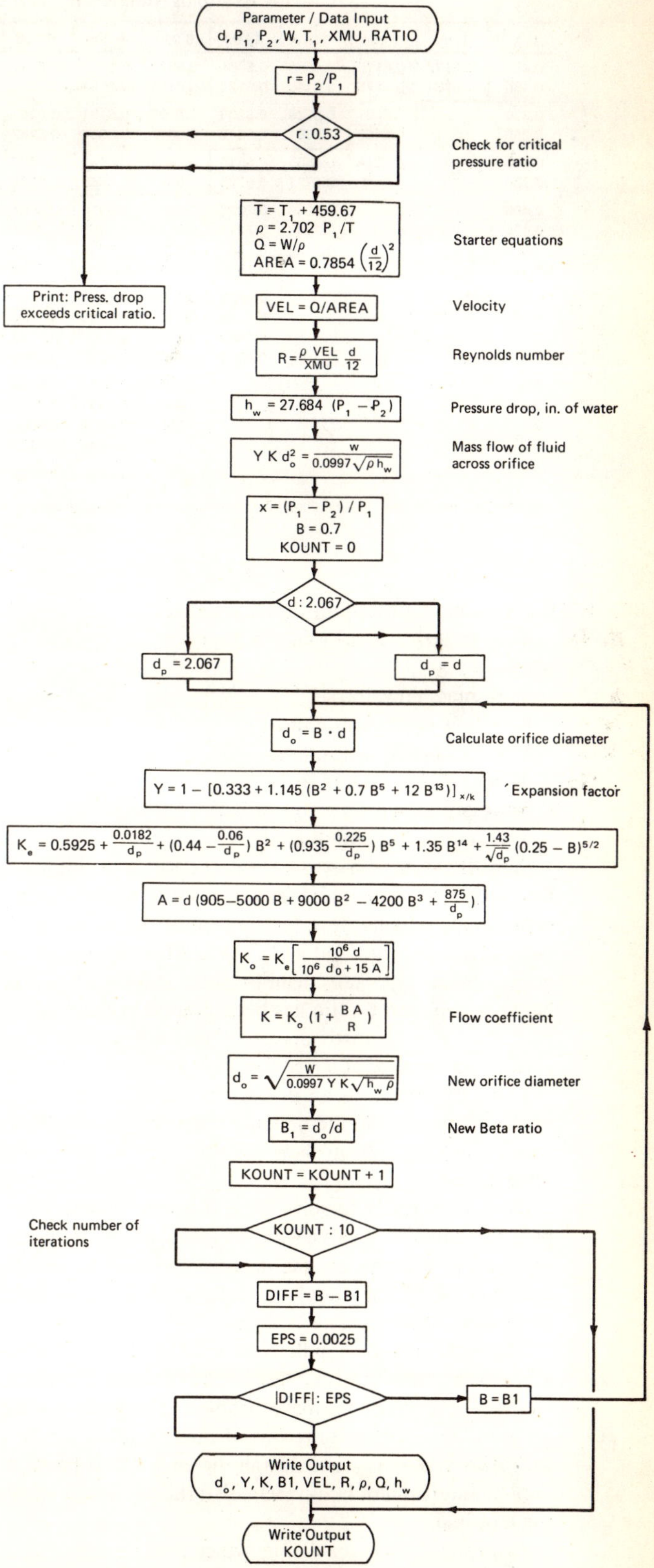

Parameter / Data Input
d, P_1, P_2, W, T_1, XMU, RATIO

$r = P_2/P_1$

$r : 0.53$ — Check for critical pressure ratio

Print: Press. drop exceeds critical ratio.

$T = T_1 + 459.67$
$\rho = 2.702\ P_1/T$
$Q = W/\rho$
$AREA = 0.7854 \left(\frac{d}{12}\right)^2$ — Starter equations

$VEL = Q/AREA$ — Velocity

$R = \frac{\rho\ VEL}{XMU}\ \frac{d}{12}$ — Reynolds number

$h_w = 27.684\ (P_1 - P_2)$ — Pressure drop, in. of water

$Y K d_o^2 = \frac{W}{0.0997\sqrt{\rho h_w}}$ — Mass flow of fluid across orifice

$x = (P_1 - P_2)/P_1$
$B = 0.7$
KOUNT = 0

$d : 2.067$

$d_p = 2.067$ | $d_p = d$

$d_o = B \cdot d$ — Calculate orifice diameter

$Y = 1 - [0.333 + 1.145\ (B^2 + 0.7\ B^5 + 12\ B^{13})]_{x/k}$ — Expansion factor

$K_e = 0.5925 + \frac{0.0182}{d_p} + (0.44 - \frac{0.06}{d_p})\ B^2 + (0.935\ \frac{0.225}{d_p})\ B^5 + 1.35\ B^{14} + \frac{1.43}{\sqrt{d_p}}\ (0.25 - B)^{5/2}$

$A = d\ (905 - 5000\ B + 9000\ B^2 - 4200\ B^3 + \frac{875}{d_p})$

$K_o = K_e \left[\frac{10^6\ d}{10^6\ d_o + 15\ A}\right]$

$K = K_o\ (1 + \frac{B\ A}{R})$ — Flow coefficient

$d_o = \sqrt{\frac{W}{0.0997\ Y\ K\sqrt{h_w\ \rho}}}$ — New orifice diameter

$B_1 = d_o/d$ — New Beta ratio

KOUNT = KOUNT + 1

KOUNT : 10 — Check number of iterations

DIFF = B − B1

EPS = 0.0025

|DIFF|: EPS

B = B1

Write Output
d_o, Y, K, B1, VEL, R, ρ, Q, h_w

Write Output
KOUNT

PRINTOUT (left) and LOGIC DIAGRAM (right) of time-shared program for calculating flow-regulating orifices—Fig. 3

Values of the Flow Coefficient, K, Velocity-of-Approach Factor Included, as a Function of the Pipe Reynolds Number, R_e, and Diameter Ratio, β, for 2-in. Pipe – Table I

β \ R_e	1,000	2,000	3,000	4,000	5,000	6,000	8,000	10,000	15,000	20,000	25,000	30,000	50,000	100,000	500,000	10^6
0.100	0.6172	0.6116	0.6096	0.6086	0.6081	0.6077	0.6072	0.6070	0.6065	0.6064	0.6063	0.6062	0.6060	0.6059	0.6058	0.6057
0.150	0.6284	0.6176	0.6140	0.6122	0.6118	0.6104	0.6095	0.6090	0.6083	0.6079	0.6077	0.6075	0.6072	0.6070	0.6069	0.6068
0.200		0.6291	0.6236	0.6209	0.6192	0.6181	0.6167	0.6159	0.6148	0.6143	0.6139	0.6137	0.6132	0.6129	0.6127	0.6126
0.250		0.6458	0.6381	0.6342	0.6319	0.6304	0.6282	0.6273	0.6257	0.6250	0.6245	0.6242	0.6236	0.6231	0.6229	0.6227
0.300		0.6666	0.6565	0.6513	0.6482	0.6461	0.6436	0.6420	0.6399	0.6389	0.6383	0.6379	0.6370	0.6364	0.6259	0.6258
0.350			0.6797	0.6728	0.6687	0.6660	0.6626	0.6605	0.6577	0.6564	0.6558	0.6550	0.6539	0.6530	0.6524	0.6523
0.400			0.7096	0.7003	0.6947	0.6910	0.6864	0.6836	0.6799	0.6781	0.6769	0.6762	0.6747	0.6736	0.6727	0.6726
0.425			0.7280	0.7173	0.7111	0.7065	0.7011	0.6979	0.6935	0.6914	0.6901	0.6892	0.6875	0.6862	0.6851	0.6850
0.450			0.7488	0.7362	0.7286	0.7236	0.7173	0.7135	0.7085	0.7060	0.7045	0.7035	0.7014	0.6999	0.6987	0.6986
0.475				0.7579	0.7491	0.7432	0.7358	0.7314	0.7255	0.7226	0.7208	0.7196	0.7172	0.7154	0.7140	0.7138
0.500				0.7829	0.7724	0.7667	0.7567	0.7515	0.7445	0.7410	0.7390	0.7375	0.7347	0.7327	0.7310	0.7308
0.525				0.8114	0.7991	0.7909	0.7806	0.7745	0.7663	0.7622	0.7597	0.7581	0.7548	0.7523	0.7503	0.7501
0.550				0.8444	0.8298	0.8200	0.8079	0.8006	0.7909	0.7860	0.7831	0.7812	0.7773	0.7744	0.7720	0.7717
0.575				0.8815	0.8645	0.8530	0.8387	0.8301	0.8187	0.8130	0.8095	0.8072	0.8027	0.7992	0.7965	0.7961
0.600					0.9044	0.8909	0.8704	0.8638	0.8503	0.8435	0.8395	0.8368	0.8314	0.8273	0.8241	0.8237
0.625					0.9502	0.9341	0.9142	0.9023	0.8863	0.8783	0.8735	0.8703	0.8639	0.8592	0.8553	0.8548
0.650					1.0025	0.9837	0.9602	0.9461	0.9273	0.9179	0.9122	0.9085	0.9009	0.8953	0.8908	0.8902
0.675						1.0446	1.0158	0.9986	0.9756	0.9641	0.9572	0.9526	0.9434	0.9366	0.9310	0.9303
0.700						1.1068	1.0742	1.0544	1.0281	1.0150	1.0071	1.0019	0.9914	0.9835	0.9772	0.9764

the mass flowrate and the orifice diameter is:

$$W = 0.0997\ Y\ K\ d_0^2\ (\rho h_w)^{1/2} \quad \text{Eq. (1)}$$

where:

W = mass flow, lb./sec.

K = flow coefficient, ratio.

Y = expansion factor, ratio.

d_0 = orifice diameter, in.

h_w = pressure drop across orifice, in. of water.

ρ = density of fluid, lb./cu.ft.

Since the desired flow is known, the Reynolds number, Re, can be calculated; and for various Res the K ratio has been tabulated against β, the ratio of orifice diameter to pipe I.D. (see e.g., Table I).[1] Also, for square-edged orifices with pipe taps handling air (ratio of specific heats = 1.4), the Y ratio has been plotted against values of r, the ratio of upstream-to-downstream pressures, at various parameters of β (see Fig. 2).[1]

Thus, by assuming an orifice diameter, the engineer obtains β; by setting the pressure drop across the orifice, he obtains r; and from these plus the Re, he can determine K and Y.

He is then ready to solve Eq. (1) for d_0, and to compare this calculated value with his assumed diameter. He repeats the assumption, the determination of K and Y, and the calculation, until the calculated d_0 is satisfactorily close to the assumed value.

This work may be handled easily with a computer program, using time sharing. Besides the obvious advantage of quickly solving the above problem, the time-sharing program gives a concise record of input and output data for future reference; and though the engineer at his desk may be interrupted, he is rarely disturbed at a time-sharing terminal.

Example: A flow-regulating orifice is to be installed in an air line, a 1.25-in. schedule 40 pipe (1.380 in. I.D.). Air flow is 0.125 lb./sec. The supply-air temperature is 225 F. The supply-air pressure and the required terminal pressure set the pressures upstream and downstream of the orifice at 25.23 psia. and 20.18 psia., respectively. What diameter orifice is required to maintain this pressure drop under these conditions of flow?

The solution by a FORTRAN program and the logic diagram for this program are shown in Fig. 3.

Input data are entered in lines 04 through 10, in the units indicated (Fig. 3). Lines 93, 94 and 115-20 show that the next-assumed orifice diameter is based on the difference between the last-assumed and the calculated values. A count is kept (line 111) of the number of iterations needed to arrive at the answer. Nearly all orifice problems are solved after three or four iterations. Occasionally, bad input data will cause a higher number of iterations. If the number exceeds 10, the program is terminated.

The program contains a check for critical pressure drop across the orifice (line 12). Since a gas will no longer expand across the orifice once this value (r_c for air equals 0.53) is exceeded, the critical flowrate through the orifice is a maximum, independent of downstream pressure. Sizing a critical-flow orifice, which requires less slide-rule time,[1, 4] is the subject of a separate computer program.

In addition to computing a required orifice diameter, the program given here may be used in cases where the orifice diameter is known but the air flowrate is unknown. A diameter can be calculated for a number of assumed flowrates, and a plot of diameter versus flowrate used to interpolate the flowrate corresponding to any given diameter for the system.

References

1. "Fluid Meters–Theory and Application," 5th ed., Amer. Soc. of Mechanical Engrs.
2. Perry, J. H., ed., "Chemical Engineers' Handbook," Third ed. (1950), McGraw-Hill, pp. 400-408.

Meet the Author

Frank G. McNulty is employed in the design engineering section of the Gas Turbine Div., General Electric Co., Schenectady, N.Y. A licensed professional engineer in the State of New York, he holds a B.S. in marine engineering (1952) from the U.S. Merchant Marine Academy, Kings Point, N.Y., and an M.M.E. degree (1960) from City College of New York.

Estimating pipe sizes for orifice runs

William T. Klapper, *Bloomfield, N.J.*

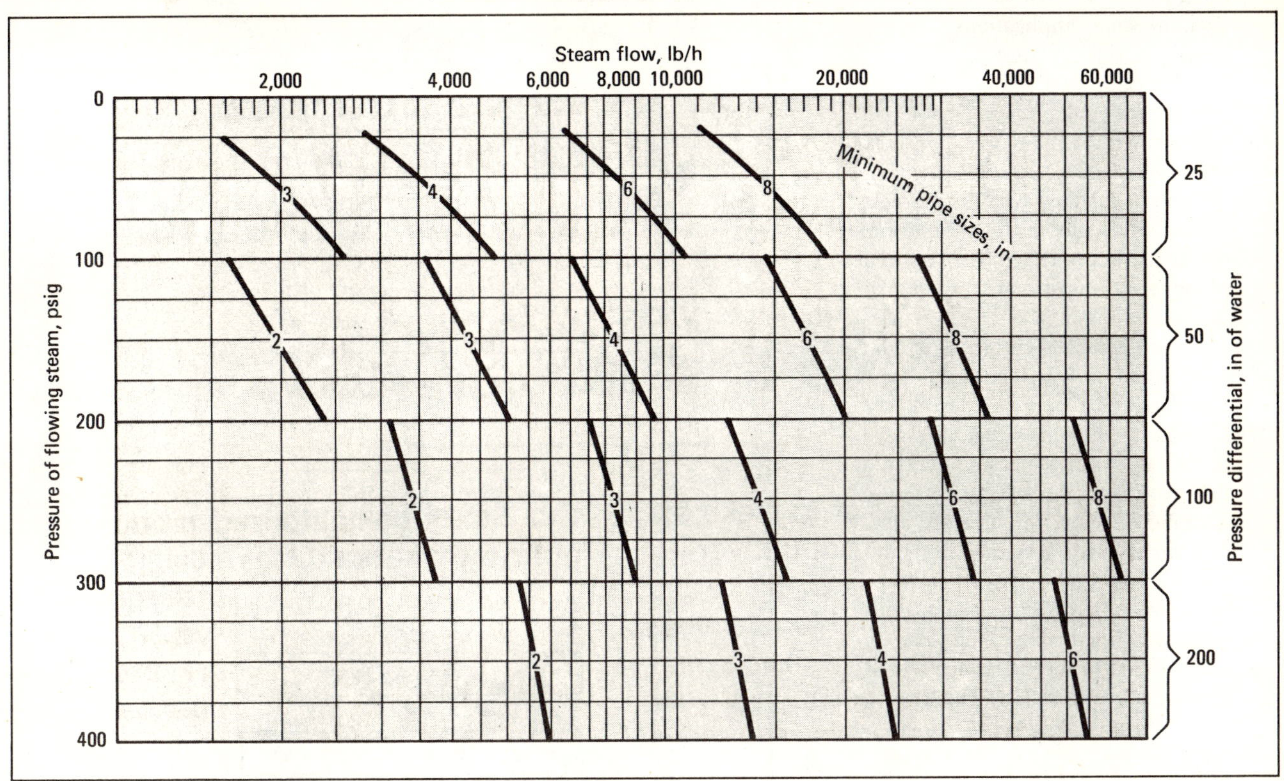

☐ This chart should simplify the iterative procedure of sizing orifice runs for measuring steam flows. In the normal design procedure, the engineer uses operating data to choose a pressure differential that will give a suitable range for control and readings, and calculates the size of the orifice. Since this calculation involves the ratio of orifice-diameter to pipe-inside-diameter, or beta ratio, the pipe sizes must be assumed before the beta ratio can be calculated.*

If the assumed pipe size of the orifice run is too small, the beta ratio may be too large for accuracy, whereas too large a pipe is uneconomical.

Four ranges of pressure differential (25, 50, 100 and 200 inches of water) are given corresponding to different line-pressure ranges, with differentials other than these reserved for capacity increase or calibration. In addition, it is assumed that:

*For discussions of the procedure, see Kern, Robert, Measuring Flow in Pipes with Orifices and Nozzles, p. 133.

Originally published August 4, 1975.

- the beta ratio is 0.7—a maximum ordinarily used for reliable metering.
- the element is a standard sharp-edged, thin-plate concentric orifice plate used with tapped orifice flanges.
- the steam is saturated.
- the pipe is schedule 40.

Also, the standard criteria, such as the required downstream and upstream runs, are assumed.

Example:

What pipe size is suitable for the orifice meter run for 300 psia steam flowing at 20,000 lb/h? A conservative design would employ a 6-in pipe with a differential pressure range of 100 in of water. This pipe size would permit up to 34,000 lb/h at 100 in or up to 48,000 lb/h at 200 in. A less expensive design would use a 4-in pipe for up to 22,000 lb/h at 200 in of pressure differential.

Designing critical flow orifices

FRANK McNULTY,
General Electric Corp.,
Schenectady, N.Y.

Sharp-edged orifices are frequently used as flow-regulating devices, and a computer program has been developed for sizing such orifices.* However, in some applications, the imposed pressures require that a critical flow orifice be used; and in these cases the conventional relations do not hold between orifice diameter, flow, and pressure.

The critical flow condition occurs when the velocity through the throat of a nozzle or orifice reaches the speed of sound at the upstream temperature. At this condition, nozzles and orifices with well-rounded entrances exhibit a maximum flow independent of the downstream pressure, whereas the square-edged orifices generally used in piping systems permit increased flow even as the downstream pressure is reduced below the critical value.

A time-shared computer program has been developed to calculate flows for sharp-edged orifices in the critical-flow region by means of a correction applied to the theoretical equation for maximum flow, which is:

$$W_{max} = A_o p_1 \sqrt{\frac{g_c k}{RT_1}\left(\frac{2}{k+1}\right)^{k+1/k-1}}$$

Assuming the gas is air, with $R = 53.3$ and $k = 1.4$, this equation simplifies to:

$$W_{max} = \frac{A_o p_1 0.5318}{\sqrt{T_1}}$$

(The constant 0.5318 for air in this equation can be replaced by the corresponding constants for use with other gases.)

This simplified equation for maximum flow is adapted to actual flow through sharp-edged orifices by a flow-coefficient, K_c, and a velocity approach factor, $1/\sqrt{1-\beta^4}$. Experimental values of K_c versus upstream-to-downstream pressure ratio are available for air, with the velocity-of-approach factor at zero.† A polynomial was fitted to these ex-

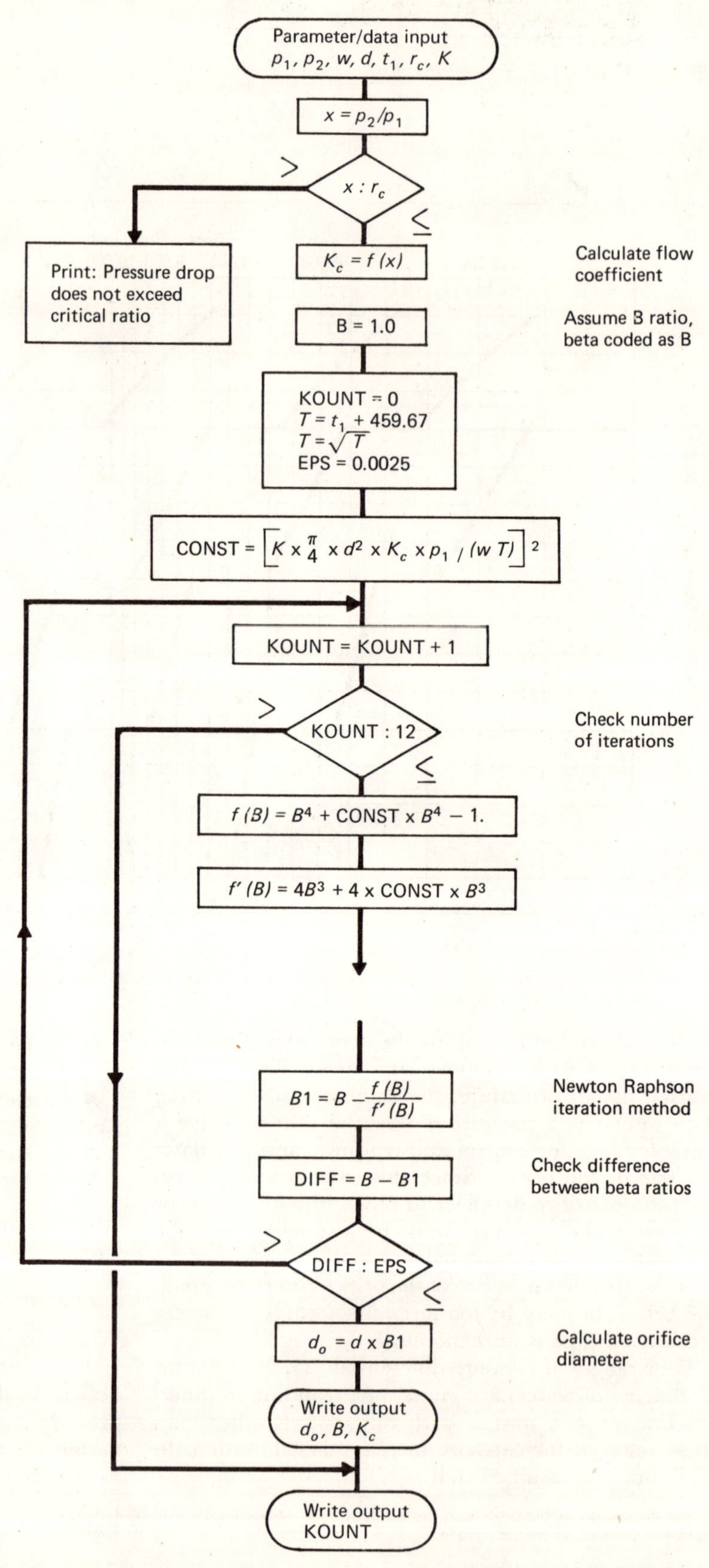

*See p. 140.

†See Perry, J. A., Jr., Critical Flow Through Sharp-Edged Orifices, *Trans. ASME*, **71**, 7, pp. 757–764.

Originally published August 5, 1974.

```
01  'A FORTRAN PROGRAM TO SIZE A SHARP EDGED CRITICAL FLOW ORIFICE.
02  'BASED ON THE MASS FLOW EQUATION WHICH IS MULTIPLIED BY
03  'A DISCHARGE COEFFICIENT AND A VELOCITY OF APPROACH FACTOR.
04  'FILL IN DATA FOR LINES 20 THRU 26 IN THE UNITS REQUIRED. FGM 7/73
20  P1=41.38                     'PRESSURE UPSTREAM OF ORIFICE, PSIA
21  P2=17.0                      "PRESSURE DOWSTREAM OF ORIFICE, PSIA
22  W=3.88                       'WEIGHT FLOW, POUNDS PER SECOND
23  D=6.137                      'PIPE INSIDE DIAM., INCHES
24  T1=310                       'AIR TEMPERATURE, DEGREES F.
25  RC=0.528                     'CRITICAL PRESSURE RATIO FOR AIR
26  XK=0.5318                    'K FACTOR FOR AIR
40  X=P2/P1
44  IF (X-RC) 10, 10, 75         'CHECK IF PRESSURE RATIO IS CRITICAL
54  10 XKC=0.835808              'CALCULATE FLOW COEFFICIENT
55  XKC=XKC+0.194516*X
58  XKC=XKC-1.02087*X*X
60  XKC=XKC+0.592086*X**3
70  B=1.0                        'ASSUME BETA RATIO
75  KOUNT=0
76  T=T1+459.67
77  T=SQRT(T)
78  EPS=0.0025
79  CONST=XK                     'GROUP PARAMETERS OF PROBLEM INTO
80  CONST=CONST*0.7854*D*D           'ONE CONSTANT TERM
82  CONST=CONST*XKC*P1
84  CONST=CONST/(W*T)
86  CONST=CONST*CONST
90  20  KOUNT=KOUNT+1
92  IF (KOUNT-12) 22, 22, 68    'CHECK NUMBER OF ITERATIONS
100 22 FB=B**4                   'USE NEWTON RAPHSON ITERATION METHOD
102 FB=FB+(FB*CONST)-1.
110 FPB=B**3
112 FPB=4.*FPB
114 FPB=FPB+(FPB*CONST)
120 A=FB/FPB
122 B1=B-A                       'NEW BETA RATIO
130 DIFF=B-B1                    'CHECK DIFFERENCE BETWEEN BETA RATIOS
132 IF (ABS(DIFF)-EPS) 40, 40, 30
140 30 B=B1
142 GO TO 20
144 40 DO=D*B1
150 PRINT 50, DO, B1
152 50 FORMAT(9X,"ORIFICE DIAM., INCHES =",F8.4,9X,"BETA =",F5.4//)
160 PRINT 55, D, T1
162 55 FORMAT(9X,"PIPE INS. DIAM., IN. =",F6.3,9X,"DEG. F=",F6.2/)
170 PRINT 60, W, XKC
172 60 FORMAT(9X,"AIR FLOW, LB/SEC =",F6.3,9X,"FLOW COEFF. =",F4.3/)
180 PRINT 65, P1, P2
182 65 FORMAT(9X,"PSIA INTO ORIFICE=", F6.3,8X,"OUT OF ORIFICE=", F6.3/)
190 68  PRINT 70, KOUNT
192 70  FORMAT (25X, "NO. OF ITERATIONS = ", 13)
197 GO TO 80
200 75 PRINT 76
202 76 FORMAT(19X,"PRESS. DROP DOES NOT EXCEED CRITICAL RATIO.")
210 80 CONTINUE
211 STOP
121 END
```

```
RUN

CRITOR    12:28     13 MON 12/31/73

          ORIFICE DIAM., INCHES = 2.7881             BETA = .4506

          PIPE INS. DIAM., IN. = 6.187               DEG. F = 310.00

          AIR FLOW, LB/SEC = 3.880                   FLOW COEFF. = .784

          PSIA INTO ORIFICE= 41.380                  OUT OF ORIFICE= 17.000

                          NO. OF ITERATIONS = 6


AT LINE NO. 211: STOP
```

perimental data for use in the program. These data were then further extended by use of the velocity-of-approach factor, $1/\sqrt{1-\beta^4}$, which approaches 1 as β approaches zero. Incorporating these two expressions in the theoretical equation gives:

$$W_{max} = \frac{K_c}{\sqrt{1-\beta^4}} \frac{A_o p_1 0.5318}{\sqrt{T_1}}$$

This is the equation used in the program. The design parameters for a given problem are grouped into one constant term and the equation is then evaluated using an assumed value of β. The program then uses the Newton-Raphson iteration method to calculate a new value of β; and the value for each iteration is compared with that from the previous iteration. When the difference between successive β values has become very small, the problem is solved, usually in 7 or less iterations. If the number of iterations exceeds 12, the problem is terminated, with only the number of iterations printed out. The engineer should recheck his input data.

Example: Air flow in a 6-in schedule 10 pipe (I.D. 6.187 in) is 3.88 lb/s at 310°F. Upstream pressure is 41.38 psia and the pressure desired downstream of the orifice is 17.0 psia. What is the orifice diameter? The input data are entered into lines 20–25, as shown in the printout, and the program run as shown in the figure. The answer is in close agreement with the test data.

Nomenclature

- A_o = Orifice area, in^2
- g_c = Conversion factor in Newton's law of motion, 32.2 ft/sec^2
- K_c = Flow coefficient, ratio
- k = Specific heat at constant pressure divided by specific heat at constant volume, ratio
- p_1 = Pressure upstream of orifice, psia
- p_2 = Pressure downstream of orifice, psia
- R = Gas constant, ft lb/lb m °R
- T_1 = Temperature upstream of orifice, °R
- t_1 = Temperature upstream of orifice, °F
- w = Mass flow, lb/s
- β = Orifice diameter divided by pipe I.D., ratio

Section VI
OTHER INSTRUMENTS FOR MEASURING FLOW

Practical Methods for Measuring Flows

Practically, flowmeters fall into two categories: those picked from a manufacturer's catalog, and those that are designed. Selection is based on required function.

W. S. Corcoran and Jesse Honeywell, Fluor Engineers and Constructors, Inc.

Because process plants are by definition manufacturing facilities in which materials flow through successive operations, flow measurement lies at the heart of process-plant design and operation.

Accurate flow measurement is essential for determining per-pass conversions and total yields of reactors; it is also required for the material balances around separation processes, such as distillation, extraction, crystallization; and it is extremely important to the testing and acceptance of materials-handling equipment such as pumps and compressors, as well of the entire unit or plant.

Consequently, the engineering design and selection of flow-measurement devices is an integral part of process-plant design knowhow. Reliable flow measurement requires sound engineering, from the selecting of the equipment through its installation, maintenance, and the interpretation and application of results. While the costs of measuring flow are less than 1% of the total plant costs, they are an important factor.

Although orifice meters are probably used in more than 80% of flow-measurement applications, it is safe to say that no single device is best for all uses. So the engineer must begin with a competent choice of the type of meter for each application. This will depend on the importance attached to the particular operation, and is based on individual judgment.

That judgment will resolve a series of inherent questions arising from considerations of accuracy, maintenance, flexibility, type of indication or control, costs, and resistance to the materials being processed. For example, will the device have adequate initial accuracy, and will it sustain that accuracy in operation under the effects of flow variation, corrosion, presence of foreign material and vibration? Will it require excessive maintenance? Is remote operation called for? Does it perform adequately for the associated functions of indicating, recording, totalizing, transmitting and controlling? And will it cost more than other choices because of necessary overrange protection, power required, special parts or pressure loss?

A mechanical approach to choosing one among the many types of devices available has been proposed by the ASME, which has classified flowmeters into two categories: quantity meters and rate meters, with the quantity meters divided into two subclasses, weighing meters and volumetric meters (Table I).

However, for a practical approach to process-plant design, the flow-meter engineer can make another classification into two types: (1) meters for which the supplier assumes responsibility for the fundamental design, and which are requisitioned according to types appearing in manufacturers' catalogs, and (2) meters for which the engineer does the fundamental calculations and assumes responsibility for the application. The first of these two classes we shall call "purchased meters," the second we shall call "engineered meters."

PURCHASED METERS

These include all the quantity meters of Table I, plus such rate meters as the turbine meter, and the variable-area meter. These meters need to be specified by the engineer in such a way that the purchasing agent will have a choice of manufacturers from which to solicit bids. Thus the engineer must differentiate between those features essential to the application and those on which the manufacturer is free to apply individual ingenuity toward better design.

In addition to the normal manufacturer's data necessary for flow measurement (Table II), the engineer will want to consider tolerance, rangeability, remote action required, flowrate, allowable pressure drop, operating conditions, the physical state of the fluid—and, based on these, will recommend a type of construction, such as:

Turbine (Closed Channel) Meter

This meter can be used almost anywhere one would use a positive-displacement meter, except for high-viscosity fluids. The price is higher than positive-displacement meters in the smaller sizes and lower in the larger sizes. Turbine meters have been used to measure ammonia, beer, gasoline, jet fuel, kerosene, crude oil, tetraethyl lead, and water. Rangeability is approximately 20 : 1; and tolerance can be within 0.1%, with special calibration.

The selection of a turbine meter should be based on demonstrated success in the given service. As for ex-

Originally published July 7, 1975.

ample, a gasoline blending system. In such a system, the turbine meters should have demonstrated an ability to handle both gasoline and lead, although lead solutions are a maintenance challenge.

Turbine meters function as a consequence of blade rotation caused by the flowing fluid. The blade tips are magnetic and induce current into a coil embedded in the meter body. Consequently, pressure ratings may be limited, due to the need for thin-walled construction.

Bearings are the most troublesome part of a turbine meter; bearing life is sometimes compromised when high rangeability is required. Ball bearings are very susceptible to damage by abrasive fines, even though a manufacturer's recommended strainer is used. In many cases, sleeve bearings will be more trouble-free, but meter rangeability will be reduced. When rotors or bearings have been damaged, the rotor requires replacement. Therefore, standardized meters with precalibrated interchangeable rotors are highly desirable.

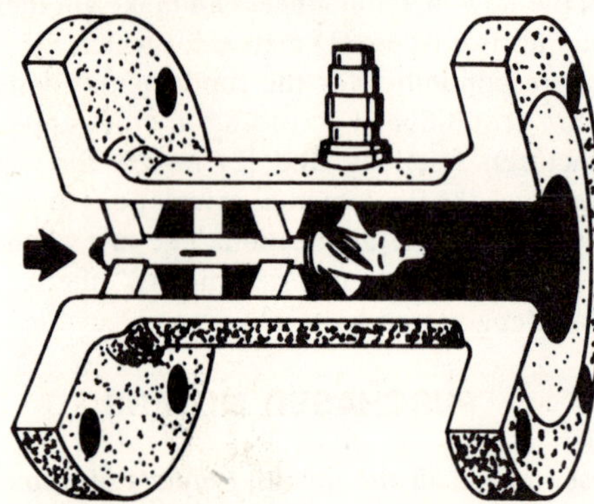

When specifying turbine meters, one manufacturer should be made responsible for both the meters and the associated system for indication and control. Also, the manufacturer should make recommendations for the filter, strainer, screen, etc., which should be strong enough to withstand full pump-differential pressure in case the element becomes plugged.

The meter should be friction-free for optimum operation. Lubricity (how the fluid affects sliding friction) cannot be directly evaluated and is only inferred from the viscosity, although some fluids with similar viscosities have different lubricity characteristics, e.g., water and jet fuel.

If the line holding a turbine meter is to be heat traced, precautions should be made against overheating the meter's signal-preamplifier, if that is mounted on the meter body. Also, a deaerator may be required, since vapor can overspeed a turbine meter designed for liquids, thus causing bearing failure. Sufficient pressure should be maintained to prevent flashing and cavitation.

Pitot Tube (Impact Tube)

These elements measure dynamic (total) pressure from which velocity is then calculated. They are installed so that the open end of a "bent" tube faces directly into the current of the moving fluid. A pressure gage at the other end measures the fluid dynamic (total) pressure. In some styles as shown in the illustration, the static

Classification of Fluid Meters and Methods of Fluid Measurement – Table I

Division/Class		Type
Quantity Meters		
Weighing		Weighers
		Tilting traps
Volumetric	Liquids	Tank
	Liquids	Reciprocating piston
	Liquids	Nutating disk
	Liquids	Sliding and rotating vanes
	Liquids, Gases	Gear and lobed impeller (rotary)
	Gases	Bellows
	Gases	Liquid sealed drum
Rate Meters		
Differential pressure		Venturi
		Flow nozzle
		Nozzle-venturi
		Critical flow nozzle
		Thick-plate rounded-edge orifice
		Thin-plate square-edged orifice
		Concentric
		Eccentric
		Segmental
		Gate or variable area
		Centrifugal
		Elbow or long-radius bend
		Turbine scroll case
		Guide vane speed ring
		Pitot tube (impact tube)
		Pitot-static tube
		Pitot-venturi
		Linear resistance or frictional
		Pipe section
		Capillary tube
		Porous plug
Area (geometric)		Gate
		Cone and float
		Slotted cylinder and piston
Velocity		
Open or closed channel		Cup (anemometer)
		Propeller
Closed channel		Turbine
Head-area		Weirs
		Flumes
Force		
Open or closed channel		Hydrometric pendulum
		Vane
Closed channel		Transverse-momentum flow meters
Thermal		
Open or closed channel		Hot wire
		Total heating
Other meters and methods		
Open or closed channel		Electromagnetic
		Tracers
		Mixtures
		Floats and screens
Closed channel		Pressure-time
		Sound velocity
		Light velocity

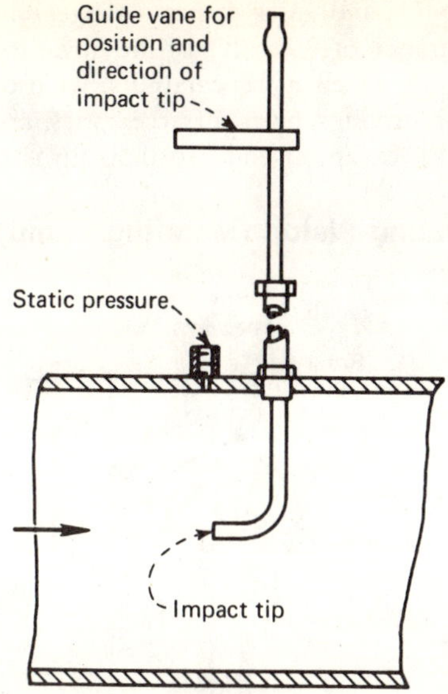

Data Necessary for Flow Measurement
Table II

Liquid	Steam	Gas	
X	X	X	Maximum chart reading and units of flow.
X	X	X	Maximum differential range and type of instrument (mercury float type, force-balance, bell, etc.).
X	X	X	Operating differential or normal flow if established.
X	X	X	Line size (Pipe schedule number or actual inside diameter).
X	X	X	Material of primary device (steel, stainless, bronze, etc.)
X		X	Reference or base temperature.
X	X	X	Flowing temperature.
X			Specific gravity of liquid at base temperature (usually 60° F.).
X			Specific gravity of liquid at flowing temperature.
*	X	X	Flowing pressure.
	X	X	Quality of steam or percentage saturation of gas.
		X	Specific gravity of gas (relative to air).
		X	Reference or base pressure.
	X	X	Barometric pressure (important at low absolute pressure).
		X	Compressibility ratio (deviation from perfect gas laws; i.e., ratio of theoretical density to actual density at flowing conditions).
X	X	X	Location of taps (flange, vena contracta, full-flow).
X	X	X	Viscosity of fluid at operating conditions†.
X	X	X	Seal liquid (if used) and its gravity.

*Needed for calculations at temperatures near critical and for high pressure.
†Always needed for precise calculations.
Note: Additional information necessary to determine outside diameter of plate: either series number and type of flange being used (300#, 400#, R.F., R.T.J., etc.) or bolt circle diameter and diameter of bolts.
Source: Spink, L.K., "Principles and Practice of Flow Meter Engineering," 9th ed., The Foxboro Co.

pressure is measured separately at the pipe wall. In other types, the dynamic pressure is measured by one tube and the static pressure is measured by openings in a second tube surrounding the first tube. The static pressure openings are at right angles to the moving fluid. Although the error is as low as 1% for laboratory style instruments with sufficient upstream straight pipe runs, the average commercial installations will give greater errors. One style of tube measures dynamic pressure at four locations and averages the result. The greatest source of error in commercial installations is the measurement of static pressure. Special considerations are required if the Reynolds number is less than 10,000. This device is easily fouled by foreign materials in the fluid. The cost is low however compared to other devices.

Bellows Meter

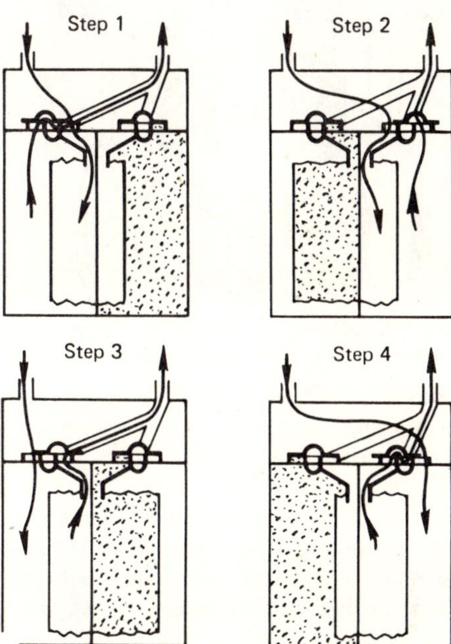

Used only for gases and primarily for custody transfer of utility gases, this device is made of flexible leather or synthetic fabrics, so that three or four chambers are successively filled and emptied in such a way that the total volume displaced from each chamber is the same. Although relatively inexpensive, the bellows meter is typically used only for totalizing services.

Rotary (Geared-Impeller) Meter

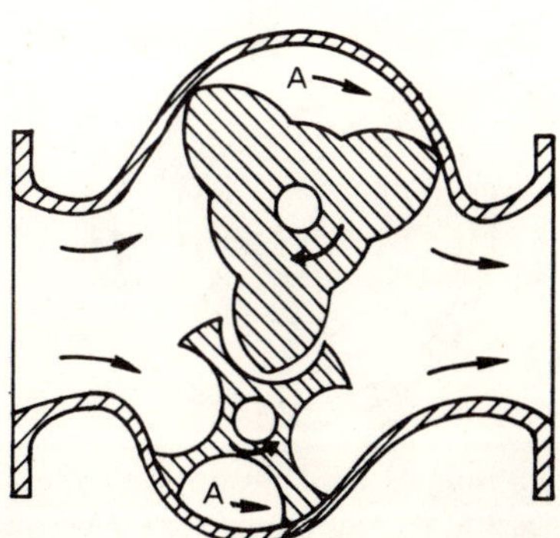

The measuring chambers of this meter are the spaces between the gear teeth and rotor lobes. Along with the sliding-vane meter, this is the most accurate positive-displacement meter that can be supplied in the complete range of sizes between 1 in and 16 in. Normally used for liquids, its rangeability of flows is a ratio of 5 : 1, with a tolerance of 0.2%.

Advantages of the rotary meter include low pressure drop and medium-contact seal area. Disadvantages include capacity limitations dependent on the maximum gear-tip speed or rpm, and difficulty of replacing the rotating element.

This meter is typically used for measuring liquids for custody transfer.

Sliding-Vane Meter

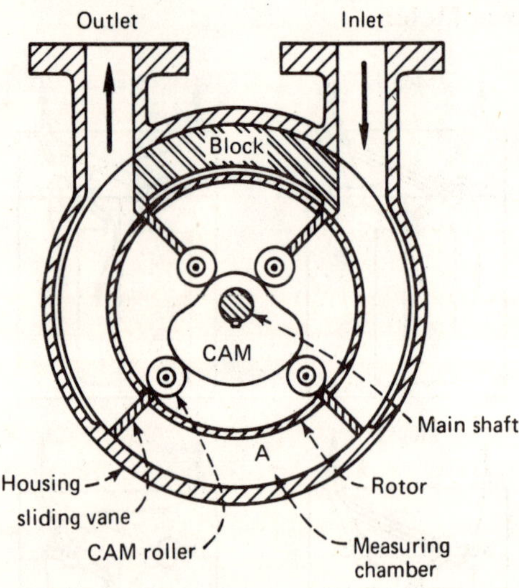

Like the rotary meter, this element separates portions of the flowing fluid by the rotation of internals. It shares with the rotary meter such advantages as low pressure drop and a medium-capacity range, as well as such disadvantage as difficult replacement of the rotating element.

This meter is typically used for the custody transfer of liquids or gases.

Rotating-Vane Meter

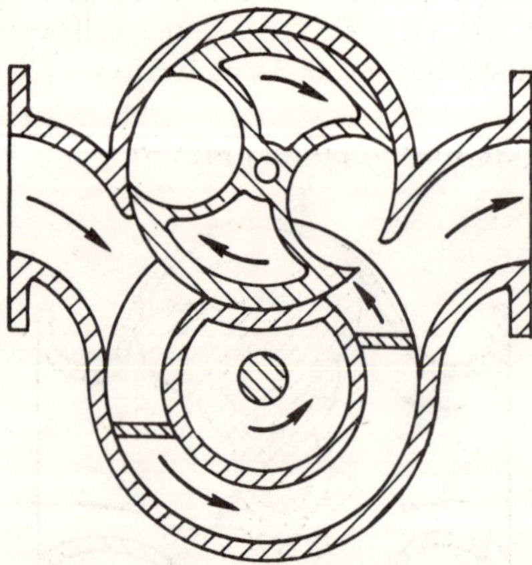

This meter isolates fluid portions by the rotation of both its vanes and its rotor. Used for the custody transfer of both liquids and gases, it has a rangeability of 5 : 1, with a tolerance of 0.2%. It has the advantages of low pressure drop, excellent repeatability, wide contact-seal area, a high driving power for accessories, and virtually no clearance, except at ends, so that slip is minimized.

Reciprocating-Piston (Metering-Pump) Meter

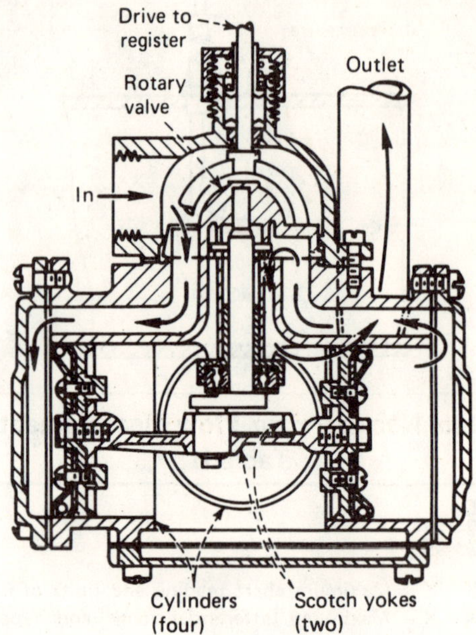

The principle of operation of this meter is akin to that of a piston pump, in which opposing pistons connected to a central crankshaft alternately fill and empty. This is the most accurate of the positive-displacement meters.

Although used only for small capacities, with available connection sizes typically 1 in or smaller, this meter has a rangeability on the order of 50 : 1, with tolerances varying from 0.1% to 1.0%. It is usually employed for metering chemicals.

Nutating-Disk Meter

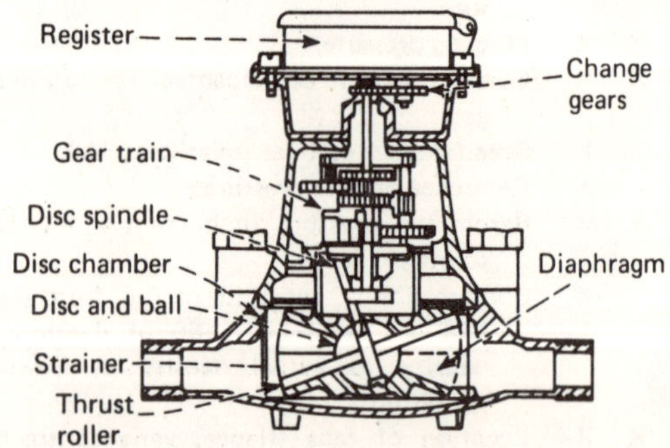

The heart of this meter is a disk surrounding a concentric ball, all fitted within a chamber between the inlet and outlet connections. As liquid passes across this disk, the disk nutates (nods); and this nutating action is transferred through a disk spindle and gear train to a register.

Widely available in various sizes, low cost, with simple construction, and relatively maintenance free, this meter has a rangeability on the order of 5 : 1, with a tolerance of approximately 1.0%. However, the device is the least accurate of the positive-displacement meters and does not meet the requirements of the National Bureau of Standards (NB5) Handbook 44, so that it is primarily used to measure domestic and industrial water or similarly inexpensive liquids.

Rotary (Oscillating) Piston Meter

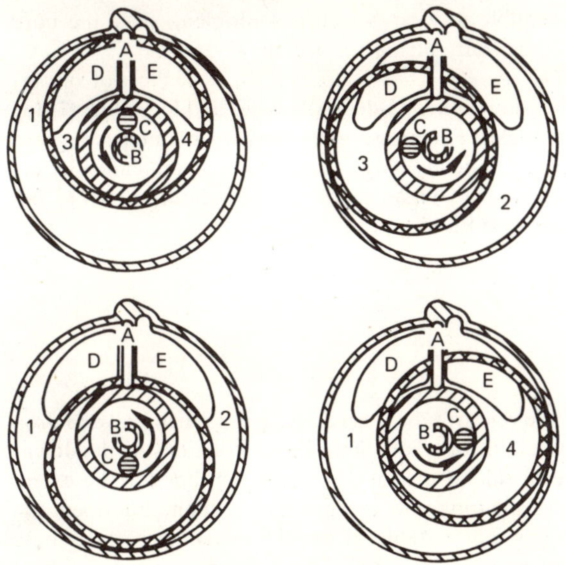

This meter consists of an eccentric cylinder rotating within a containing cylinder, so that both are alternately filled and emptied with the flowing fluid to be measured. The device is used primarily for accounting purposes at marketing terminals for viscous liquids. It has a range of up to 1,500 gpm, with a tolerance of 0.5%.

Of simple construction, this meter can supply the power necessary to operate mechanical tabulating devices.

Variable-Area Meter (Rotameter)

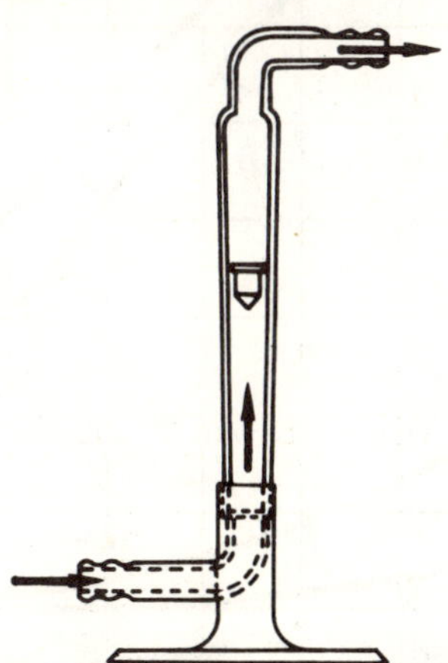

This meter, the well-known rotameter, can be thought of as a variable orifice with a constant pressure differential. Frequently used for low flows, and economical in sizes up to 3–4 in, this meter is available in many materials and can be adapted to transmit pneumatic or electrical signals. However, it becomes expensive as the size range gets larger, and the necessity for upward flow leads to awkward and expensive pipe configurations.

The rangeability of the rotameter is typically 10 : 1, with a tolerance of 2%. However, this tolerance can be reduced to 1.0% by field calibration.

ENGINEERED METERS

Engineered meters all fall into the class of differential meters in Table I, i.e., they measure a rate of flow by measuring the pressure differential across a flow restriction. Typically, the flowmeter engineer is given an operating flowrate for the fluid, the operating conditions, the properties of the fluid, and the line size. From this, he first chooses a pressure differential that will give him a legible chart reading over the expected variation in flowrate; and from this pressure differential, he calculates a restriction opening.

This calculation is not simple, however. The relationship between pressure differential and opening depends not only on the physical properties of the fluid (velocity, viscosity, density, etc.) but also on the geometry of the meter plus turbulent currents caused by the system immediately upstream of the meter.

These variables are correlated through rules of dynamic and geometric similitude. The dynamic similarity, which implies a correspondence between fluid forces, is based on dimensionless numbers. Because the only forces significant to the correlations of a restriction meter are viscosity and inertia, the Reynolds number is the most important.

Geometric similitude is provided by standardized construction of the elements. The restriction-type meter (i.e., orifice, venturi nozzle, and flow nozzle) is reasonably easy to produce to precise dimensions; the various sizes have been standardized, along with the locations of pressure taps, the flanges, and so forth; and much data have been accumulated on the standard sizes and related through coefficients applied to the sizing equation.* This means an orifice size must be assumed to get a coefficient for calculating the orifice size, so that the procedure is iterative.

The Formula

Engineered (calculated) meters all fall into the class of differential meters in Table I, i.e., they measure a pressure differential across a restriction; from this a flow rate is inferred. The data necessary for careful calculation of flowrate is listed in Table II p. 88. The engineer can calculate a primary bore with this data, using a "cookbook" approach presented in Ref. [*2*]. This method should satisfy requirements for custody transfer and plant material balances.†

The fastest method of sizing many primary elements is to use a special slide rule for the purpose. The Foxboro Co., Daniel Industries, Inc., Robinson Orifice Fitting Co., the Bristol Instruments Co. and many others, have such slide rules available.

The sliderule calculations are suitable for approximate primary-element calculation, determination of meter-run

*Design coefficients for these meters are published in tables in "Fluid Meters—Their Theory and Application," published by the Fluid Meter Research Committee of ASME.

†For a detailed discussion, see "Principles and Practice of Flow Meter Engineering," by L. K. Spink, 9th ed., published and sold by The Foxboro Co., Foxboro, Mass.

pipe diameter, and approximate material balances, but are not suitable for custody transfer calculations. Except for steam, the slide rule calculations assume Y and F_a are unity.

The formulas used in the cookbook approach of Ref. [2] and the sliderule method is based on equation I-5-36 in Ref. [8]:

$$m = 358.93\ KYd^2F_a\ \sqrt{\rho_1 h_w}$$

where:

m = flowrate, lb/hr

$K = \dfrac{C}{\sqrt{1-\beta^4}}$

$\beta = d/D$

C = primary element coefficient, a ratio

Y = expansion factor for vapors, a ratio

Y = 1 for liquids

d = bore of primary element, in

D = line size of primary element, in

F_a = thermal expansion factor (primary element), a ratio

ρ_1 = density of fluid, lb/ft^3

h_w = effective differential pressure across primary element, inches of water

Comments on the Terms

The primary-element coefficient, C, is a function of Reynolds no., beta ratio, and line size. For square edged orifices, it may be set to 0.61 as a first approximation.*

The fluid expansion factor, Y, is a function of absolute pressure of the measured fluid, pressure differential across the primary element, beta ratio, and ratio of specific heat at constant pressure to specific heat at constant volume. Y is set to unity as a first approximation.

The primary element expansion factor, F_a, accounts for the situation where the primary element will be used at a temperature different from that at which it is fabricated.

Most sliderules assume Y and F_a are unity and set $C = 0.61$. Ref. [1,2,3,5,8] describe the use of formulas for sizing primary elements and calculating flowrates.

The Engineered Meters

The foregoing formulas can be applied to sizing concentric orifices, venturi meters, flow nozzles, eccentric orifices and segmental orifices, which all may be selected for the situation according to their features:

The concentric orifice: The concentric orifice plate is the most common primary element of the industry, having the advantages of low cost, ease of fabrication, easy calibration, a rangeability of 3.5 : 1, and a relative error of 0.5%. It is used for steam, gas and liquid flow.

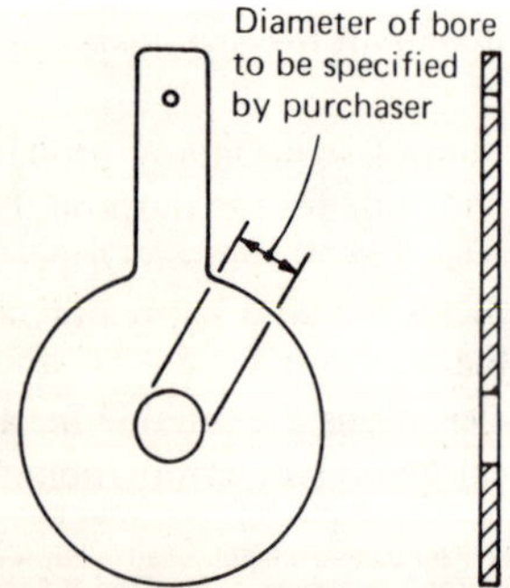

However, in some applications, the cost of the orifice flanges can be expensive. Also, this element is not good for fluids having extensive suspended solids; and it is susceptible to wear, which is a problem, since the bore must be sharp-edged for accuracy.

The venturi meter: There are two basic types of venturi tube: short cone (15-deg exit angle) and long cone (7-deg exit angle). The pressure recovery for both types is given in Fig. 1. The discharge coefficients recommended by the International Organization for Standardization are between 0.984 and 0.955 [2,8].

The venturi tube is used where not much pressure drop is available or where there are obstructions on the downstream side of the tube. It can be modified to the extent of eliminating the downstream section, but then the reading is not stable. It has the advantages of accurate measurement of clean, dirty and nonviscuous fluids at low pressure loss; is reasonably easy to install; is available in sizes ranging 4 in to 48 in. It has a rangeability of 3.5 : 1 and a tolerance of 1%.

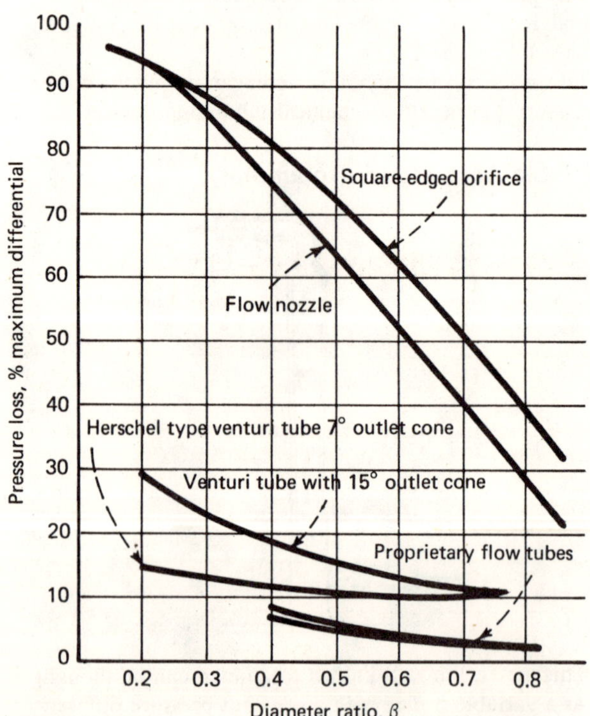

OVERALL PRESSURE LOSS by primary elements—Fig. 1

However, compared with the pitot tube, it is expensive in the larger sizes and in alloy materials.

Flow nozzles: These devices have pressure-recovery factors between orifice plates and venturis at the same

*Also, for a discussion of forumulas and pressure tap location, see Kern, Robert, Measuring Flow in Pipes with Orifices and Nozzles, *CE*, Feb. 3, 1975, p. 72.

β ratio, but closer to the orifice plate. The recommended form is the "long radius" or elliptical-inlet nozzle. They are used for measurement of high-velocity streams, and will pass 60% more flow than an orifice plate for the same line size and pressure differential. Rangeability is 3.5 : 1, with a tolerance of 1.5–2.0%. Caution: For high temperature service, the nozzle and pipe should have similar coefficients of expansion.

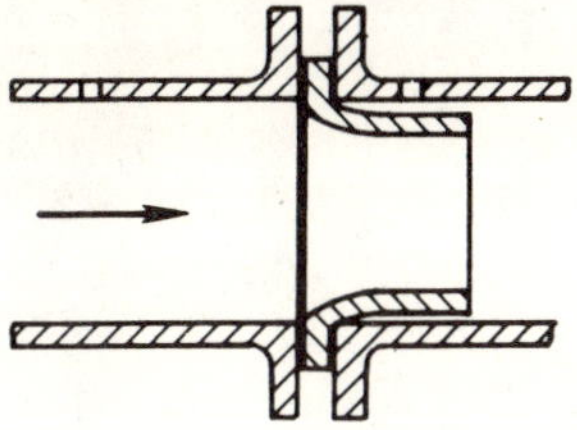

However, the flow nozzle should not be used on liquids containing large percentages of sticky solids; and their costs can be high, depending on material selection. Nozzles are made from a corrosion-resistant material. For high temperatures, stainless steel is used; and bronze may be used for temperatures below 400°F.

Eccentric orifices: These devices are sometimes used for wet steam, liquids containing solids, oils containing water, and other types of two-phase flow. Their range-

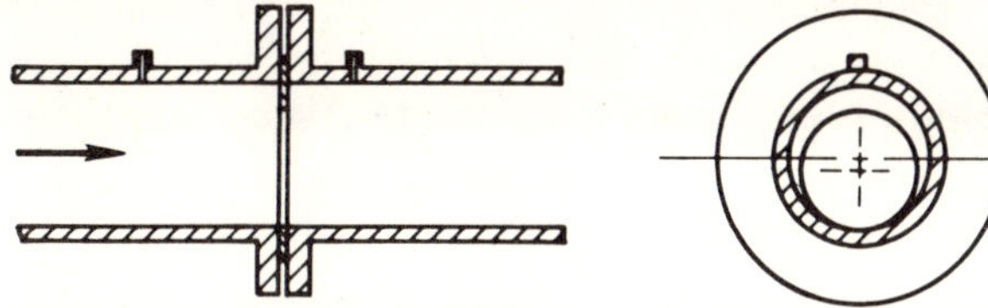

ability is about 3 : 1, with a tolerance of 1.5–2.0%. They are less accurate than the concentric orifice; they require substantial straight pipe both upstream and downstream; and the cost of the holding device can be high.

Segmental orifices: These devices have the same general types of application as the eccentric orifices, with the additional advantage that the segmental orifices do

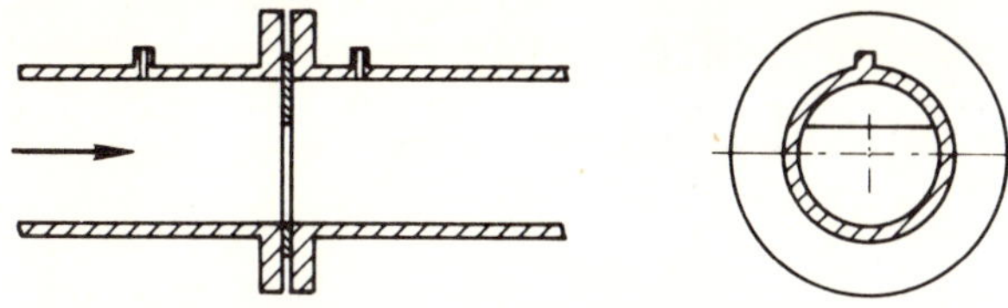

not dam solids on the upstream side of the plate. However, centering the hole of segmental orifices is critical; and these devices should be used only for large line sizes and low fluid viscosities.

Linear-resistance meters: These are a special case of the engineered meters, in that they are nonstandard. They must be calibrated to determine their coefficient.

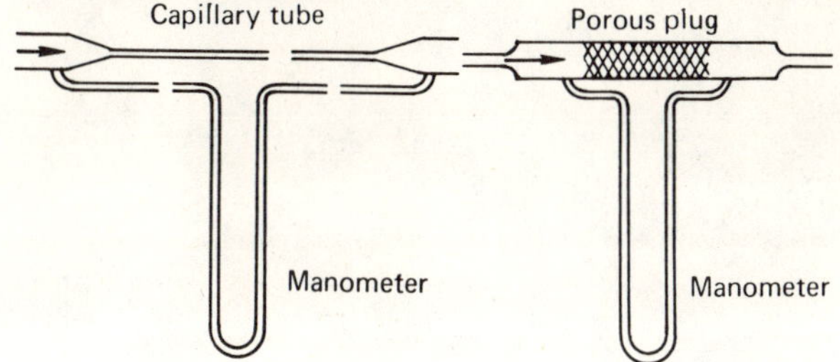

Because of the capillary-flow dimensions, flow is in the laminar range, and the relationship between pressure differential and flow is linear. These meters are subject to plugging, with consequent changes in the coefficient.

Centrifugal meters: These are another special case of the engineered meter, in that the parts, while potentially standard, have not been subject to the same tests and standardizations as the orifice-type meters. The principle

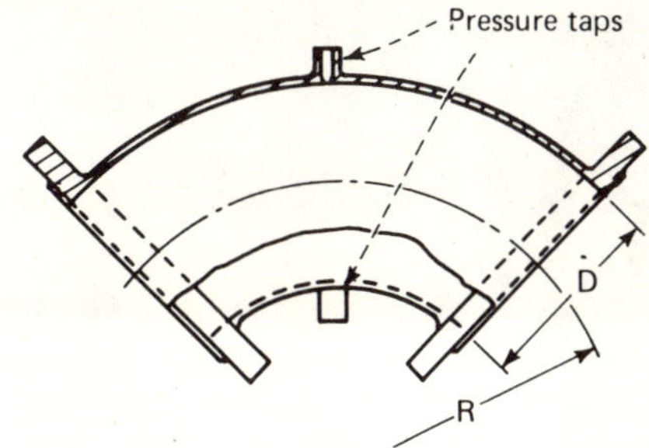

of operation is the relation between fluid velocity and the consequent difference in pressure due to centrifugal force, as measured on the inside and outside of a bend in the piping. The common pipe-elbow is a common form of this meter, which is sometimes called the "elbow meter." Rangeability is 3 : 1, with a tolerance of 1% for elements calibrated in the laboratory.

References

1. Orifice Metering of Natural Gas, Gas Measurement Committee Report No. 3, American Gas Assn., 1515 Wilson Blvd., Arlington, VA 22209.
2. Sping, L. K., Principles and Practices of Flow Meter Engineering, 9th ed., The Foxboro Co., Foxboro, Mass.
3. ASME Power Test Codes, Supplement on Instrument and Apparatus—Part 5, Chapt. 4, The American Soc. of Mech. Engrs., 345 East 47th Street, New York, N.Y. 10017.
4. National Bureau of Standards, Handbook 44, 3rd ed.
5. Shell Flow Metering Engineering Handbook, Royal-Dutch/Shell Group, 1968.
6. Daniel Orifice Flow Calculator, Daniel Industries, Inc., 8480 Beverly Blvd., Los Angeles, CA 90048.

Meet the Authors

◀ **William S. Corcoran** is a senior control systems engineer for Fluor Engineers and Constructors Inc. at 2500 South Atlantic Blvd., Los Angeles, Calif. 90040, where he has been employed since 1960. His present work includes flow measurement and control release header design and engineering analysis for computer programs. He has attended University of Southern California and UCLA.

Jesse E. Honeywell is a control systems engineer for Fluor Engineers and Constructors Inc. 2500 South Atlantic Blvd., Los Angeles, Calif. 90040. He is presently working on a coal liquefication pilot plant for the Energy Research and Development Administration. Before joining Fluor, he worked for Aeronutronics, Div. of Ford Motor Co. as an environmental tests engineer. He holds a BS in Aeronautics (1960) from Northrop Inst. of Technology in Los Angeles. ▶

Selecting sight flow indicators

Design and fabrication improvements in sight flow indicators—as well as the availability of a wider range of materials of construction to handle almost any process condition of pH, pressure or temperature—have greatly increased the dependability of these visual aids to the chemical process industries.

Roger McDonough, *Jacoby-Tarbox Corp.*

☐ Process engineers continue to include sight flow indicators (SFIs) in their systems despite the many advances in automatic instrumentation for measuring flow of liquids and gases. SFIs allow a quick and relatively accurate estimate of the rate of flow, consistency of the product, color and other vital information. They add an important human evaluation that our technological age continues to need. When properly selected and maintained, SFIs are an on-the-spot double check on more sophisticated control-board instruments.

By acquainting himself with the product lines of the various manufacturers of SFIs, an engineer can keep abreast of changes in design, fabrication, materials, styles and capacities of a standard product. Manufacturers are usually ready to offer a custom-built sight flow indicator, or a modification of a standard unit, to fill the most demanding specifications.

Types available

Generally, sight flow indicators fall into two major categories: stand-alone piping components, and sight windows for attachment to pressure vessels.

Originally published July 4, 1977.

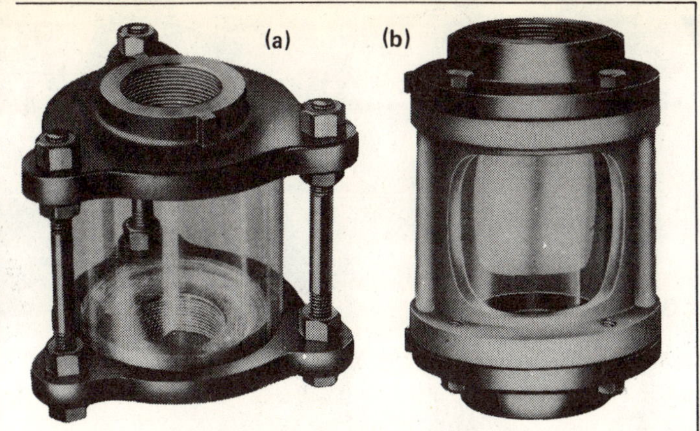

Cylindrical-type SFIs: (a) unsheathed window, (b) window with protective metal body **Fig. 1**

An examination of the SFI units offered by the several major suppliers reveals that the numerous types of cylindrical and flat-disk window units can be grouped into a few popular configurations:

Cylindrical type—These are recommended where maximum visibility is desirable, although their applications are somewhat limited because of strength considerations. They can be supplied in varying lengths and sizes, up to pipe sizes 16-in dia. An unsheathed-tube type is particularly suitable for vertical pipelines, but they can also be installed in horizontal pipes where mechanical strains in the pipelines are not present (Fig. 1a).

Those with tubes inside protective metal bodies (Fig. 1b) are suitable for vertical or horizontal lines where moderate mechanical strains may be encountered. The metal body is isolated from the solution or gas in the pipe, which contacts only the heads, glass and gasket material. Unsheathed units are fabricated with end-seal gaskets, whereas the sheathed type may also be made with stuffing-box seals. This type would be recommended for negative pressures.

Flat-disk window—These cover the widest range of applications and come in the following types:

1. Flapper—This design, with a flap visible through the sight glass (Fig. 2a) indicates changes in flowrates by the position of the flapper. It is applicable for horizontal pipelines, as well as for vertical ones with upward flow. It is frequently used in lines for gaseous products, and is best suited for indicating flow of transparent or slightly opaque solutions.
2. Ball—This design, recommended for vertical pipes with low-volume upward flow, indicates flowrate variations by the position of a fluid-suspended ball (Fig. 2b).
3. Rotary—This indicator operates successfully in horizontal or vertical positions for up or down flow (Fig. 2c). It serves well in indicating flow of dark solutions, as well as of transparent solutions and gases. The rotors are usually made of inert tetrafluoroethylene (TFE) plastic, and are visible from a distance.
4. Drip—This indicator, most often used in vertical, downward-flow pipes, is useful in distillation and similar processes where flow is low or intermittent (Fig. 2d).

There are variations of the cylindrical and the flat-window types to meet specific needs. They can be supplied with flanged ends in a variety of American National Standards Institute pressure classes, as well as screwed, socket-weld, butt-weld and silver-brazed ends. Units are available to fit standard pipe sizes from 1/8-in. through 16-in. dia. Sizes above 16 in. are generally special ones.

Although the most common system pressure is 150 psi, units rated for higher pressures are generally available from stock to handle up to 3,000 psi, in sizes up to 2 in., and from 300 to 600 psi in the larger sizes.

Special configurations

A wide variety of special configurations of SFIs are available when the more common units are not practical. Cylindrical indicators may be furnished with measuring-calibrated glasses and to serve as moisture indicators. These indicators may also be adapted for use as level controls. Flat-glass indicators are frequently specified as

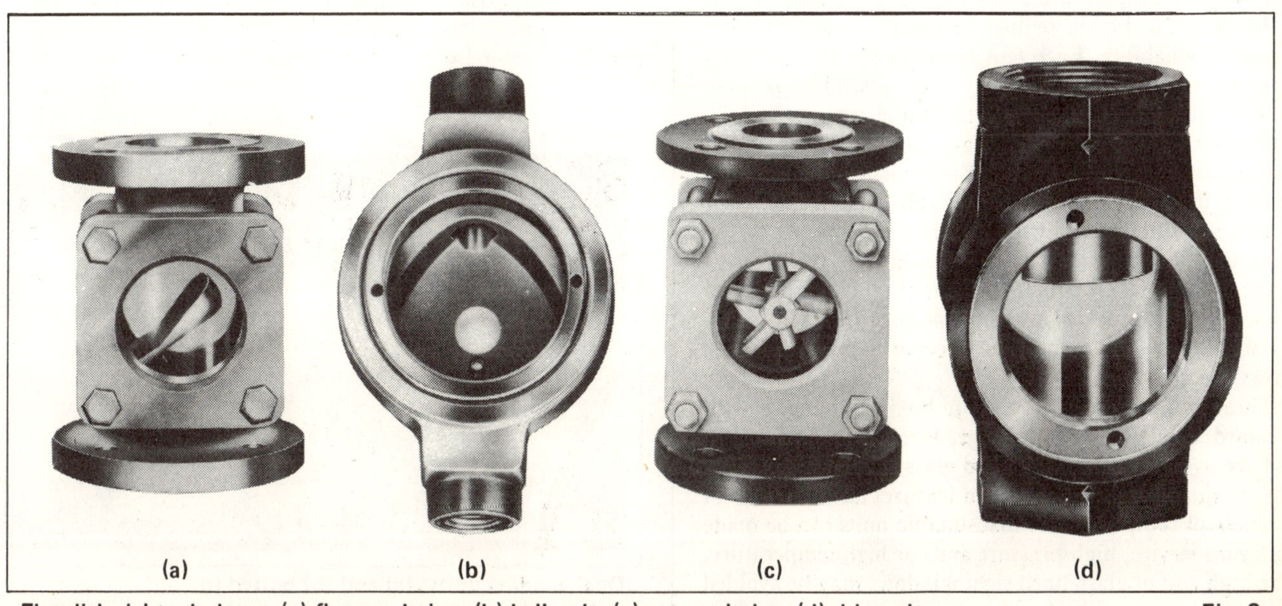

Flat-disk sight windows: (a) flapper design, (b) ball unit, (c) rotary design, (d) drip unit **Fig. 2**

jacketed models to enhance the flow of viscous materials. Angle-sight models—frequently with a drip tube, thermocouple or vent—are specified when straight-line indicators are not suitable. Three-way-flanged indicators are commonly used, as are Teflon-lined indicators.

Sanitary sight-flow indicators, in both cylindrical and flat-glass models, are available in flange, butt-weld, clamp and threaded bevel-seat types for applications where cleanliness is a major factor. They are generally available in 304 and 316 stainless steel for pipe sizes to 4 in. Thermocouple wells are frequently added to standard units, as are spray rings or nozzles as cleaning media for the inner surface of the glass, or to remove condensation.

When ambient lighting is not enough to provide visibility, illumination can be supplied with fluorescent units that are easily mounted behind the indicator. Indicators are availabile in fiberglass-reinforced-polyester moisture-proof or explosion-proof boxes with Pyrex glass, and are adjustable for the most effective illumination.

Designs for severe service

In response to increased emphasis on safety and for more precisely engineered sight flow indicators and sight windows, several manufacturers are offering more sophisticated designs to allow for more severe service and reliability.

The peripheral-compression seal type concentrates the sealing forces on the periphery of the glass disk—with adjustable resilient packing—to reduce compressive forces on the window surfaces. As temperatures increase, the glass is free to expand radially against the packing. The design makes use of a preassembled laminated lens consisting of two layers of tempered soda-lime glass (Pittsburgh's "Herculite") and an inner layer of annealed borosilicate glass (Corning's "Pyrex").

Another design uses the same laminates bonded together in a metal ferrule. This glass package is replaceable as a unit. The design strength is calculated on only a single layer of the soda-lime glass, not on the available strength of the entire laminate.

The third type provides a double lens, either of which will support the pressure loads with an adequate margin of safety, with double sealing surfaces to assure a high degree of reliability. Each lens is independently sealed by separate surfaces. The entire inner assembly is held between mounting flanges in the same manner as the previously mentioned two designs, and also has radial temperature-expansion space. Either lens is capable of holding the full line pressure. Even though the glass windows are individually replaceable, as opposed to the other two designs, the glass holder takes the compressive forces caused by bolt tightening, thus eliminating potential glass breakage due to uneven bolt torques. The dual-window design also reduces the effect of thermal gradients across the glass.

These window assemblies can be interchanged with standard windows so as to upgrade existing equipment, and are available with laminated glass, tempered borosilicate, or quartz. The construction features allow the use of a variety of components so that suitable units can be made for steam service, high pressure and/or high temperature. The high cost of glass-lined sight windows may be avoided by specifying these dual-widow units with polyvinylchloride (PVC), TFE, fiber-reinforced plastic (FRP) or rubber lining.

Sight windows in a common piping system Fig. 3

In evaluating sight windows for severe service and increased safety, be on the lookout for the Factory Mutual (FM) system approval. Each of the types mentioned above is being fabricated by manufacturers who have been granted FM approval.

Sight windows for pipes and vessels

When it is necessary to view the interior of a closed tank while reaction or mixing takes place, sight windows are available for direct attachment to the tank wall or to processing equipment. These windows are also useful in permitting visual access to large pipes.

Window assemblies for looking into tanks, pipes and vessels are of four general types: (1) those that bolt to existing flanges, (2) those that thread on the end of a pipe (Fig. 3), (3) those welded into a vessel shell or welded into a vessel or pipe, and (4) those threaded into the wall of a pipe or vessel (usually smaller sizes).

The most commonly used type has a welding-pad base that may be curved to fit the curvature of the tank or pipe wall. It may also be simply inserted into a hole through the wall and welded. Circular flat-glass sight windows are usually suitable but they are also made in rectangular and

Dual-window units: (a) and (b) bolted to nozzle necks, (c) welded into vessel Fig. 4

obround* shapes. Severe-service designs are also available for tanks, vessels and pipes, with dual and double windows and in special-glass materials (Fig. 4).

Window holders for tank-wall sights are made of carbon steel, stainless steel or brass to match specified services. Assemblies with cylindrically or spherically curved pads—or with thick welding pads for insertion in tank curvatures—are among the choices offered by vendors. When specifying a window of this type, the supplier should know the service conditions, and also whether the vessel is coded.

Materials of construction

Standard materials of construction offered by most manufacturers of SFIs will handle the great majority of products produced in the chemical process industries. Vendor catalogs of standard indicators show cast iron, bronze, carbon steel or 316 stainless steel as being the most common metals in the indicator body, although many alloys may be ordered. Neoprene seats and seals, steel bolts and nuts, and cast iron or bronze retaining rings are most frequently ordered as standard parts.

When the nature of the product in the pipeline rules out standard materials due to corrosion or high temperature, there is a host of other materials that can be used. Neoprene seats or seals serve well up to 300°F (149°C), although there are a number of commonly processed products that attack this material at temperatures below 300°F. The manufacturer can advise when other gasket materials should be used due to chemical exposure, temperature or pressure. The alternatives are asbestos, silicone, TFE, synthetic rubber such as Viton A, synthetic rubber-asbestos combinations, polytetrafluoroethylene (Teflon-Halar), other fluorocarbons, or graphite. Synthetic rubber-asbestos gaskets will withstand temperatures up to 800°F (427°C), and graphite up to 3,000°F (1,650°C).

Corrosion problems

To deal with corrosion in the metal bodies of flow indicators, there are a number of alternatives. The standard metals can be galvanized or plated with chrome, cadmium, heresite or lead. The interior can be given a protective liner of neoprene, PVC, rubber, chlorinated tetrafluoroethylene (CTFE or Kel-F), Teflon or lead.

In addition, there are the numerous corrosion resistant alloys such as Alloy 20, Hastelloys, Inconel, Monel, nickel, Ni-Resist, and titanium. There is a growing use of PVC bodies, particularly in all-plastic pipe systems, although process engineers generally prefer to use one of the alloys or a coated or lined-metal body in extremely corrosive atmospheres, because of the superior strength characteristics of the metallic units.

SFI suppliers can determine the best materials of construction, but only if they are provided with complete details of the product content, along with the pressure and temperature ranges of the process. In examining the suitability of standard materials in the construction of SFIs for some 400 commonly processed liquids and gases, it was found that cast iron, bronze, carbon steel or stainless steel would perform from excellent to good in more than 75% of the products. For exceptional cases the system designer should select his sight flow indicator in close consultation with the manufacturer.

*Flattened cylinder with parallel sides and hemispherical ends.

Since it is the visibility afforded by sight flow devices that makes them valuable, the glass windows deserve special attention in their selection and care. The suitability of various window materials for particular service conditions is best determined by the indicator manufacturer in consultation with the glass manufacturer. For instance, on occasion glass must be protected with mica or Kel-F shields to protect it from chemical attack.

There are four compositions of glass that are in common usage, with two of them also available in a strengthened (tempered) form. Each has a particular advantage.

Soda-lime glass (Herculite) is relatively inexpensive but has the poorest thermal-shock and thermal-stress resistance. However, it exhibits superior strength in its properly tempered form when thermal gradients are not present.

Borosilicate glass (Pyrex) is the most commonly used sight window material, and for good reasons. With its relatively low coefficient of expansion and good strength characteristics, it easily handles the average application. When it is strengthened through proper tempering, it can handle pressures two to four times higher than in its annealed form, up to temperatures of 500°F (260°C). In addition, it has chemical resistance superior to soda-lime glass.

Fused silica, or quartz glass, shows reasonably good strength properties and superior thermal characteristics. At continuous temperatures up to 1,832°F (1,000°C), it has about 80% of the strength of annealed borosilicate. Unfortunately, one pays for the increased thermal-shock-resistance characteristics. Continuous use of high temperatures results in a loss of strength, when compared to the tempered forms of borosilicate or soda-lime glasses. It is also more expensive.

The most expensive glass discussed here, 96% silica glass, exhibits the superior thermal properties of fused silica and the mechanical strength characteristics of borosilicates. Its relatively high cost has limited its usage to systems where there is no other alternative.

A specifying engineer must be quite certain of the properties required for his system before selecting the construction materials. By conferring with the sight-glass manufacturer, and defining his service conditions, the engineer can integrate the proper materials into standard designs to assure a high degree of reliability at the most reasonable cost.

The author

Roger McDonough is Chief Engineer at Jacoby-Tarbox Corp. (808 Nepperhan Ave., Yonkers, NY 10703). His background includes similar positions with various firms in the valve, pressure vessel, and process equipment fields. He is a mechanical engineering graduate of Farleigh Dickinson University, and is currently Ball Valve Technical Subcommittee chairman for the Valve Manufacturers Assn., and is also a representative to the Ball Valve Committee of the Manufacturers Standardization Soc.

How To Size Flowmeters

Proper installation of any flowmeter in fluid systems is essential in relation to piping layout, streamline-flow conditions, and accessibility to the device and its associated instruments and connections.

ROBERT KERN, Hoffmann-La Roche Inc.

For measuring flow in process lines, we must consider a variety of metering devices. We then select a meter, meter size and piping configuration to provide the most accurate flow-metering for the job.

Venturi Flow Meters

In principle, venturis work in the same way as orifices do. However, the permanent pressure loss across the venturi is very small; and in well-designed systems, venturis require about one-half the straight length of pipe than do orifice meters for the same accuracy. For measuring the same flowrate, venturi meters often require a smaller pipe size than do orifices. Also, venturis can handle much higher capacity ranges (10 to 1; some even 20 to 1) than orifices (4 to 1).

From the standpoint of piping design, we must resolve the following questions in order to apply venturis properly:

1. Can a calculated pipe size accommodate a venturi meter? Both the pipe and venturi are sized with the same flow data.

2. Is it possible to fit a venturi meter into a given pipe configuration without additional pipe length and fittings?

The calculation procedures yield the diameters for the inlet pipe and throat of the venturi. Manufacturers' catalogs give the overall length for a selected type and size of venturi [*1*].

Commercial Venturi Meters

The venturi meter (Fig. 1) consists of a short cylindrical section having a high-pressure connection; an inlet cone; a throat section having a low-pressure connection; and an outlet cone. As fluid moves through the throat of the venturi, its velocity increases and pressure decreases. The resulting differential pressure is proportional to the flowrate and is used for flow-metering.

The simplest venturi meter (Fig. 1a) finds use in high-temperature and high-pressure services. Compared with other venturi meters, its cost is low. It has a short overall length and high pressure-recovery characteristics. This meter can be used for slurries and for liquids containing solids. In slurry service, the pressure connections are flushed intermittently. Available sizes range from 1 to 12 in.

The standard short-form venturi tube (Fig. 1b) has a wide range of industrial applications. The low-pressure and high-pressure taps are connected to annular chambers—located around the inlet cylinder section and around the throat. Small radial holes interconnect the chambers with the inlet cylinder on the one hand, and the throat section on the other. In this way, the average pressure is sensed at the pressure taps; and hence, this venturi is not as sensitive to irregularities in the velocity distribution of the fluid. This type of venturi is usually suitable for clean liquid and gas services.

The standard long-form venturi tube (Fig. 1c) has a smaller permanent pressure loss than the short form—especially at lower throat diameter to pipe diameter ratios.

The short-form and long-form venturis operate with a wide flow range because the discharge coefficient stays constant. Because of the annular chambers at the inlet section and throat, metering accuracy is scarcely affected by upstream flow disturbances. Both are available in sizes from 1 to 48 in.

For high-pressure and high-temperature services, the previously described meters are also available as welded-

Originally published March 3, 1975.

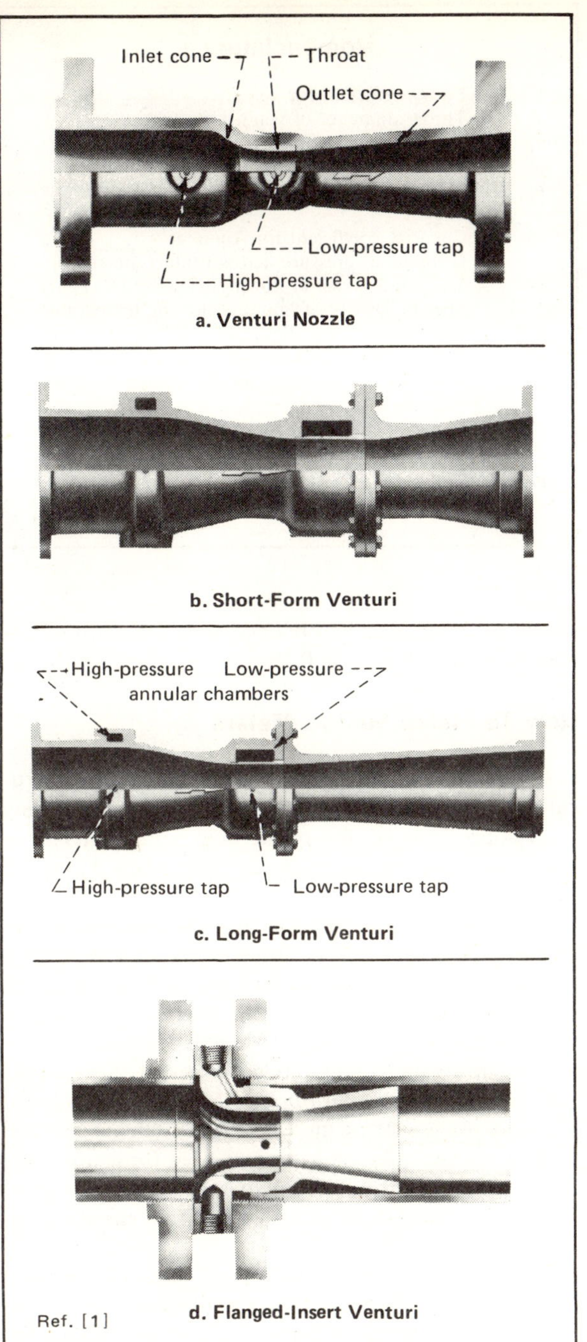

VENTURIS for various service requirements—Fig. 1

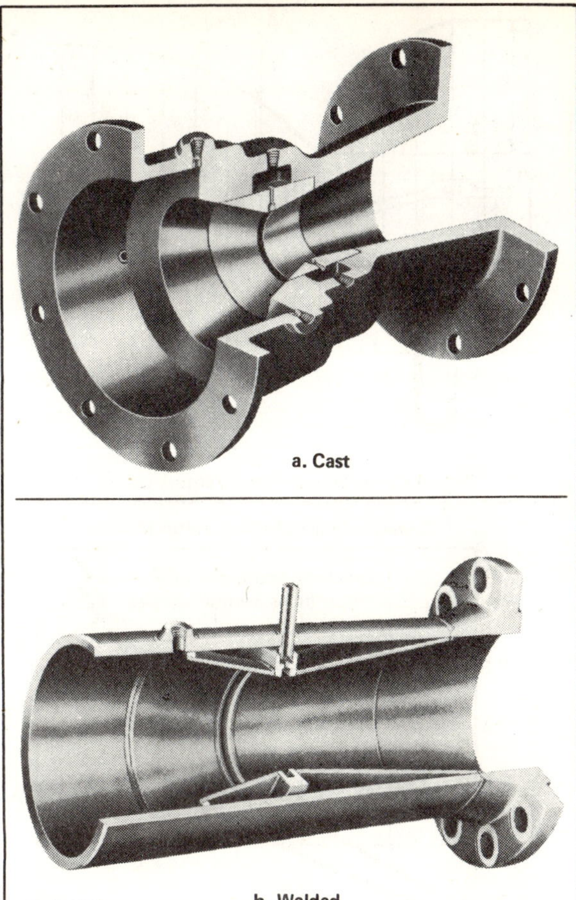

DALL flow tube has high pressure differential—Fig. 2

insert designs having flanged or beveled ends. Fig. 1d is a flanged insert nozzle with annular rings to the pressure taps. This is an economical venturi from the standpoint of capital cost and (because of high pressure recovery) utility cost. Sizes range from 4 to 42 in.

Where the lowest head loss and a high pressure differential for metering are required, the Dall flow tube is chosen. Fig. 2a shows the cast-metal version for sizes 6 to 48 in, and Fig. 2b shows the welded design. The Dall flow tube has the shortest overall length among the venturis for the same pipe size. Its installation is easy. In large sizes, a Dall tube costs less than a comparable venturi.

Sizing procedures for venturi meters (or any differential-pressure producing flow element) are identical to those for orifice calculations, as given in Part 3 of this series (*Chem. Eng.,* Feb. 3, 1975, pp. 72–75). Of course, numerical values for the flow coefficients differ, and the range of throat diameters is not as wide as that for orifices.

Flowrates and head losses across venturis are calculated from the following relations for:

Liquids at flowing temperature:

$$Q = 5.68\beta^2 C d_1^2(\sqrt{h_w}/\sqrt{S}), \text{ gpm} \quad (1)$$

$$\sqrt{h_w} = 0.176(Q\sqrt{S})/(d_1^2\beta^2 C), \text{ in}^{1/2} \quad (2)$$

Vapors or gases at flowing conditions:

$$W = 359.43\beta^2 C d_1^2\sqrt{h_w\rho}, \text{ lb/h} \quad (3)$$

$$\sqrt{h_w} = 0.00278W/(d_1^2\beta^2 C\sqrt{\rho}), \text{ in}^{1/2} \quad (4)$$

The differential pressure across venturi meters, ΔP_v, is given by:

$$\Delta P_v = (h_w/12)(62.37/144) = 0.0361h_w, \text{ psi} \quad (5)$$

A summary of sizing data is given in Fig. 3. The comparison between various venturi meters for permanent pressure loss can be obtained from the graph as shown in Fig. 3. The most economical installation from the standpoint of piping and utility costs, and the most ac-

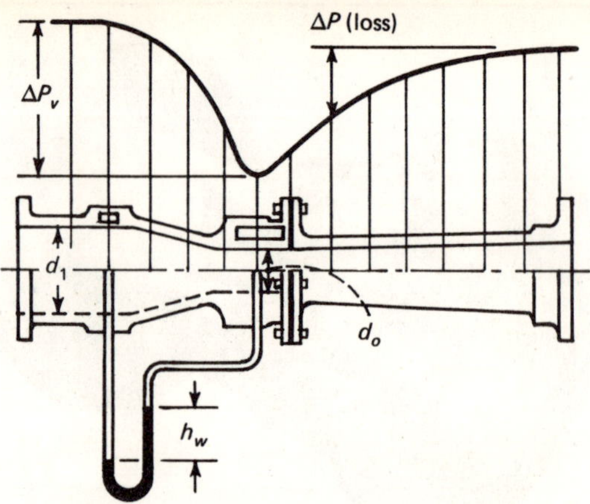

Pressure Distribution Along Venturi Tube

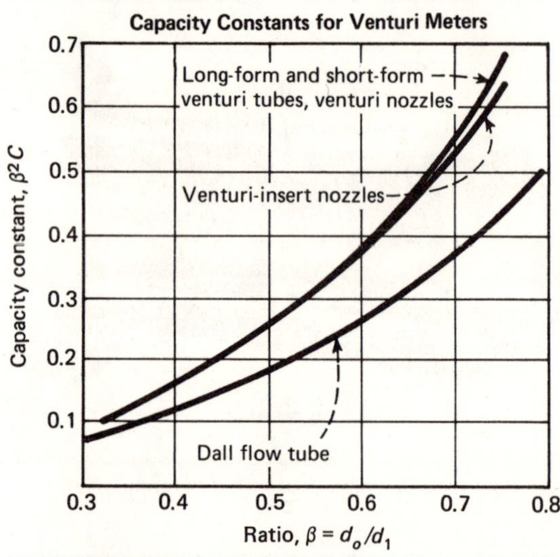

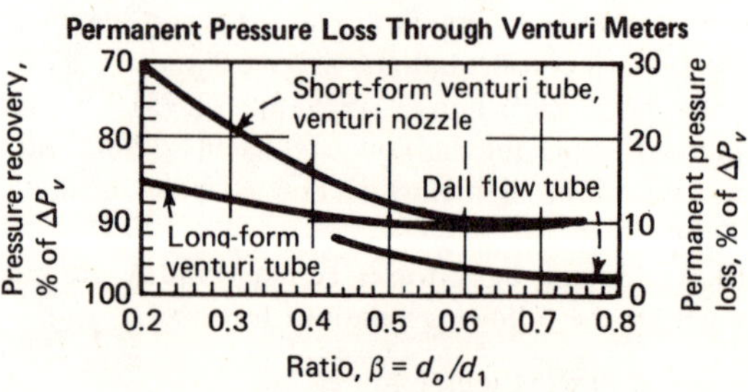

β Ratios and Capacity Constants, $\beta^2 C$, For Commercial Venturi Meters

Average value, $\beta = d_o/d_1$:	0.35	0.45	0.57	0.69	0.75
Long-form venturi, short-form venturi, venturi nozzle $\beta^2 C$	0.12	0.2	0.34	0.53	0.67
Flanged-inlet, venturi $\beta^2 C$	0.12	0.2	0.34	0.51	0.63
Dall flow tube $\beta^2 C$	—	0.148	0.24	0.364	0.445

Typical Manometer Ranges for Venturi Meters.

h_w*	20	30	40	60	80	120	160	240	320
$\sqrt{h_w}$	4.47	5.48	6.33	7.75	8.94	10.96	12.65	15.5	17.9

*In. of water

PRESSURE drop and sizing data for venturis—Fig. 3

Nomenclature

C	Flow coefficient for venturi
d_o	Throat diameter of venturi, in
d_1	Inside diameter of pipe, in.
g	Gravitational constant, 32.2 ft/s²
h_w	Manometer deflection or head loss, in of water at 60°F
$(K_g F_v)$	Capacity coefficient for Annubar
ΔP_v	Differential pressure across venturi meter, psi
Q	Volume flowrate at flowing temperature, gpm
S	Specific gravity of liquid at flowing temperature
S_{60}	Specific gravity of liquid at 60°F
W	Weight flowrate, lb/h
β	Ratio of throat diameter of venturi to inside diameter of pipe
$\beta^2 C$	Capacity coefficient for venturi
ρ	Fluid density at flowing conditions, lb/ft³
ρ_{60}	Liquid density at 60°F, lb/ft³
ρ_{60w}	Density of water at 60°F, 62.37 lb/ft³

curate metering, can be obtained with a β ratio of 0.5. The maximum β ratio is 0.75.

How To Install Venturi Meters

A venturi tube may be installed in a horizontal, vertical-upflow or -downflow, or inclined position, providing the venturi is always full of the fluid being metered. In most cases, the valved pressure taps (usually ½ in) are horizontal.

The general rule requires as much straight-run of upstream pipe as possible in order to have a symmetrical velocity profile. Venturi meters, in most installations, need less straight upstream piping than do orifices, pitot tubes or flow nozzles. Generally, with a smaller β ratio, shorter upstream piping can be provided. Specifically, with $\beta = 0.53$, a straight run equal to 10 times the inlet diameter is adequate, with $\beta = 0.63$, a straight run equal to 20 times the inlet diameter is needed for a two-plane pipe configuration. Upstream straight-run requirements for various fittings can be estimated from the diagrams shown in Fig. 4.

Straightening vanes can reduce the required upstream pipe length. For a reasonable installation, use a minimum length of two pipe diameters upstream of the inlet flange to the straightening vane, and the same length between the vane's outlet and the venturi tube's inlet. Usually the segmental type of straightening vane is chosen (see Part 3, Fig. 8, *Chem. Eng.*, Feb. 3, 1975, p. 78). Configuration of the downstream piping has no effect on metering accuracy. Reducers or elbows can be flanged to the venturi outlet. Provide a straight run of two pipe diameters if the venturi is followed by a valve.

If a noncorrosive clean fluid is being metered, the venturi meters can be buried with only the pressure-tap valves located above grade.

Large venturi meters in slurry service can have clean-out ports, vents with drains at both annular chambers, inspection openings on the outlet cone, a manhole in the piping joint just after the outlet cone, and valved purge connections in addition to the pressure-sensing taps. All

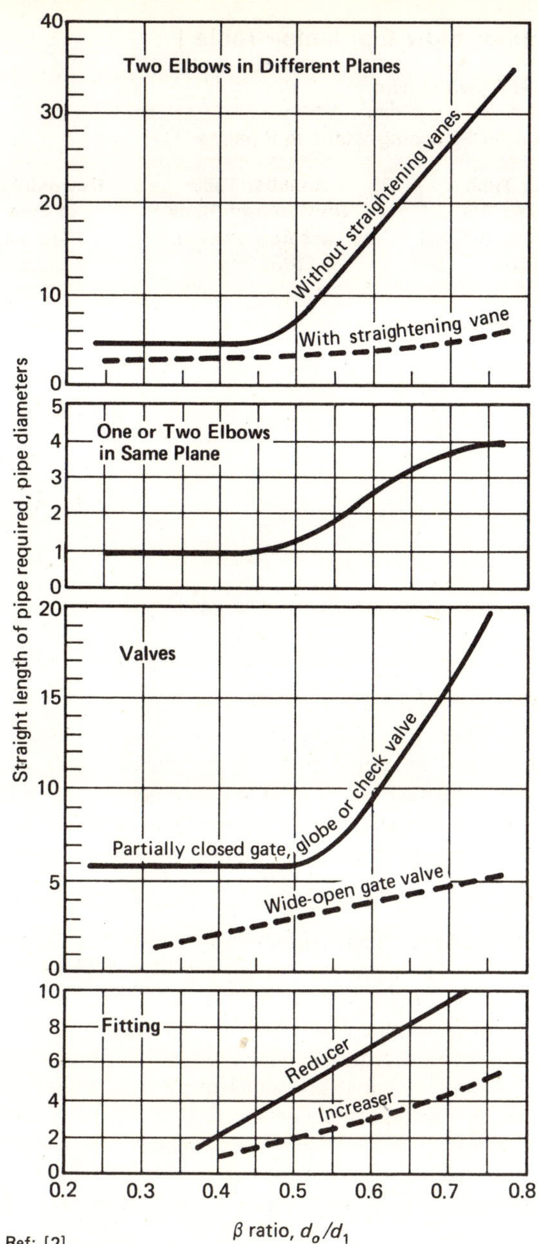

UPSTREAM straight-runs for venturi meters—Fig. 4

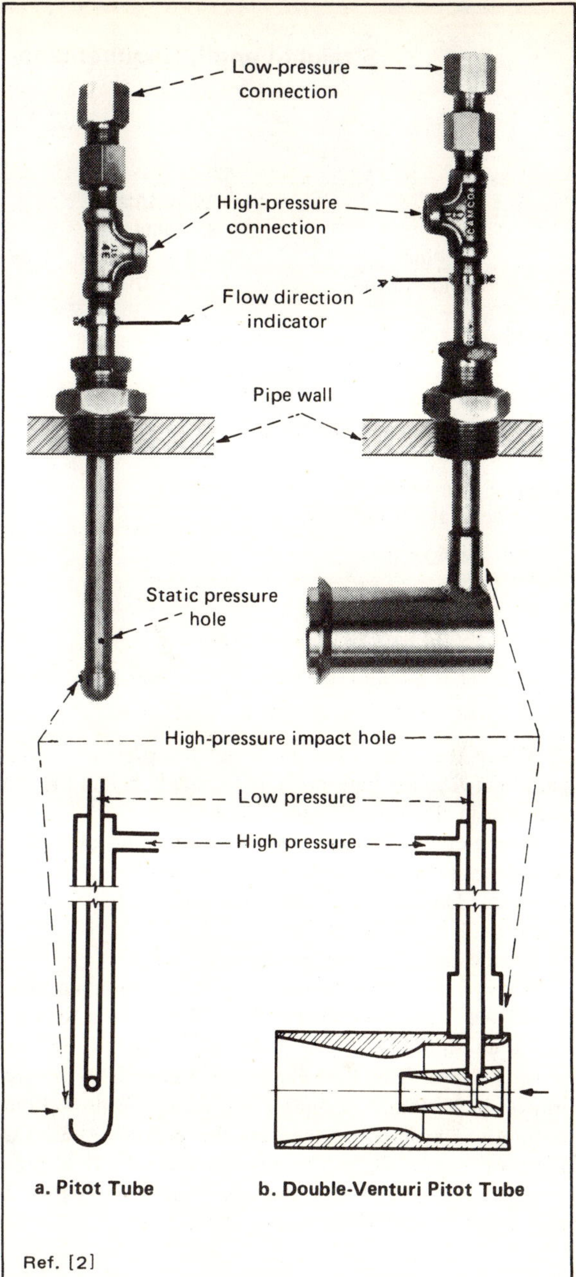

PITOT tubes measure the flow of clean fluids—Fig. 5

openings and valves should be accessible. If the pipeline is buried, an adequately sized concrete pit is provided for the venturi meter, and for instrument piping.

Pitot Tubes

The pitot tube works on the same principle as any other flow restriction. However, it is not a restriction in the pipeline but a restriction in an instrument. Because of this, pressure loss in the pipeline is negligible. The differential pressure at a conventional pitot tube (Fig. 5) is measured between the high-pressure impact hole directed against the flow, and a static hole, located at an angle of 90° or 180° to the impact hole.

Because of the single, small, impact hole, the pitot tube measures at only one point in the cross-section of a pipeline. Therefore, to obtain good measurements, the pitot tube must be precisely located at an average-velocity or maximum-velocity point and oriented in the direction of flow. A changing velocity changes the flow pattern and can result in a greater than acceptable error in measurement. Due to these conditions, we must provide the same straight length of piping as for orifice plates.

Pitot tubes are used in clean fluid service (usually in gas lines), are excellent for measuring flows having very high velocities, have a high capacity range, and are easy to install and remove.

Another version, termed a pitot-venturi tube, is also shown in Fig. 5. To the sensing tip of a pitot tube, a small venturi is added. The double-venturi arrangement,

Straight-Length Requirements for Annubar Flow Elements—Table I

	Upstream of Flow Element			
		Without Straightening Vanes in Pipeline		
Pipe Configurations	With ASME Straightening Vanes in Pipeline, Pipe Dia.	Annubar Tube (In the Same Plane as Last Approach-Turn), Pipe Dia.	Annubar Tube (In Different Plane as Last Approach-Turn), Pipe Dia.	Downstream of Flow Element, Pipe Dia.
One elbow or tee	6	7	9	3
Two elbows, or elbow and tee in same plane	8	9	14	3
Two elbows, or elbow and tee in two planes	9	19	24	4
Reducer or increaser Fully-open gate or ball valve	8	8	8	3
Partially-open valves Globe valve	9	24	24	4

Note: Control valves should be located after flow element.

shown in the illustration, increases the pressure differential between the high-pressure impact hole and the low pressure in the venturi throat [*2*].

Manometer deflections for pitot tubes and pitot-venturis are calculated in the same way as for orifice deflections. The capacity coefficient should be obtained from the manufacturers. For rough estimates, the capacity coefficient, β^2C, can be taken as 0.62.

Many of the disadvantages of the conventional pitot tube have been eliminated with an averaging pitot tube called an Annubar (Fig. 6). This device consists of two sensing tubes. The upstream tube has one to several impact holes (high-pressure side) facing the flow direction. An internal tube averages the pressure sensed at the four impact holes. The downstream tube (low-pressure side) measures the static pressure from which is subtracted the suction pressure of the flow.

The Annubar is an economical device in terms of capital and operating costs. In an unusual application, the flow element can be installed deep below grade without taking a pipeline out of service. This device is available for piping from ½ to 180 in, for pressures ranging from −30 in Hg to 2,500 psi, and temperatures to 1,200°F [*3*].

Formulas for sizing Annubars are similar to the orifice formulas. The manufacturer [*3*] provides a capacity coefficient as (K_gF_v), where K_g is a geometrical constant depending on pipe diameter, and F_v is a velocity distribution factor. For transitional and totally turbulent flow, $F_v = 0.82$. The capacity coefficient, (K_gF_v), is analogous to the orifice capacity constant, β^2C.

The Annubar has a very wide capacity range and fits in pipelines where turbulent flow exists. A change in the pipe size of the metering section is rarely necessary—and

ANNUBAR meter is an averaging pitot tube—Fig. 6

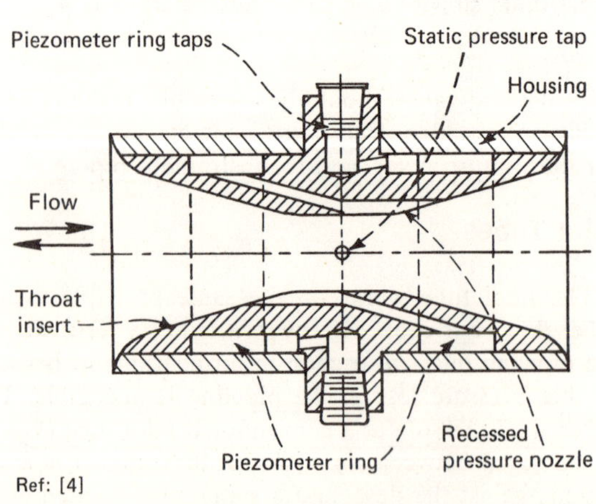

IMPACT tube handles flow in either direction—Fig. 7

then only for extremely low or extremely high flowrates. In most applications, only the operating manometer range of the instrument or control system needs to be selected. Permanent pressure loss is negligible.

Let us summarize the sizing data for Annubar pitots. The following table lists some representative values for various pipe sizes:

Pipe Size, Nominal, In	Capacity Coefficient, (K_gF_v) $F_v = 0.82$	Estimated Permanent Loss, % of h_w
½ to 1¼	0.6 to 0.62	10
1½ to 5	0.66 to 0.70	8 to 4
6 to 16	0.7 to 0.75	3 to 1
18 to 24	0.75 to 0.78	<½

Instrument deflection at flowing conditions for liquids:

$$h_w = [0.176Q\sqrt{S}/(K_gF_v)d_1^2]^2, \text{ in} \tag{6}$$

For instrument calibration at a standard 60°F liquid-flow condition, multiply right side of Eq. (6) by $(S_{60}/S)^2$.

Instrument deflection at flowing conditions for vapors and gases:

$$h_w = [0.00278W/(K_gF_v)d_1^2\sqrt{\rho}]^2, \text{ in} \tag{7}$$

Differential pressure is obtained from:

$$\Delta P = 0.0361h_w, \text{ psi} \tag{8}$$

Straight-length requirements for piping design, as recommended by the manufacturer, are given in Table I.

Impact Flow Tube

The flow-sensing tube (Fig. 7) consists of a short housing section, and a symmetrical and tapered throat section having a flow restriction in the center. The throat section contains two sets of impact nozzles. One set points upstream, the other downstream, and each is connected to an annular ring for averaging the impact pressure.

The differential pressure between the upstream and downstream openings results from the difference in impact pressures, and is a function of velocity head. Sizing of this flow tube is based on the general relationship of $v = C(2gh)^{1/2}$. Exact formulas for sizing this impact tube can be obtained from the manufacturer [*4*].

The housing and throat section of impact flow tubes are available in a wide range of metallic materials for pipe sizes ranging from 1 to 4 in. For low-pressure and low-temperature services, plastic-insert types are also available for pipe sizes ranging from 6 to 48 in.

Applications for the impact flow tube range from wind tunnels to sewer lines, and from gas to viscous flows. Flow can be in either direction. In dirty-fluid services, the impact openings can be purged.

Straight-length requirements for this device are; 6 pipe diameters upstream; 10 diameters upstream after a throttling valve; and 3 diameters downstream.

Rotameters

In rotameters, the area restriction varies in proportion to flowrate, and the pressure difference across the restric-

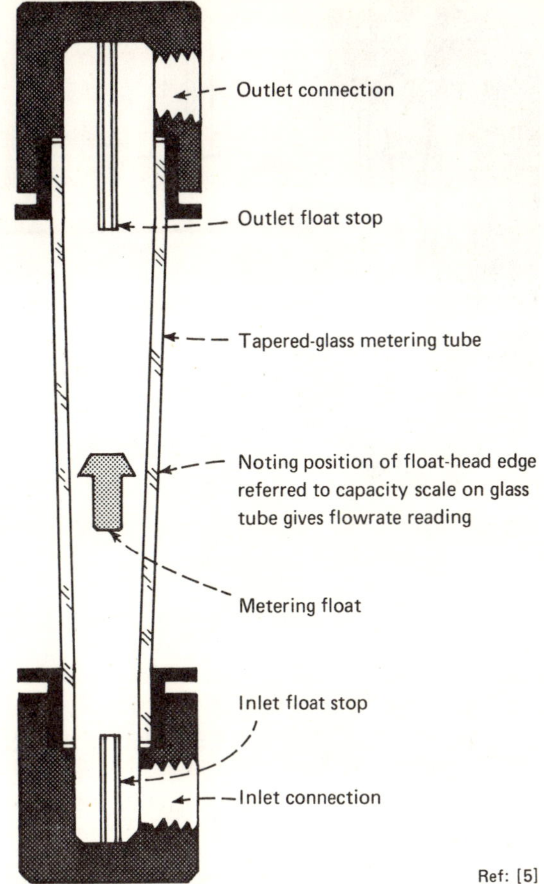

ROTAMETER has tapered metering element—Fig. 8

tion stays constant. In contrast, constant-restriction meters such as orifices and venturis have a fixed opening, and the pressure difference across the restriction becomes proportional to flow.

The rotameter consists of a tapered metering tube with a float that moves freely up and down (Fig. 8). The tube must be mounted vertically with fluid flow in an upward direction. The float will come to rest in a dynamic equilibrium when the pressure difference across the float, plus the buoyancy effect, balance the weight of the float. An increase in the flowrate causes the float to rise higher in the tube; a decrease causes it to fall.

In air and water service, the viscosity effects of the fluid on the rotameter remain practically constant. This makes possible the use of standard capacity tables for such flow streams. Standard sizing charts, tables of correction factors for any fluid, tables of correction factors for pressure and temperature, selection guides for types of rotameters, etc., are available in manufacturers' literature [*5*]. Hence, rotameter calculations are seldom made by process engineers.

In relation to piping design, pressure drop across the rotameter is negligible.

Rotameters are especially suitable for viscous liquids and very-small flowrates (less than 2 gpm, or 75 lb/h). However, a reasonable upper limit for the rotameter can be 300 gpm, and 3-in-dia. pipe. Units are available up

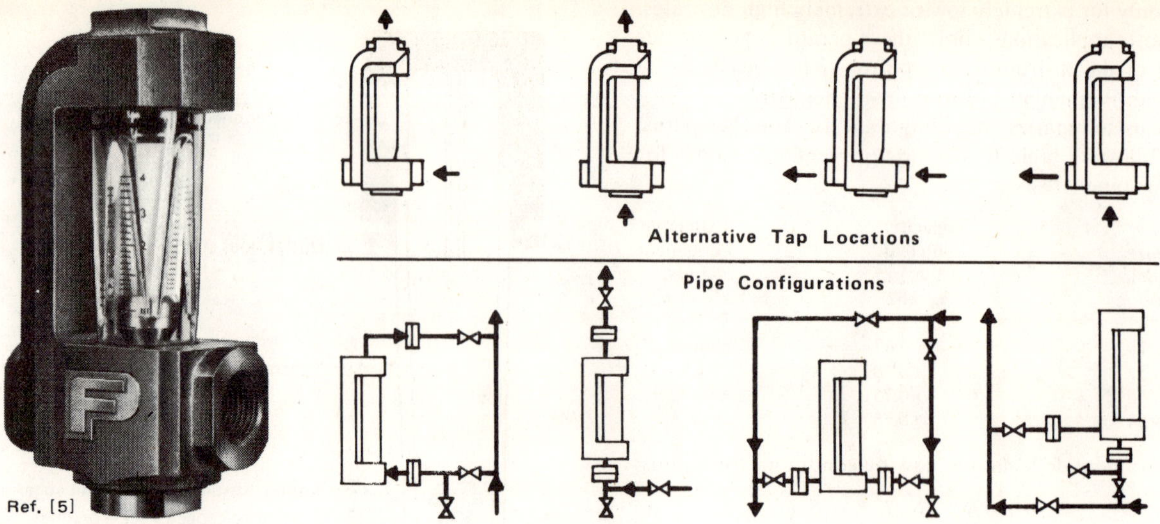

PIPING arrangements for installing a rotameter having alternative taps are simple and economical—Fig. 9

to 12 in and 4,000 gpm. The useful flow range is wide (10 to 1).

Rotameters can be used for slurries, depending on concentration of solids in the liquid, particle size and shape, density of solids relative to the carrier liquid, and degree of abrasiveness. Such applications should be reviewed with the manufacturer.

In special applications, the rotameter scale can be calibrated to show (a) fluid velocity, (b) percentage concentration in liquids in case a mixture is flowing, and (c) density or viscosity of liquid if volume flowrate and temperature can be held constant.

Piping configuration does not affect rotameter accuracy. Straight length of piping is not required. Depending on pipe configurations and rotameter design, alternative tap locations can be chosen, as shown in Fig. 9. These provide simple and economical piping arrangements.

The rotameter is usually installed between two block valves with a bypass. In clean service and with armored rotameters, a bypass globe valve is not necessary. Locate the flow-regulating globe valve to the rotameter (a) before the rotameter for liquid service, and (b) after the rotameter for gas service. Union joints in the inlet and outlet lines facilitate quick removal of the rotameter. Valves should be accessible and the rotameter scale visible from the operating aisle.

An inexpensive method for measuring large flowrates combines a rotameter with an orifice plate, as shown in Fig. 10. About 10% of the mainline flow passes through the rotameter.

Rotameter calibrations are usually nonadjustable and have differential ranges of 0–50, 0–100, 0–150, 0–200 and 0–400 in of water column. Rotameter tubes can be calibrated to show actual flowrates in the desired units. A magnetic yoke or an impedance coil added to the basic rotameter, provides the means of recording and transmitting the flow signal.

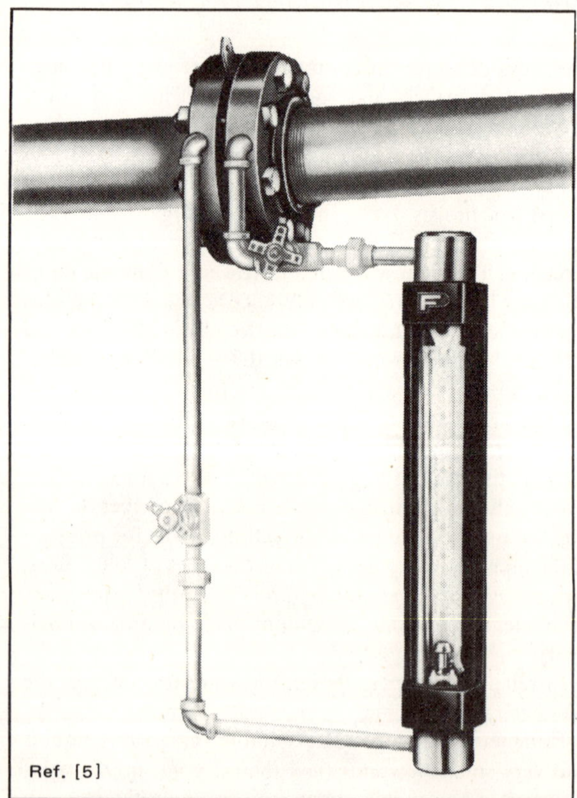

ROTAMETER-ORIFICE measures large flowrates—Fig. 10

References

1. Engineering Information on Venturi Meter Tubes, BIF Div., New York Air Brake Co., Providence, RI 02901.
2. Instructions for Pitot-Venturi Flow Element, Taylor Instrument Cos., Rochester, NY 14601.
3. "Technical Manual—Annubar," Elliot Instrument Div., Dietrich Standard Corp., Boulder, CO 80302.
4. Shea, Jr., J. A., Flow Tube Technical Paper, The Bethlehem Corp., Flow Tube Div., Bethlehem, PA 18016.
5. "Variable Area Flow Meter Handbook," Vol. I-III, Fischer & Porter Co., Warminster, PA 18974.

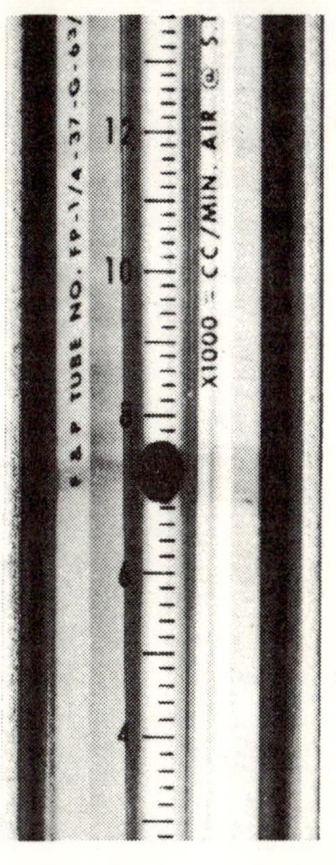

Low-Flow Measurement

JOHN YARD, Fischer & Porter Co.

Pilot-plant and other process operations involving low flows of liquids and gases present special problems in metering. Here is a look at advantages and limitations of instruments that can be used at low flowrates.

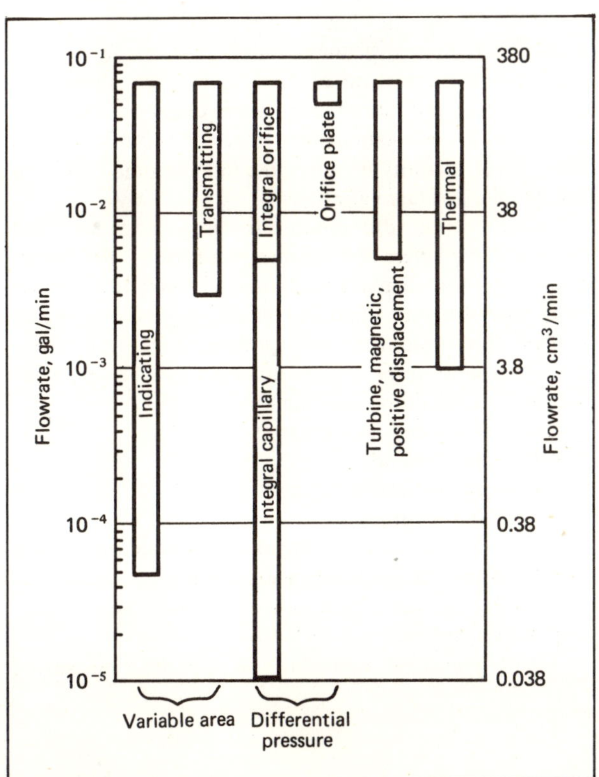

MINIMUM flow capabilities for liquid service—Fig 1a

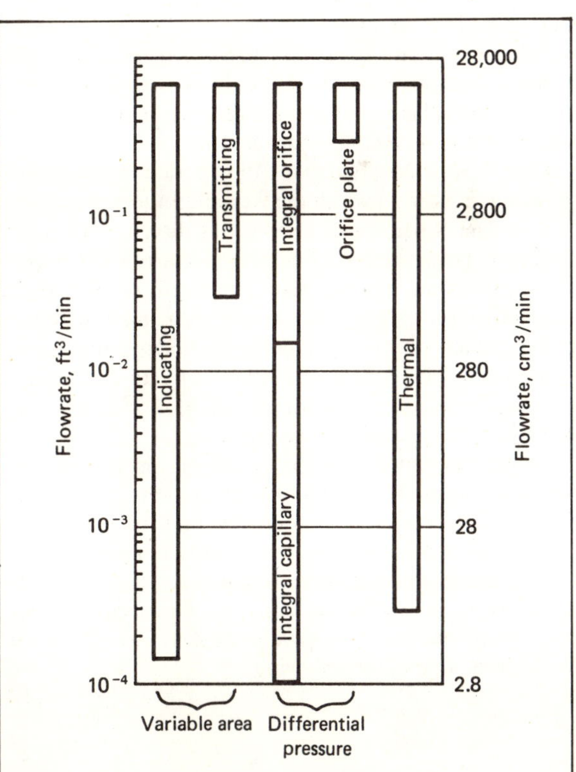

MINIMUM flow capabilities for gas service—Fig. 1b

Originally published April 15, 1974.

There is no universally accepted value below which fluid flow is called "low." One man's low flow may be another's high flow, depending on the application, the industry and the fluid being handled. However, for purposes of discussing low-flow measurement, some arbitrary limits should be established.

One possible criterion is pipe size. In most industries, fluid flow in pipes 1 in. dia. or smaller would probably be considered low. Exceptions might occur in large, single-product plants, where flow in pipes up to 4 in. dia. might be considered low.

Another definition of low flow could use flowrate as the yardstick. In this discussion, let us consider 10 gpm the upper limit of low flow. For the few flow meters that measure mass rather than volume, let us say that anything below 4,000 lb/h is in the realm of low flow.

Defining low flow of compressible fluids is more complex because pressure and temperature must be considered. For simplicity, let us revert to volumetric units and set the dividing line at 15 ft^3/min.

These arbitrary definitions of low flow include almost all types of flow-measuring devices. Among the few exceptions are ultrasonic, nuclear magnetic resonance, vortex, and angular-momentum meters. These devices are generally used to measure flowrates that would be difficult to consider "low."

Problems of Low-Flow Measurement

Perhaps the major problem in measuring low flowrates results from the mechanical limitations of meter design and operation. As meters become smaller, so do their working parts, and tolerances become increasingly important. Accuracy may be less than with similar, larger meters, while the meters themselves may be quite fragile and subject to maintenance problems.

Most flowmeters are designed to operate in the region of turbulent flow, in which the Reynolds number exceeds 10,000. The Reynolds number varies directly with pipe diameter. Thus, when fluid density, viscosity and velocity remain the same but pipe size is reduced, the Reynolds number is also reduced. As a result, with small-diameter piping, meters may be operated in the undesirable transition region where flow is neither fully turbulent nor laminar. In this range, flowmeter calibration is not predictable and repeatability may be poor.

Another potential problem in measuring low flow is purely mechanical. Most types of low-rate flowmeters employ orifices or close clearances to perform their function. These small openings can be easily plugged by particulate matter in the process stream.

The types of equipment capable of measuring flowrates within the range we have defined as low can be selected from among the following:

1. Variable area or rotameter, including
   a. Conventional round-tube type.
   b. Guided-tube type.
   c. Purgemeter.
   d. Low-density float type.
2. Differential-pressure meter, including
   a. Orifice-plate type.
   b. Integral-orifice type.
   c. Capillary type.
3. Turbine meter.
4. Magnetic meter.
5. Thermal meter.
6. Positive-displacement meter or pump.

The approximate lower measurement limits for these instruments are shown in Fig. 1a. and 1b.

Selection Factors

In addition to being suitable for measuring flow within the required range, one of the most important selection factors is whether volume or mass flowrate is desired. Some low-flow meters are volumetric, but most are hybrid designs, with output being a function of both volume flowrate and fluid density. And although flows are usually reported in volumetric units, mass flowrate is more often than not the value of interest. Thus, the influence of density and other fluid properties on the measurement cannot be ignored.

No attempt is made here to consider instrumentation for fluid-property measurement and control. Suffice it to say that the lower the flow to be measured, the more some fluid properties have a direct influence on the accuracy of flowrate instrumentation.

Low-flow meters are of two basic types: those that measure rate and those that measure quantity. Rate-type meters indicate instantaneous values of volume per unit time, while quantity meters show total volume delivered independent of rate variations.

Variable-area, differential-pressure, magnetic and thermal flowmeters are examples of rate-type instruments. Positive-displacement meters are typical of devices that measure quantity. Although it is possible to convert from volume to rate, and vice-versa, the difference in operating principle is worth considering. Other factors that should be considered are:

- Reliability and maintenance.
- Readout method.
- Lower flow limit and range.
- Cost and accuracy.
- Effect of viscosity, density and other fluid properties.
- Pressure and temperature extremes.
- Environmental conditions, such as corrosive or hazardous atmospheres.

Accuracy may be particularly important in measuring

Characteristics of Meters for Low-Rate Service

	Variable Area			Differential Pressure					Positive Displace-ment	
	5-in Scale	10-in Scale	24-in Scale	Orifice Plate	Integral Orifice	Capillary Tube	Turbine	Magnetic	Positive Displace-ment	Thermal
Service	Liquid or gas	Liquid or gas	Liquid or gas	Liquid or gas	Liquid or gas	Liquid or gas	Liquid	Liquid (electrially conductive)	Liquid	Gas preferred
Output characteristic	Linear	Linear	Linear or logarithmic	Square root	Square root	Linear	Non-linear	Linear	Linear	Non-linear (linearizing available)
Typical turndown	10:1	10:1	5:1 10:1	4:1	4:1	10:1	20:1	30:1	25:1	20:1
Typical accuracy	2% F.S.	2% F.S.	½% rate 1% rate	2½ F.S.	2% F.S.	10% F.S.	½% rate	½% F.S.	½% rate	1% F.S.
Transmission possibilities	Pneumatic or electronic	Indication only	Indication only	Pneumatic or electronic	Pneumatic or electronic	Pneumatic or electronic	Electronic (frequency)	Electronic	Electronic (frequency)	Electronic
Relative cost	Low	Low	Medium	Medium to high	Medium to high	Medium to high	Medium to high	High	Medium	High

F.S. = full scale

low rates in pilot-plant operations. Although errors after scaleup may be in the same proportion, the absolute error could be quite significant in pilot-plant work.

Now let us examine the advantages and limitations of the individual types of meters for measuring low flows.

Variable-Area Meter

The variable-area flowmeter, or rotameter, is a simple, low-cost, direct-reading indicator for measuring flow of liquids or gases. It also can be equipped with means for electronic or pneumatic transmission. Generally speaking, it is the most economical device for measuring flow of relatively clean, low-viscosity fluids.

In the variable-area flowmeter, a moving body called the "float" presents a restriction in the line. The annular area between the float and the walls of the tube forms a variable orifice. Since the float moves freely within the tube, the pressure drop across the float remains constant as the flowrate changes. The tube is designed so that the area of the annulus is proportional to the height of the float in the tube. The scale can be very nearly linear over flow ranges as broad as 12 to 1.

Rotameters tend to be self-cleaning. However, the annular area between the float and the tube may be quite small when measuring low flowrates, and the fluids should be as free as possible of entrained solids to avoid plugging the annulus or damaging the float.

Variable-area meters can be used to measure flowrates of many fluids within predictable viscosity limits. They can provide direct mass-flow measurement of low-viscosity fluids, such as light hydrocarbons or aqueous solutions, by using floats with specially selected density.

Conventional rotameters permit flow measurement as low as 0.1 cm^3/min of water or an equivalent gas flow. The ability to read the scale easily is the major factor in rotameter performance. Uncalibrated accuracy with a linear scale is ± 2% of full scale. Tubes with nonlinear tapers expand low-flow readability. For example, precision rotameters with logarithmic scales permit accuracies of 0.5%. A logarithmic scale provides greater accuracy at the lower readings, where small errors on a linear scale would result in relatively large absolute errors.

For measuring very small flows, down to 0.05 cm^3/min, variable-area meters are available with glass tubes having noncircular cross-sections, and ribs or flats to give the float proper guiding. The basis for the guided-tube design is that accurate flow measurement with a variable-area meter depends on the concentricity of the float within the tube.

Rotameter scales can be factory calibrated to show flowrates directly for a particular fluid under given operating conditions. However, a large proportion of low-flowrate meters are equipped with diameter-ratio scales in which the scale reading is based on the ratio of tube diameter to float diameter. Readings are converted to flowrates for various fluids by using flow coefficients.

For a particular float shape used with geometrically similar tubes, the flow coefficient of the meter is a function only of the Reynolds number of the fluid and the diameter ratio of the meter. The coefficient, K, is plotted as a family of curves, each representing a specific diameter ratio or scale reading as represented by the position of the float. A family of such curves plotted with the viscous-influence number, N, which is based on the Reynolds number, is shown in Fig. 2. The flow coefficient is directly convertible into flowrate from charts supplied with the meter.

It can be seen from Fig. 2 that in this case the flow coefficient is independent of viscous-influence numbers of 2 or less, and therefore is independent of fluid density and viscosity when N is less than 2. At other values of N, changes in fluid density or viscosity will produce significant errors if no correction is made.

Straight-through metal-tube rotameters are used for measuring low flows of liquid or gases at high temperatures and pressures. These instruments can be used to determine liquid flows as low as 10 cm^3/min (and equivalent gas flows) at temperatures to 500°F and pressures above 3,000 psi.

Mass-Flow Rotameters

Although variable-area meters are generally thought of as strictly volumetric instruments, their operating principle makes them adaptable to also measuring mass flow of liquids. Normally, a rotameter float is made of stainless steel having a density of 4 to 16 times that of most common liquids. A more buoyant float, with a density of about twice that of the fluid, responds to changes in fluid density so that the reading varies with mass flow even though the volumetric flowrate remains constant.

Theoretically, the relationship of float density to fluid density should remain constant. This is impossible, however, so meter indications are based upon mean fluid density. So, as long as fluid density does not vary greatly, inaccuracies will be small.

Mass-flow rotameters are extremely accurate for measuring fluids having densities that vary as much as ±10%. In cases where air entrainment causes variations in the apparent density of a fluid, the mass-flow rotameter is much more accurate than volumetric flowmeters.

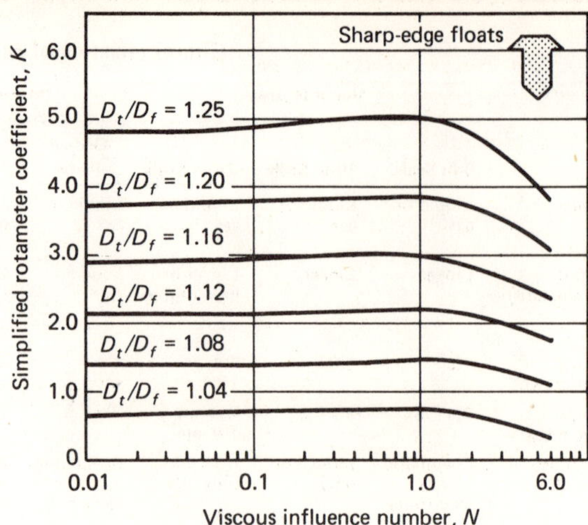

TYPICAL rotameter prediction curve—Fig. 2

Differential-Pressure Meters

Like the variable-area meter, the differential-pressure meter places a restriction in the line. However, whereas the variable-area meter maintains equal pressure across the restriction and "adjusts" the area of the orifice, the differential-pressure flowmeter has a fixed orifice. Flowrate varies as the square of the pressure drop.

An orifice meter consists of an orifice plate and a differential-pressure transmitter. The transmitter develops an output signal, either electronic or pneumatic, that is proportional to the pressure differential across the orifice. Volumetric flowrate is proportional to the square root of the pressure differential, and the output and meter scale is therefore a square-root curve.

Differential-pressure meters are simple, reliable and low cost. They are independent of line size because only the orifice need be sized. Their square-root output characteristic provides high sensitivity, and accuracy above 50% of full scale.

The sharp-edge orifice plate, one of several types, is simple, inexpensive and accurate. Although in larger pipe sizes, sharp-edge orifice-plate meters may not require calibration, in pipes below 1.5 in. dia. they must be calibrated because of the close machining tolerances involved in their manufacture. The minimum practical flowrate for this type in a 0.5-in line is about 0.05 gpm.

Integral-orifice, or shunt-tube differential-pressure, flow transmitters are used to measure lower flows. They operate in the range from 6.0 to 0.005 gpm. Integral-orifice transmitters employing capillary tubes will measure flows in the range from 40 to 0.05 cm^3/min. They are very sensitive to changes in fluid viscosity, which produce a shift in flow calibration.

Turbine Meters

Turbine meters are the best performing liquid flowmeters available today. They offer excellent repeatability with accuracies of 0.25% of rate over ranges as great as 70 to 1. Both axial and paddle-wheel type meters are produced for measuring liquid flows as low as 0.005 gpm or about 20 cm^3/min. In the smaller diameters associated with low-flow measurement, the cost of a turbine meter is comparable to that of an electronic-differential-pressure transmitter.

Turbine meters measure the quantity of fluid flowing. They provide a digital output that is proportional to fluid velocity and flowrate. Turbine meters are generally used with low-viscosity fluids, and output is linear within a given viscosity range. They can also be used with high-viscosity fluids because their output characteristic is very repeatable, although nonlinear and a function of fluid viscosity.

Liquids to be measured by a turbine meter should be clean. This is especially true in the small sizes, which have tight clearances. Turbine meters are particularly suitable with fluids having lubricating properties because the flowing liquid lubricates the bearings. They are widely used in precision batching systems, in blending, and in metering high-cost products.

Magnetic Flowmeters

Magnetic flowmeters employ electromagnetic induction to produce an alternating-current voltage that is directly proportional to the rate of flow. The signal can be converted to indicate rate or total flow. Theoretically, the magnetic flowmeter approaches the ideal because it offers no restriction in the line.

One of the primary applications for magnetic flowmeters is on corrosive fluids. Liners are available in a variety of corrosion-resistant materials. In the smaller sizes, these include vitreous enamel, fluorocarbon plastic, fiber-glass-reinforced epoxy, and polyurethane.

Other advantages of the magnetic flowmeter are: pressure loss is no greater than through an equivalent length of pipe; output is completely linear with respect to flowrate; the meter gives an electrical output signal; and results are not influenced by changes in viscosity, temperature, turbulence or conductivity above a minimum

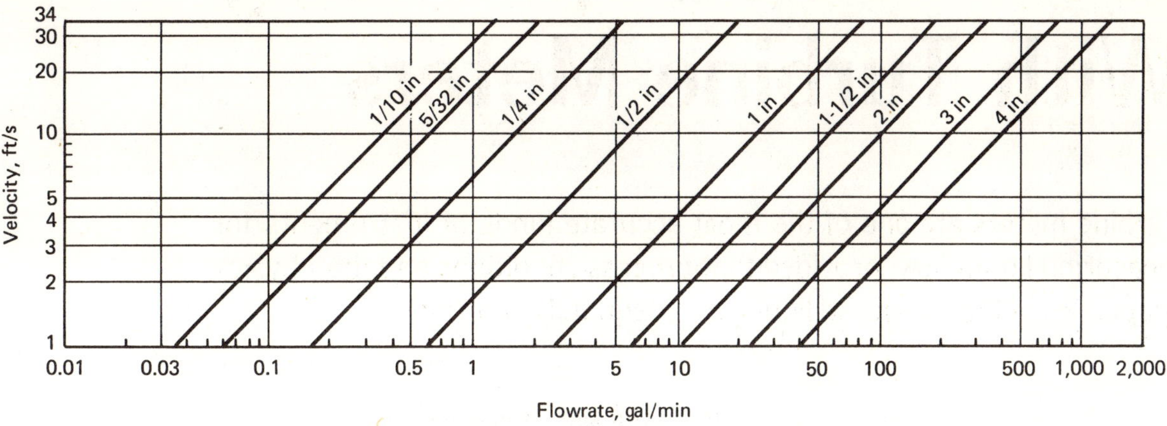

APPROXIMATE CAPACITIES for low-flow magnetic meters on lines up to 4 in. dia.—Fig. 3

threshold. In addition, the development of ultrasonic cleaning has virtually eliminated electrode fouling as a limitation in metering dirty streams.

A primary requirement of magnetic flowmeters is, of course, that the fluid be electrically conductive. An absolute minimum conductivity of 0.1 micromhos/cm is necessary. In the smaller sizes with which we are concerned, conductivity must be even higher, depending on pipe size and the distance of the receiver from the flowmeter. Magnetic flowmeters for low flowrates are relatively expensive. They also are heavy, generally requiring special piping supports.

Fig. 3 gives approximate capacities for low-flow magnetic units based on fluid velocities of from 1 to 30 ft/s. Minimum full-scale flow velocity is 1 ft/s, giving a volumetric flowrate of 0.033 gpm in a meter with 0.1-in bore.

Positive-Displacement Meters

Essentially, the positive-displacement meter divides the flow into discrete volumetric increments and then counts the increments to give the quantity delivered. Many configurations—pistons, rotating members, wobble disks, gears, etc.—are used to obtain the positive-displacement effect. In recent years, piston-type meters have been developed that are capable of measuring flows as low as 0.005 gpm.

Small positive-displacement flowmeters are designed primarily for use with low-, to medium-viscosity liquids. Better performance is achieved with liquids having lubricating properties. These meters are relatively independent of changes in fluid viscosity or density within their operating range.

Positive-displacement pumps are capable of measuring lower flows, as little as 3 cm^3/h, than any other type of meter. These operate in essentially the same manner as positive-displacement meters, except that they are active rather than passive devices. There are many types and designs. Like positive-displacement meters, pumps are relatively independent of fluid viscosity and density.

Thermal Meters

Unlike any of the meters previously discussed (with the exception of the low-density-float variable-area meter), the thermal flowmeter measures mass flow directly. Thermal meters apply heat to the fluid by various means, and measure the rate at which the fluid stream conducts the heat from the source. Heat conducted is directly proportional to the mass flowrate of the fluid.

In addition to measuring mass flow directly, other advantages of thermal flowmeters include negligible pressure drop, ability to use corrosion- and abrasion-resistant materials, and no moving parts.

Standard units are capable of measuring mass flow as low as 0.2g/min and as high as 200 lb/min. Even lower and higher rates are available on special order. Normal measuring ranges are 10 to 1.

In summary, most of the flowmeters used for larger process stream flows can be applied to low flowrates as well. Generally, the lower limit of flow-measuring capability is determined by difficulties in manufacturing the device to very tight tolerances rather than any inherent characteristic of the measuring principle. Close tolerances may also cause problems in operation and maintenance. The search for the ideal, universal flowmeter continues, with present emphasis centering on laser technology.

Meet the Author

John Yard is manager of flow products development, Fischer & Porter Co., Warminster, PA 18974. He has been in the field of fluid-flow measurement since 1952, and is currently active in the ASME standards committee on measurement of fluid flow in closed conduits and the Scientific Apparatus Makers Assn. Fluid Meters Standards Committee. Mr. Yard has a B.S. degree in chemical engineering from Lehigh University and a M.S. in engineering from Pennsylvania State University.

Accurate Flow Measurements With Turbine Meters

Turbine meters are one of the most accurate kinds of instruments for measuring liquid flow, provided that process conditions and the physical properties of the measured liquid are adequately controlled.

DONALD LYNN MAY, Monsanto Co.

When properly applied, turbine meters produce measurements that are repeatable within 0.1%, and accuracies of 0.25% in 200,000 lb. of material. These meters, however, have limitations and are not accurate in all streams.

Since the instruments are sized for a given set of process conditions—including temperature, viscosity, concentration and flowrate—the accuracy of the meter can be affected if these vary frequently. Therefore, the range of process variables in a stream should be carefully reviewed; if they cannot be controlled or limited, turbine meters must not be used.

The following paragraphs discuss factors affecting meter accuracy, and how they can be controlled to the point where turbine meters can be used to advantage.

Effect of Specific Gravity

Often, it is desirable to measure and record process flows in pounds. The formula to determine the meter totalizer dividing-factor is:

Pulses/lb. = (Pulses/gal.)/(8.33 × sp.gr.)

where pulses/gal. is a calibration number furnished by the factory for water; 8.33 is a constant; and sp.gr. is the specific gravity of the liquid at the flowing temperature.

Specific-gravity changes over the range of operating temperature must be considered. If the temperature is well controlled, or the effect of temperature on specific gravity is small enough, then the changes in specific gravity can be ignored.

Fig. 1 corresponds to a stream with a daily temperature variation of 18 C., which causes a change in specific gravity from 0.917 to 0.932. Since the water-calibration number furnished by the factory for a 1-in. meter would be approximately 935, a specific gravity of 0.917 would yield (in the above formula) a dividing factor of 122, whereas a specific gravity of 0.932 would yield a factor of 120. In a 200,000-lb. batch, this would amount to an error of 1.7%. To reduce the error, the following might be done: (1) provide temperature control; (2) for short-duration batch-processes, record the temperature at the time of batching; (3) for continuous processes, record the temperature and corresponding pounds every one to two hours.

Totalizer readings can then be converted to pounds at a common temperature. A more practical solution might be to leave the process readout as gallons, and let accounting personnel calculate and furnish operators with the amount of pounds.

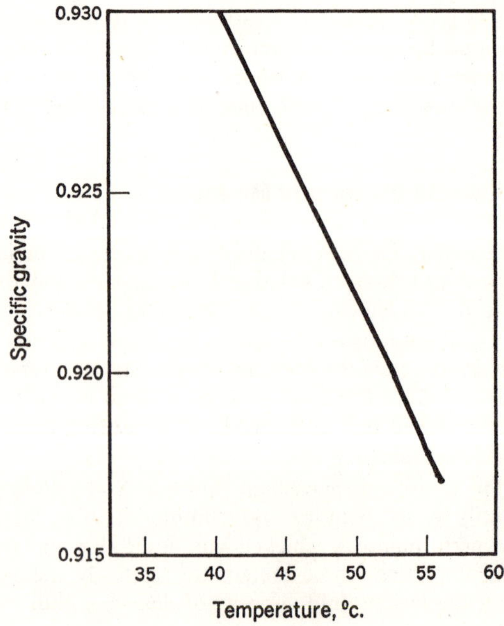

CHANGE OF SPECIFIC GRAVITY of one given liquid with temperature variations—Fig. 1

Originally published March 8, 1971.

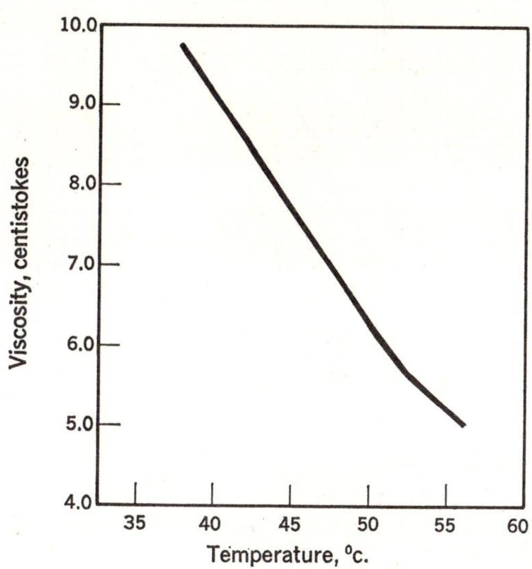

VISCOSITY CHANGE with variations in the temperature of the particular liquid being handled—Fig. 2

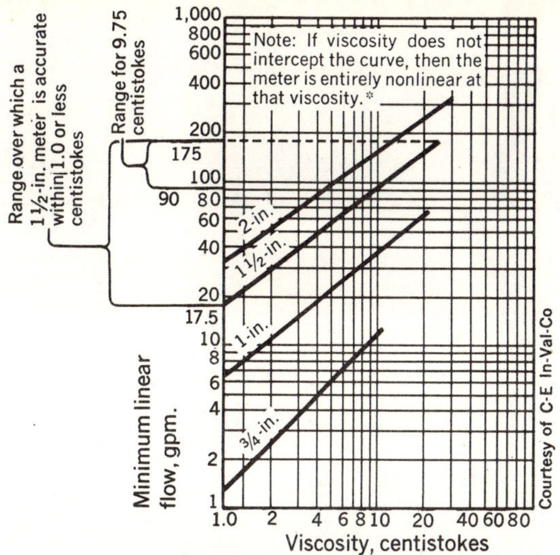

MINIMUM LINEAR FLOW vs. liquid viscosity for ¾-in. through 2-in. meters. Viscosity effects on meters vary with each instrument manufacturer—Fig. 3

Viscosity

When dealing with highly viscous materials, observe how much the viscosity changes with variations in temperature. Notice in Fig. 2 how the viscosity of the particular stream increases as the temperature decreases. Now look at Fig. 3 and note how the minimum flowrate for accurate flow is increased as the viscosity is increased.

Fig. 3 shows that, at a viscosity of 1.0 centistokes or less, a 1½-in. meter would be expected to operate accurately over a range of 17.5 to 175 gpm. However, Fig. 2 shows that the viscosity of this stream at 38 C. is 9.75 centistokes. If now a vertical line is drawn upward in Fig. 3 from 9.75 centistokes, the line will intersect the 1½-in. meter-curve at a flowrate of 90 gpm. Therefore, to stay within the rated accuracy of the meter, the flow for a 1½-in. turbine meter in this stream is limited to from 90 to 175 gpm. Operating at a flow smaller than this would produce inaccurately low meter readings.

As temperature decreases, viscosity increases. This, in turn, raises the minimum flowrate over which the meter will operate accurately. Therefore, to ensure precise flow measurement in this stream, it is necessary to establish the minimum operating temperature, and guard against the process flowing below it.

If the minimum temperature established is 50 C., the viscosity is 6.3, and the minimum flowrate 65 gpm. Although temperature controls are not required, it is necessary to keep the temperature above a certain level. Reducing the cooling, or increasing the heating to the stream, may be all that is required to maintain the rated accuracy of the meter.

Flowrate Range

Manufacturers generally calibrate turbine meters with water. Typical of the data supplied by the factory are these:

Pulses/Gal.	Flowrates, Gpm.
941.3	6.3
940.4	15.9
937.2	32.0
935.7	48.0
934.5	64.2

Since pulses/gal. average 937.8, the meter dividing factor is set at 938 pulses/gal.

The average range of error possible can be calculated as follows:

(941.3 maximum pulses/gal.) − (934.5 minimum pulses/gal.) = 6.8 pulses difference
or 6.8/2 = ±3.4 pulses
or ±(3.4 × 100)/938 = ±0.36% error

Therefore, this meter would be accurate to ±0.36% within the range 6.3 to 64.2 gpm. A standard meter from a vendor can have a range of error as high as ±0.5%.

There are two ways to decrease the range of error related to this calibration factor. One is to specify a calibration accuracy of ±0.25% when ordering this meter. Most vendors can comply with such requirements for an additional fee. The other method is to narrow the operating range of flow, and set the dividing factor for that range. For example, a dividing factor of 936 on the abovementioned meter would provide an accuracy of ±0.14% if the meter is limited to a flow range of 32.0 to 64.2 gpm.

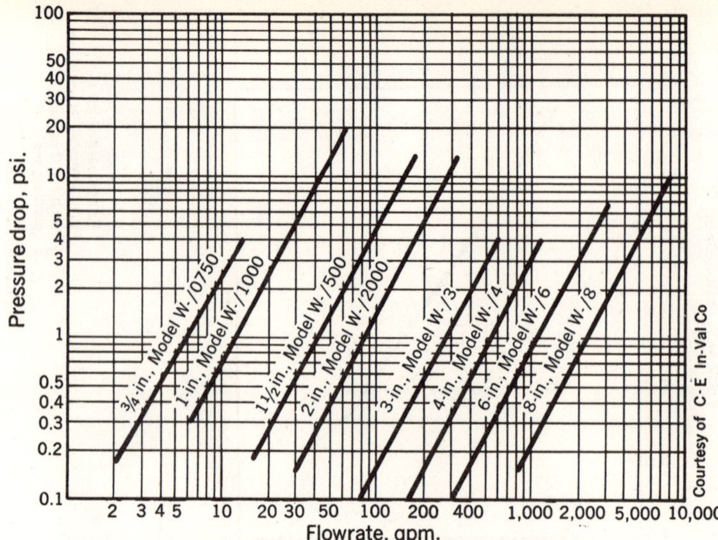

Formula for estimating pressure drop on a liquid other than water:

$$\Delta P = \mu^{0.25}\,(\text{sp.gr.})^{0.75}\Delta P_w$$

where ΔP = pressure drop of liquid, psi.; μ = viscosity of liquid, cp.; sp.gr. = specific gravity of liquid; ΔP_w = pressure drop for water, psi., as obtained from the figure.

PRESSURE DROP VALUES at different water flowrates, caused by turbine meters manufactured by C-E In-Val-Co—Fig. 4

Flashing Effects

Flashing, or the change of a liquid to a gas, can occur inside of, or downstream of a turbine meter. Although the pressure drop produced by a turbine meter is small, if the upstream pressure is already at a point close to the vaporization of the liquid, an additional pressure drop of 5 to 10 psig. may be enough to cause flashing. Fig. 4 shows a graph for determining the pressure drop caused by an In-Val-Co turbine meter in water, and lists a formula for calculating the pressure drop in other fluids.

Evidence of flashing is sometimes noted by noise, vibration or physical damage to the meter. If flashing does occur, the meter totalizer will indicate more than the actual flow. A simple way of avoiding this problem may be to raise the upstream pressure to keep the pressure downstream of the meter above the liquid's vapor pressure.

Good Installation—A Must

Proper installation is mandatory if accuracy and repeatability are to be maintained. A straight uninterrupted run of pipe the same size as the meter should be provided for a length of at least 15 to 20 pipe diameters upstream, and 10 pipe diameters downstream of the meter.

Any change in pipe I.D., nozzles, valves or fittings located closer to the meter inlet than 15 pipe diameters will cause turbulence, which results in a high meter reading. As with orifice runs, the length

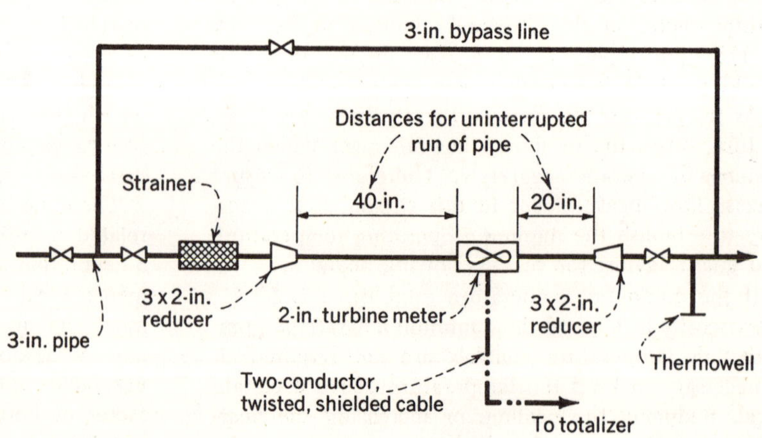

INSTALLATION DETAILS for a 2-in. turbine meter as required for installation in a 3-in. pipeline—Fig. 5

of straight pipe required can be shortened by using straightening vanes in front of the meter. If the process is to be flushed or cleaned with steam, or is not readily shut down, a bypass line is necessary. Fig. 5 shows a typical installation that meets all these requirements.

Foreign objects such as nuts and bolts, and even wrenches, sometimes get into pipelines when pumps and valves are serviced or removed. A large mesh filter or pancake strainer installed upstream of the meter will prevent such objects from damaging the meter or carrying them further downstream.

The low-voltage signals generated by a turbine meter (50 mv. to 10 v.) must be transmitted by a two-conductor twisted cable with good shielding, which should be installed in conduit or cable trays to guard against physical damage or deterioration of the insulation. The cable should not be routed in the same conduit or tray with wires used for a.c. power or control.

Since the temperature of a liquid in a vessel is not necessarily the same as it is in a pipeline leaving the vessel—difference may be caused by insulation, mixing, agitation—a temperature-measuring instrument in the tank may not suffice. A thermowell should also be installed in the pipeline near the turbine meter for temperature measurement.

Decisions must be made as to whether to install a permanent temperature monitor and, if so, how accurate a monitor. These decisions will depend upon the amount of change in viscosity and concentration caused by a change in temperature.

Operating Error

A good turbine-meter installation in a well-controlled process will not in itself always render the accuracy required, because the use given to the meter depends on the operator. He should, therefore, be made aware of the meter limitations and the consequences of going beyond them.

To minimize or eliminate chances of error, high and low-limit alarms, check valves, or automatic shutoff valves can be installed. Review the operating procedure and keep the following points in mind:

- If steam is allowed to go through the meter, the bearings will be damaged and the meter will read lower than it should.
- If any bypass valves are left open, or leak, meter readings will be lower than actual flow.
- If any conditions are present that can cause a backward flow through the line, they could cause the meter to monitor the same flow two or three times. This can occur if the meter is mounted in a low place in the line, if a tank overflows, or if the system is vented upstream for the relief of high pressure.
- If the sequence of valving is critical, it should be so noted.

This last point is illustrated by Fig. 6, which shows the valving involved around an actual truck-loading facility. Here, it was discovered that im-

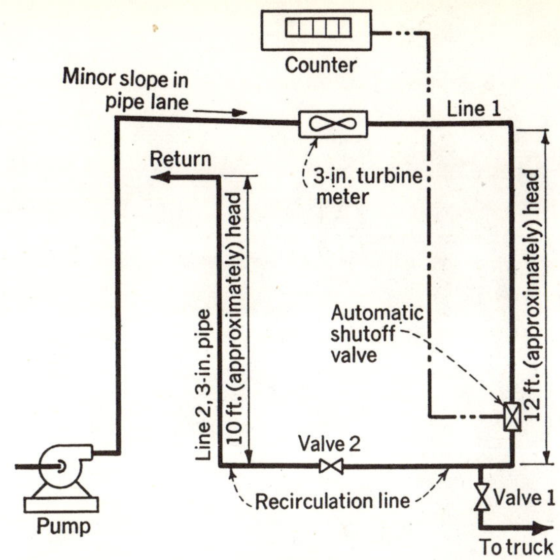

TRUCK-LOADING FACILITY with the aid of a 3-in. turbine meter placed before automatic shutoff valve—Fig. 6

proper valving sequence could cause an error of 1 to 2%, as explained in what follows:

If Valve 1 is opened with the pump off and before resetting the counter to zero, the liquid in the 3-in. line between the meter and Valve 1 will drain into the truck. Since additional flow from Line 1 upstream of the meter will also empty into the truck, about 75 to 150 lb. of liquid are loaded before the meter is started. Or: If Valve 1 is opened before the pump is started and drainage occurs, the turbine meter will be monitoring a line that is not full of liquid, causing a slightly high totalizer reading.

If Valve 1 is opened before Valve 2 is closed, the head of liquid in Line 2 will drain into the truck. The amount of liquid will depend on how soon Valve 2 is closed after Valve 1 is opened, and on the slope of Line 2 in the pipe lane. This could result in approximately 100 to 300 lb. excess fluid in the truck.

The proper sequence for this loading facility is to start the pump and fill the lines, close Valve 2, set the meter counter to zero, and then open Valve 1 to the truck.

Meet the Author

Donald Lynn May is senior instrument engineer with Monsanto Co.'s chemical plant at Pensacola, Fla. (P.O. Box 1507), where he is engaged in system engineering and instrument application for maintenance and capital projects. Previously, he worked for Brown Engineering Co. and Phillips Chemical Co., after receiving a B.S. degree in electrical engineering from Auburn University, Auburn, Ala. He is a registered professional engineer, a member of IEEE and senior member of the Instrument Soc. of America.

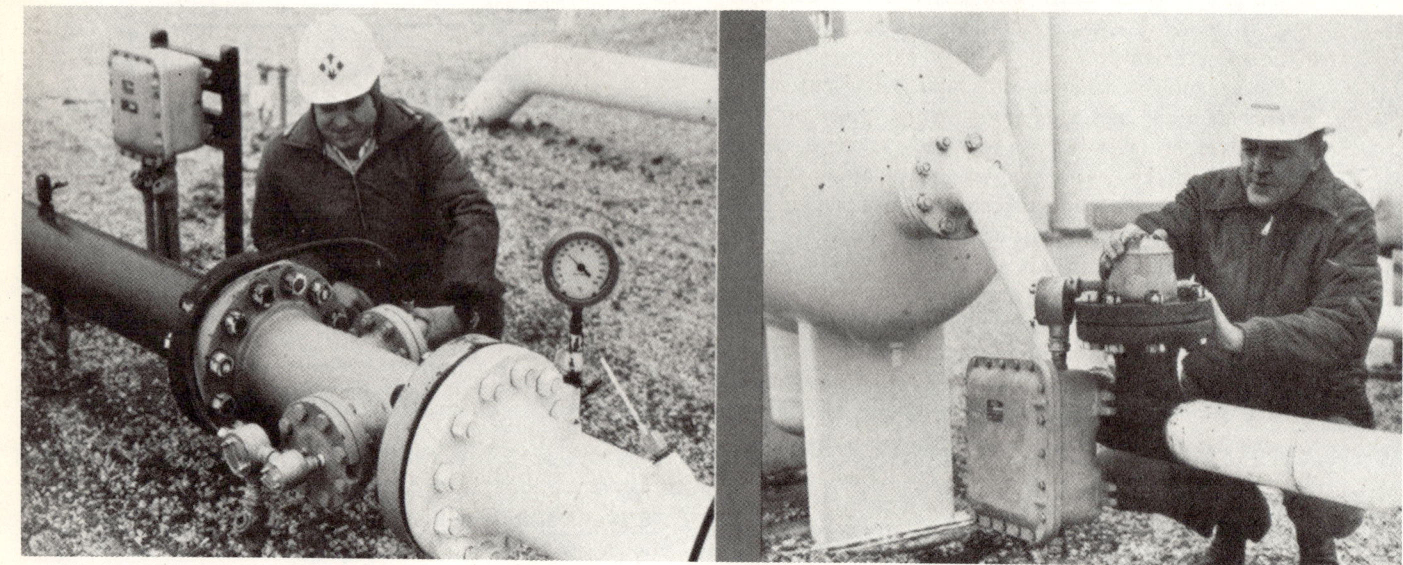

SONIC flowmeter (left) and interface detector (right) monitor petroleum products in pipelines.

Sound Velocimeters Monitor Process Streams

The velocity of sound is a function of the sonic path and the composition and temperature of the mixture. Sound velocimeters, correlating these variables, can measure certain chemical and physical properties of the solutions; or, by combining two velocimeters with the proper electronics they can measure flow.

ELLIS M. ZACHARIAS, JR., NUSonics, Inc. and DONALD W. FRANZ, HBH Associates, Inc.

Improvements in electro-acoustic transducer technology, and the emergence of electronic circuits using integrated-circuit components, have enabled sound-velocity instruments to become competitive industrial measurement devices. The addition of automatic sensitivity-control circuits allows sonic attenuation variations of over 100 to 1 to be accommodated. With these new electronics, instrument repeatabilities of 0.001% are being achieved in regular industrial use, and one user reports a stability of 0.0001%.[4]

Sound velocimeters may be used to monitor changes in physical and chemical properties of liquids. The physical properties are: density, specific gravity, bulk modulus, and compressibility (the reciprocal of bulk modulus). Chemical properties include: concentration of solutions (relative proportion of two or more liquids), solids content (relative proportion of sonic transparent solids in a liquid), and percent conversion of monomers to polymers (change in the molecular structure, and hence "elasticity" of the liquid).

Changes in viscosity may also be indirectly monitored with sound-velocity instruments as in the case of polymerization of monomers in which both the viscosity and the solids content increase. By monitoring the solids content, the changes in the viscosity of the polymerization reaction can be observed.

Originally published January 22, 1973.

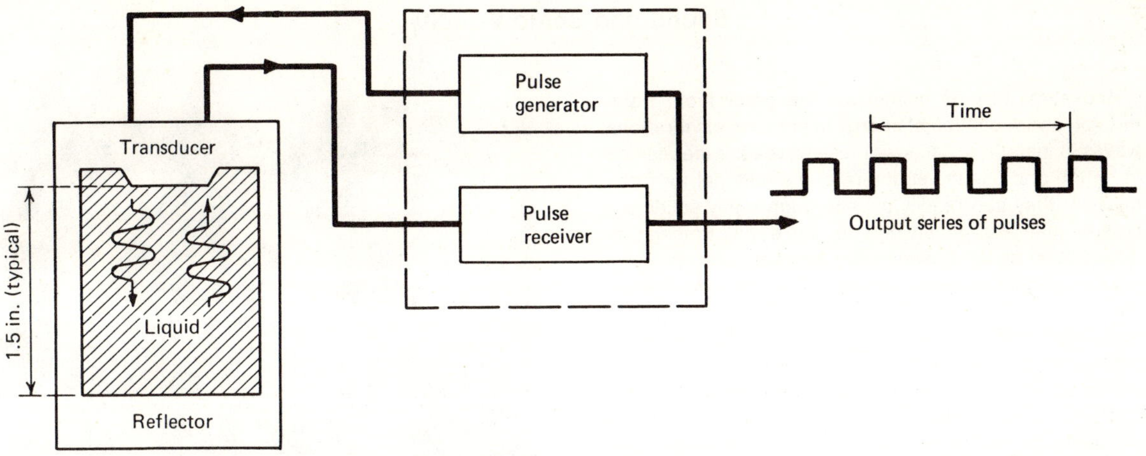

TRANSDUCER and electronic circuits form basic system for measuring sound velocity—Fig. 1

Operating Principle

To understand the operation of sound-velocity instruments, the basic "sing-around" system will be described.[3] An analyzer probe, submerged in the liquid being monitored, has an electro-acoustic transducer serving as both the transmitter and receiver, and a reflector to form a sonic path of fixed length (see Fig. 1). This sonic path along with the necessary electronics is the basic component of the sing-around circuit.

A short electrical pulse from the pulse generator excites the transducer, resulting in the emission of sonic compression waves into the liquid. These high-frequency waves (in the megahertz region) form a sonic pulse, and travel through the liquid, bouncing off the reflector and returning to the transducer that now acts as a receiver. Here the sonic pulse is converted to an electrical signal, is amplified, and threshold-detected in the pulse receiver. As soon as the pulse amplitude reaches a preset level, the threshold detector in the pulse receiver triggers the pulse generator, and the cycle is repeated. In this way, a pulse-repetition rate, or sing-around frequency, is set up.

Since the pulse-repetition rate is dependent upon the transit time of the sonic pulse through the liquid, it is a measure of the sound velocity in the liquid. It is a direct function of the:

1. Transit time of the sonic pulse through the liquid due to (a) length of the sonic path, (b) composition, and (c) temperature of the liquid.

2. Delay time of the electronic pulse in the circuitry.

The sound velocity may be determined from:

$$c = \frac{Af(1 + \alpha T)}{1 - Bf10^{-6}} \quad (1)$$

where c is sound velocity, f is frequency of the pulse-repetition rate, A is sonic-path length in meters (typically 0.086 m.), B is electonics delay time in μsec. (about 0.5 μsec.), α is coefficient of thermal expansion for the probe (causing changes in the path length of the sonic pulse), and T is temperature, deg. C.

For greater resolution, a frequency multiplier may be used to increase the pulse-repetition rate. NUSonics uses a multiplier factor of 7. Hence, the revised equation for determining the sonic velocity becomes:

$$c = \frac{Af(1 + \alpha T)}{7 - Bf10^{-6}} \quad (2)$$

For small variations of temperature and pressure, Eq. (2) may be approximated by the following simple relationship:

$$c = Kf \quad (3)$$

where K is a constant.

Typical output frequency of these units is 122,000 Hz. for a sonic velocity of 1,500 m./sec. in water.

For process monitoring, sonic instruments have proven very accurate and versatile, and have an ability to measure such diverse products as siloxane rubber and sulfuric acid without any alterations. However, to ensure that good measurements are taken, the instrument must sample a representative portion of the liquid, overcome attenuation effects that could change the sonic-velocity reading, and operate in an optimum temperature range for the particular liquid.

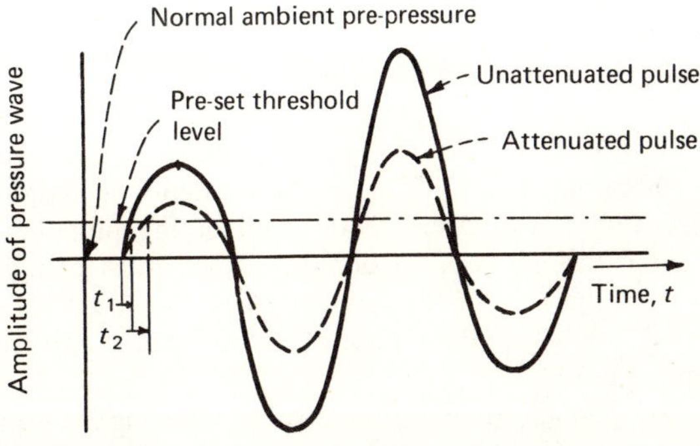

PRESSURE waves caused by sonic pulse—Fig. 2

Sound and Sonic Velocity

Acoustics may be defined as the generation, transmission and reception of energy in the form of vibrational waves in materials. Because these waves are mechanical in nature and use the structure of the material for transmission, they can readily pass through many solids and liquids that are impervious to some electromagnetic waves such as light, magnetism and gamma rays.

When atoms or molecules of a fluid or solid are displaced from their normal configuration, an internal elastic restoring force of stiffness arises. This force is easily perceived as the force created when a spring is stretched, or the increase in pressure when a fluid is compressed. It is the action of this elastic restoring force, coupled with the inertia of the system, that enables materials to oscillate and thereby generate and transmit acoustic or sound waves.

Acoustics includes the generation of sounds that are heard by us as noise or voices, due to vibrations transmitted through the air. To distinguish the part of acoustics of interest to us (i.e., the use of acoustics for industrial purposes) from the sounds and noises that we hear, we use the term sonics. Sonics is the technology of sound as applied to problems of measurement, control and processing.[1]

When a sound wave passes through a liquid, there is a small alternating increase and decrease in pressure superimposed on the static pressure. This is due to the vibrational motion of the molecules, which results in small changes in the volume occupied by the molecules. The velocity of sound in liquids is thus partly governed by the modulus of elasticity that is called the adiabatic bulk modulus, B, where:

$$B = \frac{\text{Stress}}{\text{Strain}} = \frac{\text{Force/unit-area}}{\text{Change in volume/original volume}}$$

The sonic velocity, c, also depends on the density, ρ, and may be defined as:

$$C = (B/\rho)^{1/2}$$

This velocity will vary according to the liquid's composition, temperature, and to a lesser degree the pressure.

As the sound wave propagates through the liquid, it is attenuated, and its intensity decreases due to a combination of absorption, heat loss, and scattering. Absorption is caused by viscous losses due to the relative motion between various portions of the liquid during compressions and expansions that accompany transmission of the wave. Since an incremental amount of heat is generated during compression, a small temperature gradient is set up, which causes heat to be conducted to the cooler surrounding material. However, the overall temperature of the liquid is not measurably altered with the passage of pulses.

Scattering occurs when a small amount of the sound wave is reflected by inhomogeneities in the liquid such as suspended gas bubbles, large sonically opaque particles, certain emulsion droplets, etc. In highly aerated solutions, scattering may present severe losses, limiting the use of sonic monitors.

If two liquids are mixed, the resulting mixture will have an intrinsic sound velocity made up of contributions from the sound velocities of the pure liquids. Thus, the sound velocity of the mixture will change as the relative proportion of the liquids changes, although not necessarily in relation to the amount of liquids. The effect of temperature on the sonic velocity will also be altered by changing the relative concentrations of the liquids.

Sampling and Attenuation

To ensure that a representative sample of the liquid is measured, the sonic path must extend beyond the region that is adjacent to the wall of the pipeline or reactor. In high-viscosity liquids, caution must be taken to avoid the areas of temperature gradients caused during processing, which do not represent the measured process temperature.

An adiabatic pressure wave passing through a liquid has a waveform, as sketched in Fig. 2. To prevent the electronics from spurious triggering, a threshold level is established that will be reached only when a sonic pulse is received at the transducer. This causes an inherent delay between the initiation of the wave caused by the sonic pulse and its detection in the electronics, as shown by t_1 in Fig. 2.

However, if the wave is attenuated while passing through the liquid, it maintains the same period but arrives with a reduced amplitude. Hence, the attenuated pressure wave takes a longer time to reach the threshold for triggering the pulse generator. This time delay (t_2) will affect the response of the instrument unless the condition is corrected. Automatic sensitivity-control cir-

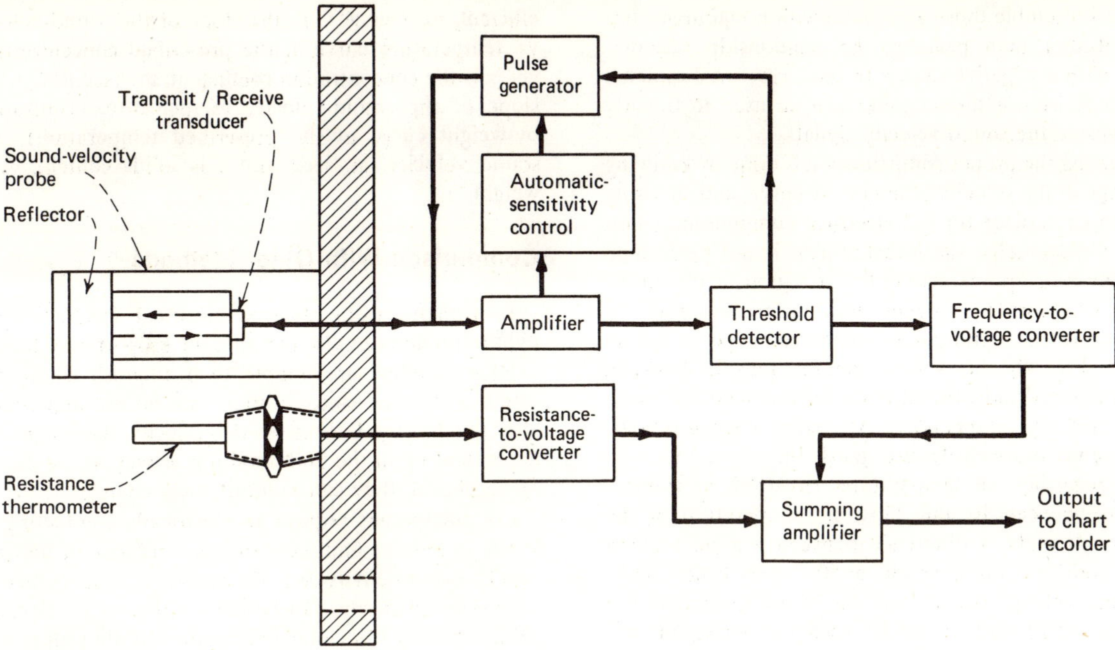

FUNCTIONAL diagram of sonic monitor for conducting and nonconducting solutions—Fig. 3

cuits detect the amplitude of the received sonic wave, and change the transmitted pulse amplitude to maintain it at the same level (see Fig. 3). Thus, the pulse generator always triggers at the same delay time, and all readings are affected equally.

Temperature Effects

Although it would be ideal to hold the process temperature constant, this may be impractical. Hence, temperature variations are monitored and a correction is applied to the measured sound velocity to offset errors. This temperature influence can be seen from a typical relationship in Fig. 4a for different concentrations of methanol in water.

The curves for the lower concentrations resemble the characteristic curve for aqueous solutions and slurries. Here, sound velocity increases with temperature to a maximum point. Beyond this, the sound velocity decreases with further rise in temperature.

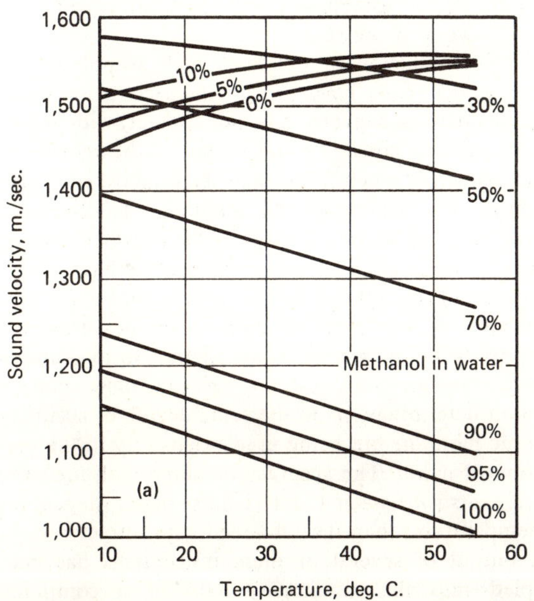

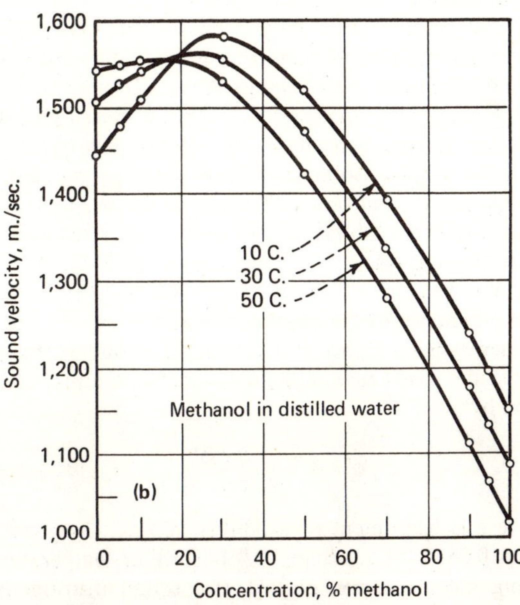

SOUND velocity in liquid mixtures is a function of concentration and temperature—Fig. 4

As the concentration of methanol increases, the curves begin to resemble those associated with nonaqueous liquids. Rather than peaking, the relationship becomes linear with a negative slope.[5] In most cases, a d.c. signal from a resistance thermometer can be used to linearly compensate the sound-velocity signal.

Selecting the proper compensation is done by knowing the slope of the velocity-temperature curve, and choosing the proper resistor for the electrical compensation network. Occasionally, the measurement is not performed in a linear or approximately linear region of the curve, and a second-order compensation may be required.

Selecting the optimum temperature range is not so simple. Fig. 4b shows how the relationship between sound velocity and concentration for methanol solutions is affected by temperature. At higher concentrations, measurements are easily distinguishable due to the steep slope, regardless of temperature. However, at concentrations between 10 and 25%, the temperature of the liquid will make a difference between a good reading and a poor one. For example, at 50 C. very little change in sound velocity occurs over this concentration range, and the instrument cannot be used for measurements. But at 10 C., a good slope exists for the same region, and accurate measurements can be taken.

Temperature effects are usually greater for nonaqueous liquids than for aqueous solutions. Only in rare circumstances will the application of sonic solution monitors be limited because of adverse temperature effects. Since process monitoring with sonic equipment is relatively new, the temperature coefficient and temperature effects on the sound-velocity-vs.-concentration curves may not be known.

Measurement Repeatability

Before the sound velocimeter can be used to measure any liquid property, the relationship between the sound velocity and the liquid property, as well as the effects of temperature, must be determined.

Measurement repeatability is dependent upon many factors, including instrument repeatability, temperature, and errors introduced by the presence of other ingredients. For concentration or solids-content measurements (determining the percentage by weight), the liquid is treated as a binary solution wherein only one ingredient is varying in concentration. (Techniques are available for separation of variables but will not be discussed here.) When two ingredients are changing together, the measurement of the ingredient of interest will be in error due to the influence of the other component.

Ignoring the presence of the changing secondary ingredients, the worst-case measurement repeatability, ϕ, is given by:

$$\phi = \frac{e + (\partial c/\partial T)\Delta T}{(\partial c/\partial s)} \tag{4}$$

where e is instrument repeatability, m./sec. (long term about 0.001% of reading); ΔT is shift in temperature during measurements, deg. C. (for sound instruments containing automatic temperature-compensation circuits, ΔT corresponds to the total equivalent circuit error and is usually less than 0.1 C.); $\partial c/\partial T$ is temperature coefficient, m./(sec.)(°C.), (the slope of the sound-velocity vs. temperature curve at the prescribed concentration); $\partial c/\partial s$ is the concentration coefficient, m./(sec.)(°C.), (the slope of the sound-velocity vs. percent-concentration-by-weight curve at the prescribed temperature); c is sound velocity, m./sec.; and s is solids content, % by weight.

Comparison With Other Methods

Instruments using density, electrical conductivity or light refraction can also measure the properties of liquids.

Densitometers are unable to distinguish changes in solids content or concentration that do not appreciably alter the density. Because sound-velocity instruments are dependent upon the bulk modulus as well as the density of the liquid, they can monitor these changes.

Conductimeters require an electrically conducting solution to operate and measure the resistance of the solution in a conductivity cell. The resistivity, and its reciprocal the conductivity, can be determined by an electrical bridge. Sound velocimeters operate equally well in conducting or nonconducting solutions.

Refractometers use a critical-angle light reflection to determine the refractive index of the liquid. Temperature changes will affect the refractive index to a greater extent than the sound velocity of a liquid, which makes the refractometer more temperature sensitive than the sonic device. While the refractometer responds to dissolved solids in solution, it cannot measure the total solids in liquids. This is possible with a sound velocimeter provided that the undissolved solids are sonically transparent.

Major Applications of Sound Meters

Sound velocimeters are being used in many applications where no other instrument has been successful. They are already established as the best instrument available in several processes.

In the production of polymers such as polystyrene, sound velocimeters have proven themselves for measuring the percent conversion (solids content) with greater accuracy than is possible with viscometers. Moreover, the measurement is much faster than the established evaporation-residue techniques. In addition, sound-velocity instruments can be used to monitor the polymer's molecular length. Generally sound velocimeters have shown accuracies in the ±0.1 to 0.3% of solids range when used for laboratory, onstream or bypass-stream measurements. Viscosity techniques, used as an alternative to evaporation-residue methods, cannot provide these end-of-reaction determinations to the same degree of accuracy.

Sonic monitors are being used extensively in the production of nylon. The accuracy obtained with the sonic instruments is between 4 and 11 times better (depending on temperature) than that obtained with refractometers. The output of several of these instruments has been coupled into the monitoring loop of a computer-controlled facility.

Monitoring sulfuric acid in the 90 to 100% concen-

Costs of Solution Monitors—Table I

Instrument Type	Cost Range
Densitometers	
Fixed-volume measurement	$1,200 to 4,000
Vibrating U-tube	$2,000 to 4,000
Radiation absorption	$3,000 to 5,000
Conductimeters	$ 800 to 1,000
Refractometers	$1,500 to 4,000
Sonic velocimeters	$2,000 to 4,000

tration range, as well as in the oleum region above 100%, has been done with densitometers and conductimeters. In the 93% region, vibrating U-tube densitometers have provided a measurement repeatability of ±0.15% concentration, whereas sound velocimeters have concentration repeatabilities of ±0.007% and better. In the 98 to 99% region, traditionally measured with conductimeters to a ±0.016% repeatability, sound velocimeters have given repeatabilities in the same order as for the 93% acid.

Sucrose is a basic chemical that can be made into a variety of products such as detergents and plastics. Refractometers have long been used for monitoring sucrose concentrations to accuracies of about ±0.08%. Sound velocimeters have proven very competitive, measuring concentrations to ±0.03%, and being less sensitive to temperature changes.

Pipeline interface detectors that use sound velocity have been able to distinguish between two different gasolines with almost identical densities. This would not be possible with densitometers. Hence, systems have been built that require dyes to be added to the gasolines. The interfaces between different products can then be detected by using colorimeters or by visual observation.

Economic Considerations

Besides the cost of the instrument, other economic factors that should be considered when comparing instruments for a particular application are installation and maintenance costs. Table I gives a quick comparison for the unit costs of the different types of instruments.[6]

Installation and maintenance costs will vary depending upon applications for each instrument. For the sound velocimeter, installation requires the welding of a blind tee-flange section or nipple onto a pipe section. The measuring probe is inserted through a flange in the section. Cost will depend upon the pipe material and pressure specifications. The only contact between the velocimeter and the liquid being monitored is the probe, which is made of corrosion-resistant materials.

Sonic Flowmeter

A sonic flowmeter is basically two sound velocimeters combined into a single instrument. It consists of three major parts: the flowtube, a transmitter, and a flow display device.

The flow tube is a pipe section, made of the same material and of the same diameter as the adjacent pipeline. Flow velocity of the liquid is measured in the flow tube. The sonic pulse is sent from the transmitting transducer diagonally across the flow to a receiving transducer, as shown in Fig. 5. As in the velocimeter, a pulse repetition rate is established that is dependent upon the transit time of the pulse through the liquid.

A similar pulse train is generated by sending pulses in the opposite direction through the liquid, using the

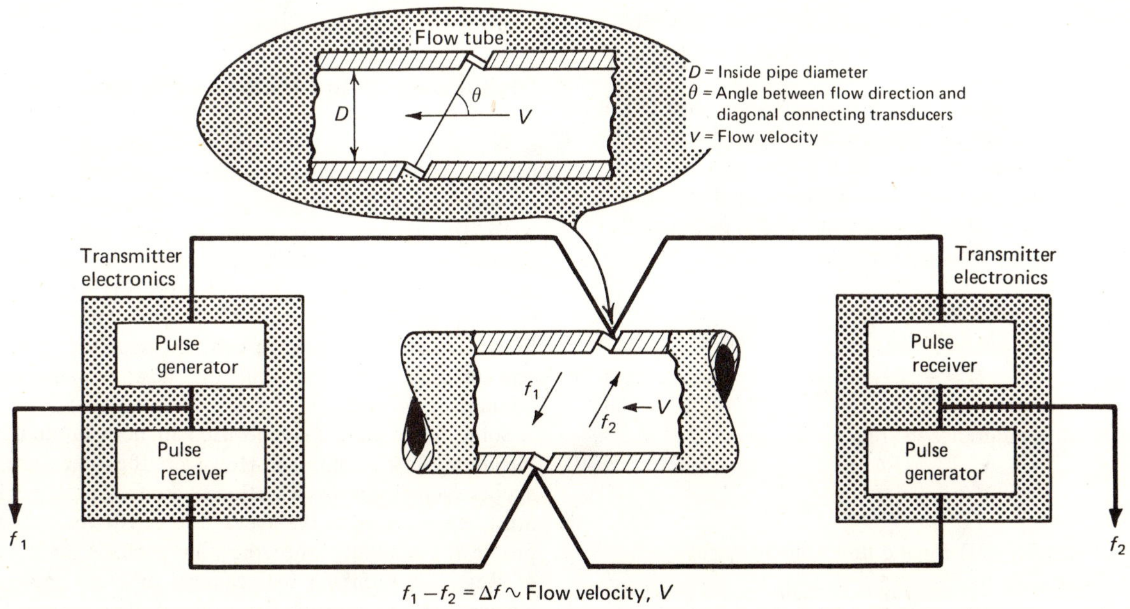

FLOWMETER combines two sound velocimeters into a single instrument for measuring flowrates of liquids—Fig. 5

same pair of transducers. Since the movement of the liquid in the pipeline will slow down pulses traveling upstream, the transit time, and hence frequency, of the upstream and downstream pulses will be different. The difference between the two frequencies is a direct measure of the liquid's flowrate. A digital pulse train, corresponding to the average flowrate, is sent from the transmitter to the flow display device.

Operating Principle of Flowmeter

Transit time for an ultrasonic pulse traveling between two electro-acoustic transducers in the flowmeter is the distance between transducers divided by the effective velocity of the sonic pulse.

In the downstream direction, the pulse travels at the sound velocity and is assisted by the flow of liquid in the pipeline. So transit time, t_1, is given by:

$$t_1 = \frac{D \operatorname{cosec} \theta}{c + V \cos \theta} \tag{5}$$

where D is inside diameter of pipe, θ is the angle between flow direction and the diagonal connecting transducers, c is velocity of sound in liquid, and V is flow velocity.

In the upstream direction, the liquid velocity tends to slow down the pulse, and the transit time, t_2, is:

$$t_2 = \frac{D \operatorname{cosec} \theta}{c - V \cos \theta} \tag{6}$$

If the instrument uses the difference in transit times to determine the liquid flow velocity, then:

$$\Delta t = t_2 - t_1 = \frac{D \operatorname{cosec} \theta \times 2\, V \cos \theta}{c^2 - V^2 \cos^2 \theta} \tag{7}$$

Solving Eq. (7) for flow velocity gives:

$$V = \frac{c^2 \tan \theta \, \Delta t}{2D} \tag{8}$$

The liquid velocity or flowrate depends upon the sound velocity, c, in the liquid, which changes with temperature, pressure, density and adiabatic bulk modulus of the liquid.

If frequencies rather than transit times are used as the basis for measurement, dependence of the reading on the sonic velocity may be eliminated.

The frequency of a pulse is the inverse of the transit time. Therefore, the downstream and upstream frequencies may be expressed as:

$$f_1 = \frac{1}{t_1} = \frac{c + V \cos \theta}{D \operatorname{cosec} \theta} \tag{9}$$

$$f_2 = \frac{1}{t_2} = \frac{c - V \cos \theta}{D \operatorname{cosec} \theta} \tag{10}$$

Taking the difference in frequencies gives:

$$\Delta f = f_1 - f_2 = \frac{2\, V \cos \theta}{D \operatorname{cosec} \theta} \tag{11}$$

and solving Eq. (11) for the flow velocity yields:

$$V = \frac{\Delta f D}{2 \sin \theta \cos \theta} = \frac{\Delta f D}{\sin 2\theta} \tag{12}$$

By using frequencies, Eq. (12) for determining the flow velocity becomes independent of the velocity of sound in the liquid.

Characteristics of Ideal Flowmeter

Sonic flowmeters exhibit many characteristics of an ideal flowmeter, such as an output proportional to the volume flowrate, applicability to any liquid that is sonically transparent, no protrusions into the flowstream. They are also accurate and have a fast response. However, the performance of different sonic flowmeters will depend upon the design and manufacture of the mechanical and acoustical parts of the system, and the circuit design, stability, and accuracy of the electronics.

Most of the immediate applications of sonic flowmeters are similar to applications now using turbine or magnetic flowmeters. All are quite linear and accurate, and magnetic meters are completely obstructionless.

Comparison With Other Flowmeters

Magnetic flowmeters require an electrically conducting liquid in order to operate. Turbine meters partially obstruct the flow of liquid and have mechanically moving parts that wear. The rated accuracy of a turbine meter generally applies only over a 10 to 1 range of flowrates. Flow velocities that exceed rated values will cause excessive pressure drops in the liquid. Most turbine meters must operate with clean, noncorrosive liquids.

When pipe size is larger than 10 to 15 in., the sonic instrument is less expensive than magnetic and turbine flowmeters. The electronics and transducer assemblies of the sonic meter do not change as the pipe diameter varies, and only the transducer-assembly mounting and the flowtube will vary. For the larger diameters, the sonic flowmeter becomes quite competitive with the venturi flowmeter, particularly in applications where the head loss incurred with a venturi is undesirable. At pipeline diameters greater than 25 to 30 in., the sonic flowmeter could be less expensive than all other flowmeters.

Digital Output and Meter Coefficient

The digital pulse-train output is similar to that of turbine meters and is compatible with the same readout devices. When the transducers are installed in the pipe walls, the average velocity in the path connecting the transducers is measured. This average varies only slightly relative to the average velocity over the entire cross-sectional area. Response of the sonic flowmeter is quite linear and the whole range can be calculated with only one reading, provided that the flow remains either laminar or turbulent.

Sonic flowmeters may be used in both laminar and turbulent flow situations. However, they measure the average velocity across a section of the flow profile, which must be converted to the average velocity over the whole profile for readout. Since the flow profile over a range of Reynolds numbers will change, different meter coefficients must be applied to different flow regions.

Fig. 6 shows the general relationship between the meter coefficient and the Reynolds number. Conditioning

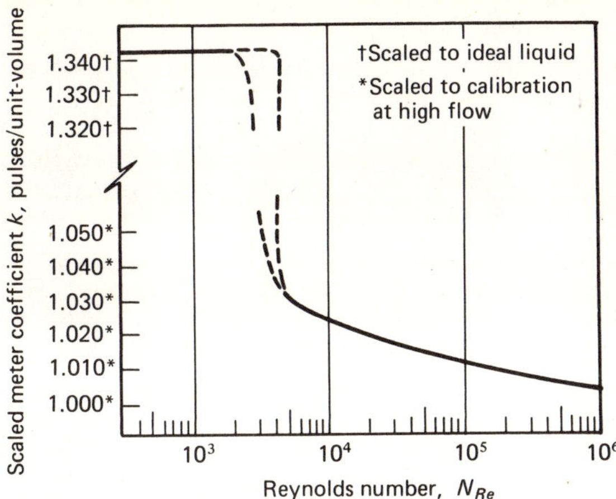

METER coefficient varies with Reynolds number—Fig. 6

can be used to adjust the approximately linear portion of the curve in the turbulent region, slightly delaying the transition. Flow straighteners are used for this conditioning. The nonlinear portion of the curve is developed from profile changes, which with a sufficiently long straight section of pipe upstream allow the flow profile to stabilize, and thereby the curve approaches the theoretical value.

Accuracies in the straight portion of turbulent flow will depend upon the use of flow straighteners. Without these accessories, accuracies of better than ±0.5% of reading are obtained. By adding flow straighteners, accuracy can be increased to between ±0.1 to ±0.2% of reading.

Along the sloping portion of the curve, where no compensation is available, accuracies will drop off to about ±1 to 3%. In the laminar region, a Newtonian flow is assumed. This results in a constant conversion factor, and an accuracy of the same magnitude as the turbulent flow without straighteners.

Low-Flow Measurements

The sonic flowmeter can handle a large range of flow velocities. Flow velocities having a minimum detectable flowrate of 0.005 ft./sec. can be measured. Below 1 ft./sec., the sonic flowmeter has an approximate uncertainty of ±0.005 ft./sec., due to noise in the electronics. This permits measurements at flowrates much smaller than 1 ft./sec., but with deteriorating accuracy as the rate approaches 0.005 ft./sec. Applying the uncertainty to a flowrate of 0.5 ft./sec. gives an expected flow error of ±1%, which is usually adequate in low flows.

Major Applications

The combination of (1) ability to measure low flowrates, (2) absence of protrusions into the flowstream, (3) ability to operate in conducting and nonconducting liquids, and (4) comparatively low cost for large pipeline diameters has made the sonic flowmeter particularly suited to water management. One installation to exemplify these characteristics is in a 42-in.-dia. glass-fiber pipeline to measure flows from 1 to 90 cu.ft./sec. This pipeline is submerged in a lake. (See photo, p. 103.)

Due to the high stability of the electronics, the sonic flowmeter exhibits very little drift. Hence, once it is calibrated, little or no change is expected in future operations, and recalibration is unnecessary. Two sonic flowmeters, calibrated in the same product, should maintain calibration within 0.1% of each other under the same conditions. This opens the field of leak detection and custody transfer to the sonic flowmeter.

Sonic monitors for solutions and sonic flowmeters are being used in applications for which they are uniquely suited. As the sound-velocity properties of more liquids and solutions are calibrated, the use of these instruments will increase. As operating experience is gained, they should find many applications in process monitoring and control.

References

1. Heuter, T. F. and Bolt, R. H., "Sonics," Wiley, New York, 1962.
2. Schaafs, W., "Landolt-Boernstein Numerical Data and Functional Relationships in Science and Technology, Group II: Atomic and Molecular Physics, Vol. 5: Molecular Acoustics," Springer-Verlag, New York, 1967.
3. Greenspan, M. and Tschiegg, C. E., Sing-Around Ultrasonic Velocimeter for Liquids, *Rev. Sci. Instr.*, **28,** 897 (1957).
4. Mathieson, J. G. and Conway, B. E., Ultrasonic Velocity in Water-Deuterium Oxide Mixtures, *Anal. Chem.*, **44,** 1517 (1972).
5. Zacharias, Jr., E. M., Process Measurement by Sound Velocimetry, *Instr. Control Systems*, Sept. 1970, p. 112.
6. Liptak, B. C., "Instrument Engineers Handbook," Vol. I, Chilton, Philadelphia, 1969.
7. Zacharias, Jr., E. M. and Parnell, Jr., R. A., Measuring the Solids Content of Foods by Sound Velocimetry, *Food Technol.*, Apr. 1972, p. 160.
8. Feil, M. F. and Zacharias, Jr., E. M., The Determination of Yeast Slurry Consistence and Wort Plato by Sonic Solution Analysis, *Brewers Dig.*, Nov. 1971, p. 76.
9. Zacharias, Jr., E. M., Sonic Monitor for Solution Analysis, *Instr. Technol.*, Sept. 1970, p. 47.

Meet the Authors

◀ **Ellis M. Zacharias, Jr.,** is president of NUSonics, Inc., 9 Keystone Place, Paramus, NJ 07652. For the past 13 years, he has been associated with the development of equipment and techniques using sound velocimetry. He has a B.S. from the U.S. Naval Academy and an M.S. in electrical engineering from Stevens Institute of Technology. He is a member of the Instrument Soc. of America, the Marine Technology Soc., and the Acoustical Soc. of America.

Donald W. Franz is director of technical publications for HBH Associates, Inc., 645 Madison Ave., New York, NY 10022. He has a B.S. and an M.S. in electrical engineering from Cornell University. He is a member of the Inst. of Electrical and Electronic Engineers, Optical Soc. of America and the American Inst. of Physics. ▶

Section VII
CONTROLLING PROCESS FLOW

Control Valves In Process Plants

For proper performance in any piping system, here are the design relations, sizing formulas and installation procedures for selecting and using control valves for fluids.

ROBERT KERN, Hoffmann - La Roche Inc.

Control valves are the basic regulatory devices in any process operation handling fluid streams. Hence, we must be thoroughly familiar with the different types of these valves and their flow characteristics. This enables us to meet process conditions, and to ensure proper installation in the fluid system.

Major Types of Control Valves

In the following brief discussions, only the general features of each control valve are given. For complete details about a specific control valve, consult the manufacturers' literature.

One major group of control valves resembles the globe valve (Fig. 1). In place of a handwheel, an actuator moves the valve stem and plug, thereby opening and closing the valve. The usual actuator is an air-operated device whose housing contains a diaphragm that separates it into two compartments. The diaphragm (and attached valve stem) is balanced in its position by a spring on one side and air pressure on the other. In flow control, the air pressure changes in response to a signal resulting from the measurement of the differential pressure across an orifice or other flow-sensing element.

The single-ported control valve (Fig. 1) finds use where tight shutoff is required in addition to flow control. The double-ported control valve (Fig. 1) has two seat rings with two plugs on a common stem. This is a higher capacity valve than the single-seated one of the same size. With hard seat rings and high temperatures, the double-seated valve cannot shut off tightly. The valve accessories, shown in Fig. 2, allow for various operating functions and conditions.

Originally published April 14, 1975.

In recent years, a second group of control valves has received wide acceptance. In these types, the actuator rotates a butterfly flap, plug or disk around its axis (Fig. 3). Size for size, these valves usually have higher capacities and less flow resistance than the contoured-plug valves. Generally, control valves with rotating axes are suitable for a wide range of flow-control applications.

Characteristics of Valve Plugs

The valve plug can be disk type, solid contoured or ported. Flow-control characteristics depend on the shape or cavities of the plug. The three basic types of plug and their flow characteristics are:

■ *Quick Opening*—A single-disk (for high temperatures) or a double-disk (for low temperatures) plug is used for total shutoff or opening. A disk-type plug has linear flow characteristics and short stem movement.

■ *Linear Flow*—A plug has linear flow characteristics when the flowrate through the valve is proportional to the lift.

■ *Equal Percentage*—A plug has equal-percentage characteristics if at any plug position, the same percentage of change in flow takes place for the same amount of plug movement. The percentage of change is related to the flowrate just before the plug is moved, as shown in Fig. 4.

Most plug characteristics are somewhere near or between those described. Manufacturers provide diagrams similar to Fig. 4 for each valve.

A plug having linear-flow characteristics is commonly specified for liquid-level control. The equal-percentage plug is used for pressure or flow control; or where only a small percentage of the overall pressure differential is

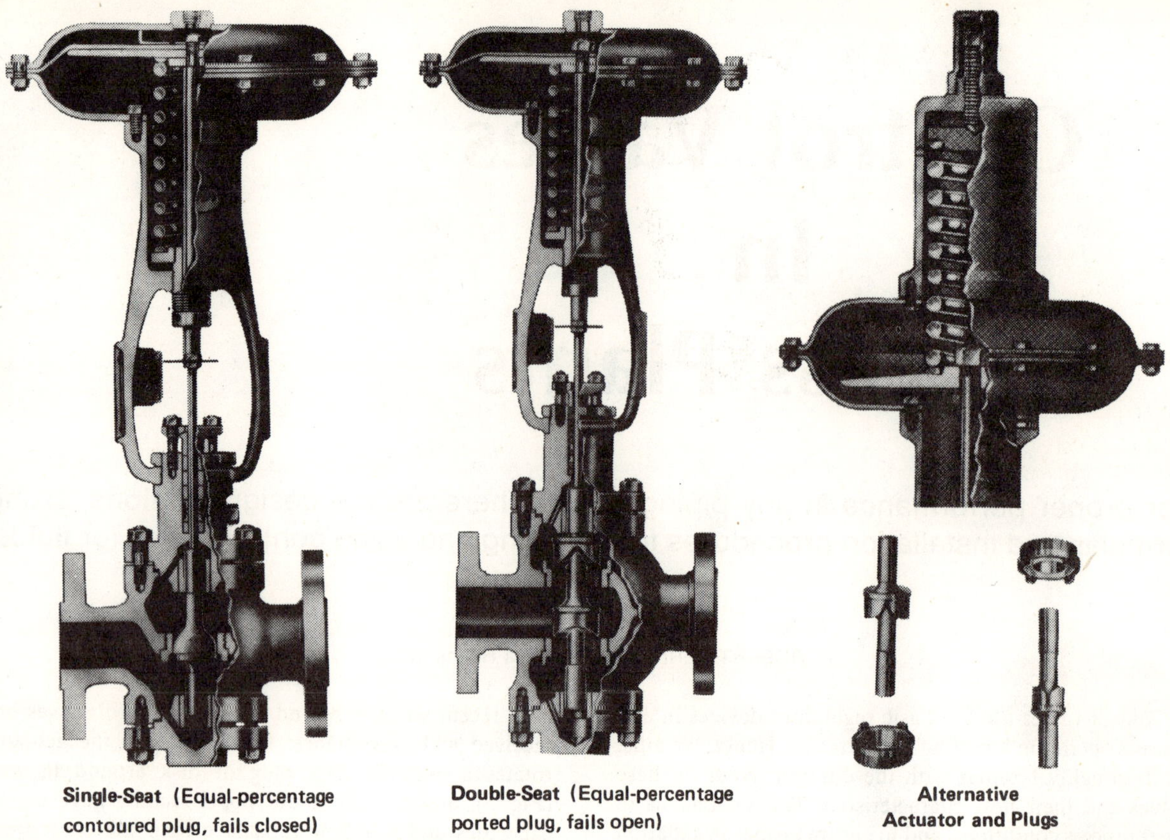

Single-Seat (Equal-percentage contoured plug, fails closed)

Double-Seat (Equal-percentage ported plug, fails open)

Alternative Actuator and Plugs

CONTROL valves handle many types of process fluids, and are actuated by air in response to a process signal—Fig. 1

available; or where pressure drop across the control valve varies greatly.

The modified parabolic-flow characteristic falls between the linear and equal-percentage characteristics. This type of plug (usually V-port) finds use where the major part of the system pressure drop is available for control.

Actuators (also called operators or valve positioners) lift the valve stem and plug above its seat, or move the plug in the seat cylinder. Butterfly or ball-type control valves have the actuators side-mounted because the actuator stem rotates the valve axle. Plug characteristics can be influenced by the linkage between actuator stem and valve axle.

The valve housing and the operator's yoke are separate pieces. Hence, after a valve is installed, the operator can be rotated around the valve stem or valve axle, relative to the valve body. This enables a convenient position to be chosen for the actuator, in order to provide access to operating points on the valve.

Hydraulic, mechanical and piston operators are also available.

Safety Requirements

Without air pressure in the pneumatic actuator, the valve can be in closed or open position. These alternative

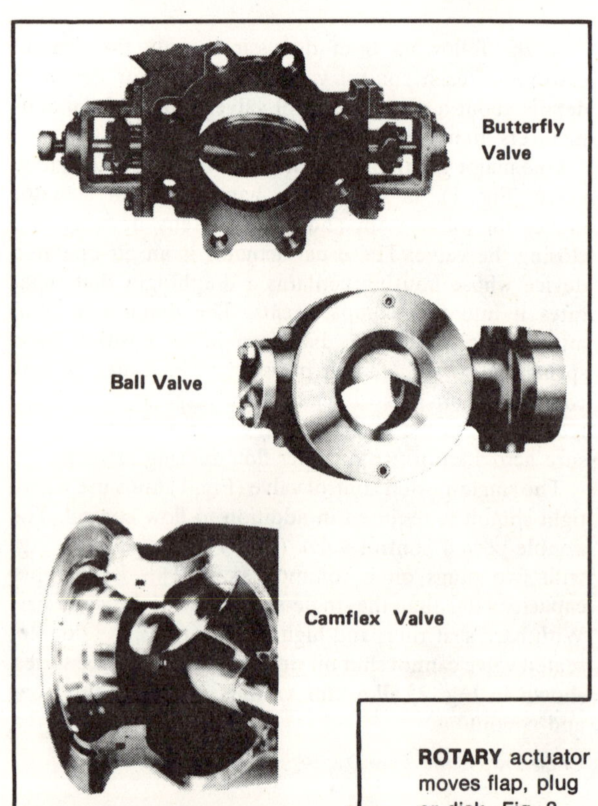

ROTARY actuator moves flap, plug or disk—Fig. 3

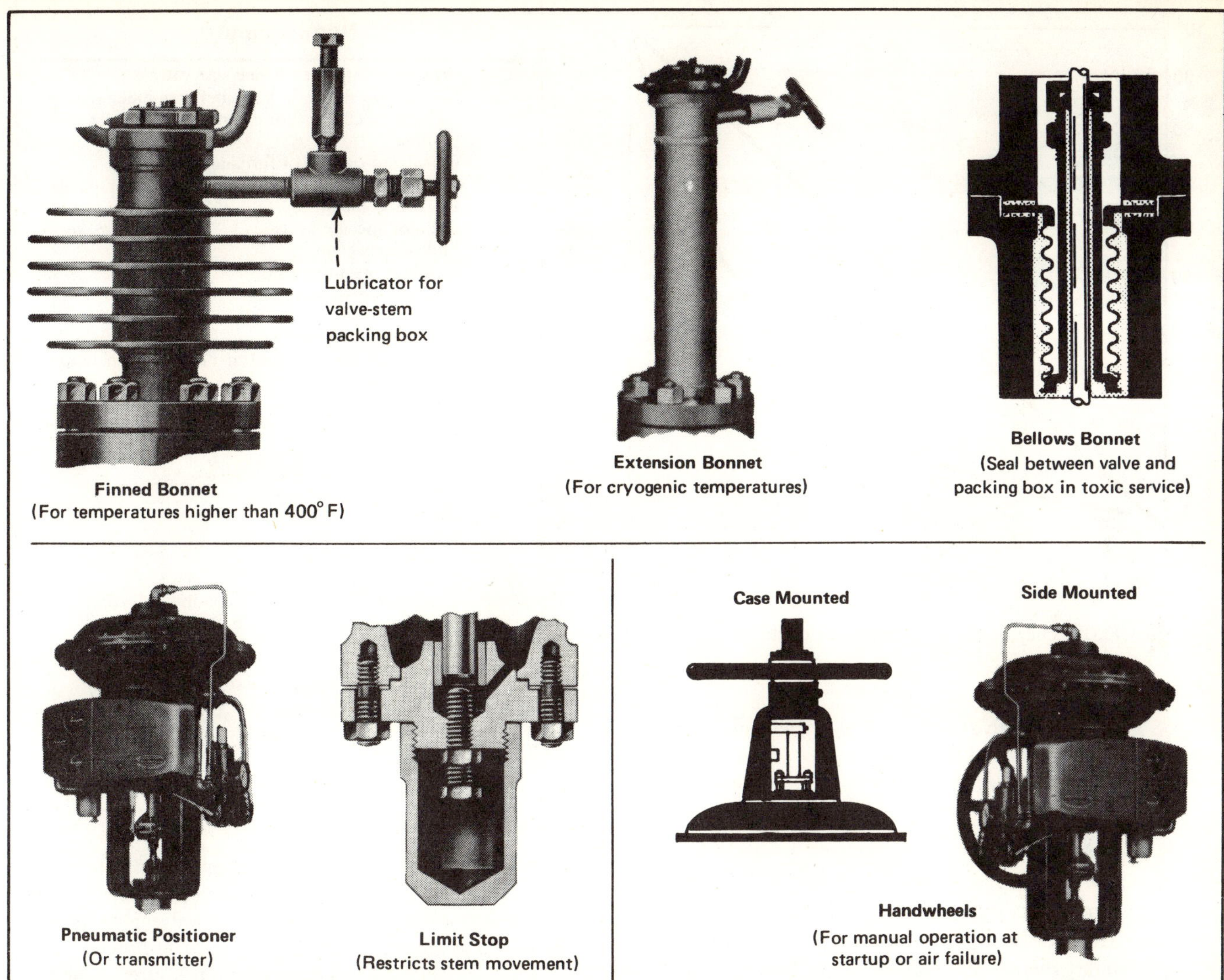

ACCESSORIES extend usefulness of control valves by providing for extreme and unusual conditions—Fig. 2

positions are accomplished by reversing the seat ring and plug, or by reversing the location of the actuator spring from below to above the diaphragm (Fig. 1).

One concern of the designer is to select valves that will fail-safe in the event of instrument-air failure. In principle, a control valve fails safe if temperature and pressure of the process system do not increase after the control valve becomes inactive.

For example, fuel-oil control valves to heater burners should fail closed. At the same time, feed to heater tubes (in most cases) should fail open to avoid overheating the furnace tubes. The feed-control valve to fractionating columns usually fails closed. Steam supply to reboiler fails closed. Reflux-drum vapor outlet and reflux pump-discharge valves fail open. Control valves in minimum-flow bypass lines at centrifugal-pump discharge lines, compressor bypass lines, and reciprocating-machine bypass lines fail open.

Reactors are protected under controlled conditions, and usually the feed-control valve fails closed. Generally, a designer of flow systems should consult process, instrumentation and equipment engineers when deciding on fail-safe positions for control valves so as to assure orderly shutdown procedures.

Capacity Coefficients of Valves

Valve flow coefficient, C_v, depends on the internal dimensions of the valve and the smoothness of surfaces. Tests made by manufacturers (using water or air at predetermined pressure difference) establish C_v values. Manufacturers give the following definition:

$$C_v = Q(\sqrt{S}/\sqrt{\Delta P})$$

C_v is a capacity index indicating the flow of 60°F water in gpm, which will pass through the completely open valve under a pressure difference of 1 psi between the inlet and outlet flanges. Obviously, if $S = 1$ and $\Delta P = 1$ psi, then $C_v = Q$.

Capacity indexes for the butterfly valve are also given at two throttling positions of the flap, in addition to the fully open position.

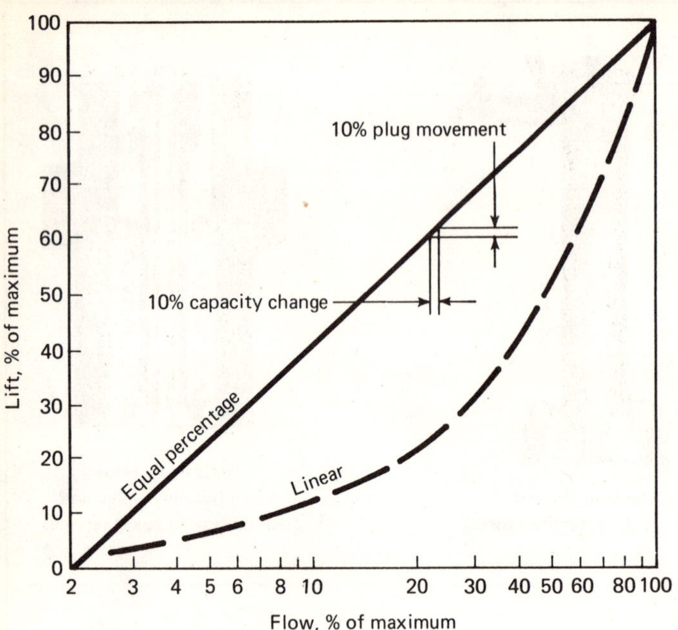

FLOW characteristics of ported or contoured plugs—Fig. 4

Nomenclature

C_f	Critical flow factor for line-size valve
C_{fr}	Critical flow factor for valve between pipe reducers
C_v	Capacity coefficient for control valve in fully open position
C_{vc}	Calculated coefficient for control valve
D/d	Ratio between larger pipe dia. to smaller pipe dia.
E	Expansion factor, ρ_{60}/ρ
k	Ratio of specific heats
M	Molecular weight
P	Absolute pressure, psia
P'	Absolute pressure, psia
P_c	Critical pressure, psia
ΔP	Differential pressure, psi
P_v	Vapor pressure of liquid at flowing temperature, psia
Q	Volume flowrate, gpm
R	Correction factor for control valve between pipe reducers
S	Specific gravity of liquid, ρ/ρ_{60w}
S_{60}	Specific gravity of liquid at 60°F
T	Absolute temperature, °R
v_s	Sonic velocity, ft/s
W	Weight flowrate, lb/h
μ	Viscosity, cp
ρ	Density of fluid at flowing condition, lb/ft^3
ρ_{60}	Density of fluid at 60°F, lb/ft^3
ρ_{60w}	Density of water at 60°F, 62.37 lb/ft^3
Subscripts	
1	Upstream condition
2	Downstream condition

Control-valve coefficients for single- and double-seated valves are given in Table I.

Calculated Flow Coefficient, C_{vc}—When sizing control valves, a flow coefficient is calculated with normal design flowrate in gpm from:

$$C_{vc} = Q(\sqrt{S}/\sqrt{\Delta P})$$

Then a valve is selected whose capacity index, C_v, exceeds C_{vc}. For a good range of control, the capacity index should fall between 1.25 to 2 times the calculated flow coefficient, or:

$$C_{vc}/C_v = 0.5 \text{ to } 0.8$$

This is an optimum range for linear and percentage-contoured plugs. Some valves have a wider optimum range. All valves will operate below and above these C_{vc}/C_v ratios, but the plug will be closer to the fully open or fully closed position. Under these conditions, we lose the important advantage of having wide flexibility in controllable flow-capacity range, and this may limit operability of the process.

High velocities across the valve orifice can wear out the plug and seat, especially if temperature is also high or when abrasive fluid is present.

Critical Flow Factor, C_f—The pressure gradient across a control valve is shown in Fig. 5. For liquids, the flow can be considered subcritical if the vapor pressure of the liquid will not get higher than the lowest pressure point across the control valve. (Vapor pressure is the pressure at which the liquid begins to vaporize at its flowing temperature. Tables of thermodynamic properties of liquids give corresponding saturated-liquid pressures and temperatures.)

If the vapor pressure falls between the ranges of A and B (see Fig. 5), vaporization or cavitation will occur in the control valve. If the vapor pressure nears the downstream pressure, P_2, cavitation can be suspected. Cavitation can cause rapid wear of valve plug and seat as well as vibration and noise. If the vapor pressure falls between upstream and downstream pressures, P_1 and P_2, vaporization can occur. In this case, there will be two-phase flow in the pipeline after the control valve. If the vapor pressure is higher than the inlet pressure, P_1, the control valve receives two-phase flow; and additional vaporization can be considered across the valve. For this condition, diameter of the downstream pipe will usually be larger than the upstream pipe.

The criteria for subcritical and critical flows in liquids are, respectively:

$$\Delta P < C_f^2(\Delta P_s) \quad (1)$$

$$\Delta P \geqq C_f^2(\Delta P_s) \quad (2)$$

where: $$\Delta P_s = P_1 - (0.96 - 0.28\sqrt{P_1/P_c})P_v \quad (3)$$

and P_c is the critical pressure, psia.

For simplicity: $\Delta P_s = P_1 - P_v$, provided that $P_v < 0.5P_1$.

The sizing formula for critical flow is:

$$C_{vc} = (Q/C_f)(\sqrt{S}/\sqrt{\Delta P_s}) \quad (4)$$

We will use a simplified version of Eq. (4) later in this article.

One example of subcritical flow is that occurring in a control valve located in the discharge line from a centrifugal pump. Critical flow can occur across a pressure-

Flow Coefficients for Control Valves—Table I

	Flow Coefficient, C_v	
Size, In	**Single-Seat***	**Double-Seat***
3/4	—	8
1	9	12
1 1/4	14	18
1 1/2	21	28
2	36	48
2 1/2	54	72
3	75	110
4	124	195
6	270	450
8	480	750
10	750	1,160
12	1,080	1,620
14	1,470	2,000
16	1,920	2,560

*These values have been obtained for Masoneilan 10,000-series (either equal-percentage or V-port) plug valves having full-capacity trim, but also apply to similar valves of other manufacturers [2].

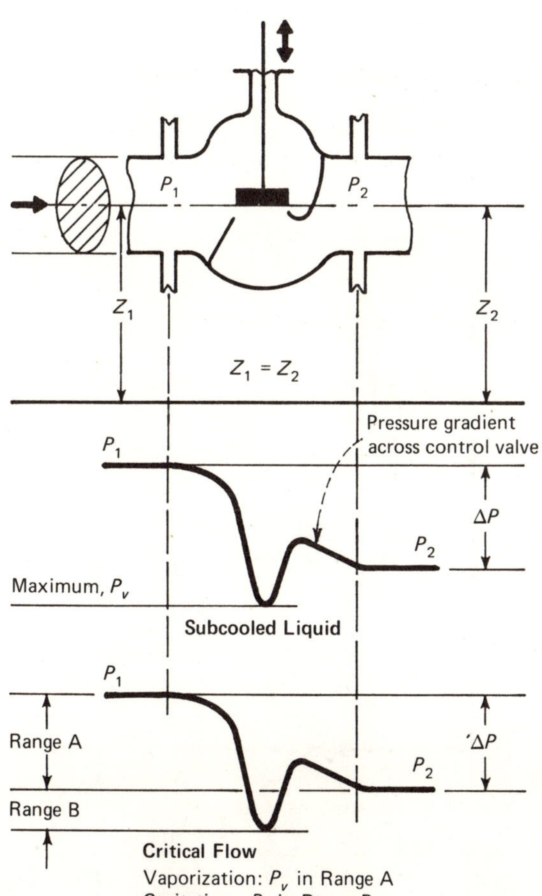

PRESSURES during liquid flow in a control valve—Fig. 5

reducing valve where the upstream liquid condition is close to the boiling point.

For gases, critical flow is assumed when gas velocity reaches the sonic velocity:

$$v_s = 68\sqrt{k(P'/\rho)},\ \text{ft/s} \tag{5}$$

Sonic velocity should be avoided because it can cause noise and vibration.

The criteria for subcritical and critical flows in gases are, respectively:

$$\Delta P < 0.5C_f^2P_1 \tag{6}$$

$$\Delta P \geqq 0.5C_f^2P_1 \tag{7}$$

Critical flow can be avoided by reducing the pressure drop across the valve, by relocating the valve in the flow system, or by choosing a valve with a high C_f value.

The critical flow factor, C_f, is a dimensionless number, which depends on the valve type [6]. C_f is the ratio between the control-valve coefficient under critical conditions and the flow coefficient as published in manufacturers' literature.

Valve Between Pipe Reducers—Flow capacity of a control valve placed between pipe reducers is slightly decreased. In subcritical flow, this is accounted for by a correction factor, R. In critical flow, the correction factor is C_{fr}, which replaces C_f in the calculations. R and C_{fr} also depend on the ratio between pipe size and valve size. C_f, C_{fr} and R have values smaller than 1. Numerical values for the valves shown in Fig. 1 are listed in Table II.

Let us now summarize a number of formulas for sizing control valves for liquid and gas services under different flow conditions [1].

Liquid Service

Subcritical Flow—For a liquid flowing well below its saturation temperature in the turbulent zone, with viscosity close to that of water, and sizes of the pipe and control valve identical, the calculated control-valve coefficient is:

$$C_{vc} = Q\sqrt{S}/\sqrt{\Delta P} \tag{8}$$

where the specific gravity, S, and flowrate, Q gpm, are taken at the flowing temperature; and $\Delta P = P_1 - P_2$.

For minimum pressure drop at the fully open plug position, C_v replaces C_{vc}:

$$\Delta P_{(min)} = (Q/C_v)^2S,\ \text{psi} \tag{9}$$

If we are interested in the pressure drop at a selected plug position between $C_{vc}/C_v = 0.5$ to 0.8, a convenient expression is:

$$\Delta P = \left[\frac{Q}{(C_{vc}/C_v)C_v}\right]^2 S,\ \text{psi} \tag{10}$$

where C_v is taken from the manufacturer's catalog, and C_{vc}/C_v is the selected plug position. (The methods of Eq. (9) and (10) can also be adapted to vapor flow.)

The calculated flow coefficient for laminar or viscous flow is:

$$C_{vc} = 0.072\sqrt[3]{(\mu Q/\Delta P)^2} \tag{11}$$

Critical Flow—If the valve and piping are the same

Correction Factors for Control-Valve Flow Coefficient—Table II

Condition	Factor	Single-Seat* Equal-Percentage	Single-Seat* V-Port	Double-Seat* Equal-Percentage	Double-Seat* V-Port
Critical flow Line size control valve	C_f	0.98† or 0.85‡	0.98	0.90	0.98
Critical flow (Control valve between pipe reducers)	C_{fr}	0.86	0.94	0.86	0.94
Subcritical flow, $D/d = 1.5$	R	0.96			
Subcritical flow, $D/d = 2$ (Control valve between pipe reducers)	R	0.94			

* These values have been obtained for Masoneilan 10,000-series plug valves having full-capacity trim, but also apply to similar valves of other manufacturers [2].
†Factor for flow to open.
‡Factor for flow to close.

size, the simplified calculated control-valve coefficient becomes:

$$C_{vc} = (Q/C_f)(\sqrt{S}/\sqrt{P_1 - P_v}) \quad (12)$$

provided $P_v \leqq 0.5P_1$.

Gas, Steam and Vapor Service

The calculated control-valve coefficient for subcritical flow will be:

$$C_{vc} = \frac{W}{11.65\sqrt{\Delta P(P_1 + P_2)\rho_1}} \quad (13)$$

where $\Delta P = P_1 - P_2$, provided that $\Delta P < 0.5C_f^2P_1$.
For critical flow when $\Delta P \geqq 0.5C_f^2P_1$:

$$C_{vc} = \frac{W}{10.13C_fP_1\sqrt{\rho_1}} \quad (14)$$

If the valve is located between pipe reducers, multiply the righthand side of Eq. (8), (11) and (13) by $(1/R)$; and Eq. (9) and (10) by $(1/R^2)$. Replace C_f with C_{fr} in Eq. (12) and (14).

These corrections can be neglected if the capacity of the selected control valve at normal flow gives a coefficient ratio, C_{vc}/C_v, well within 0.5 to 0.8. The operating position of the valve plug will perhaps not be identical to the calculated position, but this will not change valve or pipe size. Also, in sizing valves for critical flow, make sure that the plug will not operate close to its seat.

Two-Phase Flow

For well-mixed liquid and inert gas in turbulent flow with no additional vaporization, the following applies:

$$C_{vc} = \frac{W}{44.8\sqrt{\Delta P(\rho_1 + \rho_2)}} \quad (15)$$

where ρ_1 and ρ_2 are the upstream and downstream two-phase densities, respectively.

When saturated liquid enters the valve (i.e., $P_1 = P_v$), or saturated liquid and its saturated vapor flow concurrently (i.e., $P_v > P_1$), additional vaporization of the liquid can be assumed inside the control valve. For this condition:

$$C_{vc} = \frac{W}{63.3\sqrt{\Delta P\rho_1}} \quad (16)$$

where the maximum $\Delta P = 0.5C_f^2P_1$. (For calculating the densities in two-phase flow, see Part 1 of this series, *Chem. Eng.*, Dec. 23, 1974, pp. 60–61.)

Example Illustrates Computations

Let us size the control valves for handling a flow of 113,000 lb/h (348 gpm) of liquid ammonia in each of

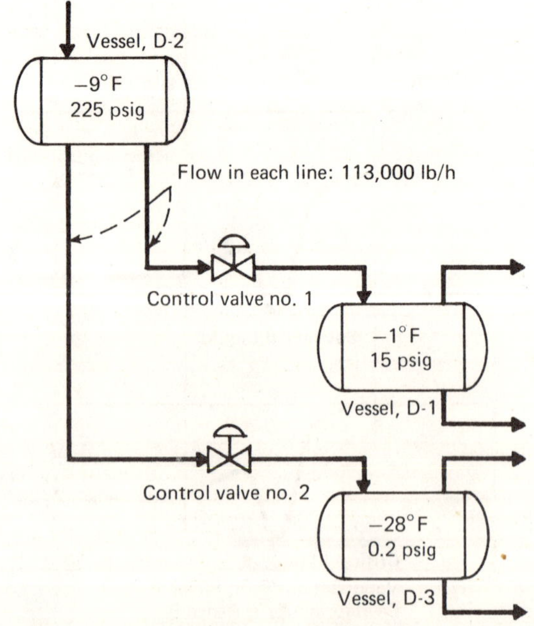

FLOW relations for sizing control valves—Fig. 6

two lines for the system sketched in Fig. 6. The three vessels are located side by side (i.e., all are at the same elevation). Physical property data for liquid ammonia are: $S_{60} = 0.615$, $E = 0.92$, and $M = 17$. Therefore, $S = 0.615 \times 0.92 = 0.566$. Pressure and temperature in each vessel, and corresponding thermodynamic properties, are:

Vessel	D-1	D-2	D-3
Temperature, °F	−1	−9	−28
Pressure, psig	15	225	0.2
Heat content, Btu/lb	41.8	33.2	12.8
Latent heat, Btu/lb	569.6	575.6	589.3

Control Valve No. 1—As liquid ammonia flows from vessel D-2 to D-1, its heat content increases, and liquid in the pipeline after the control valve is subcooled. (Pipelines before and after the control valve can be sized for liquid flow.)

Due to the large pressure difference between vessels D-2 and D-1, cavitation is possible in the control valve. Hence, we can consider the liquid to be in critical flow, and estimate the maximum vapor pressure in the valve from $P_v = 0.5P_1$, or:

$$P_v = 0.5(225 + 14.6) = 119.8 \text{ psia}$$

We will assume that a single-seat valve having a $C_f = 0.98$ (see Table II) will prove adequate by substituting the appropriate values into Eq. (12) to find:

$$C_{vc} = (348/0.98)(\sqrt{0.566}/\sqrt{239.6 - 119.8}) = 24.4$$

From Table I, we establish that a 2-in single-seat V-port control valve having a flow coefficient of 36 may be adequate. We then check the ratio $C_{vc}/C_v = 24.4/36 = 0.68$, which falls well within the desired range of 0.5 to 0.8.

The 2-in lines (before and after the control valve) are relatively short, and when handling 348 gpm will have a small pressure loss. Consequently, pipe resistance will have practically no effect on size of the control valve.

Control Valve No. 2—As liquid ammonia flows from vessel D-2 to D-3, its heat content decreases. Heat is released in the liquid, and as the liquid flows across the valve, vaporization will occur. The actual pressure drop is approximately 225 psig. This is greater than the maximum pressure drop, as determined from $\Delta P = 0.5C_f^2P_1$, or:

$$\Delta P = 0.5(0.98)^2(239.7) = 115 \text{ psia}$$

Substituting the appropriate values into Eq. (16) yields:

$$C_{vc} = \frac{113{,}000}{63.3\sqrt{115(35.2)}} = 28.0$$

We may choose a $2\frac{1}{2}$-in single-seated control valve whose C_v is 54 (Table I). The coefficient ratio will be $28.0/54 = 0.52$, which is acceptable. (Note: a low C_{vc}/C_v value was aimed for in this control valve because C_{vc} was calculated with liquid density. It is not unreasonable to use a two-phase downstream density taken at the outlet of the control valve. Flashing and vapor density are then calculated with the critical downflow pressure. A much higher C_{vc} will result, and a high C_{cv}/C_v can be accepted.)

Because of the subzero temperatures, an extension bonnet can be specified as an accessory to the control valve (see Fig. 2).

In flowing from vessel D-2 to D-3, the liquid releases $33.2 - 12.8 = 20.4$ Btu/lb of liquid, or a total of $20.4(113{,}000) = 2.3 \times 10^6$ Btu. The amount of vapor flashed with this heat is $2{,}300{,}000/582.5 = 3{,}950$ lb/h. This leaves $113{,}000 - 3{,}950 = 109{,}050$ lb/h liquid. These quantities can be used for calculating the two-phase flow resistance of downstream pipe.

Operating Conditions

Control valves are usually the same size or one size smaller than the upstream pipe size, never larger. Control valves are much smaller than line size when high pressure differentials have to be absorbed.

Control valves can accommodate a wide range of capacities and pressure differentials. Flowrates and process conditions are usually well determined for piping and components sizing. When sizing control valves, verify alternative capacities, periodically changing capacities and the related pressure differentials. Control over an extremely wide capacity range might require two control valves in parallel, one for the high flowrates, the other for the low ones. In borderline cases, or for a future increase in capacity, a larger valve body with reduced trim might be desirable.

In most instances, pressure differentials are part of the entire resistance of the piping system. Where an overall pressure differential is determined (for example, between two process vessels), one-third of the overall pressure drop can be attributed to the control valve, and two-thirds to friction losses in piping and equipment. At high pressure differentials, most of the pressure drop will be absorbed by the control valve. When pressure differentials must be minimized, the control valve should be line size, such as in steam feedlines to turbines.

Butterfly valves operate with very little pressure drop (decimals of 1 psi). They are usually suitable in compressor-discharge lines and cooling-water supply lines. However, under throttling conditions, the butterfly valve's coefficient decreases considerably. The coefficient is 50% at 72° position compared with the fully open (90°) position, and 33% at 60° setting.

At centrifugal pumps, pipe resistance in the discharge line (including that of any equipment in this stream) is usually known. An additional 25 to 50% of the discharge pipe resistance can be added for the control valve. With two control valves in series, there is double the amount of additional resistance. For a control valve installed in long discharge lines or in a system with high resistance and relatively small flow changes, the pressure drop across the valve can be 15 to 25% of total system resistance.

A control valve (except butterfly) can only regulate flow by absorbing and giving up pressure drop to the system. Economy in operation of control valves dictates lower pressure drops. However, the valve's capacity and range of control decrease rapidly with lower available pressure differentials.

Changes in the specific gravity, or inaccurate density

Flow Coefficients for Hand-Operated Throttling Valves—Table III

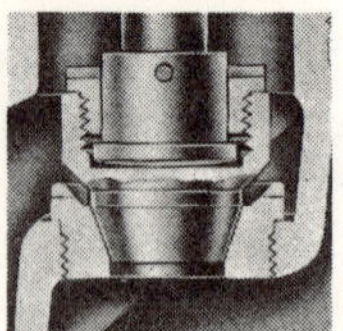
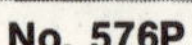
No. 576P

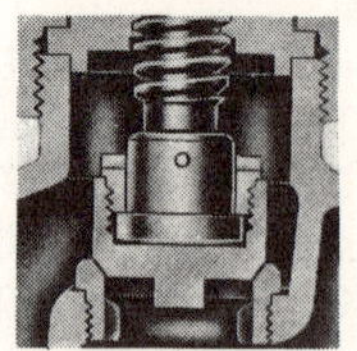
No. 556

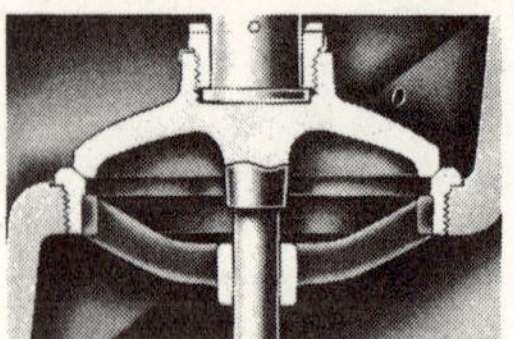
No. 1040

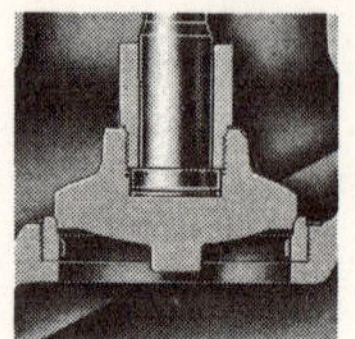
No. 1042

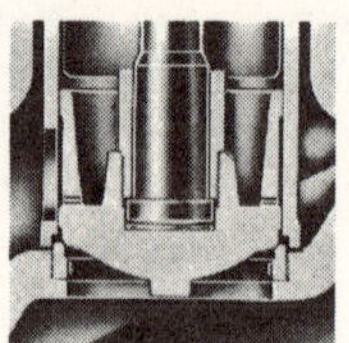
No. 1046

Jenkins Bros.

Bronze Globe Valves (Threaded)			Steel Globe Valves (Flanged)		
Size, In	Flow Coefficient, C_v For Valves No. 546P-150 Psi No. 556P-200 Psi No. 576P-300 Psi	Flow Coefficient, C_v For Valves No. 556-200 Psi No. 576-300 Psi	Size, In	Flow Coefficient, C_v For Valve No. 1040-150 Psi	Flow Coefficient, C_v For Valves No. 1042-300 Psi No. 1046-600 Psi
1/4	0.9	1.2	2	46	55
1/2	2	4.2	2 1/2	72	90
3/4	5	8.6	3	105	130
1	10	14.5	4	200	235
1 1/2	24	29.5	6	400	400
2	41	49	8	720	720

Note: Flow coefficients have been obtained for valves manufactured by Jenkins Bros., but also apply to similar valves of other manufacturers.

estimates, will have minor effect on valve capacity. These are small values—square-root functions of the calculated flow coefficient.

When critical flow occurs in the liquid, the piping after the control valve (and bypass valve) should be carefully sized. Vaporization increases pipe resistance considerably. To stay within reasonable velocities when vaporization occurs across the control valve, the downstream piping and block valve will often be larger in size than the upstream pipe size.

In some cases of saturated liquid flow, vaporization in and after the control valve can be avoided by providing a static head of liquid upstream of the valve. This should be noted on the engineering flow diagram.

At high pressures, high temperatures, or large pressure differentials, the control valve should not operate close to its seat. High velocities can wear the plug and seat. This causes inaccurate flow control, and leakage when the valve shuts off.

Bypassing the Control Valve

A bypass is usually provided for control valves smaller than 2 in., in lethal and high-viscosity services, in handling liquids containing abrasive solids, in boiler feedwater service, and in high (over 100 psi) pressure-reducing steam service.

For consistency in piping design, the flow coefficient for the bypass valve should be about the same as that for the control valve. Table III lists the flow coefficients for some of one manufacturer's globe valves. Because of various seat-and-plug designs, valve coefficients are not the same for comparable globe valves made by different manufacturers.

We find by comparing the data in Table III for globe valves with the flow coefficients for double-seated control valves in Table I that the bypass valve and control valve can be the same size. For single-seated control valves, the bypass globe valve can be one size smaller than the control valve. We can size bypass globe valves or manually operated throttling valves in the same way as control valves provided that flow coefficients are available.

Piping the Control Valve

The best position for a control valve is with the stem vertically upward. A control valve will operate in angular, horizontal or vertically downward position. Neither piping designers nor operators accept these positions. Large angle-control valves are an exception; a horizontal position for them can be most practical.

A single control valve without block valves and bypass is usually sufficient in clean-fluid service; or where parallel equipment containing control valves is installed with block valves located at pipe headers. Where dirty fluid or solid particles can be occasionally expected, a temporary or permanent strainer is installed, upstream of the control valve. Single control valves have handwheel operators.

Most piping specifications call for control valves to be

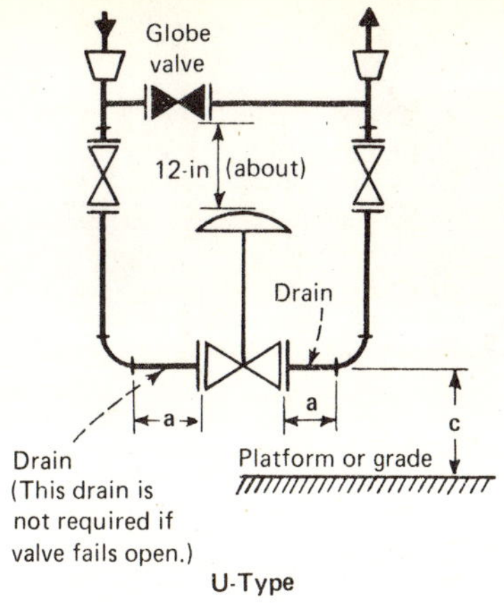

Clearance Dimensions

Pipe or Control Valve Size, In	a, In	b, In
2 to 4	9	12
6 to 8	10	15

Control Valve Size, In	c, In
2	24
3	30
4	36
6	45
8	48

Note: Clearances apply to all manifolds or bypasses.

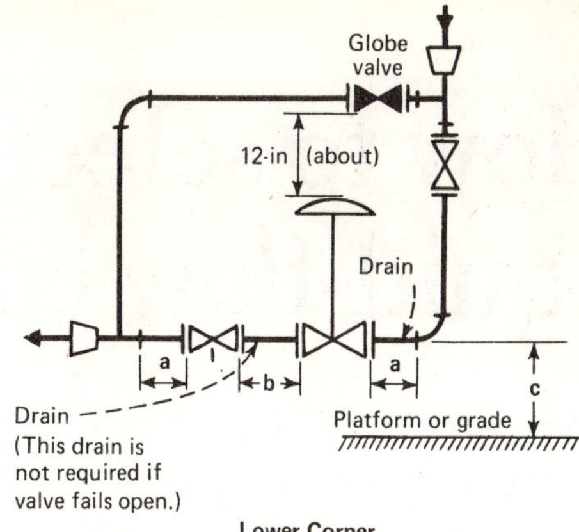

Upper corner

Looped Manifold

PLAN

Looped Horizontal Bypass

Looped Vertical Bypass

MANIFOLDS and bypasses for installing control valves into the process piping require proper clearances and drains—Fig. 7

located above grade or platform elevation, and at the edge of accessways, except for those valves that have to be located in self-draining pipelines. For example, a control valve placed in an overhead gravity-flow slurry line.

For inplace maintenance, clearance space is required below and above the valve for removing the seat, plug, actuator cover, spring and yoke. Estimated clearance requirements are shown in Fig. 7. Dimensions of control valves are given by manufacturers [*2,4*].

If flow conditions permit, manifolds for the control valve that are smaller in size than the main piping will prove economical. Typical standard manifolds are shown in Fig. 7 [*10*]. The U-type is chosen when the inlet and outlet flows approach the control valve from an elevation higher than that of the valve. The corner type is used when flow is from a high point to a low point, or the reverse. The looped-bypass type serves horizontal flows near grade. A looped-corner bypass can bring a control valve over the operating platform. For economical support, control-valve manifolds should be located near structural columns.

For pressure-relieving and draining a control-valve manifold, provide drain valves or plugs at low points. One drain point is required if the control valve fails open. Drains on each side of the control valve are needed if it fails closed. In saturated-steam flow, one or two steam traps are advisable at the low points of a pocketed control-valve manifold.

The automatic control valve is part of an instrumentation system. Sensing points for flow, pressure, temperature and level should be close to the control valve, as should the transmitter. Instrument wiring and tubing connect these elements. Air lines run from the transmitter to the diaphragm housing, and from the transmitter to the instrument-air header.

Level controllers usually have gage-glass companions. It is convenient for the plant operator to see the gage glasses from the control-valve manifold when operating the control-valve handwheel or the bypass globe valve.

References

1. "Handbook for Control Valve Sizing," Masoneilan International, Inc. Norwood, MA 02062.
2. Dimensions—Masoneilan Control Valves and Auxiliary Equipment, Masoneilan International, Inc., Norwood, MA 02062.
3. "Valve Sizing," Catalog 10, Fisher Controls Co., Marshalltown, IA 50158.
4. Fisher Control Valve Dimensions, Bulletin 1-100, Fisher Controls Co., Marshalltown, IA 50158.
5. Boger, H. W., Recent Trends in Sizing Control Valves, 23rd Annual Symposium on Instrumentation for the Process Industries, Texas A&M University, College Station, TX 77843, 1968.
6. Baumann, H. D., The Introduction of Critical Flow Factor for Valve Sizing, *ISA* (*Instr. Soc. Am.*) *Trans.*, Apr. 1963.
7. Baumann, H. D., Effect of Pipe Reducers on Valve Capacity, *Instr. Control Systems*, Dec. 1967.
8. Boger, H. W., Sizing Control Valves for Flashing Service, *Instr. Control Systems*, Jan. 1970.
9. Boger, H. W., Flow Characteristics for Control Valve Installations, *ISA* (*Instr. Soc. Am*) *J.*, Oct. 1966.
10. Hutchison, J. W. (Ed.), "ISA Handbook of Control Valves," Instrument Soc. of America, Pittsburgh, 1971.

How to select liquid-flow control valves

A chemical process plant employs numerous power-actuated throttling valves to control flow from less than a drop per minute to many thousands of gpm, and pressure drops as low as a few inches of water column to thousands of psi.

James A. Carey and *Donn Hammitt*, *Fisher Controls Co.*

☐ The range of valve applications includes a myriad of differing valve designs, from standard cage-style globe valves to massive, special designs specifically for high-pressure extremes. Choosing the right valve from the mix available might, at first, seem a bewildering task. Yet experience shows that the selection process requires close adherence to only a few fairly simple procedures.

Types of valves

Selecting the right valve for an application can be made easier by first reviewing the four basic styles of throttling-control valves: cage-style globe valves, ball valves, eccentric disk valves, and butterfly valves.

Globe valves are the "standard" of the control valve world. However, the modern cage-style valve (Fig. 1) has largely displaced the top-and-bottom-guided globe valve, either single- or double-ported. Notable in cage-style valve design is its quick-change trim consisting of the valve plug, cage, and separate seat ring.

The cage is a multi-purpose component. It provides valve plug guiding and retains the seat ring in the body, while the shape of the openings in the cage wall determines the flow characteristic of the valve. In contrast to the top-and-bottom-guided valve, the cage-guided valve gives much more stable operation. The cage offers massive valve-plug guiding for stability in the region where maximum pressure drop occurs.

The cage-style valve can meet most application requirements, since there is a wide range of trim styles from which to choose. Balanced, unbalanced, elastomer-seated, restricted or full-size trim are a few of the many options. Any trim configuration is interchangeable within a single valve body, giving the user a saving in purchase price and lowering his parts-inventory needs.

Originally published April 3, 1978.

The globe valve, typically available in sizes to 16-inch, comes in most castable alloys with screwed, flanged, or weld-end connections. It meets standard face-to-face dimensions and is available in ANSI pressure-rating designations through Class 2500.

When it is compared to other valve styles, however, some limitations of the globe valve become apparent: (1) a size limitation, normally 16 inches, (2) lower capacity when compared to an equal size line-of-sight valve such as the ball or butterfly, and at times, (3) overall expense, especially in the larger sizes.

Although many applications are likely to remain the

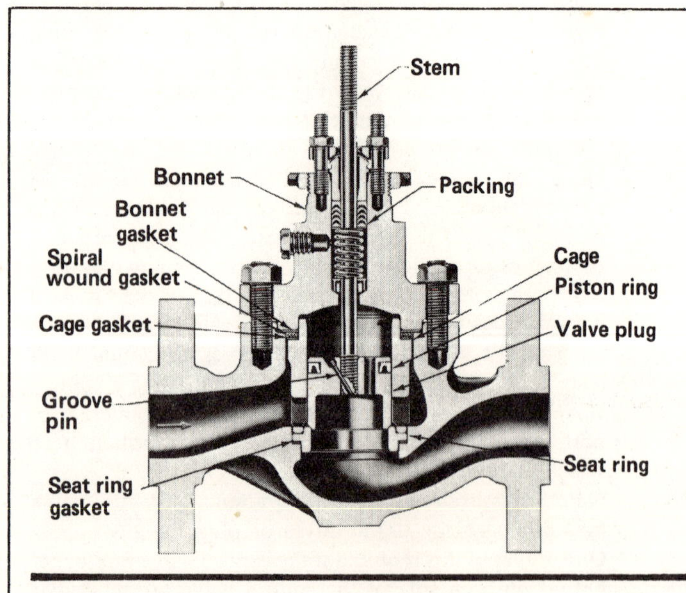

Globe valve with cage-style trim — Fig. 1

Characterized ball valve for throttling operations Fig. 2

domain of globe valves, either a ball- or butterfly-valve may prove the better choice because of higher capacity, simpler and more compact design, lighter weight, and more attractive cost.

There are several types of throttling-ball valves. In addition to the standard full ball (either full bore or reduced bore), there are designs that use a characterized, partial ball (Fig. 2). Designed originally for pulp-stock applications, these now serve in all industries. In another design, a hybrid between a ball valve and a high-performance butterfly valve, a disk pivots from the closed position to a position out of the flow stream.

Most ball valves use elastomeric or fluoroplastic ball seals, and exhibit good shutoff capabilities. Flexible metal seals are becoming standard for high temperatures, and rigid metal seals are increasingly available. The greatest progress in recent years has been made with tetrafluoroethylene (TFE) seals, which survive both relatively high temperatures and corrosive fluids.

Ball valves are classified as high-recovery valves, meaning that outlet pressure recovers to a level close to inlet pressure—higher than the pressure recovery through a globe valve. In other words, they require less pressure drop to produce a given flowrate. But ball valves are more limited in allowable pressure drop and temperature than globe valves.

ANSI Class 600, usually the top body-rating available in a ball valve, is a flangeless design that fits between line flanges. This means that tensile stresses due to piping are transmitted through the flange bolts, not the valve body.

In those applications where either a globe valve or a ball valve will serve, the ball valve can cost as little as half the price of a globe valve.

Another rotary-shaft style gaining in popularity is the high-performance butterfly valve, using a disk with its axis offset from the centerline of the valve (Fig. 3). The result is an eccentric motion of the disk as the valve strokes. An important design feature is that the disk contacts its seal for only a few degrees of rotation at valve closure. This tends to reduce seal wear and prevents permanent seal deformation. Also, because the seal does not rub against the disk when the valve is throttling, friction is low and less operating torque is required. As with a ball valve, this eccentric-design butterfly valve comes in either elastomer or metal seal construction.

This valve style is normally available in sizes through 24 inches, with ANSI Class ratings of 150, 300 or 600. It fits between line flanges, and requires very little space. The comparative costs of eccentric-disk and ball valves vary with size, but on the average an eccentric disk costs less. The cost differential becomes significant at sizes above 6 inches.

The fourth basic type of throttling-control valve, the standard butterfly valve (Fig. 4), is the most economical of all designs on a cost-per-flow-capacity basis.

For general applications not requiring low leakage rates, butterfly valves with swing-through disks are recommended. These valves can handle high inlet pressures and pressure drops over a wide range of

A high-performance rotary-shaft valve Fig. 3

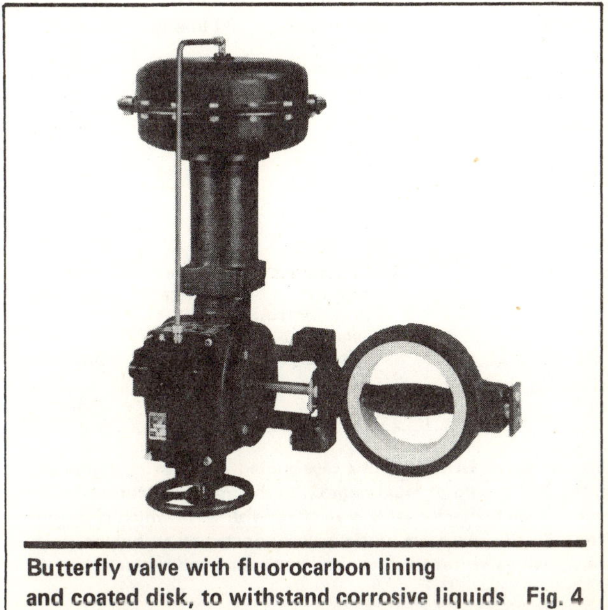

Butterfly valve with fluorocarbon lining and coated disk, to withstand corrosive liquids Fig. 4

temperatures. The swing-through design, with clearance between the body-bore and disk, allows differential thermal expansion of the disk and body without seizing of the disk in the body.

Where low leakage rates are required, piston-ring designs can provide rates approximately one-third those of valves with swing-through disks. They can control high inlet pressures over a wide range of temperatures. The piston ring, retained on the disk periphery, also allows differential thermal expansion of the disk and body.

Valves with either an elastomer T-ring seat or a fully lined body-bore can provide tight shutoff. Also, in fully lined butterfly valves the liner provides total isolation of the body from the process fluid. This permits use of low-cost body materials for corrosive-fluid service.

Butterfly-valve sizes and styles cover a wide range of flow rates, inlet pressures, pressure drops, and shutoff requirements. Sizes range above 100 inches, with body ratings from ANSI Class 125 through Class 2500.

Table I shows principal characteristics of popular designs for globe and rotary-shaft valves.

Pressure considerations

The four major types of control valves discussed above account for the majority of control-valve applications. To choose a specific construction from the options obtainable first requires a look at the pressure conditions the valve must withstand.

Most commercially available control valves meet the standards of the American National Standards Institute (ANSI). These codes establish a pressure and temperature rating for a valve body, based on its material of construction. The first step then in the selection process is to determine which valve style or styles meet the pressure rating of the application. Some valves do not have pressure ratings above ANSI Class 600, while others are not available in relatively low pressure ratings, such as ANSI Class 125.

Selection of the best valve configuration for a particular use depends also on the maximum pressure drop expected. Although a quick study of manufacturers' literature will reveal precise pressure-drop capabilities, it is generally safe to expect that globe valves will control higher pressure-drops than rotary-shaft valves. Cage-guided globe valves serve successfully for pressure drops as high as 50,000 psi. Cage-guided designs are generally used where no other type of valve-plug guiding is satisfactory. Other globe valves also give relatively good pressure-drop capabilities, although some designs are intended for applications where pressure drops do not exceed 100 psi.

The pressure-drop capability of ball, butterfly, and other rotary-shaft valves depends largely on (1) the support of the throttling member, (2) shaft-bearing sizes and locations, and (3) the sealing-member design. Consequently, pressure drop capabilities vary considerably among common rotary-shaft valves. For example, trunnion-mounted ball valves can handle very high pressure drops, while various designs of floating-ball

Pertinent characteristics of common nonproprietary control valves **Table I**

	Globe valves				Rotary-shaft valves			
Characteristic	Cage guided	Split body	Flangeless	Top-and-bottom guided	Lined butterfly	High performance butterfly†	Characterized ball valve¶	"Floating" ball valve**
ANSI pressure classes	125 to 2,500	150 to 600	150 to 300	125 to 2,500	125 to 300	150 to 600	125 to 600	125 to 300
Usual size range, in.	1/2 to 16	1/2 to 8	1/2 to 4	1/2 to 24	2 to 24	2 to 42	1 to 24	1/2 to 12
Temperature range, °F	Cryogenic to 1,500	−100 to 450	−100 to 450	50 to 800	−20 to 300	Cryogenic to 1,500	Cryogenic to 1,000	−50 to 400
Pressure drop, psi	to 14,000	300	740	3,000	200	1,440	up to 1,440	up to 720
Capacity relative to 3-in.	100*	75	50	90	250	200	240	350
Purchase cost relative to 3-in.	100*	70	70	110	50	65	80	80
Relative maintenance cost	Low	Moderate	Moderate	High	Highest	High	High	High
Flow characteristic	Any	Any	Any	Any	Equal percentage	Linear	Modified equal %	Modified equa
Rangeability‡	30 to 50:1	30 to 50:1	30 to 50:1	30 to 50:1	100:1	75:1	200 to 300:1	100:1
Typical application	Steam letdown; high ΔP flow control; general application	Corrosive service	Corrosive service	Flow control; general applications	Low ΔP general service; corrosives; HVAC applications	General service; steam; water; hydrocarbons; natural gas dehydration	General service; pulp and erosive slurries; dry chlorine service; superheated steam	General servic hydrocarb water; satu rated stea
Usage trend	Level	Down	Down	Down, sharply	Up	Up	Up	Up

Legend
*Values are compared with the globe cage-guided valve having a base of 100.
‡Rangeability is the ratio of maximum C_V to the minimum controllable C_V
†High performance butterfly valve is an offset disk valve which is trunnion supported and suitable for high pressure drop.
¶Characterized ball valve has a partial ball supported by trunnions.
**"Floating" ball valve has a full or solid ball supported by seals.

Suggested service temperature limits for elastomeric valve components **Table II**

Material	Limits, °F
Natural rubber	–60 to +160
Neoprene	–40 to +175
Nitrile	–20 to +200
Polyurethane	–40 to +200
Butyl	–20 to +300
Ethylene propylene	–40 to +300
Tetrafluoroethylene (TFE)	0 to +450
Silicone	–65 to +400

valves are usually limited to drops of 300 to 400 psi or less. Special butterfly-valve designs accommodate pressure drops well in excess of 1000 psi, but the common rubber-lined butterfly valve ordinarily operates up to 100 or 200 psi.

Flowing fluid

The flowing fluid must enter into selection of the valve configuration. For example, fibrous or erosive slurries are difficult to handle at best; specifying valves designed for these services give optimum control and service life.

Another example is corrosive service. Exotic alloys to withstand corrosive liquids are often not readily available in cast form; therefore, TFE-lined bodies or valves made from bar stock might provide the answer. Fortunately, most applications involve relatively noncorrosive fluids at reasonable pressures and temperatures. Therefore, cast iron and cast carbon steel are the most commonly used valve-body materials. For further information, see the article "Corrosion and Piping Materials in the CPI in this Deskbook.

Temperature limitations

Temperature is another major operational consideration. Valves that rely on elastomeric materials for seals or for bearing surfaces cannot handle temperatures much higher than 300 or 400°F. Temperature ranges in Table II suggest limits within which elastomers will function adequately. Dynamic forces imposed on the materials also enter into consideration, since, in many cases, tear strength and other physical properties decrease rapidly as temperature increases.

Cryogenic service demands valves with low cooldown mass, short face-to-face dimensions for ease of installation, and a way to keep the valve packing out of the cold-box area. Extremely high-temperature service usually means a special body and valve-trim configuration and special materials of construction.

Shutoff capability

Besides temperature, other secondary criteria help determine the valve trim—the so-called "innards" of the valve. Often neglected, but very important to the selection of an economical control-valve package, is the degree to which the valve must shut off. This degree of shutoff can range from a relatively large amount of seat leakage in the closed position (such as one might expect in a swing-through butterfly valve) to leakages of less than 1 bubble per minute (the standard for soft-seated globe control valves). ANSI standard B16-104 defines the industry-recognized standard levels of leakage for control valves.

Stringent shutoff specifications usually add to the cost of a valve. Assume a 4-inch globe valve operating at temperatures around 600°F, with a pressure drop in the shutoff position of 1,000 psi. If leakages of 0.5%, or 0.1% of maximum valve flow, can be tolerated, the net price might be around $1,500, complete with a pneumatic actuator. If, on the other hand, the leakage has to be held at the level of 1 or 2 bubbles per minute, the price could easily exceed $3,000. Know what leak rates can be tolerated, and then specify to that level. Otherwise, expect to pay for oversized actuators or special valve construction.

Flow characteristics

Although thoroughly explored in innumerable technical discussions, valve flow characteristics remains one of the least-understood factors in selecting valve trim. The three most common characteristics—linear, equal-percentage, and quick-opening—are inherent; i.e., they measure the inherent capacity of a valve with a constant pressure drop across it as the valve plug is traveled.

- The linear characteristic is self-explanatory; the valve capacity varies linearly with the travel of the valve plug.
- The quick-opening characteristic is linear for the first portion of the travel, and very little gain in capacity is experienced thereafter. This is the type of characteristic that is found in a poppet type valve where no throttling surface is exposed to the fluid flow to modulate capacity in the early stages of control valve travel.
- An equal-percentage characteristic increases flow capacity by the same percentage for each equal increment of travel. For example, assume capacity increases by 10% in the first 0.1-inch of travel; it will increase another 10% in the next 0.1-inch, and so on until fully opened. This characteristic exhibits a straight line on a semilogarithmic plot; it is exponential, not linear.

As its main objective, valve-flow characterization can show how the gain of the valve varies to compensate for changes in process gain with changing loads. The gain of the valve indicates the sensitivity of its output (flow) to changes in the input (valve travel). A high-gain valve exhibits a large change in flow for a small change in the valve position.

Thus, (1) a quick-opening characteristic has a high gain in the initial travel of the valve plug and then a low gain in the upper portions of the travel, (2) the linear valve characteristic maintains a constant gain throughout the entire travel, and (3) the equal-percentage and related modified-parabolic characteristics show low valve gains in the low-travel regions but, as the travel increases, valve gain increases sharply.

The flow characteristics discussed so far are described as inherent, and are those observed when pressure drop across the valve is constant. Flow through the valve, however, is influenced by more than just the flow area. Variations in pressure drop across the valve also cause changes in the fluid flow, even at constant valve area. In laboratory testing, holding pressure drop constant produces flow that is strictly a function of valve travel. In actual applications, the flow/travel relationship that results is referred to as the installed flow characteristic—the characteristic obtained in service where pressure drop varies with flow and with other system changes.

Pressure drop across the valve also influences the amount of flow change that occurs as a result of a travel change. Consider a linear valve in a system where pressure drop across the valve increases with load flow. At low load flows in correspondingly low travels, the pressure drop will also be low. As the load flow increases, the pressure drop also increases. Plotting flow-data points shows that, even though the inherent flow characteristic of this valve is linear, the installed flow characteristic more nearly approaches equal-percentage.

Conversely, if the pressure drop changes with flow in such a manner that, as the load flow increases, the pressure drop across the valve decreases, then the installed flow characteristic will appear more like a quick-opening characteristic. To establish the necessary installed characteristic requires a dynamic analysis of the system, so that the valve gain will compensate properly for gain variations in the process.

Reliable guidelines that help in selecting the proper valve flow characteristic are tabulated in Table III. The optimum method for selecting a characteristic involves a dynamic analysis of the control system. For those wishing to pursue the subject farther, more information may be found in the bibliography at the end of this article.

If a good dynamic analysis of the system is available, or if there is sufficient time to make one, the valve characteristic that is best suited to a particular control loop can be selected without reliance on the guidelines listed in Table 3. However, without a dynamic analysis, trouble can usually be avoided by following the previous suggestions, although optimum control may not be reached.

Long experience and numerous analyses show that it is better to pick an equal-percentage characteristic when in doubt. Using a linear characteristic where an equal-percentage one would have been better often leads to an unstable system. However, the reverse situation seldom causes instability. Fortunately, an equal-percentage characteristic is inherent in most of the common ball and butterfly valves.

A recently revived theory from the late 1950s holds that a "universal" characteristic exists between linear and equal-percentage. The fit of a universal characteristic to a flow system depends on (1) the percentage of total system pressure-drop designated for the valve, (2) the range in load experienced, (3) the gain of the control loop as load varies, and (4) the change in total system loss as a function of load. It is possible that this characteristic suits some applications, but in most instances it is not likely to give control as good as that provided by a properly selected linear or equal-percentage characteristic.

Flow-characteristic selection guidelines — **Table III**

Control factor	Conditions encountered	Characteristic to use
Liquid level	Pressure drop increases by 2 to1 ratio, or more, as flow increases	Quick-opening
	Any other	Linear
Pressure	Liquid	Equal-percentage
	Compressible fluid:	
	Fast system—low volume downstream (generally, less than 10 ft of pipe); pressure increases quickly.	Equal-percentage
	Slow system—generally, more than 100 ft of pipe downstream	Linear
	If pressure drop varies by 5-to-1 ratio, or more, for either a fast or slow system	Equal-percentage
Flow	Measuring element in series with valve	Linear
	Measuring element in bypass:	
	Linear measuring device	Linear
	Square-root measuring device	Equal-percentage
	Flow range small; pressure-drop at valve changes greatly	Equal-percentage

Rangeability of control valves

The word rangeability, used alone, is not definitive. Any discussion of it must address itself to inherent rangeability and installed rangeability.

Inherent rangeability can best be defined as "the ratio of maximum to minimum flow-coefficients between which the gain of the valve does not exceed the specified gain by some stated ratio." This definition must, of course, be further qualified by the limitation that in no case can the inherent rangeability exceed 100 divided by the percent of leakage.

As an example, if a double-ported or balanced type of valve with a leakage of approximately 0.5% could throttle down to the point of seat contact, the inherent rangeability would be 200. If the valve had tight shutoff (zero leakage), and again could throttle down to the point of seat contact, the inherent rangeability would be infinite. Unfortunately, it is not possible to achieve accurate control down to the point of seat contact, due to the necessary tolerances between parts. (This has a significant effect on the rangeability as the valve plug approaches the seat.)

The point of minimum controllable flow most commonly comes at the point of clearance flow, a flow that would be achieved through the dimensional clearance between the valve plug and cage. Since the clearance flow can be as much as 2% of the total flow, this would indicate an inherent rangeability of 50.

Installed rangeability can be defined as "the ratio of maximum to minimum flow over which the valve, as installed, will provide satisfactory control." Most people feel that the installed rangeability of a valve can be established from the inherent characteristic curve plus a knowledge of the range of actual service conditions. This is rarely the case and, in fact, applies only where the limiting factor is leakage flow.

The following points help make a reasonably accurate estimate of installed rangeability:

- The required valve gain over the intended flow range.
- The inherent flow characteristic of the valve.
- The maximum permissible ratio of actual to required gain.
- Stem force characteristics in the low-flow region.
- Actuator stiffness.

At this point, it should be obvious that there are some serious limitations in trying to estimate installed rangeability. To do this with any degree of confidence requires considerable time, along with access to information not readily available. Although a complete rangeability analysis can seldom be justified, and comprehensive guidelines are not feasible, there are a few rules that may help in avoiding rangeability problems:

1. Whenever possible, select the flow characteristic that best matches system requirements.
2. Do not expect installed rangeability to exceed that calculated from leakage flow.
3. Operating a globe-style balanced valve below 0.050 inches lift requires that spring rate (pounds per inch) of the actuator be substantially greater than pressure drop (psi), or negative gradient problems may possibly arise.

Packing materials

Unless otherwise specified, a control valve comes equipped with packing the manufacturer considers standard for the application. Sometimes the manufacturer's past experience and knowledge of the strong and weak points of available packing materials leads him to recommend a packing other than that specified.

Basic packing materials in widespread use today include asbestos, graphite, and TFE. Asbestos is low in cost and stable to about 800°F, but tends to give a high-friction seal.

Graphite packings may also have higher friction. Graphite is stable to 3,000°F in nonoxidizing service, but practical limits are 740°F for oxidizing service and 1,200°F for nonoxidizing service. It has high thermal conductivity and long service life, but requires higher gland pressures for leak-free operation. Graphite is impervious to a wide range of liquids, including sulfuric acid (up to 95% acid at 160°F), sodium hydroxide at all concentrations, and almost all inorganic compounds. Strong oxidizers, however, are not always compatible. Sodium chlorite, sodium hypochlorite, bromine, chlorine, and iodine will attack graphite packings above room temperature.

TFE materials, more chemically inert, are suitable for use with strong oxidizers. They also have very low friction properties. They require a smooth stem finish (4 to 6 micro-inches RMS) and will leak if stem or packing surface is damaged. Temperatures should range between −40 and 450°F.

To produce optimum properties, various materials can be mixed with each other. Two good examples of materials that are not suitable as packing when used alone, but which result in better packing when used as impregnants, are soft metals and fiber glass. Some soft metals provide better retainment and can provide cathodic protection in systems where stem or packing-box corrosion is a problem. Substances such as molybdenum disulfide strengthen packing materials, and are often used in TFE packings.

To achieve the most sensitive actuator response to an input signal requires the lowest possible packing friction without leakage. Accordingly, packing suppliers offer a number of packing compositions designed for low friction performance:

- Braided asbestos impregnated with TFE, molded into split rings.
- Asbestos fiber, lead wool, flake graphite, metal particles, and neoprene binder.
- Braided, preshrunk Teflon-fiber yarn impregnated with TFE.
- Molded rings of TFE, glass fibers, and compounding material.
- Square-braided blue African asbestos impregnated with TFE.

All packings require some periodic maintenance. Although spring-loaded TFE packing requires less maintenance than others, slight leakage should be expected. "Zero leakage" calls for a bellows seal with leak-off connection and backup packing. Usually this expense cannot be warranted except for hazardous fluids or extremely precious materials.

Lubricants can be used to help reduce friction. Lantern-ring spacers strategically placed on the stem between packing rings allow lubricant access to the stem; lubricant can be added at any time. Lubricant composition is optional, but it must be compatible with both controlled fluid and packing material. For moderately high and low temperatures, silicone fluids are widely used. Above 500°F, however, silicones "coke," or oxidize, and often become more of a problem than an aid in obtaining low friction or a tight seal.

Valve sizing

With the valve configuration and flow characteristic selected, the control valve can be sized. Sizing is exactly what the name implies; it is the technique of determining the valve size *best* suited to control the process. Obviously, a valve too small will not pass the required amount of flow. A valve too large will be unnecessarily expensive and may create trim wear and control problems at very low increments of travel.

Several sizing techniques are available, some in a form designed for calculation by hand-held programmable calculators. This permits convenient calculation of required valve capacity, based on maximum and minimum flow rates of the system and the pressure drop at each condition. For liquids, this capacity is expressed by the coefficient C_v, numerically equal to the number of U.S. gallons of water at 60°F that will flow through the valve in one minute under a 1 psi differential. C_g and C_s coefficients were developed for gases and steam to correct for compressibility effects and critical flow limitations at increased pressure drops.

(Some manufacturers do not use the C_g and C_s coefficients, but employ a conversion method utilizing the liquid sizing coefficient C_v.)

With the manufacturer's sizing technique, it takes only a few minutes to determine the flow coefficient, check for flashing or cavitating conditions in the case of liquid service, or critical flow for gases, and then select a valve with the proper capacity from a sizing chart. Most manufacturers have a logical progression of capacities to aid in selecting valve sizes. Restricted trim is available for applications where future demands for capacities might be greater than today's, or where the designer wishes to avoid a situation which could see an 8-inch line swaged down to a 2-inch valve.

Cavitation

There are three categories of liquid flow through a control valve: cavitating, noncavitating, and flashing. Each can be of concern to the valve specifier, because of their potential for mechanical valve damage and excessive valve noise.

For noncavitating and flashing liquid flow, laboratory tests and field experience show that noise levels are quite low, and so are generally not a problem. Also, proper valve design and material selection can eliminate mechanical damage from these two threats. (As an example, erosion damage caused by flashing liquid can be reduced or eliminated by specifying erosion-resistant valve body and trim materials.)

Cavitating flow, on the other hand, can cause considerable noise, and cavitation damage can make a valve useless. Cavitation must, therefore, receive major consideration in selecting valves for liquid flow.

Cavitation refers to the formation and subsequent collapse of vapor bubbles in the flowing liquid stream. As the liquid flows through the orifice of a control valve, its velocity increases while its static pressure decreases. In many applications, the increase in velocity causes pressure in the control valve to drop below the vapor pressure of the liquid, and vapor bubbles are formed. As the fluid moves downstream within the control valve into a larger flow area, the velocity decreases, with resulting pressure recovery. When the static pressure exceeds the vapor pressure of the liquid, implosions of the vapor bubbles occur, generating extremely high-pressure shock waves that hammer against the valve outlet and piping. Pressures in these collapsing cavities, reported as high as 500,000 psi in magnitude, can cause severe and rapid damage to the valve and piping.

The equation used to determine the maximum pressure drop effective in producing flow through globe valves can also be used to indicate when significant cavitation can occur. Minor cavitation will occur at a slightly lower pressure differential than that predicted by the equation, but should produce negligible damage in most instances. This formula is:

$$\Delta P = K_m(P_1 - r_c P_v)$$

where ΔP = Maximum allowable pressure drop; K_m = Valve-recovered coefficient; P_1 = Inlet pressure (psia); r_c = Critical pressure ratio; P_v = Vapor pressure of the fluid at the inlet condition (psia).

For high-recovery rotary valves, the equation becomes

$$\Delta P_c = K_c(P_1 - P_v),$$

where K_c is a cavitation index.

Solving cavitation problems begins with either controlling the cavitation process or, ideally, eliminating cavitation altogether. Before "anti-cavitation" valve trims appeared, several techniques served—and some still are serving—with various degrees of success.

Erosion-resistant body material, along with hardened valve trim, became a common technique. A sacrificial member, either in the valve itself or in the downstream piping, was another attempt to extend valve or piping life. An example is the "flow-down" angle valve, with a hardened liner in the downstream part of the valve body. The success of this method depends on the energy levels involved in pressure drop and flowrate. If both are high, the sacrificial member may soon disappear.

Another approach to preventing cavitation involves sharing the overall pressure drop, either between valves in series or between one valve and a breakdown orifice device or capillary tubing. If the series method is the choice, complicated control problems may occur, and the initial equipment cost may be prohibitive. The drawback of the downstream-orifice method is that it applies to only a narrow range of flow, and therefore has little application in throttling control.

Strategic location of the valve along the flow path can give effective and economical cavitation control. For example, locating the valve next to a tank and allowing it to flash into the tank eliminates cavitation problems that could occur if the valve were installed elsewhere in the system.

All these methods have limited applicability or limited design life—perhaps both. In the final analysis, the requirement is usually for a single valve/trim combination that will either control the location of cavitation and prevent damage, or eliminate the potential for cavitation altogether.

Pressure drop and flowrate are usually the chief factors in choosing the best valve/trim combination where cavitation is possible. The higher these two factors, the more sophisticated must be the valve/trim package.

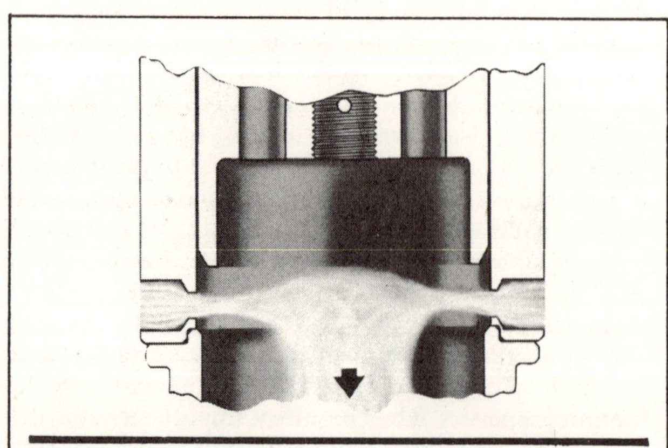

Cage-trim holes form a flow cushion that prevents cavitating liquids from contacting metal surfaces Fig. 5

Criticality of the application and economics quickly enter the picture, so it is advantageous to have several valve/trim styles available, so the best and most economical trim can be matched to the system application.

For applications with low pressure drops and relatively low flowrates, a cage with knife-edge orifices in the cage wall (Fig. 5) was one of the first valve/trim answers. As the valve plug moves away from the seat, more orifices open for the fluid. Each hole's jet aims at the center, to meet the diametrically opposite jet. A fluid cushion surrounds the central core of cavitating liquid, preventing contact of imploding bubbles with metal seating surfaces.

For applications with high pressure drops (1,000 to 3,000 psi), the high energy levels call for a valve/trim combination that eliminates cavitation. If a single control valve serves, the valve trim must "absorb" the pressure drop in stages, so that flowing pressure remains above the vapor pressure of the liquid. Concentric, cylindrical sections, each containing specially drilled orifices, are designed into valves with this style trim (Fig. 6). In operation, each section stages the pressure drop, and the number of stages required depends on the inlet pressure and the total pressure drop across the valve.

This cage design can also be characterized for applications where pressure drop across the valve decreases with increasing valve-plug travel. The characterization gives pressure-staging plus cavitation-prevention at low plug-lifts where it is needed. The number of stages becomes progressively lower as the pressure drop becomes smaller at high travel. Many other designs can prevent cavitation. Most work on some variation of a multiple-stage or tortuous-path principle to break the overall pressure drop into many smaller ones.

Excessive valve noise

Pressures, flows, and temperatures of many processes have increased over the years with a resulting increase in the operating noise level of valves controlling compressible flow. The response by control-valve manufacturers has been to develop noise-reducing valve trims and high-capacity valve configurations.

Understanding the major sources of control-valve noise is the first step to understanding today's noise-abatement technology.

One noise source, the mechanical vibration of valve components, results from either random pressure fluctuations within the valve body, or fluid impingement upon the movable or flexible parts. A more common type of vibration, the lateral movement of a valve plug against its guiding surfaces, produces sound with a frequency of less than 1,500 Hz, often described as a metallic rattling.

Another type of vibration is a valve component resonating at its natural frequency. This generates a single-pitched tone at a frequency of 3,000 to 7,000 Hz. Valve components susceptible to natural-frequency vibration are contoured valve plugs with hollow skirts, and flexible members such as the metal seal ring of a ball valve.

With both types of vibration, noise is usually a secondary concern, since it may be a warning of valve

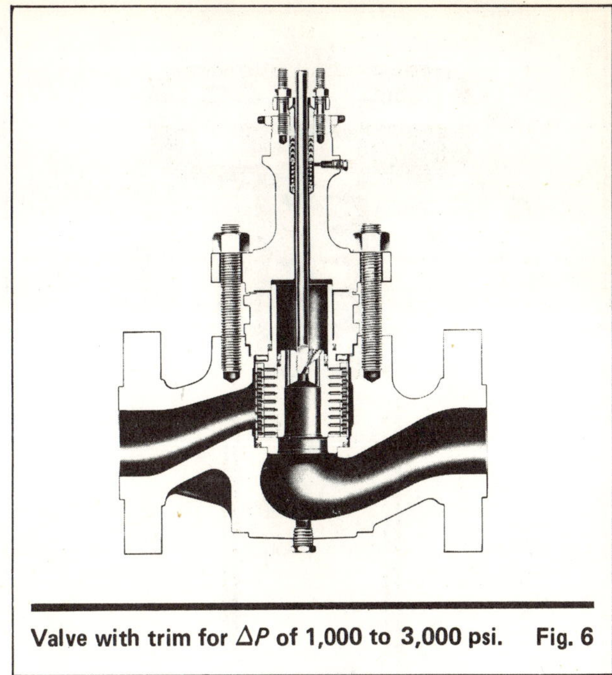

Valve with trim for ΔP of 1,000 to 3,000 psi. Fig. 6

failure. Resonant vibration produces high levels of stress which can fatigue the part. Fortunately, most modern control valves have been designed very conservatively—sturdy, rugged, and well guided—and they suppress vibration to assure a long service life. Vibration-caused noise is a minor consideration today.

Actuator

Usually, the final step in valve specification is selecting the valve actuator. An integral part of every automatic control loop, it provides the muscle, or motive force, required to position the final control element. And, since stability and operability of the loop hinge on satisfactory actuator performance, the actuator must be able to take control of the many varying static and dynamic forces created by the valve.

For the many available valve styles, there are four basic actuator types suitable for throttling-control: 1. Spring-and-diaphragm; 2. Pneumatic piston; 3. Electric motor; 4. Hydraulic or electro-hydraulic.

Diaphragm actuators

The pneumatic spring-and-diaphragm actuator, a very common, extremely simple design (Fig. 7), offers low cost and high reliability. Diaphragm actuators normally operate with air-supply ranges of 3 to 15 psi or 6 to 30 psi. Because of this, they are often suitable for throttling service using direct instrument signals. Available designs offer either adjustable springs or wide spring selections, for tailoring the actuator to the application. Spring-and-diaphragm actuators have few moving parts that can contribute to failure and, therefore, offer extreme reliability. If they fail, maintenance is simple.

The overwhelming advantage of spring-and-diaphragm actuators is the ever-present provision for fail-safe action. As air is loaded on the actuator casing, the diaphragm moves the valve and compresses the spring.

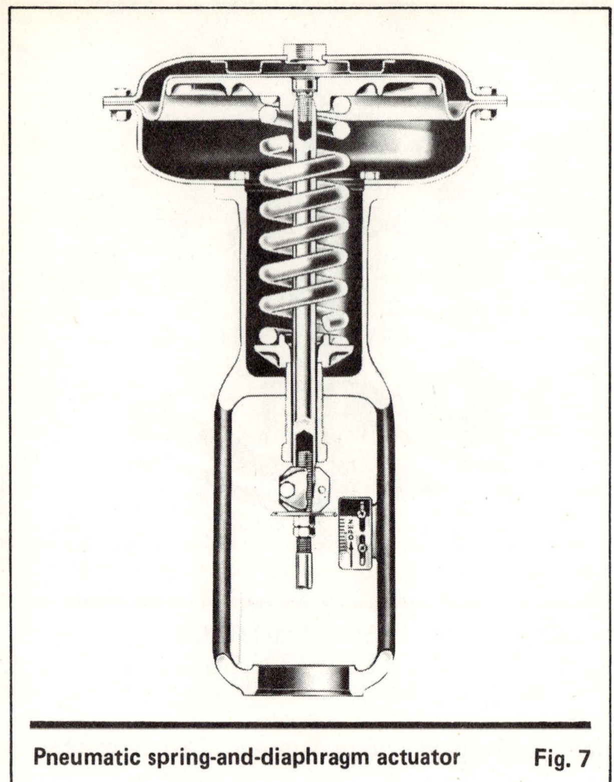

Pneumatic spring-and-diaphragm actuator Fig. 7

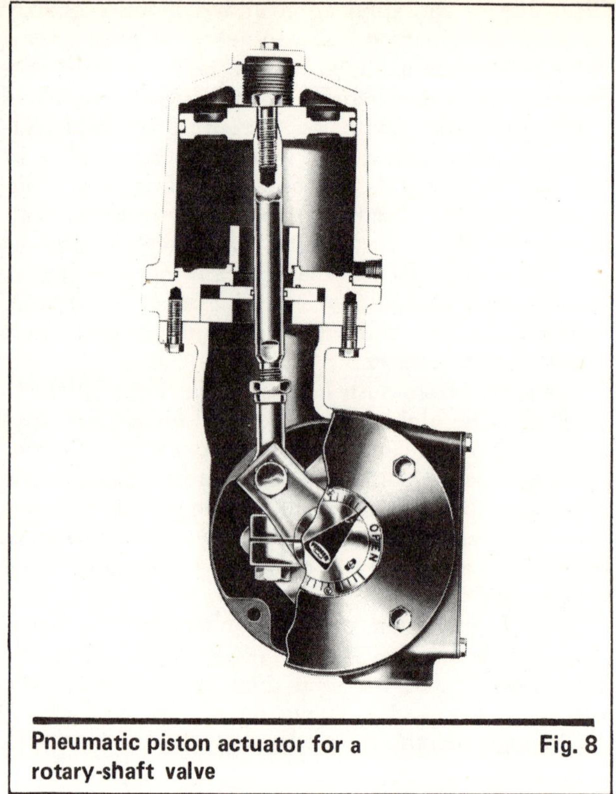

Pneumatic piston actuator for a rotary-shaft valve Fig. 8

Stored energy in the spring acts to move the valve back to its original position as air is removed from the casing. In the event of signal-pressure loss to the instrument or actuator, the spring moves the valve to its initial (fail-safe) position. Spring-and-diaphragm actuator designs offer either fail-open or fail-closed action on loss of signal pressure.

The primary drawback of a spring-and-diaphragm actuator is its relatively limited capability. Much of the thrust created by the diaphragm is taken up by the spring and does not result in output. The spring-and-diaphragm actuator ceases to be cost-effective for requirements in excess of approximately 2,000 lb thrust or 5,000 in.-lb torque. Unless there are severe overriding factors, use of diaphragm actuators above this level can be costly. It is simply not economical to build and use diaphragm actuators in this thrust range because the size, weight, and cost grow out of proportion.

Piston actuators

Higher-force requirements than those provided by spring-and-diaphragm actuators, call for one of the other three types previously mentioned. Pneumatic-piston actuators provide the next economical force-output range for operation of automatic control valves. Piston actuators normally work with supply pressures between 50 and 150 psi. Although some feature spring return, this construction has limited capabilities.

Piston actuators used in throttling service must be furnished with double-acting positioners which will simultaneously load and unload opposite sides of the pistons (Fig. 8), causing travel toward the lower-pressure side. The positioner senses the motion of the piston, and when the required position is reached, the positioner equalizes the opposing pressures on the piston, creating equilibrium.

The pneumatic-piston actuator is an excellent choice when compact, high-thrust units are required. It can also serve very effectively where varying service conditions require a wide range of output forces. Constructed primarily of metallic parts with few elastomers, piston actuators adapt readily to situations involving high ambient temperatures or humidities.

The main disadvantages of piston actuators are the high supply pressures required, the requirement for positioners when used in throttling service, and the lack of built-in fail-safe systems. As mentioned, piston actuators can be equipped with spring-return options, but addition of springs limits the construction to much the same force outputs as the diaphragm actuator. The only alternatives to springs are pneumatic-trip systems to move the piston actuator to its fail-safe condition. While these systems prove highly reliable, they add to overall system complexity, maintenance and cost.

Other available double-acting, high-pressure pneumatic actuators use vanes or rubber bladders to create the output thrust or torque directly.

Electric actuators

Electric-motor actuators, useful in many process situations, usually consist of motors with gear trains, and are available in a wide range of torque outputs. They offer a prime advantage in remote installations where no other power source is available.

Electric actuators are economical for normal application in smaller size ranges only. Larger units generally operate slowly and weigh considerably more than their pneumatic counterparts. At present, no economically

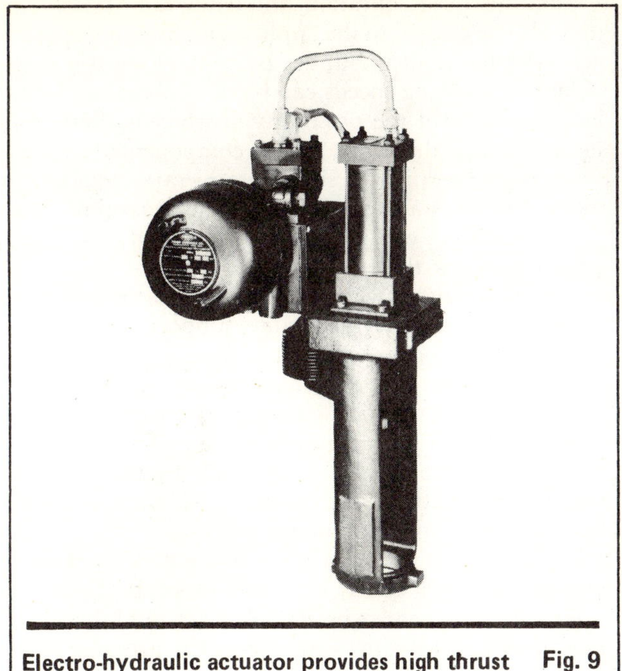

Electro-hydraulic actuator provides high thrust Fig. 9

feasible high-thrust electric actuators are available with fail-safe action other than lock-in-last-position. Throttling versions of electric-motor actuators have limitations in both capability and availability. In continuous, closed-loop applications requiring frequent changes to control valve position, the electric actuator may not be suitable, primarily because of its limited duty cycle.

Hydraulic and electro-hydraulic actuators

Electro-hydraulic actuators (Fig. 9) are electric actuators in which a motor pumps oil at high pressures to a piston, which in turn creates the output force. An electro-hydraulic actuator is an excellent choice for throttling service, because of its high stiffness (resistance to changing valve-body forces) and its compatibility with analog signals. Most electro-hydraulic actuators can provide high-thrust outputs (often up to 10,000 lbs). They are, however, handicapped by high initial cost, complexity, and size.

Hydraulic actuators, essentially the same as the electro-hydraulics, differ principally in gaining their power from an external pumping unit. A central hydraulic pumping unit can supply hydraulic fluid under considerable pressure, sometimes up to 3,000 psi. Actuator control is accomplished through a servo amplifier and a system of hydraulic valves. This system can provide the ultimate in actuator performance: exceptional stiffness, fast stroking speeds, very high thrust (often up to 50,000 lb), and excellent dynamic-response characteristics. But the price tag for all this performance is quite high.

Actuator selection

Selecting an actuator for power-operated valves requires looking at both performance and economics. The control valve can perform its function only as well as the actuator withstands the forces exerted on it. Also, an actuator can represent a significant portion of the total package price, especially when used with a small control valve. Careful selection, though, can lead to significant savings. As mentioned before, the wide range of actuator types and sizes might make the selection process seem highly complex. It is not. With a few simple rules in mind, knowledge of fundamental process needs can make the selection process very simple.

Several major actuator characteristics can be of help in making a selection:

Power source—The available source of power at the location of the valve can point directly to what type of actuator to choose. Typically, valve actuators are powered either by compressed air or electricity. (Some designs, however, use water pressure, hydraulic fluid, or even pipeline pressure.) The majority of actuators now use compressed air, and operate at supply pressures from 15 psi to 200 psi.

Since most plants have readily available electricity and compressed air, actuator selection depends on the convenience and cost of furnishing either power source to the actuator location. Other considerations include reliability and maintenance requirements of the power system (and their effects on subsequent valve operation), and provision for backup power to critical plant loops.

Fail-safe characteristics—Although the overall reliability of power sources is high, many processes demand specific valve actions if the power source fails. Fail-safe systems incorporated into many actuator designs automatically shut down the process, thereby preventing possible loss of product in case of power failure. Some systems store energy either mechanically in springs or pneumatically in volume tanks or hydraulic accumulators. Failure of actuator power triggers fail-safe systems to drive the valves to the required position and then hold this position until resumption of normal operation.

Available designs allow choice of valve-failure mode: failing open, failing closed, or holding in the last position. Many actuator systems, such as the spring-and-diaphragm, incorporate failure modes at no extra cost. In other designs, they may be optional.

Actuator capability—An actuator must have sufficient torque or thrust for the specific application. In some cases, torque requirements can dictate actuator type as well as power-supply requirements. For instance, large valves which require a high amount of torque, or thrust, may be limited to electric or electro-hydraulic actuators, because of unavailability of pneumatic actuators with sufficient capability. Conversely, electro-hydraulic or hydraulic actuators would represent a poor choice for a valve with very low force requirement.

Matching actuator capability with valve-body requirements is best left to the control-valve manufacturer. Although the sizing is not difficult, the great variety of designs on the market and the ready availability of vendor expertise (normally at no cost) make detailed knowledge of the procedures unnecessary.

Valve positioners

Pneumatic valve positioners transduce an instrument signal to a valve position, rather than using the air signal to act on the actuator directly. Several studies have shown that positioners are often used where pneumatic amplifiers (volume boosters) would be a better choice. Further, in many cases, better control can be achieved through use of neither an amplifier nor a positioner.

The principal reasons for selecting either a positioner or a volume booster are:

- A split range is required.
- A maximum loading pressure greater than the instrument signal is desired.
- The best possible control is desired. (Examples might include fast recovery from disturbances, or minimization of overshoot.)

Selection of positioner or pneumatic amplifier is loosely related to process dynamics, but not to valve size, unbalance, packing friction, or transmission-line length:

Process type	Positioner or pneumatic amplifier
"Slow" processes Most thermal systems, reactors, liquid-level control, and some low-pressure large-volume gas processes.	Fast positioner, no input restriction. Large-port 1-to-1 pneumatic amplifier may be included in positioner loop for the large actuators.
"Fast" processes Liquid-pressure, small volume gas-pressure and flow processes.	Standard ($\frac{1}{8}$-in. port) or high-capacity pneumatic amplifier commensurate with actuator size and stroking-speed requirements. Pressure gain of amplifier may be 1 or higher, depending on actuator-loading pressure. Positioner generally not recommended. If used, input restriction should be on the smaller actuators, and may or may not be desirable on larger-sized actuators.

Control functions

The functions of actuator control most clearly define the options available on the actuator-selection process. They include signal type, signal range, ambient temperature, vibration level, operating speed and frequency, and the quality of control required.

Signal groups are generally grouped as being either two-position (on/off) or analog (throttling). On/off actuators are controlled by two-position electric, electro-pneumatic or pneumatic switches. This is the simplest type of automatic control, and therefore the least restricted by actuator hardware.

Throttling actuators have considerably higher technological demands put on them, from the standpoints of both compatibility and performance. The throttling actuator receives its input from electronic or pneumatic instruments which measure the control-process variable. The actuator must then move the final control element accurately and timely, in response to the instrument signal, to ensure effective control. Compatibility with instrument signals is inherent in many types of actuators, and it can be obtained with add-on equipment.

Stroking speed, vibration, and temperature resistance may also be critical to the application. Stroking speed is generally not critical, but flexibility to adjust it is desirable. Fast stroking speeds can be detrimental on liquid loops, because of the possibility of water hammer and its subsequent damage to valve components.

In many cases it is desirable to operate the actuator manually for startup or emergency situations. Most power actuators can be equipped with optional handwheels to allow this.

Actuator weight, when combined with the weight of the control valve, may necessitate support structures for the assembly. Proper selection of compact, light-weight actuators can eliminate this unnecessary expense. Often, economics is an overriding factor in actuator selection. The total cost of an actuator includes not only its initial cost but also the operating and maintenance expenses over its total lifetime. These costs vary widely, but easily lend themselves to qualitative judgments.

A simple actuator with few moving parts is easier to service and will generally cause fewer problems; in addition, maintenance personnel understand them, and can work comfortably with them. An actuator made specifically for a control valve eliminates the chance for a costly mismatch. An actuator manufactured by the valve vendor and shipped with the valve will eliminate separate charges for mounting, and assures easier coordination of spare-parts procurement. Inventory of spare parts, a significant hidden cost, can be minimized by selecting actuators designed with commonality of parts in mind.

Spring-and-diaphragm actuators generally cost less than piston actuators of comparable quality. Part of this saving results from using instrument output-air directly, eliminating the need for positioners or boosters, in many cases. The inherent provision for fail-safe action in the diaphragm actuator is also an important consideration.

If a diaphragm actuator cannot be used for the application, the next best choice would be a pneumatic-piston actuator. This type offers a good combination of high thrust and relatively low initial cost, and the simplicity and maintainability characteristic of pneumatic actuators.

In choosing the type of actuator, the fundamental requirement is to understand the application. Knowing control signal, operating mode, power source available, thrust or torque required, manual-operation requirements, and fail-safe position can make decisions easier. Consider the simplicity, maintainability, and lifetime cost of actuator selected. Safety is another factor not to be overlooked.

The spring-and-diaphragm actuator is the most popular, versatile, and economical type. Consider it first. If the inherent limitations eliminate it, then consider pistons, and then electrics or electro-hydraulics, keeping in mind the capabilities and limitations of each (Table IV).

Installation tips

The consideration of line size vs. valve size involves primarily the strength of the valve in relation to the strength of the adjacent piping. The two rules of thumb

Summary of actuator advantages and disadvantages Table IV

Pro	Con
Spring-and-diaphragm	
Low cost	Limited torque availability
Simplicity	Limited temperature range
Inherent failsafe action	Inflexibility to changing service conditions
Low supply-pressure requirement	
Adjustability	
Maintainability	
Ability to throttle sans positioner	
Fast stroking speeds possible	
Pneumatic piston	
High torque capability	Failsafe requires accessories
Compactness	Positioner required for throttling
Light weight	Higher cost
Adaptable to high ambient temperatures	High supply-pressure requirement
Adaptable to varying valve-torque requirements	
Fast stroking-speed possible	
Relatively high actuator stiffness	
Electric motor	
Compact	High cost/torque ratio
Suitable for remote applications	Lack of failsafe action
	Limited throttling ability
	Slow stroking-speed
	Lack of adjustability
Electro-hydraulic/hydraulic	
High torque	High cost
Very high actuator stiffness	Complexity
Excellent throttling stiffness	Large size and weight
Fast stroking-speed	Failsafe action requires accessories

commonly used in this regard are: valve size not less than one half pipe size; and valve size not less than two sizes below line size. Either rule may be used with the assurance of safety.

Control valves are often installed with little thought as to how the piping arrangement immediately adjacent to the valve will affect its performance. Process-control engineers carefully size each valve and specify suitable valve-plug characteristics, but then the actual installation layout is turned over to piping designers who may give little thought to the control functions of the valve. Allowance for pressure drop through the adjacent block valves and piping is often neglected.

It is regrettable that in many instances the care used in specifying control valves in the engineering stages of a project, and the efforts of manufacturers to provide precise valve characteristics and capacities, are partially nullified by poor installation practices. The following general recommendation will help to assure optimum control-valve performance:

1. Avoid arrangements which amount to a nozzle-effect into the inlet of a control valve, as this will alter the valve characteristic adversely.

2. Avoid close-coupled inlet-block valves reduced from line size, since they affect both characteristics and capacity.

3. If a block-and-bypass arrangement must be used around a control valve, the valve is best placed in a straight-through section. Avoid the tortuous manifolds sometimes provided for control valves to achieve accessibility. They may have an adverse affect on valve characteristic and capacity.

4. Avoid sharp turns close to the valve inlet. Straight-line sections into and out of the control valve should, as closely as possible, simulate the piping required to originally establish flow capacity and characteristic test data. These test data are based on straight sections of approximately 5 nominal pipe diameters downstream of the valve and 12 diameters upstream for reducers and expanders; 13 pipe diameters upstream to the outlet of a tank; 18 pipe diameters upstream for elbow fittings oriented in the same plane; and 30 pipe diameters upstream for elbows oriented in a different plane.

While these recommendations tend to be ideal in nature, following them, and taking care in selecting the piping arrangements around the control valve, can help in getting the best performance from valve selection.

References

1. Control Valve Handbook, 2nd ed., Fisher Controls Co., Marshalltown, Iowa, 1977.
2. Coughanowr, L. B., and Koppel, L. D., "Process System Analysis and Control," McGraw-Hill, New York, 1965.
3. Harriot, P., "Process Control," McGraw-Hill, New York, 1964.
4. Lloyd, S. G., and Anderson, G. D., "Industrial Process Control," Fisher Controls Co., Marshalltown, Iowa, 1971.
5. Lloyd, S. G., Generalized Control Theory, TM-8, Fisher Controls Co., Marshalltown, Iowa.
6. Murrill, P. W., "Automatic Control of Processes," International Textbook Co., Scranton, Pa., 1967.
7. Schuder, C. B., The Dynamics of Level and Pressure Control, TM-7, Fisher Controls Co., Marshalltown, Iowa.
8. Schuder, C. B., "Control Valve Characteristics," *Instr. Control Syst.*, Mar. 1967.

The authors

James (Jack) A. Carey is assistant manager of the Chemical Sales Group at Fisher Controls Co., Marshalltown, IA 50158. He has also held positions in the refining, petrochemical, and pulp and paper sales areas. He holds a B.S. in industrial technology from Texas A&M University.

Donn Hammitt is Product Manager, Control Valves, within the marketing organization of Fisher Controls Co., where he has had both marketing and valve-design experience. He holds a B.S.M.E. degree from Iowa State University and an M.S.M.E. degree from the University of Nebraska. He is the Fisher representative to the Valve Manufacturers Assn.

Specifying Control Valves

This article presents a basic review of control valves, useful to beginning as well as seasoned engineers. Some theory is presented, but only to illustrate concepts.

SANFORD CHALFIN, Fluor Corp.

Control valves are the most important element in the control system of a chemical process plant. Many new and improved types of control valves are available, making it timely to review the state of control-valve design and application.

The proper selection of a control valve requires knowledge of: the process, the user's design criteria, cost, availability, delivery time, manufacturer's engineering services, and spare parts. The specifying engineer must also consider plant standardization requirements and field-maintenance practices. Furthermore, an early selection of vendors can assure the timely delivery of special valve types and materials when required, and can be helpful in valve sizing, since some factors necessary for sizing are peculiar to particular manufacturers.

For these reasons, chemical engineers need to have an understanding of up-to-date features of control valves, and of the techniques for sizing and selecting them, even when the services of an instrument engineer are available for actual calculations and selection.

Technical Considerations

A control valve is a variable orifice used to regulate the flow of a process fluid in accordance with the requirements of the process. There are three features important to this use: *capacity, characteristics and rangeability*.

The *capacity,* is commonly measured as C_v, which is defined as the number of U.S. gallons per minute of water at 60°F that will flow through the valve with 1-psi pressure drop at a stated pressure and percent of rated travel (see Table I).

The *characteristic* is the relationship of the change in

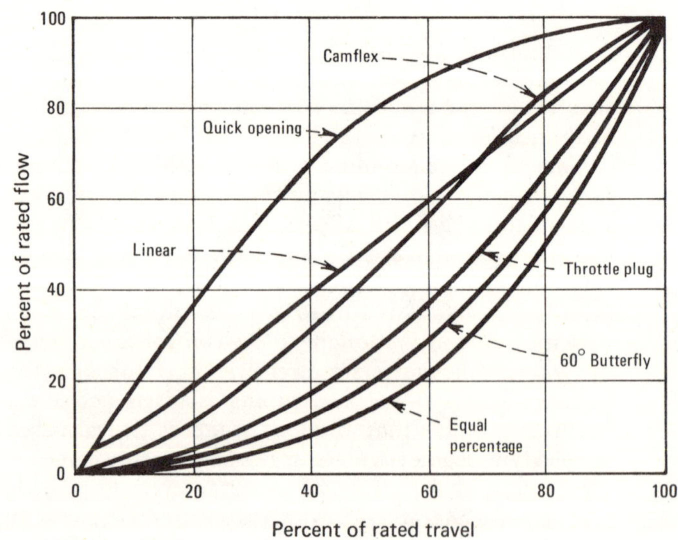

CHARACTERISTIC CURVES of common valves—Fig. 1

Originally published October 14, 1974.

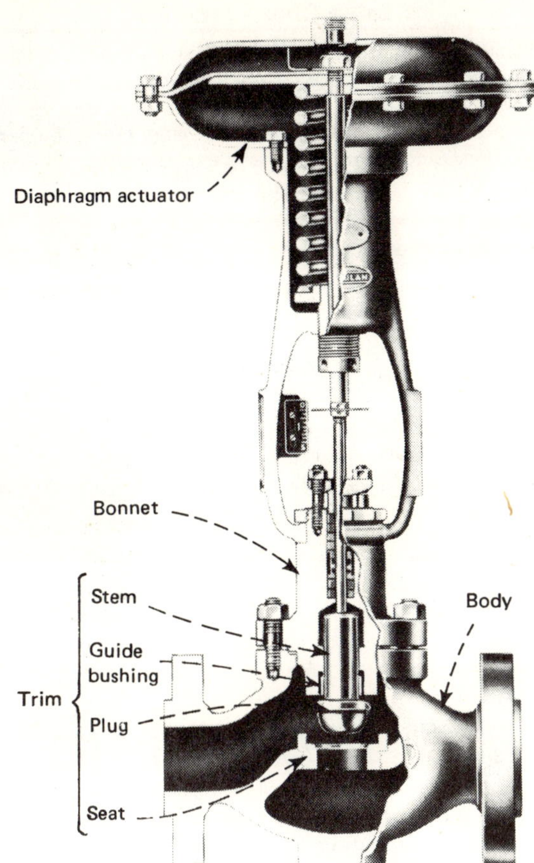

BASIC STRUCTURE of a control valve—Fig. 2

valve opening (plug or vane position) to the change in flow through the valve. The most frequently used characteristics are those of: equal percentage, linear, and quick opening (see Fig. 1).

Rangeability is the ratio of the range of flows, high to low, through which the control valve can give stable control.

In addition to a valve's features, there are two other technical considerations: structure and type.

Most control valves have certain structural features in common. (Fig. 2). These include: a body, capable of withstanding the pressure, temperature, corrosion, abrasion and other properties of the process fluid. A trim, with seat, a plug or vane, a stem or shaft to position the plug or vane, guides for stem or plug, and bushings, all of which act to determine the valve's characteristic. A bonnet, or other pressure-sealing structure, located on the valve body, subject to the same pressure and temperature conditions as the valve, and containing those stem-sealing parts, that enable the stem to open or close the valve without leakage of process fluid. And finally, an actuator, normally an air-operated diaphragm or piston, which translates a signal from the valve controller into stem or vane motion. Springs oppose the force of air on the diaphragm or piston to hold the plug or vane in position against the forces of fluid flow.

Relative Flow Capacities of Control Valves – Table I

Valve Type	C_d*	C_dF_p†	C_dF_L**
Double-seat globe	12	11	11
Single-seat top-guided globe	11.5	10.8	10
Single-seat split body	12	11.3	10
Sliding gate	6-12	6-11	na
Single-seat top-entry cage	13.5	12.5	11.5
Eccentric rotating plug (Camflex)	14	13	12
60° open butterfly	18	15.5	12
Single-seat Y valve (300 & 600 lb)	19	16.5	14
Saunders type (unlined)	20	17	na
Saunders type (lined)	15	13.5	na
Throttling (characterized) ball	25	20	15
Single-seat streamlined angle (flow-to-close)	26	20	13
90° open butterfly (average)	32	21.5	18

Note: This table may serve as a rough guide only since actual flow capacities differ between manufacturer's products and individual valve sizes. (Source: ISA "Handbook of Control Valves" Page 17)

* Valve flow coefficient $C_v = C_d \times d^2$ (d = valve dia., in.)

† C_v/d^2 of valve when installed between pipe reducers (pipe dia. 2 x valve dia.)

** C_v/d^2 of valve when undergoing critical (choked) flow conditions.

The types of control valve most often used are: globe, with various kinds of trim; butterfly; eccentric rotating plug; ball; angle; three way; and velocity control. A wide variety of more-specialized types exist. The various types are compared in Table 2, which can help the specifying engineer narrow his choice to a minimum of types before making a detailed selection.

Control valves may be sized by formula, slide rule, or computer. All methods find the C_v; or for gas and steam, some find the C_g or C_s, respectively. Given these criteria, a valve suitable to the process conditions and with the required capacity can be selected from a manufacturer's catalog. Let us now consider the problems related to liquids, then, gas and steam.

Sizing Liquid-Control Valves

For a liquid, the following formula is widely used:

$$C_v = \frac{Q_1}{\sqrt{\Delta P/G_f}}$$

Where: C_v = valve capacity, gpm/psi; Q_1 = flow, gpm; ΔP = pressure drop across valve body, psi; G_f = Liquid specific gravity at the flowing temperature.

This formula seems simple enough at first glance. A little closer look at some terms of the equation reveals special problems:

Q_1, *flow:* Generally, a process has a required design or normal flow. Sizing the valve with this flow as the maximum would be unwise, since it would leave no margin for flexibility. A better procedure is to size the valve for a maximum flow that is some percentage above the normal flow. Most users take the greater of the following:

$$Q_{(sizing)} = 1.3\ Q_{(normal)}$$

or

$$Q_{(sizing)} = 1.1\ Q_{(maximum)}$$

where only a maximum flow is given.

The reason why designers use these factors is to build some "fat" into the valve. At a wide-open position, the slightest upset otherwise would not be compensated for by further opening of the valve and would cause the system to go out of control.

Δp, pressure drop: Three questions become important: Is the pressure drop critical? Is the Δp across the valve enough to give good system control? What is the maximum Δp across the valve at any condition?

If the pressure drop is critical, flow through the valve would be limited by flashing of the liquid into vapor, which would choke the flow and cause cavitation, erosion and noise [*1.1,1.2,1.3,1.4,1.5*]. This flashing occurs when the fluid passing through the valve trim reaches such a high velocity that its pressure goes below the vapor pressure of the liquid, which then turns to vapor.

Under these conditions and a given upstream pressure, an increase in the Δp across the valve will not produce an increase in flow, and Δp, for sizing is limited to this critical (allowable) value. A little further downstream, the fluid velocity decreases, and the pressure recovers, collapsing the vapor bubbles, with ensuing cavitation, erosion and noise. Streamlined valves normally have a high pressure recovery. Manufacturers have developed recovery factors, K_m or C_1, for each type of valve to take into account the valve's tendency to choke the flow [*10.1,10.2,10.3,11.1*]. The lower the K_m factor, the higher the possibility for flashing or cavitation (Table III). The engineer can check for the allowable Δp as follows:

$$\Delta P_{(allow.)} = K_m(P_1 - r_c P_v)$$

Where: K_m = valve recovery coefficient; P_1 = body inlet pressure, psia; r_c = liquid critical-pressure ratio; P_v = vapor pressure of liquid at flowing temperature.

Another equally good check for critical Δp uses the factor C_f, which, like K_m, is a recovery coefficient based on valve geometry.

Now, check $\Delta P_{(allow.)}$ against $\Delta P_{(normal)}$:

If $\Delta P_{(allow.)}$ is $< \Delta P_{(normal)}$, use $\Delta P_{(allow.)}$ as $\Delta P_{(sizing)}$.

If $\Delta P_{(allow.)}$ is $> \Delta P_{(normal)}$, use $\Delta P_{(normal)}$ as $\Delta P_{(sizing)}$.

Next, check for cavitation and flashing:

If $P_2/P_v > 1$, and $\Delta P_{(normal)} > \Delta P_{(allow.)}$, cavitation results.

If $P_2/P_v < 1$, and $\Delta P_{(normal)} > \Delta P_{(allow.)}$, flashing results.

Where: P_2 = valve downstream pressure, psi;

P_v = vapor pressure of liquid, psi.

To avoid either of these conditions, the engineer should consider: (1) selecting a new valve with a higher K_m, (2) changing the pipe size, or (3) designing for a lower $\Delta P_{(normal)}$. If flashing cannot be avoided, locate the valve so that it flashes into a vessel. If either flashing or cavitation cannot be avoided, select the valve type and trim materials that can withstand these conditions. A number of valve-sizing factors are related to each other, as shown in Table IV.

The correlation between these factors is not exact, since they differ from size to size for the same type of valve by a given manufacturer, and from manufacturer to manufacturer for the same type and size of valve. Nevertheless, the correlations exist, and one manufacturer's factors can be used to size another's valves within those limits. Best results, of course, are obtained when the sizing coefficient is used for the manufacturer's valve for which it was developed.

The second question about pressure drop: Is the Δp across the valve enough to provide good system control? The "system" is the complex of pipe, elbows, valves, heat exchangers, etc., from the source to the destination of the flow-motivating pressure. Unless the Δp across the valve is above a certain percentage of the total dynamic pressure-loss in the system, the valve will not provide good control. For pumped systems, $\Delta P_{(normal)} \geqq 1/3\ \Delta P_{(system)}$ or 15 psi, whichever is greater, where $\Delta P_{(system)}$ includes the Δp across the valve. For other systems (e.g., vessel to vessel, compressor to its own suction, etc.), the characteristics of the system pressure-variation must be taken into account [*2.1*].

Finally, the design engineer must consider the maximum Δp across the valve at any condition in order to size valve-operators, bodies and shafts.

The above calculations could be accomplished by using a manufacturer's slide-rule [*10.4,10.5*] to get the C_v and select the valve. Such a slide-rule fully takes into account the sizing factors cited above but does not deal with other considerations for which special corrections must be made, such as viscosity, pipe reducer effects and noise:

For viscosity, a rule of thumb is [*10.1*]:

If viscosity $\leqq$ 20 centistokes, no correction need be made.

If viscosity $>$ 20 centistokes, find C_{vc}, as:

$$C_{vc} = C_v F_v$$

Valve Characteristics

Equal Percentage: Denotes a percentage change in flow equal to the percentage change in lift or rotation. Valves having this characteristic are more sensitive to plug- or vane-position change as the valve nears the wide-open position. Since this is normally the operating position, most control valves are equal percentage.

Linear: Denotes equal change of flow for equal plug or vane change at all points.

Quick Opening: Denotes a very rapid change in flow, (in some cases as much as 80%) in the first 10% of plug or vane travel.

Other valve characteristics can be produced by varying the shapes and relationships of valve plugs and seats, or by cam or linkage arrangements in valve positioners. The chemical engineer must select the valve characteristic whose sensitivity matches the needs of the process.

Comparison of Control Valve Types – Table II

Type	General	Advantages	Disadvantages
Globe, single port	Widely availabe in smaller sizes; standard sizes to 16 in; all characteristics abailable; standard ratings 1,500 psig or 2,500 psig and 450° F; useful in most services; gradually being replaced by cage-trim valves.	Low pressure recovery; standard face-to-face dimensions; reduced trim available; operation can be reversed by reversing trim; tight shutoff.	High cost/C_V; may require larger operator; sizes above 4 in may not be competitive (check vs. butterfly, etc.); limited rangeability; screwed seats corrode, hard to remove; poor performance on slurries.
Globe, double port	Same as single port.	Lower unbalanced forces, requiring smaller operators than single port; same as single port, except no tight shutoff.	Limited rangeability, but slightly better than single port; sizes above 4 in may not be competitive (check vs. butterfly, etc.); no tight shutoff.
Globe, cage, guided, balanced	Good availability; all characteristics available; standard sizes to 6 in; standard ratings to 2,500 psig, 450° F; hardened and noise-reducing trim.	Quick inline change of trim; standard face-to-face dimensions; higher capacities, less cost/C_V, much higher rangeability than globe; more guiding surface provides smoother plug motion; low-noise trim available.	Trim not reversible; tight shutoff requires special trim; low-noise trim reduces capacity.
Globe, cage, guided, unblanaced	Same as balanced.	Same as balanced, plus tight shutoff; low-noise trim available.	Trim not reversible; low-noise trim reduces capacity.
Butterfly, standard vane	Widely available in larger sizes and flangeless wafer-style bodies; standard sizes to 36 in, 2,500 psig; special sizes to 60 in, 2,500 psig. normal characteristic approaches equal percentage; other characteristics available with characterized positioners.	Very low cost/C_V; fair rangeability; high capacity; good control at low ΔP; lightweight, small; fewer parts to maintain.	Control limited to 60° open; high pressure-recovery, susceptible to choked flow, cavitation and noise; tight shutoff requires special linings; temperature rating limited to composition of linings (where required); high torque or high shutoff pressures require oversized shafts and operators; too-fast closing can cause water hammer; must be installed with shaft horizontal.
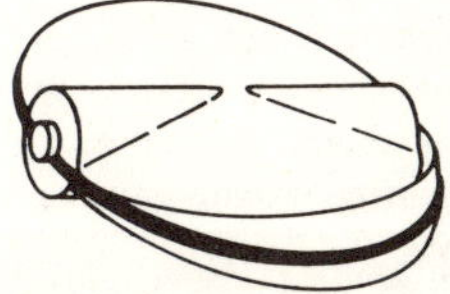 Butterfly, fishtail and low-torque vanes	Vanes designed to reduce torque.	Same as standard vane plus: good control to 90° open; increased capacity; smaller shafts and operators.	Sames as standard vane, except control to 90° and lower torque permits smaller shafts and operators than standard vane.

Comparison of Control Valve Types – Table II (continued)

Type	General	Advantages	Disadvantages
Eccentric rotating plug (Camflex)	Standard sized to 12 in-600 psig; flangeless for inclusion between standard ANSI/flanges, 150,300 and 600 psig; standard temperature range; −320 to +750° F; equal-percentage and linear characteristics available.	High rangeability; competitive cost/C_V; small size/C_V; low weight/C_V; good capacity under average conditions; high capacity under critical-flow condition; tight shutoff; handwheel standard.	Must remove from line for maintenance; pressure limited to 600 psig.
Ball valve	Equal-percentage characteristics; flangeless for inclusion between standard ANSI/flanges;	Handles slurries and fibrous materials; very high capacity; tight shutoff; low cost; high rangeability.	High pressure-recovery, choked flow, and cavitation; temperature limited by seat materials; side-thrust produces bearing wear; remove from line for maintenance; handwheel cost high.
Characterized ball valve	Standard sizes to 2 in-600 psig ANSI; standard sizes to 12 in-300 psig ANSI; standard sizes to 24 in-150 psig ANSI; all characteristics; flangeless for inclusion between standard ANSI flanges.	Low cost; very high rangeability; high capacity; handless slurries, fibrous and high-viscosity fluids.	High pressure-recovery; choked flow and cavitation; no reduced trim; needs large piston operators, positioners and high air pressure; additional engineering of manifolds; nonstandard face to face; long through bolts; remove from line for maintenance; handwheel cost high.
Angle valve	Used for special applications: coking hydrocarbons, erosive catalysts.	Very high capacity; prevents valve body erosion; self-draining.	High cost; high pressure-recovery; cavitation, noise and downstream erosion; C_V's somewhat unpredicatable. special piping required.
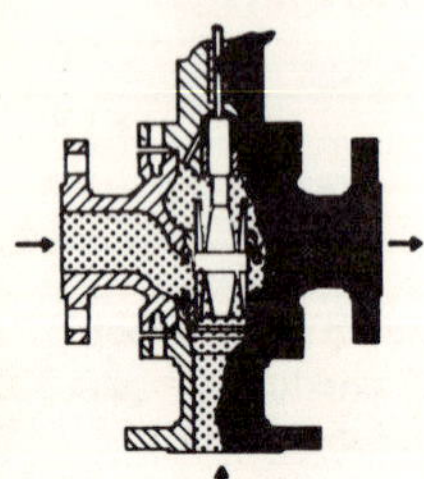 Three-way combining	Special application: temperature control around heat exchanger, etc.	Available all sizes.	High cost (check vs. two butterfly valves); special operator sizing required: linear characteristics only; limited rangeability: high stress on body where stream temperatures vary widely.

Comparison of Control Valve Types – Table II (continued)

Type	General	Advantages	Disavantages
Self drag	Special applications: Labyrinth stacked disk dissipates pressure and reduces fluid velocity; available in several styles; angle, Y and straight-through.	Handles extremely high Δp's; handles severe services well; cuts noise, vibration and cavitation at source; tight shutoff available; eliminates silencers, baffle-plates and insulation as noise reducers.	High valve cost; check for total system cost vs. valves requiring silencers, baffles, diffusers, etc., to perform same function; long delivery; special piping for angle construction; becomes easily plugged in dirty service; requires strainers, especially at startup.
Whisper I	Multiple slotted orifice.	Trim fits many standard cage trim bodies; cuts down noise and vibration.	Often requires "path" acessories for noise reduction.
Other velocity-control valves	A variety of multistage, multiple-orifice, tortuous-path valves now coming on the market.	Cuts down noise and vibration; useful in severe Δp service; limits cavitation and erosion.	Special applications; hugh cost; often require "path" noise accessories; special piping; tendency to plug easily; need special computation of C_V and rangeability.
Saunders	Widely used for erosive and corrosive service with appropriate liners; sizes to 14 in; useful in water tower service.	Tigh shut-off; simple construction; liner keeps fluid off working parts; self cleaning; high capacity.	Extremely limited line pressures (50-150 psig); operating temperature limited to linings used; linings subject to wear; high maintenance on linings; control fair up to 50% open, otherwise not predictable.
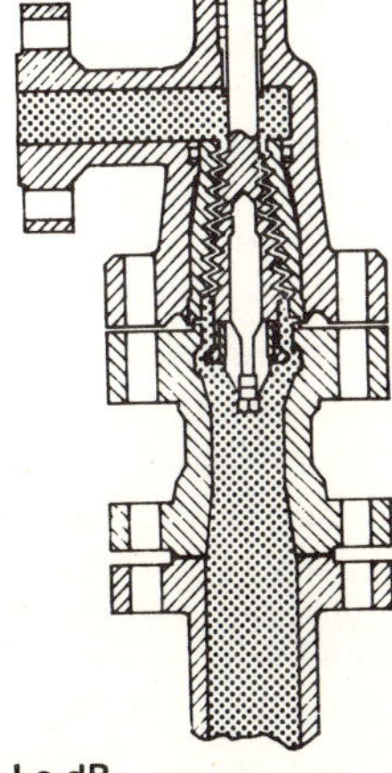 Lo-dB	Available in several styles: 1. angle style with expanding labyrinth plug; 2. bar-stock angle body with multistep guided plug; 3. standard body with multistep.	Useful in high ΔP service; often cuts noise at source; limits, and sometimes eliminates, cavitation, erosion and vibration; High rangeability; tight shutoff.	Linear characteristics only; special piping for angle construction; high cost of valve must be checked for total system cost; often requires downstream expansion plates as silencers; downstream plates may limit valve rangeability.

Where: $C_{vc} = C_v$ corrected for viscous flow; F_v = viscosity correction factor.

To compensate for pipe reducer effects [*4.1,4.2*], some manufacturers publish modified C_v values to adjust for the loss of pressure drop due to swaged pipe reducers. Where such published data are not available, the engineer must check and correct the C_v, if required. A rule of thumb is:

If $K_m > 0.75$ and valve-body size/pipe size < 0.75, no correction is required. Otherwise:

$$C_{vr} = C_{vc}/F_p$$

Where: C_{vr} = required C_v; F_p = pipe-swage correction factor.

Federal legislation (the Walsh-Healey Act, Title 41) sets maximum noise levels to which plant personnel may be exposed. Control valves sometimes exceed these levels. Valves in severe services—high pressure drops, high pressure recovery, high sonic velocity, etc.—should be checked for noise in excess of allowable [*7.1,7.3,7.5,7.6, 7.7,7.8,7.9*].

The development of a labyrinth multistage-disk valve plug has been an important contribution to the reduction of noise [*7.5*]. This plug dissipates pressure and reduces velocity by taking the fluid through a tortuous path. Since noise is often only a symptom of much more serious cavitation, erosion and vibration problems, the valve meets an important need. Services that used to chew up valves in a matter of days can now be handled with valve life extending to months and even years.

Many manufacturers are offering valves with a variety of multistage, multi-orifice and multi-path trims. Those that can completely handle the service solve the problem at its source. Others require auxiliary devices such as diffusers, silencers, multiple-restriction orifices, heavier pipe walls, and insulation. This type of treatment (path control) is less desirable than source control but is often suitable.

If the Δp is high, the engineer should follow step-by step procedures for determining sound-pressure level, and taking corrective action [*10.1, 10.2*]. He can then size his valve for liquid service.

Sizing Gas- and Steam-Control Valves

As with liquids, gas- and steam-flow through a valve can reach a critical-flow condition [*1.3,1.6*]. This critical flow develops when the fluid reaches sonic velocity at the narrowest point (*vena contracta*) downstream of the valve trim. At this condition, no increase in the Δp across the valve can further increase the flow *at a given upstream pressure.* As with liquid flow, pressure recovers from its low point at the *vena contracta.* The amount of recovery depends on the configuration of the valve (see Table II) and the downstream velocities thus influence the selection of valve type.

In the past, industry used formulas that failed to take into consideration the compressible nature of gas at low pressure drop, the transition to incompressible flow at critical pressure drop, and the phenomenon of high pressure-recovery. The actual capacity of valves sized by those formulas was far below the predicted capacity. New empirical formulas, based on test data, have now been developed, with critical-flow factors C_1 [*1.6*], C_f [*1.1*], and Y/[*1.3*]. The formulas using C_1 are given here. Formulas for C_f and Y may provide equally good results (Table IV).

Consideration of the critical-flow condition is included in the following empirical formulas:

$$C_g = \frac{Q_s}{\sqrt{\frac{520}{GT}}\,P_1 \sin\left[\frac{3417}{C_1}\sqrt{\frac{\Delta P}{P_1}}\right]_{\text{deg}}} \quad \text{(Gas)}$$

$$C_g = \frac{Q_s}{1.06\sqrt{d_1 P_1}\,\sin\left[\frac{3417}{C_1}\sqrt{\frac{\Delta P}{P_1}}\right]_{\text{deg}}} \quad \text{(Steam and vapor)}$$

$$C_g = \frac{Q_s(1 + 0.00065T_{sh})}{P_1 \sin\left[\frac{3417}{C_1}\sqrt{\frac{\Delta P}{P_1}}\right]_{\text{deg}}} \quad \text{(Steam under 1,000 psig)}$$

Where:

Q_s = sizing flow, std ft³/h for gas, lb/h for vapor and steam.
C_g = gas, steam or vapor capacity coefficient.
C_s = vapor or steam capacity coefficient.
G = specific gravity = $\frac{\text{mol wt}}{29}$.
T = temperature, °R.
P_1 = inlet pressure, psia.
ΔP = normal ΔP across valve, psi.
C_1 = C_g/C_v, valve recovery coefficient found in manufacturer's tables. Where not published, use values in Table IV.
d_1 = density of steam or vapor, lb/ft³.
T_{sh} = degrees of superheat, °F.

To use these formulas:

First, find the Q_s.

For maximum load, $Q_s = Q_{(\text{max})} \times 1.1$*

For normal load, $Q_s = {}_{(\text{norm})} \times 1.3$*

Then select a preliminary C_1 by making a tentative valve-type selection. (See Table II and III or consult a manufacturer's catalog.)

If using the longhand method, calculate the sine function in the formula. If equal or greater than 90 deg, critical flow has been reached. Calculate a value of Δp that will reduce the sine function to sine 90 deg = 1, and thus cancel it out. This is the maximum allowable Δp for sizing for the selected C_1.

You can now calculate C_g or C_s by use of the formula, or by using a manufacturer's slide rule, which also takes into account the selected C_1 [*10.4*]. Use of this slide rule is recommended over the calculation method because it is faster. Other slide rules similarly make correction for valve geometry, critical flow and high pressure recovery [*10.5*]. With all slide rules, care should be taken to use the specified units. For example, temperature on one slide rule is in °F, not °R, and P_1, for convenience, is in psig, not psia [*10.4*].

If C_g or C_s is not listed by the manufacturer, convert to C_v (Table IV).

Valve selection next involves choosing the body and trim size from a manufacturer's catalog. The selected C_g,

*When both normal and maximum conditions are given, a calculation of sizing coefficient C_g or C_s should be made for each condition, using Q_s, to determine maximum required C_g or C_s. Calculation of C_g or C_s may, in some cases, not be the governing condition, where valve Δp at normal and maximum loads are significantly different.

Average Valve-Recovery Coefficients, K_m and C_l – Table III

(For use only if not available from manufacturer.)

Type of Valve	K_m	C_l
Cage-trim globes:		
Unbalanced	0.8	33
Balanced	0.70	33
Butterfly:		
Fishtail	0.43	16
Conventional	0.55	24.7
Ball:		
Vee-ball, modified-ball, etc.	0.40	22
Full-area ball	0.30	
Conventional globe:		
Single and double port (full port)	0.75	35
Single and double port (reduced port)	0.65	35
Three way	0.75	
Angle:		
Flow tends to open (standard body)	0.85	
Flow tends to close (standard body)	0.50	
Flow tends to close (venturi outlet)	0.20	
Camflex:		
Flow tends to close	0.72	24.9
Flow tends to open	0.46	31.1
Split body	0.80	35

Correlations of Control-Valve Coefficients – Table IV

$C_l = 36.59\, C_f$	$C_s = 1.83\, C_f C_v$
$C_l = 36.59 \sqrt{K_m}$	$C_v = 19.99\, C_s/C_l$
$C_g = C_l C_v$	$C_g = 19.99\, C_s$
$K_m = C_f^2 = F_L^2$	$C_l = 39.9 \sqrt{X_T}$

Values of K_m calculated from C_f agree within 10% of published data of K_m.

Values of C_l calculated from K_m are within 21% of published data of C_l.

C_s (or C_v) must be greater than the required C_g, C_s (or C_v). Next, check the C_1 of the selected valve against the average valve used for calculation. If there is a significant difference, recalculate, using the actual C_1 of the selected valve.

When checking outlet velocity in ft/sec [*6.1, 6.2*] be sure not to exceed Mach 0.25 in saturated steam, because this causes erosion and abrasion on valve parts. Outlet velocity exceeding 0.3 × fluid sonic velocity is highly likely to result in noise and vibration.

An inlet velocity greater than 300 ft/s for valves 2 in and less in diameter, or greater than 200 ft/s for valves over 2 in will, produce serious wear on valve body and trim. To check this velocity, use the formula, [*6.1,6.2*] substituting inlet for outlet conditions. Take corrective action to reduce velocity if the limits are exceeded. This usually requires a change in pipe size. Finally, check for noise, as in liquid valves.

A number of computer programs for control-valve sizing and selection have been developed. These eliminate longhand or slide-rule calculations. Unfortunately, the software is not available to everyone.

Choosing the Appropriate Valve Characteristic

The relationship of change in valve opening to flow can be assumed as either installed or inherent. The characteristics referred to below are installed. Installed characteristics take into account that dynamic pressure losses in the pipe system will increase as flow increases, reducing the pressure drop available to the valve, whereas inherent characteristics assume a constant pressure drop across the valve.

Ideally, the installed valve characteristic should produce the most linear system control. Thus, if a system is linear, a linear valve will maintain that status. If the system is nonlinear, an equal percentage or other characteristic can compensate for this nonlinearity, as follows:

Use equal percentage: (1) for flow control where measurement is nonlinear, (2) for flow, temperature or level control where the measurement is linear but the $\Delta p_{(sizing)}$ is less than $0.3\Delta p_{(system)}$ and (3) for pressure control where the process changes faster than the valve.

Use linear characteristics: (1) for flow, temperature or level where the measurement is linear and where $\Delta p_{(sizing)}$ is greater than $0.3\Delta p_{(system)}$, and (2) for three-way valves, and (3) two-way valves used in three-way services.

Use quick-opening: (1) for most on-off applications, and (2) for direct-connected valves, such as pump governors, back-pressure regulators and high-capacity reducing regulators (1 in and larger).

Check For Required Rangeability Factor

Valve rangeability is determined by the formula:

$$\underset{\text{(required)}}{\text{Rangeability factor}} = \frac{C_v \text{ selected}}{C_v \text{ at minimum required flow}}$$

Check manufacturers' catalogs for C_v versus lift to see whether the valve can meet the following:

$$\underset{\text{(required)}}{\text{Rangeability factor}} = \frac{C_v \text{ of selected valve at 100\% lift}}{C_v \text{ of selected valve at 10\% lift}}$$

If the catalog does not publish C_v vs lift relationships, use Table V for an approximate value. The "Rangeability factor" may differ widely from the "inherent rangeability" claimed in manufacturer's literature.

These formulas give rangeability factors that are on the safe side. Manufacturer's data will often give higher rangeability factors than listed in Table V. In borderline cases, consult the manufacturer. If the required rangeability factor is greater than either the computed value or the value listed in Table V, reconsider operating requirements, recheck the valve rangeability with the manufacturer, select a new valve or, as a last resort, use two valves in parallel.

Valve-Actuator Sizing

Sizing of value actuators is highly specific to each application and dependent on manufacturer's data not always available to the sizing engineer.

An actuator must generate enough force to open or close a valve, overcoming the spring force, stem-friction force, and unbalanced fluid forces on the plug area, applying force sufficient to seat the plug to the required leakage rate. In rotary-action valves, (butterfly, ball, camflex), it must also overcome the highest torque developed in the course of rotation. These forces are influenced by: flow direction (whether flow tends to open or close the valve), stem size (which can detract from seat area), and effective diaphragm area (which changes in the course of the valve stroke). Since the unbalanced forces also change in the course of the valve stroke, selection of an operator is almost unique to the type, style and action of the chosen valve, and no general formula can be used for valve-operator sizing.

For all these reasons, the wisest course is for the chemical engineer to work closely with the valve manufacturer on actuator sizing, using the manufacturer's suggested sizes. All manufacturers will supply calculations of their operator sizing, if required. Some have manuals devoted to actuator sizing, with formulas and valve constants for each particular valve type and action [*8.5*].

One example of dozens of possible formulas is presented to illustrate the problem. This is a case where flow tends to open the valve.

$$A_d = \frac{\Delta P A_s + P_2 A_{st}}{P_m - S_m - 2}$$

Where:

A_d = final effective diaphragm area, in^2.
A_s = valve seat area, in^2.
P_2 = downstream pressure, psig.
P_1 = upstream pressure, psig.
$\Delta P = P_1 - P_2$, psi.
A_{st} = stem area, in^2.
S_m = maximum spring force in closed position, psi.
P_m = maximum air pressure used for control, psi.
C = friction constant, psi = 2 psi.

Additional diaphragm area must be added for the degree of seat-tightness required. Where force requirements make the diaphragm-actuator-size excessive, piston actuators may be used. Rolling diaphragm actuators (Camflex), electrohydraulic actuators, high-performance servo-actuators, and electro-mechanical operators are also available. For further discussion of actuators, see the reference list.

Block Valves and Bypass Valves

Block and bypass valves are installed around a control valve to enable the plant to function if the control loop or control valve fails. Their cost is high, particularly in larger sizes. Because users differ widely, engineers from the operating company should be consulted. In general, the following considerations apply:

- The effect of the bypass manifold on the control valve's properties.
- The critical nature of the service.
- The type of valve trim, the availability of spare trim or parts, and the speed with which the trim can be replaced. (Obviously, quick change of trim while the valve remains in line has great advantage. How long the valve can be out of service without affecting overall throughput is of prime importance.)
- The scheduled regularity of plant shutdowns viewed against the probability of valve failure.
- Whether or not blocks and bypasses are used. Handwheels should be used for manual control during control-loop failure when blocks and bypasses are absent. Handwheels, preferably, should be side-mounted on globe-type valves and yoke-mounted on butterfly and ball valves. Some people, for fire and other emergencies, prefer handwheels that always close the valve when turned clockwise. These handwheels are special, and care should be taken to include this in the specifications.
- Usually, in valves 2-in and less, it is more economical to install blocks and bypasses than handwheels. (Some valve makes have handwheels as standard.)

Where blocks and bypasses must be used, the following is good practice: *Block valves* should normally be full-port gate valves. When control valves are smaller than line size, block valves should be one size smaller than line size, but in no case be smaller than the control valves. *Bypass valves* should normally have a capacity at least equal to the calculated or required C_v of the control valve

Approximate Rangebility Factors for Different Valve Types – Table V

Valve Type	Size	Trim Characteristic: Linear	Equal Percentage
Cage trim (balanced plug)	1½", smaller	9	20
	2", larger	9	30
Cage trim (nonbalanced plug)	1½", smaller	9	20
	2", larger	8	30
Camflex style (Masoneilan)	–	–	50
V-ball	All sizes	9*	120
Butterfly-(conventional)	All sizes	6*	20
Fishtail (low torque, etc.)	All sizes	6*	35
Split body	1½", smaller	9	20
	2", larger	9	40
Angle body	–	9	20
Standard globe - Single port	1½", smaller	15	22
Double port	2", larger	10	15
3-way globe (combining or diverting)	All sizes	6	–

* Using characterized positioner.

but not greater than twice the selected C_v of the control valve. Globe valves should normally be used as bypass valves up through 4 in size. However, if the capacity requirement dictates larger than a 4-in bypass valve, gate valves may be used.

In special applications, such as high pressure drops, flashing service, and fast filling and emptying, block and bypass valves are sized by process calculations. When gate valves are used as bypass valves, their minimum size should be two sizes under line size. This requirement should override the capacity requirement. In cases where the predicted noise level in the control valve is a problem, the noise level in the bypass valve should also be considered.

Valve Positioners and Booster Amplifiers

These are special elements added to the control valve to assure that it achieves the position requested by the controller. Criteria differ more on the use of valve positioners than on any subject related to control valves. One useful guide, backed by both theoretical and test data, recommends that a booster and/or positioner be used when a control signal is split-ranged between two or more control valves or when it must be amplified to obtain sufficient actuator thrust [*9.1*]. Positioners or boosters are also recommended for better control, with minimum over-shoot from system disturbances and the fastest possible recovery. (For example, this may be as a result of stringent process requirements, high control-valve stem friction or long transmission lines.) If the preceding situations do not apply to a given application, neither positioner nor booster is usually necessary.

Generally, when a system is slow, the choice is to use a valve positioner. (Slow systems usually involve thermal or level processes, most mixing processes and a few large gas processes.) When the system is fast, the choice is to use a booster. (Fast systems are usually those involving liquid or gas pressure and most flow measurement.)

Although this short article necessarily has omissions—body materials, trim materials, accessories such as cooling fins, lockup devices, valve speed controllers, limit switches, etc.— the reader should have enough material to make the first steps in control-valve selection.

Bibliography

1.0 Critical-Flow Effects in Control Valves.

1.1 Baumann, H. D., The Introduction of a Critical Flow Factor for Valve Sizing. ISA Transaction, Vol. 2, No. 2, Apr. 1963.

1.2 Driskel, L. R., Practical Guide to Control Valve Sizing, Instrumentation Technology, 1967.

1.3 Boger, H. W., Recent Trends in Sizing Control Valves, Texas A & M 23rd Annual Symposium for the Process Industries, Jan. 1968.

1.4 English, M. L., An Introduction to the Fluid Principles Used in Control Valve Sizing, ISA Final Control Elements Symposium 1970, Paper No. 3.4.

1.5 Stiles, G. F., Development of a Valve Sizing Relationship for Flashing and Cavitating Flow, ISA Final Control Elements Symposium 1970, Paper No. 3.4.

1.6 Buresh, J. F., and Schuder, C. B., Development of a Universal Gas Sizing Equation for Control Valves, Fisher Controls Paper No. TM-15.

2.0 Control Valve Systems.

2.1 Moore, R. W., Allocation of Control Valve Pressure Drop, ISA Final Control Elements Symposium 1970, Paper No. 6.1.

3.0 Viscosity Effects.

3.1 Stiles, G. F., and Sheldon, C. W., Liquid Viscosity Effects on Control Valve Sizing, ISA Final Control Elements Symposium 1970, Paper No. 3.2.

4.0 Effects of Reducers.

4.1 Baumann, H. D., Effect of Pipe Reducers on Valve Capacity, Instruments and Control Systems, Dec. 1968.

4.2 Technical Paper No. 410. Crane Co., 1957.

5.0 Control Valve Characteristics

5.1 Moore, Ralph, L., Flow Characteristics of Valves, ISA Handbook of Control Valves, 1971.

5.2 Wing, P. Jr., Control Valve Characteristics. Process Instruments and Control Handbook, by Douglas M. Considine, edit., McGraw-Hill Book Co.

5.3 Schuder, C. B. Control Valve Characteristics, Instruments and Control Systems, March 1967.

5.4 Boger, H. W., Flow Characteristics for Control Valve Installations, ISA Journal, Oct. 1966.

6.0 Velocity Limits

6.1 Wing, P. Jr., Why Use Velocity Limits in Selecting Valves, ISA Journal, Aug. 1964.

6.2 Baumann, H. D., Why Limit Outlet Velocities in Reducing Valves, Instruments and Control Systems, Sept. 1965.

7.0 Control Valve Noise,

7.1 Arant, J. B. Special Control Reduce Noise and Vibration. Chem. Eng. Mar. 6, 1972.

7.2 Arant, J. B., Control Valve Cuts Noise and Pressure Drop, Chem. Eng. May 22, 1971.

7.3 Allen, E., Noise Control in Valves and Regulators, Fisher Controls, Paper NO. TM-24.

7.4 Schuder, C. B., Coping with Control Valve Noise *Chem. Eng.* Oct. 19, 1970.

7.5 Self, R. E., Why Velocity Control? Control Components, Inc.

7.6 Baumann, H. D., On the Prediction of Aerodynamically Created Sound Pressure Level in Control Valves, Masoneilan International Technical Paper 1970.

7.7 Baumann, H. D., Control Valve Noise—Cause & Cure, *Chem. Eng.* May 17, 1971.

7.8 Lighthill, M. J., On Soung Generated Aerodynamically II, Turbulence As A Source of Sound, Proceedings of The Royal Society of London, Vol. 222, Series A (1954).

7.9 Butler, Paul, Valve Noise, Live With It, Muffle It, or Cure It, *Proc. Eng.* (England) Nov. 1973.

8.0 Actuators

8.1 O'Connor, J. Jr., Control Valve Actuators, Chapter 7 ISA Handbook of Control Valves Ed. by J. W. Hutchison, edit.

8.2 Emery, J. T., Control Valve Operating Ranges, Honeywell Controls.

8.3 Muller, J. T., Selecting Springs for Control Valves, Instruments and Control Systems, Oct. 1966.

8.4 Ives, R. P. Guide to the Section of Control Valve Actuators, ISA Conference, New York, 1966, Paper No. 11. 4-3-66.

8.5 Luther, Fred J., Proper Sizing of Diaphragm Actuators for Control Valve Service TM-25 Fisher Controls Company Marshalltown, Iowa.

9.0 Valve Positioners and Booster Amplifiers.

9.1 Lloyd, S. G. Guidelines for the Use of Valve Positioners and Booster Amplifiers, ISA Conference, New York, 1968 Paper No. 68-922.

10.0 Manufacturers Catalogues and Technical Manuals.

10.1 Fisher Controls, Marshalltown, Iowa, Catalog 10.

10.2 Masoneilan International, Inc., Norwood, Mass. "Masoneilan Handbook for Control Valve Sizing." Third Edition, 1971.

10.3 Hammel Dahl, A Unit of IT & T, Warwick, R. I. "Hammel Dahl Technical Manual, General Engineering Data." Vol. 1.

10.4 Fisher Controls Company "Universal Valve Sizing Rule, Instruction Manual."

10.5 Masoneilan, "Valve Slide Rule Instructions".

11.0 Standards

11.1 Instrument Society of America. ISA-S39.1 1972 Standard Control Valve Sizing for Incompressible Fluids.

11.2 Instrument Society of America. ISA-S39.3 1973 Standard Control Valve Sizing Equations for Compressible Fluids.

Meet the Author

Sanford "Sandy" Chalfin is a senior instrument engineer at Fluor Corp., 2500 S. Atlantic Blvd., Los Angeles, CA 90040, with which he has been employed since 1966. His responsibilities with Fluor include those of lead instrument engineer and as acting chief instrument engineer of Fluor Netherlands. Prior to joining Fluor, he worked for Bechtel Corp. (1956–66). A member of ISA he is past chairman of the R.P. 5.2 Digital Control Logic Committee. He holds a B.Sc. in chemistry (1933) from New York's City College, where he graduated Phi Beta Kappa.

Applying Ratio Control To Chemical Processing

Calculation methods, equipment hookups and flow diagrams clearly explain how two or more fluid streams can be readily proportioned in a simple fixed or variable ratio, or by ratio computation with an analog computer.

J. B. ARANT, E. I. du Pont de Nemours & Co.

Ratio control is a special form of cascade control in which a secondary variable is held in some proportion to a primary variable. These variables may be any type of measurement function, but are most commonly applied to flow.

For example, flow ratio may be used to blend two or more fluid streams, regulate material balance in a fractionating column, or regulate heat input to a column in proportion to a varying feed rate. Flow ratio may be used as an aid in overcoming some undesirable control aspects in difficult programs, such as pH control.

Regardless of the reason for the ratio-control application, the proper solution can usually be obtained from one of three systems:

- Simple, fixed ratio.
- Simple, variable ratio.
- Ratio computation.

A basic description of each with representative examples and control diagrams will be outlined later in this article.

Ranges for ratio systems may be very narrow or very wide. However, there is not unlimited freedom in selecting this range. Two of the factors involved may often require compromises to be made. These are (1) limitation in the range adjustment of the ratio hardware, and (2) maximum and minimum flowrates required under various operating conditions and load changes (process turndown). As such, the actual system ratio is a function of the orifice-flow ratios plus the ratio-equipment range. This restraint on the system ratio will be seen more clearly in the following examples.

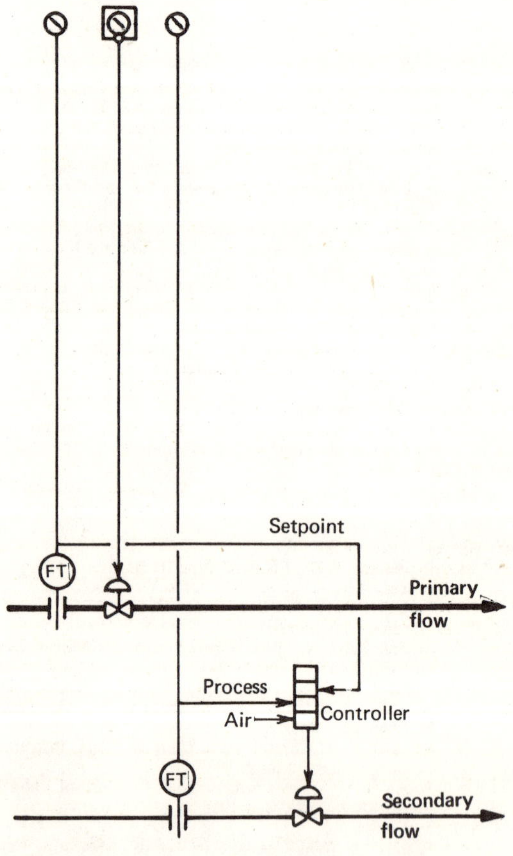

Simple Fixed Ratio

Often, it is desired to ratio one flow to another in an invariable or rarely changed ratio. The primary flow may be "wild" (uncontrolled), manually set, or under auto-

Primary flow on remote manual set.
Secondary flow follows primary from zero to maximum.

FIXED ratio: secondary flow follows primary—Fig. 1

Originally published September 18, 1972.

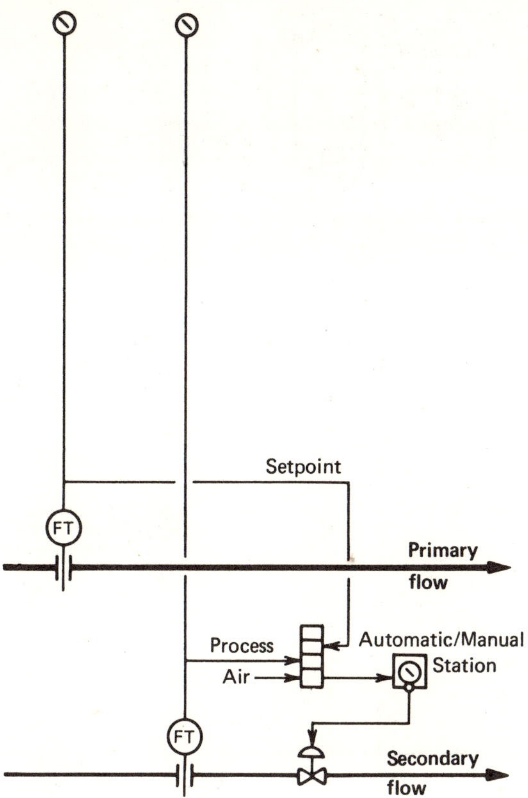

Primary flow uncontrolled.
Secondary flow follows primary from zero to maximum.
Provision for local automatic to manual switching control of secondary control valve.

AUTOMATIC/manual control of secondary valve—Fig. 2

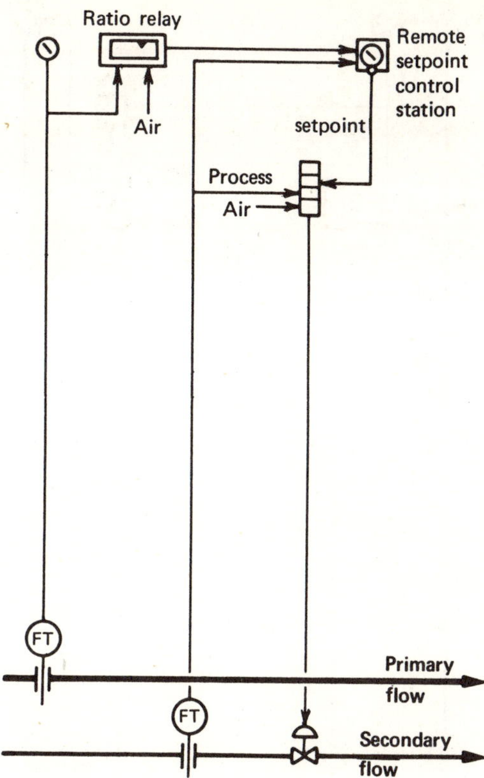

Primary flow uncontrolled.
Secondary flow follows primary through ratio relay output to setpoint on control station.
Secondary flow can be operated on ratio from primary on secondary automatic-flow control only, or on automatic to manual switching.

RATIO relay provides system flexibility—Fig. 3

matic flow control. As this flow changes, the secondary flow is to follow in a fixed proportion. This is done very simply by calculating the orifice-flow ranges to be in the desired ratio. The rate of flow in each line is then measured by a suitable device (such as a differential-pressure unit) that is coupled to a flow transmitter.

One flow-transmitter output goes to the setpoint chamber of a controller, and the other flow-transmitter output goes to the process input chamber of the controller (Fig. 1). The controller will then operate the control valve in the secondary flow until the two transmitter signals are equal, and the exact ratio maintained. Either linear or square-root signals may be used, as long as they are the same type for a given ratio-control system. If the ratio must be revised for any reason, one or both orifice plates can be changed.

If it is necessary to be able to manipulate the secondary flow manually at times, then a simple modification to the control circuit can be made by incorporating a local automatic/manual station between the controller's output and the control valve on the secondary flow. (Fig. 2). An example would be in a water-treatment plant where it is desired to ratio fresh water from a river to well water in a 25/1 ratio. If the maximum fresh-water demand is 4,000 gpm., then we require 160 gpm. of well water. Our actual orifice calibration would probably be 5,000 gpm. of fresh and 200 gpm. of well water (5,000/200 = 25/1).

Simple Variable Ratio

Where simple ratio control is desired, but ratio changes are made often enough to make orifice changing undesirable, then the system can be made more flexible by using a ratio relay or substation (Fig. 3). This relay is in effect a multiplier-bias system. That is, it can be adjusted so that the outgoing signal is either larger or smaller than the incoming signal. Over the range of the relay, this has the same effect as changing orifice plates in the fixed-ratio system.

The relay range depends upon whether the flow system is linear or square root. Wider rangeability is obtained with linear signals. For example, a typical ratio-relay range for square-root signals is 0.55 to 1.7, while for linear signals the range is 0.3 to 3.0. The scales and adjustments on these devices are relatively coarse, so fine adjustment requires calculation of the ratio from the in-

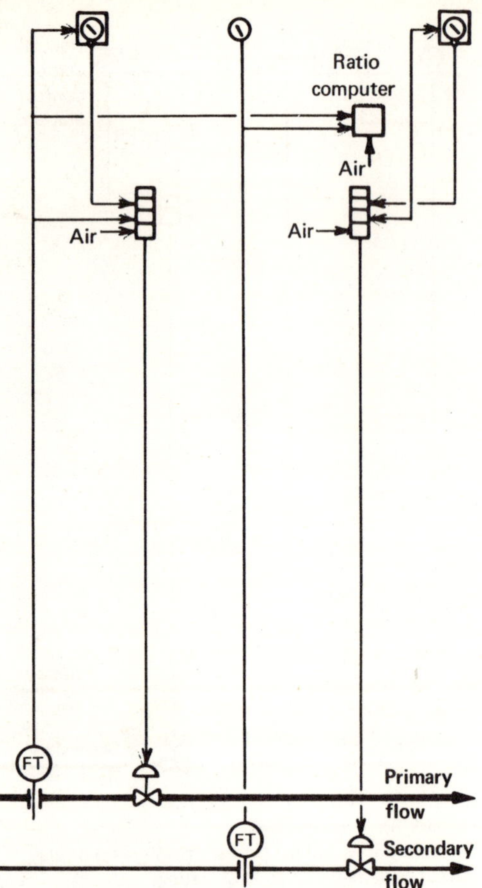

COMPUTER maintains specific ratio settings—Fig. 4

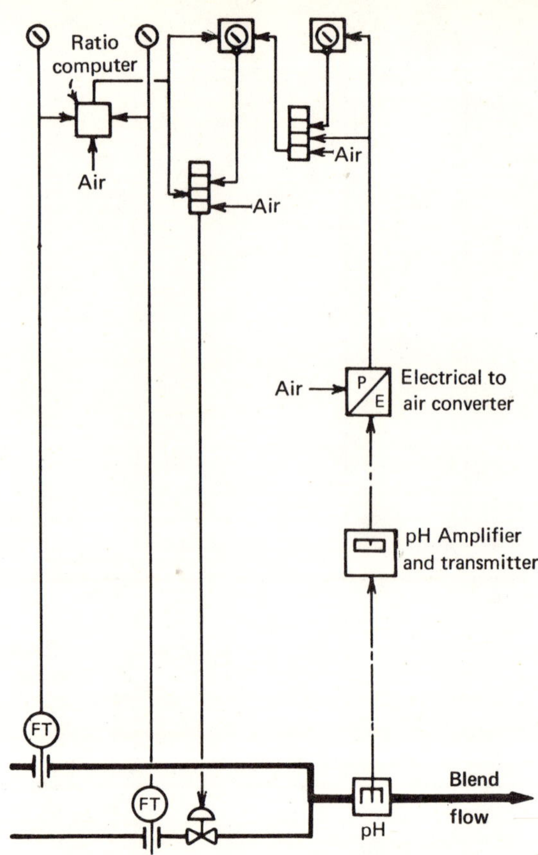

pH establishes setpoint for ratio control—Fig. 5

dividual flow readings and corrections to the relay setting.

Let us take our previous example and carry it one step further to illustrate this method. Our incoming water rate was essentially 5,000 gpm. maximum.

It is desired to initiate treatment by adding a 10% (by weight) lime slurry to the water. Also, the feedrate for the lime slurry must be proportional to the water rate at any given time. The proportion must be changed from time to time depending upon water temperature, chemical balance, and other factors. It is desirable to have a lime dosage for this water treatment ranging from 180 lb./100,000 gal. (0.00180 lb./gal.) to 425 lb./100,000 gal. (0.00425 lb./gal.). Ratio-relay range on square-root signals is 0.55 to 1.7.

A 10% lime slurry contains 0.8345 lb./gal. Therefore, the dosage range is:

$$0.00180/0.8345 = 0.00216 \text{ gal./gal.}$$
$$0.00425/0.8345 = 0.00510 \text{ gal./gal.}$$

In order to use our ratio-relay range, the orifice ratio should be about midway between these two values, or 0.00363 gal./gal. This works out to be about 18 gpm. of 10% lime slurry. 18/5,000 = 0.00360. Now, our overall system ratio can be calculated for actual scale calibration:

Minimum Dosage:

$$\frac{0.00360(0.55)(0.8345)}{1} = 0.00165 \text{ lb. lime/gal. water}$$

Maximum Dosage:

$$\frac{0.00360(1.7)(0.8345)}{1} = 0.00510 \text{ lb. lime/gal. water}$$

Therefore, the minimum dosage is 165 lb. lime/100,000 gal. water, and the maximum is 510 lb. lime/100,000 gal. These values bracket our desired dosage adjustment of 180 to 425 lb./100,000 gal.

Ratio Computation

In some applications, it is desirable or necessary to know the actual flow ratio at all times. Generally, this occurs where process requirements demand more elaborate control systems, or the actual ratio must be used for alarm or interlock purposes. Here, the two flow-transmitter signals are fed to an analog computer and the output is the actual ratio.

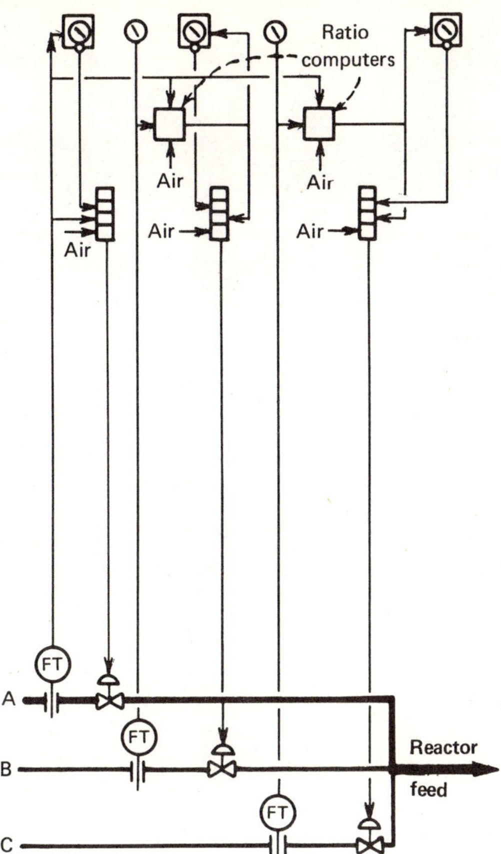

Stream A on automatic flow control.
Streams B and C ratioed to stream A by ratio computer control.

RATIO computers for multiple-feedstreams—Fig. 6

Since the computer is in effect a transmitter, the dial length in an indicating or recording control station allows for highly accurate and easily set specific ratio settings (Fig. 4). It also allows for multiple stream blending with wide rangeability and flexibility, as well as more complex systems where the ratio is cascaded to another control system. With the ratio-computation system, the secondary flow is always controlled by a full, standard control loop. The primary flow may be wild (i.e., uncontrolled), manually set, or under automatic flow control.

A good example of computer control of ratio (to aid in overcoming undesirable problems) is the control of pH by blending two streams where the load demand is subject to change (Fig. 5). pH is a highly nonlinear process measurement. Some pH systems have very steep, almost discontinuous pH changes with small changes in one of the streams. The lag time of the pH-measurement system coupled with system load changes often results in a disastrous control situation. Since the approximate ratio of one flow to the other to produce a given pH can be calculated (assuming fixed ion-concentration), it is possible to absorb gross load changes by ratio flow control very quickly. Then, pH control has only the burden of trimming the ratio control point for small flow inaccuracies or changes in ion concentration. In this way, an uncontrollable system may be improved manyfold to an acceptable control level.

Another example where ratio-computation control is of great value is for controlling multiple-stream feeds to some catalytic-reactor processes (Fig. 6). In many such systems, it is necessary to proportion the feedstreams in a very precise ratio to each other. Further, the ratio range may be critical and be very narrow because too high a ratio is costly of raw materials and too low a ratio can lead to severe catalyst damage. Yet, the ratio setting must be changed dependent upon catalyst condition and feedstock quality.

An example of a three-feedstream system calculation is as follows:

Stream	Orifice Range
A	500,000 std. cu. ft./hr.
B	35,000 lb./hr.
C	35,000 lb./hr.

Desired feed ratios:

B/A = 69/1,000 to 74/1,000 lb./1,000 cu.ft.

C/A = 64/1,000 to 76/1,000 lb./1,000 cu.ft.

Base orifice ranges for the secondary and primary flows, B/A and C/A, are both:

$$35{,}000/500{,}000 = 70 \text{ lb.}/1{,}000 \text{ cu.ft.}$$

Since our overall range limits for both systems are from 64 lb./1,000 cu.ft. to 76 lb./1,000 cu.ft., the ratio-computer range would be:

$$\text{Minimum: } 64/70 = 0.915$$

$$\text{Maximum: } 76/70 = 1.085$$

Therefore, we will select an actual ratio for the computer calibration range of 0.9 to 1.1, and the overall system range will be:

$$0.9(70) = 63 \text{ lb.}/1{,}000 \text{ cu.ft. (minimum)}$$

$$1.1(70) = 77 \text{ lb.}/1{,}000 \text{ cu.ft. (maximum)}$$

While the above example appears straightforward, some changes were required in the orifice ratios and in the ratio range adjustment of the computer to give the desired overall system ratio. This is an example of a very narrow range, high-gain, or tight system on ratio. As such, it is usually necessary to incorporate inverse-derivative action into the circuit for stability under upsets or load changes.

In conclusion, ratio control is applicable to a variety of simple to complex process-control requirements. The price of the hardware is modest and, in view of the benefits, gives a lot of control flexibility for the money.

Meet the Author

J. B. Arant is a senior engineer in the Design Div., Engineering Dept. of E. I. du Pont de Nemours & Co., Wilmington, DE 19898. He is presently responsible for instrument design for plant projects and is involved with special control-valve problems. He joined Du Pont in 1950 at the Sabine River Works in Orange, Tex., as a division engineer in the instrument department. He has a B.S. in chemical engineering from the University of Texas and is a registered professional engineer in Texas.

Section VIII
CONTROL VALVE PERFORMANCE

How to improve online control-valve performance

Reliability, fail-safe operation and economics are the essential criteria for making the final decision in selecting optimum control valves.

E. Ross Forman, *United Engineers & Constructors Inc.*

☐ The control valve is the only controlled variable-restriction in the control loop. In effect, it is a variable-area orifice that removes discrete amounts of energy from the system in order to control the process. Fig. 1 shows the place of the valve in the loop.

While pumps, reactors, boilers and compressors introduce pressure rise into a process system, the valve deliberately introduces pressure drop. Unlike equipment, such as a turbine, that generates useful work and ultimately electrical energy from a pressure drop, the control valve accomplishes control of the process variable by consuming energy. Since 30 to 50% of the total dynamic system drop may be expended in the control valve, there is always concern about its functioning reliably under continual stress.

There are more points of potential failure in a valve than in any other device in the control loop. As can be seen from the typical valve body (Fig. 2), the interior can wear due to erosion and corrosion. Furthermore, the constant need for the valve to moderate against the process medium causes packing wear, bellows fatigue, and wear of accessories such as positioners. Aging can cause diaphragms to fail. Frequent cycling to open and close the valve during operation can wear out auxiliary devices such as limit switches and solenoid valves (Fig. 3). Under such conditions, it is amazing that the control valve is as reliable as it is.

Failure analysis

There are a number of reasons why it may be necessary to account for the failure of a valve despite the most well-written specification and the finest-manufactured and best-tested valve that can be obtained.

One reason is the pure economic penalty of not having a key valve available for operation, whether its function is to control or to isolate the process. If the valve is in a crucial part of the process where a serious interruption in production can be caused by its failure, it means a loss of income. This loss is easily calculated by the hour or day, and can be postulated ahead of time. This is increasingly important in most plants because of the increasing size of processes. Major sources of income come from the larger facilities, so that any downtime is significant.

Another reason may involve the release of hazardous or toxic materials in the plant environment.

Regardless of the reason, valve failure cannot be tolerated in most circumstances, and the design phase must include appropriate analysis.

Obviously, the best way to preclude the possibility of failure is to have redundant valves. This may mean spare valves that duplicate the entire function of the original ones, or perhaps just the duplication of the components most likely to fail, such as limit switches. The task is to determine whether such redundancy is required and how much there should be. This is done by failure analysis of the process in the design phase, since no amount of quality control, field testing, or maintenance can compensate for a deficient design.

Unfortunately, it is usually easier to conceive of how a process will work than how it will fail. The human mind is basically success-oriented and looks for positive results. Conceiving negative solutions is difficult, and until recently was considered an art rather than a science. By applying proven statistical methods, the selection of a design can be success-oriented and also have a low incidence of failure that will be adequate for the plant's objective.

Reliability and availability

The term "reliability" is often confused with "availability." A valve is reliable when it works. In other words, it is a measure of the time stability of its performance (e.g., it will operate 997 h out of 1,000 h).

Reliability is defined as "the characteristic of an item expressed by the probability that it will perform a required function under stated conditions for a stated period of time." This definition of reliability mentions

Originally published June 5, 1978.

only the time of operation and not the time that the item may not operate. For example, the valve involved in a failure may be in a hazardous area, and not accessible until the unit is shut down for maintenance. In this case, the valve must be reliable for this time period, no matter how long that may be. At the end of this period, the plant is shut down and repairs can be made at will, so that no further consideration is necessary.

However, let us suppose that the valve is accessible for inspection, test and repair during plant operation. Then, a more meaningful measure of the integrity of the valve will be "availability" [defined by the IEEE (Institute of Electrical and Electronics Engineers) as "the characteristic of an item expressed by the probability that it will be operational at a randomly selected future instant of time."]

When designing the system, it is necessary to (a) define the success required for the valve and its system, and (b) establish reliability goals for the valve.

Success of the valve and its system must be defined as rigorously as possible to assure that valve operation takes place as planned. A complete appreciation of the system must exist for it to be properly defined. There may be many criteria for success, and all of them will influence the number and arrangement of the valves in the process. The definition must include the environmental conditions, how long the system must function, the number of cycles of operation involved, and any other data. The following is an example of such a "definition":

"The valve and actuator must operate satisfactorily during accident and post-accident conditions at any time during the design life of the equipment. Accident conditions are: temperature = 310°F, pressure = 62 psig, relative humidity = 100%, atmosphere = saturated steam plus nitrogen. Temperature and pressure duration time is 10 h. The valve shall be capable of actuating 250 cycles/yr (full open to closed, and return)."

Reliability goals are based on several factors. In the nuclear industry, the "single failure" criterion is often used as the primary goal. Simply stated, the valve should fulfill its success definition in the event of failure of a single component. In numerical terms, the data should provide the expected system reliability or availability. Such goals are arrived at by:

1. *Risk acceptance*—What is the highest risk that can be accepted in return for the benefits of the system as designed? Risk is defined as the product of the probability of failure and the consequences of that failure. Consequences may be measured in terms of lost profits, injuries, or the release of toxic or hazardous substances.

2. *Grandfather systems*—Past experience on existing systems that have had a good history of reliability and availability.

3. *Industry-standard goals*—This is often a tentative objective based on current knowledge.

Examples illustrate analysis methods

The most convenient method for explaining the studies made on reliability and availability is based on a system where two valves are required for isolating the process fluid. For instance: How many valves are re-

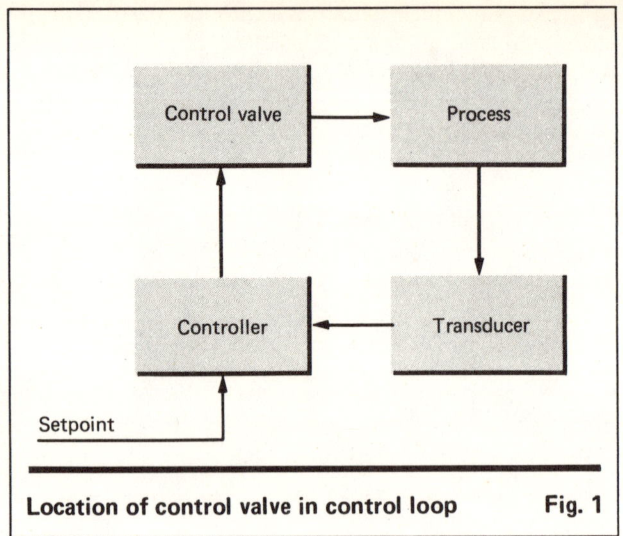

Location of control valve in control loop **Fig. 1**

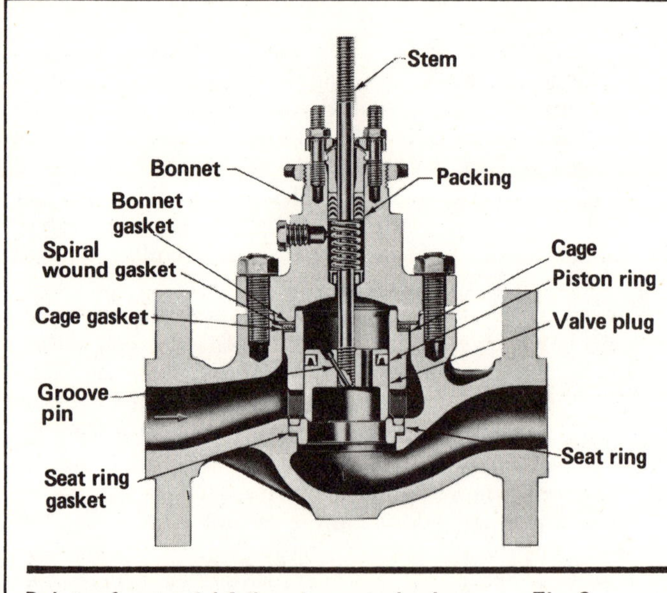

Points of potential failure in control valves **Fig. 2**

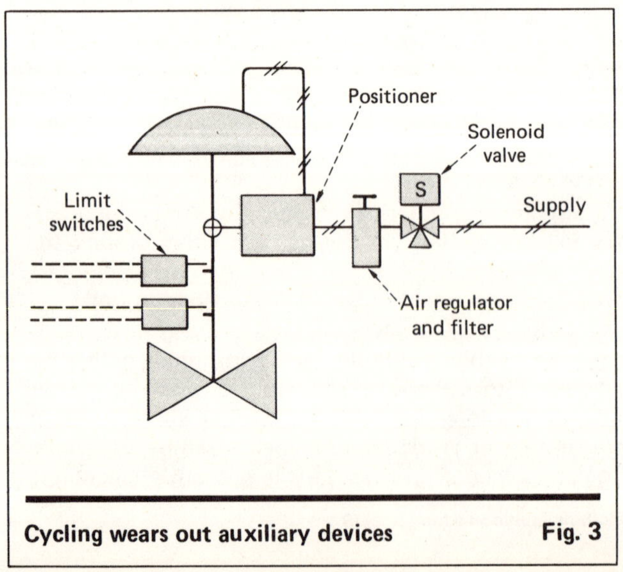

Cycling wears out auxiliary devices **Fig. 3**

quired that will meet the demands of the system, so that it will be available as much as possible?

Let us consider two valves in series, as shown in Fig. 4. The probability of successful operation of the system (i.e., closing the two valves) is:

$$P_g = P_1 \times P_2 \tag{1}$$

where P_g = probability that system will work, P_1 = probability that Valve 1 will work, and P_2 = probability that Valve 2 will work.

If the probability of each valve working is 90%, then substituting the appropriate values into Eq. (1) yields:

$$P_g = 0.90 \times 0.90 = 0.81$$

Hence, the system would have a probability of working only 81% of the time.

If an additional valve, P_3, is added in parallel to valves P_1 and P_2, and has a probability of 90% (Fig. 4), the probability of successful operation becomes:

$$P_g' = P_g + (1 - P_g)(P_3) \tag{2}$$

Therefore, the probability of working for this system becomes 98.1%—a substantial improvement over the original design. At this point in the design, a decision must be made whether to add redundant valves or to leave the arrangement as is. The answer depends upon the system goals. (It should be stressed that the probability for valves is considerably higher than shown in the examples.)

Arrangement	Failure rate, λ_T	Mean time between failures, h	Mean downtime, θ_T, h	Probability of operation, %
One valve	0.01	100	10	90
Two valves in series	0.02	50	10	81
Two valves in series with one in parallel	0.004	250	5	98.1
Two valves in parallel	0.002	500	5	99

Redundant valves can increase system reliability **Fig. 4**

Availability analysis

In design work, we must not only analyze for failure but also for the time required to fix a failed valve so that the process can operate again. Therefore, our analysis must consider the time that it takes to get the valve repaired and back online.

For this analysis, both the failure rate ($\lambda_1, \lambda_2, \ldots$) and the mean repair time ($\theta_1, \theta_2, \ldots$) are used. For our examples, failure rates are assumed to be constant, and the repair is started as soon as failure occurs.

For two valves in series:

$$\lambda_T = \lambda_1 + \lambda_2 \tag{3}$$

$$\theta_T = \frac{\lambda_1\theta_1 + \lambda_2\theta_2}{\lambda_1 + \lambda_2} \tag{4}$$

For two valves in parallel:

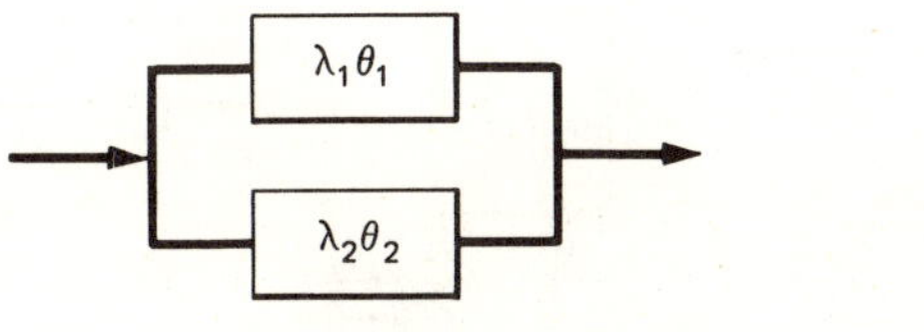

$$\lambda_T = (\lambda_1\lambda_2)(\theta_1 + \theta_2) \tag{5}$$

$$\theta_T = \frac{\theta_1\theta_2}{\theta_1 + \theta_2} \tag{6}$$

For our discussion, let us assume that each valve has a failure rate of 1% of its operating time, and that 10 h are needed to repair it. Thus, in a 100-h cycle, the valve could fail once and it would take 10 h to repair.

For the two valves in series, we find λ_T and θ_T by substituting into Eq. (3) and (4), respectively:

$$\lambda_T = 0.01 + 0.01 = 0.02$$

$$1/\lambda_T = 1/0.02 = 50 \text{ h}$$

$$\theta_T = \frac{(0.01)(10) + (0.01)(10)}{0.02} = 10 \text{ h}$$

Two valves in series result in a mean time between failure of 50 h, and a mean downtime of 10 h.

If the two valves are in parallel, the mean time between failure becomes 500 h, a considerable improvement. The mean downtime becomes 5 h.

For the system of two valves in series with one valve in parallel (Fig. 4), the mean time between failure becomes 250 h, and the mean downtime is 5 h.

These examples show how redundant valves can increase the success of a system. Each case must be analyzed on the specific objective of the system in order to ensure that excess equipment is not installed.

Fail-safe considerations

Valves have an additional function aside from their control assignment, in that they must fail-safe to meet operating and safety requirements. Fail-safe is defined in this discussion as the position of the valve following the loss of the operating medium, whether it be air, electronic or electrical.

The fail-safe operation depends on the type of process. In chemical processes, the essential objectives upon failure of the operating medium are to shut off the feed, eliminate the source of heat energy, and reduce operat-

Ratios of control-valve costs for 6-in. globe body **Table I**

Body construction and rating	Single-seated,* top-guided	Double-seated,† top- and bottom-guided
Iron, 125 psi, standard construction, 150 psi ΔP shutoff	1.1	1.0 (base)
Steel, 150 psi, standard construction, 150 psi ΔP shutoff	1.7	1.6
Steel, 300 psi. standard construction, 150 psi ΔP shutoff	1.7	1.6
Chrome-moly steel, 300 psi. Construction for 900° F, Extension finned-bonnet Stellited seat joint Stellited guide bushings Stellited plug-guide posts	2.9	3.0
Steel, 600 psi, 1,200 psi ΔP, 60° F, 0.5% C_V Maximum leakage Stellited seat-joint	2.1	1.8
Steel, 600 psi, 1,200 psi ΔP, 60° F, tight shutoff Stellited seat joint High-pressure piston actuator (90 psi) Not fail-safe	2.1	–
Chrome-moly steel, 600 psi, 1,200 psi ΔP, 600° F Pilot-operated, balanced design, reliable fail-safe action	2.8	–

*With diaphragm actuator of 200 in.^{2}
†With diaphragm actuator of 145 in.^{2}
Source: Hutchison, J.W., ed., "ISA Handbook of Control Valves," 2nd ed., Instrument Society of America, Pittsburgh, 1976.

ing pressure. Failure of the control valve to accomplish these objectives could cause chemical reactions to increase in rate, with resulting increases of heat and pressure. This would eventually cause loss of valuable products, or send wastes through the relief system, and possibly damage equipment by means of burnout, hot spots, coking or rupture. There is also the possibility of injury to personnel.

Fig. 5 shows various examples of fail-safe action. Note that a symbol is used that clearly shows the fail-safe position of the valve. This symbol is used on system descriptions and loop diagrams, so that the intent of the design is fully understood.

Specification techniques

While there can be no compromise with the engineering of the valve, other considerations can influence the final cost. Some of these factors must be evaluated

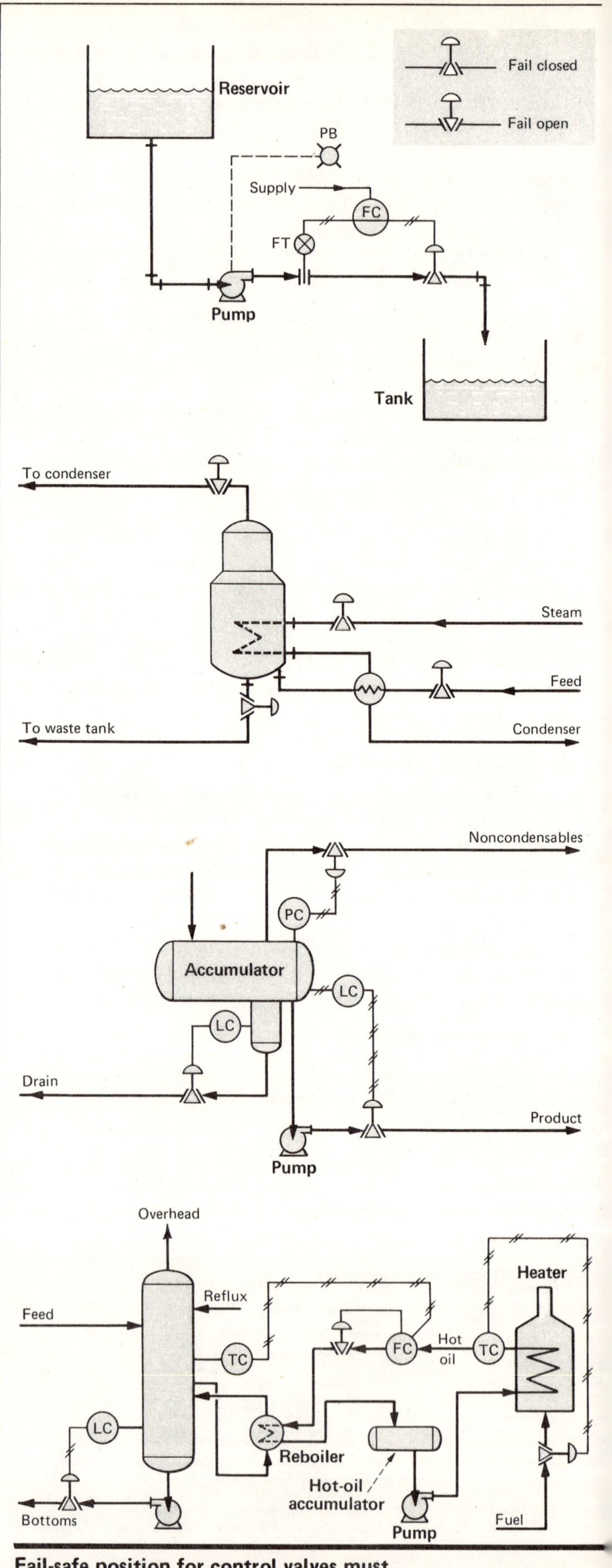

Fail-safe position for control valves must be clearly shown in flow diagram **Fig. 5**

Distribution of control valves by size in a chemical plant — **Table II**

Size range, in.	Amount, % of total
1 1/2 and under	65
2 and under	83
3 and under	91
4 and under	96

during the specification phase if the optimum valve is to be obtained.

The price of the valve is related to the type, size, construction and accessories. Valves can be purchased for less than $100 to practically any amount—in some cases, $250,000. However, in most applications, valves cost less than $2,000. Table I shows cost-ratio data for a typical 6-in. globe valve, and the changes in price as it becomes more complex due to a change in service. Obviously, the valve must not be overspecified if the cost is going to match its use.

Often overlooked is the need for standardization of spare parts and maintenance. These may override first cost of the control valves. This is particularly true for the single plant, where it would be expensive to stock parts for a large number of different-type valves. When stocking (inventory) is done on a companywide basis, a more flexible policy on unique installations may be followed.

Budgets are prepared long before individual valves are sized, so that list prices are used for costing the valves. However, some procedures can be followed to get the lowest price at the time of bidding, which again may occur long before the final sizes are selected.

Competitive bidding is effective if there is an adequate specification and a full bid-analysis done. Furthermore, a vendor that wants the order will bid competitively even on small orders. Also, allow the vendor to suggest alternative constructions that best fit his line and still meet engineering requirements for the valves.

A good practice is to lump as many valves together in the bid package. This will result in quantity discounts, where available. During the early phase of a project, an inquiry and order may have to be placed on a preliminary basis in order to ensure the availability of items that require long delivery times. This means that a bid-package must be written and later amended. While this sounds like bid-guessing, there are successful techniques to use. Base prices and discounts can be established with the vendor, along with unit prices for add-ons and deletions of positioners, limit switches, trim and special materials. This is even more feasible if data on previous plant use of such valves are available. For example, Table II shows the distribution of valve sizes in chemical processing facilities. The same information can be developed for any type of plant.

At the time of bidding, consideration should be given to obtaining prices for spare parts and for valves—in suitable quantities (say, 10, 20 or 30)—in addition to those stated in the specification. Pricing based on such possible future purchases could have an influence in the selection of the vendor, and will prevent a low bidder from making it up on extras. A certain percentage of any large order should be for spare valves, in order to prevent delays at the plant when emergency replacements are required due to design changes or failures.

Generally, pricing will be competitive unless many unique features such as exotic trim, nonstandard materials, fail-safe capacity tanks, solenoid shutoff valves and valve-position switches are required. Then, the vendor may no longer be able to use mass-production techniques that save money.

What about the low bidder?

The bid analysis should fully explore all facets of the bid to make sure that the low bid is valid. To ensure that the vendor understands all of the requirements of the specification, a vendor review meeting should be held to go over all engineering requirements in detail. Also, the shop can be visited to make sure that the vendor has the manufacturing capacity for the order.

Ideally, a maintenance history would exist on some other installation so that the durability of the product can be evaluated. A complete analysis might indicate the selection of someone other than the low bidder if the facts proved that other costs really have made his low bid a high.

Valve delivery

While the delivery of a valve would seem to be automatic if a reputable bidder has been selected, there are some factors to consider. A standard valve can take six to eight weeks for delivery if it is in normal production. Special requirements such as exotic materials and the nuclear code "N" stamp can result in a delivery date of over a year. In fact, the valve can easily be the critical item in the construction schedule.

One of the first things to do is to make sure that the vendor can meet the delivery, as promised, before the order is placed. Shop loads can be checked. If the order is large, the purchaser's expeditor can follow the order through the vendor's shop to make sure the parts are not diverted to a larger order for another customer. No reputable vendor minds such expediting. In fact, substitutions can often be made that ensure faster delivery because of close coordination between the purchaser and vendor.

Lastly, once the specification is issued, it should not be changed by the purchaser in any way if the delivery dates are to be met.

The author

E. Ross Forman is supervisor of the power division for United Engineers & Constructors Inc., 30 S. 17 St., Philadelphia, PA 19101. He joined United in 1971 and is responsible for all phases of instrumentation and controls. Previously, he was employed for 12 years by Catalytic, Inc. as chief instrument engineer. He has a B.S. and an M.B.A. from Drexel University, and is a member of ASME and Instrument Soc. of America. He is a professional engineer in N.J., Pa. and N.C.

How To Estimate Pressure Drop Across Liquid-Control Valves

Engineers doing preliminary designs for equipment procurement must often estimate pressure drops before control valves are finally chosen. Here is a technique that is easy to use and that gives good results.

HANS D. BAUMANN, Masoneilan International, Inc.*

Pumps are usually ordered early in an engineering project—before the control valves have been specified. But in order to make an intelligent pump selection, the engineer must be able to estimate the pressure drop across the control valves in the pumped fluid's path.

Various rules-of-thumb have been proposed for this purpose, such as assigning 33% of the dynamic pipe-friction loss, or selecting a 15-psi minimum drop [*1*].

While these methods work reasonably well, they can lead to the selection of an unnecessarily large pump, thereby wasting energy in the form of excess horsepower. This article suggests an alternative method—the selection of valve pressure-drop as a function of two parameters that should already be known to the process engineer: the pipe velocity at maximum design flow; and the maximum pipe frictional loss and pump droop (reduction in pump head between zero and maximum design flow).

The total-dynamic-head change determines the relative valve capacity and, therefore, the valve style (if one wants to stay within certain pressure-drop ratios as far as installed characteristic and efficiency are concerned), while the maximum pipe velocity sets the pressure-drop across this valve. The accompanying graphs and tables make the parameters easy to determine.

Line Velocity Determines Valve Pressure-Drop

Determining maximum line-velocity has to be done early in order to specify the pipe diameter for the known maximum design-flow volume that the system will handle, and in order to calculate the pressure drop across pipe and fittings.

There seems to be no "universal" rule. While the tendency in the past 20 years has gradually been to increase liquid line-velocity for the sake of low installation cost, the reversal of this trend has recently been seen due to concern for hydrodynamic noise. For example, an 8-in Schedule-40 pipe will radiate above 90 dbA at 3 ft from the pipe, due to normal turbulence in elbows, fittings, etc., if the line velocity exceeds about 24 ft/s. As a result, one major engineering company has made a general rule not to exceed 7 ft/s as maximum line-velocity for liquids.

I believe a more flexible rule would be in order, assigning higher velocity to smaller piping systems and vice versa, since the sound pressure level seems to vary as 20 times the log of pipe diameter; i.e., for a given velocity, a 10-in pipe will tend to be 20 db louder than a 1-in pipe with the same wall thickness.

Here are some other suggested limits currently in use: Crane Technical Paper #410 recommends for boiler-

Typical C_v/D^2 Values (Nonchoked conditions) at 90% Open — Table I

Valve Type	0.4 Factor Reduced Trim*	Pipe/Valve Diameter Ratio 1	1.5	2
Single-seated globe	4	10	4.2	2.3
Split-body valve	4.3	10.8	4.6	2.5
Double-seated globe and eccentric rotary plug (Camflex)	5	12.6	5.4	3
High-capacity cage 1 to 3-in. Butterfly, 60 deg open	6.4	16	7	3.4
Y-style and angle, flow-to-close	7.2	18	7.2	3.9
Contoured ball		25	8.8	4.5
Butterfly, full-open low-torque type		32	12	5.2

*For line-size valves

*Since writing this article, the author has left Masoneilan to become a consultant—see "Meet the Author" at the end of the article.
This article is based on a paper given at the 29th Annual Symposium on Instrumentation for the Process Industries, sponsored by the Dept. of Chemical Engineering, Texas A&M University, College Station, Tex., Jan. 16–18, 1974.

Originally published April 29, 1974.

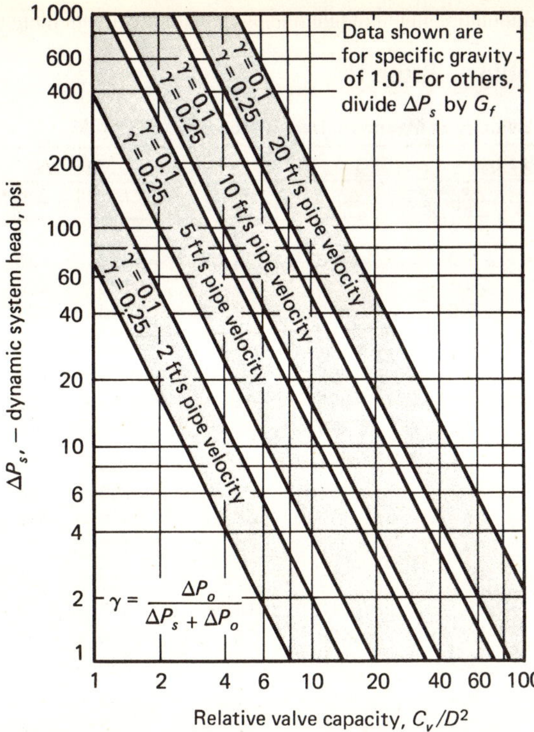

RELATIVE control-valve capacity determination—Fig. 1

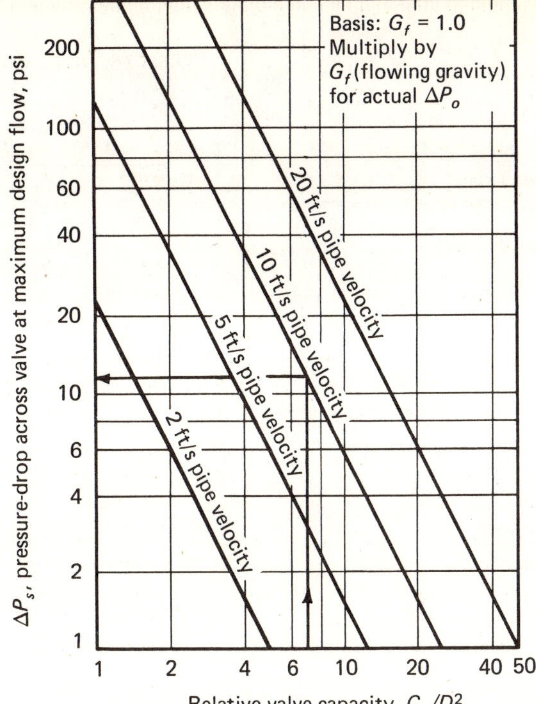

VALVE pressure-drop at maximum design flow—Fig. 2

feed application, 8 to 15 ft/s; pump suction, 4 to 7 ft/s; general service, 4 to 10 ft/s; and city water, up to 7 ft/s. Major engineering contractors specializing in paper mills allow pipe velocities up to 20 ft/s, while a well-known pump company suggests 3 ft/s maximum for suction and 5 ft/s maximum for discharge velocity.

The selected maximum pipe-velocity indirectly determines the pressure drop across the control valve once a specific valve type is selected. Rather than be concerned with the actual valve diameter and trim size, I feel that it is better at an early stage of the project to select only a style of valve with a known "relative valve capacity" (taken here as $C_v/(\text{pipe dia.})^2$) in order to relate the valve capacity to the pipe size rather than the body size. The valve type may logically be chosen because of the restrictions of the system (see Fig. 1 and Table I) or may be selected on the basis of experience with similar systems. The use of such a "relative valve capacity," C_v/D^2, leaves the instrument engineer full choice later on to select line-size valves, reduced-trim valves, or full-ported valves installed between reducers.

The aforementioned relationship between line velocity, U, and valve pressure-drop, ΔP_o, (at maximum design flow) can be expressed as follows:

$$\Delta P_o = G_f V^2/C_v^2 \quad \text{(psi)} \tag{1}$$

wherein $V = 60 \times \text{pipe area (ft}^2) \times U \times k^*$ (gpm) (2)

ΔP_o could therefore be expressed as:

$$\Delta P_o = \frac{6U^2 G_f}{(C_v/D^2)^2} \quad \text{(psi)} \tag{3}$$

For detailed development of this equation, (see box at left). Fig. 2 shows the results, indicating the dependency of valve pressure-drop on the square of the maximum pipe velocity and also the square of the relative valve capacity.

Development of Eq. 3

From the basic C_v equation (see ISA Standard SP39.1):

$$\Delta P_o = \frac{G_f V^2}{C_v^2} \text{ (psi)}$$

$$V = 60(D^2\pi/4)(Uk/144)$$

$$V = 2.45\, U D^2 \text{ (gpm)}$$

Therefore: $$\Delta P_o = \frac{G_f\, 2.45^2\, U^2 D^4}{C_v^2}$$

and $$\Delta P_o = \frac{6\, U^2 G_f}{(C_v/D^2)^2} = \text{(psi)}$$

Concern for Installed-Valve Characteristic

From the data in Fig. 2, one could quite easily select the valve pressure-drop, assuming a C_v/D^2 for a favorite style of valve from the information given in Table I [*2*]. Incidentally, the information in Table I incorporates a safety factor of 10% suggested by R. Moore [*1*], allowing for general manufacturing tolerances between the stated catalog C_v and a given valve's actual flow-capacity.

While this simplified method works, a better argument would be to select the relative valve capacity (and,

*k = 7.48 gal/ft³, D = pipe dia., in.

Pressure-Drop Selection Table — Table II

Valve Type	D/d	2 Ft/s ΔP_s*	2 Ft/s ΔP_o*	5 Ft/s ΔP_s*	5 Ft/s ΔP_o*	10 Ft/s ΔP_s*	10 Ft/s ΔP_o*	20 Ft/s ΔP_s*	20 Ft/s ΔP_o*
		Pipe Velocity at Maximum Design Flow							
Single-seat globe	1	0.8	0.24	14	1.5	54	6	216	24
	1.5	12	1.3	77	8.5	308	34	1224	136
	2	42	4.7	260	29	1050	116	4180	465
Ditto with 0.4 factor trim →	1	14	1.5	86	9.5	342	38	1370	152
Split-body valve	1	1.8	0.2	12	1.3	47	5.2	189	21
	1.5	10	1.1	63	7	252	28	1010	112
	2	34	3.8	216	24	865	96	3460	385
Ditto with 0.4 factor trim →	1	10	1.1	63	7	252	28	1010	112
Double-seated globe and eccentric rotary plug valve (Camflex)	1	1.4	0.16	9	1	36	4	144	16
	1.5	7.5	0.83	47	5.2	189	21	755	84
	2	24	2.7	153	17	612	68	2450	272
Both with 0.4 factor trim →	1	23	2.5	54	6	216	24	865	96
High-capacity cage valve 1-3 in. size, butterfly 60-deg. open	1	0.9	0.10	5.4	0.6	22	2.4	87	9.6
	1.5	4.5	0.50	28	3.1	112	12.4	450	50
	2	19	2.1	117	13	468	52	1870	208
Cage valve with 0.4 factor trim →	1	5.4	0.60	33	3.7	1.35	15	540	60
Y-style and angle valve, flow-to-close	1	0.5	0.06	3.6	0.4	14	1.6	58	6.4
	1.5	4.3	0.48	27	3	108	12	432	48
	2	14	1.6	90	10	360	40	1440	160
Both with 0.4 factor trim →	1	4.2	0.47	26	2.9	108	12	432	48
Contoured ball	1	0.4	0.05	2.7	0.3	11	1.2	43	4.8
	1.5	2.7	0.30	17	1.9	69	7.6	270	30
	2	12	1.3	72	8	288	32	1150	128
Butterfly, full open	1	0.2	0.02	1.2	0.14	5.4	0.6	22	2.4
	1.5	1.4	0.16	9	1	36	4	144	16
	2	7.7	0.86	49	5.4	198	22	792	88

*Figures denote Max. ΔP_S, (i.e., droop in pump cahracteristic and pipe friction losses for $\gamma = 0.1$) and corresponding valve pressure drop ΔP_O.

thereby, indirectly the type of valve) from two other criteria, namely:

a. Conserving pump horsepower.

b. Keeping the system gain as linear as possible by limiting the distortion of the inherent valve characteristic.

If one assumes as an arbitrary limitation that under (a) the efficiency of the system will be detrimentally affected if the control-valve pressure drop exceeds 25% ($\gamma = 0.25$) of the total dynamic head (pipe friction plus pump droop plus valve pressure-drop)—and, on the other hand, assumes that the installed valve characteristic (even with a percentage plug) becomes too nonlinear if the valve pressure-drop is less than 10%* of the dynamic head loss ($\gamma = 0.10$)—then the choice of available C_v/D^2 becomes very narrow indeed, as can be seen from Fig. 1. For example, the calculated pipe-friction loss in a given piping system with fluid having a specific gravity of 1 comes to 50 psi. The droop in pump characteristic is estimated to be 28 psi, which gives a ΔP_s of 78 psi. Assuming a pipe velocity of 10 ft/s, one has a choice from Fig. 1 between a "relative valve capacity" of 8 C_v/D^2 (limitation of installed characteristic), or a C_v/D^2 of 4.7 (valve

*Except for rotary valves with high inherent rangeability (> 100:1), where a limit of 5% is suggested.

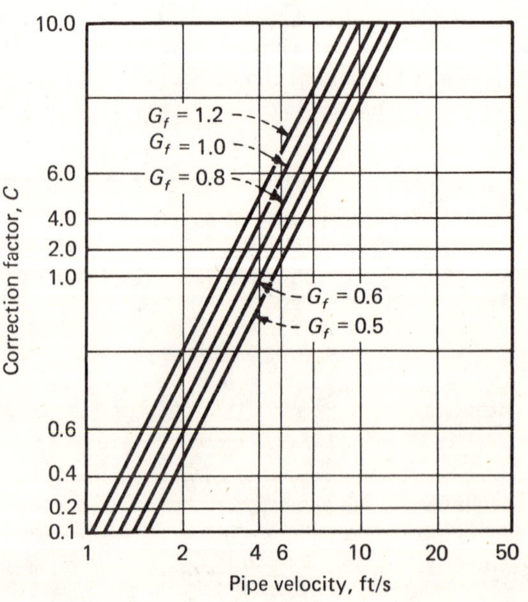

CORRECTION factors for velocity and sp. gr.—Fig. 3

Nomenclature

C	Correction factor (see Fig. 3)
C_v	U.S. gal/min per 1-psi pressure-drop across valve
D	Pipe diameter, in.
G_f	Specific gravity under flowing conditions (water at 60°F = 1)
k	Conversion factor, 7.48 gal/ft^3
P_o	Pressure drop across valve at maximum design flow (see Fig. 7), psi
P_s	Pressure drop caused by pipe friction and drop in pump head (see Fig. 7), psi
S	Actuator signal
U	Pipe velocity at maximum design flow, ft/s
V	Flowrate = $(60\pi D^2/4)(Uk/144) = 2.45\ UD^2$, gal/min
γ	Ratio of valve pressure-drop at maximum design flow to valve pressure-drop near zero flow

pressure-drop exceeds 25% of system drop below this capacity). From Table I, we have a choice between:

Camflex valve or double-seated globe with 0.4 factor trim. $C_v/D^2 = 5$	Camflex valve or double-seated globe with 1/1.5 line size. $C_v/D^2 = 5.4$
High-capacity cage valve, 0.4 factor trim. $C_v/D^2 = 6.4$	High-capacity cage valve, 1/1.5 line size $C_v/D^2 = 7$
60-deg., open butterfly, 1/1.5 line size. $C_v/D^2 = 7$	Y style or angle valve, 0.4 factor trim or 1/1.5 line size. $C_v/D^2 = 7.2$
Contoured ball valve, ½ line size. $C_v/D^2 = 4.5$	Full-open low-torque butterfly valve, ½ line size. $C_v/D^2 = 5.2$

Assuming a 60-deg open butterfly valve is selected (installed between reducers and giving a C_v/D^2 of 7), and turning to Fig. 2, the valve pressure-drop is found to be 12 psi, which can now be added to the pump specification data (minimum pump head = $\Delta P_s + \Delta P_o$ = 78 + 12 = 90 psi).

The data from Fig. 1 and 2 are tabulated directly against valve types in Table II. Data shown represent the *maximum* permissible dynamic system drop, ΔP_s, in a given system having a chosen valve style, in order to allow for an acceptable installed flow-characteristic. For a given maximum pipe-velocity, select the preferred valve type that shows a ΔP_s figure *larger* than the one calculated for your system. However, the next available ΔP_s figure should not exceed 4 times the actual ΔP_s. If it does, then either a different style of valve has to be selected or a valve pressure-drop of more than 25% of ΔP_s has to be tolerated.

For velocities other than shown in Table II, or for specific gravities other than 1, divide your actual ΔP_s by correction factor C (obtained from Fig. 3), in order to find the usable ΔP_s for Table II under the 10-ft/s velocity column. After finding the corresponding ΔP_o, multiply this number again with correction factor C, in order to obtain the actual ΔP_o applicable to your given velocity and specific gravity conditions.

Example: $G_f = 0.8$, velocity = 9 ft/s

Estimated ΔP_s = 80 psi, preferred valve—single-seat globe.

1. Correction factor from Fig. 3 for U = 9 ft/s and G_f of 0.8 = 0.65.
2. ΔP_s for use in Table II = 80/0.65 = 123 psi.
3. Table II gives next highest ΔP_s of 308 psi for globe valve installed between reducers 1.5/1 pipe/valve-dia. ratio and the 10-ft/s reference velocity; the corresponding ΔP_o = 34 psi.
4. Actual $\Delta P_o = 34 \times C = 34 \times 0.65 = 22.1$ psi.
5. Specify minimum pump head to be 80 psi (ΔP_s) + 22.1 psi (ΔP_o) plus static head if any.

The ΔP_s given in Table II is purposely limited to a maximum dynamic head loss for a γ ratio of 0.1. This really reflects the main criterion that the process engineer should shoot for, namely: What is the lowest possible pressure drop the valve will tolerate (in order to do a satisfactory control job) for the sake of reduced pump size and conservation of pumping horsepower? Modern control valves have no mechanical requirements that call for a certain minimum, internal pressure-drop for reasons of stability, hysteresis or otherwise. Therefore, the only real need for such minimum pressure-drop is the maintenance of a reasonable "installed characteristic" (the gain of the final control element in regard to the control system). The gain of the control is defined as the slope of the installed valve characteristic.

In most systems, the ideal control-valve gain should be constant—i.e., a linear relationship should exist between actuator signal and system flow. Only in loops where the transmitter input-signal is nonlinear, such as from orifice-type meters, is there a need for nonlinear control-valve gain. Nevertheless, with the limited choice of available control-valve characteristics, too great a distortion even here can be anticipated if the γ ratio (valve drop at full flow divided by the total system pressure-drop) is less than 0.1.

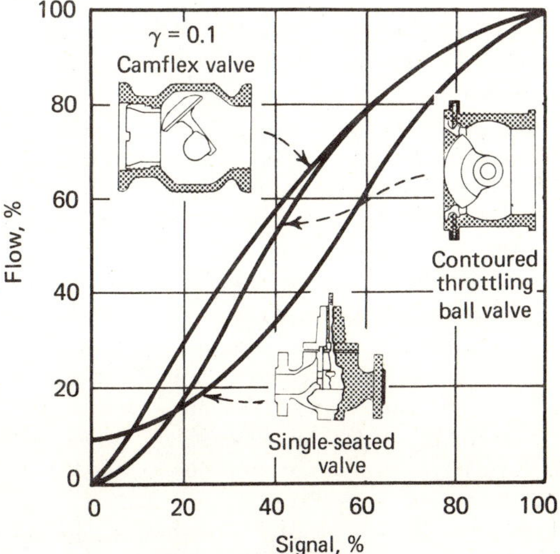

INSTALLED CHARACTERISTIC for three typical valve styles, all passing identical flowing quantities with equal percentage characteristic—Fig. 4

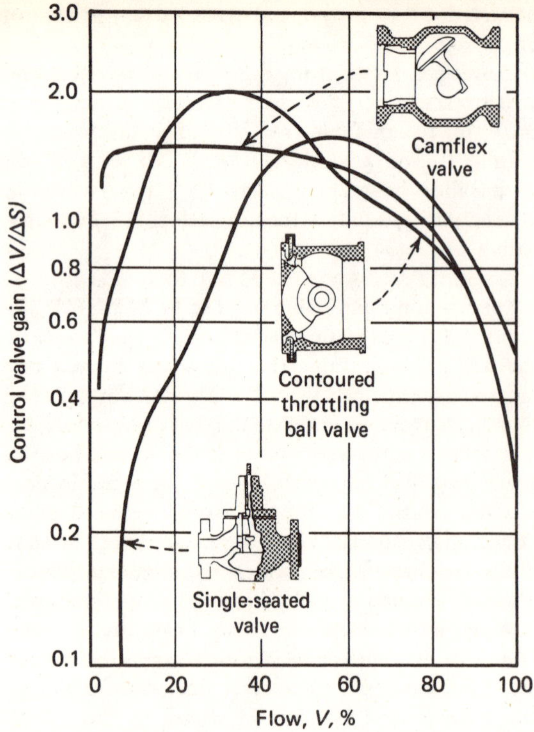

INSTALLED GAIN for three valve styles all having equal % inherent characteristic and identical design flow—Fig. 5

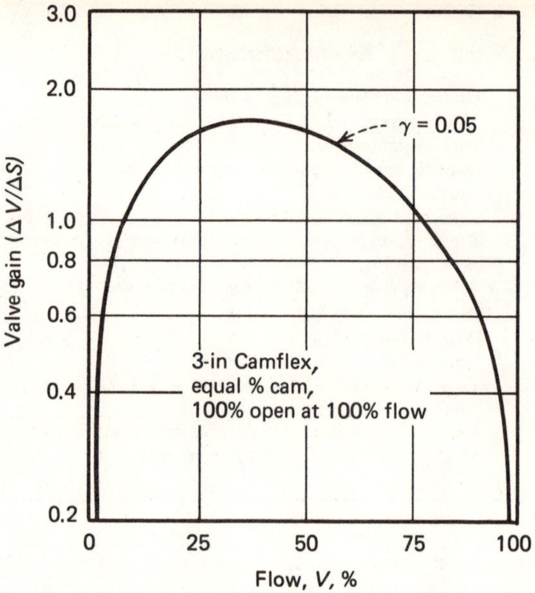

INSTALLED GAIN characteristic of 3-in Camflex valve with ratio of only 0.05 (pressure drop wide open only 5% of drop near closed position)—Fig. 6

To illustrate this point, actual installed characteristics are shown for three commonly used valve styles in Fig. 4. The valves chosen are a single-seated globe valve, at 90% open; a rotary valve with eccentric plug, 75% open; and a throttling contoured ball-valve installed between reducers 1/1.5 line size. All have an equal-percentage characteristic and are used for identical flow capacities.

As can be seen, the inherent limitation in rangeability of the globe-style control valve is very apparent and control is impossible below 10% of flow. On the other hand, modern rotary style valves do have both a superior installed characteristic and an actual usable rangeability to around 1% of system flow (due to their high *inherent* rangeability).

Perhaps a more meaningful presentation is to plot "gain" against "percent of flow" (data shown in Fig. 5). If one arbitrarily assumes that a given control system would yield stable and acceptable control with a control-valve gain varying between 0.5 and 2 (i.e., ± 2:1), then the eccentric rotating plug-valve could operate from near zero to 95% flow, the contoured ball valve from between 2% to 95% flow,* and the globe valve only from 21% to 99% flow. The data represented here indicate that, depending on the particular installation, a γ ratio of 0.1 might already be too low for typical globe valves, while modern rotary control-valves still give acceptable gain characteristics with γ ratios even down to 0.05—i.e., valve pressure drops of only 5% of total system head-change (see Fig. 6).

*The ball valve is somewhat handicapped by the added distortion of the "inherent characteristic" by adjacent pipe reducers, typical of all "high C_v" valves.

Working Example

As an example: In a typical depropanizer application, a pump has to circulate propane into the reflux tower at an elevation equivalent to 20 psi; the maximum drop through pipe and fittings at the maximum design flowrate amounts to 18 psi; the droop in pump head is estimated to be 10 psi. The total dynamic system loss, ΔP_s, is therefore 28 psi. The pipe velocity is calculated at 10 ft/s. Since the specific gravity is 0.5, enter 0.5 × 28 psi = 14 psi ΔP_s on the left-hand side of the graph of Fig. 1 and find that for 10 ft/s we could select C_v/D^2 between 11 and 19, while staying within the recommended limits. Since this is a relatively small installation, a single-seated globe valve is chosen.

Using a C_v/D^2 of 10 from Table I, we find, from Fig. 2, a pressure drop of 6 psi for water. Multiply by a sp. gr. of 0.5 to give an actual drop of 3 psi for the control valve. The total pump head that has to be selected is now 28-psi dynamic loss + 3-psi valve drop + 20-psi elevation, making the minimum required head 51 psi.

Including some safety factor, the next standard pump size to choose has a head of 60 psi, which then should be specified. Note that the process engineer should *add* this extra head to the valve pressure-drop in his specifications, which would then be used by the instrument engineer to select the final valve. The *actual* pressure-drop for sizing purposes (as seen by the instrument engineer for final selection) would be 60 psi minus an assumed safety factor of 10% (6 psi) and minus 48-psi dynamic loss and elevation, which equals 6 psi. This increase in actual pressure drop changes the required valve capacity by the square root of the ratio of original-to-actual pressure-drop. The required capacity therefore is 10 C_v/D^2 multiplied by $\sqrt{3/6}$ or 7.1 C_v/D^2. This means that either

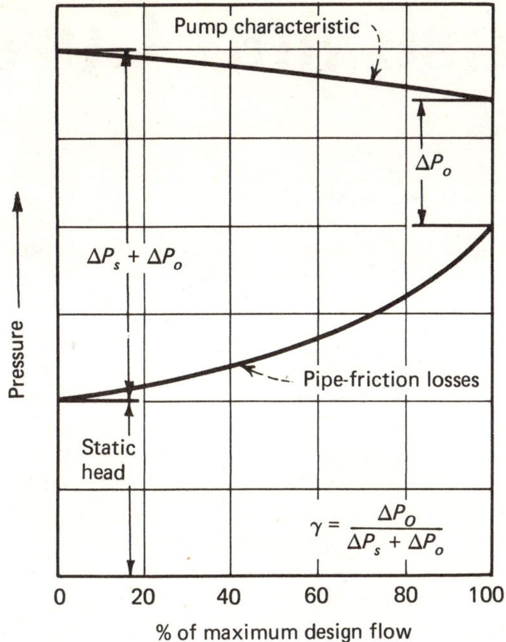

HEAD-CHANGE schematic for typical pumping system—Fig. 7

the originally selected valve will operate at 70% stroke, or a different valve style (smaller valve size) will have to be selected.

The most desirable inherent flow-characteristic is dependent upon the γ ratio. In this case, the γ ratio $= \Delta P_o/(\Delta P_s + \Delta P_o) = 6/34 = 0.177$, indicating that an equal-percentage characteristic is most desirable.

It should be noted that safety factors and "round off" on pump head always *increase* the final valve drop. This is one more reason to prudently select a γ ratio as low as possible (i.e., small valve-pressure-drop to begin with).

Economic Considerations

As explained in the foregoing, the limitation for selecting the lowest possible valve pressure-drop (ΔP_o) is concern for the installed characteristic. The upper limit is really controlled by the desire for reasonable operating expenses (pump horsepower). Simon and Roby [5] have investigated the relationship between pump operating expenses and valve size (valve ΔP) quite thoroughly, coming to the conclusion that the wide-open pressure-drop across the valve (ΔP_o) should be limited to 15 psi for pumping systems of up to 47 hp.* Above 47 hp, the drop should be lower. While this study, conducted in 1960, did not take into account the availability of newer rotary-style control valves, it should still be fairly up to date. In order to understand the economics, one must realize that it costs \$0.08 per year for each gallon pumped against 1-psi pressure-drop (based on 60% pump-motor efficiency and \$0.012 per kW, specific gravity = 1).

Assuming a system with 500-gpm average flow, the yearly cost for 10-psi pressure-drop is \$400; the savings realized over a period of 2 years would pay for the *installed cost* difference between a 4-in and 3-in control valve (10-psi ΔP_o vs. 20-psi ΔP_o).

The suggested upper limit for ΔP_o of 25% system drop is quite arbitrary and, as one can see from the above, might already be too high for pumping systems above 60-psi-head pressure. The range between $\gamma = 0.1$ and $\gamma = 0.25$ was chosen primarily because it covers roughly the average difference in C_v/D^2 between successive valve sizes. Referring to Fig. 1, one should always start from the $\gamma = 0.1$ line and move to the left to choose the next available valve size (C_v/D^2).

*At less than 240-psi head.

Summary

To find the applicable control-valve pressure-drop for pump-selection purposes, select the basic valve style (and relative flow capacity) from the known dynamic system head (pipe friction and pump droop). Select a valve capacity near $\gamma = 0.1$ (except for conventional globe style and angle valves if the desired *installed* rangeability or "turndown" should be larger than 5:1). This results in a relatively *low* valve pressure-drop. N.B. added pump safety factors always increase final valve pressure-drop.

After selection of the valve style, the pressure drop is given as a function of the known pipe velocity and, when added to the system losses plus elevation, can determine the required pump head.

This recommendation will lead to a more economical selection of control-valve pressure-drop, particularly when utilizing the advantage in wide rangeability of modern rotary-type control valves, which work well with a valve-to-system-drop ratio down to 0.05. Finally:

■ The calculations show that not all valve styles are interchangeably applicable for given flow conditions if one desires to stay within γ ratios between 0.25 and 0.01.

■ Conventional control valves are not very suitable for conserving pump horsepower, since their poor inherent rangeability allows only a limited "turndown" range due to drastic gain changes at low γ ratios.

References

1. Moore, R. W., Allocation of Valve Pressure Drop, Paper 6.1, First ISA Final Control Elements Symposium, Wilmington, Del., Apr. 1970.
2. Liptak, Bela G., "Instrument Engineers' Handbook, Vol. II, Process Control," Chilton Book Co., Philadelphia, p. 44.
3. Boger, H. W., The Effect of Installed Flow Characteristic on Control Valve Gain, ISA Paper 68-920, Oct. 1969.
4. Buckley, P. S., Selection of Optimum Final Element Characteristics, Proceedings of 5th National Chemical & Petroleum Instrumentation Symposium, Wilmington, Del., May 1964.
5. Roby, M. A., Simon, H., Pump and Control Valve Design Determined by Minimum Cost, ISA Paper 3-SF60, May 1960.

Meet the Author

Hans D. Baumann, 29 Villa Dr., Foxboro, MA 02035, is an independent engineering consultant specializing in automatic control systems, control valves and regulators. At the time of writing this article, he was Vice President—Technology for Masoneilan International, Inc., where he headed all U.S. and worldwide engineering activities, including R&D.

He is a member of American Soc. of Mechanical Engineers and Instrument Soc. of America, and is a registered professional engineer.

Control-Valve Noise: Cause and Cure

HANS D. BAUMANN, Masoneilan International, Inc.

As part of the present great interest in bettering the environment, a new attitude has developed toward control valves. To paraphrase a familiar saying: "Control valves should be seen but not heard."

The Federal Government has added legal backing to this attitude by setting a 90-db. noise time for equipment operating in the vicinity of plant personnel (for 8-hr. exposure). As a result, process and contracting engineers alike are setting up standards to ensure adherence to this law. The obvious question to put to the equipment manufacturer is, "How noisy will the valve be in our plant?"

In order to answer this question and be able to guarantee the specific noise level, control-valve manufacturers have had to study the problem of valve noise in depth (Fig. 1).

VALVE being tested in acoustical chamber—Fig. 1

The first requirement is for a method to predict the anticipated noise level under operating conditions. The equally important second question is, "What can we do to reduce the noise level if excessive?"

One must distinguish the three distinct and different noise phenomena emanating from a control valve:

1. Noise induced by mechanical vibration of the trim.
2. Noise produced by cavitating liquids.
3. Noise caused during aerodynamic throttling.

It is quite important that these three noise sources be understood so far as their generating mechanism is concerned. Only then can an effective improvement or cure be made. Luckily, mechanical-vibration noise seldom happens simultaneously with cavitation and aerodynamic noise. However, if this happens, the cure of one is usually the cure of the other.

DISPLACEMENT of guide posts of stainless-steel valve caused by severe resonant vibration—Fig. 2

Originally published May 17, 1971.

Noise Produced by Mechanical Vibration

Two mechanisms are involved: The first is mechanical vibration, induced by pulsation of the fluid passing through the valve. The frequency is usually low, i.e., between 50 and 500-Hz. However, if this turbulence-induced vibration of the valve trim approaches the natural frequency of the plug-stem combination, then we have the second mechanism—resonance. This resonance occurring at frequencies between 2,000 and 7,000-Hz. is most harmful, since it can lead to fatigue failure of the valve stem or plug post and can even displace solid stainless-steel parts by fractions of an inch (see Fig. 2).

The only beneficial aspect of this vibratory noise is that it warns operators that a mechanical failure is in the offing.

Luckily, the phenomenon has become less common since the introduction of top-guided single-seated valves, since they have (as a rule) less clearance in the guide bushings; also, the lower weight of a single-seated plug increases the natural frequency of the trim, making it less susceptible to fluid-induced vibration.

Possible cures for this type of noise include reduction of guided clearances, and increase in stem size (a 40% increase in stem diameter doubles the undamped natural frequency of the valve trim). Another attempt to cure this problem can be made by changing the flow or pressure conditions to which the valve is subjected. Quite often, a simple reversal of the flow direction through the valve sufficiently alters the flow pattern to shift inducing frequencies away from the trim excitation range.

Noise Produced by Cavitating Liquids

Cavitation noise should never be heard in a well-designed process plant. Hardly anything destroys a valve trim as surely as a cavitating liquid. With the introduction of special valve trims having very little pressure recovery and special valves having multiple velocity-headloss trim, there is seldom an excuse to have cavitation in a throttling valve (except, perhaps, for some rather large valve sizes where no anticavitation trim is available as yet).

With the present availability of good engineering data, it is possible to predict quite accurately whether or not a selected valve will cavitate under a given process condition. One such equation introduced by me in 1962 allows the prediction of a critical pressure drop at which a given value will cavitate,[1] i.e.,

$$\Delta p_{crit.} = C_f^2 (p_1 - p_v). \qquad (1)$$

If the operating pressure drop in the plant exceeds $\Delta p_{crit.}$ then the valve is cavitating. In that case, one should solve for the required Critical Flow Factor, C_f, and then select a valve from the manufacturer's catalog that has a C_f factor equal to or higher than the one calculated by the following equation:

$$C_f = \sqrt{(p_1 - p_2)/(p_1 - p_v)} \qquad (2)$$

Even though cavitation can be avoided in almost all cases, there is still interest in the prediction of cavitating noise. Our laboratory investigations indicate noise to be a function of: the amount of decrease in downstream pressure beyond the pressure that causes incipient cavitation, and the difference between downstream pressure and vapor pressure. The peak in cavitation noise can be expected where these two variables are nearly equal; i.e., the noise decreases as the difference between p_2 (actual) and p_2 (incipient cavitation) approaches zero, and if the difference between the outlet pressure and the vapor pressure approaches zero. The latter is understandable since the process of *cavitation* is converted into a process of *flashing*. The relationship just described is best illustrated in Fig. 3. Note a gradual increase in noise due to a normal fluid-borne turbulence up to the $\Delta p_{crit.}$, the point of full cavitation. The noise value reaches a peak when the excess pressure head and what we might call "cavitation energy"—i.e., the difference between p_2 and p_v—hold equal sway. Finally, the sound pressure level goes down again as the cavitation energy approaches zero with further decrease in p_2. It is of interest that this relatively small angle-valve ($C_f = 0.55$) with a C_v of only 4.25, an inlet pressure of 240 psia. and a pressure drop of only 89 psia. already exceeds the 90-db. limit of the Walsh-Healey Act.

To satisfy demands for an empirical equation to predict cavitation noise, I suggest:

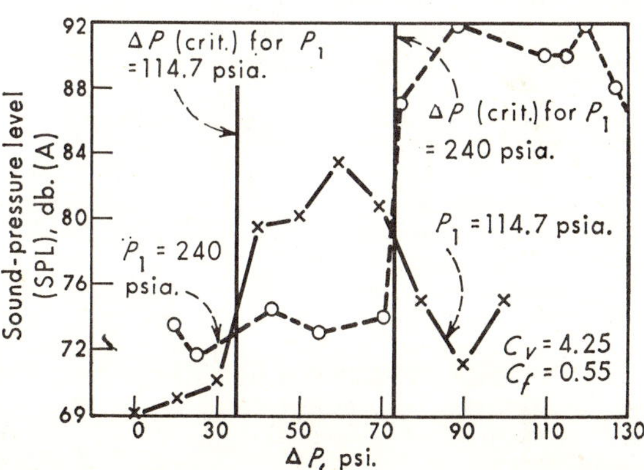

CAVITATION noise of single-seated valve at two inlet pressures (water at 60 F.)—Fig. 3

$$SPL = 10 \log (C_v C_f) + 8 \log (p_{2(crit.)} - p_2) + 20 \log (p_2 - p_v) + 33 \text{ (in dbA.)} \quad (3)$$

Where: $p_{2(crit.)} = p_1 - C_f^2 (p_1 - p_v)$

Note this equation is reasonably accurate only for water, and when using Schedule 40 downstream pipe. We expect that further research will enable extension of this equation to other liquids and let us predict the effect of differences in pipe-wall thickness.

Using this equation, one is able to predict the noise of a 6-in. ball valve reducing 100-psia. water to 25 psia. as follows:

Given: $C_v = 1{,}000$, $T = 70$F., $C_f = 0.7$, $P_v = 0.5$ psia. Hence, $p_{2crit.} = 100 - 0.7^2 (100 - 0.5) = 51.2$ psia.

$$SPL = 10 \log (700) + 8 \log (51.2 - 25) + 20 \log (25 - 0.5) + 33 = 28.5 + 11.3 + 27.8 + 33 = 100.6 \text{ db.}$$

Aerodynamic Noise

This is the most important form of acoustical annoyance so far as control valves or pressure-reducing valves are concerned. Aerodynamic noise is a byproduct of the reconversion of kinetic energy through turbulence into heat downstream of the throttling orifice. These are two basic contributory factors. One is the terminating shock front of a supersonic jet generating from the vena contracta of the valve orifice (at higher-than-critical pressure drop). The second comes from the general turbulence of the fluid boundary and is effective above, as well as below, choked flow in the valve orifice.

Apparent Noise-Producing Orifices—Table I

Valve Type	n Factor
Control ball-valve	1.0
Camflex Single-port globe valve Angle valve Butterfly valve	1.4
Double-port valve 4-port cage valve	2.2

Unfortunately, there is no way to avoid aerodynamic noise, since we have not as yet invented a valve that can reduce pressure without causing turbulence. However, the degree of noise generation can be affected by various parameters, as will be discussed later.

The important question facing an instrument engineer when laying out a new plant is: Which control valves will be noisy, (i.e., exceed the 90 db. usually considered the upper limit)? We published in Sept., 1969, a practical equation[2] that allows the prediction of the sound-pressure level for any style of valve under any given pressure condition. This equation has been further improved and converted into a graphical form for simpler handling.[2]

The general theory behind this equation has recently been made public[3] and I would like to repeat it here in broad terms in order to promote a better understanding as to which parameters affect the sound pressure level in a valve and how they may be favorably altered.

Nomenclature

C_f	Critical-flow factor of control valve, a ratio
C_v	Flow coefficient, gpm./(psi.)$^{1/2}$ (for water, 60 F.)
f	Frequency, cycles/sec. (Hz.)
g	Gravitational constant, ft./sec.^{2}
L_T	Transmission loss, db.
m	Weight of pipe wall, lb./sq.ft.
N	Strouhal number, a ratio
n	Number of apparent frequency-producing orifices (see Table I)
P	Sound pressure, lb./sq.ft.
p	Static pressure, psia.
R_{cv}	Critical-pressure ratio, p_1/p_2, at Mach 1
r	Radial distance to noise source, ft.
S_g	Gas-property correction factor, db. (see Table II)
SPL	Sound-pressure level, db. (Ref., 2×10^{-4} microbar)
V	Velocity, ft./sec.
W	Power, ft.-lb./sec.
X	Fraction of mechanical power conversion ($p_1 - p_2/0.47\, p_1$) limit to 1.
η	Acoustical efficiency, a ratio
ρ	Mean density lb./cu.ft.
Subscripts	
a =	Acoustical
i =	At vena contracta
m =	Mechanical
o =	downstream
s =	Sonic
v =	Vapor
1 =	Inlet
2 =	Outlet

Theory of Aerodynamic Sound

As mentioned before, the sound pressure measured in the proximity of a throttling control valve is a result of pressure waves in the atmosphere—the pressure waves' root mean square (rms.) values are expressed in microbars. The sound-pressure level (SPL) in decibels is equivalent to $20 \times \log_{10}$ of the ratio between the rms. value and the reference value (taken as 2×10^{-4} microbars).

The acoustical power that generates these pressure waves is created by the supersonic shock front in a jet, and by turbulent boundary layers within the valve. It is directly related to the amount of mechanical energy converted in the valve. This then makes the SPL value a direct function of mass flow or C_v, since the latter is an expression of flow capacity.

The acoustical efficiency factor η describing the ratio between acoustical power and mechanical power is a function of the valve style and the pressure ratio across the valve. Here is an expression for the mechanical power converted in a valve:

$$W_m = [\rho_i v^3 \pi (2.3 \times 10^{-4}) C_v C_f]/8g \quad \text{(in ft.-lb/sec.)} \qquad (4)$$

(For the mathematical development of this and other equations see Ref. 3.)

Multiplying the mechanical power by an acoustical efficiency factor, η, will now yield the acoustical power:

$$W_a = W_m \times \eta \qquad (5)$$

Fig. 4 shows semi-empirical, acoustical efficiency factors plotted against pressure ratios across the valve. The acoustical efficiency varies with the pressure recovery characteristic of a particular valve as expressed by the Critical Flow Factor, C_f (see Table III). It shows, for example, that a streamlined angle valve, flow-to-close ($C_f = 0.5$) will produce relatively higher noise than a V-ported globe valve ($C_f = 0.95$) when below the critical-pressure ratio of approximately 2 to 1, but considerably less noise above this ratio.

Having found the acoustical power, a conversion into sound pressure can be made as follows:

$$P = (W_a \rho_o V_s / 4\pi r^2 g)^{1/2} \quad \text{(in lb./sq.ft.)} \qquad (6)$$

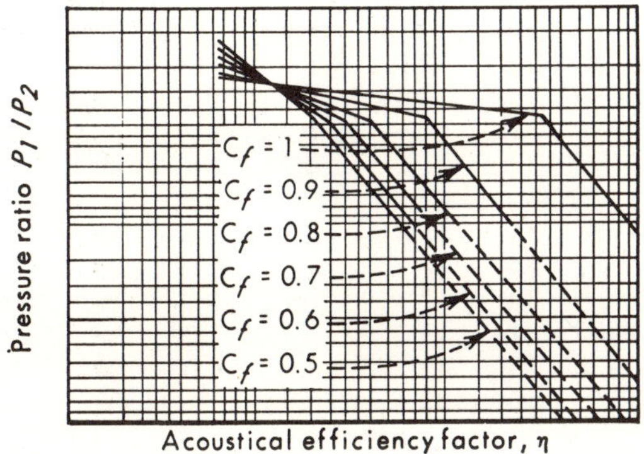

ACOUSTICAL efficiency factors of control valves for various critical-flow factors, C_f—Fig. 4

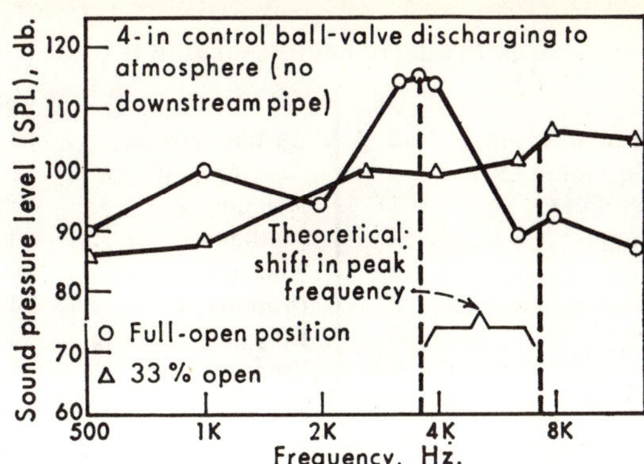

$$\text{Peak freq. } (\Delta) = \text{peak freq. } (o) \sqrt{\frac{C_v\ C_f\ (o)}{C_v\ C_f\ (\Delta)}} = 3{,}500 \sqrt{\frac{258}{65}} = 7{,}000$$

PEAK NOISE frequency shift with shift in apparent valve orifice diameter—Fig. 6

Finally, the transmission loss must be known, i.e., the amount of sound power attenuated by the downstream pipe wall. Assuming that the predominant frequency of the valve noise coming from the orifice is below the coincidence frequency or "ring frequency" of the pipe (see Ref. 4 and Fig. 5), one can relate the pipe wall attenuation to that of a flat panel, i.e.:

$$\text{Transmission loss, } L_T = 17 \log (mf) - 36 \qquad (7)$$

wher $fe = NVn/0.015\ \sqrt{C_v C_f}$.

The Strouhal number, N seems to vary between 0.1 and 0.2, depending on the pressure ratio across the vena contracta. The frequency is also a function of the apparent orifice diameter, expressed as $(1/n)$ $0.015\ \sqrt{C_v C_f}$ (see Fig. 6), wherein n is taken as a modifier to convert an irregular orifice into an equivalent circular shape (see Table I). Note also that the attenuation is a function of 17 log pipe-wall density, m, (see Fig. 7). Finally, a correction factor has to be used to make the basic equations (strictly

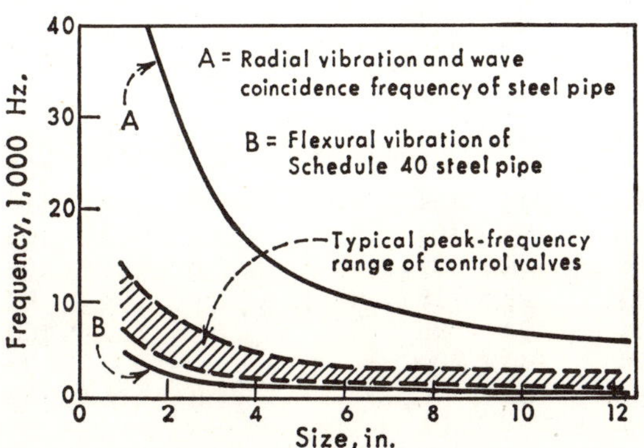

PEAK frequencies (typical) of control valves as related to vibration of adjacent piping—Fig. 5

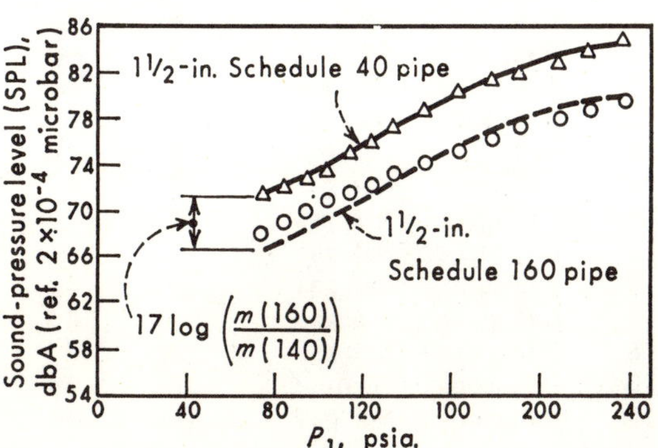

TRANSMISSION LOSS effect of pipe wall thickness. Switch to heavy pipe reduced noise by 4.5 db.—Fig. 7

Gas-Property Factor—Table II

	S_g, db.		S_g, db.
Saturated steam...	−2	Carbon dioxide.....	+1
Superheated steam	−3	Carbon monoxide..	0
Natural gas........	−1	Helium.............	−6.5
Hydrogen..........	−10	Methane...........	−1
Oxygen.............	+0.5	Nitrogen...........	0
Ammonia..........	−2	Propane...........	+1
Air.................	0	Ethylene..........	−1
Acetylene..........	−1	Ethane.............	−1

applicable only for air) suitable for other gases. This is done with the use of a gas-property factor S_g listed in Table II. Thus the final equation for the aerodynamic sound-pressure level at 3 ft. from the valve outlet and pipe wall becomes:

$$SPL = 10 \log_{10} (X \eta 10^{11} C_v C_f p_1 p_2) - L_T + S_g \quad (8)$$

This equation has been found quite accurate despite the rather complex nature of the subject. Table III shows calculated vs. tested data on valve sizes 1 to 12 in., and a wide range of pressures (up to 4,000 psi.).

What To Do About Aerodynamic Noise

One problem with the SPL equation is that in the planning stages an engineer may not know the exact style and type of valve that will be used in a given application. Since this is essentially a filtering-out stage, i.e., for isolating critical applications, a simple rule-of-thumb would be very handy. One such rule is not to worry about a valve (handling compressible media at critical pressure drop) as long as the product of absolute inlet pressure in psia. multiplied by C_v is below 1,000. All valves falling

Comparison of Calculated SPL With Laboratory and Field Test Data—Table III

Valve Size, In.	Type*	Fluid	Pipe Size, In.	Pipe Schedule	n	P_1	P_2	C_vC_f	C_f	$C_{FR/R}$†	SPL Calc.	SPL Test
1	Y style	Air	1	40	1.4	145	17.5	3.04	0.65		81.1	83
1	Y style	Air	1	40	1.4	24.7	14.7	3.04	0.65		56.8	57
1	Y style	Air	1	40	1.4	115	19.0	3.04	0.65		78.0	80
1	Y style	Sat. steam	1½	40	1.4	44.7	15.7	2.9		0.65	63.7	65
1½	S.P. globe	Air	1½	40	1.4	145	15.7	2.40	0.50		78.0	78.8
2	4-port cage	Sat. steam	2	40	2.2	71.7	21.7	31.6	0.95		89.9	89.5
2	4-port cage	Sat. steam	2	40	2.2	111	32.2	31.6	0.95		94.1	96
2	Angle	Nat. gas	10	160	1.4	4,000	2,300	25.5		0.50	110.6	111
2	Regulator	Sat. steam	8	40	1.4	170	34.7	11.8		1.0	98.3	96
2	D.P. globe	Ind. gas	4	80	2.2	530	80	14		1.0	106.6	105
2	D.P. globe	Ind. gas	4	80	2.2	530	80	17		1.0	108.1	110
3	Regulator	Sat. steam	8	40	1.4	170	34.7	19.3		1.0	99.7	102
3	D.P. globe	HC vapor	6	40	2.2	241	72	48		0.95	102	103
4	Ball	Air	4	40	1.0	150	50	180	0.76		108.7	107
4	Butterfly	Air	4	40	1.4	150	50	.180	0.62		105.3	103
4	8-port cage	Air	4	40	2.5	150	50	180	0.90		105	110
4	Ball	Air	4	40	1.0	100	72	336	0.60		103	100
4	D.P. globe	Air	4	40	2.2	150	50	180	0.86		104.6	107
4	D.P. globe	Fuel gas	8	40	2.2	255	90	97		0.89	104.5	108
4	Ball	Air	8	40	1.0	615	115	108		0.84	121.2	120
4	Ball	Air	8	40	1.0	615	315	108		0.84	113	114
4	Ball	Air	8	40	1.0	150	50	180		0.79	109.1	108
4	Ball	Air	8	40	1.0	615	75	108		0.84	123.2	122
6	S.P. globe	Nat. gas	8	40	1.4	305	91	76.7		0.80	104.4	106
6	D.P. globe	Nat. gas	8	40	2.2	270	90	100		0.90	104	105
6	Ball	Nat. gas	8	40	1.0	650	325	244		0.76	116	110
6	D.P. globe	Nat. gas	8	40	2.2	635	295	70		0.9	101	105
6	D.P. globe	Nat. gas	8	40	2.2	665	340	191		0.9	110	110
6	D.P. globe	Nat. gas	8	80	2.2	250	105	101		1.0	98.3	102
8	Regulator	Nat. gas	12	40	3.0	365	165	177		1.0	101.8	101
8	D.P. globe	Fuel gas	10	40	2.2	190	22.7	216		0.5	105.9	98
8	D.P. globe	Nat. gas	10	40	2.2	200	23	225		0.55	104.5	100
10	D.P. globe	Steam	10	40	2.2	245	130	800	0.95		110	107
12	D.P. globe	Nat. gas	18	40	2.2	165	45	.367		1.0	114.4	122

* S.P. = single port; D.P. = double port.
† Use $C_{FR/R}$ instead of C_f when valve is installed between reducers.
The weight, *m*, of pipe reducers is based on valve-size pipe wall. Calculated values include 3 db. for hemispherical radiation.

above this limit should be calculated and treated properly once the final specifications become available.

An important requirement in noise control is to keep the valve-outlet velocity below a certain limit, depending on the type and size of the valve, in order to prevent the occurrence of a secondary noise source that might be even worse than that produced by the valve itself. This is particularly important with valves having special "low noise" trim. To demonstrate the importance, Fig. 8 gives the SPL values of air flowing through various pipe sizes as calculated from equations suggested by Heitner.[5] This example draws attention to this often overlooked aspect and points out the need for proper valve and pipe dimensioning.

If the calculated SPL value of a reducing valve under maximum load exceeds the stated limit by only 5 to 10 db., then the following simple cures are applicable:

- Increase the pipe-wall thickness downstream (doubling the wall thickness will decrease SPL by 5 db.).
- Use acoustical insulation downstream. This will reduce the SPL number between 5 and 10 db. per inch of insulation, depending on the density of the insulating material.

If the valve noise is more than 10 db. above the selected limit, then one must choose a different approach such as the use of downstream, in-line silencers. These devices generally attenuate between 10 and 20 db., depending on the frequency range. Care should be taken to mount the silencer directly adjacent to the valve body and to make sure that the valve outlet velocity is below sonic; otherwise the silencer will act as a pressure-reducing device for which it is not suitable. One should also be aware that part of the valve throttling noise will be transmitted upstream and reradiated through the upstream pipe. This means that the valve that would have a calculated untreated SPL of 100 db. could

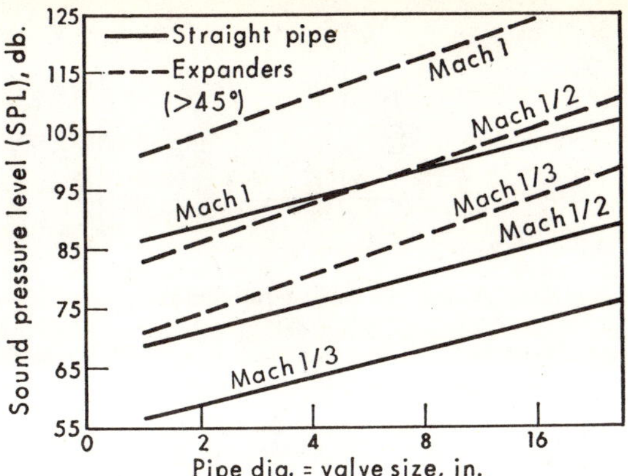

AERODYNAMIC NOISE generated by pipe velocity, including effect of downstream expanders—Fig. 8

be attenuated by a silencer to 80 db. downstream. However, the system would still radiate noise at 90 db. upstream, cancelling most of the silencer's beneficial effects.

Another approach usually recommended by my company is the use of noise-reducing expansion plates[6] downstream of a valve. The primary function of these plates is not to attenuate the valve noise, but to absorb some of the pressure reduction over the whole system. This way, the pressure ratio across the control valve can be kept below critical, with the associated benefit of having lower acoustical power generation. Noise-reducing Lo-db plates can be used if the product of absolute inlet pressure (p_1) times C_v is below 5,000 when used downstream of conventional control valves. This limit does not exist when placed downstream of multiple-step reduction valves and conventional valves used for *on-off applications* only. One major benefit of these plates is the reduction of the valve-outlet velocity to safe levels by the buildup in static pressure.

The use of multiple plates is recommended where

Sound Pressure Levels of Regular Vs. Low-Noise Valves—Table IV

Lo-db Size	Valve Type	X†	Volume, Std. Cu. Ft./Hr.	Fluid	P_1 Psia.	P_2 Psia.	Pipe Schedule	Sound-Pressure Level Lo-db Valve Calcul.*	Lo-db Valve Tested	Standard Valve
1½ × 1½	S. Seat	2	31,180	air	240	17.5	40	79.4	78.5	87
2 × 2½	S. Seat	6	1.4×10^6	gas	3,725	1,151	160	78.5	81	107
2 × 2½	S. Seat	6	$0.5 \times 10_9$	gas	4,000	1,060	160	70.5	67	96
2 × 4	S. Seat	4	2.7×10^6	gas	4,000	1,060	80	89.3	88	112
2 × 4	S. Seat	4	4.1×10^6	gas	4,000	1,060	160	89	94	110
3 × 6	S. Seat	4	2.7×10^6	gas	3,850	1,060	80	87	83	111
3 × 6	S. Seat	4	3.4×10^6	gas	3,850	1,060	80	88.8	85	113
3 × 6	S. Seat	2	4.1×10^6	gas	2,300	1,060	80	94.2	96	112
4 × 4	D. Seat	2 × 2	350,000	air	215	115	40	79.2	80	91.2
6 × 6	D. Seat	2 × 2	360,000	gas	282	138	40	75	79	90

* $SPL = 18.5 \log (C_v C_f) + 28 \log (P_1) - 13 \log (P_2) + 20 \log (4/X) + 28 - Sa + Sg$ (in dbA) † X = trim area ratio

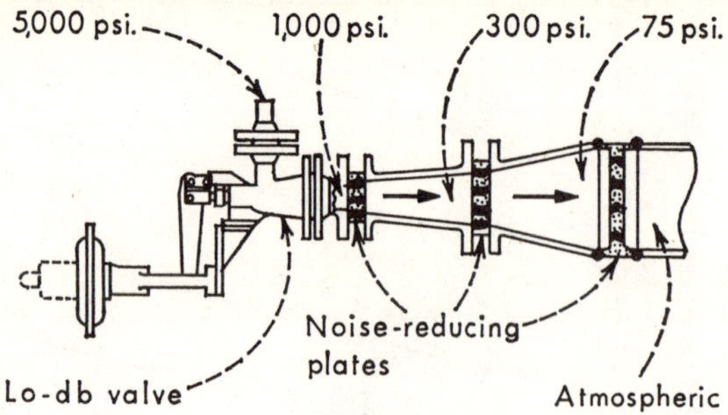

LOW-NOISE valve installation reducing from pressure of 5,000 psi. down to atmospheric—Fig. 9

the ratio between upstream pressure and ultimate downstream pressure is more than 10 to 1. Fig. 9 shows typical layout of a valve reducing from 5,000 psi. to atmospheric. Here the valve could reduce the pressure from 5,000 to 1,000 psia., the first plate from 1,000 psi. to 300 psi., the second plate from 300 psi. to 75 psi., and the last plate from 75 psi. to atmospheric. Note that these plates are installed in expanders whose increasing flow areas match the given pressure level and keep the velocity gradient through the whole system nearly constant.

Analyzing Eq. 4 and 6 will yield the conclusion that the most effective way to combat aerodynamic sound in a valve is by reducing the throttling velocity. This is indeed so, since SPL varies nearly to the 8th power of throttling velocity. Several valve styles now on the market make use of this beneficial effect, notably the Self Drag valve marketed by Control Components, Inc., and the Lo-db valve introduced by Masoneilan in 1967.[7] The latter valve uses a multistep cone trim that forces the fluid to make repeated sharp turns, each of which has a high-velocity head-loss coefficient, producing a nearly "constant adiabatic" pressure drop across the valve. The high specific head-loss change per stage means low velocity. Hence, noise level is substantially below that of conventional angle or globe valves (Table IV).

Unfortunately, these valves are more costly than their predecessors, owing to the more-expensive trim configuration and the valve size, which as a rule is double that of a conventional valve. The reason for the increased valve size is, of course, to keep the fluid velocity within limits. No such restrictions are placed on a conventional valve (when disregarding the sound pressure level).

The comment has often been made that if a valve is installed in a remote area without operating personnel being constantly present, the noise problem can safely be disregarded. This can be at times a dangerous simplification, since noise is nothing but audible *vibration*. The vibration itself can cause considerable mechanical damage. Damage to pressure gages and valve- or pipe-mounted instruments is one of the many detrimental effects. Of more-drastic consequence is that even 2,500 lb. ANSI flange bolting can vibrate loose, as I have witnessed myself. Fig. 10 shows typical acceleration values measured downstream of a 2-in. high-pressure angle valve "flow-to-close," reducing gas from 4,000 psi. to 2,300 psi. Notice that after the Lo-db valve was installed the pipe acceleration level decreased from 125 db. to 103 db. (14g. to 1g. acceleration level). The original acceleration level of 14g. was quite severe and compares to that experienced when holding a riveting hammer.[8]

In any case, elimination of vibration is an important aspect that should not be overlooked. Money spent for safety and for the reduction of maintenance costs, is money well spent.

Note: The acoustical-test results presented in this article, with the exception of Fig. 6 and 10, refer to the A-weighted scale (dbA), i.e., the frequency range corresponding roughly to the sensitivity of the human ear. Such overall frequency-weighted data are often called "Sound Levels" and usually exceed the specific SPL at a given peak frequency by 1 to 3 db.

References

1. Baumann, H. D., The Introduction of a Critical Flow Factor for Valve Sizing, *ISA Transactions*, 2, No. 2 (Apr. 1963) Paper presented at ISA Conference, Oct. 1962.
2. "Masoneilan Handbook for Control Valve Sizing," Masoneilan International, Inc., Norwood, Mass.
3. Baumann, H. D., On The Prediction of Aerodynamically Created Sound Pressure Level of Control Valves, ASME Paper WA/FE-28, Dec. 1970.
4. Sawley, R. J., While, P. H., Energy Transmission in Piping Systems and in Relation to Noise Control, ASME Paper 70-WA/PET-3, Dec. 1970.
5. Heitner, J., How to Estimate Plant Noises, *Hydrocarbon Process.*, Dec. 1968.
6. Hynes, K. M., The Development of a Low Noise Constant Area Throttling Device, ISA Paper 839-70, Oct. 1970.
7. Baumann, H. D., Multistep Valve Design Cuts Throttling Noise, *Instrum. Tech.*, Oct. 1969.
8. Harris, C. M., Crede, C. E., "Shock and Vibration Handbook," Vol. 3, pp. 44-42 and 47-53, McGraw-Hill, New York, 1961.

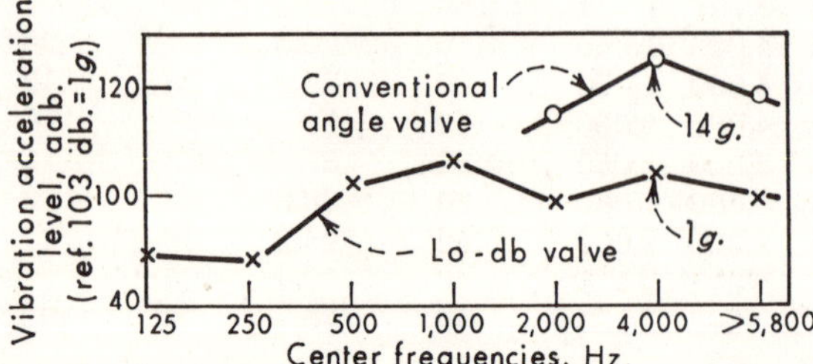

VIBRATION acceleration levels when reducing 4-million cu.ft./hr. gas from 4,000 to 2,300 psi.—Fig. 10

Meet the Author

Hans D. Baumann is Vice-President Engineering for Masoneilan International, Inc., Norwood, Mass. 02602. He received his undergraduate education in Germany, and has done graduate work at Western Reserve University, Cleveland and Northeastern University, Boston. He is a member of the American Soc. of Mechanical Engineers, and a senior member of the Instrument Soc. of America. He is a registered professional engineer in Illinois and Pennsylvania.

Special Control Valves Reduce Noise and Vibration

High-velocity fluid flow through conventional control valves causes erosion, vibration and noise in the valve and pipeline. Here is an analysis of how to handle these problems, and an evaluation of some new control valves and other devices that can reduce these problems or eliminate them at the source.

J. B. ARANT, E. I. du Pont de Nemours & Co.

The conventional plug-and-orifice control valve is subject to severe maintenance and operational problems when forced to operate under conditions of high fluid velocity. This is also true of other valve configurations, such as ball and butterfly.

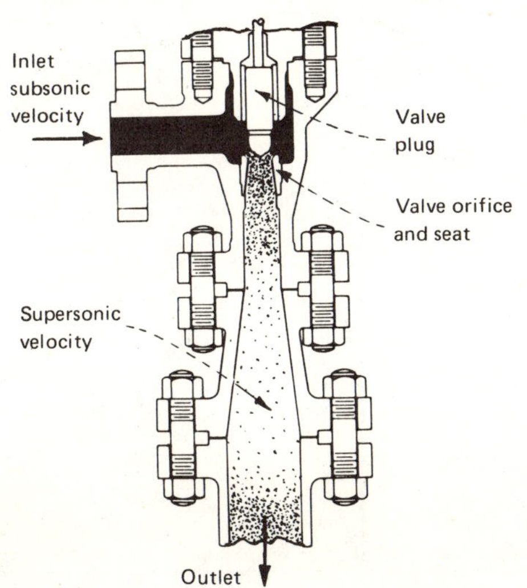

ANGLE valve, conventional, has supersonic flow—Fig. 1

Originally published March 6, 1972.

High velocity is the big problem in control valves, such as the angle valve shown in Fig. 1. It results from the conversion of pressure head to velocity head as the fluid undergoes a pressure drop while flowing through the valve orifice.

In liquid services, the problem arises when the velocity through the valve trim exceeds about 300 ft./sec. In gas or vapor services, it arises when the velocity exceeds about Mach 0.4. At these velocities, one or a combination of the following problems occurs: (1) erosion, (2) erratic control, (3) vibration, (4) shutoff leakage and (5) noise. Let us examine each of these problems in detail in order to define their parameters.

Erosion

Erosion can result from liquid flow, by high-velocity "washing" of the metal surface whereby the protective film is continuously removed and the metal wears away. It also occurs in fluids containing solids particles such as scale, catalyst fines and liquid droplets, whereby the metal surface is abraded away.

In some cases, a special type of surface damage is caused by cavitation. While the exact mechanism of cavitation is complex and not completely understood, the following is generally accepted:

As the pressure head of a liquid is converted to velocity head through the valve trim, it will reach a zone of maximum velocity and minimum pressure called the vena contracta. If this minimum pressure is below the vapor pressure or gas-solution pressure of the liquid, localized flashing and bubble formation will occur. If this were as far as the phenomenon would go, there would be no problem because the bubbles do not cause damage. However, since pressure recovery occurs downstream in the valve, the bubbles quickly collapse in an implosion. The resultant energy dissipation can impose concentrated surface stresses on the order of 100,000 psi. or more. These act like a myriad of small, high-energy hammer blows that will chip metal particles from the surface to form the easily recognized "sand blasted" or "termite hole" damage. Even hardened-trim materials cannot withstand these stresses for long.

One type of valve design that has been used for cavitation situations is the "swiss cheese" or "flash flow" trim. While extremely helpful, this trim does not always solve the problem.

Generally, clean dry gases do not cause erosion problems, but the presence of liquid droplets or solids can cause severe erosion in a relatively short time. This is highly accelerated when the valve is leaking or operating near the seat. Such erosion is called wiredrawing.

Steam is particularly notorious for wiredrawing since it is often in the saturated or wet condition. Even dry, superheated steam is highly erosive, though we would expect it to behave like any dry gas. The explanation for this is a reverse situation to cavitation, with liquid droplets formed in the vapor (Fig. 2).

Here, inlet pressure, P_1, drops in isentropic throttling to the pressure in the valve throat, P_3, where it is below the saturation temperature. This causes water droplets to form, and creates the same erosion situation as does wet steam. As pressure recovery takes place, the water returns to the vapor state, and dry steam leaves the valve at pressure P_2. Though reduction in steam pressure seems to be isenthalpic, actually it is not. The high velocities generated, the water droplets formed and the high fluid temperature, all combine to play havoc with the valve trim. It is not uncommon for a valve used in reducing 1,500 psig. steam to the 300 to 500-psig. level to be destroyed internally in a matter of days.

Erratic Control

Erratic control can result from excessive velocity, which may be due to one or a combination of factors. Extremely high velocities will produce turbulence that causes erratic side-loadings on the stem and plug. Changes in the relationship between velocity and static pressure across the plug face can seriously alter the force-balance relationship in the vertical direction. These situations can:

1. Alter the concentric plug-to-seat relationship, which will vary the flow characteristic.
2. Cause high-frequency vibration and wear, which will destroy the inner-valve components.
3. Create vertical instability in the valve operator, which will produce vertical oscillations.
4. Cause variable flashing or cavitation in liquids, which will give poor control.

Vibration

Vibration is often the result of high velocity both within the valve and in the piping. Such vibration will cause valve failure and, worse, may lead to piping failure.

Pipe vibrations, downstream of a high-drop gas-control valve, with a measured frequency on the order of 2,500 Hz. and an amplitude of 100 mils (1 mil = 0.001 in.) have been personally observed. Needless to say, this caused line failure. Attempts to dampen the vibration with an overwrap of lead sheeting resulted in destruction of the sheeting in less than 24 hr. Correction required complete redesign of the line's configuration, an increase in wall thickness of the pipe, bracing of the pipe, and lead-sheet wrap. The correct answer would have been to reduce or eliminate the high-velocity energy from the valve, which was the vibration source.

Shutoff Leakage

Loss of shutoff capability is the end-result of several of the problems previously mentioned. Erosion, by whatever means, will destroy the seating and shutoff surfaces. Vibration may peen and distort the plug and seat. In many cases, it is a cannibalizing situation, since the more the valve leaks, the worse it leaks. And, high velocity often limits the ability to use a soft elastomer

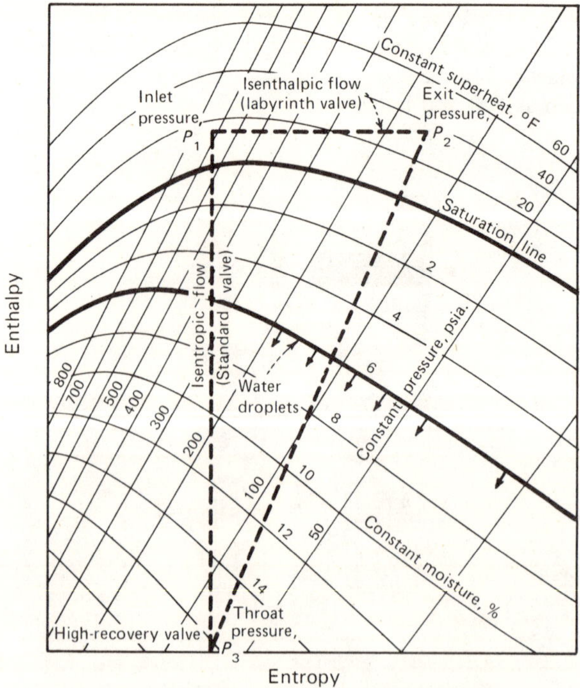

ISENTROPIC throttling creates water droplets—Fig. 2

or Teflon® seat to effect tight shutoff because of inability to retain these seals mechanically.

Noise in Control Valves

Noise and velocity go together. There are three main sources of control-valve noise: mechanical vibration, cavitation, and aerodynamic action.

Mechanical vibration is caused by turbulence generated by velocity and/or large massflows, and is usually unpredictable. It is most common with double-seated valves, but single-seated valves are not immune. Noise levels are usually low and in the 50 to 1,500-Hz. range. Frequencies above this are usually due to excitation of the valve's internals at natural resonance. This resonance is often very noisy and may be up in the 90-dbA range, with a frequency up to 7,000 Hz. This may be acceptable as noise, but is very likely to lead to unacceptable valve failures.

Cavitation can be noisy depending upon the degree present. Cavitation noise will increase with increasing pressure drop to some maximum value and then decrease to some lower magnitude. This is due to the progressive reduction in the downstream pressure recovery, which causes bubble collapse. Equations for predicting cavitation and for calculating the attendant noise are available in one vendor's valve handbook.[1] Noise levels during cavitation are generally in the 90 to 100-dbA range, but can be higher.

Aerodynamic noise is the commonest and worst source of valve noise. It is caused by fluid turbulence and shock waves due to high velocity or massflow of gas, and can be readily predicted as to occurrence and magnitude.[1,2]

While Mach 1.0 velocity (or sonic flow) guarantees noise, high-noise levels can be generated when Mach number is as low as 0.4 and there are relatively low pressure drops where large massflows are handled (Fig. 3).

However, the noise that affects us most is that generated by sonic or near-sonic flow. As an example of the tremendous power encountered, noise levels up to 140 dbA have been measured near valves. Natural-gas letdown from 4,000 psig. to 1,000 psig. generates about 5,000 w. of sound pressure level. About 4,950 w. will be absorbed by the valve body and pipe wall. The 50 w. escaping will yield about 110 dbA at 3-ft. distance. If the letdown had been a free jet or vent, the sound pressure level would have been greater than 150 dbA. For comparison, the noise level of a jet aircraft as it leaves the terminal is about 117 dbA.

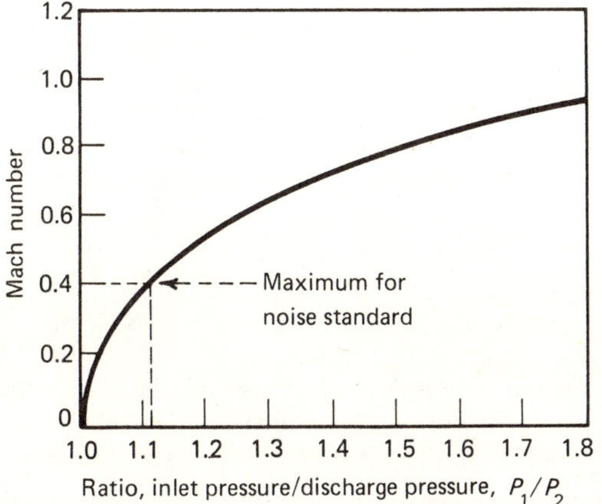

SONIC or near-sonic flow causes noise—Fig. 3

Anyone who has been around high-noise-level valves knows how intolerable this can be. Temporary symptoms may be dizziness, nausea, headache and deafness. Prolonged exposure can give permanent hearing damage. Even at a distance, such noise can be extremely wearing to the nerves and temper, and can cause self-generated safety hazards.

Noise pollution is drawing more attention from federal, state and local authorities, and increasingly stringent regulations concern not only the worker exposed to noise, but also the effects of noise on nearby residential areas. It behooves industry to recognize the impact of this trend and to bend every reasonable effort to eliminate or alleviate such pollution.

Valve noise is one of the major offenders in many plants. Its elimination is possible with current technology and is a big step in meeting noise standards. A guideline that is becoming accepted as a valve-noise standard is that the sound pressure level should be equal to, or less than, 85 dbA when measured 3 ft. from the source. While this is not exactly quiet, it is acceptable for hearing conversation. This criterion normally requires that gas or vapor velocity be limited to less than Mach 0.4 and, preferably, no more than Mach 0.3.

Much work has been done in predicting aerodynamic noise, and appropriate information is readily available.[1,2] A quick and approximate way to start is with the equation: $P_1C_v = X$, where P_1 is the upstream pressure, psia., and C_v is the valve-capacity coefficient at maximum service flowrate. If X equals or is less than 1,000, the maximum noise level is about 90 dbA. If X has a value of more than 1,000, more-definitive calculations are required in order to assess the problem. If the noise prediction is 90 to 100 dbA, there are several approaches possible. These are based primarily on path treatment rather than on source treatment.

1. Evaluate effect of distance. Maybe the valve's location on top of a structure or a pipe bridge indicates there would be no problem for personnel at their normal working locations. This can be checked by using the equation for reduction in sound pressure level (SPL):

$$SPL_{distance} = SPL_{source} - 10 \log\left(\frac{\text{distance in feet}}{3}\right)$$

2. Increase wall thickness. Doubling the pipe wall thickness reduces escaping sound by 5 dbA. Adding one inch or more of acoustical insulation will reduce sound level another 10 dbA. However, keep in mind that sound travels down a pipeline, and increasing the pipe wall thickness and/or applying acoustical insulation may be an expensive solution.

3. Use a silencer. This can also be expensive and is not always effective. Inlet velocity must be subsonic and the silencer should not be forced to act as a pressure

reducer. The silencer itself can act as a sound propagator if the shell-wall thickness is too light and vibrates.

4. Try to limit exit velocity at valve so that it is equal to or less than Mach 0.3. Even if we expand to a larger diameter of pipe, exit turbulence will concentrate at the expander and generate noise. It is a common fallacy that we can eliminate noise by reducing the exit velocity at the valve by pipe enlargement.

If the noise-prediction calculation is equal to or greater than 100 dbA, then we have a problem of the worst order. The only recourse is to go to source treatment.

Resolving the Problems

Until recently, all we could do was to install one of a variety of hardened-trim materials, strengthen the stems and guides, and hope for the best. We might try using an angle-, Y-, or S-body configuration, or a different trim type, or reversing the body, etc. All we were doing was to try to live with the problem and stretch out the service life of the valve. If any of our changes increased valve life, we counted it as a success and it passed from the realm of an engineering problem to a maintenance one. It then became a way of life, accepted as such, and often we denied we had a problem anymore.

In the last three or four years, some new control-valve designs and other devices have come on the market. None is an absolute cure-all, although one comes close.

In general, any special design can be more expensive than a standard control valve by a factor of two to five. For larger valves or for higher temperatures and pressures, the cost difference between custom-built conventional valves and the special designs may be very modest. There are many benefits to be gained from special designs, such as reduced maintenance costs, increased service life, improved control, and reduced noise levels. We have many cases where the entire cost of the special valve was absorbed by elimination of vent mufflers or line silencers or support steel, by reduced piping complexity, and vastly reduced space requirements. The point is, you can not balance valve cost against valve cost.

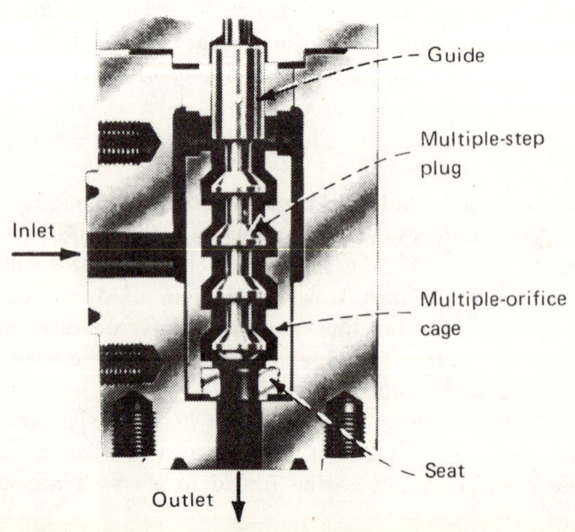

STAGED-TRIM valve distributes pressure drop—Fig. 4

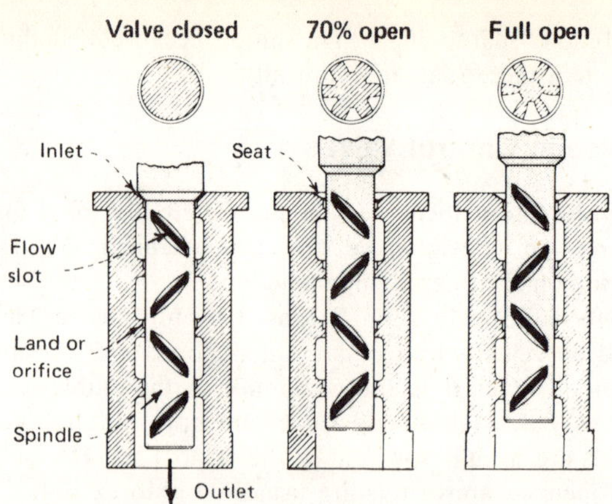

MILLED slots on plug provide staging—Fig. 5

Multiple-Stage Design

The first design is the staged trim (Fig. 4), which is a plug-and-orifice valve in a series or multistage configuration. There are several varieties available from various manufacturers for liquid service. One manufacturer also offers a version for gas or vapor service.

The virtue of the staged trim is that instead of taking the total pressure drop in one "fell swoop," it is distributed across the several plugs and orifices, or stages. These vary from two to a maximum of about six.

For best results, the pressure drop per stage should be limited to about 600 to 700 psi. Under these conditions, the velocity through each stage is considerably lowered. However, each stage still acts as a high-recovery valve, which generally means higher than desirable throat (or trim) velocities due to the efficiency of the nozzle-type trim. Therefore, staged valves are not immune to erosion or cavitation, although these are usually reduced in magnitude. Such valves are generally limited in size to 2 or 3 in. maximum, using pneumatic operators. This is due to power limitations as a result of their unbalanced or semibalanced design. They can be subject to vertical instability in the larger sizes. Flow coefficients, C_v, are on the order of one-third to one-half those of an equivalent-size standard control valve, and turndown, or flow-control rangeability, is much lower. These valves are large and massive for their pipe size and capacity, due to body-design requirements to contain the inner valve.

One manufacturer makes an interesting variation of the staged trim for liquids (Fig. 5). In this design, the spindle or plug is a straight cylinder, and the flow passages are obtained by milled slots or flutes that are cut at a 45-deg. angle to the spindle's vertical axis. As the spindle moves out of its corresponding land or orifice in each stage, its projected slot area is increasingly exposed and the flow-capacity area developed (Fig. 5). Further, the slots in each stage are angled 90 deg. to the preceding and following stages, so that flow direction from stage to stage is reversed. This aids in dissipating additional energy due to increased turbulence from the flow reversals. The efficiency of this reversal in making a low-recovery valve decreases as the valve plug moves toward the closed position, and the valve then behaves as a conventional staged-trim valve. In general, it has the same basic limitations as other staged-trim valves.

These various liquid-service staged-trim valves are commonly applied in high-pressure-drop services such as

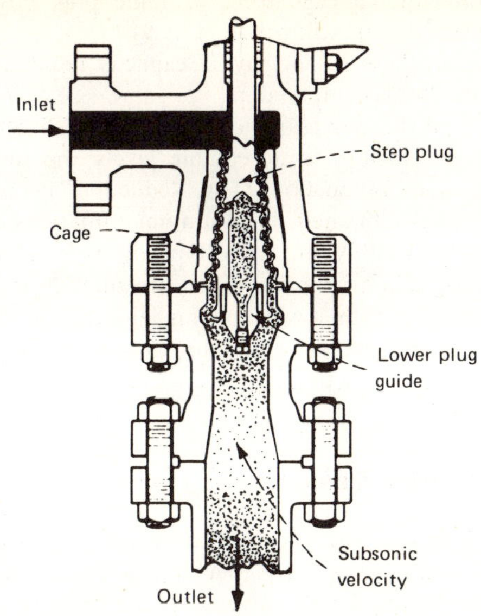

STAGED-TRIM valve for gas or vapor service—Fig. 6

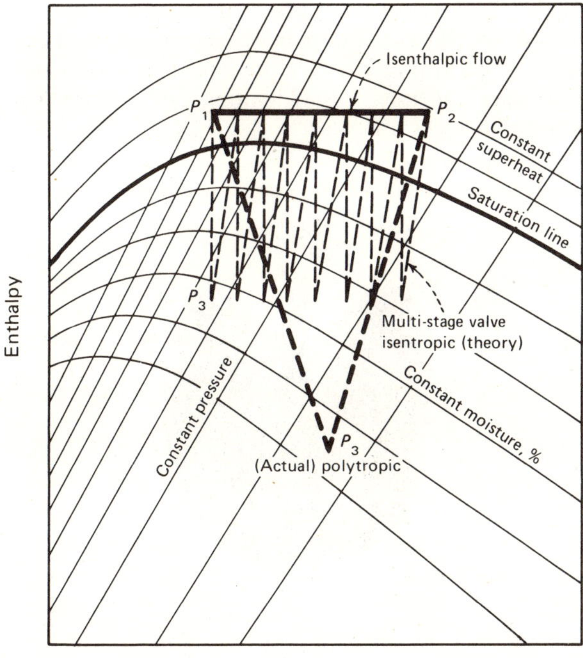

THROTTLING may be polytropic with Lanco trim—Fig. 7

boiler-feedwater-pump bypass and recirculation, desuperheater spray water, boiler-feedwater startup, and oil-separator levels. They can be considered for other services within their operating capabilities and size availability.

Only one staged-trim valve for gas and vapor service is manufactured (Fig. 6). This is the "Lo-Db," sometimes referred to as the "Lanco valve." Though in use in Europe, to the writer's knowledge none is in service in the U.S. The main differences between this valve and regular liquid-service staged-trim valves are:

1. The step plug has a Christmas-tree shape. The circumference and thus the flow area is gradually increased toward the discharge. This helps to compensate for the increase in specific volume as the pressure decreases, and helps to limit the increase in velocity buildup.

2. The step plug and mating-body lands have a semi-labyrinth pattern that provides some changes in fluid direction (similar to the liquid valve described previously). This can result in additional velocity-head loss.

The manufacturer states that the valve uses the principle of adiabatic flow with friction. The throttling process itself is adiabatic. The unique difference in a staged-trim vs. a plug-and-orifice trim is the degree of isenthalpic throttling achieved. The desired isenthalpic effect is that obtained in a long friction tube, like a capillary, but this is impractical. A short tube can be used if it is made to turn often and quickly.

A close look at the Lanco trim raises a question as to its efficiency as a friction device. It is probably more effective, in this respect, in the first portion of the plug's travel; and after that, it is doubtful if the fluid even knows the turns are there. What probably exists is multi-stage throttling, with each stage being isentropic initially. As the valve opens and the turns become less effective, the valve probably becomes polytropic (Fig. 7). In-service pictures have shown it heavily coated with frost and ice, which tends to confirm this analysis. The valve's effectiveness in noise reduction is more probably due to the geometry of the passage size coupled with limitation in the velocity buildup that is inherent in the trim shape.

These staged-trim valves are even larger and more massive for their pipe size and capacity than the liquid-service versions, due to the body-design requirement to contain the inner-valve geometry. To be effective, they are limited to a maximum pressure reduction of 4 to 1. Above this ratio, it is necessary to add restrictors in series to impose enough backpressure to limit the ratio to 4 to 1, or less. The necessity for fixed restrictors in this situation imposes limitations on flow turndown. This is because the restrictors lose their effectiveness as the flow decreases and quickly absorb available pressure drop as flow increases.

The valve has a very short stroke, and even though it has a partial hydrostatic-balance system, it must be operated with a ratio-multiplication lever system to obtain the necessary power and position sensitivity.

The manufacturer of the Lo-Db valve also supplies a unique device called the "Lo-Db Expansion Plate." This is a constant-area special-restriction orifice device that will handle part of the pressure drop in the system. One or more units in series can be used as needed (Fig. 8).

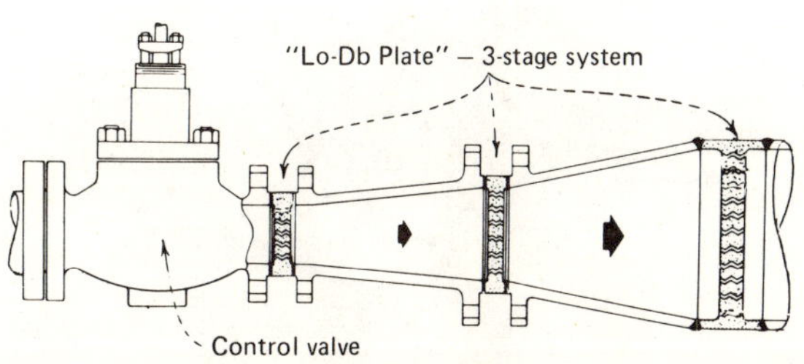

EXPANSION plates are multiple-orifice devices—Fig. 8

Such devices have a multiple-path geometry with some change in the flow-direction path. They not only act as restrictors, but also produce much lower noise levels than a larger single-path equivalent restrictor. Sound pressure level can be reduced as much as 20 dbA. The devices are effective only over a limited flow range. The vendor's guideline is that if $P_1C_v \leq 5{,}000$, these can be evaluated for throttling services. Above 5,000, they should only be used for on-off services such as free vents or system dumping. They can be an economical answer to some noise-throttling problems, either in conjunction with standard plug-and-orifice valves or to extend the pressure-ratio range of the Lo-Db valve.

Multiple-Path Design

The second design available is a conventional valve fitted with a special trim called the "Whisper Trim." This trim is for use with gas or vapor, and was developed by experimental and empirical means. The trim operates on the principle that a lower noise level is generated by a given massflow passing through a number of small areas rather than one large area. In other words, it functions in a manner similar to the Lo-Db plate except the multiple-path geometry is different, and the design allows flow modulation. The size and shape of the flow slots used for the multiple-path system have been developed by experimental testing (Fig. 9). The trim fits into the body like a conventional cage trim, with the plug position exposing the slot-area flowpath (Fig. 9).

In general, these valves have a capacity reduction of about 30 to 40% compared to the same size standard valve. It may be necessary to use oversized bodies to reduce exit velocities to acceptable levels and to gain optimum noise reduction. The reduction in sound pressure level obtained in optimum situations is a maximum of 20 dbA, but more typically is 10–15 dbA. This occurs at approximate sonic pressure drop. It is less effective at both lower and higher pressure ratios. Therefore, these valves should be considered only for moderate-noise-reduction and near-sonic pressure-drop applications.

Labyrinth-Disk Design

The third design is perhaps the most unique of all—a new approach to valve design, technology and operation. Called the "Drag Trim" valve, it is the most versatile valve currently available for severe-service and noise-reduction applications. It can be used for liquid, gas, vapor, or a combination of these, and is essentially immune to erosion, cavitation, wiredrawing, vibration and shutoff leakage. It does not generate noise and can be designed to give any level desired. It is a constant-impedance and true-velocity-limiting valve with actual isenthalpic letdown flow.

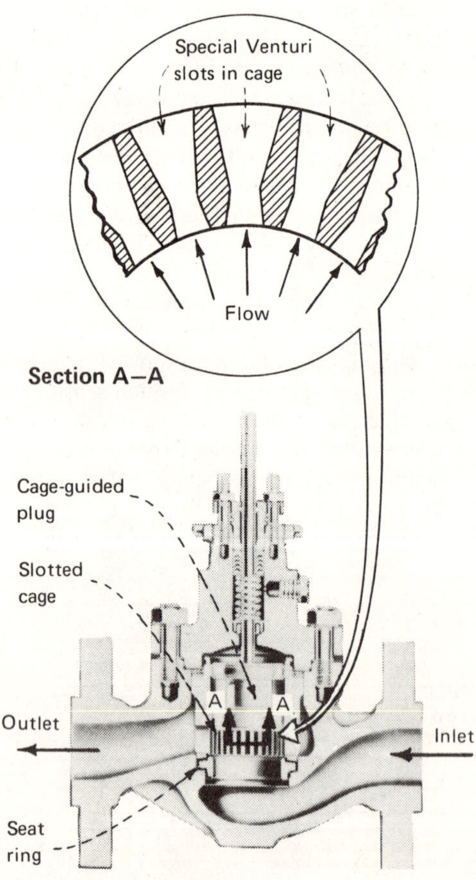

MULTIPLE paths generate lower noise levels—Fig. 9

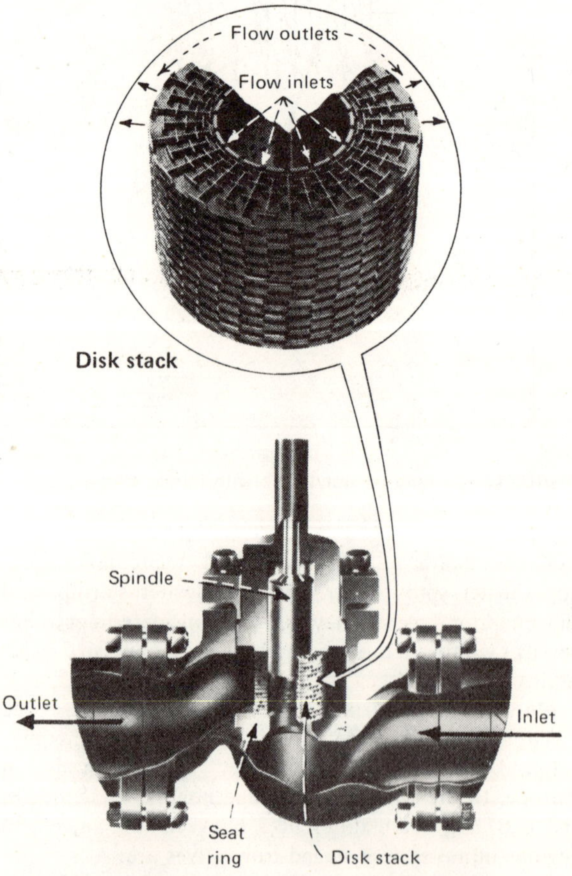

LABYRINTH design provides isenthalpic flow—Fig. 10

Basically, this valve consists of a stack of round disks with a hole in the center and a cylindrical spindle or plug in the hole. The individual disks are designed with a number of inlet passages, and each passage is a labyrinth that is actually a short friction tube consisting of a number of right-angle turns in series (Fig. 10). The number and shape of these turns are calculated and designed to be sufficient to dissipate the total required valve-drop at a velocity roughly equal to that at the inlet. Each disk has a specific flow capacity, so the required number of disks are made up in a "stack" to give the desired flow.

In general, the inlet area is about equivalent to the inlet-pipe area. Stack-discharge area is that necessary to handle the required flow, specific volume and desired exit-velocity.

The complete disk-stack is assembled in the valve body, similar to a conventional cage trim (Fig. 10). The spindle or plug positioning by the operator determines the exposed flow-area. This is inherently a linear-characteristic valve. It is relatively easy to characterize to other requirements by altering the disk-stack design and the configuration of the disks in the stack, or by using a cam positioner, or both.

In liquid service, the passage dimensions are the same from inlet to exit. In liquid flow with flashing, two-phase flow, gas flow and vapor flow, the passages increase in flow cross-sectional area after each turn. This accommodates the changes in specific volume while maintaining complete velocity control independent of pressure ratios. Another feature is the ability to provide continuous tight shutoff with the elimination of seat erosion. Bubble-tight shutoff up to about 500 F. with ultrahigh differential pressure can be obtained with an encapsulated Teflon® seat. Other designs are available for higher temperatures.

While the Lo-Db staged-trim design is limited to maximum pressures ratios on the order of 4 to 1 to achieve noise reduction, the Drag Trim valve is virtually unlimited in this respect. Services are in operation at 400 to 1, with valve-noise levels well under 85 dbA at a 3-ft. distance. Further, the design concept lends itself to a relatively efficient body configuration for the inner valve at any given pipe size and capacity. Since the disk-stack design has a greater height compared to the diameter, the long spindle stroke and ability to characterize disks can give turndowns of 100 to 1, or better.

However, utopia is not here yet. These valves are sensitive to line junk such as weld beads, slag, scale and rock. The disk stack does look like a good filter but it is not very susceptible to plugging from normal solids such as catalyst fines, dusts, slurry particles, etc. Generally, anything that can go in a passage will go through. In most cases, the passage size is larger than we might expect.

Line junk can cause jamming in the close-fitting spindle in the stack, where the flow direction is from the center hole to the disk's outside diameter. This orientation is required for flashing liquids, two-phase flow, gas, and vapor. For liquids, this problem is essentially eliminated because the flow direction is from the disk's outside diameter to the center hole. Here, the stack acts as its own filter and will tolerate a reasonable amount of debris without much effect on capacity or operation because there are two or three entrances into each disk passage. In any case, screens of adequate size and strength installed ahead of the valve are highly recommended. Of course, there may be situations where the system is known to be clean and free of line junk.

These valves have been used in the services previously mentioned, and in such tough applications as urea, diamine, ammonia, methanol and polyethylene letdown. Other applications include: high-pressure, steam, hot-process-gas, and natural-gas letdown and venting; and light-duty services such as moderate-pressure natural gas, steam and air at high massflow rates.

At this time, no other design can operate under high-velocity situations with such a degree of immunity to noise and destructive internal vibrations. For example, in a high-pressure liquid-letdown service, a custom-design angle-pattern plug-and-orifice valve with tungsten carbide trim has a service life on the order of 250 hr. A Drag Trim valve in the same service has had better than 6,000 hr. of operation to date with no problems.

Conclusion

There are several special-control valve designs or special devices for the tough, severe-service applications and noise problems. Each has its areas of usefulness and its limitations. Selection will depend upon a number of factors such as type of fluid, static and pressure-drop conditions, temperature, corrosion, and relative costs to benefits. All of these devices being specialized designs require more engineering evaluation of the application, the mechanical design and the metallurgy than is usually the case with conventional control valves that are used in many processes.

Illustration Credits

The illustrations in this article were contributed by the following firms: Fig. 10, Control Components Inc., Los Alamitos, CA 90720; Fig. 9, Fisher Controls Co., Marshalltown, IA 50158; Fig. 1, 4, 6, 8, Masoneilan International, Inc., Norwood, MA 02062; Fig. 5, Yarway Corp., Blue Bell, PA 19422.

References

1. Bulletin No. 340E, Masoneilan International, Inc., Norwood, MA 02062.
2. Bulletin TM-24, Fisher Controls Co., Marshalltown, IA 50158.

Meet the Author

J. B. Arant is a senior instrument specialist in the Design Div., Engineering Dept. of E. I. du Pont de Nemours & Co., Wilmington, DE 19898. He is presently responsible for instrument design for plant projects and is involved with special control-valve problems. He joined Du Pont in 1950 at the Sabine River Works in Orange, Tex., as a division engineer in the instrument department. He has a B.S. in chemical engineering from the University of Texas and is a registered professional engineer in Texas.

Section IX
LEVEL, PRESSURE AND TEMPERATURE INSTRUMENTS

Sighting in on level instruments

A full range of level instruments—from capacitance probes to gage glasses—are described, their advantages and limitations cited, and costs noted. Also discussed are control modes, and calibration and mounting methods.

Leonard M. Wallace, *Fluor Engineers and Constructors, Inc.*

☐ The many applications for level instruments have resulted in a wide variety from which to choose. Proper selection depends on such things as: application, cost, reliability, serviceability, safety, the caliber of technicians, delivery time and dependability, quality of service from the local representative, the additional amount of parts to be stocked, and standardization policy.

Selection also depends on the type of vessel, what is to be measured (i.e., height of solids, granules, or liquids, or a liquid/liquid or liquid/foam interface), process conditions (i.e., pressure, temperature, specific gravity, boiling point, viscosity and pour point), what the instrument is to accomplish (monitor, on-off or modulating control, or alarm), and whether the signal is to be electronic or pneumatic.

For a level instrument to help smooth out the process by providing a fairly constant flowrate downstream, its vessel must have adequate surge capacity. Surge time depends on flowrate and control-valve size. The following are typical surge times:

Type of service	Full volume surge time, min.
Feed from storage	5-10
Primary feeding secondary column	10-20
Column bottoms to storage	1-2
Reflux accumulation	5-10
Knock-out drum to drain	1-1½

Surge rates vary depending on the type of service and the composition of the feed. A feed of fixed composition needs less surge time than a variable one.

The instrument range required depends on the type of service, surge time required, vessel diameter and maximum flowrate. Although the range selected may be sufficient to cover the required surge time, the rate of level change may be too fast for proper control. Because of this, vessel diameter is important to proper control. As a rule of thumb, the ratio of maximum flowrate (in^3/s) to liquid surface area (in^2) should be less than 4.5 in/s [2]. If the level changes at any faster rate, increasing the range can cause instability, because the level will be changing faster than the instrument or control valve can respond.

The rate of change in vertical vessels is, of course, uniform. In horizontal vessels, it varies and is greatest at the extremes of the range. The control valve will be wide open, and the flowrate highest, at the top of the range when the flow is out of the vessel; and at the bottom when the flow is into it. In a vertical vessel, then, the rate of change $=$ gal/min $\times$ $4.90/D^2$, with D in inches. If the vessel is horizontal, the rate of change$=$gal/min $\times$ $3.85/W \times L$, with width and length in inches. If the rate is faster than 4.5 in/s, the diameter of the vessel should be larger.

A control valve should always operate from fully closed to fully open faster than the full-volume surge time; otherwise, it will cycle. A typical operating time for a 2-in valve is 15 s. The valve will not cycle as long as the surge time is greater than 15 s and the level rate of change is less than 4.5 in/s.

Because most calculated ranges are not standard, the next-greater range is selected. However, one must be careful not to widen the range beyond critical low- and high-level limits. The low limit is generally taken as the minimum level for proper pump net-positive suction head, and the upper limit must be below seal pans or vapor outlet lines. The process and instrument engineers should work closely together to establish the proper range.

With this done, and other variables determined, the choice of instrument is narrowed. Selection then depends on capital expense, maintenance cost and reliability. Capital expense includes such costs as those for the instrument, accessories (fittings, valves, pipings and tubing), installation and calibration. The table affords some idea of what is involved for a variety of instruments.

Originally published February 16, 1976.

Instrument summary—from level controllers to level indicators

Type	General	Applications	Advantages	Disadvantages	Limitations	Cost range
Displacer transmitters and controllers	Available in all materials; body flanges per ASA ratings; top or side mounted or furnished with a cage; electronic or pneumatic	All fluids and wide range of sp. gr; wide pressure and temperature range; interface service	Easy to put into service and adjust; reliable, relatively maintenance free; wide range of applications	Range usually limited to 120 in, but mostly to 72 in; long cage assemblies hard to handle and must be plumbed; cage takes up space; mounting cost high; high pressure can crush displacer	Ranges from 14 in to 120 in; to 850°F and 2,500 psig	\$500 – \$2,000
Displacer level switches	Same as ball float switch except for solid displacer	Same as ball float switches; also where a wide level differential required between starting and stopping equipment, usually by tandem float arrangement	Simple, reliable and low cost; wide ranges	Only for on-off services	Range limited to length of cable; generally to 30 ft; to 500°F and 10,000 psig	\$150 – \$400
Ball floats, direct actuated	Available in variety of materials; ball can be mounted inside vessel or in external cage	Low pressure vessels not requiring local or remote indications and non-critical service	Low cost, simple, relatively maintenance free; measures level directly	Limited; narrow range; dificult to mount because of linkages; high pressure can crush ball; limited pressure drop across valve; internal type requires downtime for maintenance	Limited range of about 30 in; to 600 psig	\$200 – \$1,200
Ball float, pilot operated	Basically same as direct actuated, except pneumatic pilot on float housing allows controlling a remote diaphragm operated control valve	Same as direct actuated, except valve can have higher pressure drop; internal float type also used on viscous service by adding weight to ball	Relatively simple and reliable; pilot can operate control valve with high unbalance forces	Limited applications; narrow range; high pressure can crush ball	Limited to about 30 in; to 600 psig	\$300 – \$1,400
Ball float level switches	Mounted internally or in external cage; can be furnished in variety of materials, with single or multi-switch, and fins for high temperature	Low to high-pressure vessels to activate alarms, start and stop equipment for all types of fluids; for interface service by adding weight to ball	Relatively simple, reliable, and maintenance free; low cost	Only on-off service; high pressure can crush ball; limited range	To 2,000 psig, 750°F; down to .35 sp. gr.; differential range to 132 in with special floats	\$150 – \$400
Hydrostatic head, pressure	Can be provided as pressure gage, switch or transmitter; last, pneumatic or electronic	Open or vented tanks by static or bubbler system; for corrosive liquids, lead line can be sealed	Simple, reliable, low cost, easy to maintain; wide range	Only for vented tank unless pressure compensated; must have seal or purge with many fluids	Unlimited range and temperature because pressure lead is static	\$50 – \$500
Hydrostatic head, differential pressure	Generally furnished as transmitters or local differential indicators; transmitters available for pneumatic or electronic signals; local differential pressure switches are also available	Low to high pressure vessels for all ranges over 80 in, but can go down to 5 in; for interface services when sp. gr. difference and measured range are large enough	Simple installation; wide range, for variety of fluids and high pressures	Need sealing fluid or purge, calibration more complicated than for displacer instrument	To 6,000 psig and 200 ft; unlimited temperature because leg is static	\$100 – \$600
Capacitance	Wide selection of probes for non-conducting or conducting liquids and granular solids; provided as a switch or for continuous control	Silos, bins and low to high pressure vessels to measure powders, grains, pellets, sewage, food products, coal, sawdust, hydrocarbon products and other liquids and granular solids	One vessel connection normally required; simple installation; can be mounted in cage assembly; inexpensive	Hard to calibrate unless in cage assembly; probe can become coated and change calibration; dielectric or temperature change causes error in output	To 100 ft; to 500°F and 1,000 psig; special probes available to 1,000°F for non-conducting materials	\$150 – \$800

External instrument mounting is preferred, unless the instrument must be mounted internally for process reasons, or unless the vessel can be taken out of service to calibrate the instrument.

A backup instrument may be necessary when a gage glass is not permitted on a vessel. Some means of determining level should be included other than the one instrument.

Instrument mounting methods

An instrument's sensing element can be mounted internally, or in an external cage connected directly to the vessel or on a standpipe (Fig. 1). Elements are generally mounted internally for highly viscous fluids, batch operations, or when the vessel can be taken out of service without shutting down the process.

A displacer and a top-mounted float usually require a stilling well to prevent erratic operation (Fig. 2). The well should have holes drilled in it, or a slot along its length, to keep materials from stagnating (in interface service this is necessary). This type of installation reduces cost but can increase maintenance problems.

An external cage assembly is used whenever possible because it lets the instrument be taken out of service for maintenance or calibration. The cost is higher because of the additional isolation valves and fittings. Some of the vessel connections may be eliminated by installing a standpipe.

A standpipe reduces vessel connection cost. Rather than install several different instruments to a vessel,

Instrument summary—from level controllers to level indicators (continued)

Type	General	Applications	Advantages	Disadvantages	Limitations	Cost range
Conductivity	Good selection of probes from one to three elements can be set at various depths; can actuate alarms or other electrical devices	Any liquid that conducts; for oil-water interface, water level alarms on boiler	Simple; low cost; trouble free; easy installation; low maintenance; no moving parts	Limited to on-off control and to non-hazardous liquids	Approximately 60 in, 600°F, 500 psig	\$100 – \$300
Nuclear	Available as a level transmitter or switch; wide range of source materials available	Low to high pressure vessels; coking towers or where coking is problem; on catylist hoppers and other services in which element should not contact material	Neither source nor detector contacts material; can be used on solids or liquids; no moving parts	Difficult to calibrate and to get linear signal unless source size and location are exactly correct; limited range; potential safety hazard	Unlimited pressure and temperature; range limited to number of sources and length of detector	\$800 – \$3,000
Sonic	Variety of probes available; provided as single, dual and gap type; continuous control output signal or on-off switch	Elevators, silos, and bins for chemicals, pellets, powders, paper, coal, liquids and interface service	Low cost; simple installation; accurate, no moving parts; probe does not have to contact material	Variations in density cause errors	Standard probes to 3,000 psig; –100°F to 300°F; special probes from –450°F to 1,000 °F at 1,000 psig	\$130 – \$1,500
Tank gage, float	Indication can be local board, ground reading head, or remote console; electric switch contacts available, as are variety of float and tape materials	Liquid storage vessels below 50 psig; tank signals multiplexed to central console	Gage board is simple, relatively low cost, can be seen from distance; ground reading head can operate at higher pressure than gage board; transmitters or switches can be mounted on unit	Gage boards limited to few in pressure; local indication; ground reading head local readout cannot be seen unless person at head; temperature must be measured; tape causes mechanical problems	To 140 ft. 50 psig; high accuracy	\$500 and up
Tank gage, strain gage transducer	Readout in total weight of content but can be liquid height by adding temperature measurement	All liquid storage vessels, low to high pressure; signals can be multiplexed to central console	No moving parts; simple installation; no temperature measurement required	Purging or seals required	Unlimited pressure and temperature	\$3,000 and up
Visual gage, tubular	Readily available; also in variety of materials, and with special glass protectors; furnished with gage cocks and guard rods	Low pressure vessels containing non-flammable or toxic fluids	Low cost and simple to install; externally piped and valved for online repair	Glass easily broken; limited applications	72 in; to 350 psig	\$70 – \$200
Visual gage, reflex	Readily available; furnished with gage cocks; avialable with heating or cooling jackets and frost preventive window; purging possible	All liquids; from low pressure to high pressure vessels	Low cost, simple to install, reliable; easily seen from distance; externally piped (except pad type)	Glass can be broken, but more durable than tubular; liquid cannot be seen; not for interface service or with dark, viscous liquids	5-section glass; 69 in visibility; 3,000 psig at 100°F, or 1,700 psig at 750°F	\$50 – \$1,000
Visual gage, transparent	Readily available; generally furnished with gage cocks and illuminators; available with heating or cooling jackets, and in welding pad gages; purging possible	Interface service or when fluid color or clarity visibility desired; low and high pressure	Low cost; simple to install, reliable; see interface level; externally piped and valved for online repair	Glass can be broken, but more durable than tubular; needs electric power with illuminators; liquid level harder to see than with reflex type; with pad type, vessel must be down for maintenance	5–section 69 in visibility; 2,500 psig at 100°F and 1,420 psig at 750°F; special gages for up to 10,000 psig	\$1,200
Visual gage, armored	Available for top or side mounting in variety of materials; steam jackets and alarm switches also available	Low to high pressure vessels containing toxic and flammable fluids; also for interface service	Safe because there is no glass; level easily seen from distance; alarm switch can be furnished and can be externally piped	Requires larger vessel connection than standard gage glass; cannot see actual liquid	Standard is 6 in to 36 in; multi-units available to 10 ft, to 10,000 psig, and to 600°F	\$600 –\$1,400

they are connected to a standpipe, which requires only two vessel connections. Although more common on horizontal vessels, it can also be used on vertical ones. Connections to this piece of pipe are much less expensive than they would be if made to a vessel.

Instrument calibration

Ease of calibration is important in instrument selection. Generally, an instrument should be bench calibrated before installation. It should also be possible to calibrate an instrument in place, rather than removing and taking it to the shop. Therefore, some means of calibration should be provided. The main advantage of an external cage assembly is that is permits isolation from the vessel for calibration in place.

The simpler the calibration equipment, the better the results. Water, for instance, can be used to calibrate and set a displacer instrument; and a simple squeeze bulb for setting d/p cells. Radiation instruments, on the other hand, require sheets of lead equal to the absorption rate of the material to properly calibrate them.

The backup instrument, calibrated before put into service, can be used to calibrate the operating instrument by varying the level and making appropriate adjustments. Usually, it is not possible to run the level from zero to 100% during operations. As long as the instrument is linear, however, it can be calibrated by changing the level (only a small amount, so as to not upset the process) with the following equations:

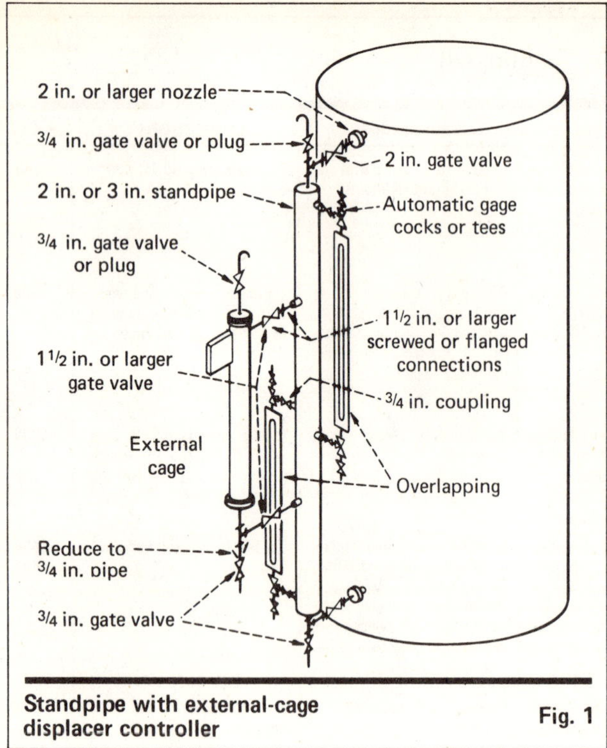

Standpipe with external-cage displacer controller Fig. 1

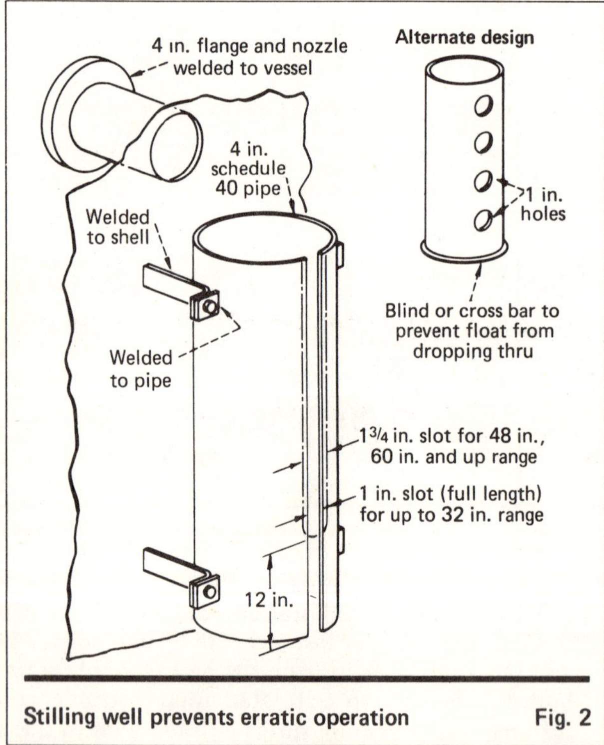

Stilling well prevents erratic operation Fig. 2

If the instrument is a 4–20-mA electronic, then output (mA) = [16 mA × *d*/range(in)] + 4 mA; if it is a 3–15-psi pneumatic, then output (psi) = [12 psi × *d*/range(in)] + 3 psi (*d* = distance from bottom of range to the measured level, in).

If the range is 32 in, and *d* = 10 in, the output should be 9 mA. If it is not, the zero should be adjusted to get the 9-mA output. If the upper level is at 20 in, the output should be 14 mA. If it is less, the span should be increased; if more, decreased. The operation should be repeated between any two points until the instrument is calibrated. The wider the two levels, the more accurate the results.

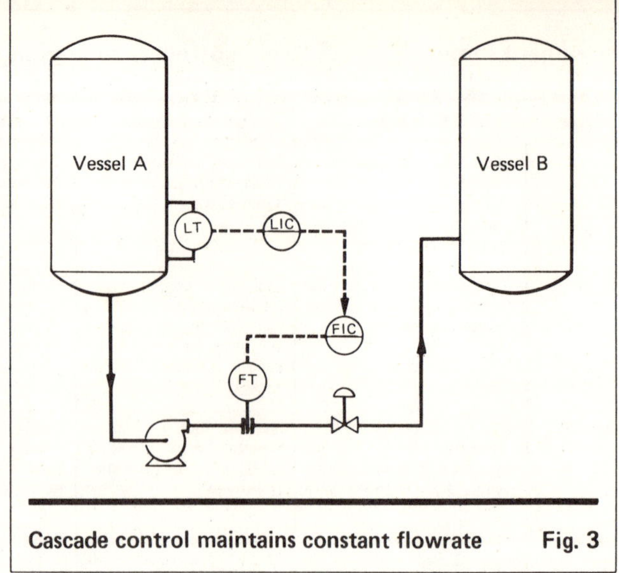

Cascade control maintains constant flowrate Fig. 3

Level-control modes

The type of control needed should first be determined. If on-off control is all that is necessary (such as for makeup to a storage tank), do not install a more-elaborate instrument.

Proportional control—This alone is required in most level applications. Usually, the level can be allowed to fluctuate, depending on load conditions. The larger the surge space, the smoother the flow from the vessel. However, cost usually limits surge capacity. The least capacity is that which prevents cycling. Optimum capacity depends on where the liquid is being discharged. If discharged to a tank or sewer, or if the flow is a small one that combines with other flows, minimum surge is adequate. If, however, one vessel discharges into a second that requires a fairly constant feedrate, a wide range is desirable.

A proportional controller can be mounted locally or on the main control panel. The problem with the latter is that the setpoint will seldom be at the process level because of load changes. Although this really does not do any harm, it can upset the operator. If he sees the level deviate considerably from the setpoint, he may put the instrument on manual or readjust the setpoint. In such instances, the instrument man will generally narrow the proportional band, and so reduce the surge capacity. Because this will cause wider fluctuation downstream, it may be better to install a local level controller, or to add manual or automatic reset.

Proportional plus reset—Reset rate can be added to bring the level back to setpoint after a load change; however, the reset rate should be very slow (5 to 20 min). This way, the surge capacity is still used, but the level will return gradually to the setpoint after a load change. In most cases, reset is used to maintain a fixed level, such as that in a reboiler whose tubes are 100% covered.

Differential-gap controller—This is generally used on knock-out drums in which liquid accumulates slowly, and so a proportional controller is not required because the flow is small. The small flowrate could cause a valve with a small port to become plugged. With the gap controller, however, the level is allowed to rise until the upper limit is reached. At this point, the valve, which can be oversized to prevent plugging, is fully opened. The level then drops to the lower limit. At this point, the output changes and the valve closes.

Cascade control—If the pressure should change in vessel A or B of Fig. 3, the flowrate will change. Small suction or discharge pressure-changes of a centrifugal pump can alter the flowrate considerably. If a fairly constant feedrate to vessel B is desired, the flow controller should be set by a level controller. The flow controller will maintain a flowrate based on the setpoint level signal; and if the flowrate starts to change because of a pressure change in either vessel, the output of the flow controller will vary to open or close the control valve.

Displacer-type instruments

Displacer level instruments use a cylindrically shaped float whose length corresponds to the specified measurement range (Fig. 4). In service, the actual movement of the float is small. The buoyant force exerted on the float (resulting from the liquid displaced by the partially or wholly submerged float) causes it to operate the instrument's pilot mechanism. This mechanism may be an integral part of either a level signal transmitter or a controller whose output signal positions a remotely mounted control valve.

The buoyant force varies directly with the density of the fluid. Consequently, the instrument must have a compensating specific-gravity adjustment to permit calibrating the transmitter or controller for the measured fluid conditions.

Displacer transmitters are available with either pneumatic or electronic output signals for operation of remote controllers, indicators and alarms. The transmitter output increases in direct proportion to the height of the liquid level in the vessel.

The sensitivity of displacer instruments is sufficiently high to permit inferential detection of an interface between two immiscible fluids having different specific gravities. In such applications, the float operates totally submerged, with the lighter fluid surrounding the upper portion of the float and the heavier fluid covering the lower part.

In some cases, the displacer float assembly is installed inside the process vessel through a flanged opening, with only the pilot mechanism housing outside. Most process conditions permit the displacer float assembly to be housed in a float cage that is externally mounted and piped to vessel connections. This arrangement can be valved in such a manner that the instrument can be isolated and serviced or replaced while the vessel remains in operation.

Although an externally mounted displacer instrument is more conveniently serviced, it has some disadvantages, such as the need for it and the connecting piping to be heated either when the fluid is viscous, or when the vessel operating temperature is high (so that the specific gravity of the fluid in the float cage will equal, or approach, that inside the vessel).

When both water vapor and hydrocarbons are present (such as in steam stripper columns), there is a danger of droplets of condensed water vapor falling onto the surface of hot hydrocarbon liquid inside the external float chamber. The resultant boiling can cause

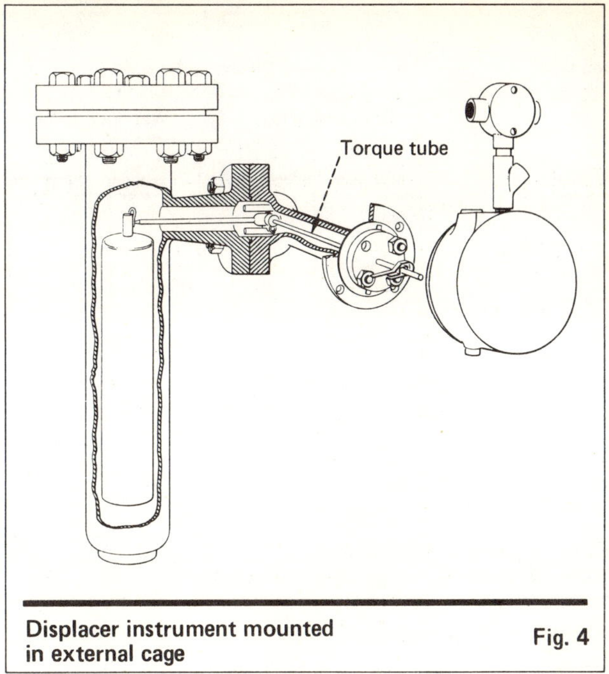

Displacer instrument mounted in external cage — Fig. 4

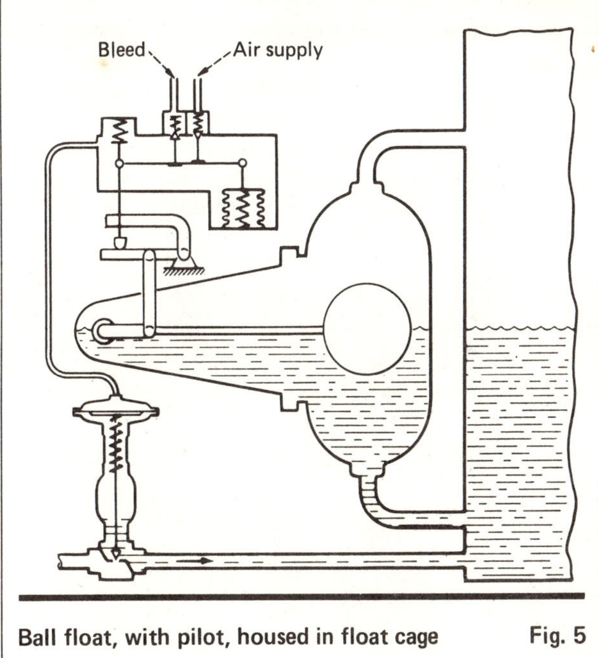

Ball float, with pilot, housed in float cage — Fig. 5

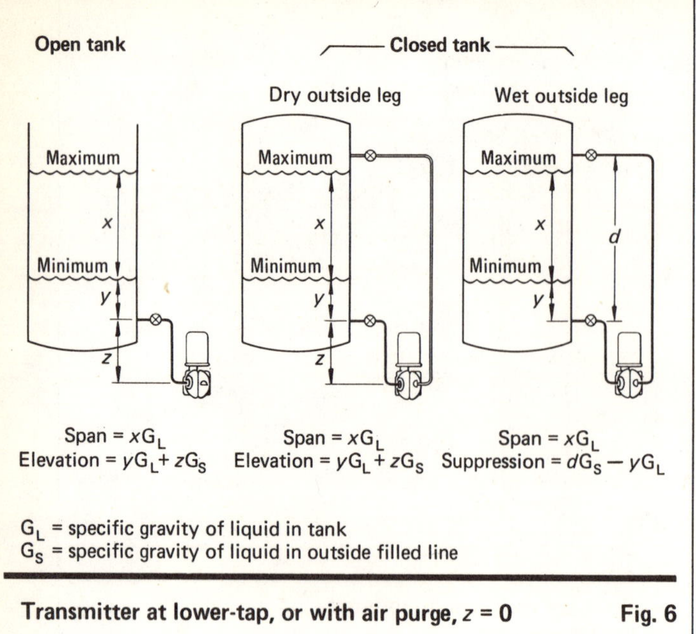

Transmitter at lower-tap, or with air purge, $z = 0$ Fig. 6

severe surging of the float, and unwarranted instrument response. A continuous flow of purge gas into the vessel through the upper-level instrument connection may be necessary to keep significant amounts of water vapor from entering the float chamber.

Some applications may require introduction of a continuous flow of purge liquid to dilute or limit the amount of vessel fluid that enters the external float chamber. Also, a flushing liquid may need to be piped to the external displacer-float assembly to enable periodic flushing of the float chamber and its piping. Fluids that have a tendency to coat metal parts or to solidify, or that contain suspended materials that settle out, could be handled in this manner.

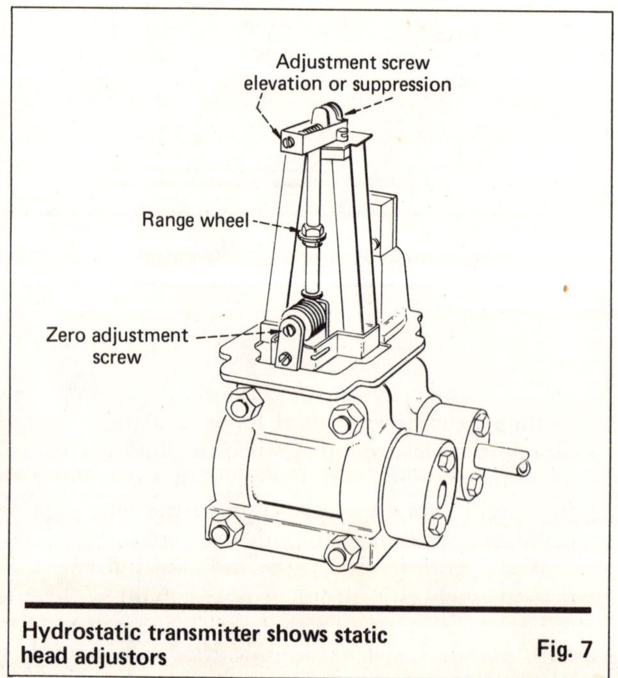

Hydrostatic transmitter shows static head adjustors Fig. 7

The cost of mounting a displacer instrument externally is usually greater than installing it inside the vessel. A decision must be made as to whether this cost is worth being able to service the instrument without shutting down the process.

Ball-float level instrument

In this instrument, a ball float rides on the surface of the liquid. As with the displacer instrument, the ball float may be housed in a float cage for external mounting (Fig. 5). Used less frequently than the displacer type, the ball float has measurement ranges that are narrower because of practical limits to float-arm length. The ball float may have to be installed inside the vessel for coking services, and when the fluid is waxy or contains suspended solids.

Ball floats may be designed to provide sufficient power to directly operate a control valve, which may be connected to the float arm by mechanical linkage. Such directly operated controllers are usually restricted to noncritical utilities services, when the vessel is at atmospheric condition.

Hydrostatic-head instruments

Other conventional instruments for inferential liquid-level measurement are actuated by the hydrostatic head of liquid inside the process vessel. From a knowledge of the operating specific gravity of the vessel liquid, the hydrostatic pressure may be readily converted into units of liquid height above a reference elevation inside the process vessel.

On vessels that operate at atmospheric pressure, a simple pressure gage, transmitter, or controller may be connected to the side of the vessel, near the bottom, to sense this hydrostatic pressure.

On vessels that operate at elevated pressures, a differential-pressure instrument is frequently used. Its low-pressure side is connected to the vapor-space section of the vessel. This is necessary in order to compensate for vessel pressure. The high-pressure side is connected to the lower section of the vessel to sense the hydrostatic head of the content above a reference elevation. The instrument is usually mounted below, or at the same elevation as, the lower-vessel connection (Fig. 6).

In addition to a lower installed cost (compared to the valved, external displacer instrument), these pressure sensing instruments offer other advantages over the displacer float type:

1. When the vessel fluid is a slurry (or of similar nature) that would settle in the external piping. (Pressure sensing leads often need to be continuously purged with a clean fluid.)

2. When agitation of the vessel liquid would cause a float-operated device to bounce.

3. When the required range becomes excessively long for a displacer float. (Although available in longer lengths, displacer floats are often restricted to ranges between 5 and 8 ft. Depending on the nature of the process, a differential-pressure instrument is sometimes used down to ranges of 20 in or lower.)

When the instrument is located below the lower-vessel connection, the low-pressure lead line (usually called the outer leg of the instrument) is filled with liquid (often the same as that inside the vessel). Some applications require that the low-pressure lead line be kept filled with a sealing fluid. In either case, the differential-pressure instrument must be equipped with an adjustment to compensate for the static head of liquid in the low-pressure lead line (Fig. 7).

Differential-pressure instruments for level measurement are usually transmitter types, which provide a continuous output signal that varies linearly with the height of fluid throughout the calibrated range. This signal (pneumatic or electronic) operates remote receiver-controllers, indicators, alarms, etc.

The transmitter is also available with a corrosion-resistant sensing diaphragm. This type is usually mounted on a flanged nozzle at the bottom side of the process vessel, often without valving. Vessel contents are in direct contact with the outside surface of the diaphragm. This type has particular merit when vessel contents either are corrosive or would solidify in external lead lines.

These instruments are also suitable for measuring an interface between two immiscible liquids. For such service, the range of measurement selected must fall within practical minimum sensitivity limitations of the instrument. This is a function of the difference in the specific gravities of the two liquids.

Ball-float and displacer switches

Ball-float level switches are used on all types of vented and pressure vessels, for high and low alarms, starting and stopping pumps, and shutting down compressors when the liquid level in the suction drum becomes too high.

The ball actuates a switch by one of several methods—one of the most popular being a magnetic sleeve attached to the ball by a rod, either directly or via a pivot. The sleeve moves up and down inside a pressure-tight, nonmagnetic tube. When the level rises, a magnet on the outside of the enclosing tube is attracted to the sleeve magnet, which causes the outside magnet to swing against the enclosing tube. This movement activates a single-pole, double-throw switch (which can be either a mercury bottle, microswitch or a pneumatic switch), to allow actuation on a rising or lowering of the level (Fig. 8).

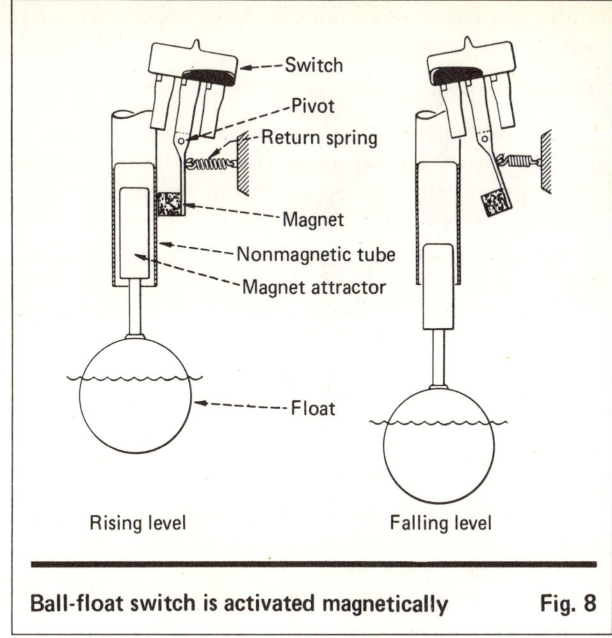

Ball-float switch is activated magnetically Fig. 8

To get high pressures, the ball can be pressured up with a gas; and cooling fins can be added to keep the switch head cool when high temperatures are involved.

Displacer switches can be used in some of the same services as ball-float switches. However, the displacer switch is most common where wide ranges are required. It is frequenctly used to start and stop one or more sump pumps. It can be adjusted to be located at any desired distance on a cable. Stilling wells are provided when conditions are turbulent. Displacer switches are available in stainless steel, porcelain and other mate-

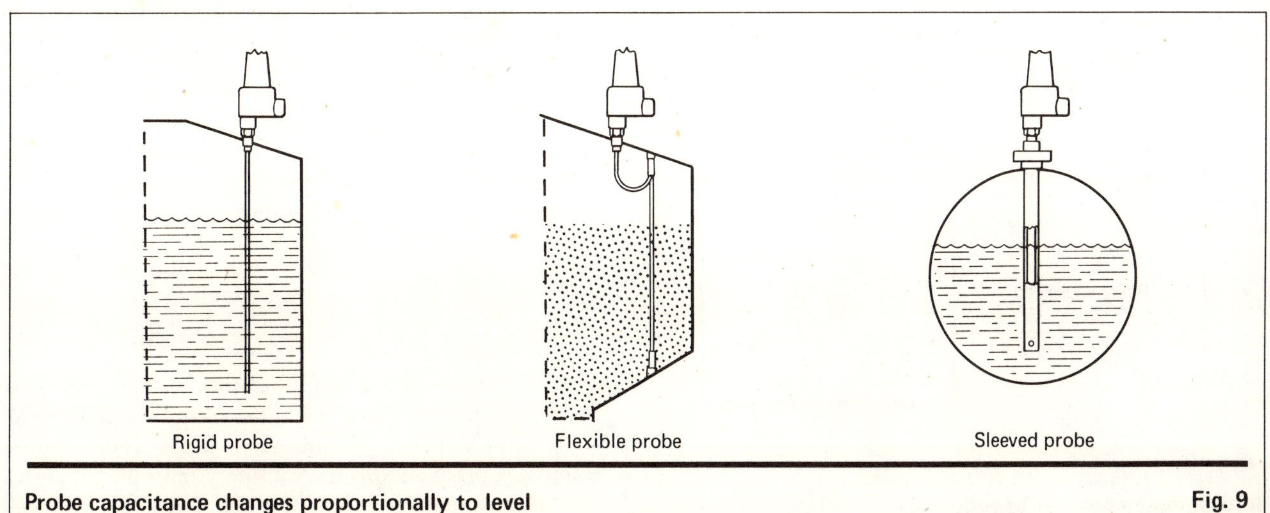

Probe capacitance changes proportionally to level Fig. 9

rials. The cable also comes in materials to meet many service requirements.

Capacitance instruments

Material surrounding a special probe alters the probe's capacitance in proportion to the level. All capacitance instruments depend on this principle to measure and control liquid or granular-solid levels.

The capacitance-probe instrument works with nonconducting materials, or with conducting materials that do not coat the probe. Probes are available insulated or uninsulated. The former are used for conducting materials and liquids.

The information needed to select the right probe includes:

- Whether the probe should be sheathed or unsheathed.
- Inactive capacitance of the probe.
- Nominal length of the probe.
- Length of probe that will be covered by liquid.
- Capacitance when vessel is empty.
- Capacitance span.
- Dielectric constant of the material.

There are three basic types of probes: the rigid, the flexible, and the sleeved (or sheathed) (Fig. 9).

The rigid probe is normally Teflon insulated, and thereby suitable for liquids or granular solids, whether conductive or nonconductive. A bare (noninsulated) probe is used with nonconductive materials having low dielectric constants.

The flexible probe is designed for applications that have a range wider than 12 ft. Because the electrode is a flexible wire, it can be coiled, which simplifies shipping and installation. The flexible probe is generally used on bins, silos and hoppers. The sleeved (or sheathed) probe's design is similar to that of the rigid probe, except that a metal tube surrounds the electrode. This provides linear capacitance change without the probe being affected by the shape of the vessel wall. This probe is used only with liquids.

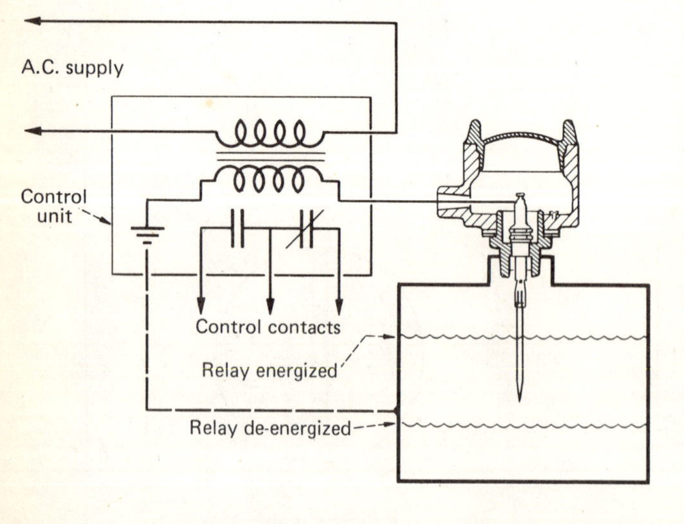

Rising level completes circuit to vessel wall **Fig. 10**

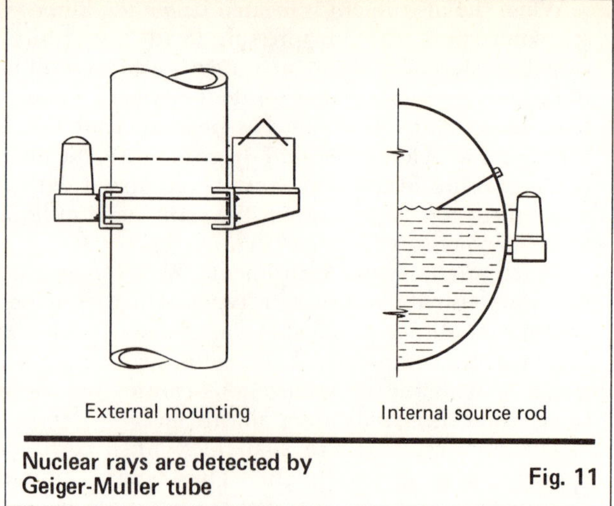

Nuclear rays are detected by Geiger-Muller tube **Fig. 11**

Conductance level switch

The conductance level detector operates on the principle that when a conductive material rises and touches a probe electrode an electrical path will be completed to the vessel wall. The shorting of the probe causes a relay in the control unit to be actuated (Fig. 10). The contacts on the relay can activate alarms, or operate valves, pumps and electrical equipment. These probes generally are used at pressures below 500 psig and temperatures to 600°F; however, special probes can serve at higher operating conditions.

Nuclear level instruments

Nuclear level instruments measure by gamma rays that are detected by a Geiger-Mueller tube. A radioactive source, such as radium or cobalt 60, is placed so that the vessel contents are between the source and the tube (Fig. 11). When the vessel is empty, the count rate is high; it decreases as the level rises.

The strength of the source at the tube depends on the density or thickness of the material in the vessel, the distance between the source and the tube, and the thickness of the vessel wall and insulation. The range of the instrument is limited by the size of the source. Too large a source can be dangerous and impractical to handle. Multiple sources are used to measure a wide range.

A source is held in a container of lead, housed in steel. The housing is locked to prevent tampering with the source. A special cover is opened and closed to turn the radiation gamma rays on or off. When the vessel's diameter is large, the source must be located in a well inside the vessel.

The design of the source container, the size of the sources, and their handling must comply with state and the U.S. Nuclear Regulatory Commission requirements. A plant should have a safety man who knows all the requirements and who can perform safety checks.

Nuclear instruments are used where other types of

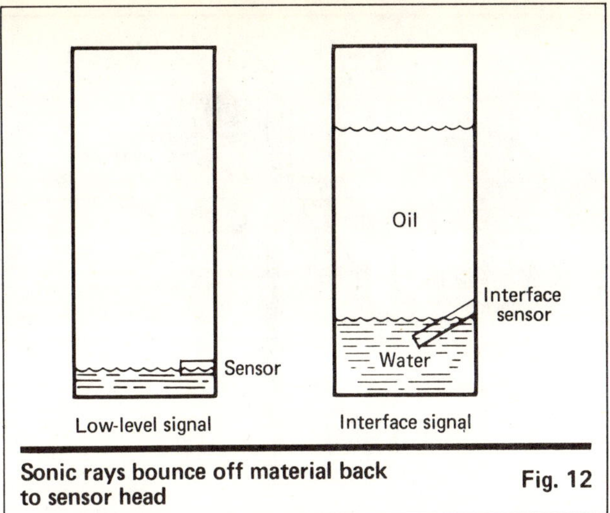

Sonic rays bounce off material back to sensor head Fig. 12

internal or external instruments cannot be, such as on coking towers, vacuum towers, and high-pressure or high-temperature vessels.

Sonic and ultrasonic instruments

These function similarly to naval sonar. An ultrasonic wave above 20 kHz, transmitted from a transducer-sensor head, bounces off the material being measured back to the sensor head. The instrument measures the time it takes for the wave pulse to leave and return (as an echo) to the sensor head. This time is proportional to the distance from the head to the material. The instrument is calibrated to convert time measurement into distance. It can function as a level switch or continuous level monitor (Fig. 12).

The range of the instrument is from a few inches up to 100 ft. The transducers are provided generally with a threaded or flanged connection. Some applications are in measuring liquid, solid, slurry and powder levels, and liquid-liquid and liquid-foam interfaces.

Measuring level by weight

Many times, fairly accurate measurements of the level of toxic or dangerous contents are needed, and the only vessel connections permitted are those to the process, with none allowed for instruments or venting.

One means of measuring such contents is by weight. From one to four load cells can be placed under the vessel—the number depending on the accuracy desired, the weight of the tank and the tank contents. Inlet and outlet lines must be flexible so that forces are not exerted on the vessel.

There are two types of load cells, one the hydraulic, the other the strain gage. Either can be equipped with a local readout or transmitter, or both. The dead weight of the vessel can be suppressed or tared out, and the indicator or recorder can be calibrated in any units, such as percent, feet or weight.

Tank gages

The most common tank gage is the float type (Fig. 13). The float is attached to a metal tape having holes exactly spaced, and the tape runs over pulleys at the top of the tank and down to a ground reading-head, where it passes over a spiked wheel (counterdrive sprocket). The excess tape is coiled up on a storage sheave. As the level changes, the tape is pulled out or taken up. The counterdrive sprocket turns over a set of digits that read out in feet, or in other units.

A transmitter can be attached to this assembly, and a remote selector may be used to choose any of several tanks. This system reads level very exactly. However, because a temperature change will alter the specific gravity of tank contents, the temperature of each tank must also be read out to establish the tank's true content. These gages require considerable maintenance.

Another type of tank gage simply measures head pressure by means of a strain gage rather than a standard d/p cell. These gages are generally calibrated to read out in pounds of liquid, but readouts in gallons and barrels are possible.

This type of gage has been used in the chemical industry, but not extensively for the storage of petroleum products. However, there is no reason that it should not be used more extensively, as it is simple and requires little maintenance.

Tubular gage-glasses

These gage-glasses are generally used on low-pressure vessels that do not contain toxic or flammable liquids. The glass is Pyrex, of ⅝ in or ¾ in dia. and up to 72 in long. A special gage-cock with stuffing box assures a positive seal and prevents stress on the glass when the packing nut is tightened. The body is generally of bronze, but it can be of other materials. Guard rods can eliminate glass breakage, and a special wire-glass or plastic shield can prevent injury and damage from exploding glass (Fig. 14a).

Reflex gage-glasses

These gage-glasses are for high-pressure and -temperature services. They can be read from a distance.

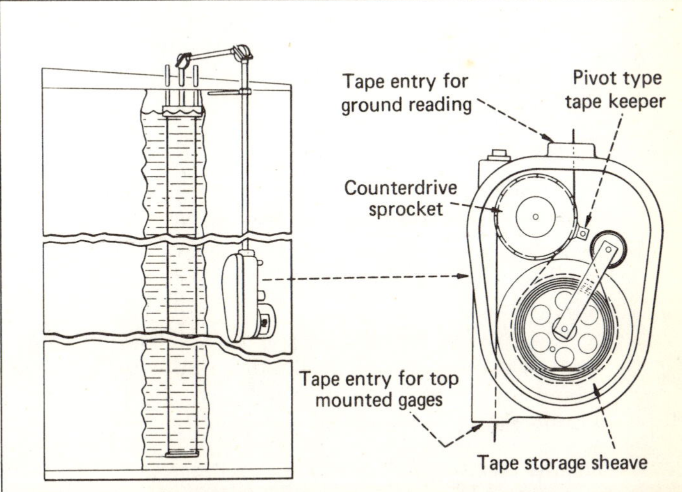

Float tape is stored on sheave that is run by motor Fig. 13

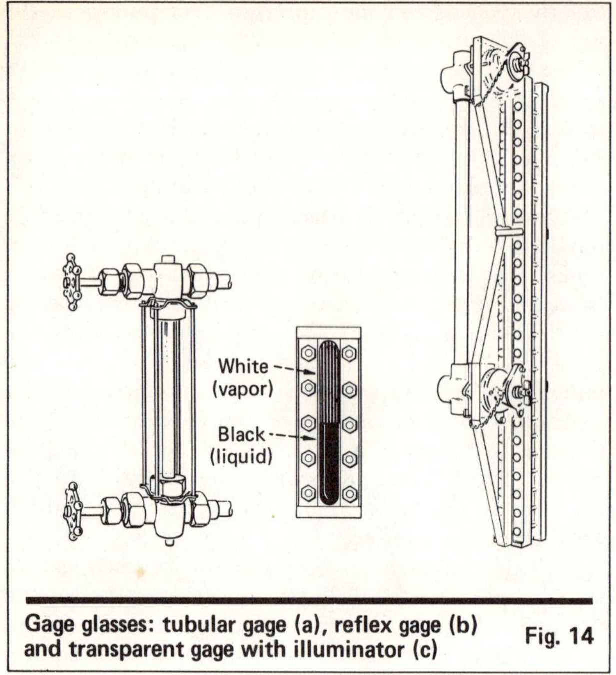

Gage glasses: tubular gage (a), reflex gage (b) and transparent gage with illuminator (c) Fig. 14

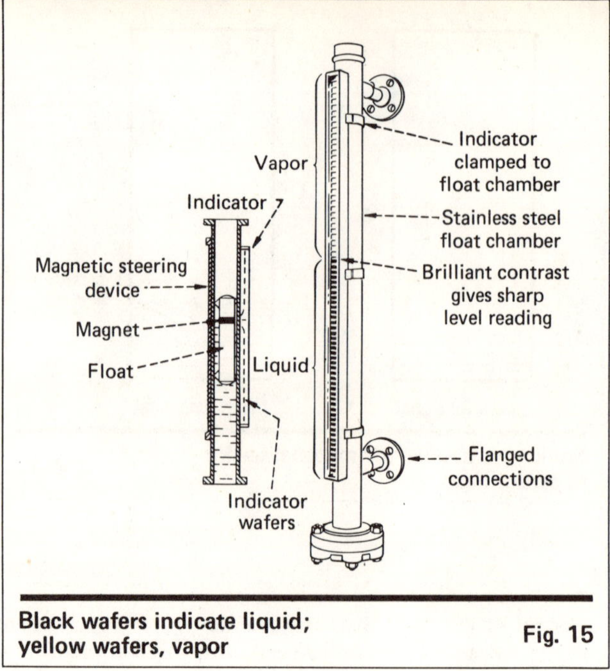

Black wafers indicate liquid; yellow wafers, vapor Fig. 15

Grooved facets cut at an angle on the inner surface of the glass allow light to pass through that part of the glass that is covered with liquid, and to reflect that part of the glass having vapor or gas behind it. A distinct white color represents the vapor, and black the liquid (Fig 14b). Because of this, the gage-glass cannot be used to detect interface levels; also, the color of the fluid cannot be seen.

Transparent gage-glasses

These gage-glasses reveal the color or clarity of the liquid, and show an interface. Glass at the front and rear of the chamber permits a clear view through it. An illuminator, of plastic with a light in the center, is mounted on the entire back of the gage. This diffuses the light evenly over the entire length of gage-glass (Fig. 14c).

For high-pressure steam service (over 250 psig), a mica shield can be mounted between the chamber and the glass. This prevents erosion of the glass by steam and water.

Gage-glass valves

To isolate the gage-glass from the vessel for servicing, some prefer a standard gate-valve, others a specially designed valve with ball checks that will force the ball against a seat if the glass breaks, thereby preventing loss of fluid.

Gage-valves are available with a variety of important features, including: floating-shank or spherical-vessel union, handwheel or quick-closing lever, union connection, and renewable seats.

Armored gage

For services that involve dangerous or toxic fluids and when glass breakage cannot be tolerated, the armored (magnetic) gage is suitable (Fig. 15). At the back of the gage, a magnetic strip positions a float that contains an actuating magnet. Magnetized wafers—which form the indicator, and are yellow on one side and black on the other—are rotated 180 deg as the float passes. The wafers remain in this position until the float magnet turns the wafers in the opposite direction. Liquid is shown by black wafers, and vapor by yellow ones.

The gage comes in several styles for different mountings. It can be mounted on the side or top of a vessel and is available in lengths up to 10 ft. The standard models are used for up to 400 psig and 450°F. Special models can be furnished that can operate up to 10,000 psig and 600°F.

References

1. American Petroleum Institute, Manual on Installation of Refinery Instruments and Control Systems, API-RP 550, March 1965.
2. Anderson, G. D., "Guidelines for Selection of Liquid Control Equipment," Fisher Controls Co., Paper No. TM-26.
3. Considine, D. M., "Process Instruments and Controls Handbook," McGraw-Hill, 1957.
4. Rasmussen, E. J., Alarm and Shutdown Devices Protect Process Equipment, *Chem. Eng*, May 12, 1975, p. 74.
5. Ryan, J. B., Pressure Control, *Chem. Eng.*, Feb. 3, 1975, p. 63.

The Author

Leonard M. Wallace is Principal Instrument Engineer for Fluor Engineers and Constructors (2500 S. Atlantic Blvd., Los Angeles, CA 90040). He has been lead instrument engineer for many major refinery projects in the U.S. and overseas, and has also supervised field-construction personnel in the installation, calibration and checking of instruments. He graduated from Harbor College's instrumentation and automation course and received a diploma in automation and industrial electronic engineering technology from Capital Radio Engineering Institute.

How To Select Liquid-Level Instruments

Selecting the appropriate liquid-level measuring device involves matching the requirements of a vessel with the capabilities of an instrument. Here is the information necessary to make this match.

VICTOR N. LAWFORD, ITT Barton

The liquid contents of process tanks and vessels are measured with a variety of instruments, which can be described as visual types, float-operated systems, hydrostatic gages, and electrical devices. Satisfactory performance can often be obtained from many of these. However, each has its limitations; and even when a number of arrangements are satisfactory, the costs will usually eliminate all but one or two.

Thus the chemical engineer who must specify a liquid-level measurement system is faced with a myriad of possibilities, from which the best instrument must be selected based on the specific application.

Selection by Tank Type

The type of tank or vessel used to hold the liquid usually determines the type of gage to be installed. When considered from this point of view, vessels can be placed into one of ten categories, each of which poses its own special requirements and restrictions. These categories are described as: buried, vented, pressurized, elevated, cryogenic, high temperature, CO_2, chlorine, interface, and vacuum.

Buried Tanks

These usually contain oil or gasoline at atmospheric pressure. The dipstick (Fig. 1) is the most common gage. It is simply a rod or scale that is inserted into the tank and withdrawn, so that the wetted surface of the scale indicates the height of the liquid. When the tank does not have access holes, a sounding tape, which is a flexible dipstick, is guided through a pipe to the tank's bottom.

The hook gage (Fig. 2), another variation of the dipstick, measures the distance from the surface to the top of the tank. The hook is immersed in the liquid and raised until the point is just starting to break the surface. This gage is generally used with open tanks, such as water tanks.

Also, hydrostatic types of gage, such as the air-purge system, are feasible. An air pipe is immersed in the liquid to the bottom of the tank, with air-flow adjusted so that slow bubbling occurs. The air pressure, which is equal to the hydrostatic pressure at the bottom of the tank, is then calibrated in tank units (Fig. 3). When compressed purge-air is not available, liquid-filled hydrostatic systems and differential-pressure gages may be used.

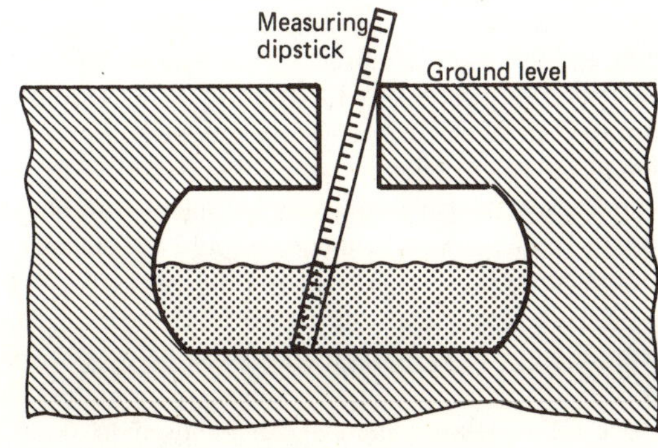

DIPSTICKS measure wetted length—Fig. 1

Originally published October 15, 1973.

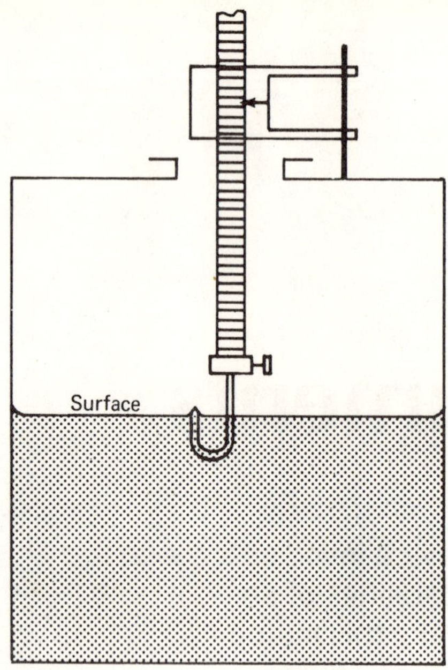

HOOK GAGES measure distance from top—Fig. 2

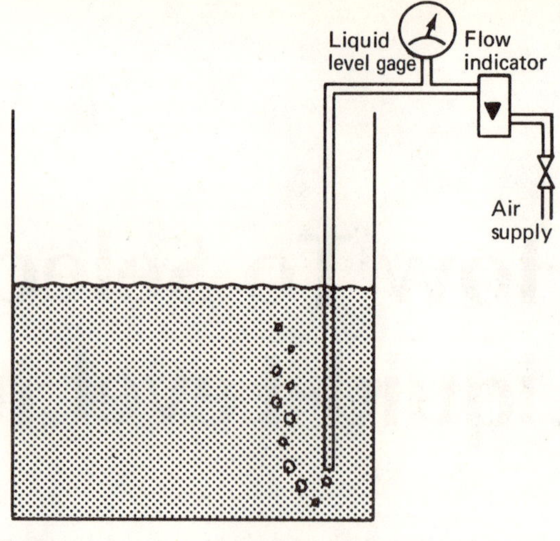

AIR BUBBLERS give pneumatic readings—Fig. 3

Vented Tanks

Any of the methods are suitable, but float gages and hydrostatic types are most common.

Float gages may be direct-connected, magnetic coupled, or hydraulic operated, depending on the requirements. The direct-connected float, probably the oldest and most widely used gage for automatic level control, is simply a buoyant sphere that rests on top of the liquid. For direct level gaging, this float is connected by a tape or cable (Fig. 4) to an indicator outside the tank wall; it may also be connected by a rod or lever to a controller that operates a valve.

The magnetic-coupled float (Fig. 5) has only one moving part in the tank, i.e., the float resting on the surface of the liquid. This float is guided along a sealed tube, in which there is a tape attached to a pulley. Magnets in the float and on the tape keep the parts coupled, while a take-up spring eliminates backlash.

A hydraulic float is connected to a bellows, so that its motion, expanding or compressing the bellows, generates varying hydraulic pressure for transfer to a remote indicator or controller. Such a system may be compensated for temperature or pressure changes in the tank (Fig. 6).

With hydrostatic gages, the range of the instrument depends on the height to be measured and the density of the liquid, since these gages measure the pressure exerted by the liquid rather than the actual level. Manometers, pressure meters, or diaphragm gages can be used from 1 ft. to 50 ft. of liquid. For higher ranges, manometers and Bourdon-type pressure meters can be used to directly measure the pressure at the bottom of the tank.

Also, the differential-pressure diaphragm gage (DP unit) is suitable. One side of the gage is exposed to the pressure from a reference column of liquid (Fig. 7); and the DP unit thus measures only the difference in pressure between that due to the tank level and that due to the constant reference. Otherwise, a diaphragm gage may be exposed directly to the pressure at the bottom, with pneumatic pressure from an instrument air system balancing this pressure (Fig. 8).

The low-pressure sight glass or gage glass is also suitable for visually indicating levels in vented tanks. The simplest and most direct method for level measurement, it consists of a transparent tube mounted on the side of the tank and connected at top and bottom (Fig. 9).

Variable-resistance gages (Fig. 12) employ an electrical-resistance strip extending from top to bottom of the tank. Although the resistance element is sealed from the liquid, the windings are shorted out as the liquid rises, thus indicating the level.

Pressurized Tanks

The pressure in these tanks usually determines the selection of liquid-level gage. Of the visual types, only the high pressure gage glass—two flat pieces of gage glass mounted in steel housing—is suitable above 200 psig. Float gages are also usually suitable to several hundred psig., although the vessel design will often preclude the bulky float mechanisms.

Displacer gages and DP units are used to pressures of 3,000 psig. and 6,000 psig., respectively. These two instruments measure liquid level by the hydrostatic principle; pressure is a function of height and density of tank fluids, both liquid and vapor phases. The displacer is a metal rod that is suspended from a spring mechanism. Hydrostatic pressure, acting on the bottom area of the rod, develops a force (equal to the bouyancy) that lifts the rod and thus operates the gage. The DP unit is mounted externally with one side connected to the pressure at the bottom of the tank and the other side to the top of the tank. A reference column is used to prevent condensate from forming in the piping.

Also, electrical gages are suitable. These electrical gages include capacitance probes, sonic systems, weight-

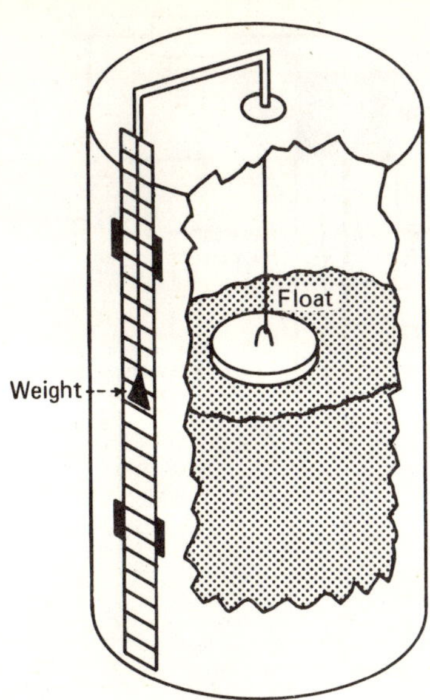

FLOATS give direct level readings—Fig. 4

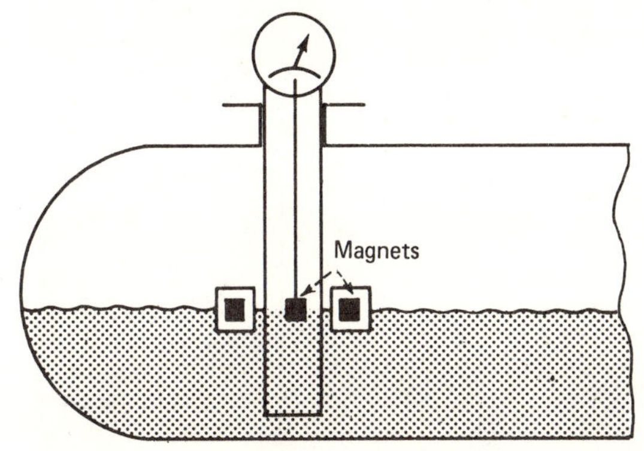

MAGNETIC-COUPLED FLOATS have only one moving part exposed inside of the containing vessel—Fig. 5

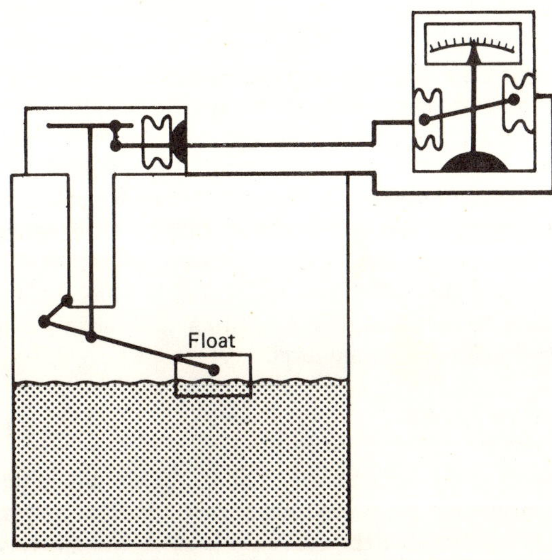

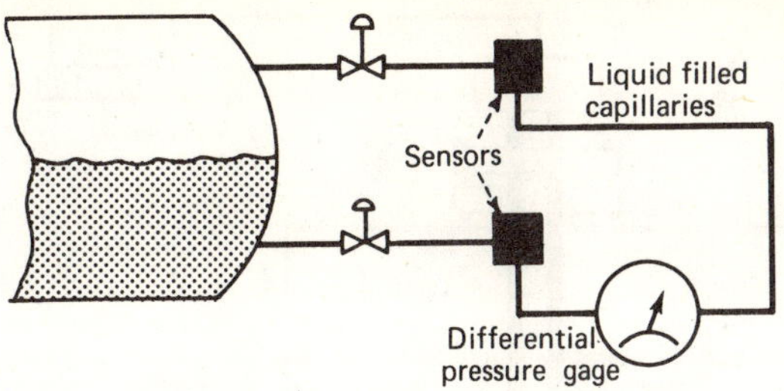

DP GAGES measure the difference between the level inside the tank and a reference-column level—Fig. 7

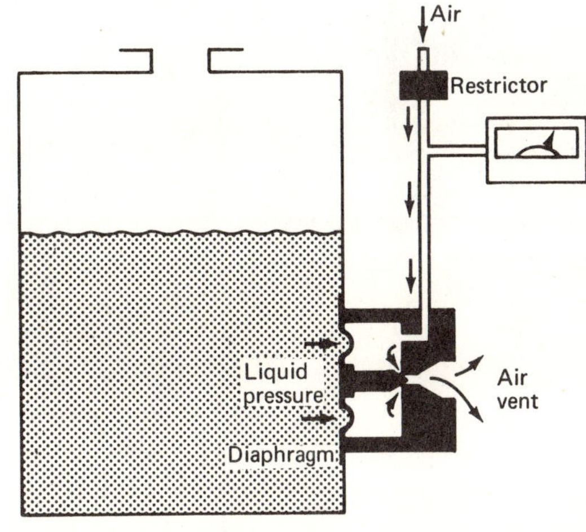

FLUSH DIAPHRAGMS balance hydrostatic pressure of the liquid in the tank with pneumatic air pressure—Fig. 8

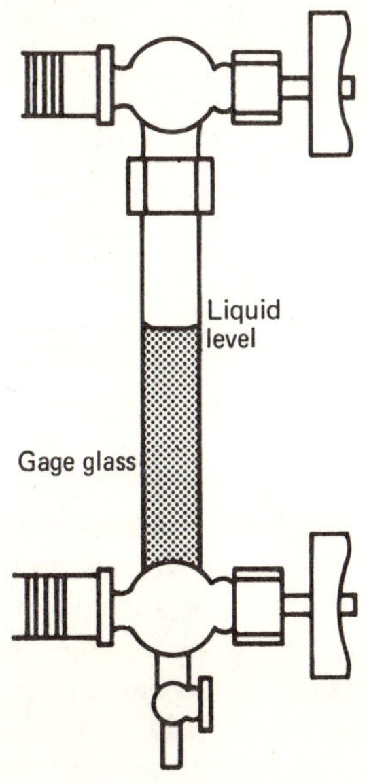

GAGE GLASSES give direct readings—Fig. 9

← **HYDRAULIC FLOATS** operate a bellows—Fig. 6

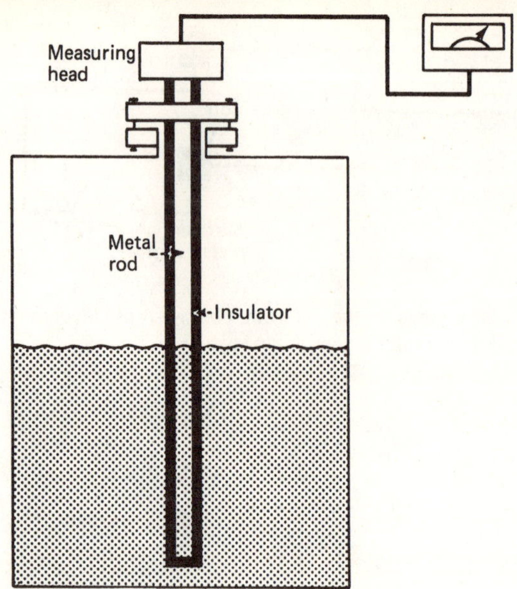

CAPACITANCE PROBES use the bed of liquid in the tank as part of the circuit—Fig. 10

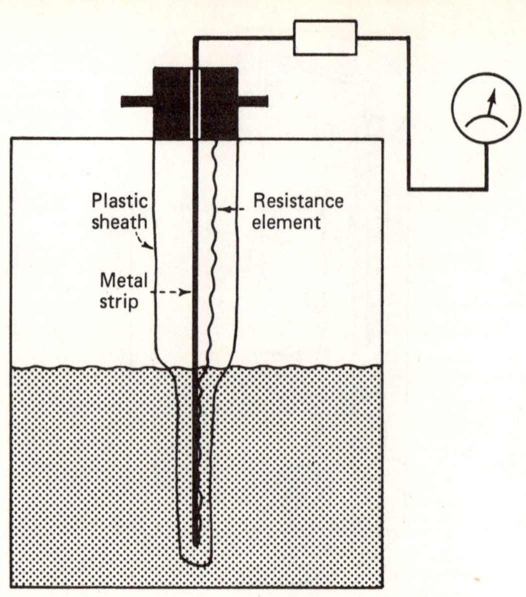

VARIABLE RESISTANCE GAGES have electrical windings that are shorted out by the liquid pressure—Fig. 12

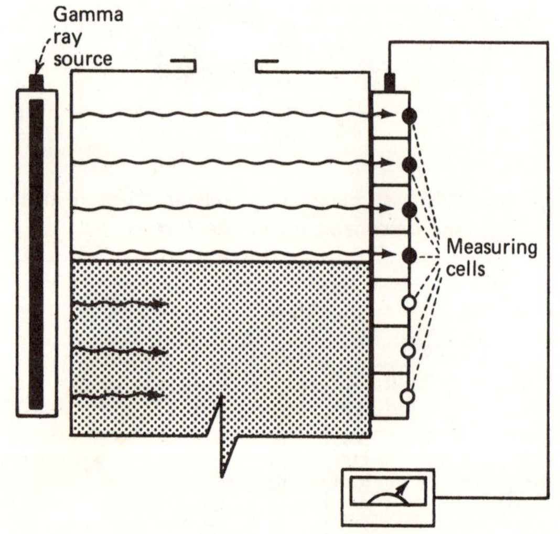

RADIATION SYSTEMS measure the resistance to passage of the liquid mass—Fig. 11

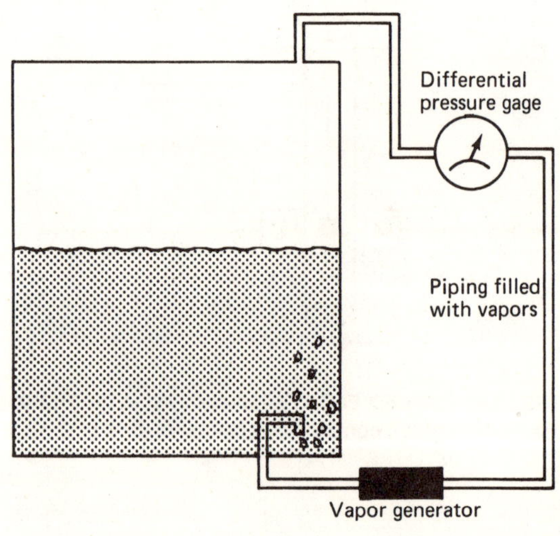

DP GAGES can read hydrostatic pressure exerted against a small flow of the liquid vaporized outside the tank—Fig. 13

measuring load cells, and radiation systems. The capacitance probe extends from top to bottom of the tank (Fig. 10). It is a metal rod covered with an insulator, if the liquid is a good conductor. It is a combination of a metal rod inside a metal tube, if the liquid is dielectric. In the first case, the rising conductive liquid forms an increasing capacitance around the rod; and in the second, the liquid rising between rod and tube forms the capacitance.

With the sonic systems, a sound generator located in the top of the tank transmits pulses toward the liquid surface, as a pickup transducer receives reflected sound waves and a control unit computes the distance.

With the radiation systems (Fig. 11) a source of gamma rays is located in a vertical column on one side of the vessel, while measuring cells are located on the other. Since the transmission of rays is inversely proportional to the mass in the vessel and the walls are constant, the radiation received by the measuring cells is inversely proportional to the liquid level.

Elevated Tanks

This type of tank usually requires hydraulic transmission systems, since these experience almost no limitations in tank size, height above the ground, and pressure. The water level in a 20-ft.-high tank atop a 100-ft.-high

Advantages and Disadvantages of Liquid-Level Measuring Methods – Table I

	Instrument	Advantages	Disadvantages
Visual	Gage glass	Low cost for low pressure; no moving parts; measures direct level; accurate.	Local reading only; requires cleaning; may not be applicable on dirty liquids; breakage may be dangerous; sealing problem; expensive on high-pressure systems; inaccurate at high temperatures.
	Dipstick	Low cost; direct reading; any liquids; simple; accurate.	Local reading only; vented tanks only; best for shallow tanks; dangerous; clumsy.
	Dip probe (rotary)	Low cost; simple; direct reading; accurate. Used on propane tanks.	Local reading only; pressurized tanks only; potentially dangerous.
	Hook gage	Low cost; simple; direct reading; accurate.	Local reading only; open tanks only.
Float	Direct-connected float	Measures direct level; low cost for open tanks; adaptable to transmitters; no limit to height of tank; any liquid specific gravity.	Moving parts exposed to fluids; cables or tapes can break; obstruction in the tank may interfere; limited in pressure ratings; float must be kept clean.
	Magnet-coupled float	Suitable for pressure or vacuum tanks; accurate in deep tanks; independent of specific gravity.	Float may jam on tube; long unsupported column may be damaged by liquid surges.
	Hydraulic-operated float	Remote (250 ft.) indicator gage; gage may be elevated above tank; suitable for closed tanks.	Complicated installation/calibration; internal clearances required; moving parts in liquids.
Hydrostatic	Direct-connected pressure gage	Flexible; accessible from outside the tank.	Limited to open tanks; affected by variations in specific gravity.
	Bubbler pneumatic pressure-gage	Location is flexible; elevation practically unlimited; may be used for corrosive fluids, severe temperatures, dangerous fluids, and with suspended mater.	Requires an equalizer line and relief valve on closed tanks; distances are limited; piping leaks can cause errors; pressurized tanks cannot use this system; source of purge air required. May contaminate liquid.
	Sealed reference column to bellows or diaphragm gage	Purge air flow not required; liquid in reference column need not be replaced; remote, continuous indication; large dials possible without auxiliary power; transmission possible; accurate; sensitive; reliable; no moving parts in tank; adaptable to variety of tanks; unaffected by operating variables; safe; no contamination; easily installed.	High cost for small or vented tanks.
	Displacer gage	Measures liquid level, specific gravity, or interface; low cost in small sizes; good sensitivity; used in both vented and pressurized tanks.	Moving parts in liquids; build-up of solids may impair accuracy; signal must be amplified; must extend full length of liquid to be measured; extra-large diameters required for interface measurements.
	Flush diaphram gage	No probes, etc., in tank; simple installation, usually flange mounted; easily cleaned; can be corrosion resistant.	Open tanks only.
	Diaphragm gage with low-pressure side connected to vapor space of tank	Suitable for pressurized tanks; purge air not required; remote, continuous indication; transmission possible; no moving parts in tank; safe; easily installed; no contamination; accurate; reliable; unaffected by pressure fluctuations.	Vapors from tank can condense in connecting tube; exposed to corrosion from tank contents.
Electrical	Capacitance probes	No moving parts; light-weight equipment; good corrosion resistance; easily cleaned.	Temperature effects the dielectric constants; contaminants may adhere to probe.
	Weighing	Safe (liquid always enclosed); measures mass; no moving parts.	Expensive; winds cause errors; batch process only.
	Sonic systems	Best for all tanks; any type of liquid, slurry, etc.	Sensitive to density of fluids.
	Radiation	No contact with liquid; simple installation.	Expensive; license needed.
	Variable resistance	Measure any liquid or solids; easily installed and calibrated.	Pressure limitations; float or sheath susceptible to damage.

tower, for example, could be measured by a DP unit attached to the water pipe at ground level, with a single reference column running from the gage to the top of the tank 120 ft. above. Since the water in the pipe would counter-balance the first 100 ft. in the tube, the DP gage would be measuring only 20 ft. of head; and with a potable liquid sealed in the reference column, it would never freeze or need refilling.

Cryogenic Tanks and LNG Tanks

At temperatures of -150 F. and below, a pneumatic system with a DP unit (Fig. 13) is usually preferred. When a small amount of the cryogenic liquid is vaporized in the tubing connecting the gage to the bottom of the tank, the pressure of this gas equals the hydrostatic pressure at the bottom of the tank. Tank pressure is

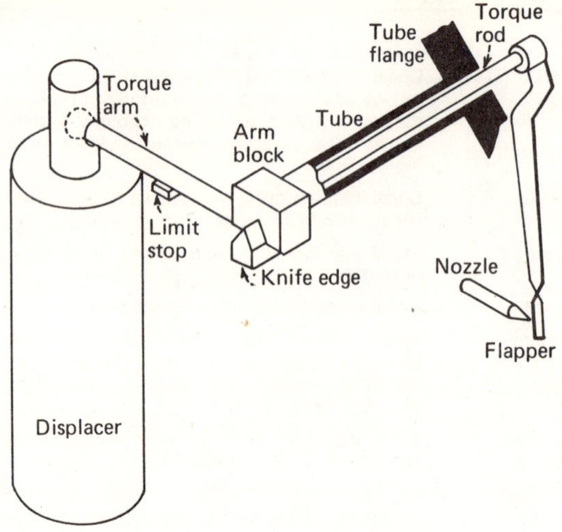

DISPLACER GAGES measure buoyant forces—Fig. 14

counter-balanced by connecting the low-pressure port of the DP unit to the vapor space in the top of the tank.

The storage of LNG in underground caverns holds interesting possibilities for such systems. In this application, the liquid-level devices will need to transmit information to instruments several hundred feet above the surface of the liquid.

High Temperature (Boilers)

High temperature tanks usually contain high pressures as well. These conditions reduce the selection of liquid-level gages to two types: the gage glass and the DP gage.

Gage glasses are required on steam boilers, but they exhibit a continuous error typical of high-temperature services. Because of a lower temperature in the gage, its water is more dense and therefore at a lower level than the connected level of saturated water inside the boiler. A boiler operating at 2,650 psig. and 677 F., for example, might have a 10-in. glass operating at 450 F.; and the reading in this glass will be 20% low, when the boiler is at normal level.

DP gages will also be affected by similar density differences between the hot liquid in the tank and the reference column. However, these gages may be precalibrated for true readings at the operating temperature, or compensating techniques may be used to accurately indicate the actual level over a wide range of conditions.

Carbon Dioxide Tanks

These present a special case of cryogenic tanks, because they are at higher pressures, up to 300 psig., and because the ASME code specifies a vapor space of approximately 15%, so that the tank is considered full when 85% of the volume is reached. The most popular level-measuring system is the DP gage, with the tank-liquid generating gas for the high-pressure side of the diaphragm.

Chlorine Tanks

Chlorine is an archetype of difficult-to-measure liquids. The tanks are under pressure; the environment is corrosive; and safety requirements tend to prevent connections on the sides or bottom of the tank. Gage glasses are sometimes used, but require frequent cleaning. Weight measure is also used, but the tanks are invariably out-of-doors, where wind effects prevent accurate weighing. Gamma-ray gages would be suitable; but the shape of the tanks (horizontal cylinders) causes a nonlinear relation between volume and liquid height. DP gages with liquid-filled tubes have been successfully installed through top openings and at gage glass connections.

Interface Levels

The interface between two immiscible liquids is usually measured by float devices or hydrostatic gages. The floats are designed to sink in the lighter fluid but float in the heavier fluid. The buoyancy (displacer) type of hydrostatic gage (Fig. 14) may require large-diameter displacers in order to develop sufficient force to operate the transmitter.

However, the DP gage can operate on these applications without any limitations. With a 10-ft.-high tank, for example, the hydrostatic pressure at the bottom would be 4.33 psig. when full of water, and about 3.48 psig. when full of kerosene. This gives the DP gage 0.85 psi. maximum differential. This differential, equivalent to about 24 in. of water, is easily measured by sensitive, high-accuracy DP gages.

Vacuum Tanks

Gage glasses are suitable for visual indication, but cannot transmit signals. Bubbler systems are not suitable. Floats are seldom used, because the fittings may leak. Displacers and DP gages are the most popular systems.

In conclusion, instruments for continuous measurement of liquid level are available with a wide choice of operating principles. For every tank and for each liquid, at least one instrument, and probably many, will meet the requirements. At the present time, most level-measure-systems use floats, displacers, or DP gages. But new devices appear on the market continually, with the trend toward electrical operation and external mounting. A summary of current instruments, with their advantages and disadvantages, is given in Table I.

Meet the Author

Victor N. Lawford is Manager of Product Engineering at ITT Barton Process Instruments and Controls, 580 Monterey Pass Rd., Monterey Park, CA 91754. An author of several papers on liquid-level measurement and holder of five patents in instrumentation, he is a member of the Instrument Soc. of America and the American Soc. of Mechanical Engineers. He holds an MS degree in mechanical engineering from California Institute of Technology.

Pressure Control

Here, for the novice, is an introduction and an incentive toward furthur study, for the seasoned instrument engineer, a stimulant to further simplify descriptions of pressure instrumentation, and for other engineers, a help toward overcoming the uncomprehending acceptance of instrumentation so prevalent in the industry.

J. B. RYAN, Fluor Corp.

Pressure measurement and control serves three purposes: the protection of equipment, the protection of personnel, and the production of on-spec product.

The average applications engineer is not often asked to design pressure instruments. Many instrument companies have perfected instruments that are proven, reliable, relatively inexpensive, and often available in quantity as off-the-shelf items. Thus the instrument engineer's function becomes one of choosing.

This means that the instrument applications engineer must be concerned with fundamentals of pressure measurement and control. These are most easily understood in terms of functional parts, i.e., of pressure-sensitive elements, transmitters, and controllers.

Pressure-Sensitive Elements

Pressure instruments are usually categorized according to their pressure-sensitive element and range (Fig. 1). However, all of the elements now used were initiated by the development of the Bourdon tube.

Eugene Bourdon, a French engineer who died in 1884, invented a pressure-sensitive element that has since carried his name and has served as the basis for a steady accretion of additional developments. He bent a piece of good quality tubing into the shape of a horseshoe, sealed one end, and connected the other end to a source of fluid pressure.

When pressure was applied, the tube would flex, as if trying to straighten out. Since the elastic limit of the metal was not exceeded, the tube would return to its original form when the pressure was released; and the amount of flexure was proportional to the increase or decrease in pressure. If the open end of the tube were fixed, and the closed end linked to a pointer, it could be made to indicate the pressure.

The stage was thus set for pressure measurement. Different wall thicknesses and various alloys could be used to make Bourdon tubes for various pressure ranges. Clockwork-type fittings, such as a pinion gear, a quadrant, a pointer, a dial, and a suitable hairspring to take up slack from the gears, could be used to provide precision. And a set of calibration attachments completes the parts necessary to a pressure gage (Fig. 2).

The Bourdon tube has been a very important milestone toward control of process pressures previously considered unsafe. Before a pressure can be safely controlled, it must be measured, and this gage provided the necessary reliability for such measurement.

However, the Bourdon tube provides only a limited distance of travel at low pressures. This led to development of the *helix*, an extended Bourdon tube in the form of a helical coil, which gives additional movement of the closed travelling end (Fig. 3). Such additional movement is used to power transmitters, recorders, or controllers. Even at high pressures, it gives protection against overpressure and less hysteresis.

Usually the helical element can handle pressures ranging about 10:1 and be capable of precise measurement over continuous fluctuations. Accuracy is normally ±1% of the calibrated span. Such elements are available in various materials provided by various manufacturers.

The *spiral* element permits more compact housings than the helix, but retains extra travel not available with the Bourdon tube. Thus, it does not require linkages to amplify the movement of the indicator (Fig. 4).

The same principle of elastic metal deformed under pressure has been used for the *bellows* type of element (Fig. 5), which is suited to lower pressures. Originally

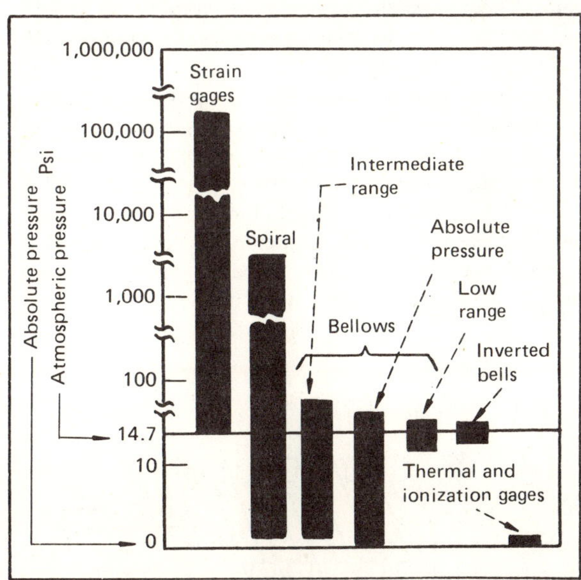

NORMAL PRESSURE for elements and gages—Fig. 1

Originally published February 3, 1975.

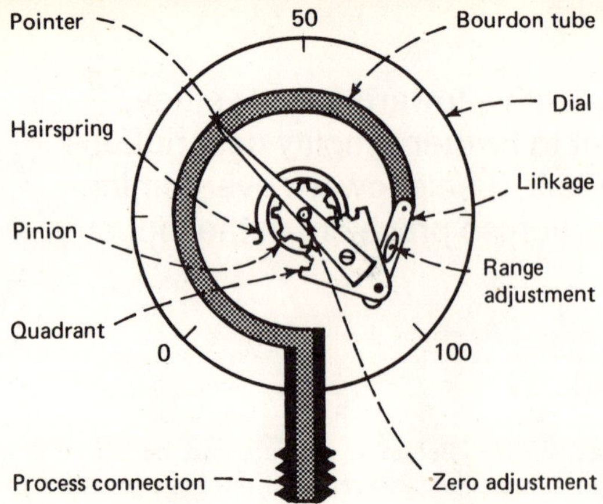

BOURDON TUBE type pressure gage—Fig. 2

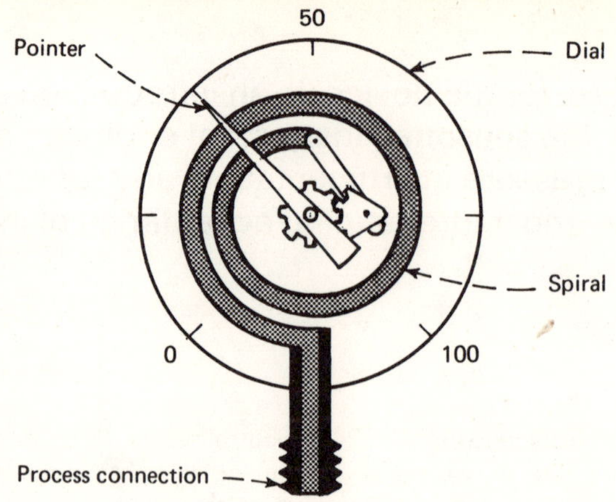

SPIRAL TUBE type pressure gage—Fig. 4

beryllium was used because it is quite rugged and elastic. Other materials such as stainless steel, nickel, and brass are now used extensively. In time, a bellows element tends to become work-hardened and needs replacement. Thus bellows-type instruments require more maintenance than other pressure devices. Normally, a return spring is integral with a bellows assembly (Fig. 10).

The Transmitter

Although mechanical linkages from element to needle can translate pressure to a reading in a gage mounted at the equipment, they cannot transmit that information on to a control room or a remote location. Nor are they powerful enough to move a valve to regulate the pressure. Some sort of system is needed to transport the measured signals and to convert them to strong forces.

Two such systems have evolved—pneumatic and electronic. The pneumatic system usually employs air at 3-15 psig pressure. The electronic system usually employs 4-20 milliamps of direct current.

Pneumatic transmitters translate the position of a pressure element into air pressure that varies in strict proportion to the element position, so that this instrument air can be piped to other components as a measured signal.

Within the transmitter (Fig. 6) the main flow of instrument air passes through a pneumatic pilot valve that regulates the signal pressure, while a sidestream passes through a restriction orifice to a nozzle. A connection transmits the pressure in this nozzle to one side of the diaphragm actuator for the pilot valve, and a flapper over the nozzle outlet lets air escape according to the position of the primary element.

Since air flow to the nozzle is restricted, a change in element and flapper positions causes a simultaneous change in nozzle-air pressure. This in turn changes the diaphragm pressure and the pilot valve position to put the signal air at a different pressure level. The signal air is stabilized at this new pressure by means of a feedback bellows that acts to oppose the continued action of the flapper. And the instrument can be calibrated at this feedback balance point (Fig. 6).

The system in Fig. 6 is known as a force-balance-type transmitter. There are also motion-balance transmitters. In either type, the process pressure is balanced against the spring action of the pressure-sensitive element, whether it be bellows, Bourdon tube, spiral, helix, etc. A spiral or helical element would be used where accuracy could be a problem because of loss of motion.

Calibration adjustments for rangeability relate the pressure-sensitive element to the fulcrum of the force

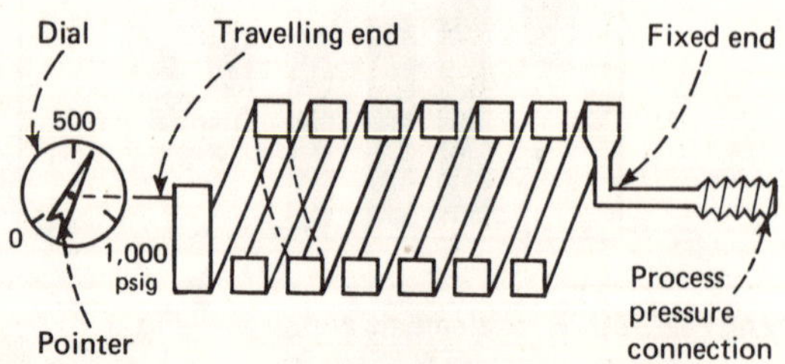

HELICAL TUBE type pressure gage—Fig. 3

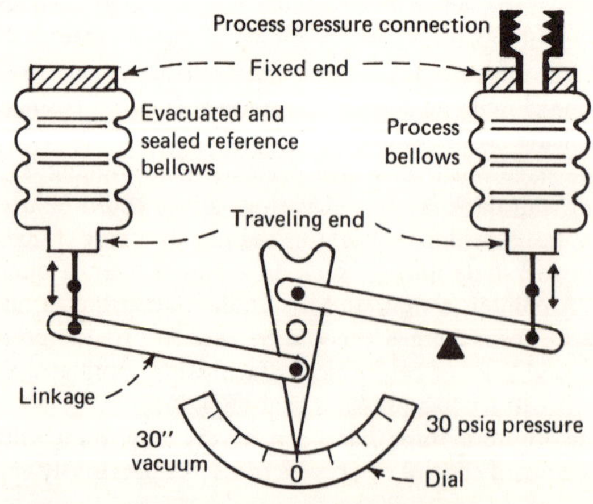

ABSOLUTE-PRESSURE-BELLOWS indicator—Fig. 5

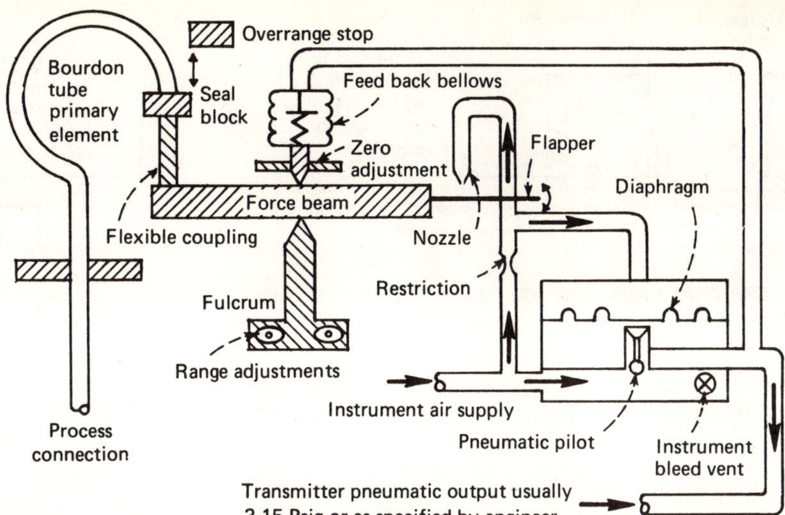

BOURDON TUBE primary element activating a force-balance type of pneumatic transmitter—Fig. 6

beam, for a corresponding relationship between the flapper and the air output of the transmitter. This latter relation depends on the coordination between the feedback bellows and the pilot diaphragm. An increase in output air pressure is felt by the feedback bellows which forces the flapper back to its throttling position.

Normally, the output air from the transmitter is fed to a controller, which in turn actuates a final control component, such as a valve. The normal calibrated range of these components would be 3 psig for 0% pressure and 15 psig for 100% pressure, so that midrange would be 9 psig. Similarly, electronic transmitters (Fig. 7) and controllers normally use an input of 4 milliamps (mA) and an output of 20 mA, so that 4 mA is 0, 12 mA is 50% and 20 mA is 100% of the range.

Similar relationships exist for instruments having ranges of 6-30 psig or 10-50 mA.

Differential-pressure elements, normally in force-balance-type transmitters, are used in addition to the direct-acting elements to measure pressure, as well as flow and liquid level. They sense a difference in pressure between two sources acting on opposite sides of a diaphragm (Fig. 8). Great care should be exercised in specifying them. They are available in exotic metals, as well as most of the common materials compatible with process fluids and conditions.

Because of their many uses, differential-pressure elements command a unique position in pressure instrumentation. Within the transmitter, a primary force-balance beam passes through a seal and fulcrum integral in the diaphragm housing to either a pneumatic or electronic system. Movement of the diaphragm is thus transmitted to the secondary beam, and the changing output from the pilot is the measured variable.

Still other pressure-sensing and transmitting devices (i.e., slack diaphragm detectors, pressure repeaters and

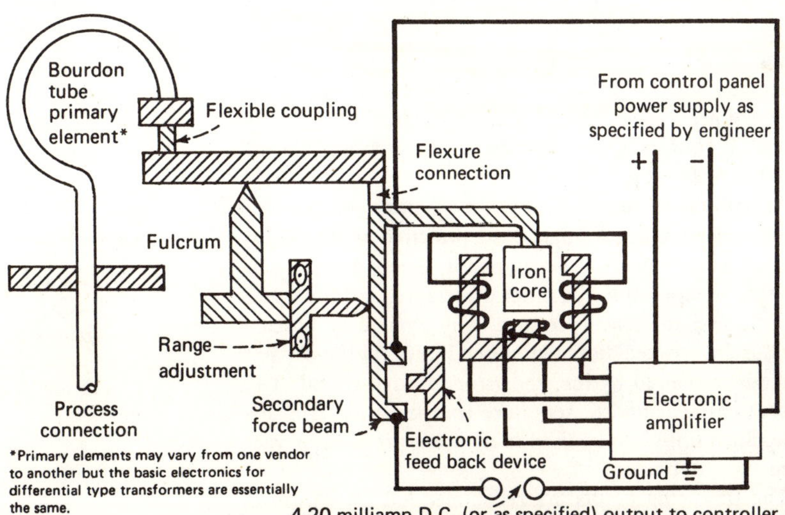

FORCE-BALANCE type electronic transmitter using a differential type of transformer—Fig. 7

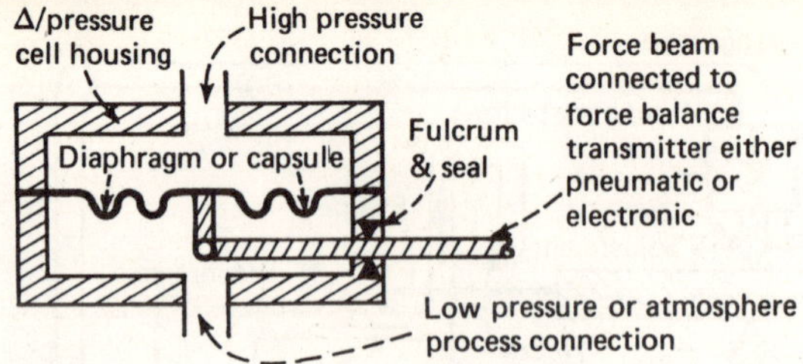

DIFFERENTIAL PRESSURE type element—Fig. 8

boosters, absolute pressure sensors, opposed bellows type detectors, fused-quartz helical pressure sensors, inverted bell manometer type sensors, strain gages, U-tube manometers, inclined manometers, etc.) are available and described in detail by their manufacturers' literature; but the foregoing principles of operation are generally applicable.

All such devices do not have equal electronic/pneumatic possibilities. The strain gage, for example, is available in an electronic version only (Fig. 9). It employs a wire resister, which is stretched like a violin string, thereby varying the electrical resistance of the wire according to the pressure. Its readout is translated through an electronic bridge circuit into a signal which can do the work (i.e., measure, indicate, record, or control). Such an instrument is expensive and should be applied only where very high pressures warrant its use (See Fig. 1). Nevertheless, it is very accurate if properly applied.

Also, the repeatability of the bellows can be extended, when fitted with a positive return spring (Fig. 10). In an electronic transmitter (Fig. 10), this element can be used when pivoted across a fulcrum, so that a change in pressure changes the position of a noble-metal contactor in a variable potentiometer. The resistance signal is fed to a wheatstone bridge circuit. Readout devices can take the form of an indicator, a recorder, or a controller. Accuracy is about ±2%.

Where the process warrants higher accuracy, a capacitance-type pressure transmitter with a differential-pressure element should be used (Fig. 11). Such devices are expensive and very sensitive to the dielectric properties of the process, so that they should be limited to dielectric processes and to dry noncondensible gases. However, they are accurate to ±0.2% of their range, and with the proper diaphragms, can handle pressures from 3 in of water to 5,000 psig.

As with other electric or electronic instruments, the developers have taken a single basic principle with predictable characteristics and developed it. One side of a diaphragm is exposed to the process pressure, and the other to a known reference pressure. Any change in process pressure moves the diaphragm; and in Fig. 11 the capacitance is measured by the wheatstone bridge circuit. The output is linear and temperature compensation is available. Such units can be used to measure pressure or vacuum.

The following points are recommended for choosing pressure components:

1. *Transmitters*: For suppressed-range control, transmitters should normally be nonindicating force-balance type, piped in parallel with an indicating pressure gage. For full-range control, or for indicating or recording service only, transmitters may be motion balance, indicating type, or nonindicating type, piped in parallel with an indicating pressure gage. Transmitter output should be 3-15 psig where pneumatic instruments are used and 4-20 or 10-50 mA direct current where electronic instruments are used.

2. *Controllers*: For applications where transmission to remote receivers is not required, controllers should be nonindicating, pilot-operated type, with the pressure-sensitive element and controller mounted in a weatherproof case suitable for installation by the valve or remotely. Direct-connected pressure gages should be piped in parallel with local indication. An adjustable proportional band of at least 2-20% and convenient control point setting should be provided. Output should be 3-15 or 6-30 psig as is normally required for the final control element.

3. *Elements*: Except where the process requires special material, process-measuring elements should be AISI

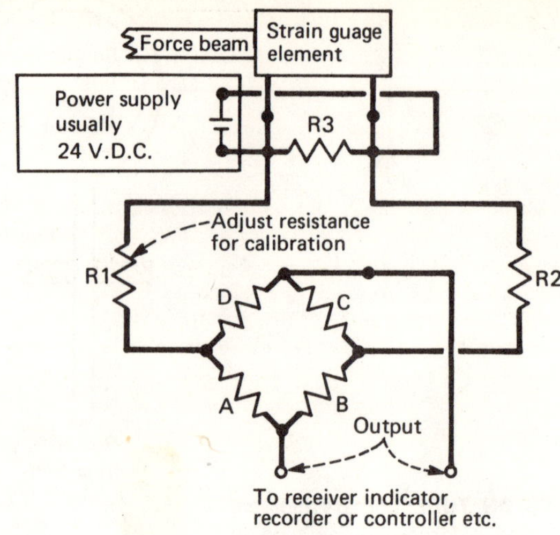

STRAIN GAGE with electronic circuit—Fig. 9

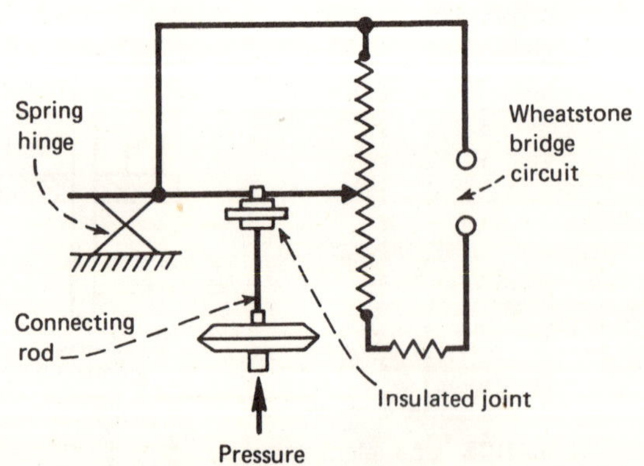

RESISTANCE TYPE of transmitter—Fig. 10

Pressure Instruments for Temperature

Some pressure instruments are used to determine other properties, particularly temperature. The sensing elements of temperature instruments are usually modifications based on the expansion and contraction of a liquid or gas with temperature, and the system is essentially a pressure gage fitted with a readout device. The gage is connected by a small-bore capillary tube to the temperature-sensing device, filled with a suitable fluid, and sealed. If the gage is a bellows, for example, the change in temperature would act like a change in the pressure to expand or contract the bellows; and the movement due to fluid in a large bulb could be used to actuate a direct-connected control valve. Such self-actuated temperature control valves are in widespread use in heating systems for storage tanks and other rough applications, where a few degrees, plus or minus, is not acute.

Type 316 stainless steel or "Ni-Span C." Bronze or steel may be used for utility services.

4. Differential pressure instruments should be liquid-filled manometers, bellows, or force-balance type, according to requirements.

5. Control modes should normally be of adjustable proportional control with automatic reset. Local-mounted pilot-type controllers may be limited-proportional type. Derivative response should be provided only when required by process conditions.

Pressure Controllers

Before selecting a pressure controller from the wide variety available, the engineer must know the purpose and preferred mode of operation for the entire control circuit, including transmitter, controller and valve. Normally the transmitter output signals the controller, which in turn positions the valve.

A pressure controller should include an adjustment for proportional control in addition to other calibrational adjustments, so as to obtain proportional action over a wider or narrower range according to the needs of the process. Since the advent of electronic controllers, this proportional action is sometimes called "gain." Another desirable feature, called "rate" or "derivative" action is used to smooth out undesirable pressure peaks that cannot be corrected by proportional action only.

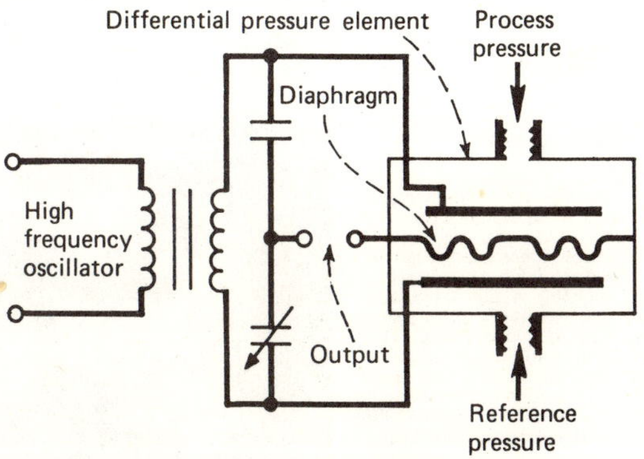

TWIN CAPACITANCE plate type transmitter—Fig. 11

These adjustments are critical to the operation of a control circuit and should not be tampered with by unqualified operating personnel. Such tampering is responsible for a large proportion of the instrument troubles experienced during the startup of new plants.

The operating principles of controllers resemble those of transmitters, whether pneumatic or electronic. Whereas the transmitter receives input data in a Bourdon tube or similar sensing element, the controller receives the transmitted signal in a bellows or electronic bridge, which has an adjustable setpoint. When the transmitted signal differs from this setpoint, the controller generates an output signal to the valve for corrective action, which is supposed to return the transmitter signal to the setpoint.

The choice between pneumatic or electronic control should depend largely on the plant's management, who will know the qualifications of operating and maintenance personnel. Sometimes plant managers convert an existing unit from pneumatic to electronic control, because technicians have gained exposure to electronic circuitry from work in television and space programs. By themselves, the components would rarely command precedence one over the other; and when the choice is made by an engineering contractor, it is based largely on economics, availability, and the area where a new plant is to be located.

The following points might be helpful in choosing between the two:

Pneumatic	Electronic
Less expensive	More expensive
Small plants	Large plants
Short transmission	Long transmission
Simple process	Complex process
Slow response	Fast response
Greater dependability	More maintenance
No computer capability without transducer	Adapted to computers
Intrinsically safe	Added cost for safe components
Analog signals not available with digital	Digital signals; also available with analog

Installation

The engineer should also be aware of accessories, such as snubbers and blowout backs, as well as good practice regarding such items as dirty services and instrument position.

Snubbers: Pressure snubbers on reciprocating pump discharges and other pressure-cycling streams tend to smooth out the peaks and thus give longer life to an instrument. They should always be considered for pressure transmitters and direct-connected pressure controllers.

The wear on the quadrant of a pressure gage, for example, usually occurs in the center portion, and the pinion gear also wears. If such wear is left unchecked, the hairspring can no longer take up slack movement, with resulting loss of accuracy. In recent years, several manufacturers have filled the cavity in the pressure-gage housing with a nonfreezing liquid to absorb the shock of cycling pressures as a means of smoothing out peaks.

Blowout backs: These are provided on some models of pressure gages, so that a rupture directs the process fluid out the back end of the gage, and not into the face of a bystander in front of the unit. However, if the proper metal and ranges are specified for the pressure-sensing device, chances of the element rupturing are considerably lessened.

Dirty or corrosive services: The chemistry of the process should always be evaluated before specifying any pressure device, so as to choose the proper type of element for the service. Suppose, for instance, that the pressure from a 1,200 psig crude-feed supply line is let down to 50 psig for a distillation unit, with the crude containing particles of sand and traces of water, but not enough to warrant a strainer or filter on the upstream side of the transmitter. A measuring element other than a Bourdon tube, spiral or helix should be chosen, since such elements present a collection chamber for foreign matter.

Water or other foreign matter can cause corrosion, thereby limiting the choice of metal to stainless steel or Monel. Diaphragms have been developed by various instrument vendors for such service. These work reasonably well but are limited to the ranges shown in Fig. 1. Because of the increased surface area of a diaphragm, relative to that of Bourdon-tube type elements, the latter requires less maintenance and offers longer life.

Position: A pressure instrument applied to piping and vessels should be mounted in an upright position, so that particles carried in the process stream will drop out of the sensing element before they become lodged there. Also, water can act like particles, when it is carried in an immiscible lighter liquid such as hydrocarbons. In time such impurities can cause an unnecessary shutdown of the unit. Regardless of the instrument selected, periodic maintenance for such dirty service is necessary, and perhaps a redundant pressure control circuit is warranted.

Usually the housing of the capsule diaphragm is complete with a blowdown valve, and the controller in that particular loop will have been placed on manual control before the maintenance operation is undertaken.

Calibration

All pressure devices require proper calibration, and a number of standard instruments have been developed for this. The first and most essential is the simple glass-tube manometer. No instrument shop would be complete without one. Generally known as "U" tubes because of the shape of their glass tubes, these instruments are partially filled with water or mercury, and one end fitted to a variable pressure source. The pressure can be read in the difference of liquid levels on a scale located between the legs of the U tube.

The component to be tested is usually connected in parallel with the manometer, and a reference pressure applied to give the component's 0%, 50% and 100% points, with the readings checked on the manometer. Other points can also be checked, especially the known operating point.

Inclined manometers are used for calibrating instruments such as draft gages, where the U-tube is limited by its sensitivity. Still other limitations relate to the length or size of glass manometer that can be built. If the desired calibration range lies outside that of the standard water-filled manometer, a mercury-filled unit might be used. But if the range lies still outside that of such heavier solutions, the most logical standard calibration device would be a "dead weight" tester. No well-equipped instrument shop should be without a dead weight tester.

This device functions by measuring hydraulic pressure of oil in a chamber with dead weights loaded onto a vertical piston that rests down against the pressurized liquid. A connection puts the instrument to be calibrated against the same liquid pressure. In use, the instrument is connected, weights are loaded, and a hand pump used to bring liquid from a builtin reservoir into the chamber until the piston can just lift the weights. Then these are checked against the indication on the instrument.

Dead weight testers are certified by a governmental bureau of standards and must not be altered in any way. Their care and maintenance is of paramount importance. Before using such sensitive calibration equipment, the instrument to be tested should be thoroughly cleaned, especially if it has been exposed to corrosive chemicals or dirty substances.

Although normally a first task for new technicians, calibrating pressure gages should not be taken lightly as a menial task; every time an instrument technician or mechanic says a given instrument is calibrated, that person has taken responsibility for the safety of fellow workers and the safe operation of a process.

In conclusion, it should be noted that safety is a primary function of pressure control. Consequently, a few extra dollars are always justified for a more reliable installation. It follows that only properly qualified and well-trained technical people should be used in its application.

References

1. Moore, J. F., others, "Process Control and Guide to Process Instrument Elements," *Chem. Eng.*, June 2, 1969, pp. 94-164, also *Chem. Eng.* reprint no. 68.
2. Chalfin, Sanford, "Specifying Control Valves," *Chem. Eng.*, Oct. 14, 1974, pp. 105-114.
3. Harvey, Glenn F., edit. "Standards and Practices for Instrumentation," 3rd edit., Instrument Soc. of America, 400 Stanwix St., Pittsburgh, PA 15222.
4. Liptak, Bela G., ed., "Instrument Engineers' Handbook, Vol. I, Process Measurement," Chilton Book Co., New York

Meet the Author

J. B. (Bernie) Ryan is an instrument engineer at Fluor Corp., 2500 S. Atlantic Blvd., Los Angeles, CA 90040, where he has been employed since 1965, working on major contracts including a current assignment to the Alaska Pipeline. Before joining Fluor, he worked for Du Pont Co. of Canada, in Sarnia, Ontario. A senior member of the Instrument Soc. of America, he holds a diploma in Industrial Instrumentation (1958) from McGill University in Montreal.

Response of temperature-measuring elements

The response time of thermal elements varies according to the type of element, whether it is bare or in a thermowell, and the medium in which it is immersed. Understanding the response time is critical in selecting proper elements for a system.

Paul W. Kardos, Merck Chemical Manufacturing Div.

☐ Temperature is a controlled variable in many industrial processes, where it is a critical factor in meeting product specifications, as well as in maximizing production efficiency and safeguarding the plant's processing equipment.

In some systems, such as those for temperature control of outside oil-storage tanks, the ambient load conditions change very gradually and the response time of the temperature sensor is not important. However, when the load on the temperature-control system can change very rapidly, as in many exothermic reactions, the speed of response of the temperature sensor becomes an important factor in maintaining good temperature control. Tests of thermal elements show a surprisingly wide variation in response times, depending both on the type of element and the environment of the temperature sensor.

Temperature control often involves heat transfer into the contents of a tank and is characterized by a long first-order time lag. If temperature is sensed instantly and there are no other significant lags, then temperature control is very easy with a zero-proportional-band (or "on-off") control. However, if the response of the temperature measurement is also a first-order lag, then the control loop has two first-order lags in series, producing a second-order system. If τ_1 and τ_2 are the time constants of the slower and faster lags respectively, then the degree of difficulty of regulation is proportional to the ratio τ_2/τ_1, with the most difficult case being $\tau_2 = \tau_1$. In addition, the natural period of the system varies as τ_2 only. Since τ_2 is almost always the lag of the thermal element, reduction of this lag will improve both the proportional-band required for control and the speed of response. In the two-capacity process, derivative control action, theoretically, could completely cancel out the faster lag, making the process appear to be first-order and controllable with "on-off" control. Although the best derivative devices will reduce the secondary lag τ_2 only to another lag equal to $\tau_2/10$, this is still a sizable improvement [*1*].

Most tank temperature-control systems have additional capacity lags such as the heat capacity of the contents of the jacket (or coils) and jacket wall; they also have the dead time of coolant delivery and circulation in the jacket. Thus a more-common temperature control loop can be characterized as multi-capacity plus dead time. Derivative action is not nearly as effective here as it is in the two-capacity system, due to the presence of dead time, but is still useful in compensating for the capacity lags [*2*].

Improved control of temperature can be achieved by

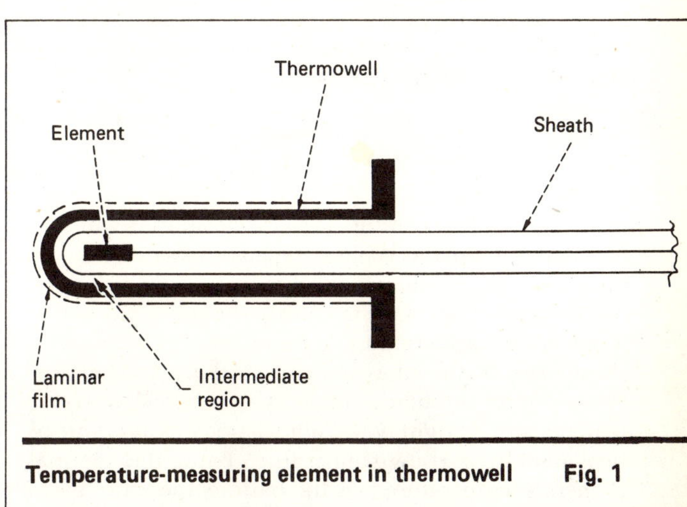

Temperature-measuring element in thermowell **Fig. 1**

Originally published August 29, 1977.

Response times of bare temperature-measuring elements **Table I**

Bare elements	Time for 63.2% response, s: Experimental	From literature
Thermocouples		
1/8-in, sheathed and grounded	0.48	0.34*
Beaded	0.52	
1/4-in., sheathed, insulated		4.5*
1/4-in., sheathed, grounded		1.7*
1/4-in., sheathed, exposed loop		0.09*
Resistance temperature detectors		
1/16-in., dia.		0.8†
1/8-in. dia.		1.2†
1/4-in. dia.	6	5.5‡
1/4-in. dia., dual element		8‡
Nitrogen-filled bulb		
3/8-in. dia.	2	2‖
Mercury-filled bulb		
1/4-in. dia.		1.6‖
3/8-in. dia.		2.5‖
3/4-in. dia.		6.5‖

Literature sources:
*Thermoelectric (boiling water)
†Rosemount (oil at 3 ft/s)
‡Rosemount (water at 3 ft/s)
‖Taylor (agitated water)

Nomenclature

A Area of heat transfer
C Heat capacity of thermal element
C_p Heat capacity of flowing fluid
D Diameter of element
E Activation energy in Arrhenius equation
G Mass velocity of flowing fluid
h Film heat-transfer coefficient
k Thermal conductivity
M Mass of thermal element
R Universal gas constant
T Temperature of thermal element
T_o Initial temperature of element before step change
T_u Temperature of surrounding fluid, or ultimate temperature
t Time
x Thickness
Z Proportionality constant in Arrhenius equation
μ Viscosity
τ Time constant
exp Exponent on the base e

Subscripts

e Referring to thermal element
i Referring to intermediate region between well and thermal element
w Referring to thermowell

minimizing the secondary lags and dead time in the control system. Both jacket capacity-lag and dead-time can be reduced by reducing the volume of the jacket, by using cascade control of coolant temperature, and by maximizing the coolant flow with a jacket recirculation pump. For example, in one particularly hot reaction, cooling was achieved by injecting liquid nitrogen directly into the reactor contents, thus eliminating nearly all secondary lag. Knowledge of the factors that affect the response time of thermal elements will permit the selection of a temperature sensor with minimum contribution to the secondary lag time.

In most exothermic reactions the rate-constant is dependent on temperature according to the Arrhenius equation:

$$\text{Rate} = Z \exp\left(\frac{-E}{RT}\right)$$

At a certain temperature the heat-generation rate will be greater than the cooling capacity, resulting in an uncontrollable exotherm, or runaway reaction. In many reactions it is important to control the exotherm before it reaches the temperature of no-return. With a rapidly rising temperature it is imperative that the chemical operators or automatic safety system know what the temperature is "right now," not what it was a minute or two ago. In a rising-temperature bath, the thermal element's temperature lags the bath temperature by as much as one time constant. Thus, a fast temperature-sensor is needed for this type of reaction.

The batch-wise reaction of phenol and formaldehyde, to produce bakelite resin, is subject to runaway reactions. During the initial phase of the reaction, the operators closely watch the temperature trajectory, to determine if an uncontrollable exotherm is underway (possibly due to an overcharge of raw material). For 30 or more years, the standard temperature sensor for this process was a custom-fabricated thermocouple. A hole was drilled in the tip of a stainless steel thermowell, thermocouple wires were inserted through the hole, from the inside, and silver-soldered to the outside of the well. The resulting device was a mechanically strong element with fast response, but it could not be removed from the vessel for servicing owing to the hardening of the phenolic resin on the threads of the well. With great effort, an old thermowell was recently removed and replaced with a standard filled-bulb system in a thermowell. The time lag of the filled bulb in the viscous resin was of the magnitude of a few minutes and was completely unsatisfactory, halting production. The soldered-thermocouple device was quickly returned and production then resumed.

Derivation of first-order response

To understand the factors that affect response time, it is necessary to go through the mathematical derivation of the response of a thermal element. The response for a general element as shown in Fig. 1 is derived from the unsteady-state energy balance.

$$\text{Input rate} - \text{output rate} = \text{accumulation} \quad (1)$$

If the input rate is of the basic form, driving force divided by resistance, and the output rate is zero, then:

$$\text{Driving force/resistance} = \text{accumulation} \quad (2)$$

The overall resistance is the sum of the following individual resistances:

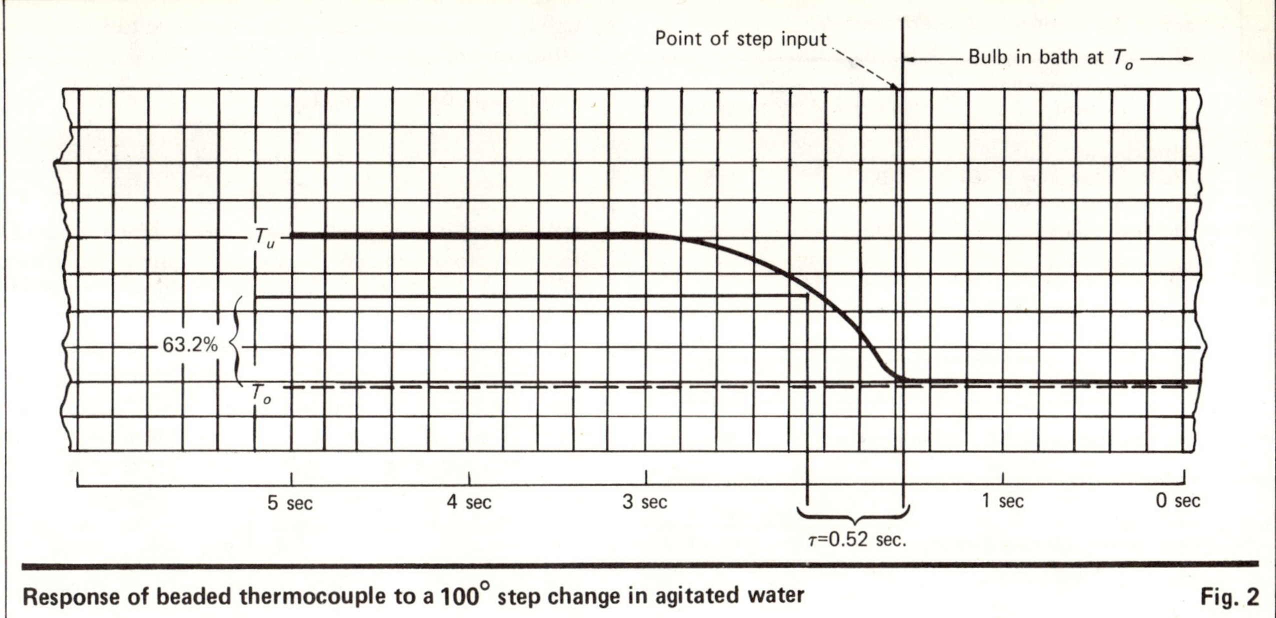

Response of beaded thermocouple to a 100° step change in agitated water **Fig. 2**

Outside Film Resistance—In turbulent flow around any surface, there always exists a thin laminar layer at the surface through which heat must flow by conduction. The coefficient of heat transfer for this layer is defined as h, with the units of Btu/(h) (ft^2) (°F). Outside film resistance is $1/h$.

Thermowell resistance—This is x_w/k_w.

Intermediate resistance—Between the well and the thermal element is a layer of fluid with resistance of x_i/k_i.

Element resistance—The thermal element has a cover or sheath with a thermal resistance of x_e/k_e.

An assumption must be made that the entire thermal capacity of the system is lumped together, having a mass of M with heat capacity C, and that the thermal element is at the same temperature, T, as this mass. The driving force of Eq. (2) is the temperature difference between T_u and T, where T_u is the temperature of the turbulent fluid around the well. Eq. (2) now becomes:

$$\frac{T_u - T}{\frac{1}{h} + \frac{x_w}{k_w} + \frac{x_i}{k_i} + \frac{x_e}{k_e}} = \frac{MC}{A}\frac{dT}{dt} \qquad (3)$$

The solution of Eq. (3) for a step input, with initial conditions $T = T_o$ at $t = 0$ is:

$$T = T_u - (T_u - T_o)\exp\frac{-At}{MC\left(\frac{1}{h} + \frac{x_w}{k_w} + \frac{x_i}{k_i} + \frac{x_e}{k_e}\right)} \qquad (4)$$

Define:

$$\tau = \frac{MC}{A}\left(\frac{1}{h} + \frac{x_w}{k_w} + \frac{x_i}{k_i} + \frac{x_e}{k_e}\right) \qquad (5)$$

Then Eq. (4) becomes:

$$T = T_u - (T_u - T_o)\,e^{-t/\tau} \qquad (6)$$

The response time of a thermal element is determined by measuring the response of the element to a step-change in temperature. At an elapsed time equal to one

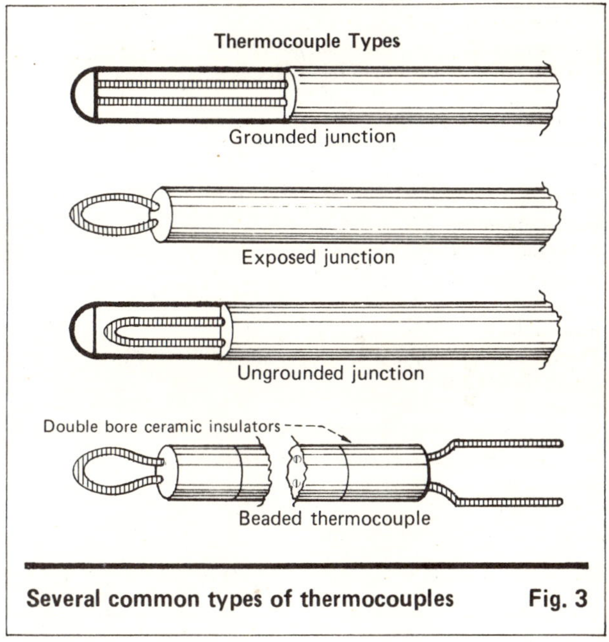

Several common types of thermocouples **Fig. 3**

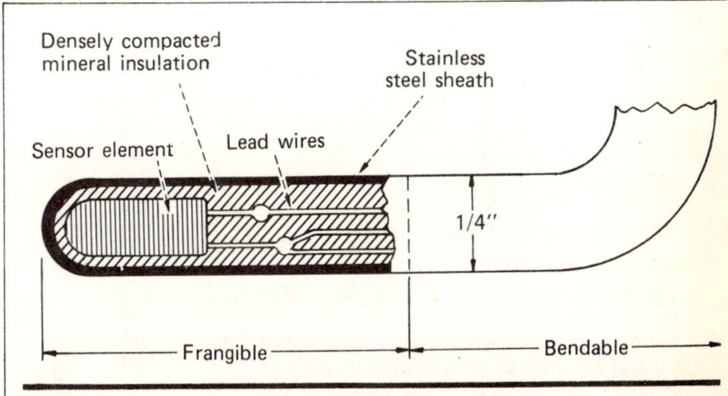

Schematic of a resistance temperature detector **Fig. 4**

Effect of the laminar film coefficient, *h* **Table II**

Temperature-measuring element	Time constant, s In water	In light oil
Thermocouples		
In Hastelloy alloy B well	93	125
In stainless steel well	19	37
In Teflon tube	22	28
Filled bulbs		
Bare	2	5.5
In stainless steel well	17	46

Effect of a thermowell on the time constant **Table III**

Temperature-measuring elements	Time constant, s Water	Light oil
In thermowells of Type 316 stainless		
Thermocouple soldered in well*	8	
Beaded thermocouple; well, 1/8-in.-thick tip, 1/4-in. I.D.	19	
Thermocouple 1/8-in., sheathed, grounded; same well as above	12	35
Thermocouple 1/8-in., sheathed, grounded; well, 3/8-in.-thick tip, 3/4-in. I.D.	23	
Filled bulb in 3/8-in. I.D. well	17†	46
Hastelloy alloy B thermowell, 1/4-in.-thick tip		
1/4-in. RTD in 3/4-in. I.D. well	105	
Beaded thermocouple	93	
Thermocouple, 1/8-in., sheathed, grounded	28	
Glass-lined thermowell		
Filled bulb in same I.D. well	65‡	
3/8-in. filled bulb in 3/4-in. well		355‖
RTD	195	
Beaded thermocouple		155
Thermocouple, 1/8-in., sheathed, grounded	100	
Tantalum thermowell		
Filled bulb	33‡	

Notes:
*Special element used in phenolic-resin process
†Boiling water
‡Pfaudler data
‖Does not include 2 min dead time

Effect of heat-transfer fluid on time constant **Table IV**

Temperature-measuring elements	Time constant, s Water	Light oil
Filled bulb in 3/8-in. I.D. stainless steel well, dry	17*	46
Ditto with mineral oil covering bulb	9*	38
1/2-in. RTD in 3/8-in. Teflon tube, dry	55	
Ditto with mineral oil in tube	41	
1/8-in. sheathed, grounded thermocouple in 3/8-in.-thick well	23	
Ditto with dab of silicone grease	23	
(Effect of well I.D.)		
1/4-in. RTD in Hastelloy alloy B well, 3/8-in. I.D. and 1-in. O.D.	105	
1/4-in. RTD in Hastelloy alloy C well, 1/4-in. I.D., and 3/4-in. O.D. with silicone grease	24	

*Water boiling

time constant ($t = \tau$), the temperature of the element will have reached 63.2% of the distance to the ultimate temperature.

From Eq. (6)
$$\begin{aligned} T &= T_u - (T_u - T_o)\,e^{-1} \\ &= T_u - 0.368(T_u - T_o) \\ &= T_o + 0.632(T_u - T_o) \end{aligned}$$

After another time constant has passed, 63.2% of the remaining distance will have been reached, and so forth. At four time constants, the element temperature will have traversed 98% of the original temperature difference.

Fig. 2 shows the actual temperature response, on a high-speed recorder, of a beaded thermocouple when taken from a cold bath of water (0°C) to a hot-water bath (near 100°C), stirred at approximately 1 ft./sec. For the bare thermocouple, Eq. (5) reduces to:

$$\tau = \frac{MC}{A}\left(\frac{1}{h}\right)$$

which tests out to be very small, at 0.52 s, as shown in Fig. 2.

The response time of a bare mercury-filled bulb is one of the few that can be readily calculated from Eq. (5), which, because there is no thermowell, reduces to:

$$\tau = \frac{MC}{A}\left(\frac{1}{h} + \frac{x_e}{k_e}\right) \tag{7}$$

For a ⅜-in.-dia. by 3-in.-long bulb, the required constants are: M = bulb volume × density of mercury = 0.162 lb; C = specific heat of mercury = 0.033 Btu/(lb)(°F); $A = 0.0245$ ft^2; $x_e = 0.0026$ ft (estimated); $k_e = 9.4$ Btu/(ft)(h)(°F) for stainless steel.

For heating- and cooling-liquids flowing perpendicular to single cylinders, the following equation is used for the laminar film coefficient [3]:

$$h = \frac{k}{D}\left[0.35 + 0.56\left(\frac{DG}{\mu}\right)^{0.52}\right]\left(\frac{C_p\mu}{k}\right)^{0.3} \tag{8}$$

For 99°C water flowing at 0.5 ft/s:

$k = 0.393$ Btu/(ft)(h)(°F)
$D = 3/8$ in. $= 0.03125$ ft
$G = 0.5$ ft/s × 62 lb/ft^3 = 31 lb/(ft^2)(s) = 111,600 lb/(ft^2)(h)
$\mu = 0.000192$ lb/(ft)(s) = 0.6912 (lb)/(ft)(h)
$C_p = 1.007$ Btu/(lb)(°F).

Plugging these values into Eq. (8) yields:

$$h = 709 \text{ Btu/(ft}^2\text{)(h)(°F)}$$

From Eq. (7) and converting to units of seconds, the time constant of a filled bulb is calculated:

$$\tau = 1.3 \text{ s}$$

Experimental response times

Eq. (5) is cumbersome, if not impossible, to use in determining any but the simplest of thermal systems. It is much easier to test the thermal element in question than to try to track down all the factors in the equation. Still, the equation is useful in interpreting the results of thermal element testing and understanding why some sensors behave the way they do.

The response times of bare elements, tested experimentally, are shown in Table I. All response times in this article are for a step-change in temperature, for a low-velocity water system, except where noted. This is the most common system used for measuring response times.

As the elements get larger, the M/A ratio in Eq. (5) gets larger and the response times suffer accordingly. For the RTD (resistance temperature detector) and the sheathed and insulated thermocouple, the element resistance, x_e/k_e term of Eq. (5) becomes significant.

The effect of the laminar film coefficient, h, for water and light oil is shown in Table II for various configurations. The different times show the importance of using a standard test medium when comparing response times. Some typical ranges of h are shown below [4]:

Medium	Range of h, Btu/(ft²) (h) (°F)
Boiling water	300–9,000
Water, hot or cold	50–3,000
Oils, hot or cold	10–300
Air	0.2–10

The large range of the value of h for water is due mainly to the effect of fluid velocity. For water, h increases only about 70% due to temperature effects from 0° C to 99° C from Eq. (8)—but increases almost five-fold if velocity is raised from 0.5 ft/s to 10 ft/s.

Literature of the Conax Corp. gives the following guidelines for estimating the effects of media velocities on time constants:

H_2O at 15 ft/s = $\frac{1}{4}$ × (time constant of device in still water)

Air at 10 ft/s = 4 × (time constant of device in still water)

Still air = 20 × (time constant of device in still water)

The effect of adding a thermowell is shown in Table III, for various element types. The thermowell resistance term, x_w/k_w, is now significant, and effective mass M is increased. Thermal conductivities of well and intermediate materials are as follows:

Material	Conductivity, (Btu) (in.) / (ft²) (h) (°F)
Carbon steel	360
Tantalum	377
316 stainless steel	113
Hastelloy alloy B	72
Hastelloy alloy C	61
Glass	7.5
Teflon	1.4
Light heat-transfer oil	0.91
Air	0.19

For the glass-lined well with the well I.D. mismatched to the bulb element (see Table III), the area of heat transfer between the well and bulb (A_e) is very small, with disastrous consequences to the thermal response. The assumption of a lumped mass at constant temperature is no longer valid. The response is now first-order plus considerable dead time.

The influence of the fluid in the intermediate region between the well and the thermal element is shown in Table IV. For the grounded thermocouple pressed tightly against the well, the x_i/k_i term does not exist because $x_i = 0$. Addition of heat-transfer grease has no effect.

Time constants of some special elements — Table V

Element	Time constant in water, s
Thermocouple in 1/4-in Teflon tube	22
1/4-in. RTD in 3/8-in. Teflon tube	41
Thermocouple embedded in glass of baffle	17*
Resistoflex Fluoroflex thermowell with tantalum cup and grounded thermocouple, and dab of silicone grease	36
Same as above with RTD	73

*Pfaudler data

Filled bulbs are normally slightly bent by the manufacturer so that they are force-fit into a well to maximize contact with the well. However, there are always some voids along the bulb, where it does not touch the well. Wetting the bulb with a heat transfer fluid will increase heat transfer, as shown in Table IV.

The miscellaneous devices shown in Table V are special elements used for specific applications. A bare thermocouple is inserted in a $\frac{1}{4}$-in. Teflon tube for corrosion resistance and fast thermal response. The tube must be strapped to an external device, such as a glass-lined baffle, for support.

The RTD in a $\frac{3}{8}$-in. Teflon tube has some advantages over the thermocouple. The RTD is easier to support and can be serviced externally without having to go inside a tank to install or remove it. However, the thermal resistances, of both the RTD's sheath and insulation, degrade the response time.

The response times listed in this article will seldom, if ever, be exactly the same response times found in industrial processes, due to the profound effect of media physical properties and turbulence. Instead, the response times listed may be used to compare and allow a choice of different temperature-sensor configurations on an equivalent basis.

References

1. Shinskey, F. G., "Process Control Systems," McGraw-Hill, New York, N.Y., 1967, pp. 29–30.
2. Ibid., p. 34.
3. McCabe, W. L., and Smith, J. C., "Unit Operations of Chemical Engineering," McGraw-Hill, New York, N.Y., 1956, p. 459.
4. Ibid., p. 439.

The author

Paul W. Kardos is an Engineering Associate, Instrument Engineering, for Merck Chemical Manufacturing Div., Merck & Co., Rahway, NJ 07065. He has previously worked as a project engineer for Union Carbide Corp., and was a job shopper for Hoffmann-LaRoche Inc. He holds two chemical engineering degrees—a B.S. from Clarkson College of Technology and an M.S. from Northwestern University. He is an associate member of AIChE and a senior member of the Instrument Soc. of America.

Section X
PROBLEM SOLVING, MONITORING AND MAINTENANCE

Simple solutions to control problems
Estimating instrument accuracy
Plant instrument zeroing is now a function of process computers
Instrument arrangements for ease of maintenance and convenient operation

Simple solutions to control problems

Take some simple devices such as impulse relays and pilot-operated regulators, combine them with a little control theory and some ingenuity, and you can solve problems that might otherwise require complex instrumentation.

R. L. Martin, *Tex-A-Mation Engineering, Inc.*

☐ The automatic-process-control field is in the midst of an accelerating technological change and development. Witness, for example, the expansion of computer control—on the one hand the large systems capable of intricate optimization of large and complex processes; on the other, the beginnings of an explosion in the application of microcomputers. Similar developments are taking place in plant-stream analyzer systems, as well as in control and display systems.

In the midst of all this, it is easy to become enraptured with the power and promise of the new technologies, sometimes at the expense of overlooking the availability of relatively simple solutions to many process control problems. This article illustrates some of these simple solutions to control problems through use of (1) instrument computing relays, (2) a little control-system theory and, most of all, (3) ingenuity aimed toward a simple solution.

The examples used in this article are not necessarily new but have been selected to stimulate the imagination of engineers toward practical applications of instrument relays. While such solutions may not be glamorous, they are practical and impressive in terms of cost/benefit ratios.

Before proceeding, it should be recognized that there exists a great deal of bad experience with instrument computing relays. In many of these cases, the lack of success can be traced to improper engineering application—particularly the lack of proper consideration of rangeability and commissioning problems. For example, summing relays are sometimes used to combine two signals for control purposes, with the result that the primary controller can no longer stroke the control value through its entire range. Such a simple misapplication also creates commissioning and operational problems that are due to the signal bias introduced by the relay.

Look at the impulse relay

One of the most useful, versatile and underutilized devices available is the impulse relay. Because it is an unusually versatile and powerful device, let us review its functional operation.

The essence of the impulse relay is shown in electrical

This article was originally presented as a paper to the 33rd Annual Instrumentation Symposium for the Process Industries, Jan. 18–20, 1978, Texas A&M University, College Station, Tex.

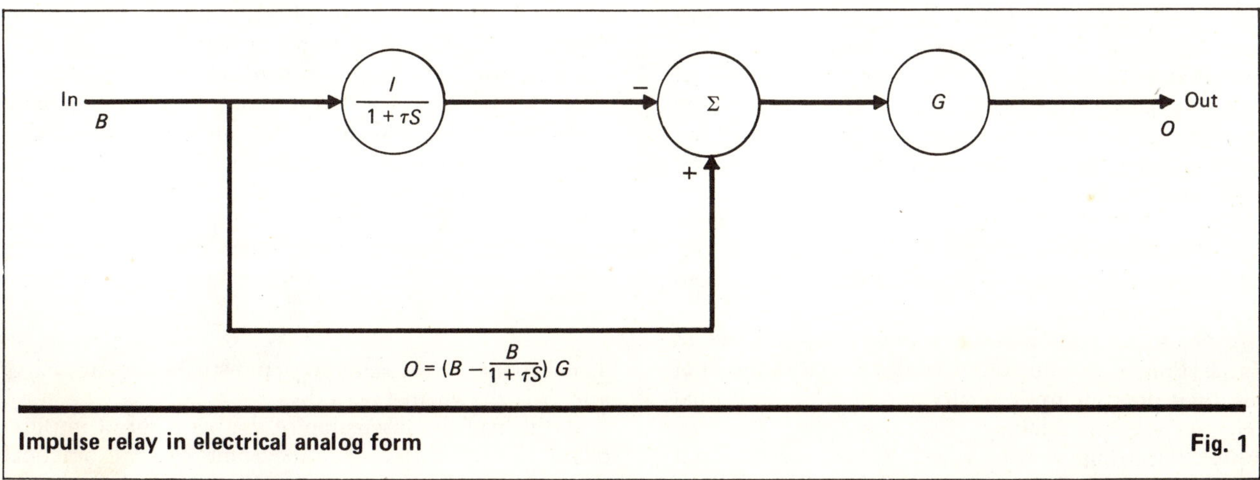

Impulse relay in electrical analog form **Fig. 1**

Originally published May 22, 1978.

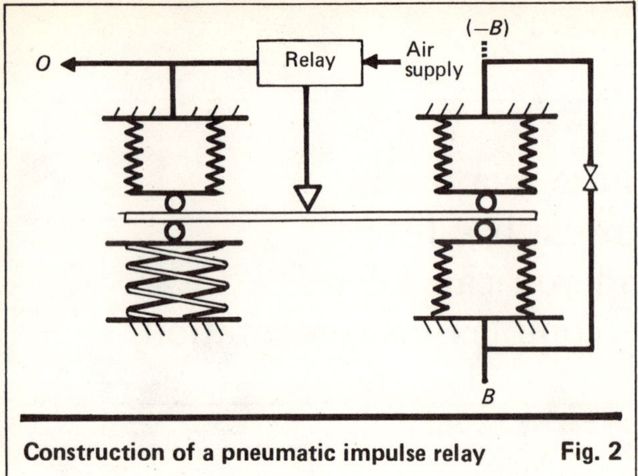

Construction of a pneumatic impulse relay **Fig. 2**

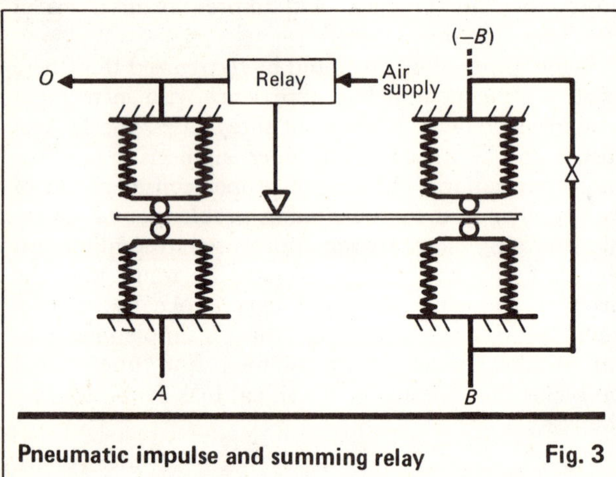

Pneumatic impulse and summing relay **Fig. 3**

Instrumentation symbols

FC	Flow controller
FR	Flow recorder
FT	Flow transmitter
I/P	Current-to-pressure converter
LC	Level controller
PC	Pressure controller
PdC	Pressure-differential controller
TC	Temperature controller
TS	Temperature switch

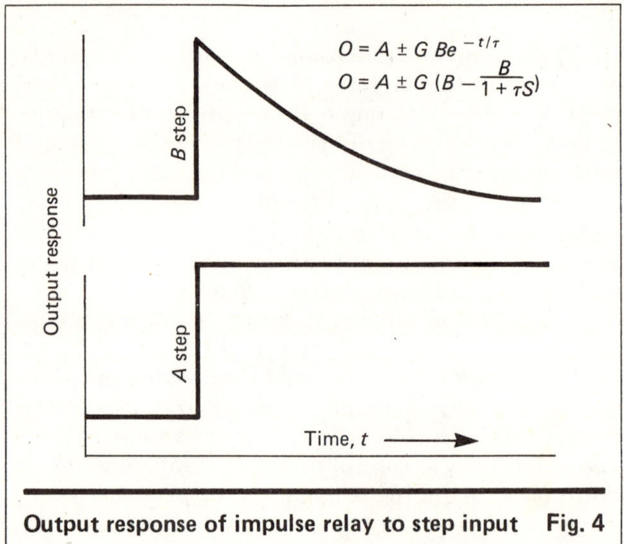

Output response of impulse relay to step input **Fig. 4**

analog form in Fig. 1. The operation of the relay is governed by the formula

$$O = \left(B - \frac{B}{1 + \tau S}\right)G$$

where: B = input value, O = output value, G = multiplier gain, S = Laplace operator, τ = time constant. Consider a step-type change in the input value, B. Immediately following the change, the value of $B/(1 + \tau S)$ is zero, hence $0 = B \times G$. However, as time passes, $B/(1 + \tau S)$ approaches B as a first-order lag function, and eventually the function $B - [B/(1 + \tau S)]$ decays to zero. Adjustment of the gain, G, of the multiplication function provides adjustment of the amplitude of the impulse, while adjustment of the time constant, τ, sets the speed of decay of the impulse, and hence its duration.

A pneumatic version of the impulse relay is shown in Fig. 2. (Pneumatic illustrations will be used in much of this discussion due to simplicity and because most final elements are pneumatic.) The principles, however, apply to electronic systems as well. An increase in input pressure, B, raises the right end of the sensing beam, increasing nozzle back-pressure, which in turn increases the output pressure and restores the beam to a null position. Following the initial pulse, pressure B leaks through the needle valve into the upper bellows (a resistance-capacitance, first-order lag), creating a decaying action. Eventually, the pressure equalizes in both bellows, at which time the pulse function has decayed to zero. In this pneumatic version, the gain may be adjusted by moving the sensing nozzle along the beam; the duration is adjusted by opening or closing the needle valve. Duration of the pulse may be reversed by inputting the signal to the upper bellows (as indicated by the dotted line) instead of the lower bellows.

Neither of the above two versions of the impulse relay is of significant practical use. To render the relay useful requires combination with a summing relay. This can be accomplished as shown in Fig. 3 by replacing the spring with a bellows into which input A can be introduced. The operation of this relay is governed by:

$$O = A + G\left(B - \frac{B}{1 + \tau S}\right)$$

The response of the relay to step disturbances in each A and B is shown in Fig. 4.

With this brief discussion of the behavior of impulse relays, we can now examine some typical practical applications. (Some commercial impulse relays do not include a multiplication function, so practical application will require addition of a second relay in the B input-line, to multiply B by a selectable constant.)

Fig. 5 illustrates a simple feedforward application on

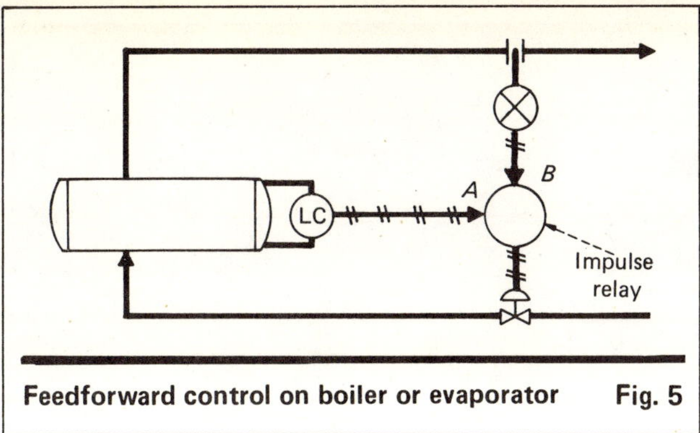

Feedforward control on boiler or evaporator Fig. 5

a small process-evaporator or boiler. The basic control provides for level control on the drum to adjust the liquid-makeup valve. The level controller is normally set with a wide proportional band to provide an averaging-type level control. Without the impulse relay, a sudden change or disturbance in the vapor withdrawal rate will cause a large change in level in the drum. (Of course, the reset action of the level controller will eventually eliminate the offset.)

Performance can be improved by the addition of the impulse relay, which provides an immediate signal to the control valve, hence providing a feedforward signal from the vapor flowrate to the level-control valve. Proper selection of the gain and time-constant of the relay will practically eliminate any effect of the disturbance on the drum level.

The above simple example can be used to summarize several important points regarding the application and advantages of the impulse relay:

- The impulse relay provides a means of combining both measurement and control signals.
- The time constant of the impulse relay should normally be set to a higher value than the effective reset rate of the primary controller. This permits the reset function of the controller to track, and compensate for, the decay of the pulse function.
- Insertion of the relay into the circuit does not restrict the ability of the primary controller to stroke the valve through its entire range, except during the transient period. Note that the use of a summing relay in place of the impulse relay would permanently restrict the primary controller in this regard.
- For commissioning purposes, the time constant of the impulse relay may easily be set to 0, effectively eliminating the feedforward feature. Thus, the level-control will function as if the relay were absent on both manual as well as automatic modes. On more complex applications, this feature becomes the difference between a workable and an unusable system.

The simple application in Fig. 5 can be extended to a variety of level-control configurations by providing a feedforward of the disturbance-variable measurement to the control valve, thereby reducing the deviation of the controlled (level) variable.

A waste-gas-fired furnace

Fig. 6 shows an application of the impulse relay, along with other standard relays, to provide feedforward control of a furnace fired by plant waste gases having widely varying requirements for fuel/air ratios. In this example, the blend-gas fuel supplies are measured by flow transmitters. The multiplication relay is required to scale the relative air requirements of the two blend-gas measurements. The high-select relay chooses the greater of the two air requirements, which in turn generates the impulse signal through the impulse relay. In operation, the oxygen measurement operates in a feedback path, through the combustion control system, to adjust the combustion air supplied to the furnace.

In this example, either of the blend-gases requires a significantly higher air/fuel ratio than does the normal fuel gas. Also, changes in blend-gas supply occur quite suddenly. A sudden supply increase to the furnace requires additional combustion air immediately to prevent starving the furnace. The response of the oxygen control to demand disturbances is too slow (due to sample lags, etc.) to permit adequate response to sudden increases in blend-gas supplies. Decreases in blend-gas do not cause serious problems, since they incur only excess combustion air (a safe condition), and only for the length of time required for the oxygen analyzer to take corrective action.

With the addition of the impulse feedforward circuit, the pulse caused by the increase in blend-gas supply acts to increase the combustion air immediately. The relay gain-factor is set to provide the correct amount of air increase. The relay time-constant must be set to a value greater than the reset time in the oxygen controller. This permits the oxygen analyzer to compensate for the decay of the impulse signal.

A graphical representation of the functioning of these signals in reaction to a step disturbance in blend-gas supply is shown in Fig. 7. Proper adjustment of the impulse-relay gain and time-constant in relation to the analyzer response-time can approximate a step response to a step disturbance, as shown. It is recognized that this response might give a perfectionist the "rigors," but we are concerned with the practical aspects of controlling the disturbance, and this system reduces its magnitude to an acceptable level.

In adjusting the relay, the relay time-constant must be greater than the response time of the base control to permit the base controller to compensate for the gradual decay of the impulse. Even though the impulse signal has decayed to zero, the correct combustion-air is maintained by the oxygen controller. In effect, the impulse relay has made the initial required change for a long enough time to allow the primary controller to track and hold the steady-state requirement.

Stabilizing a refrigeration compressor

Another very interesting application of the impulse relay is to compensate for a fundamental process problem. Fig. 8 shows a basic two-stage refrigeration compressor system. When the system was placed in service, an instability was present that was traced to mechanical problems in the condenser, accumulator, and associated piping. The mechanical configuration permitted condensate to accumulate in the exchanger and its piping to the point where it would slug-discharge into an overly small accumulator. This overloaded the accu-

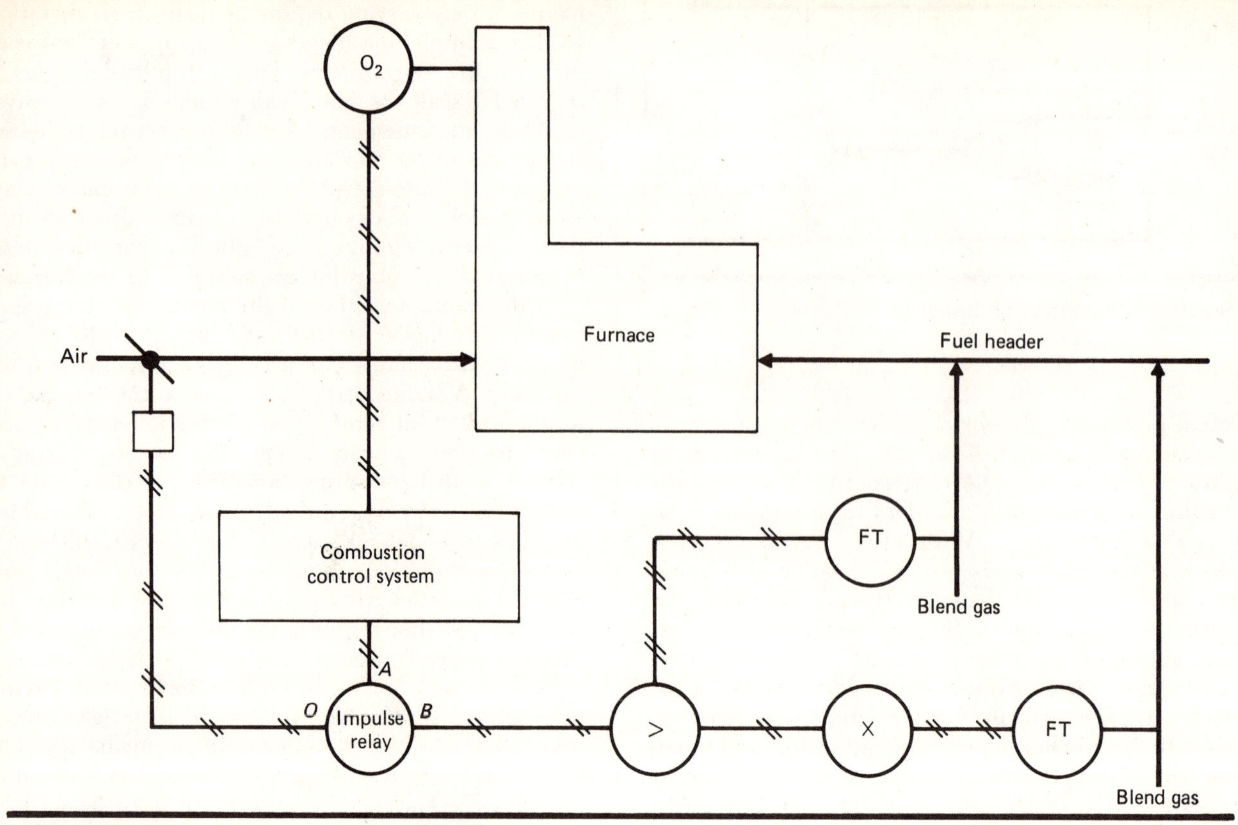

Feedforward control on waste-gas-fired furnace **Fig. 6**

mulator, which could be relieved only by recycling refrigerant back to the compressor through the interstage drum. As a result, the entire compressor system would break into a slow, undamped oscillation, driving the compressor into an intolerable condition of surge. Further, the oscillation could not be tuned out by any adjustment of the controls.

While the basic problem could have been solved by redesign and rebuilding of the condenser and accumulator system, this obviously would have been very expensive in terms of both mechanical costs and lost production time. An alternative was to examine means of controlling the system through instrumentation.

Application of a little frequency-response knowledge theorized that the system must not be far from stability, since the oscillation amplitude built up slowly over a number of cycles. While the exact shapes of the frequency-response curves were unknown, it was further theorized that they must look something like Fig. 9, at least in the area of the critical 180-deg phase-crossover

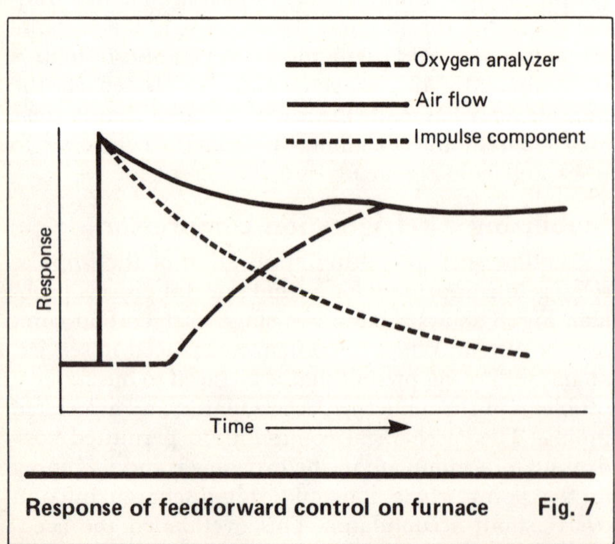

Response of feedforward control on furnace **Fig. 7**

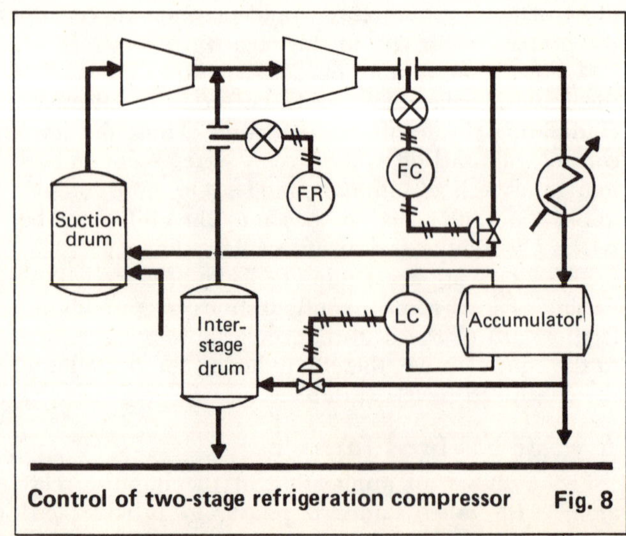

Control of two-stage refrigeration compressor **Fig. 8**

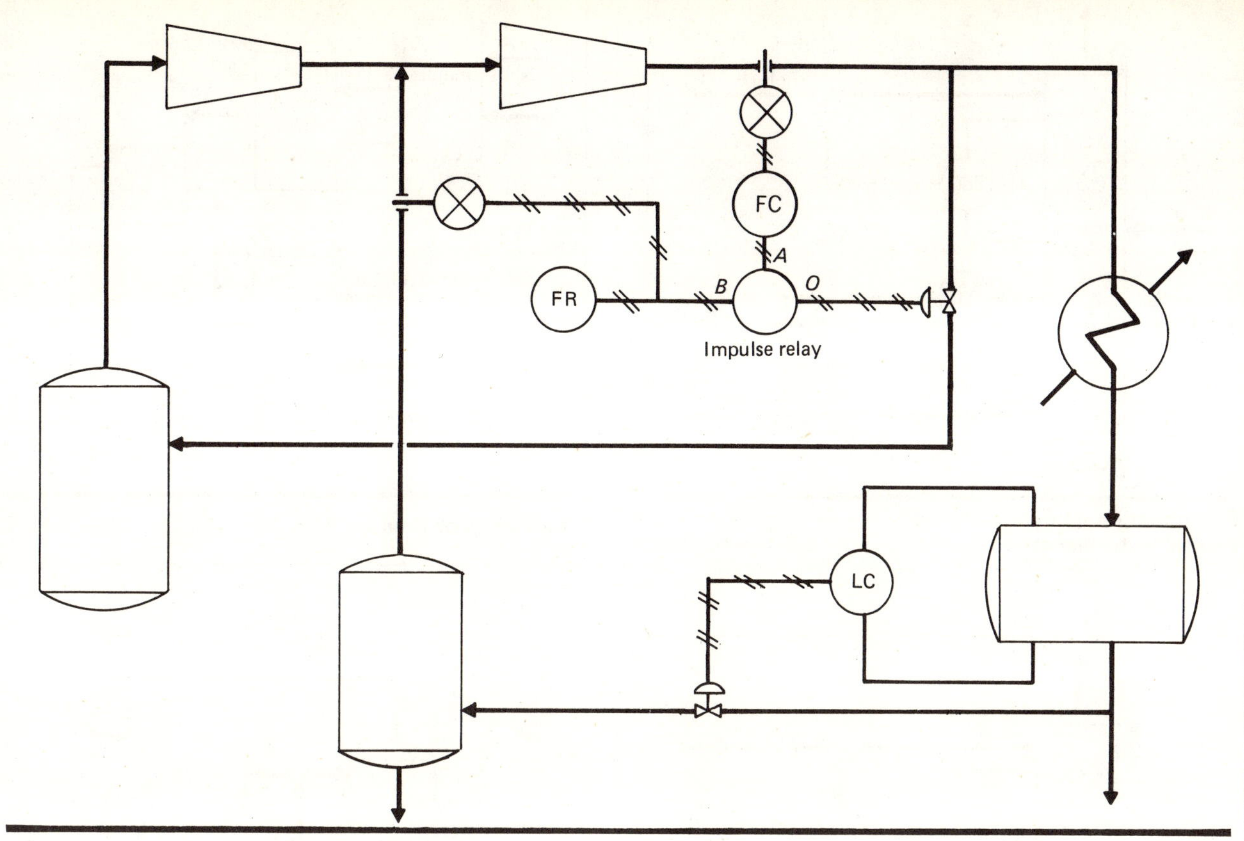

Addition of impulse relay to compressor control **Fig. 10**

point. If this were true, even a minor displacement of the phase curve toward the right might convert the system from an undamped to a damped (hence stable) system, and the problem would be solved.

Observation of the system revealed that the interstage flow-measurement led the compressor discharge-flow by some 15 to 20 s. Coupling this interstage flow-measurement signal into the low-flow bypass control circuit in a feedforward manner would result in a phase advance. It was hoped that this advance would lift the phase curve of Fig. 9 sufficiently to cross the 180-deg phase shift at a magnitude ratio of less than unity, and thereby create the transition from instability to stability. The impulse relay was an excellent device to implement the plan, as shown in Fig. 10.

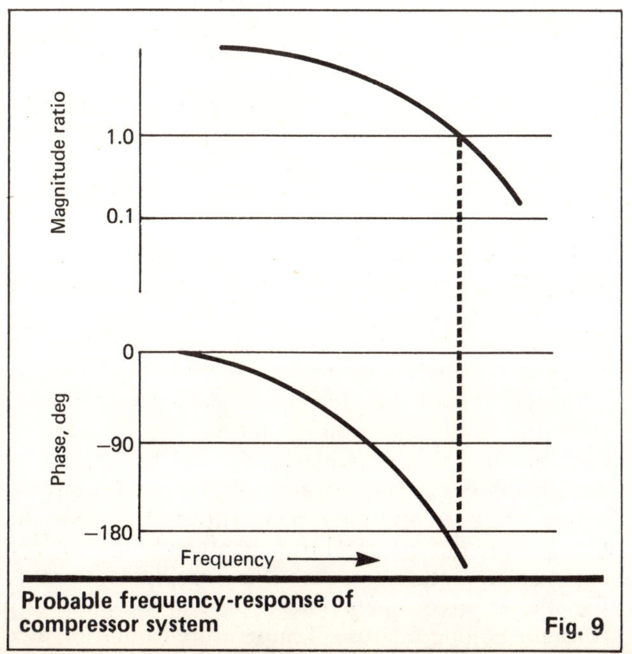

Probable frequency-response of compressor system **Fig. 9**

The impulse relay was installed, did its job, and the compressor system was stabilized. The $500 cost of installation of an impulse relay saved several hundreds of thousands of dollars in mechanical-change costs and lost production time.

Control of fractionators

Fractionation columns pose interesting instrumentation challenges because of the complexity of maintaining material balance, heat balance, and product quality simultaneously. Although it is impossible to generalize completely on fractionation control, many fractionators can be represented by the basics shown in Fig. 11. Heat balance in particular is difficult to maintain because of vast nonlinearity in response to heat input at the bottom, as compared with reflux at the top of the column. However, for every change in heat input at the bottom, a change will be required in the reflux to maintain heat balance. Conversely, every change in reflux necessitates a change in heat input. In the usual situation, the manner in which each of the two ends of the column

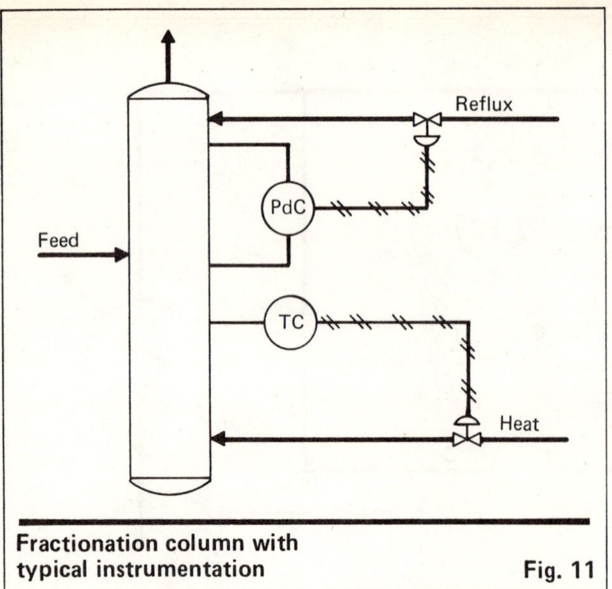

Fractionation column with typical instrumentation **Fig. 11**

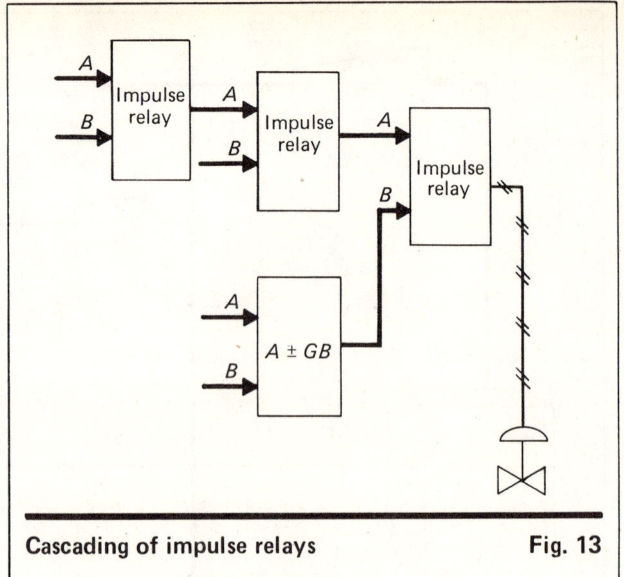

Cascading of impulse relays **Fig. 13**

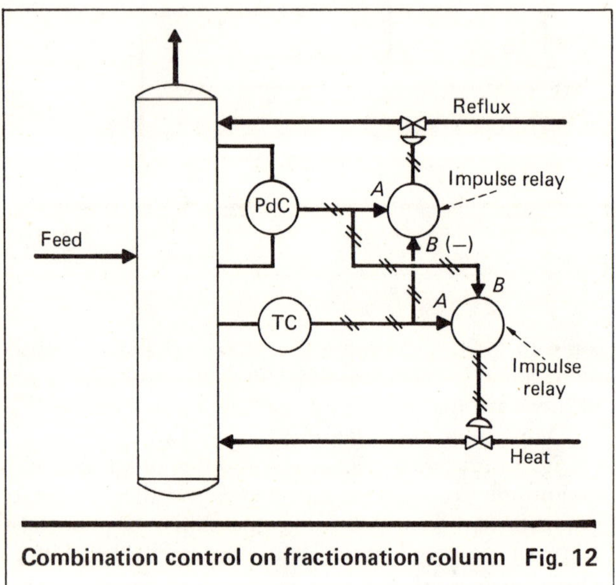

Combination control on fractionation column Fig. 12

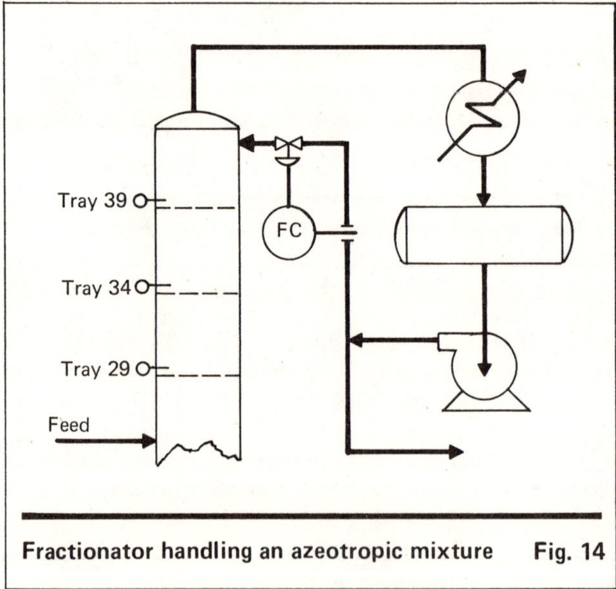

Fractionator handling an azeotropic mixture **Fig. 14**

"learns" what has happened at the other end is after the effect has worked its way, tray by tray, up or down through the tower, with delay lags up to an hour, or even more.

The response of the tower can be materially improved by supplementing the slow "internal communication" with external communication through instrument relays, as shown in Fig. 12. A change in reflux, for example, can now be communicated to the heat-input control immediately, and adjustment of the heat input immediately made. Likewise, a change in heat input can be communicated directly to the reflux control, and compensating corrective action initiated immediately. Here again, impulse relays are excellent devices for implementing the external communication scheme. They provide the needed communication, while at the same time permitting the primary controllers to stroke their respective valves through their full ranges without restriction. The system can also be commissioned by first setting both time constants to zero, placing the controllers on automatic, and then setting the time constants to their proper values.

I have witnessed this scheme attempted with ordinary summing relays substituted for the impulse relays. Not only was commissioning next to impossible, but the output of the controllers was limited to an unrealistically small range, e.g., 2 to 3 psi, the exact value dependent on the constants selected. The practical value of the impulse relays is demonstrated by the difference.

Impulse relays also may be used to combine more than two signals into a single output. Fig. 13 indicates how the use of multiple relays could be applied to combine five separate signals into one control output.

Departing from impulse-relay applications, consider the operation of a typical fractionator dealing with an azeotropic mixture, as shown in simplified form in Fig. 14. An azeotrope separates only to the point of a constant-boiling mixture. Temperature cannot be used

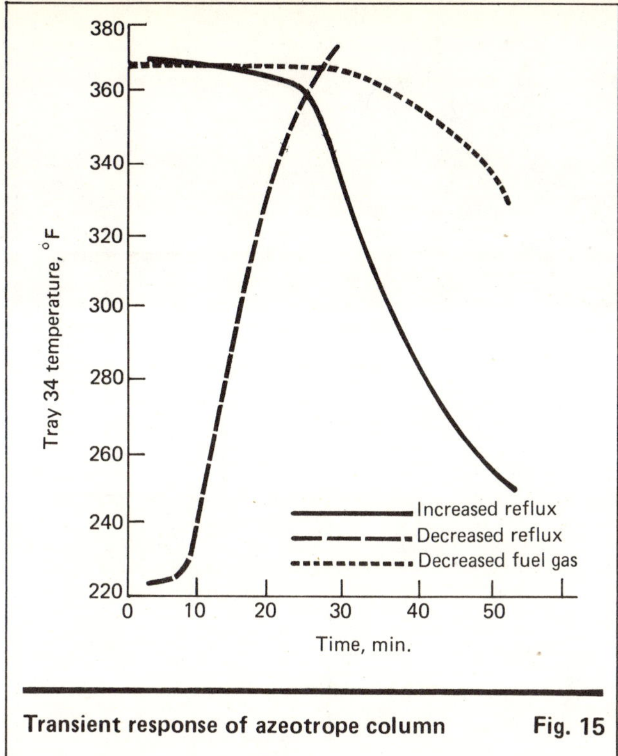

Transient response of azeotrope column **Fig. 15**

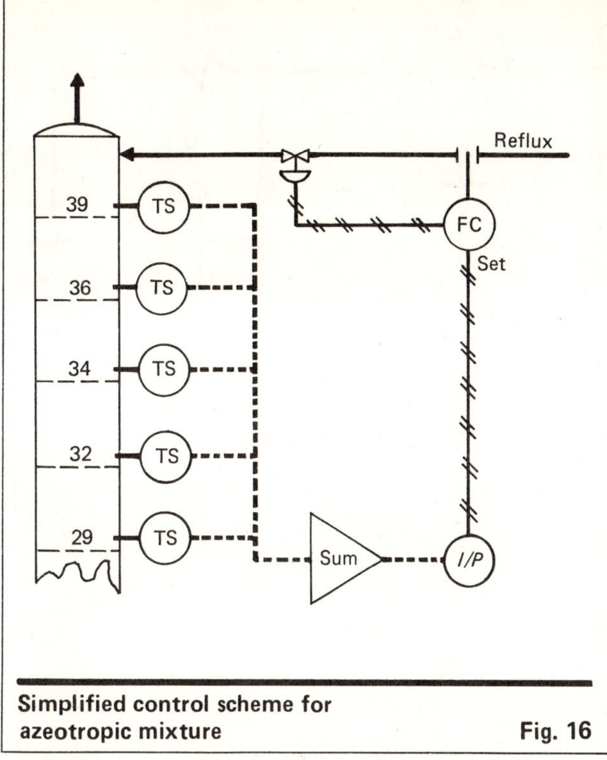

Simplified control scheme for azeotropic mixture **Fig. 16**

directly as a measure of the performance in the constant-boiling range.

Observation of the operation of such a tower shows a sharp temperature break below the azeotrope, with one constant temperature above the transition. This effect is shown in the transient response of a tower in Fig. 15. The three step-response curves were the result of very small changes in the indicated variables. If carried to completion, the responses assume a bistable nature (flip-flop)—i.e., the response goes to either of two fixed conditions, independent of the size of the disturbance. Attempts to control this on-off system with a proportional-and-integral controller is hopeless. The nonlinearity of the upscale and downscale responses of Fig. 15 further complicate the picture.

Automatic control of a system of this type can be accomplished if a means can be devised to create a process transfer-function that is either a proportional or integral function of the control variable. In this case, the level of the "interface" is an inverse measure of the integral of the reflux rate. While level measurement, per se, is not applicable, measurement of the temperature on several trays can indicate the location of the "interface" level as a sharp temperature break.

This approach was implemented as shown in Fig. 16. Five temperature switches were installed between the 29th and 39th trays. All five were set to trip at identical temperatures selected to be midway in the response range of Fig. 15. As the interface rises, each of the five switches is activated in sequence. The digital-temperature signals are then converted to a proportional analog output through a summing amplifier, which in turn sets the reflux flowrate.

Additional detail of the system is shown in Fig. 17. As each of the five temperature switches closes in sequence, the summing amplifier increases the analog output by 20%. With all switches open (interface below Tray 29), the output is zero. The output increases in steps of 20% as each switch closes sequentially due to the rising interface. The pneumatic summing relay allows for (1) placing the system on manual, (2) providing an easy means of adjusting the system gain for control tuning purposes, and (3) providing a means of biasing the reflux-control setpoint to the desired value.

In real application, the summing amplifier resistors were selected as equal values to create even, 20% output steps. It is obvious that a nonlinear response can be easily designed by selecting unequal resistances.

Analyzer control systems offer splendid opportunities for significant performance improvement. A typical system is shown in block-diagram form in Fig. 18. Most analyzers exhibit significant dead times, either in the form of sample-system transport lag or inside the analyzer mechanism itself (such as in a chromatograph), or in both. Further, the dead time is in the feedback path. Such systems can benefit significantly by utilizing a sampled-data controller, in which a dead time is generated in the controller equal to the dead time in the analysis.

Such a sampled-data controller can be easily achieved by installing a three-way valve in the controller circuit, as shown in Fig. 19. The three-way valve is operated periodically by a timer set at a cycle-time equal to the system dead time. When the valve is positioned to connect the two inputs to the summing junction, the controller sees zero error and hence is dead—i.e., takes no corrective action. When the valve connects the feedback signal to the negative summer junction, the controller generates a control output as a function of the error.

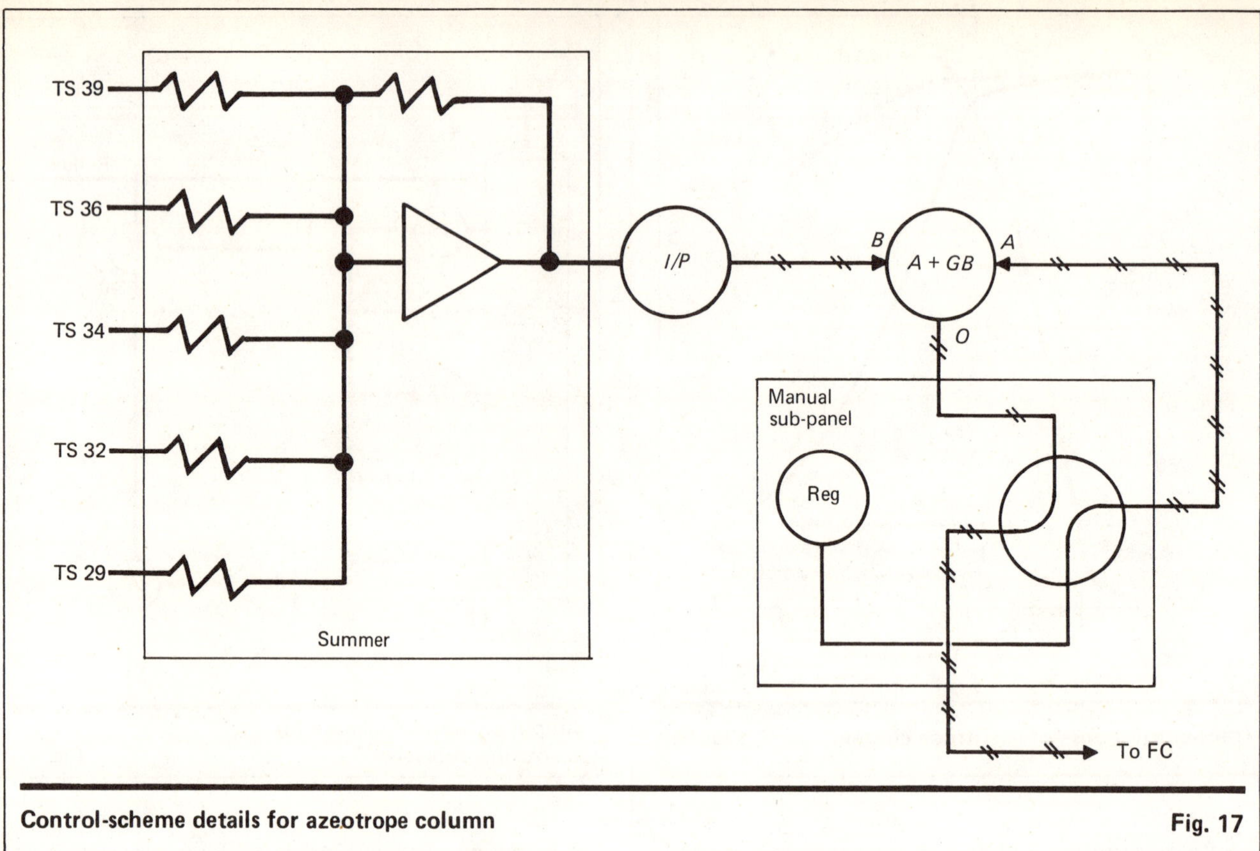

Control-scheme details for azeotrope column — Fig. 17

The valve is operated by a timer (usually a part of the analyzer). The timer should be set to activate the controller only for a short period, and only when new analysis information is available. In the case of the chromatograph, this is when the new analysis is outputted. For a continuous-type measurement, such as pH, the timer should have a cycle approximately equal to the total system lag. In operation, the system reads a new piece of data and takes full corrective action based on that information.

The controller should then be deactivated and no further action should be taken until the results of the change have been effected and measured. Only then should further corrective action be made and only if an offset exists.

A somewhat different type of problem involves the unstable control system—i.e., one that oscillates continuously. A common example is a furnace fuel-gas supply system with a pressure-regulated pilot-supply valve having the pressure-sensing tap located remote from the valve. (Fig. 20). It is not unusual for such a system to oscillate when the main control valve is closed. The oscillation comes about because the lag of the process piping is approximately equal to the lag of the control valve.

One's first impulse may be to relocate the pressure tap to a point near the valve. This would speed up the valve response and thus stabilize the system. It would also produce an undesirable droop in the pressure at the burners, due to the pressure drop in the supply line. An

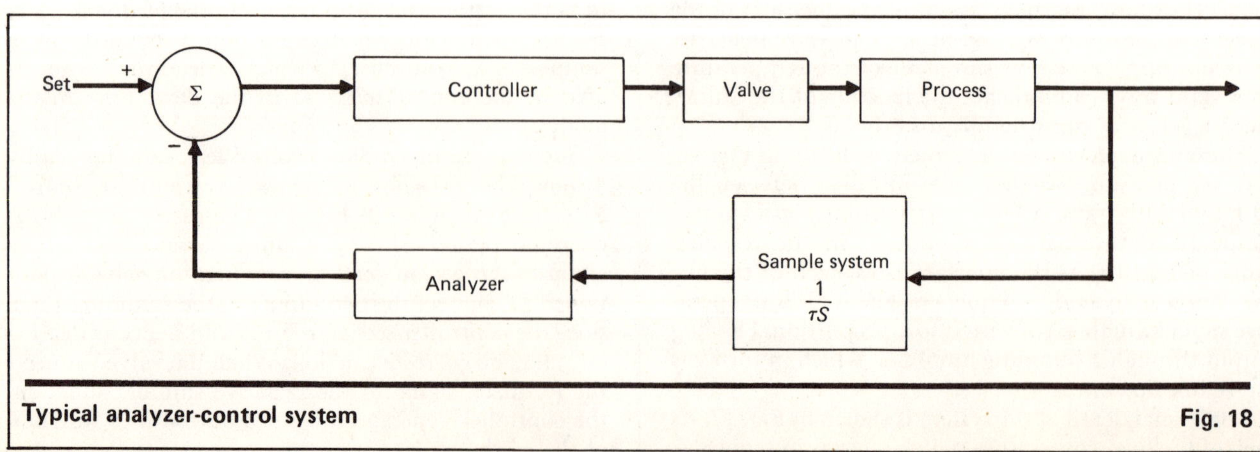

Typical analyzer-control system — Fig. 18

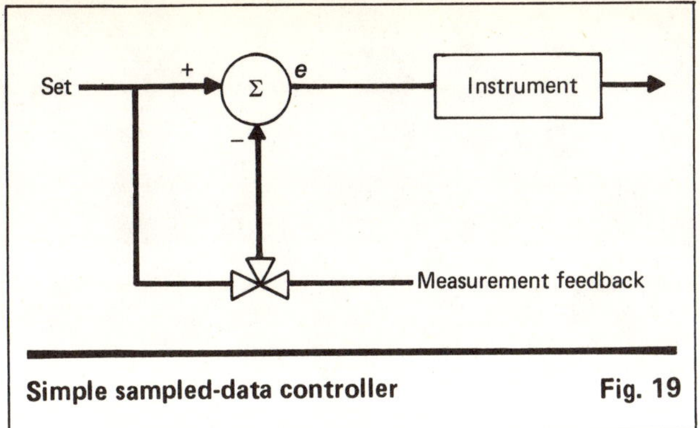

Simple sampled-data controller Fig. 19

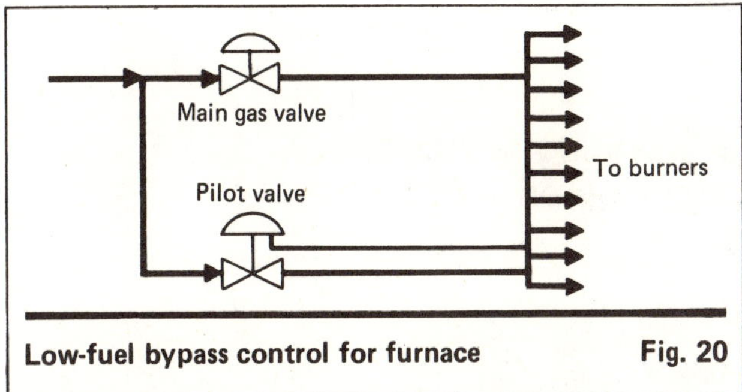

Low-fuel bypass control for furnace Fig. 20

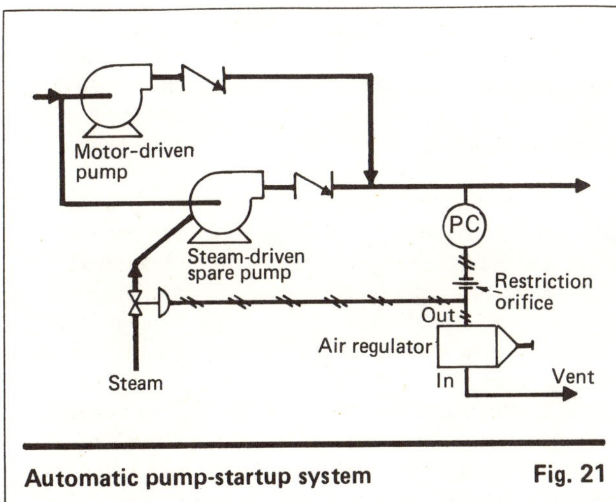

Automatic pump-startup system Fig. 21

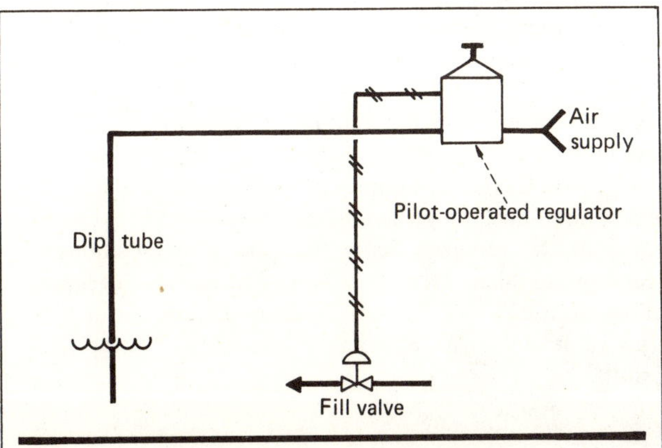

Pilot-operated regulator used for level control Fig. 22

effective alternative is to go the other direction and slow down the valve by using a restriction orifice in the pressure-sensing lead. The important criterion is to separate the magnitude of the two nearly equal lags. To gain stability, *increasing* a lag can also be effective.

Starting up steam-driven pumps

Have you ever needed an inexpensive, foolproof automatic startup for a steam-driven pump? A simple non-pilot-type air-reducing regulator, hooked up as shown in Fig. 21, can do the job. A drop in the discharge pressure of the motor-driven pump is sensed by the pressure controller, dropping its output pressure. As this pressure falls below the setting on the air regulator, the regulator trips, opening the steam-supply valve.

Once the regulator opens, it will remain open until reset, keeping the steam valve open even after the pump output pressure has recovered. Reset is effected merely by holding a finger over the vent port until the pressure in the control circuit exceeds the regulator setting. (If this seems too uncomplicated, you could instead put a spring-return, normally open valve in the vent line.) The restriction orifice is inserted to reduce instrument air consumption during the tripped state.

Level control in open tanks

Probably the most underused of the simple devices is the pilot-operated pressure regulator employed as a level controller on open tanks (such as cooling-tower basins, etc.). All that is required is a bleed-type pilot-operated air pressure regulator in which the air bleed can be piped to a dip tube. Such regulators are available commercially with typical fixed spans of from 1 to 7 in. of water (depending upon the pilot). The pressure spring adjustment provides bias as may be desired.

The pilot bleed is piped to a dip pipe inside the cooling-tower basin, as in Fig. 22. The bleed serves both as the dip-tube purge supply and also as the sensing input to the regulator. The system has the advantage of no moving parts to corrode in the cooling tower. The regulator can be mounted at a distance from the tower.

References

1. Buckley, P. S., "Feedforward Summer" Has an "Impulse" Type Feedforward Signal Injection Method, Twenty-Fifth Annual Symposium on Instrumentation for the Process Industries—1970, p. 6.
2. Warnock, J. D., Interaction Compensator (Circuits), Twenty-Seventh Annual Symposium on Instrumentation for the Process Industries—1972, p. 66.
3. Gracey, J. O., Analog Sampled Data System—Application to a Sulfur Recovery Process With Deadtime, Thirtieth Annual Symposium on Instrumentation for the Process Industries—1975, p. 39.
4. Martin, R. L., and Webber, W. O., Combination Control of Fractionation Columns, Texas A&M Eighteenth Annual Symposium on Instrumentation for the Process Industries—1963, p. 67.

The author

Robert L. Martin is president and cofounder of Tex-A-Mation Engineering, Inc., P. O. Box 1355, LaPorte, TX 77571. He is responsible for the direction and operation of a service corporation specializing in the field of process-instrumentation and automation. He holds a B.S. in chemical engineering from the University of Texas and is actively involved in the Instrument Soc. of America, for which he has served as national vice-president, Houston Section president and (currently) chairman of the admissions committee.

A Simplified Approach to . . .

Estimating Instrument Accuracy

A faulty scale could mean expensive errors in weighing raw materials. Here is how to apply statistical methods to determine errors of accuracy and precision in measuring devices.

BRUCE L. HARSHE, Dayton Power and Light Co.

In many process plants, bulk raw materials make up the largest operating expense. Since millions of dollars per month may be involved, even small errors in weighing can result in substantial economic losses.

Over a period of time, we have developed a relatively simple procedure for evaluating the performance of weighing devices. This work concerns a belt scale used for weighing coal at our power plant, although the technique applies to any instrument or instrument system.

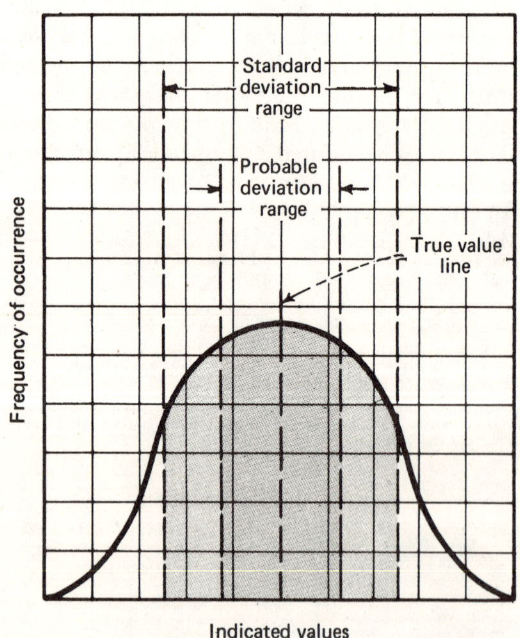

PRECISION ERROR gives bell-shaped curve—Fig. 1

Originally published March 18, 1974.

Types of Error

First, it should be realized that all measuring devices have some degree of error. The question then becomes: Is the error small enough so that it can be tolerated?

There are two types of error—accuracy and precision [*1*]. Accuracy is predictable and may be thought of as a constant-value bias or an error that is proportional to the true value. An example of accuracy error would be a bend in a pressure-gage needle. Once the error caused by the bend were determined—for instance 10 psi low—it would be a simple matter to obtain the true value by adding the amount of error to all future readings.

Precision, the second type of error, is that which introduces uncertainty into the results. Referring again to a pressure gage, friction in the gearing and linkages between the Bourdon tube and the needle will cause an erratic error. A plot of gage reading vs. frequency of occurrence gives the familiar bell-shaped curve shown in Fig. 1.

Fig. 1 shows that sometimes the pressure gage will read a little high and sometimes a little low, with the average value of the readings being the true value if enough readings are taken. Deviation from the true value (degree of uncertainty) will be in a random pattern about the true value unless an external influence is affecting the results.

Standard and Probable Deviation

To express the likelihood of a reading falling within a given range, the normal distribution curve in Fig. 1 is used to define two of the most common precision-error bands. The first is the standard deviation, which is the error band that includes 68% of all the data. The second is the probable deviation, which takes in 50% of all the data.

Usually when the term "accuracy" is quoted it refers to the standard deviation of an instrument, although this is not always the case. Some manufacturers use the "20-to-1 odds" method. This error band is twice the standard deviation and encompasses 95% of the data, hence the name 20-to-1. Still another error band, three times the width of the standard deviation, includes 99.7% of the data [*2*].

When discussing instrument accuracy, the question "What percentage of the readings will fall within a given error band?" is perfectly valid and necessary to determine the instrument's suitability for the application.

Of course, both accuracy error and precision error are possible and in fact probable. A pressure gage could have a bent needle (accuracy error) and friction in the linkage (precision error). Assume that a number of readings are taken with a gage attached to a constant pressure source, with the gage disconnected between readings. If the pressure is a constant 30 psi and if the gage exhibits only precision error, the data could be expected to fall as in the upper portion of Fig. 2, with the average being 30 psi (dotted line). If the gage had only an accuracy error of 4 psi, the data would appear as the solid line in Fig. 2. And in the case of both accuracy and precision error, the data would be scattered, as in the lower portion of Fig. 2, with the average value being 26 psi.

The preceding discussion assumes that the data are not being affected by an external influence, such as a leaky Bourdon tube in the pressure gage. Thus it is necessary to determine whether the data do indeed fall in a normal distribution pattern, that is, in a random fashion free of outside influence.

Analyzing Scale Data

We are now ready to look at the data from the instrument in question, the coal scale. We know that coal-barge draft readings taken before and after unloading will yield "true values" because this method has proven very accurate when checked against weights from certified scales. Data from our actual scale are analyzed by dividing the indicated values by the true values. The results (see table) are arranged in descending order, along with the percentage of values at and below each value.

To test distribution, the data are plotted on probability paper. (Such paper can be made easily by dividing the ordinate into 17 equal segments, as shown in Fig. 3.) If

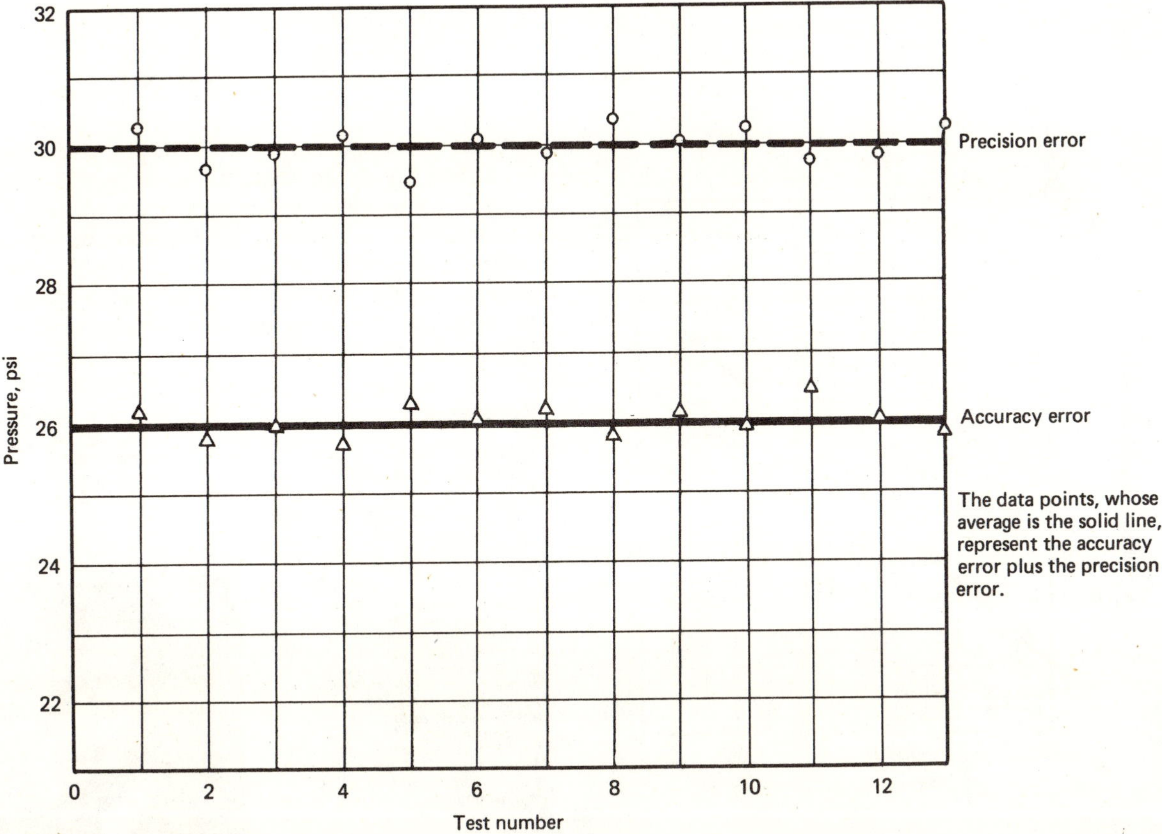

RESULTS from errors of accuracy and precision in readings taken from pressure gage—Fig. 2

Typical Data From Coal Scale

	Data	% At and Below Each Value
	0.989	100.00
	0.985	92.86
	0.982	85.71
	0.979	
	0.979	
	0.979	78.57
	0.976	57.14
	0.975	50.0
	0.972	
	0.972	
	0.972	42.86
	0.967	21.43
	0.965	14.29
	0.962	7.14
Average	0.975	

the data produce a straight line, and if the line passes through or close to the point where 50% and the average value of the abscissa meet, then the data are representative of normal distribution. If, on the other hand, the data result in a curved line, or if the line does not cross the 50% line at the average value, then an outside influence is affecting the system.

When an outside influence is indicated, the system should be investigated further for flaws. However, continuing the data analysis to obtain standard and probable deviations will yield at least a ballpark accuracy figure. Naturally, the closer the data approach normal distribution, the more accurate the error-band determinations will be.

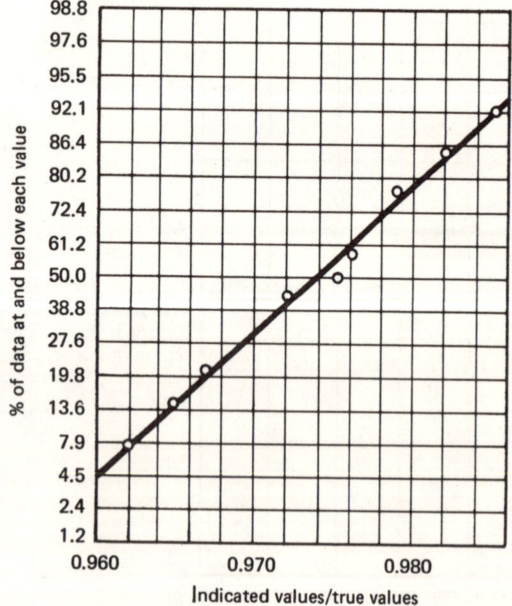

NORMAL DISTRIBUTION is indicated by straight-line relationship of typical data from coal scale—Fig. 3

Assigning Error Values

Once it has been established that the data vary in a normal distribution pattern, values for the errors can be assigned. The accuracy error is the average deviation of the data from the true values, whether a constant value or a constant percentage of the true value. In other words, it is the consistent error that can be corrected—equal to 0.975 in the example presented in the table. Since this is the ratio of indicated value to true value, dividing the indicated scale weights by the average error factor of 0.975 should yield the true values.

Now, the uncertainty, or deviation from the average, can be expressed:

$$s = \sqrt{\frac{\Sigma x^2}{n}}$$

where s standard deviation
x deviation from average value (0.975)
n number of data points

$$s = \sqrt{\frac{754 \times 10^{-6}}{14}}$$

$$s = \pm 0.0073$$

Since the average value in this example is 0.975, the standard deviation of ± 0.0073 is equal to $\pm 0.75\%$.

The standard deviation expressed by the above equation is the most frequent method of describing accuracy. It means that 68% of all readings will fall within this band, and 95% of readings will be within twice the band.

The probable deviation, ϕ—the error band within which 50% of the data will fall—is equal to 0.477 divided by the index of precision, h.

$$h = 0.707/s$$
$$h = 0.707/\pm 0.0073 = \pm 96.85$$
$$\phi = 0.477/\pm 96.85 = \pm 0.0049$$

With the average value of 0.975, the probable deviation of ± 0.005 is equal to $\pm 0.52\%$.

Once the precision and accuracy errors are known, the engineer can either modify the instrument calibration to eliminate the bias of accuracy error or simply correct the readings for the known error. Little can be done with the precision error of the instrument except to adjust the system to minimize its effects.

References

1. Schenck, H., "Theories of Engineering Experimentation," McGraw-Hill, New York, 1968.
2. Doebelin, E. O., "Measurement Systems, Application and Design," McGraw-Hill, New York, 1966.

Meet the Author

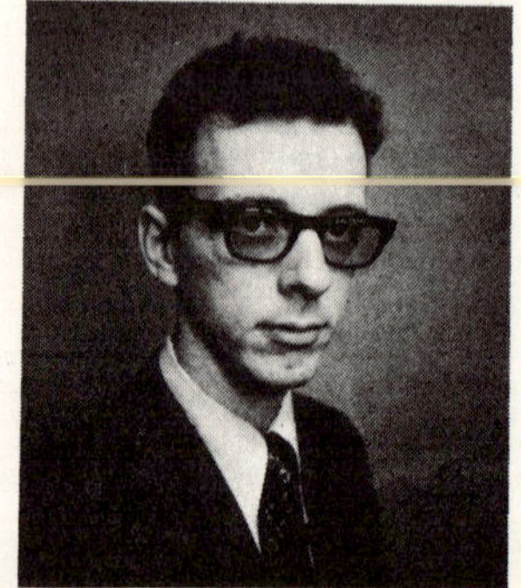

Bruce L. Harshe is assistant plant engineer at the J. M. Stuart Station of Dayton Power and Light Co., P.O. Box 158, Aberdeen, OH 45101. He joined the company in 1969 to assist on startup of the Stuart Station and conduct performance testing of new equipment. He assumed his present position in 1971. Mr. Harshe has a bachelor of mechanical engineering degree from Ohio State University and is working toward an M.S. in nuclear engineering at the University of Cincinnati.

Plant instrument zeroing is now a function of process computers

The control of auto-zeroing of instrument transmitters by computers increases input accuracy. Here, the author describes how this can be done with hydrocarbon streams.

***A. F. St. Pierre**, Cities Service Oil Co.*

☐ The quality of the work output of a computer depends on the accuracy of its input. While programming logic can reject input data that is obviously incorrect, most believable data will be processed. A process-control computer at Cities Service's Lake Charles operations has been extended to control zeroing of selected instrument transmitters to improve the accuracy of the input.

How auto-zeroing is performed

"Auto-zeroing" (or zeroing) of instrument transmitters is accomplished by permitting the computer to control a manifold that equalizes the pressure on each side of a static or differential-pressure transmitter cell. The equalized pressure results in a "zero output" signal. Periodically, a zero-offset error is obtained by the computer to correct transmission of process measurements. Should the offset error be beyond preestablished limits, the computer is programmed to alarm and request instrument maintenance.

Process flows, pressures and other variables measured by plant instruments are transmitted to recorders and to the computer as 1–5 mA electrical signals, which span the range of the measurement. A 1-mA signal is equivalent to 0% of full scale for the process variable being measured.

When a measurement is in doubt, the first thing usually checked by instrument technicians is the signal transmitted at 0% of scale. If the transmitter is zeroed, recalibration normally is not required. Thus, periodic rezeroing of a meter transmitter increases the confidence level of a measurement.

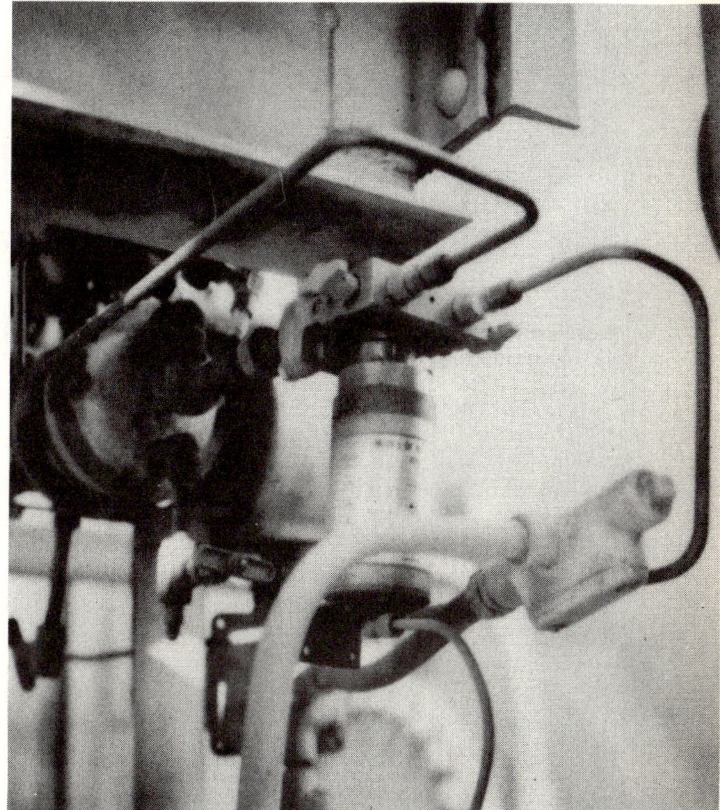

Computer zeroes flow transmitter by controlling pneumatically actuated ball-valve manifold in foreground

Advantages of zeroing instruments

Zero offset readings correct process measurements, upon which costly computer-control action may hinge. For example, hydrocarbon feedrates to the cracking furnaces in Cities Service's ethylene unit are often limited by the effluent gas pressure at certain points in the gas-compression and treating system. Since feedrates to the furnaces are controlled by the computer, accuracy and reliability in measuring effluent constraints are important to production rates. Zero-offset corrections here

Originally published November 10, 1975.

Pressure transmitter is zeroed by computer through control of ball-valve manifold behind board

may prevent some unnecessary feedrate cutback or flaring caused by excessive feedrates.

Need for frequent zeroing

Another example of the need for frequent zeroing of transmitters was found in the measurement of a bidirectional liquid-propane stream flowing between the propane unit and propane storage. This stream varied from no flow to full scale, and was measured with a single orifice and a differential-pressure transmitter zeroed at midscale.

Because the flow of this stream was near zero for a considerable part of the time as the flow changed direction, accuracy of the measurement suffered greatly when it was not corrected for zero-offset error. The nonlinear relationship of the differential pressure and the flow was responsible for the extreme inaccuracy as the flow approached zero. In this case, periodic zeroing of the transmitter by the computer—with subsequent flow corrections—improved the material balance being performed by the computer.

Data collected over a period of four months showed the zero-offset error of the two transmitters to be within the manufacturer's stated specifications. For instance, drift was less than 0.5% over a one-month period under normal operating conditions.

However, we have experienced appreciable sudden zero-shifting in a number of our transmitters over the years. This variance is assumed to be associated with mechanical shock and vibration from surrounding equipment, and long-time wear in the mechanical linkage of the instrument.

Thus, regular and frequent checking—by a computer, as in this particular case—can prevent long periods of inaccurate measurements. Also eliminated is any error introduced when the transmitter output is manually reset to zero.

How zeroed transmitters work

When zeroing a differential-pressure transmitter, a pneumatically actuated ball-valve manifold (which permits remote control) will block the low-pressure side first, and then open the high-pressure port to both sides of the transmitter.

In the case of the static-pressure transmitter, a ball valve turns to vent the transmitter to the atmosphere through snubbers for the zero reading. These two devices have logged over 13,000 and 2,500 trouble-free cycles, respectively.

Neither of the streams serviced revealed any limitations regarding the ball-valve manifolds. The propane stream flow was at 140 psig and 20–75°F. The furnace-effluent-flow conditions were at 245 psig and 200°F. There is no reason why this particular type of equipment, if temperature and pressure limits are correctly adhered to, would not function properly in other similar hydrocarbon streams.

A computer program runs periodically to obtain and update the zero-offset value that zeroes the transmitter. About 2 s are required for the valve to operate, when using about 100 ft of ¼-in metal tubing between the solenoid and the valve. The differential-pressure cell needs an additional 4–5 s to stabilize, while the static transmitter—because of restriction by snubbers—takes a little longer.

At this point the computer reads the transmitter signal several times and stores an average reading to be used for the zero-offset correction. When the readings are completed, the solenoid is deenergized by the computer, and the manifold valve is spring-returned to the measurement position.

Other computer programs

Another computer program—usually referred to as the "Analog Scan Program"—cycles once per minute to read all analog input values, and applies zero-offset corrections where available. Proper interprogram communication prevents the process variable from being measured during the zeroing procedure.

Notice that there are benefits to be derived from remote zeroing of transmitters, even without the use of computers. When zero is called for, either by an operator or a computer, the recorder is zeroed and the total system is checked. The zero signal traced by the recorder can be used to correct the process-flow trace. Safety and convenience are additional advantages to be gained by remote zeroing, when the transmitter is located in very remote or hazardous areas.

The author

A. F. St. Pierre is a process engineer with Cities Service Co. (Box 1562, Lake Charles, LA 70601) in the field of automation development, particularly with digital process-control computers in petrochemical operations. Previously, he was a petrochemical laboratory chemist with the same company. He holds a B.S. degree in chemistry and zoology from Louisiana State University, and has presented various technical papers at area seminars.

Instrument arrangements for ease of maintenance and convenient operation

Fluid flow, liquid level, pressure and temperature will be used to illustrate the principles of graphic design when planning instrumentation for meeting process conditions and the needs of plant operators and maintenance crews.

Robert Kern, Hoffmann - La Roche Inc.

☐ From a layout and piping-design standpoint, instruments are just as important as the structural steel, vessels, piping, or any other component of plant design.

Layout and piping designers make detailed nozzle-orientation studies when arranging instrument connections to vessels. Beyond this, plant designers and project engineers normally pay little attention to all of the details for instrumentation- and electrical-hardware arrangements during the graphic plant-design phase of a project. And overall graphic design rarely goes farther than general routing diagrams for field erection of instrument and electrical systems.

Further, installation personnel often place electric cables, instrument tubing, and components of instrument or electrical hardware in access space intended for other purposes and in locations of questionable functionality. The result is often a poor installation—especially when plant-design practicality interfaces with plant-operation requirements.

Let us review instrumentation in terms of layout design.

Process control

Instruments give information about the physical state and properties of fluids in vessels, piping and process equipment. They indicate, record and control process variables such as flow, liquid level, pressure and temperature. Each instrument has a connection to the piping or vessels for sensing the physical state or condition of the fluid. The sensing element is connected mechanically, pneumatically or electrically (or a combination of these) to an instrument that indicates, records or transmits the physical condition for observation, control and operation.

The final link in the instrument circuit is the control valve (or final control element) that operates according to commands from the sensing elements through the transmitting and controlling devices. In short, automatic-control functions proceed as follows:

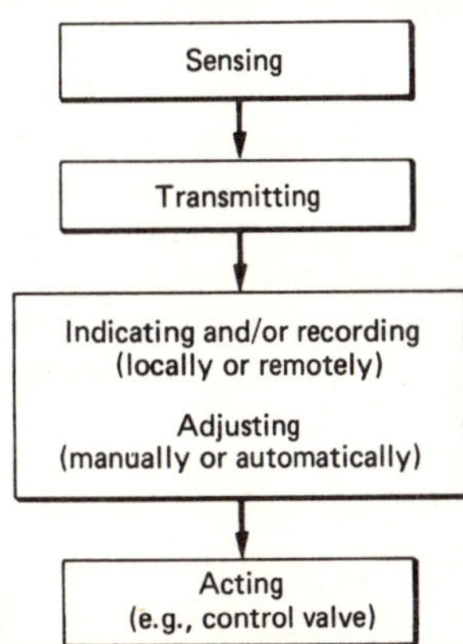

Among these functions, elements for sensing and acting have hardware built into pipelines or other equipment. The transmitting element is usually grouped with the sensing and acting functions. Indicating and recording instruments can be grouped with the remaining functions, or can be mounted on a local

Originally published April 10, 1978.

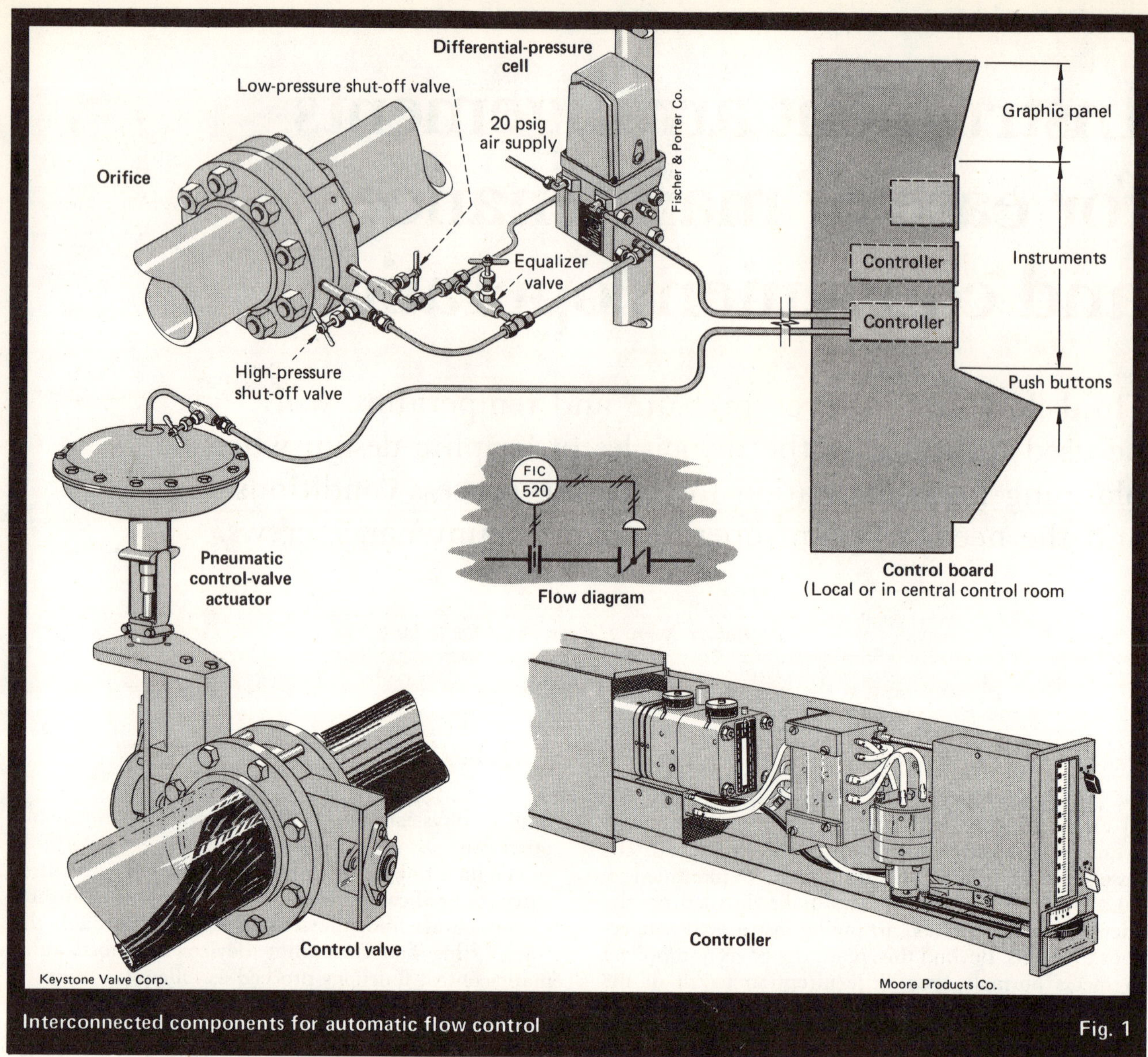

Interconnected components for automatic flow control Fig. 1

panel, or remotely on a panel in the central control room.

Plant-layout and piping designers must be sure that (1) the instrumentation hardware for sensing, transmitting, indicating and acting elements is functionally installed and will operate as intended by the pipe-systems, instrumentation-systems and process designers, and that (2) these components are visible and accessible to plant operators and maintenance crews.

Instrument components and control systems come in endless varieties. However, graphic-design principles can be explained via some basic installations:

Flow control

Fig. 1 shows the interconnected hardware for automatic flow control. An orifice plate in the pipeline creates variations in the differential pressure across the orifice, due to changes in flowrate. Instrument-air tubing connects the transmitter (differential-pressure cell) to the controller/recorder and, in turn, sends an air signal to the pneumatic actuator of the control valve. Instead of pneumatic tubing, a converter can change pneumatic-pressure signals to electrical ones. This will not change the graphic configuration of the hardware. The supporting steel will carry electric cables instead of air tubing.

Flow metering and components sizing have been covered in previous articles [*1, 2* and *3*]. Let us now examine the graphic-design aspects for the arrangement of Fig. 1.

Orifice—In liquid service, an orifice can only operate well if it flows completely flooded. This is also true for all other flow-sensing elements, such as flow nozzles, venturi-pitot tubes, impact tubes and rotameters. The flooded condition is prevalent in pump-discharge and pressurized lines. To measure flow in gravity-flow lines, a static head must be provided in front of the orifice line in order to overcome orifice and control-valve resistances. The termination point of the pipeline, following the orifice straight-run, should be higher than the orifice-line elevation. In short, gravity-flow orifice lines are horizontal, and they must be pocketed.

Practical alternatives for positioning flow-sensing elements in piping systems. **Table I**
(Positions represented by thick arrows are preferred)

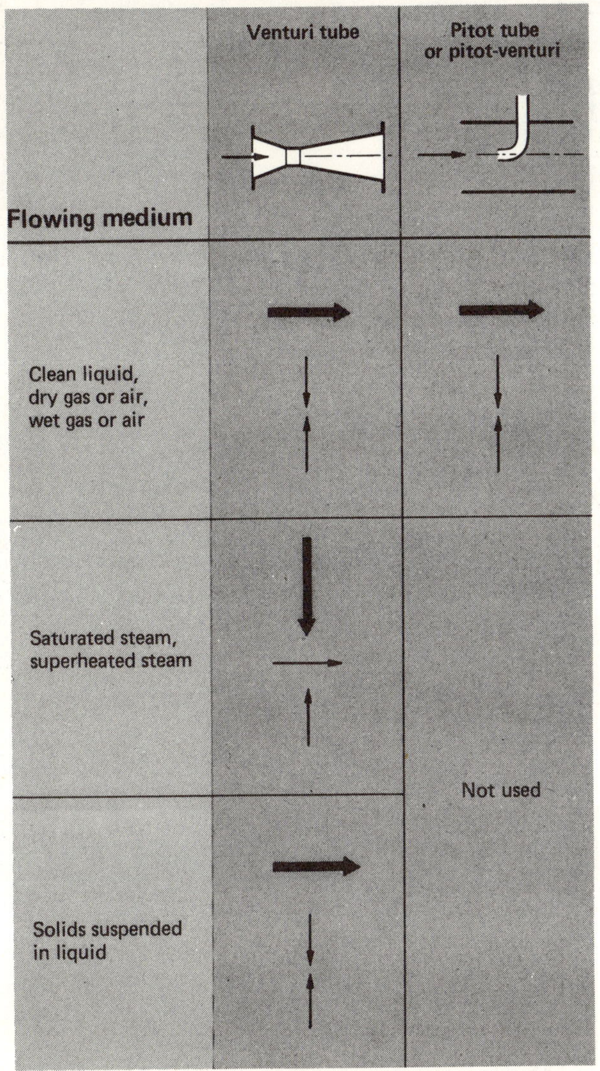

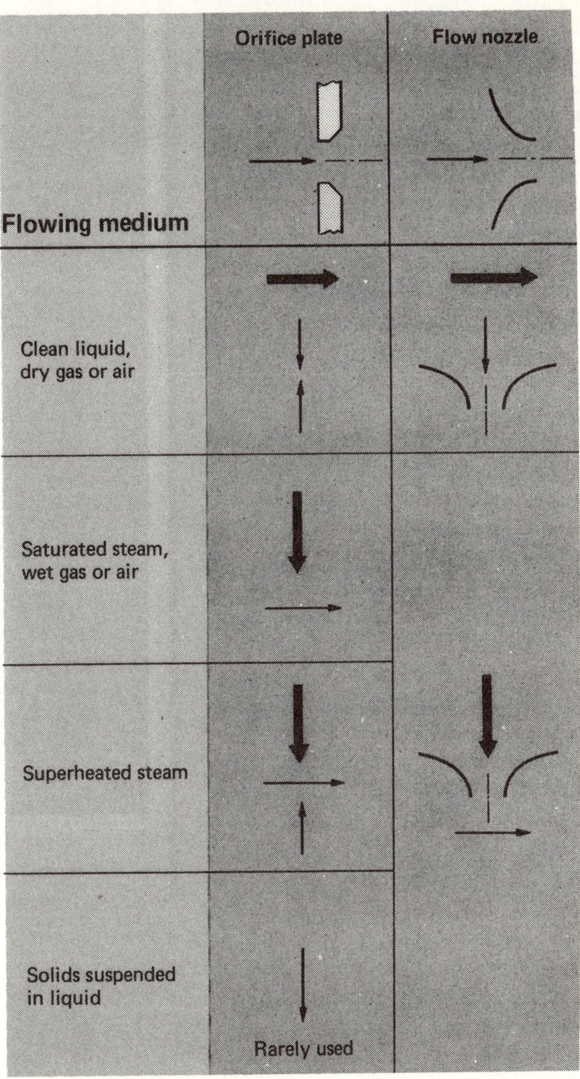

Orifices in pressurized piping can be vertical, horizontal or slanted. Table I shows some practical alternatives for positioning flow-sensing elements in various services.

Two-phase flow cannot be accurately measured with differential-pressure sensing elements.

Transmitter—The differential-pressure transmitter (Fig. 2a) consists of a diaphragm, an attached pivoted linkage, and pressure relay. The diaphragm divides the housing into high-pressure and low-pressure sides. Each side is connected to the corresponding orifice tap. The diaphragm deflects in proportion to the orifice pressure differential. The pivoted linkage (attached to the diaphragm) transmits the motion to an air-nozzle plate, thus opening and closing the air-supply nozzle. This changes the air pressure in the relay, which sends an output signal to the controller, and an air signal from the controller to the control-valve actuator. A feedback bellows in the transmitter balances pressure fluctuations.

Occasionally, the transmitter housing has to be removed to make adjustments. For this purpose, clearance must be given above the differential-pressure cell. The preferred arrangement is close to the orifice [*1*]. Often, differential-pressure cells have to be remote from the orifice. Where necessary, drip legs, sediment chambers and air chambers (as specified by the hydraulic designer) are provided between the orifice and differential-pressure cell (Fig. 2b). These separators should be close to the differential-pressure cell, and connections to them should be self-draining.

The most economical support for differential-pressure cells is on existing structures or walls. Manufacturers of these cells have versatile attachment and support designs. Local supports can be provided with tubular stanchions, flanged to the floor. These should not be placed in accessways. Electric-starter supports can be combined with those for differential-pressure cells.

Controller—The controller, often combining an indicator and recorder, is usually built into an instrument panel board that is mounted locally, or remotely in a central control room. For a board placed in a control room (Fig. 1), considerable space is required, e.g., 3-ft access space behind the board and a minimum of 4 ft in front of it. Long control boards need 6- to 10-ft frontal space for convenient observation by the operators.

Control valves—These are located close to the flow-metering devices. Extensive literature is available on control valves [*3*].

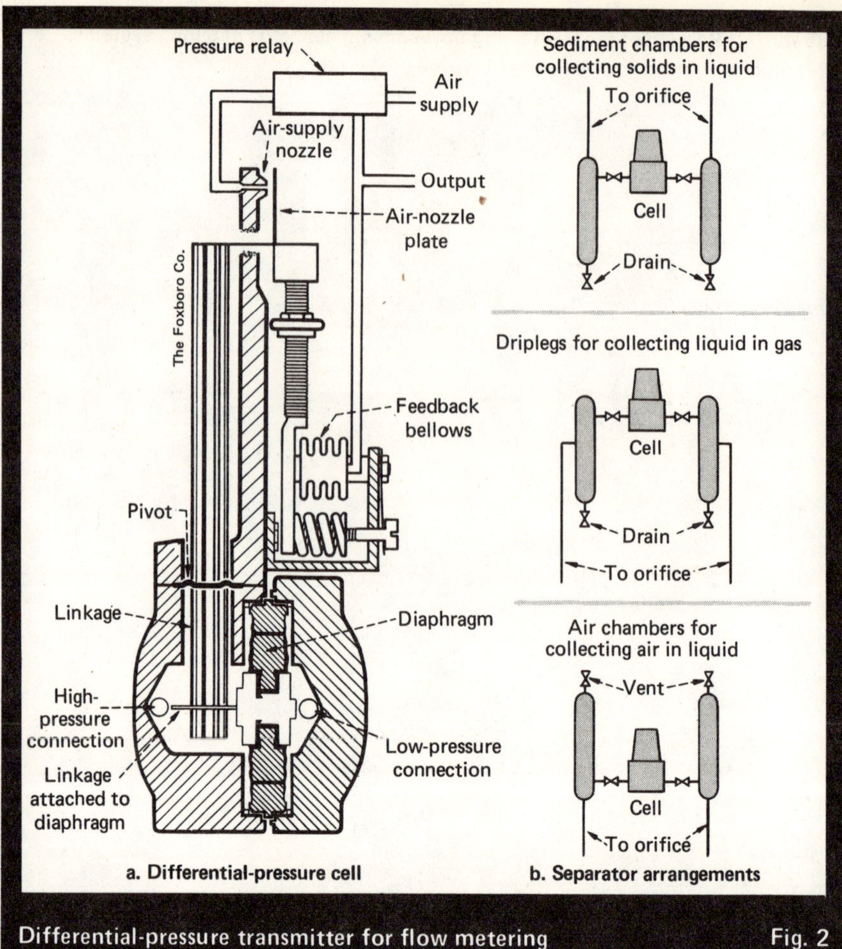

a. Differential-pressure cell

b. Separator arrangements

Differential-pressure transmitter for flow metering — Fig. 2

Liquid-level control

Level instruments are located on process vessels and storage tanks. The level range and elevations for instrument-nozzle connections are determined by the instrument specialist. Orientation of nozzles, access to isolating and drain valves, clearances and access to instrument boxes are usually the responsibility of piping designers.

Displacer-type level controllers are the most widely used instruments in process plants. The principle of this instrument is shown in Fig. 3a. As the liquid level rises in the container, the buoyant effect on the immersed displacer increases and its apparent weight decreases. This weight is held in balance by a torsion bar. At the pivot, an indicator points to a calibrated scale. With no liquid in the chamber, the displacer force will be heaviest and the torsional deformation will be the largest. Fully immersed, the displacer force will be smallest; and the torsional deformation, the least. The controlling range for the liquid level depends on the length of the displacer.

The arrangement of Fig. 3a (with much refinement) is placed in a housing. Piping connections provide the means for process liquid to flow from a vessel through the displacer chamber. Fig. 3b shows an externally mounted, displacer-type level controller with integral transmitter head.

Fig. 3c to 3e show construction details, alternatives and dimensions that affect vessel-nozzle orientation, clearances, and access to the level controller, transmitter, and isolating valves.

For convenient access to the valve and displacer cover, any angular location can be used for the level controller around a vertical vessel, as shown in Fig. 3c. The transmitter can be ordered with the case mounting to right or left (Fig. 3d). Clearances should be given for swinging the cover open. For hot liquids, a finned torque-tube extension is specified. For subzero temperatures, a plain torque-arm extension is used. These extensions increase the distance from the displacer centerline to the transmitter cover (Fig. 3e).

For convenient orientation, the torque-mechanism housing can be rotated relative to the displacer chamber in increments of flange-bolt spacing (Fig. 3c).

The two connecting nozzles on the displacer chamber can be at the side, or at top and bottom. Isolating and drain valves can be accessible from a ladder; platform access to the transmitter is preferred.

The displacer can be placed directly in a vessel such as a storage tank. This is an advantage if the level-sensing range is wide. (Consequently, the displacer becomes very long. Installation and removal clearances must be provided for this version.)

Controller and control-valve arrangements coming after the transmitter are similar to those shown for the flow-control circuit of Fig. 1. The gage glass on the vessel (covering the range of level controller) should be reasonably visible from the related control-valve bypass, so that the operator can observe the glass while turning the bypass-valve handwheel.

There are a number of variations for level sensing [*4*]. The differential-pressure type is a mercury-filled U-tube in which a varying liquid head in one leg is balanced by a constant static head of liquid in the other leg. A mercury float with mechanical linkage transmits the level variations to an indicator or recorder.

The internal (or external) ball-type level indicator/controller is used in vessels containing viscous liquids. A floating ball is placed in the vessel through a nozzle 3 to 10 in. dia. The ball is installed and removed through the nozzle. Hence, clearances must be given in addition to the appropriate nozzle size. The ball is mechanically

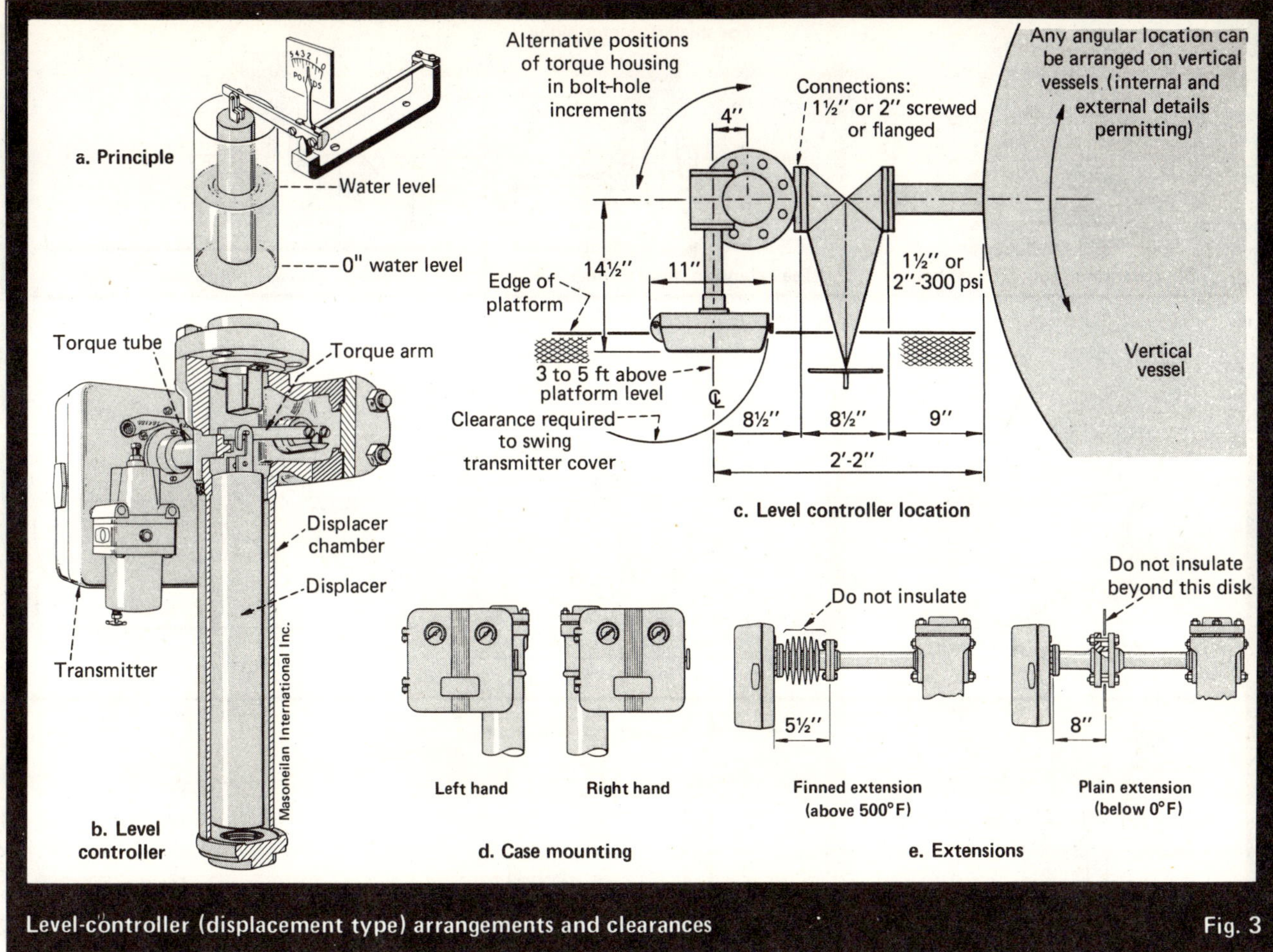

Level-controller (displacement type) arrangements and clearances Fig. 3

linked to a transmitter that converts the signal to an air pressure. An air line is connected to the control-valve actuator.

Temperature measurement

Local or control-room indicators and automatic-control hardware for pressure and temperature are usually not as elaborate as flow and level controls. Applications can range between a control system (similar to Fig. 1) and a local dial indicator. In the following discussion, we will cover only how the sensing elements and local indicators are installed, because much of the information on the hardware for achieving automatic control is repetitive.

Temperature indicators come in great variety, with straight and angular heads so that they can be oriented in a visible position. These indicators are bimetallic or bulb types.

Thermocouples provide the vast majority of temperature measurements in process plants. When two dissimilar metals are joined at one end and heated, a small continuous voltage is generated, which is proportional to the temperature of the thermocouple.

Resistance thermometers measure temperature by changes in the resistance of a nickel or platinum wire. As the temperature increases, so does the resistance of the wire. This resistance is proportional to the temperature change and is detected by an appropriate electrical-bridge circuit. The resulting signal is transmitted to indicating, recording and controlling equipment, either locally or remotely.

Temperature instruments are inserted with or without thermowells (protective tubing) into process furnaces, vessels and piping. Most temperature connections are threaded, and made of high-temperature alloys. When frequent removal is required for cleaning, a 1-in. ball-joint connection is used. In specified services, 1½-in. ball-joint or flanged connections are also installed. The pressure rating for these connections is usually 6,000 psi.

Often, special provisions have to be made in piping for installing thermowells (Fig. 4). Usually, there is no problem in inserting thermowells in large-diameter piping. In overhead lines, reboiler drawoff and return lines, and generally vessel inlet and outlet lines, the temperature points should be as close to the vessel as possible.

Smaller lines may require angular (45°) thermowell connections if the inserted length exceeds the internal diameter of the pipe. A flanged tee can be provided for a longer thermowell. Elbow connections permit the insertion of any length of thermowell. It is impossible to insert thermowells in small-diameter piping (¾ to 2 in.) without restricting the flow. In this piping, a local increase in line size will be necessary.

For long thermowells or thermocouples, a vertical

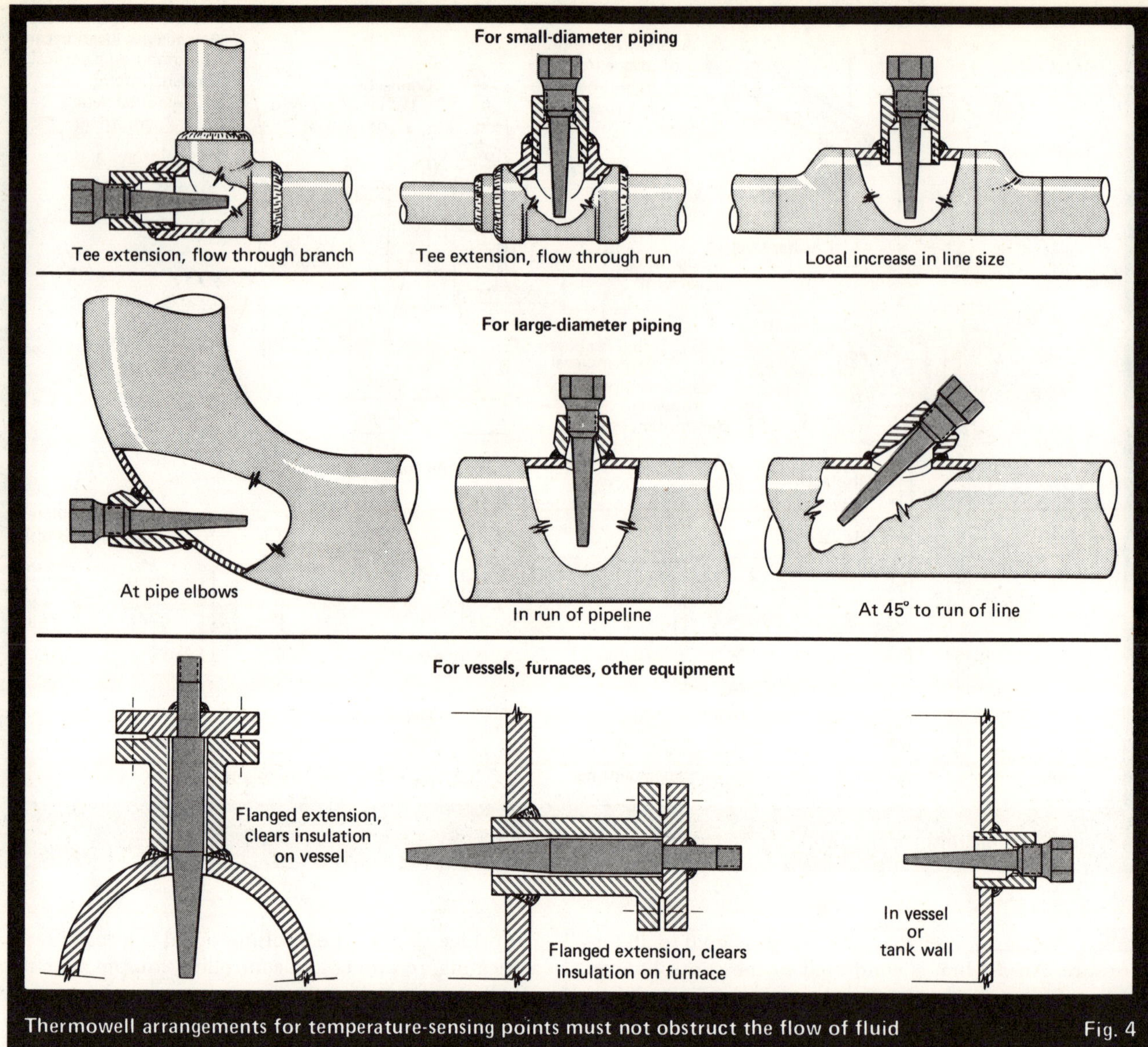

Thermowell arrangements for temperature-sensing points must not obstruct the flow of fluid Fig. 4

arrangement is preferred. A horizontal instrument might need extra support inside the vessel.

The piping designer must find practical locations for all temperature connections in order to satisfy the following: (a) instrument location, as shown on the piping-and-instrumentation (P&I) or the process-control diagram, (b) internal clearances for the length of thermowell, and (c) external clearances for removal and access to the connection.

Dial-type thermometers should be visible from the access space. Temperature instruments are delicate; they should not be located near manhole covers because of possible damage when opening the cover.

Temperature instruments in vessels and piping must give representative readings. For this reason, they are usually placed in the liquid space in vessels, close to outlet nozzles, (Fig. 5a), or at the bottom of the downcomer in a distillation column, with a radial or hillside connection depending on internal vessel dimensions and length of thermocouple. When required, temperature connections are close to inlet and outlet at pumps, exchangers and control valves (Fig. 5b). These temperature locations are downstream of pressure connections, upstream of additive-injection points.

Thermowells should be placed well ahead of orifice runs, or just after the straight length of orifice piping, in order to avoid flow disturbances (Fig. 5c). Where two streams of differing temperatures meet, a minimum of 10 pipe diameters should be allowed in order to mix the flow and even out differences in temperatures before sensing them.

Pressure measurement

Pressure taps are placed in the vapor space of vessels (Fig. 6a); on pump-suction and pump-discharge nozzles; at a minimum distance in piping upflow to exchangers, control valves and orifices (Fig. 6b and 6c).

After a restriction in a pipeline, velocity increases and pressure locally decreases. So for representative readings, a length equal to five pipe diameters is recommended after the restriction or flow disturbances, for locating the pressure point at valves, pipe junctions,

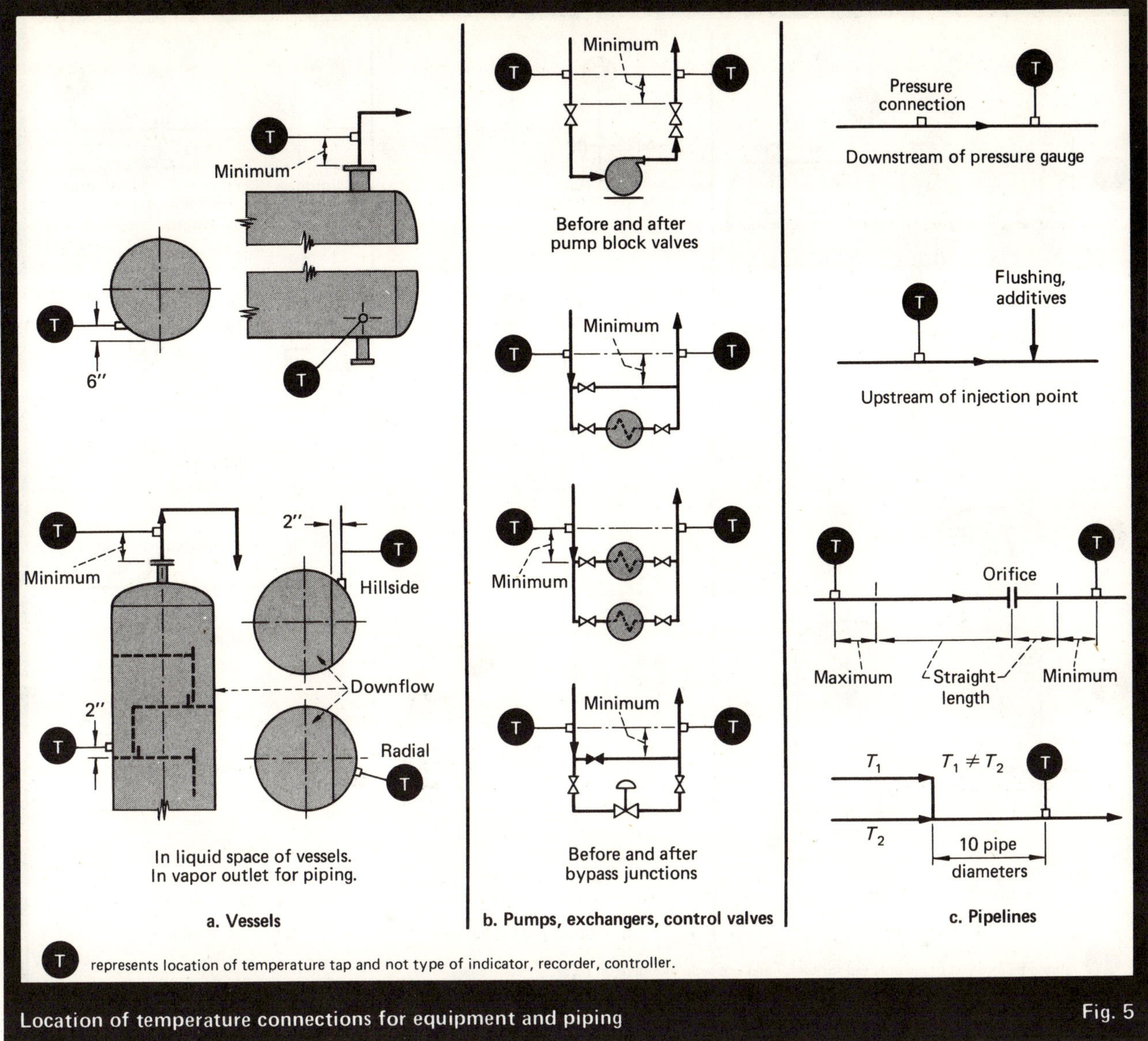

Location of temperature connections for equipment and piping — Fig. 5

throttling valves, and thermowells (Fig. 6b and 6c).

Pressure taps to piping are threaded (3/4 in.) or flanged connections. A valve usually separates the sensing instrument from its piping connection. Local indicators come in straight, angular or movable heads. These should be visible from access aisles. Temporary access by ladder for such pressure sensors and valves is usually acceptable.

Flushing connections are added between the pipe coupling and the isolating block valve when coking, scaling, polymerization or slurry can plug up the pressure tap.

Instrument location and layout

All instrumentation hardware in vessels and piping is provided by layout designers. They are responsible for the visibility of indicators and recorders, and access to sensing points, operating points and control valves. For this work, they need: (a) P&I flow diagrams that show the hydraulic- and instrumentation-systems design; (b) vessel sketches that show location and elevation of instrument connections; (c) physical details and dimensions of instrumentation components, obtained from manufacturers' catalogs or certified drawings.

Orientation and exact locations for instrument nozzles on vessels have to be arranged for at the very early stages of design, after the above-listed information has been released. Vessels have to be ordered, and design finalized, as soon as possible.

Instrument locations in piping are produced during detailed piping design, or at later stages when a piping model is built.

Transmitter and electric-starter supports and locations should be arranged during the planning phase of design [*5*]. For this purpose, a meeting is held by layout, instrument and electrical designers.

For construction, a plot plan or any suitable drawing or model is marked up for final location of supports, transmitters, electric starters and junction boxes. Floor attachments are dimensioned, elevations are given for locations on existing steel, additional steel requirements are indicated and, most important, orientation of in-

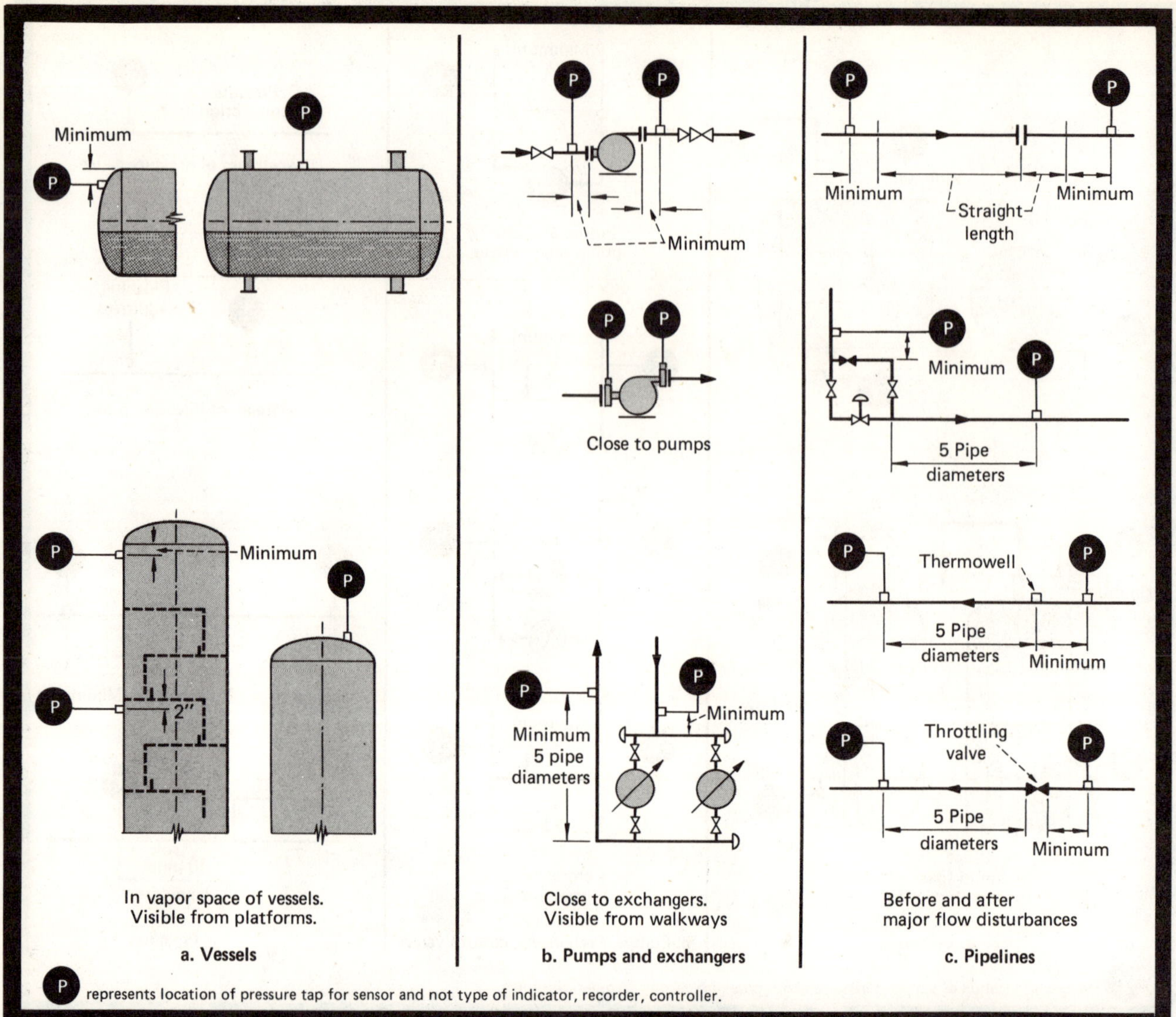

Minimum
Minimum
Minimum
2"
In vapor space of vessels.
Visible from platforms.
a. Vessels

Minimum
Close to pumps
Minimum
Minimum
5 pipe
diameters
Close to exchangers.
Visible from walkways
b. Pumps and exchangers

Minimum
Straight length
Minimum
Minimum
5 Pipe diameters
Thermowell
5 Pipe diameters
Minimum
Throttling valve
5 Pipe diameters
Minimum
Before and after
major flow disturbances
c. Pipelines

P represents location of pressure tap for sensor and not type of indicator, recorder, controller.

Location of pressure connections for equipment and piping **Fig. 6**

strumentation and electric hardware is shown. Operation and maintenance should be convenient from access aisles.

Consideration should be given to the location of analyzers. Space is required for the often-bulky instrument, storage compartments, wash basin, portable equipment, and containers. Sample connections usually have floor or platform access and double-valving arrangements. The sample cooler is connected closely to the sample point. Because operators' attention is frequent, good access to these analyzers and accessory components must be provided.

Operators usually review P&Is from an instrumentation standpoint. The same careful review should be given to the physical locations of all instruments. While such a review can be a manager's activity, an effective model review can be given by the operator who will observe the instruments and turn the valve handwheels.

The next article of this CE REFRESHER will appear in the issue of May 8, 1978, and will deal with plot plans for process units.

References

1. Kern, R., Measuring Flow in Pipes With Orifices and Nozzles, Fig. 7, *Chem. Eng.,* Feb. 3, 1975, p. 72.
2. Kern, R., How To Size Flowmeters, *Chem. Eng.,* Mar. 3, 1975, p. 161.
3. Kern, R., Control Valves in Process Plants, *Chem. Eng.,* Apr. 14, 1975, p. 85.
4. Liptak, B. G., "Instrument Engineers' Handbook," Vol. I, pp. 23–170, Chilton Book Co., Philadelphia, 1969.
5. Kern, R., How To Manage Plant Design To Obtain Minimum Cost, *Chem. Eng.,* May 23, 1977, pp. 130–136.

The author

Robert Kern is head of the plant-layout section in the corporate engineering department of Hoffmann-La Roche Inc., Nutley, NJ 07110. He is a specialist in hydraulic-systems design, plant layout, piping design and economy. He is the author of a number of articles in these fields, and has taught several courses for the design of process piping, plant layout, graphic piping and flow systems, both in the U.S. and South America. Previously, he was associated with M. M. Kellogg Co. in England and the U.S. Mr. Kern has an M.S. in mechanical engineering from the Technical University of Budapest, and is a member of AIChE.

Section XI
INSTRUMENT ANALYZERS

Online process analyzers

Better onstream availability and analytical reliability create greater acceptance of online analyzers. Combined with computers, these analyzers improve profitability via optimization of raw-materials, utilities and process-equipment usage.

Victor C. Utterback, *Shell Chemical Co.*

☐ Online process analyzers provide just about any analysis desired. System complexity can range from a simple pH monitor to computer-controlled analyzer systems. For field installations, process analyzers perform surprisingly well under plant conditions. When properly applied and maintained, they produce an information signal that can be applied in closed-loop or computer control. Our intent here will be to discuss the more commonly used devices and some of their applications, and to present some facets of sampling systems.

Sample systems for online analyzers

Sample-preparation and delivery systems for online analyzers have a design scope as large as the instruments they feed. The simplest system is the insertion of a sensor directly into the process stream. Sample-system complexity increases when the sample is extracted and a delivery system required. Real time usually becomes a factor in such sampling systems for process analyzers. Therefore, the volume of the system between process and analyzer becomes critical.

With increased application of process analyzers, reduced-volume devices and sample-handling equipment have been developed. Special devices such as low-flow pressure regulators, small-volume flow regulators, low-volume switching and self-purging valves, bypass filters, coalescing filters, and small-volume inline filters are now readily available. These devices have been increasingly made with corrosion-resistant materials.

Availability of specially packaged tubing having efficient heat tracing, either steam or electric, and corrosion-resistant material have simplified transportation of vapors and gases. Instrument manufacturers have developed off-the-shelf integral sample-preparation systems for multiple-stream, calibration-fluid-insertion and many other special sample-handling requirements.

A device having particular value in special-analyzer systems requiring cooling is the vortex tube. Vortex tubes provide a very convenient coolant for knockout pots, reflux condensers, analyzer temperature control, etc. Compressed air serves as both the operating medium and source of coolant.

Permeable membranes have been used to a small extent for sampling, and could become more widely adopted.

Off-the-shelf sample-preparation systems account to a large extent for major advances in the capability of onstream analyzers.

Measuring pH and transmitting the signal

The pH meter has been simplified and improved by hardware modernization. So-called permanent-reference electrodes have been developed that can be used at elevated pressures without special precautions needed. In addition, these electrodes do not require solution reservoirs. A pH measuring system has also been developed that uses two glass electrodes instead of the conventional glass/calomel system. This all-glass system is easily temperature compensated and has a low-impedance output, which eliminates the need for close coupling of electrodes to the transmitter for long-distance signal transmission.

Another electrode development has been to incorporate a 1:1 amplifier with the glass electrode, either directly attached or in the electrode holder. This also provides a low-impedance signal for long-distance transmission. Several pH systems incorporate ultrasonic vibrators in order to reduce coating problems at the electrodes.

A rather simple pH meter now available combines a blind pH amplifier with an air or milliamp transducer. The instrument is easily field-installed with no more trouble than for a conventional E/P (voltage to pneumatic air) or E/I (voltage to current) transducer. It is convenient for remote control of pH in applications such as cooling-water towers or effluent streams.

Originally published June 21, 1976.

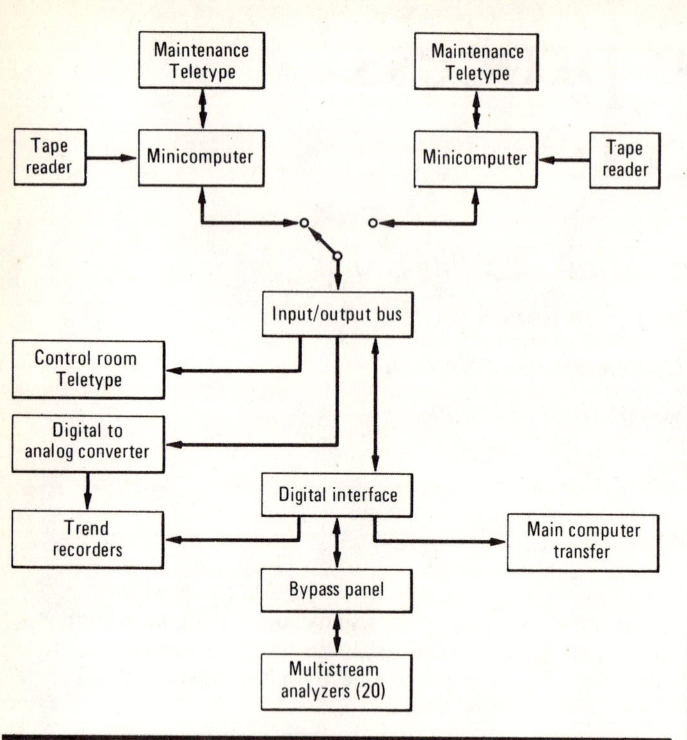

Process analyzers are under direct computer control.

Oxygen content of emission gases

Many applications for oxygen analyzers require the gas sample to be pretreated in order to prevent condensibles from reaching the measuring element. Removing the condensibles gives an analysis that provides an empirical relationship. The information is usable but makes absolute relationships very difficult to determine.

A relatively new oxygen analyzer makes monitoring and control of emission sources such as stack gases easier. This analyzer uses a zirconium-oxide sensor operating at elevated temperatures. Thus, an analysis of the total sample is achieved, including water and other noncondensibles. The analytical probe of such analyzers can be inserted directly into the hot stack gases, or a sample can be extracted and transported to the analyzer.

However, care must be taken to keep the gas sample above its dew point for delivery to the sensor. This analyzer is particularly valuable in simplifying the overall job of source monitoring to meet EPA requirements. Care must be exercised in using the zirconium-oxide analyzer because it is a possible ignition source. Also any combustibles that are present will combine with oxygen in the sample to give oxygen readings that are too low. Therefore, the instrument's application is limited to determining oxygen in inert gases.

Spectrophotometers

Spectrophotometers find broad use as process analyzers because they provide a continuous analytical output, and can be used for closed-loop control or action initiators. Spectrophotometers operate in the ultraviolet, visible, near-infrared and infrared regions. Generally, these analyzers are nondispersive, and depend on various means of filtration to provide more or less monochromatic wavelengths of light for measuring and reference purposes. The light is tailored to the analysis by one or a combination of the following: optical filters, reference-gas mixtures (sealed in a cell), and sensitized detectors. In ultraviolet units, an energy source emitting a limited number of desired wavelengths is combined with optical filters to obtain analytical and reference spectra.

Since physical and chemical characteristics tend to be shared by compounds of similar nature, the spectrophotometer is best suited for simple mixtures to analyze for a component that has a relatively different nature or reaction characteristic compared to the other components of the mixture.

Analysis with ultraviolet radiation

An ingenious application for ultraviolet spectrophotometers concerns the control of a Klaus sulfur-recovery plant. One such system combines two split-beam photometers that use optical filters to produce light of specific wavelengths in order to provide sensitivity for the analysis of sulfur dioxide and hydrogen sulfide. The output of each photometer is sent to recorders, and to an analog-computational device. The output of the analog computer is proportional to the ratio of hydrogen sulfide and sulfur dioxide, and is transmitted to a ratio controller that sets the air feed to the reactors.

Analyses using visible light

Visible-light applications generally fall into two categories: (1) turbidity or opacity, and (2) colorimetry, via product color or color-producing reactions.

Opacity monitors or turbidimeters are normally *in situ* devices that look through the process stream rather than extract a sample. Historically, this type of installation has caused servicing problems. However, present designs have decreased the maintenance problem to some extent. One of the popular applications is monitoring stack-gas opacity of power generators.

Colorimetric procedures are becoming more widely used in monitoring or closed-loop control. Most of this activity is in water-quality management. The instruments are required to provide analyses that range from parts-per-billion silica in high-purity water, to hundreds-of-parts-per-million silica in effluent waste water. A variety of colorimetric analyzers are commercially available.

Near-infrared techniques

The near-infrared spectrophotometers are nominally single, chopped-beam photometers using optical filters in the light chopper to create a measuring wavelength and a reference wavelength. Though the near-infrared technique is less widely used than others, it does a good job in determining water in some products. Water contents analyzed by this procedure range from parts per million to percentages. Normally, this analyzer is applied to relatively pure streams—principally, one component, with water being the main impurity.

Infrared-radiation analyzers

Infrared analyzers use different methods to obtain the optimum frequency for a specific analysis. These instruments are normally double-beam, nondispersive analyzers, with one beam functioning as a reference frequency and the other as a measuring frequency. Spectral wavelengths for analysis are achieved by:

1. Cells filled with selected components in the reference-light and measuring-light paths, which permit only selected frequencies to pass through the sample cell and reach the detector.
2. Optical filters that can be used to provide the correct light frequencies for the desired analysis.
3. A combination of the above.

The detectors for each of these procedures provide the final spectral selectivity and sensitivity by being filled with the component being analyzed. Several manufacturers supply infrared analyzers having well-developed analytical schemes. Best results are obtained with these analyzers by avoiding complex mixtures. The speed of response and reliability is adequate for process control or action initiation.

In one application, an infrared analyzer controls the methane concentration in demethanator column bottoms. In another, an infrared unit monitors carbon monoxide in furnace stack-gas.

A system that incorporates dispersive infrared analyzers coupled with a computer to achieve rapid multicomponent analysis, has been developed. However, the advantage of these analyzers compared to other types of online analyzers has been marginal.

Mass spectrometry

At various times, mass spectrometry has been investigated for online analysis. However, installation problems in the chemical process industries have inhibited this technique for online use.

The recently developed combination of a quadrapole mass spectrometer and a computer appears to have some potential for online analysis. Currently, quadrapole spectrometers monitor and analyze for contaminants in controlled atmospheres such as in semiconductor manufacturing or clean rooms. The quadrapole mass spectrometer is presently limited to nonhazardous locations.

Moisture content of process streams

Various types of online water analyzers have been developed. Some will work at any pressure from ambient to 5,000 psi in vapor or liquid. Others are restricted to gases at atmospheric pressure. They provide a continuous signal that can be used in closed-loop control or as action initiators. In applying these analyzers, considerable care must be exercised in choosing an analyzer that will be compatible with the media in which it will operate. Most of the detectors in these analyzers can be easily destroyed or affected by oil, acidic or basic materials, some solvents or polymerizable compounds.

Several of the moisture analyzers employ hygroscopic crystals that adsorb or desorb water, based on the partial pressure of the water vapor in the stream. Another type detects water content by the reaction of water and phosphorus pentoxide—developing a signal that is proportional to the partial pressure of the water. Water concentrations that can be measured range from less than 1 ppm to thousands of ppm.

Process liquid chromatography

Process liquid chromatographs are just becoming available as online instruments. Their application and analytical techniques are similar to gas chromatographs. Online liquid chromatographs will provide improved accuracy and easier analysis for mixtures that are currently analyzed with difficulty by other methods. Mixtures of high-boiling components or temperature-sensitive compounds can be handled by this procedure fairly easily. Though considerable laboratory experience with liquid chromatography has been developed, the online version is still in its infancy.

Analyzers based on liquid chromatography will be batch type, with automatically controlled repetitive cycles. Probably a fair degree of "dead time" between analyses will be involved, on the order of tens of minutes. Therefore, use in a control loop will require some type of cascade-control or calculated-control output.

Gas chromatography

The online gas chromatograph is probably the most widely used analyzer for continuous analysis of complex mixtures in the chemical process industries. This analytical technique can separate rather complex mixtures into individual components, and measure the concentration of these components.

Gas chromatography, actually gas-liquid or gas-solid chromatography, is a nonspecific technique. One must already know the components and their concentrations in a mixture in order to develop an appropriate application. Once this has been done, a continuous surveillance program must be maintained to assure analysis accuracy.

Component concentrations provided by gas-chromatograph analyzers range from parts per million to tens of percent. The sensitivity of the analysis is determined by the type of detectors used with the chromatograph. The most common are: thermal-conductivity detectors, normally used for measuring concentrations above 500 ppm in a mixture; and flame-ionization detectors (FID), used for detecting organic compounds at concentrations down to parts-per-million levels.

Other detectors such as helium ionization are available for special applications.

Since the online chromatograph is a programmed cyclic analyzer, special handling is required in order to use it in closed-loop control. A high-speed gas chromatograph, capable of performing analyses every 10 to 60 s, is now available. This higher rate has made closed-loop concentration control easier.

Application of these analyzers ranges from the analysis of a finished product to measuring organic contaminants in water. For example, analysis of the impurities (in the 1 to 1,000-ppm range) in high-purity ethylene manufacture is used to control reactors and distillation equipment. The requirements for the speed and continuous composition data necessary to produce high-purity ethylene preclude reliance on laboratory analysis, temperature and pressure measurements, or mass balance control. Another application involves the analysis of feed and effluent streams around a process for material balance and yield control. The information is used in both feedforward and feedback control loops.

Computer-analyzer systems

Dedicated minicomputers now control analyzers and compute the desired information. A typical example is a system combining 20 gas chromatographs and a digital computer. The process analyzers are under direct computer control and do not have conventional programmers as backup. Instead, a redundant computer and manual switchover provide backup in case the online computer fails. The accompanying flowchart shows a schematic of the system. The offline computer for the analyzer offers faster backup than conventional analyzer-programmers.

Complete programming and startup of the offline computer-and-analyzer system can be done in approximately 30 min. This is considerably faster than the time required to bring a single conventional programmer and analyzer online. The offline computer can also be used to modify and update the program, to train operators, and to schedule computer maintenance.

Using the computer as a programmer greatly expands the type of applications. Any desired chromatographic analysis can be produced. The computer system can be programmed for one component or for total stream analysis. The calculation procedures include the choice of baseline-correction technique to accommodate the desired analytical procedure.

Techniques for multiple columns, multiple-sample injection and multiple detectors for a single analysis are practical. One such application incorporates flame-ionization and thermal-conductivity detectors plus double-sample inject and triple-column techniques to provide a total analysis for an eight-component mixture having component concentrations ranging from parts per million to 60%. The computer sees this as one analyzer.

As shown in the flowchart, analytical results are transferred to a larger plant-computer. The combination of these two systems provides a powerful tool for process-control schemes. The main computer can use the analytical data in any desired control algorithm for computer control. Also, the analyses can be combined with other data for material-balance calculations.

Maintenance, calibration and capability

From a maintenance standpoint, the computer-chromatograph system is easier to maintain than the equivalent system that has conventional programmers. Since all programmer tasks and communication routines are centrally generated, surveillance of the total system is a job for about two men.

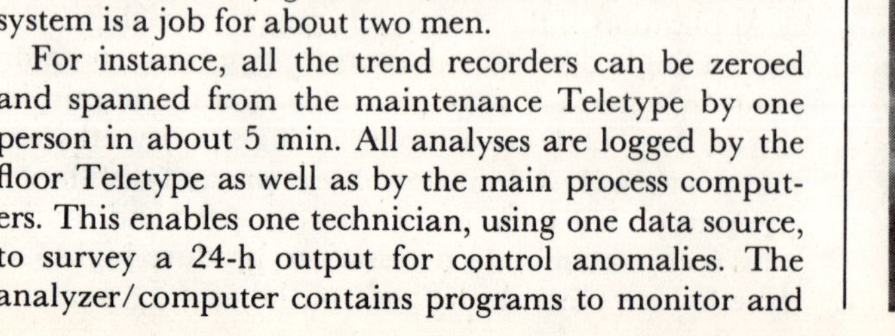

For instance, all the trend recorders can be zeroed and spanned from the maintenance Teletype by one person in about 5 min. All analyses are logged by the floor Teletype as well as by the main process computers. This enables one technician, using one data source, to survey a 24-h output for control anomalies. The analyzer/computer contains programs to monitor and alarm analytical or system failures. Common failures flagged out include: out-of-specification concentrations, shifting elution of components from the chromatograph column, and improper sample size.

This system offers online analysis capability that is at least equivalent to results from a quality-control laboratory. Further, it has the advantage of performing at a frequency completely impractical for a quality-control laboratory. Thus, high-quality analytical data are immediately available at the process site for maximizing production.

Online control of processes

The coupling of online analyzers with computers has broadened the scope of such analyzers for process control. Computers for this service vary from small, dedicated analog or digital units to very large multipurpose machines. The computer combines data from several sources to provide a control signal based on changes in composition or changes in a combination of composition and process variables. The computer/analyzer combination is a good way to obtain compositional process control when the online analyzer is a repetitive, batch-type device. However, all online analyzers will fit this scheme. Analyzer/computer systems are being used to control chemical reactors (batch or continuous), complex distillations, and optimization routines.

Past reluctance to use online analyzers for process control has been due to their poor onstream and analytical reliability as well as lack of operator understanding. Several factors are responsible for the increasing acceptance of online analyzers:

1. Improved onstream availability and analytical reliability.
2. Trained operators.
3. Increased value for optimizing the use of raw materials, utilities, and process equipment.

Improved onstream availability and reliability results mainly from a well-designed maintenance program. Basic to this program is a technically competent support group that implements a preventive maintenance routine. The group should include technicians whose sole duty is maintenance of analyzer systems. The technicians must be supported and supervised by a technical staff that understands laboratory procedures in chemistry, instrument-engineering practices, and probably has some familiarity with computers. This support group must also familiarize the operators with online analyzers and systems.

The author

Victor C. Utterback is a staff engineer at Shell Chemical Co., P. O. Box 2633, Deer Park, TX 77536, where he specializes in the application, installation and maintenance of online analyzers. Since joining Shell as a chemist in the quality-control laboratory and pilot-plant operation, he has spent more than 24 years on analyzer developments and process-control applications. He has a B.S. in chemistry from the University of California at Los Angeles.

Guidelines for Selecting Online Process Analyzers

Proper choice of an analyzer depends largely on the nature of the properties to be measured in the process stream. Equally important, however, is assurance that the sample stream remain representative of the process fluid. Here are the advantages, limitations and applications of typical instruments.

RICHARD A. FOSTER, Fluor Engineers and Constructors, Inc.

To determine the composition of a process stream, or the concentration of some component in that stream, we can use a process-stream analyzer. This article describes some of the available analyzers and their applications.

An analyzer is a device that measures the concentration of the component of interest in a process stream. Often, we can correlate the measurement of a single physical property to the quantity of a given substance in the stream. Many instruments that provide a readout of the physical property only are commonly referred to as analyzers. This group of instruments includes those that measure boiling range, pH, specific gravity, density, viscosity, refractive index and optical density. Usually, these instruments can be calibrated to read out a given component directly, provided the composition of the other components in the stream is sufficiently uniform.

We justify the installation of any analyzer on the same basis as any other type of capital equipment. We must expect an advantage or a monetary payout from such installation. All of the advantages of an analyzer arise from the immediate knowledge of stream purity, stream composition, or a particular physical property. Some of these advantages are:

- Fewer off-specification streams or products.
- Minimum product giveaway caused by exceeding specifications.
- Reduction in laboratory analyses, and thereby savings in costs.
- Use of the analyzer's output signal in a control loop.
- Optimization of lineout time for intermittent units.
- Examining progress of catalyst regeneration.
- Monitoring effluents to meet regulatory requirements of government agencies.

We must also consider both the cost and maintenance of an analyzer installation. Obviously, these will depend on the type of analyzer chosen.

We have witnessed the development of many new types of analyzers. Some instruments that formerly were used only for laboratory procedures have been adapted for online process applications. Other instruments have been developed primarily for online uses, while others were designed for both laboratory and process applications. As a result, several instruments may often yield the same or similar information. Typical applications of some analyzers are shown in Table I. Let us examine the details for some of the analyzers that are available for online service.

Gas Chromatography

The most versatile analyzer is the gas chromatograph. Unlike most other analyzers, the components of a sample are separated from each other, and their concentrations measured.

Chromatography basically involves a two-phase system. The stationary phase is a surface-active solid or a liquid supported on a divided solid. The stationary phase is packed in tubing, which forms the chromatograph column. The column's length may vary from a few inches to over 100 ft, depending on the separation. The diameter of the tubing may vary from capillary to about ⅜ in. The moving phase, in this case, is the carrier gas, usually hydrogen or helium.

A sample of suitable size is injected into the carrier gas just before it passes into the column. The components of the sample establish equilibria between the two phases as the carrier gas moves the sample through the column. If the equilibrium coefficients of the components differ sufficiently, the components will move through the column at different rates, thereby becoming separated. The carrier gas then passes into the detector, which provides an electrical signal. This signal may be recorded, normally

Originally published March 17, 1975.

Selected Online Process Analyzers and Their Applications – Table I

Analyzer	Sample	Approximate Range	Remarks
Gas chromatograph	Gas, vapor or volatile liquid		Can be specially modified for various applications. For example: traces of pesticides in foodstuffs; ethyl alcohol, esters, higher alcohols in beverages.
Thermal conductivity detector		0.1 to 100%	
Ionization detector		Ppm to 100%	
Nondispersive infrared	Usually gas or vapor Occasionally liquids	0.1 to 100%	Lower ranges are obtained by increasing cell length or pressures.
Fourier-transform infrared	Liquid, vapor or gas	0.1 to 100%	Will analyze a number of components in a single sample. By repeated scans, the lower limits can be extended to parts per billion.
Thermal conductivity	Gas	0.1 to 100%	Normally used for hydrogen determination; also a chromatograph detector.
Ultraviolet adsorption	Liquids	0.01 to 100%	Used for chemicals contianing a chromophore that absorbs ultraviolet, such as aromatic nucleii or conjugated dienes.
Colorimeters	Usually liquid	0.1 to 100%	Range depends on color density. Used as a detector in automated chemical analyzers. Often used to monitor color specifications of a process.
Physical property	Liquids	0.01 to 100%	Solids or solutes in solution.
Density			
Specific gravity			
Refractive index			
Automated chemical	Liquids	0.01 to 100%	Dynamic range depends on normality of titrating solution. Much-lower ranges may be obtained with some colorimetric analyses.

as a differential plot. The signal may also be digitized for computer interface, or it may be fed to a peak-picking and holding amplifier for an analog or trend record, or for providing a signal for a control loop.

Many of the online chromatographs contain series or parallel columns to effect the more difficult separations or to eliminate unwanted components from the analysis. Column technology is well developed for most separations, and suitable application data are available from manufacturers.

Although the gas chromatograph appears to solve many analytical problems, it possesses certain disadvantages:

- Relatively high capital cost.
- Expert maintenance requirements.
- Discontinuous batch analysis.
- Accuracy limited by calibration techniques.
- Possibility of error due to interference from unknown components.

Of course, some of these disadvantages can apply to other types of analyzers as well.

The installation of a chromatograph must be carefully planned. In many cases, the investment for shelters, sampling systems and mounting exceeds the cost of the analyzer itself. Generally, the analyzer section must be completely protected from the weather, and still be reasonably close to the sample point. Fig. 1 shows a typical shelter for a chromatograph, and Fig. 2 shows the interior of the shelter with the analyzer mounted, wired and piped.

In most installations, the control unit of the chromatograph is mounted in the main control room, with electrical cables to and from the analyzer. Fig. 3 illustrates a panel-mounted control device for the chromatograph and the bar-graph recorder. The instrument between the two units is an H_2S analyzer.

When only one component in a stream is of interest, it is good to choose an analyzer that depends on the correlation of a property of the stream with the concentration. We must choose a property that differs sufficiently from the remainder of the stream so that interference is eliminated as completely as possible. Ideally, if we have a two-component system with perfect mixing, we have a choice of several instruments. For example, a single determination of specific gravity, density, refractive index or thermal conductivity suffices because the value of the selected property varies in a simple proportion to the properties of the pure components.

Infrared Analyzers

Nondispersive infrared analyzers have been successfully applied to many vapor or gas streams, and to some liquid streams. These instruments consist of an infrared radiation source, beam chopper, reference cell, sample cell and detector. The detector consists of a condenser microphone that is usually filled with the gas or vapor of the component to be determined. Whenever any of this material is present in the sample cell, the energy is absorbed by the sample rather than by the gas in the detector. Thus, an imbalance in pressure is produced, which is picked up by the condenser microphone. The resulting electrical signal is then amplified (linearized, if necessary) and may be used for recording, controlling or indicating.

Infrared analysis cannot be applied to monoatomic gases or symmetrical diatomic gases, which do not possess infrared absorption spectra. It is sometimes successful with gases such as acetylene that have a few narrow absorption bands, but only if the background absorption does not mask the peaks.

Infrared analysis is most successful where a particular compound has a chemical structure that differs from the background material. A good example is the determination of isobutane in a stream containing *n*-butane and propane. If we tried to analyze for *n*-hexane in a mixture of *n*-pentane and *n*-heptane, we would find this method of analysis unsatisfactory. Materials other than hydrocarbons can also be analyzed, such as carbon monoxide, carbon dioxide, oxides of sulfur, and oxides of nitrogen.

In general, the infrared analyzer gives excellent results, and the output is fast and continuous. The process stream to be analyzed must be carefully dried to remove water vapor in order to prevent damage to the optical components that may be fabricated from salt, potassium bromide, or other water-soluble materials.

The Fourier-transform infrared analyzer has been adapted for online work. An interference pattern is obtained while scanning the sample, and a dedicated minicomputer reconstructs the absorption spectrum. Extremely fine resolution and sensitivity are claimed. This instrument can analyze a number of different components in a single sample if their absorption spectra are sufficiently varied. By using multiple scans and averaging, extremely high sensitivity has been obtained. For example, one instrument has been used in the measurement of nickel carbonyl at a molar concentration of less than 1 ppb.

Monochromatic Analyzers

Most ultraviolet analyzers, colorimeters and nephelometers use filters or monochromators to attain a fairly narrow band of energy. Absorbance of the various bands can be correlated with the concentration of a particular compound. Organic chemicals with aromatic nuclei or conjugated olefinic bonds are chromophores that strongly absorb ultraviolet energy. Since some displacement of the absorption bands occurs with chemical substitution, it is often possible to choose an absorption band that will minimize interference from other constitutents.

Colorimeters differ from ultraviolet instruments in that they operate in the visible-light range. Again, either filters or monochromators may be used to isolate the desired radiation band, but optical filters are more com-

OUTDOOR shelter protects online chromatograph—Fig. 1

INTERIOR of shelter shows the sample-conditioning system and cabinet for gas chromatograph—Fig. 2

CENTRAL panel contains control unit and bar-graph recorder for the online process chromatograph—Fig. 3

mon. Often, only the color density is measured, but an analysis can be performed if the component of interest is colored.

Nephelometers are widely used to determine smoke density, or cloudiness in liquids. With these instruments, either the transmitted energy or scattered energy may be measured. Since particulate matter is being determined, white light is adequate. The readout is calibrated in arbitrary units, optical density, or light transmission, which can be correlated to particle concentration.

In any instrument that depends on the absorption of radiant energy, the output is essentially nonlinear. Energy absorption follows the Beer-Lambert law:

$$I_t/I_o = e^{-kcd} \qquad (1)$$

where I_o is incident energy; I_t is transmitted energy; k is absorptivity, a constant for each constituent at a defined wavelength; c is concentration of the component of interest; and d is distance of absorbing path or cell length.

Eq. (1) simplifies to:

$$\ln(I_o/I_t) = kcd = A \qquad (2)$$

where A is absorbance or optical density.

Unless linearizing circuits are installed in the instrument, charts or scales should be calibrated in absorbance or optical-density units so that the readout will be proportional to the concentration. Some departures from the Beer-Lambert law have been noted that can be caused by stray light or wavelength bands of a finite width. These problems are due to the limitations of the instrument. For best results, each instrument should be calibrated over several concentrations for the compound of interest.

Catalytic Oxidation Devices

Instruments using catalytic oxidation chiefly detect nonspecific explosive mixtures and combustibles—usually in stack gases and ambient air. The detector consists of two wire filaments connected in a Wheatstone bridge circuit. The active filament is coated or imbedded in a catalyst while the reference filament does not have the catalyst. When combustibles or explosives are present, the active filament becomes hotter than the reference filament, which increases its electrical resistance. This, in turn, unbalances the bridge circuit, providing an electrical readout to activate an alarm, to indicate, or to record and control.

Specific analyses have also been successfully done by instruments of this type. When excess hydrogen is added to a sample of stack gas, the readout can be calibrated to give the percentage of oxygen. This application is widely used to control the excess air in boilers. Specific analyses for hydrogen sulfide and carbon monoxide are possible by modifying the catalyst and by controlling filament temperature.

Catalytic-combustion analyzers will fail to operate if the oxygen content of the gas mixture is insufficient. This occurs in the presence of an excess amount of combustibles, or with an inert atmosphere. We can overcome this difficulty by using a flow-through cell and adding a known amount of air or oxygen to accomplish the desired oxidation. This is difficult to do in an instrument having a diffusion-type sensor.

Electrochemical Analyzers

The principal applications of electrochemical analyzers are with aqueous solutions. The analyzers are of two types: (1) those requiring electrical energy to be supplied to the electrodes, and (2) those generating an electrical potential or current created by electrochemical action of the sample at the electrodes. Examples of the former are: coulometric, conductivity and polarographic analyzers; and of the latter, oxidation-reduction potential (ORP), pH, and analyzers based on the fuel-cell principle.

The electrolytic-conductivity analyzer is a nonspecific device because it responds to any electrolyte. It is widely used to monitor boiler feedwater, aqueous effluents, and streams in desalting plants. The area of the electrodes is designed to accommodate the electrolyte concentration, and to keep the resulting current within the operating range of the instrument. Alternating current is applied to the electrodes to prevent polarization.

The electrolytic analyzer for water vapor in gases or vapors consists of tubing that is coated on the inside with phosphorus pentoxide. Two platinum wires are imbedded in the phosphorus pentoxide, which has a very high affinity for water. A potential sufficiently high to electrolyze water is applied across the electrodes, and the resulting measured current is proportional to the water vapor present. No calibration is needed, since the current is a measure of the coulomb value for electrolyzing water. This instrument is useful in the low parts-per-million range.

Other substances that electrolyze near the potential for water will interfere. For example, hydrogen sulfide not only shows up as water but the elemental sulfur formed will foul the analyzer tube. Fouling may also occur from substances that react with or are polymerized by phosphoric acid. For this reason, olefins and other reactive gaseous materials cannot be analyzed for their water content by this means. The analyzer has a tendency to give high readings for water vapor in streams that contain high concentrations of hydrogen. This problem has been largely overcome by careful control of the flow and the length of the path through the analyzer, or by using electrodes that do not catalyze the recombination of hydrogen and oxygen.

Acidity or alkalinity of aqueous solutions has been determined by the glass electrode referenced against a suitable half-cell, usually mercury-calomel or silver - silver chloride. A potential, related to the hydrogen ion concentration, appears on the glass electrode and is measured by a high-impedance amplifier to give a readout in millivolts or pH units. The determination of pH does not give the concentration of acid or alkali but rather their activity. As a consequence, other means must be used to determine the concentration of acid or alkali, except in extremely dilute solutions when the activity approaches unity.

Various metallic and modified glass electrodes have been developed that respond to the activity of ions other than hydrogen. Some of the ions, successfully measured, include: sodium, potassium, cupric, silver, sulfide, chloride and fluoride. These electrodes are useful for monitoring aqueous effluents.

Oxidation-reduction potential (ORP) may be measured with pH-type equipment by using an inert metallic electrode instead of the glass electrode. The measurement is not specific to any particular ion but does indicate the activities of oxidants or reductants. The device has been used to monitor a chemical reaction where oxidation or reduction occur.

Coulometric titrators have been applied to a limited extent to determine the concentration of sulfide or sulfite.

Thermal-Conductivity Analyzers

The thermal-conductivity cell has wide use as the detector for gas chromatographs. It is very sensitive when hydrogen or helium is the carrier gas. With other carrier gases, peak inversion may occur when the thermal conductivities of some eluted components are greater than the thermal conductivity of the carrier gas. By itself, the device may also be used to determine a given gas in a mixture when the background gas is reasonably constant in composition. As such, the greatest application is in measuring hydrogen concentration or impurities in hydrogen.

The thermal conductivity cell consists of a metal block, drilled to accommodate four filaments or thermistors. The block also has openings so that sample gas may diffuse into two of the filament chambers, and a reference gas in the other two. The block is temperature controlled to prevent ambient conditions from affecting the cell. The filament or thermistors are connected in a bridge circuit, and any unbalance current may be read directly by a potentiometric recorder or by a sensitive meter, without additional amplification.

If the sample gas has a lower thermal conductivity than the reference gas, the sample filaments will become warmer, thereby increasing in resistance. This will cause an electrical imbalance in the bridge circuit and provide an output signal.

Sometimes, only two ports are used—one for the sample gas, one for the reference gas. The remainder of the bridge contains resistors. This arrangement is only half as sensitive, but may be adequate. Since thermistors have a negative resistance-temperature relationship, a current-limiting device should be specified when they are used. This prevents runaway overheating that can destroy the thermistors.

Specific-Gravity and Density Analyzers

A device for monitoring the carbon dioxide content of stack gas (and other gas mixtures) measures the specific gravity of the stream. The most-common type consists of two mechanically driven rotating impellers, and two nonrotating impellers. These are coupled by a linkage so that the torque generated by the drag of one impeller opposes the drag of the other. A small flow of sample gas passes through one impeller while air flows through the reference impeller. Specific gravity of the sample gas is indicated by the balance point produced by the two nonrotating impellers. Other instruments are available that operate on the balanced-flow principle or that measure buoyancy. All yield similar information. Often, we can obtain the same analysis with either a thermal-conductivity cell or a gas specific-gravity analyzer. Hence, we can choose the most suitable device.

Liquid density can be determined by several devices. The balanced "U" tube often finds application where temperatures and pressures permit. A modification of this instrument has a vibrating "U" tube for higher temperatures and pressures. The damped oscillation of the tube in the liquid is electrically sensed and compared to the reference frequency, and converted to a readout signal.

A displacer can measure density, provided the effects of liquid flow past it are minimized. A differential bubbler, consisting of two dip tubes of unequal length, can be easily applied to liquids in a tank or vessel. Liquid or air, slowly bled into the dip tubes, produces a differential pressure across the tubes. The readout is in head of liquid, which is proportional to density. Instead of density units, the readout may be calibrated directly as percent of solute in a solution, or other appropriate units.

Property-Inspection Analyzers

Property-inspection analyzers directly read a property that is a specification of a product or plant stream. These analyzers are automated versions of laboratory appa-

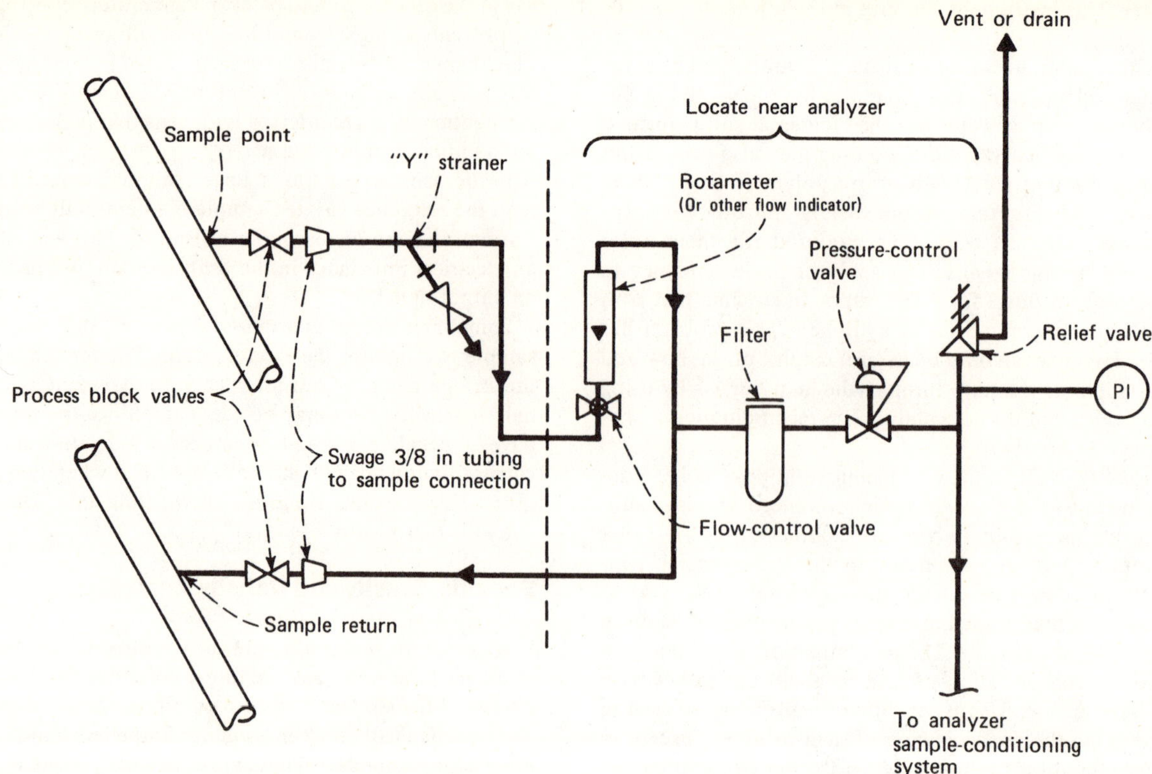

SAMPLE LOOP for liquids, gases or vapors circulates process stream from high-pressure entry to low-pressure return—Fig. 4

ratus, and check certain properties as required by inspection tests. Instruments for measuring the properties of color, density and specific gravity have been previously described. Other online analyzers perform checks on initial boiling point, final boiling point, viscosity, melt index, flash point, etc. Oil refineries use these instruments to prevent specification limits from being exceeded and to prevent giveaway of product quality.

Chemical Analyzers

Both wet- and dry-chemical analyzers have been applied to online service. The dry type contains a roll of chemically impregnated paper that becomes colored when exposed to the compound of interest. One such analyzer determines the hydrogen sulfide content of air or gases. In this case, the paper, impregnated with lead acetate, darkens on exposure to hydrogen sulfide. The degree of darkening can be measured photometrically and correlated with the concentration of hydrogen sulfide.

Wet-chemical analyzers are applicable mainly to aqueous solutions and find wide use in the control of water pollution. In general, they perform any analysis that can be done by chemical titration. When no physical property can be correlated with the composition of a stream, chemical analysis may provide an answer. Chemical analyzers serve with acid-base, oxidation-reduction, precipitation, and color-forming reactions. The end-point is usually detected colorimetrically. Other methods for determining end-points can also be used. Reagents may be added volumetrically, or generated electrolytically, as with coulometric titration. All such analyzers require routine maintenance to replenish reagents and clean the system.

Miscellaneous Analyzers

A number of other analyzers, not classified in the preceding groups, have online applications. For instance, mass spectrometry proposed for online analysis is difficult to justify economically. An oxygen analyzer employs oxygen's paramagnetic properties successfully, since the only interfering gases are some oxides of nitrogen. Refractometers and polarimeters have been adapted for on line work. Nuclear-radiation absorption has been applied for measuring density, level and thickness. Nuclear radiation and X rays have shown promise for determining sulfur in petroleum, but some elements—especially metals—interfere.

Sampling Systems

The proper choice of a sample-conditioning system is nearly as important as the choice of analyzer. Many analyzers fail to give the expected results because of basic problems with sampling methods.

The sample system is composed of two parts: (1) the

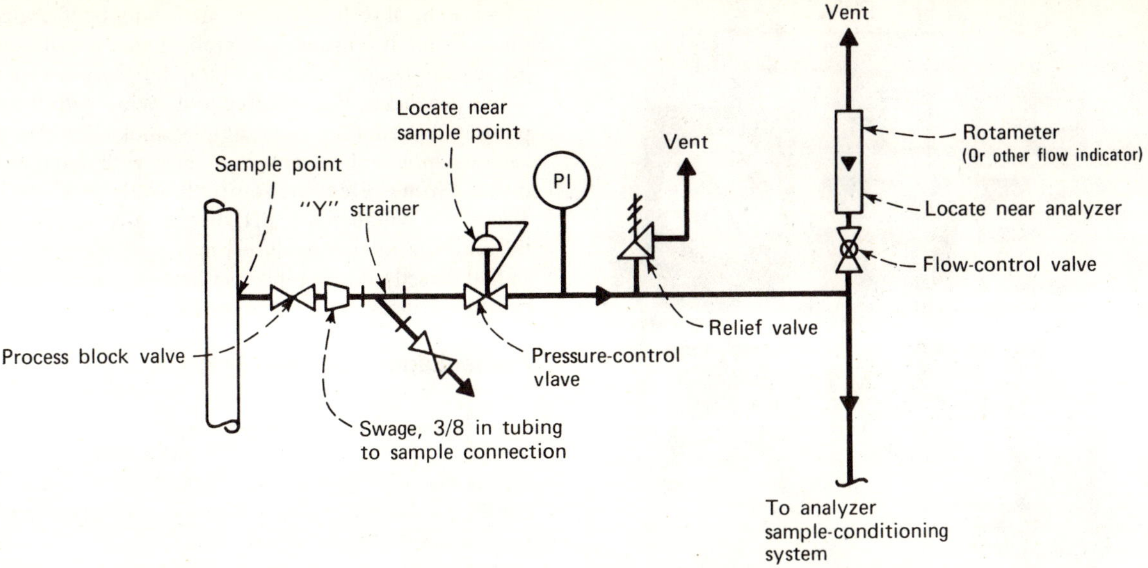

SAMPLE LINE is suitable for vapors and gases that may be safely vented to the atmosphere—Fig. 5

sample loop or sample line from the process connection to the analyzer location, and (2) the sample-conditioning system that is usually mounted near or in the same cabinet with the analyzer.

When we sample a stream from a long distance, the possibility of time lag in the sample lines must be considered. The amount of lag that can be tolerated depends upon the analyzer's application. If we use the analyzer as a component of a control loop, the sampling lag time should be as small as possible. On the other hand, if the analyzer is used for a periodic check on the process stream, lags of several minutes can be tolerated between sampling and analysis.

Fig. 4 shows a typical sampling loop, where the sample is circulated from a high-pressure point in the process to a low-pressure point. Modifications can be made to this loop to better adapt it to various conditions at the sampling point. For example, if we have a high-pressure gas or vapor and sufficient pressure drop, we may reduce the pressure at or near the sample tap. This minimizes the

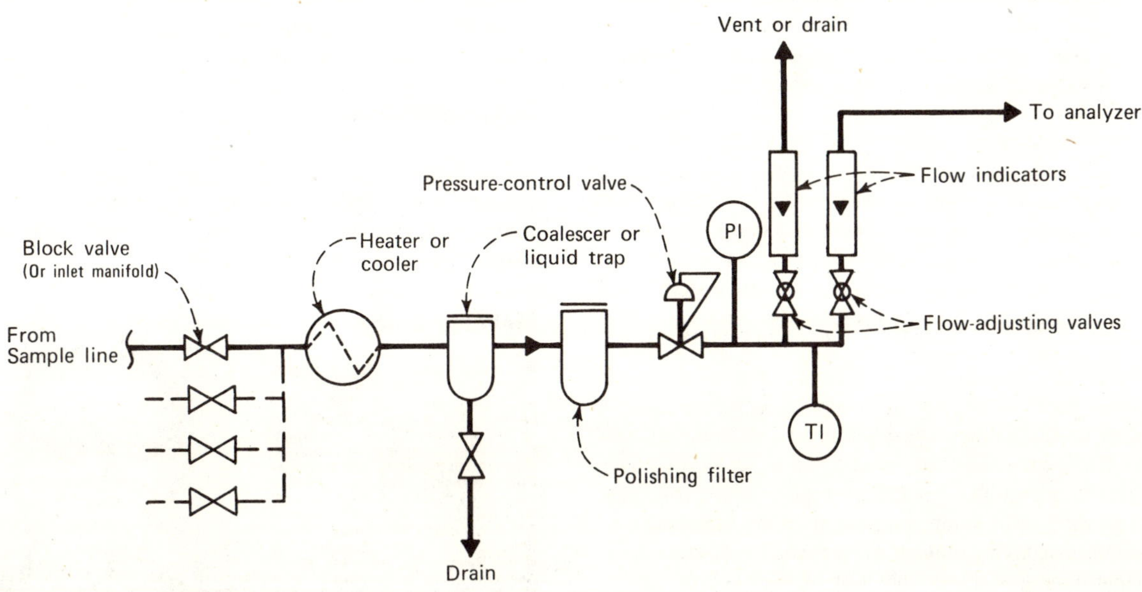

SAMPLE-CONDITIONING system is usually an integral part of a process analyzer for single or multiple streams—Fig. 6

PACKAGED analyzer systems are ready for use—Fig. 7

quantity of material circulating in the sample loop, and the sampling lag is still sufficiently reduced.

If a sample can be safely vented, we can use the sample line shown in Fig. 5. This is a simpler system than the circulating loop, but should not be used on streams that are valuable, or on those that are objectionable when released to the surroundings. Again, we may eliminate some components if they are not required. For example, clean streams may not require filters; low-pressure streams may not require pressure reduction or a relief valve.

The sample-conditioning system is usually an integral part of the analyzer as furnished by the manufacturer. The purpose of sample conditioning is to ensure that temperature, pressure and flow of the sample meet the requirements of the analyzer. A typical system is shown in Fig. 6. Again, some components of this system need not be included if not required for the analyzer. This should be checked with the manufacturer of the analyzer.

Significantly changing the composition of the sample stream presents the greatest pitfall in any sampling system. This can occur in several ways: a vapor-phase sample may partially condense, a liquid sample may partially vaporize, or some component of the sampling system may selectively remove a compound of interest.

Heat tracing of the sample loop or sample line may be necessary when sampling a gas or vapor, while cooling may be necessary with a liquid. In some cases, desiccants have been used to dry a sample, but polar compounds may also be absorbed. If moisture cannot be tolerated by the analyzer, it is usually preferable to use a coalescer for liquid samples, or a cooler and trap for gases or vapors.

In a few cases, it is possible to dispense with the sampling system entirely. For example, smoke-density analyzers are mounted directly in a stack with suitable windows. Some electrochemical analyzers such as electrolytic conductivity, pH, oxidation-reduction potential, and specific-ion analyzers may have their probes installed directly into the process line—provided that temperature or pressure is not excessive.

Prepackaged Analyzer Systems

With increasing applications of online analyzers in the chemical process industries, we encounter certain difficulties that can lead to user dissatisfaction. These difficulties involve the choice of the type of analyzer and the installation of the analyzer system.

A number of control-panel fabricators, specialty contractors, and most analyzer manufacturers can now furnish a complete package, with the analyzers mounted, wired and piped. Except for the taps at the sampling point, the sample system can also be provided. This service can include selection of the analyzers, rack or panel mounting of the components, testing of their operation, field checkout, and placing the system in operation.

A total-responsibility arrangement can be attractive when neither manpower nor time is available to completely specify the required instruments, or when field labor is not capable of making a satisfactory installation. When the complete package is provided, only the sample lines to the process need be connected. Electrical power and signal connections are relatively simple to install. Thus, the complete-system approach is often economically justified. A typical system to monitor the condition of effluent water is shown in Fig. 7. Any size installation may be put together because the system concept is not confined to large walk-in enclosures. Local cabinets or racks with rain shelters are available if these meet the requirements of the analyzer.

Acknowledgements

The following firms kindly supplied illustrative material for this article: Beckman Instruments, Inc., Fullerton, CA 92634 (Fig. 1, 2 and 3); Delphi Industries, South El Monte, CA 91733 (Fig. 7).

Meet the Author

Richard A. Foster is a Senior Instrument Engineer with Fluor Engineers and Constructors, Inc., 2500 S. Atlantic Blvd., Los Angeles, CA 90040. He is Chairman of Fluor's Analytical Instrument Committee, and does instrument engineering for projects. Prior to joining Fluor, he was employed by Beckman Instruments, Inc., and Shell Oil Co. He holds a B.S. in chemistry from Loras College (Iowa), with further courses at the Naval Reserve School, U.S. Naval Academy, and a certificate from UCLA's Graduate School of Business Administration. He is a senior member of the Instrument Soc. of America.

Optical instruments monitor liquid-borne solids

By passing a beam of light through a process or effluent stream and measuring changes in the liquid's optical properties, these instruments can detect the count and size distribution of suspended particles.

Alvin Lieberman, Royco Instruments, Inc.

☐ Various optical instruments have been developed for measuring the number and size of particles suspended in various fluids. Such instruments have been devised for analyzing both gas and liquid streams; however, this article will be limited to those units used to determine the characteristics of particles suspended in liquids. Since many design variations exist, this discussion will not attempt to cover all of them; instead, attention will be given to several optical devices that illustrate general principles.

Optical instruments offer several advantages. They can produce real-time data; they can handle a broad range of liquids; they can operate online (within specified concentration limits); they can measure over a fairly wide dynamic size range; they can gather a large amount of reproducible data quickly; and they can be operated by relatively unskilled personnel.

However, certain limitations on optical instruments should also be kept in mind. They take size data based on optical properties that must be referred to a calibration base; they yield a size distribution based on particle population, rather than particle mass; their concentration restrictions may necessitate sample dilution; and they are subject to bias if air bubbles or immiscible liquids are present.

How they work

Commercially available optical instruments fall into two general classes. In the first, an assemblage of particles is scanned to reveal information on overall concentration. With more-sophisticated instruments of this type, spatial filtering of scattered light permits determination of mean diameter and variance of particle size distributions.

In the second class of instruments, the optical system views a volume so small that individual particles in a flow are observed and each one provides an indication of size. Both concentration and size data can be derived, since the instrument reports the number of particles over several size ranges measured in a given volume of liquid.

Originally published December 18, 1978.

Photometers

The operating principle of these devices is quite simple. A light beam is directed through the sample. The total amount of light scattered at specific angles, or absorbed by the opaque particles, can be related directly to the quantity of particulate material in the liquid. Empirical calibration to a turbidity unit scale is normally used to define the quantity of suspended solids, although the degree of scattering is dependent on the size, shape and concentration of the particles.

Online photometric analyzers are used primarily to measure turbidity and low concentrations of solids in process liquids or water lines. Back-scattering systems are used to determine concentrations in slurries con-

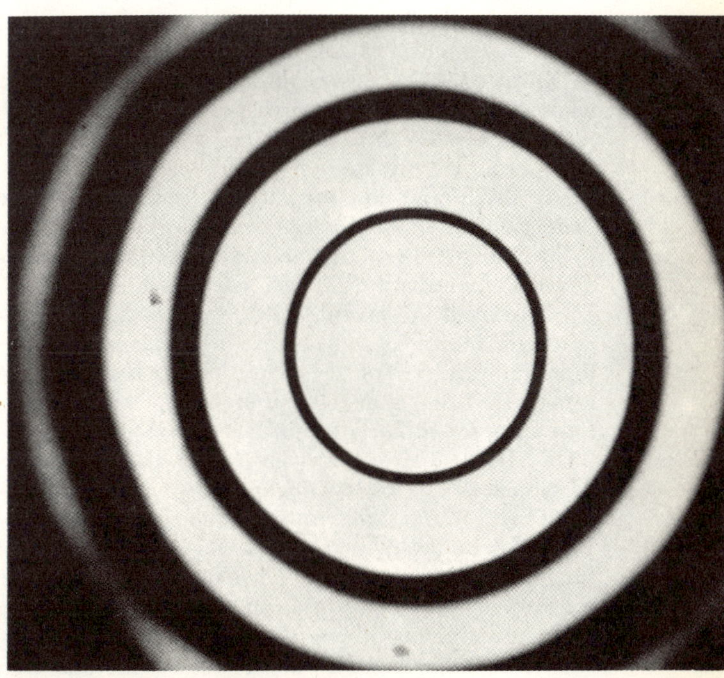

Laser beam diffracted by suspended particles produces typical Fraunhofer pattern

taining 10–15% solids. Angular scattering systems can be used to monitor solids contents up to 50,000 ppm. Absorption analyzers, which handle lower concentrations (to approximately 1,000 ppm) are used for monitoring water treatment systems and process lines.

Depolarization measurements

Under development by the Precision Products Div. of Badger Meter, Inc. (Tulsa, Okla.) is a method for determining suspended particle concentration, based on the depolarization of scattered light. A depolarization ratio is determined by measuring the intensity of scattered light passing through a polarizing prism. Measurements are made along the optical axes both perpendicular and parallel to the plane of incident polarized light. The ratio of intensities represents the depolarization ratio.

This ratio can be related to concentration, since multiple scattering increases with the number of particles present. Primary scattered light becomes increasingly polarized with multiple scattering.

Reports indicate that this method gives good response to solids concentrations in waste liquids from a few ppm to 5,000 ppm, with no undue bias from particle size or dissolved materials. Solids buildup on the optical windows is not believed to be a problem.

Fraunhofer diffraction analyzers

The French company Industrielle des Lasers Cilas (Marcoussis) offers a particle-size analyzer (Model 227) that measures the intensity of Fraunhofer diffraction patterns produced by passing He-Ne laser light through a suspension. Although the pattern diffracted through polydisperse material is monotonic, smaller particles diffract the laser beam over a larger angle than do larger particles. The overall energy level at any point varies with concentration.

Diffracted light passes through a rotating mask, whose window diameters increase toward the periphery. A photodetector then determines the levels of light energies at various radii from the laser beam's axis. From this information, the particle distribution can be deduced using a preprogrammed matrix calculation.

Size measurements can be made from 1 to 128 microns. Cumulative weights are calculated at 1, 2, 4, 8, 16, 32, 64 and 128 microns. Data reported by the manufacturer indicate good results with cement and alumina samples. Sample concentrations should be in the range of 0.1 to 1 g in 500 mL of liquid. Measurement time of approximately 5 min is required.

The Leeds & Northrup Co. (North Wales, Pa.) has developed a line of low-angle light-scattering instruments (Microtrac) that operate on the principle of Fraunhofer diffraction, using laser light sources, optical filters, and microprocessors. These instruments measure particles in either liquid or air suspension, and may also be used online for process monitoring.

Although the intensity of the total light scattered (photo) by particles is proportional to the square of their diameters, the angle of diffraction is inversely proportional to their diameters. By placing a specially shaped spatial filter in the diffraction plane of the collecting lens, one can relate light fluxes to scattering angles; these fluxes are related to particle volume.

The spatial filter is rotated to extract signals serially. (Collected light can be related to the second, third or fourth power of the particle diameter.) Once these signals are detected and converted to digital form, a microprocessor computes the distribution of particle volume and surface area. Output data are presented in the form of the particle volume in each of 13 size bands. Information is also given on the percentage of particles passing each size interval, as well as their mean diameters, surface areas and relative total volume. Measurements are made over two size ranges: 1.9 to 176 microns and 3.3 to 300 microns.

The product line has recently been extended to include a low-cost monitor that provides a continuous measure of solids concentration in wastewater and other industrial streams. This instrument measures true volumetric concentration, which can be converted to mass concentration with a one-time, single-point gravimetric calibration.

Similar instruments are being used in refining operations and chemical processes, including the manufacture of powdered resins, catalysts and a variety of powdered and slurried chemicals.

Lab methods detect particle groups

This brief summary of some operating laboratory methods is restricted to those used to characterize particle assemblages in liquids. Holographic analyzers and automated image analyzers are treated later as single-particle devices.

1. Bagchi and Vold [*1*] describe a method for determining the average particle size of coarse suspensions from measurements of apparent specific turbidity. They show that the turbidity is inversely proportional to the average radius from about 12 to 50 μm for materials whose turbidity is independent of the wavelengths of light. The method has been used with materials in concentrations of 1–10 g/L.

2. Dobbins and Jizmagian [*2*] describe two methods for finding mean size by measurement of optical cross-section. In the first method, when concentration is known, the relationship between mean scattering cross-section and volume-surface mean diameter is fixed. In the second method, both the volume-surface mean diameter and concentration can be measured if transmittances are measured at two wavelengths. Data were obtained for mean diameters ranging from 1–5 μm.

3. Groves, Yalabik and Tempel [*3*] discuss the operation of a centrifugal photosedimentometer using laser light. A transparent hollow disk is rotated, causing particles to settle under the influence of centrifugal force. The relationship between sedimentation time and optical density at any point along the radius is a characteristic of the particle size distribution.

A data logger is used to record light-transmission level, location along the radius, and settling time. This enables computer treatment of information to yield relatively rapid descriptions of particle size data. The method can size particles from 5–10 μm to less than 0.1 μm, and can handle sample concentrations up to one percent.

4. Jordan, Fryer and Hemmen [*4*] use a hydro-

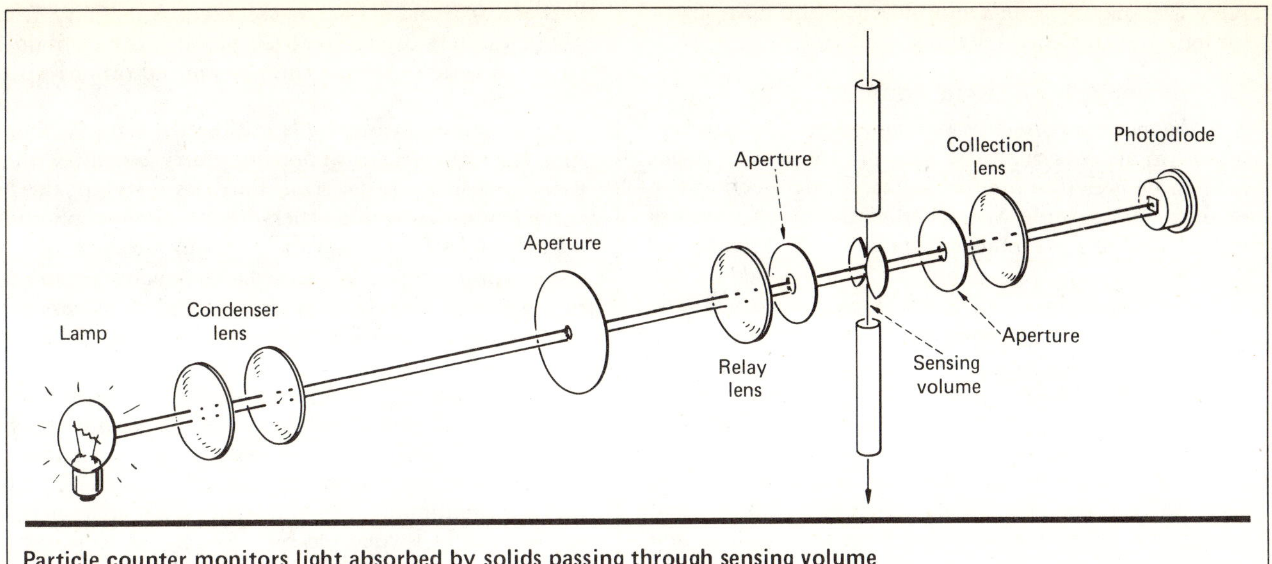

Particle counter monitors light absorbed by solids passing through sensing volume

photometer to find particle distribution in the size range from 2–50 μm via a sedimentation method. Optical density at a fixed point in a gravity settling column is measured as a function of time and is correlated to mass loading.

5. Robillard, Patitsas and Kaye [5] describe the use of Mie scattering measurements at two wavelengths to determine the mean diameter and refractive index of particulate material. The method requires comparison of experimental data with computer-generated curves derived from Mie theory, followed by selection of best-fit parameters. Size distribution is inferred from the intensity ratio of the scattering extrema. This work follows and extends earlier work by T. P. Wallace [14]. It is not applicable to multimodal distribution.

Nonimaging, single-particle counters

Among these are three optical single-particle counters commercially available in the U.S. All pass a liquid sample through an illuminated sensing volume; the amount of light detected depends on the degree of absorption or scattering as the particles traverse the sensing volume. The sensing volume is small enough so that single particles are detected, even when they occur in relatively high concentrations. From the amplitude of the light pulse, particle size information can be derived. By sorting and counting the light pulses, the instrument can ascertain the distribution and concentration of particles.

Climet Instrument Co. (Redlands, Calif.) manufactures a liquid particle analyzer (Model Cl 220), which detects the white light scattered by individual particles as they traverse the sensing zone. Scattered light is collected at scattering angles from 15 to 105 deg. Suspended particles—sized in terms of their equivalent optical diameter as referred to latex spheres—are detected and can be resolved between 2 and 200 μm. A sample flowrate of up to 500mL/min can be used, and by varying the ratio of sample fluid to filtered recirculating fluid, one can measure particle populations between 1 and 10^8 per milliliter.

HIAC Instruments Div. of Pacific Scientific Co. (Montclair, Calif.) manufactures a design that observes a telecentric white-light beam and detects reduction in light level caused by particle passage. The firm makes a series of sensors that can detect particles from 1 μm to a maximum of 9 mm. Each sensor has a dynamic range reported to be 60 : 1, based on the smallest detectable particle (as referred to an equivalent optical diameter for a latex sphere) and the minimum dimension of the internal passageway. The smallest sensor handles particles 1 to 60 μm, in concentrations up to some 30,000/mL at a flowrate of 4–6 mL/min. The largest one handles particles to 9 mm in concentrations up to 0.2/mL at a flowrate of 100 L/min.

Royco Instruments, Inc. (Menlo Park, Calif.) manufactures two instrument packages. One observes a focused beam and detects reduction in white-light level caused by particle passage. Two flow cells are available with this package, both of which can detect particles of less than 2 μm with a dynamic range of at least 100 to 1. Sizing is based on calibration with either latex spheres or the AC Fine-Test Dust [15]. The smallest cell has a concentration capability of 12,000/mL at flowrates to 35 mL/min; the larger cell can handle concentrations of 3,000/mL at flowrates to 150 mL/min.

The second package detects near-forward scattering produced by particles passing through a laser beam. The same flow cells can be used with this package, permitting identical concentration and flowrate capabilities. Minimum size sensitivity is 0.5 μm, and a dynamic range of 50 : 1 is achieved.

Spectrex Corp. (Redwood City, Calif.) markets a device that uses light scattered from a laser beam to detect particles larger than approximately 5 μm to 100 μm in concentrations to 1,000/mL within a glass container. The laser beam can scan a volume of 10 cc in 15 s. Since out-of-focus particles and the bottle walls are not detected by the collection optics, only particles within the in-focus point for both the illumination and collection optics are reported.

A microprocessor attachment automates the complete size analysis, prints and plots size and mass distribution, and provides a settling scan.

Imaging single-particle counters

Image analysis systems. Many manufacturers produce image analysis systems. These are basically video-camera devices that are focused on a microscope stage or on a photograph. Signals from the video tube are transmitted to a computer processor for data reduction. A wide range of information can be developed by the computer, including length and area size data, shape factors, and statistical analyses.

Differential scattering systems. Science Spectrum, Inc. (Santa Barbara, Calif.) has developed instruments that can locate a single particle and retain it electrostatically in position so that differential scattering intensity can be recorded as a function of scattering angle. In this way, particle composition and shape information can be retrieved by matching measured data with those in a computer file. Particles as small as 1 to 2 μm can be studied, but the measurement time is on the order of seconds per particle. Data produced include average size, size distribution, shape, and other particle suspension characteristics.

Lab methods for analyzing single particles

Numerous techniques for characterizing single particles have been reported in the literature. Included among these are:

1. Bartholdi et al. [*6*] describe a scanning system in which a particle is illuminated by an argon-ion laser beam as it passes through a flow chamber. Scattering between 7.5 and 21.5 deg is measured by a 128-element photodiode array. Measurement is initiated by a delay signal once the particle has passed through a Coulter orifice. Approximately 500 microseconds per particle are required to record a complete scatter pattern for particles larger than approximately 10 μm.

2. Breitmeyer and Sambandam [*7*] detail an inline holography analysis, where holograms are made of blood cells in saline solutions to determine orientation and cell shape. An argon-ion laser was used and photographs were observed for data reduction.

3. Eisert et al. [*8*] discuss a microphotometer that uses a flow-through system for orienting nonspherical cells along the flowaxis. A focused laser beam at the axis of the hydrodynamically focused flowpath is arranged so that its diameter is smaller than that of the cell. If constant flow velocity is maintained, the pulse width will indicate the cell length in the flow system. Sizing from approximately 5 μm to 300-μm lengths at rates up to 50,000/s is reported.

4. Kaye [*9*] describes the use of a low-angle laser-light-scattering instrument. Scattered light is measured at angles as low as 1.5 deg from volumes of approximately 10 microliters, illuminated with a 5-mW He-Ne laser. Particles as small as 0.1 μm can be detected. The sample flowrate is extremely low, but concentration capabilities are high.

5. Mullaney et al. [*10*] have employed a forward angle photometer. Hydrodynamic focusing maintains a sample stream of 50-μm dia. that is illuminated by a He-Ne laser beam of 100-μm dia. Light scattered at between 0.5 and 2 deg is collected. Sensitivity from approximately 3 μm up to 20 μm has been obtained with a flowrate of 1 mL/min at concentrations up to 50,000/mL.

6. Ricci and Cooper [*11*] outline the use of a flying spot laser that scans a flowing slurry stream with a two-dimensional raster scan. Particles interrupt the focused beam that would otherwise illuminate a photodiode. Particles from 5 μm to 1,000 μm are measured in concentrations up to 2%, in streams flowing at rates up to 5 gal/min. The laser beam, reduced to 10-μm dia., scans a 3-cm distance at a frequency of 500 Hz and a velocity of 50 m/s.

7. Salzman et al. [*12*] describe a flow system, multiangle light-scattering instrument. Particles are characterized by their light-scatter patterns. A He-Ne laser is focused at the center of the sample stream. As each particle is illuminated by the beam, a 250-μsec pulse of scattered light is collected by a 32-photodetector array laid out in a concentric pattern that encompasses scattering angles from 0.3 to 20 deg. The scatter pattern is then stored in a computer for analysis by a mathematical clustering algorithm.

8. Uzgiris and Kaplan [*13*] describe application of a laser Doppler velocimeter to determination of electrophoretic mobility distributions. A standard LDV system was used to observe the frequency shift in scattered laser light due to applied voltage on the suspension of particles in a sample. Mobilities ranging from 1–5 μm/s/V/cm were recorded.

Material handling requirements

Batch systems. Since essentially all of the optical devices use flow systems, it is necessary to transfer material from a container through the instrument. One may assume that the necessary preliminary steps of adequate particle dispersion and uniform mixing have been completed. Since the instruments measure particles as local nonhomogeneities in the liquid, it is necessary to avoid artifact introduction. This requires elimination of bubbles, inclusions of immiscible liquids, and contamination from airborne dust or surface debris on the container. In addition, if concentration data are required, then the need for careful volumetric control is obvious.

For the single-particle counting devices, if the maximum recommended concentration is exceeded, coincidence errors can occur. In this case, it may be necessary to dilute the sample. If dilution is required, then aside from the normal care in maintaining good volumetric control, dilution ratio selection must be chosen high enough to avoid coincidence errors, but not so high as to reduce concentration of large particles in typical distributions to a level that does not permit accumulation of statistically significant data.

If a suitable dilution ratio cannot be found for all size ranges, then it may be necessary to measure small particles at a high dilution ratio, and large particles at a low ratio. For the most part, dilution is a manual procedure; however, there are many commercially available automatic dilutors manufactured for clinical chemistry procedures. These typically dilute by a factor ranging from 10 to 500, with high precision and accuracy. Diluted samples of 10–20 ml are typically produced; the concentrated material is aspirated in microliter quantities and then dispensed along with clean diluent in milliliter quantities.

Inline systems. If an optical particle counter is connected inline, then the problems of solids dispersion and mixing, sedimentation during sample storage, careless handling, etc., associated with container preparations are eliminated. However, a number of other problems are present. First, it is necessary that sample stream flowrate be controlled; an isokinetic sample probe may be required; a means of ensuring gas bubble elimination may be needed; an inline dilution system may be required.

Sample stream flowrates can be controlled by using pressure-compensated flow control valves with suitable protection to avoid orifice clogging, or by using a metering pump; both are downstream of the particle counter. Bubble control can be attained to some extent either by maintaining system pressure so high as to ensure that all gases remain in solution in the liquid or by incorporating a low-pressure vacuum deaerator immediately upstream of the particle counter inlet.

At present, no commercially available inline dilution system is available for dispersion sampling. Some one-of-a-kind developments have been made. These are basically modifications of batch dilutors. Clean diluent is pumped at a fixed flowrate into a mixing chamber, where a smaller sample flow is mixed with it. The diluted sample is then fed to the particle chamber. These systems are limited by the available quantity of clean diluent, particle losses in the mixing chamber, and flow measurement stability.

Some preliminary work has been reported on fixed-ratio sample-diluent flow systems feeding into motionless mixers for minimum particle loss before presentation to the optical instrument. Clean diluent is supplied by recirculating the diluted stream through a suitable filter.

Data development

Data production. Optical instruments have a number of common features in terms of the data they produce. In each device, the measurements are dependent upon the optical properties of the particulate material being different from those of the substrate; the primary data output is related to a projected area for the particles; a truncated size measurement is always made, with the low end varying with instrument sensitivity. These comments apply to both multiple- and single-particle analyzers in which size information is produced.

Multiple-particle optical instruments will usually produce data describing a particle-size number distribution. With judicious data-processing circuitry, either differential or cumulative distributions can be defined. Single-particle optical instruments will produce data based on the number of particles in several size ranges.

Depending on the resolution of the instrument, the width of each range can be made small enough so that essentially all particles in the range can be represented as having the mean diameter of the range. In this way, mathematical manipulation can be performed to convert from a number base to a volume or an area base with minimum error.

Data processing. The conversion from number base to volume or area base includes certain assumptions that must be kept in mind if comparison with data obtained by other means is desired. First, the size description is based on calibration to an idealized basis; next, a limited range of shape factors is assumed; next, uniform particle composition in the sample is assumed. Once these assumptions are accepted, conversion from optical diameters to aerodynamic diameters or sieve data can be accomplished for a range of materials. In some cases, empirical calibration may be required.

All of the optical instruments produce data that can be easily transmitted to a computer for storage and/or processing. The data can be a series of d.c. signal levels, or parallel or serial output digital pulses. Part or all of the data processing system can be included with the instrument.

Data can be provided that indicate cumulative or differential particle-size distribution curves, mean diameters with standard deviations, total quantity of particles per unit volume of liquid, and number of particles in several particle size ranges. With this easily acquired and easily processed data, application of the instrument output through a computer or microprocessor to a control function for process lines can be accomplished with relative ease. It is necessary to select a particular size parameter or ratio of sizes as the dependent variable in a control function, define an optimum operating range, and choose the process parameter that controls the variable of concern.

References

1. Bagchi, P., and Vold, R.D., *J. Colloid Science,* Vol. 53, No. 2, 1975, p. 194.
2. Dobbins, R.A., and Jizmagian, G.S., *J. Opt. Soc. of America,* Vol. 56, No. 10, 1966, p. 1,351.
3. Groves, M.J., Yalabik, H., and Tempel, J.A., *Powder Technology,* Vol. 11, No. 3, 1975, p. 245.
4. Jordan, C.F., Fryer, G.E., and Hemmen, E.H., *J. Sed. Petrology,* Vol. 41, No. 2, 1971, p. 489.
5. Robillard, P., Patitsas, A.J., and Kaye, B.H., *Powder Technology,* Vol. 10, No. 6, 1974, p. 307.
6. Bartholdi, M., et al., Optics Letters, Vol. 1, No. 6, 1977, p. 223.
7. Breitmeyer, M., and Sambandam, M.K., *J. Assn. Adv. Med. Instrum.,* Vol. 6, No. 6, 1972, p. 365.
8. Eisert, W.G., et al., *Rev. Sci. Instr.,* Vol. 48, No. 8, 1975, p. 1,021.
9. Kaye, W., *J. Colloid Int. Sci.,* Vol. 44, No. 2, 1973, p. 384.
10. Mullaney, P.F., et al., *Rev. Sci. Instr.,* Vol. 40, No. 8, 1969, p. 1,029.
11. Ricci, R.J., and Cooper, H.R., *ISA Trans.,* Vol. 9, No. 1, 1970, p. 28.
12. Salzman, G.C., et al., *Clin. Chem.,* Vol. 21, No. 9, 1975, p. 1,297.
13. Uzgiris, E.E., and Kaplan, J.H., *J. Colloid Interface Sci.,* Vol. 55, No. 1, 1976, p. 148.
14. Wallace, T.P., and Kratohirl, *J. Polymer Sci.,* Vol. 8, Pt. A-2, 1970, p. 1,425.
15. American National Standards Inst. Calibration Method, ANSI B93.28, 1973.

The author

Alvin Lieberman is vice-president of Royco Instruments, Inc., 141 Jefferson Drive, Menlo Park, CA 94025, which makes optical instruments for particle measurement in gases and liquids, and electrical-resistance blood cell counters. He holds B.S. and M.S. degrees in chemical engineering from Illinois Institute of Technology. He has served on the Technical Advisory Committee for the Illinois Air Pollution Control Board, and on the National Academy of Engineering Ad Hoc Panel on Abatement of Particulate Emissions from Stationary Sources.

Section XII
PROCESS CONTROL BY COMPUTER

Fill-in-the-forms computer language for process control
Networking computers for process control
Combined analog and digital control
Computer control in the design of a grass roots plant
Computer control of fractionation plants

Fill-In-the-Forms Computer Languages For Process Control

Several languages available for programming process-control computers involve merely filling in blanks on special forms, and do not even require the ability to use high-level languages such as FORTRAN.

CECIL L. SMITH, Louisiana State University

A major obstacle in using computers in process control has been the development and preparation of applications software, i.e., programming. For this reason, considerable emphasis has been placed on use of languages such as FORTRAN or BASIC, as opposed to assembly language. But although significant savings in time and effort are realized by using languages such as FORTRAN, software development has continued as a major expense.

In most projects, one or more individuals—control engineers—are responsible for the functional design of the control system. While some companies teach control engineers enough about software so that they can do the applications programming, use of even high-level languages such as FORTRAN requires too much time to be completely acceptable. Other companies have software specialists write the programs, but the communications problem between control engineers and software specialists reduces the effectiveness of such an approach.

The Fill-In-the-Forms Concept

Several attempts have been made to develop control-oriented languages through which the control engineer can quickly obtain the desired control functions from the computer system. To date, the most practical of these languages have been the fill-in-the-forms* languages. To use these, the control engineer need only fill in certain parameters in blanks on specially prepared forms. From these parameters, the system ascertains the functions to be performed.

Fig. 1 illustrates a form used in the IMPAC system developed by Foxboro Co. IMPAC is a block-oriented system, and this particular form is for a SCAN block, specifying the reading of Point 55 on the input multiplexer at a 5-s sampling interval. The reading is conditioned (e.g., compensated for a nonzero bias), smoothed, and the resulting value stored. This value is available to other blocks by referring to the name F2011 assigned to this block. High and low limits are specified, and Program Number 25 is to be called if either limit is violated. In fact, all necessary operations associated with the reading of this value may be obtained by inserting the appropriate information in the form, thus greatly reducing the programming effort.

The control system can be designed by specifying the proper combination of blocks, such as the one in Fig. 1. To illustrate, consider the cascade control system shown in Fig. 2. Assume that the fill-in-the-forms system supports both a block capable of performing the three-mode control calculation and a block, serving as an output module, that provides the setpoint to a conventional flow controller. The entire system can be specified by completing the forms associated with the five blocks in Fig. 3.

Not all fill-in-the-forms systems are inherently block-oriented. However, most control designers use schematic or block-oriented diagrams to express the configuration of the control system. Consequently, to be effective, a fill-in-the-forms system must support the operations necessary to implement the functions designated by at least all commonly used components of these diagrammatic representations of the control system.

The Fill-In-the-Forms System

The fill-in-the-forms system consists of two rather-distinct parts:

1. A forms-processor (sometimes referred to as a compiler) that constructs the data base from the information on the forms.

2. An interpreter that performs the functions specified on the forms.

Both of these parts will now be examined:

The fact that the system creates its own data base, of itself, reduces the software development costs. Fortunately, it is not necessary to be familiar with the detailed structure of the data base in order to use a fill-in-the-forms system. The data base consists of several items of different types, including the following:

1. Fixed data, such as the function code for each block,

* Also known as "fill-in-the-blanks."

Originally published March 3, 1975.

Foxboro Corp.

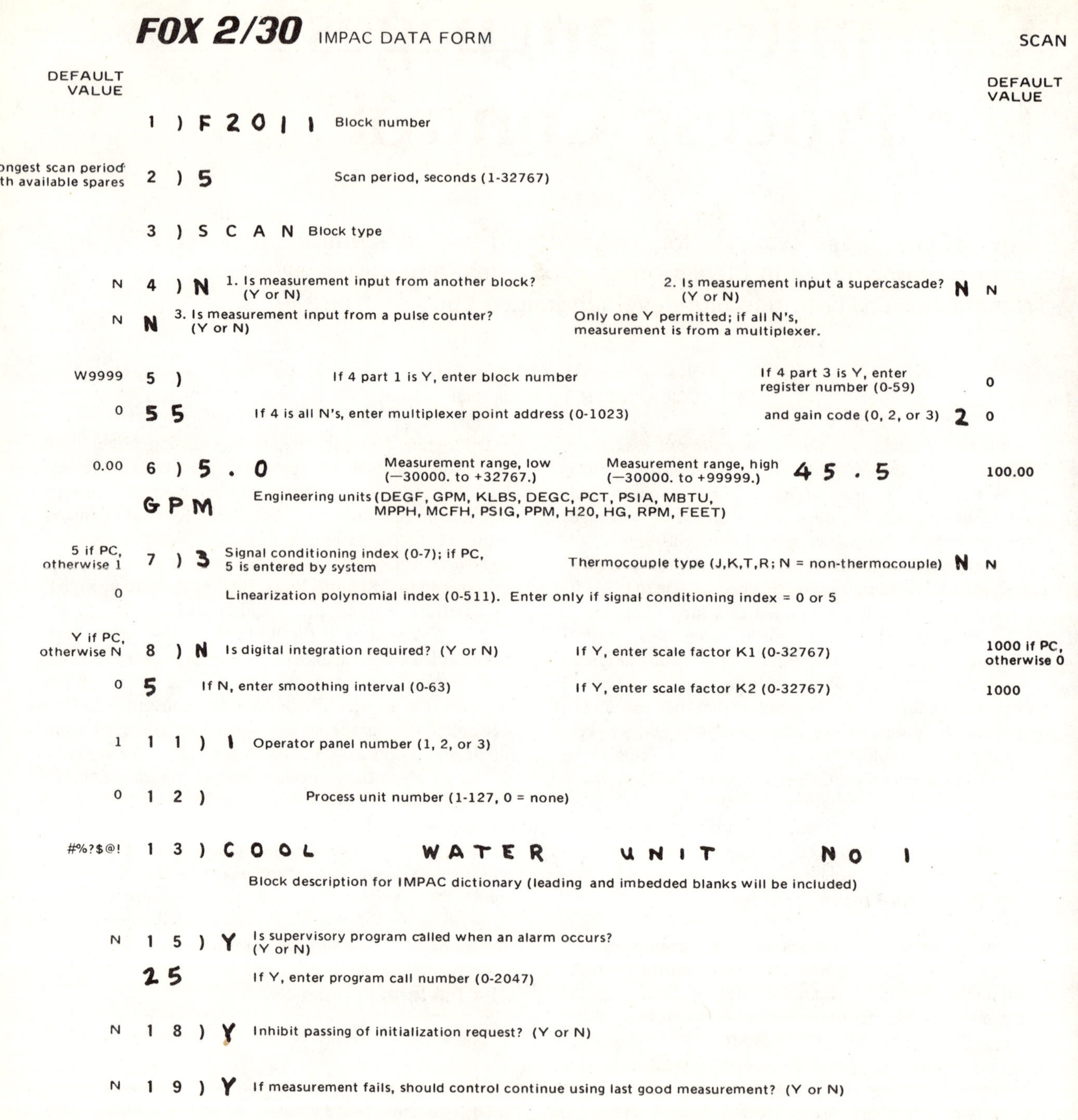

FOX 2/30 IMPAC DATA FORM — SCAN

DEFAULT VALUE	Entry	Description	Entry	DEFAULT VALUE
	1) F 2 0 1 1	Block number		
Longest scan period with available spares	2) 5	Scan period, seconds (1-32767)		
	3) S C A N	Block type		
N	4) N	1. Is measurement input from another block? (Y or N) / 2. Is measurement input a supercascade? (Y or N)	N	N
N	N	3. Is measurement input from a pulse counter? (Y or N) / Only one Y permitted; if all N's, measurement is from a multiplexer.		
W9999	5)	If 4 part 1 is Y, enter block number / If 4 part 3 is Y, enter register number (0-59)		0
0	5 5	If 4 is all N's, enter multiplexer point address (0-1023) / and gain code (0, 2, or 3)	2	0
0.00	6) 5 . 0	Measurement range, low (—30000. to +32767.) / Measurement range, high (—30000. to +99999.)	4 5 . 5	100.00
	G P M	Engineering units (DEGF, GPM, KLBS, DEGC, PCT, PSIA, MBTU, MPPH, MCFH, PSIG, PPM, H20, HG, RPM, FEET)		
5 if PC, otherwise 1	7) 3	Signal conditioning index (0-7); if PC, 5 is entered by system / Thermocouple type (J,K,T,R; N = non-thermocouple)	N	N
0		Linearization polynomial index (0-511). Enter only if signal conditioning index = 0 or 5		
Y if PC, otherwise N	8) N	Is digital integration required? (Y or N) / If Y, enter scale factor K1 (0-32767)		1000 if PC, otherwise 0
0	5	If N, enter smoothing interval (0-63) / If Y, enter scale factor K2 (0-32767)		1000
1	1 1) 1	Operator panel number (1, 2, or 3)		
0	1 2)	Process unit number (1-127, 0 = none)		
#%?$@!	1 3) C O O L WATER UNIT NO 1	Block description for IMPAC dictionary (leading and imbedded blanks will be included)		
N	1 5) Y	Is supervisory program called when an alarm occurs? (Y or N)		
	2 5	If Y, enter program call number (0-2047)		
N	1 8) Y	Inhibit passing of initialization request? (Y or N)		
N	1 9) Y	If measurement fails, should control continue using last good measurement? (Y or N)		

"FILL-IN-THE-FORMS" languages use forms like this one, which give compiler the needed parameters—Fig. 1

the parameters specified (alarm limits, multiplexer address, etc.), and other similar data. The operator may display and/or change some of these data (e.g., alarm limits) through the operator's console.

2. Variable data available internally to the block only. An example is the two previous error readings for the PID control calculation.

3. Variable data available to any block needing them. For example, the value of the flow (F2011) resulting from the block specified in Fig. 1 is available to any other block needing this value.

Depending upon the system, other information (such as a cross-reference table between user-defined names and corresponding data-base addresses) may be included.

In the initial generation of the system, maximum limits are placed on the size of the data base or parts thereof, and sufficient area is reserved in high-speed memory, and/or the disk (or drum), to accommodate the system.

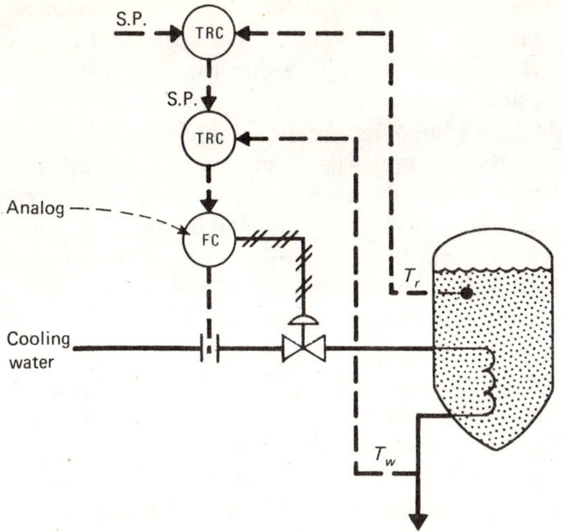

CONTROL SYSTEM specifications flowchart—Fig. 2

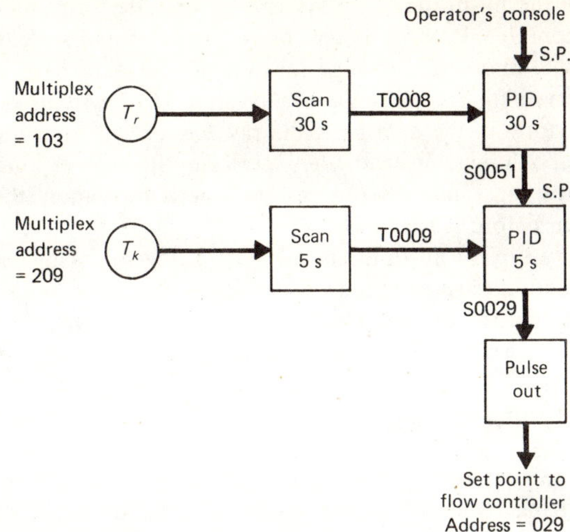

CONTROL SYSTEM schematic for a reactor—Fig. 3

The data on the forms in Fig. 1 are punched either on cards or paper tape for entry into the system, where these data are read by the forms processor and the data base generated.

After the initial data-base has been configured, most systems permit changes in the block configuration while the system is online, typically from the console typewriter. The only restriction is that any blocks affected by the changes must be taken off-scan or off-control while the changes are being made. Blocks can be added until the limits imposed at initial system generation are reached. Thus, the control engineer can readily experiment with different control concepts.

Documentation

Another problem with software development is preparing documentation. For this reason, most fill-in-the-forms systems support various types of report generators; so, at least to some extent, the systems are self-documenting.

The actual control calculations are performed by a master-type program running in an interpretive mode. This program scans the data base periodically, performing the operations required at that time by each block. The necessary logic for performing such operations is included in this program, as are the routines for smoothing, linearizing, signal conditioning, etc.

Documentation for the fill-in-the-forms system describes the various capabilities provided. In order to better suit his needs, the user may extend the capabilities of the interpreter in some respects, but with varying degrees of difficulty, depending on the language chosen.

In computer-control systems, the fill-in-the-forms interpreter usually runs as a task under the operating system. While most control functions can be accomplished by means of a fill-in-the-forms system, most processes require a limited number of control functions that are not readily supported via such a system. However, if, for example, 90% of the required control functions can be obtained via the fill-in-the-forms system, the custom programming required for the remaining 10% becomes a manageable problem.

To provide these additional functions, communication with the fill-in-the-forms system is usually essential. For this purpose, routines are normally provided to retrieve information from the data base and/or to store information in it. One approach is to implement these routines as subroutines callable either from FORTRAN or from assembly language. Alternatively, the user-written routines can be subroutines called by the fill-in-the-forms system, which passes the necessary data when a subroutine is called.

Fill-In-the-Forms System Has Drawbacks

Although the use of the fill-in-the-forms approach results in considerable savings in software development costs, certain penalties are incurred. First, since the control program is running in the interpretive mode, execution time will be slower than for custom-prepared software (written in either FORTRAN or assembly language). In addition, the programs tend to be large, requiring additional memory as well as a disk.

Input Scan

As indicated earlier, one of the functions performed by the interpreter is the input scan. The available systems generally permit input of three types of signals:

1. Analog signals.
2. Digital signals in the form of pulses.
3. Digital signals in the form of contact closures (on-off signals).

Each of these will be treated separately.

Analog signals—Common examples of analog measurements include outputs from thermocouples, pressure transmitters, flowmeters, etc. In practice, these signals must usually be compensated for offset and amplifier drift, then linearized, smoothed or integrated, and compared to the high and low limits for alarm purposes.

In the fill-in-the-forms systems, a separate form must be completed for each point on the analog input system. In this manner, a storage location in the data base is reserved for receiving the input value. Most systems do not store, in the data base, the raw value read from the input scanner; instead, they perform the conversions, smoothing, and other necessary operations upon the value before it is stored.

In a typical fill-in-the-forms system, the following information must be specified for each analog input point:

1. Multiplexer address.
2. Scan frequency (usually one of several permitted values).
3. Amplifier gain (usually one of several permitted values).
4. Signal conditioning and linearization (usually by referring, by number, either to the equation to be applied or to a routine for performing the necessary calculations).
5. Digital integration (if used, the equation is normally of the type $y_n = y_{n-1} + kx_{n-1}$, where x is the input value, y is the integrator output, and k is a constant whose value must be specified).
6. Smoothing (exponential smoothing is generally used, and a filter time-constant or equivalent smoothing index must be specified).
7. High and low alarm limits.
8. Action to be taken in case of alarm (message to be typed or special routine, usually user-written, to be invoked).

The various parameters for each of these functions must generally be specified on the fill-in-the-blanks form. However, the process operator normally can (through the operator's console) place a point on or off scan, change the alarm limits, initialize the integrator at a desired starting value, and perform similar functions.

Pulse Inputs—Devices such as turbine meters, tachometers and totalizers output a train of pulses. Although the computer can be programmed to count the pulses, the most common approach employs a pulse counter, thereby requiring that the computer only read the value in the counting register of the pulse counter. In one approach, the register is automatically cleared each time it is read, the total in the register then being the number of counts received since the last time the device was read. In the case of a tachometer, the count is a measure of speed; in the case of a turbine or swirl meter, it is a measure of flow.

For this type of input, the following information must typically be specified:

1. Device (pulse counter) address.
2. Scan frequency.
3. Conversion to engineering units.
4. Smoothing.
5. High and low alarm limits.
6. Action to be taken in case of alarm.

This information suffices to permit the system to regularly read a raw value from the pulse counter, and then to store the processed value in a storage location in the data base.

Digital Inputs—Devices such as limit-switches, contacts, or relays may be in only one of two possible states—open or closed. In most process situations, one of these states will indicate normal conditions, e.g., reactor pressure below an alarm condition. The other state will indicate a condition that calls either for some response by the system or an action by the operator.

On most computers, the digital input modules provide the interface between the computer and the devices in the plant. Normally, each digital input module accommodates the same number of digital inputs as the number of bits in a word. For example, for a computer with a 16-bit word-length, the digital input module would accommodate 16 digital inputs. The address of a digital input is the address of the digital input module followed by a bit address e.g., (0 to 15) to designate the specific contact on the module.

Depending upon the particular system, the digital inputs may be scanned at a selected frequency, at a preset frequency, or on demand. The system normally looks for changes in the states of the contacts. When a change is detected, it indicates either an abnormal or a normal condition. For the abnormal condition, the message to be displayed to the operator and/or the program to be executed must be specified.

In most processes, certain conditions demand the fastest possible response. Digital signals relating to these conditions are fed to the computer's interrupt bus, which permits the processor to respond immediately rather than waiting until the next scan of the digital inputs. Programs to service these interrupts are generally supported directly by the operating system, but provision is made for these programs to access the data base of the fill-in-the-forms system.

Process Outputs

The typical process-control system permits four types of outputs: pulse trains, variable-width pulses, analog outputs, and digital outputs (contact closures).

The pulse-train output is a sequence of fixed-width, fixed-amplitude pulses used to drive stepper motors. The most common application of such an output is in supervisory control, to change the setpoint of a primary controller. In this case the following must be provided: the hardware address of the controller on the pulse-output system, the address of the auto/manual switch indicator on the digital-input system and, if desirable, the address of the deviation indicator and/or computer acknowledge light.

The variable-width pulse is normally used in DDC (direct digital control) to drive valve motors. For outputs to these devices, the hardware address must be provided.

Analog outputs are generated by D/A (digital-to-analog) converters, and are used to drive analog devices such as trend recorders. Normally, only the output address is required.

Digital outputs are used to open or close field contacts such as relays for starting or stopping motors. Digital-output modules similar to the digital-input modules form the interface between the computer and the process. The output address is composed of the digital-output module address, and a bit address within that module to specify a specific digital output.

Control Functions

The functions for control purposes differ considerably from system to system in terms of configuration and specification, but the overall operations performed are very similar. The following is a rather minimal example of functions required for control purposes. Most systems currently available offer a more expanded complement.

PID Controller—This function is used to implement the PID (proportional, integral, derivative) control law commonly used in the chemical process industries. The specifications must include the source of the setpoint and the source of the feedback variable. Generally, the setpoint may be specified by the operator or obtained from the data base, thereby permitting cascade or supervisory control. Other information required includes the sampling time and the tuning parameters (gain, reset time, and derivative time). Some systems accommodate variations in the basic control law, including basing the proportional and/or derivative modes on the feedback variable instead of the error signal, using error-squared ($e|e|$) instead of error, and permitting only reset action to be used (floating control).

Add/Subtract—General equations of this type follow the form:

$$Y = \frac{k_1X_1 \pm (k_2X_2 + b)}{k_3}$$

where Y = output; X_1,X_2 = inputs; k_1,k_2 = scaling factors; k_3 = weighting factor; and b = bias.

This equation can be used to add, to subtract, to bias a reading, to compute a weighted average, and to implement other similar functions.

Multiply/Divide—The generalized equations of this type permit multiplications and divisions according to:

$$Y = \frac{kX_1X_2}{X_3}$$

where Y = block output; X_1,X_2,X_3 = block inputs; and k = gain.

By selectively specifying one or two of the three inputs as being unity, one can perform multiplication, division, ratio control (multiplication by a constant), or taking the reciprocal of X_3. One system, PROSPRO, combines multiplication and addition into a single equation, with exponentiation also included.

Lead/Lag—The equations used to approximate the conventional lead-lag transfer function are of this form:

$$Y(s) = k\frac{\tau_1 s + 1}{\tau_2 s + 1}X(s)$$

where Y = output; X = input; k = gain; τ_1 = lead-time constant; and τ_2 = lag-time constants.

In addition to k, τ_1 and τ_2, the sample interval must be specified.

Dead Time—The delaying of signals—primarily for dynamic compensation in feedforward control systems, or for use in deadtime compensation algorithms—is described by the equation:

$$Y(t) = kX(t - \theta)$$

or

$$Y(s) = ke^{-\theta s}X(s)$$

where Y = output; X = input; k = gain; and θ = delay time.

In addition, the sample interval must be specified.

On/Off—This function compares two inputs and sets the output according to the equation:

$$Y = \begin{cases} b_1 \text{ if } X_1 > X_2 \\ b_2 \text{ if } X_1 < X_2 \end{cases}$$

where Y = outputs; X_1,X_2 = inputs; and b_1,b_2 = output values.

To prevent excessive chatter when the two inputs are nearly equal, a deadband is normally incorporated. This function can be used either to emulate the conventional on-off controller, or as a single-stage cutoff in batch-sequencing operations.

Auto-Select—For this function, the output is simply the larger (or smaller) of the two inputs.

Setpoint Ramp—Ramping setpoint changes, at a predetermined rate, are frequently required in batch control.

Switch—This function is used to select between two inputs (or two outputs), depending upon the status of a two-state device.

Available Implementations

In most situations where commercially available products are being categorized, a few products fall clearly into a defined category, but if the definition is changed slightly, other systems can be included. The three most well-known systems of the fill-in-the-forms type are IBM's PROSPRO, Foxboro's IMPAC, and Honeywell's BICEPS. However, in the case of the last, the forms are essentially used only to specify the inputs and outputs, which also permit the data base to be generated. The control functions are then implemented by a language derived from FORTRAN.

Other systems come close to this. For example, Fisher's DC^2 system permits the input/output operations to be readily specified by techniques other than by the use of forms, and then permits control functions to be implemented by a language derived from BASIC.

The following is a brief summary of PROSPRO, IMPAC and BICEPS:

PROSPRO—Available from IBM for its 1800 computer, PROSPRO provides the following functions:

1. Process I/O.
2. Definition and maintenance of data files.
3. Both supervisory and DDC control.
4. Definition of nonstandard control algorithms.
5. Computer/operator interface.

With PROSPRO, the control strategy can be modified online.

Using PROSPRO, the information required to formulate the data acquisition and control functions is supplied via:

1. *Variable Information for Supervisory Control* form supplies the necessary information to produce the disk-resident record that is needed to direct the processing of each variable.
2. *Adjustment Information for Supervisory Control* form specifies the data necessary to effect supervisory control action.
3. *General Equation, Supervisory* form is used to spec-

ify special arithmetic computations needed to extend or tailor the monitoring, control and regulation strategy without burdening the user with programming in assembly language or FORTRAN.

4. *General Action, Supervisory* form permits specification of: a combination of actions including logic decisions, a series of action steps (such as: read contact point, operate contact, type message, etc), execution of a special program—or some combination of such actions.

5. *DDC General Block* form specifies DDC control calculations and actions to be taken. The associated processor program is written to obtain fast processing but yet to be compact so that it can be core resident.

6. *Variable Information for DDC* form specifies a 16-word core-resident data block that contains all of the information necessary to perform basic DDC.

7. *DDC-Variable-Associated Data Block* form extends the 16-word DDC data block to provide for additional storage of information associated with the processing of certain DDC variables.

As indicated by these forms, PROSPRO provides for both DDC and supervisory functions. The main distinction is that the DDC-related programs and data are intended for frequent processing and thus are core resident. Supervisory-related programs and data are primarily disk resident.

For the System/7, IBM supports a fill-in-the-forms system called PCP/7 (Process Control Program for System/7). Processing options supported by PCP/7 are measurement only, measurement with limit checking, five-mode algorithm, supervisory control algorithm, ratio control algorithm, setpoint adjustment, output tracking, weighted sum, material integration, cold-junction compensation, arithmetic check, ADC check, DAC check, and three user-defined algorithms.

IMPAC—Written by Foxboro for the FOX 2 process control computer, IMPAC incorporates a "fill-in-the-blanks" method to allow the user to perform supervisory control, process optimization, generation of reports, and performance-level calculations. The system provides the following functions:

1. Scan and control.
2. Data-base generation.
3. Supervisory control.

Using IMPAC, the user can add to or modify the data base online, which includes modifications in the control strategy.

The following data forms are used in IMPAC programming:

1. Setpoint ramp.
2. Reference input switch.
3. Contact point.
4. Scan.
5. Setpoint control.
6. Lead/lag or deadtime.
7. Ratio.
8. Bias.
9. Weighted average.
10. Parabolic.
11. Controller (P_eID, P_bID, E^2, I)
12. Auto-select.
13. On/off.
14. Two-stage cutoff.
15. Multiply/divide.
16. Measurement input switch.
17. Output switch.
18. Data.

From these forms, the IMPAC compiler (or forms processor) generates the data base, which specifies the control actions to be performed, the computations required, and other similar information.

BICEPS—Originally developed by General Electric for the GE 4020 control computer, BICEPS incorporates the following functions:

1. All input/output functions, including filtering, units conversion, limit checks, etc.
2. Generation and maintenance of instruction and data files.
3. Communication between operator and computer.
4. Special loop-processing, conversions, and similar special functions via the BICEPS Programming Language (BPL), a FORTRAN-like language.

Using BICEPS, new loops can be added, existing loops modified, and BPL programs compiled, while the system is online.

The BPL language is a simplified, control-oriented version of FORTRAN. The language includes most of the arithmetic features of FORTRAN, the logical IF, the capability to retrieve values of process variables from the data base, and some special subroutines for lead/lag, status change, etc. Provision is made to test programs by suppressing of storing operations until the program is debugged. BPL is implemented via a compiler rather than an interpreter and, consequently, execution is faster.

With the availability of a language like BPL, there is no need for a large number of special forms to specify control calculations. Instead, BICEPS requires only two forms—one that identifies the process variables, and a second that provides the information required for control action.

While BICEPS is intended primarily for supervisory functions, a DDC package is available that can be implemented along with BICEPS on a 4020 computer.

Acknowledgment

This article is based on a chapter written by Cecil L. Smith for a book on minicomputer systems edited by Duncan A. Mellichamp, Dept. of Chemical Engineering, U. of California at Santa Barbara.

References

1. IMPAC User's Guide, 73002DA, Foxboro Co., c1972.
2. BICEPS Application Manual, GET-6064A, General Electric Co., c1970.
3. PROSPRO II (TSX/1800) PROcess Systems PROgrams Application Description Manual, GH20-0718-0, International Business Machines Corp., c1970.
4. System/7 Process Control Program (PCP/7) Operations Guide, SH19-7026-0, International Business Machines Corp., c1974.

Meet the Author

Cecil L. Smith is Professor and Chairman, Dept. of Computer Science, and Professor, Dept. of Chemical Engineering, Louisiana State University, Baton Rouge, LA 70803, where his interests lie mainly in the areas of digital control and simulation of industrial processes. He holds B.S., M.S. and Ph.D. degrees in chemical engineering and is the recipient of the 1972 Donald P. Eckman Award for outstanding contributions to the field of automatic control. He is a member of AIChE, Instrument Soc. of America, and Assn. for Computing Machinery, and is a registered professional engineer in Louisiana.

Networking computers for process control

In chemical process plants, computer networks are consolidating management information and supervision. Such applications extend beyond the use of even the most powerful digital devices for plant operating functions. Such networks will gain in importance as organizations become larger and more centralized, and as broad-scale optimization becomes increasingly critical to economic survival.

Leonard A. Bizarro, Digital Equipment Corp.

☐ The ability of computers to communicate with each other is well known and is being exploited increasingly in commerce and industry. Networks of interacting computers facilitate the exchange of information freely through an organization, while offering economical and efficient use of expensive data-handling resources by sharing with remote stations.

NETWORK CONCEPTS

A computer network includes the hardware and software needed to perform application-oriented tasks, as well as those required to automatically implement reliable communications. The network is a collection of nodes having different functions and amounts of computing capability; these nodes are interconnected by data links, using various signal-transmission media and techniques.

Function of nodes

In a completely generalized network, every node is a full-power computer, able to function independently of any master or host machine. Each node, accordingly, has a complement of hardware and software such as memory units, process input-output devices, operator and programmer terminals, and executive and application programs. In addition, each autonomous node includes communication hardware and software, which permit exchanges of stored data, programs, and computing resources with other machines that may be included in the network.

Specialized networks normally include nodes having less than full computational power. For example, remote-data entry terminals, analog-to-digital and digital-to-analog converters, dedicated controllers, and alarm annunciators might serve as nodes in a process control system. In some cases, these devices would operate directly through specified host computers. However, the devices could include sufficient intelligence of their own to function independently, and to interact with several other nodes through their own communication interfaces. Remote input-output modules, with internal engineering units that perform conversion or linearization, multiplexing, and error-checking functions, are an example.

Configurations

Networks may be arranged in numerous ways, as suggested in Fig. 1, so structures can be built to meet diverse application needs. Any of the data links could be implemented with two or more redundant channels when the reliability of a particular path is considered essential.

Meshed loops, as shown in Fig. 1(a), are most general, and are suggested when traffic can be expected among any combination of nodes. Star networks, shown in Fig. 1(b), might be more appropriate when one machine is to serve as host, for resource sharing or supervision. Ring structures, shown in Fig. 1(c), can be developed when all functions are essentially parallel, but outputs from one unit are needed as inputs at other points. Multi-drop or party-line communications—Fig. 1(d)—are also used, and can be cost-effective when remote terminals report only to the host computer, when abnormal conditions exist.

Originally published December 6, 1976.

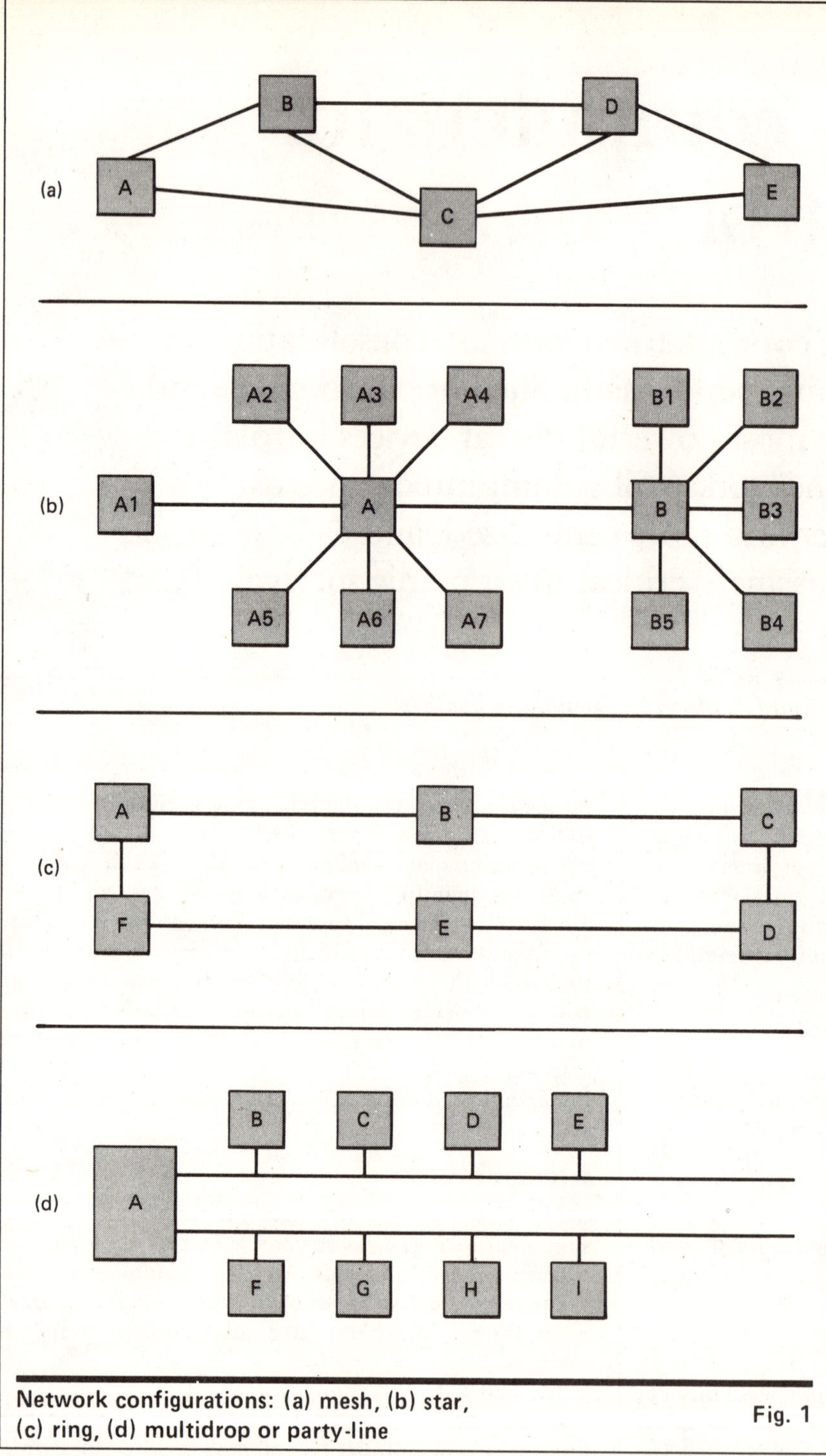

Network configurations: (a) mesh, (b) star, (c) ring, (d) multidrop or party-line

Fig. 1

Architecture

Ideally, network functions should be invisible to system designers and operators, in the same sense that the internal structure of a single computer is not apparent to those that use it. This is accomplished by means of network architecture, comprising the hardware and software necessary to interface computers of various types to common data channels, create links among nodes, exchange various types of information existing in different formats, share devices and files, and control execution of programs in remote machines.

A four-level architecture is typical. At the lowest level, *interface elements* operate communication devices such as telephone models. *Physical links* perform coding, sequencing, error detection and recovery, synchronization, and other transmission-management tasks. *Logical links* provide multiplexing, demultiplexing, message routing, transmission acknowledgment, and similar network-control services. *Dialog functions* convert messages into meaningful exchanges at either end of the communications line.

All instructions or commands are also provided as part of the network architecture, in order to permit programmers to implement the various communications functions. The command structure must also include link, file and control instructions of the types suggested in the accompanying table.

Node function

Each node in a network can be optimized to perform specific computational functions. Several broad classes of such functions are important as building blocks for individual applications:

Host processing involves executing programs requested from remote computers and terminals, to make resources and files available to distributed users. Star networks, such as shown in Fig. 1(b), are often used, with the hosts at the center. The host computers typically have large core and disk memories, and support sophisticated operating systems.

Front-end communications processing relieves the burden of multiplexing, coding polling and other message-handling tasks from host machines. This frees significant memory and time on the hosts, greatly extending the capacity for more-complex computations. A minicomputer tied directly to the host machine is normally used for this function.

Remote concentration is similar to front-end processing, except that the functions are implemented at a distance from the host. Concentrators can also be used in networks where no computer acts as a host, but where nodes are organized into distinct groups. As an example, nodes A and B in Fig. 1(b) might be intended primarily to concentrate communications from the computers to which they are directly interfaced. Primary incentives are to provide economy through replacing several slow data links with just a single high-speed line.

Remote job entry enables users to submit batch or production jobs directly to a host, and to receive output through an onsite terminal. The terminal normally consists of a card reader and line printer.

Remote computing places minicomputers onsite for direct user interactions, to perform nonfile tasks, such as routine calculations. The link to the host is established automatically, when needed for access to the data base or high-capacity peripherals. This network function reduces the burden on host machines, making it possible for more users to take advantage of the extensive central resources.

Message switching involves use of minicomputer nodes to accumulate and retransmit communications from remote points. This alternative to route switching provides functions such as assembly and disassembly of messages, polling and addressing terminals, controlling lines, checking and correcting errors, performing code and speed conversions, and adding time and date identifications to messages.

NETWORK APPLICATIONS

Control systems incorporating multiple interconnected computers are already widely used in the chemical process industries. Newly emerging network disciplines simplify to quite an extent the development, operation and modification of all of these systems and therefore greatly broaden the scope of most practical applications.

Communications management

Some applications involve distinct or geographically isolated processing areas or plants, which must be coordinated. In such cases, there are economic incentives for computer networks to perform communications management. As examples, data concentrators permit one high-speed line to replace many slower channels, and message-switching nodes can optimize routings to increase data throughout and circumvent lines out of service.

Resource sharing

Monitoring and control system effectiveness can frequently be upgraded by adding computational resources to augment online functions. The resources might include such hardware as mass storage for data or auxiliary programs, core memory to support online background program development, high-speed printers for report generation, or graphic displays for operator interfacing. All these various resources may also include such software as multiprogramming executives, complex optimization algorithms, and rather extensive system diagnostics.

Advanced resources tend to be expensive, and are generally underutilized unless shared among many system nodes. The resources may be concentrated at a host machine or placed where they will be most often used, and in either case accessed through the network when needed for remote tasks. For example, process input-output hardware and complex feedforward model logic might be dedicated to particular control computers, while disk-storage-containing optimization routines and facilities for operator interactions are in a more centralized machine.

Overview of typical network commands

Basic commands

Connect (3, B, Y)—When issued by Program X at Node A, this requests a connection with Program Y in Node B. If Program Y accepts the connection, the link will be made and subsequently will be called Connection 3.

Send (3, Data)—When issued by Program X, this indicates that the specified data should be transmitted to the program at the other end of Connection 3.

Receive (3, Buffer)—When issued by Program X, this indicates that Program X will accept data being transmitted by the program at the other end of Connection 3. No data are normally moved over Connection 3 until Program Y has indicated it has data to send, and Program X has indicated it has a buffer to receive the data.

Disconnect—When issued by Program X or Y, this constitutes a request to terminate the link.

File commands

Open (4, C, New [100, 100]—This causes *File New*, stored in Record 100, 100 on a computer at Node C, to be opened as logical Unit 4 within Program X at Node A. The same command would be used to reach remote devices, such as printers or card readers, rather than files.

Get (4, Buffer)—This causes data to be transferred from the remote file or device designated as Unit 4 to the buffer of Node A.

Put (4, Data)—This causes data to be transferred from a buffer at Node A to the remote file or device designated as Unit 4.

Close (4)—This causes logical Unit 4 to be closed, and the link to the remote system to be terminated.

Delete (C, Old [100, 100]—This causes the file Old, stored under Record 100, 100 on the computer at Node C, to be removed from the file directory.

Control commands

Run (B, Z)—This requests the execution of Program Z on Node B.

Abort (B, Z)—This causes the execution of Program Z on Node B to be terminated.

Distributed processing

Another important application of computer networks is distributed processing. This permits a job to be segmented according to any appropriate criteria, so the computation can be performed in several machines simultaneously. One advantage is in reducing hardware and software complexity at each node. This

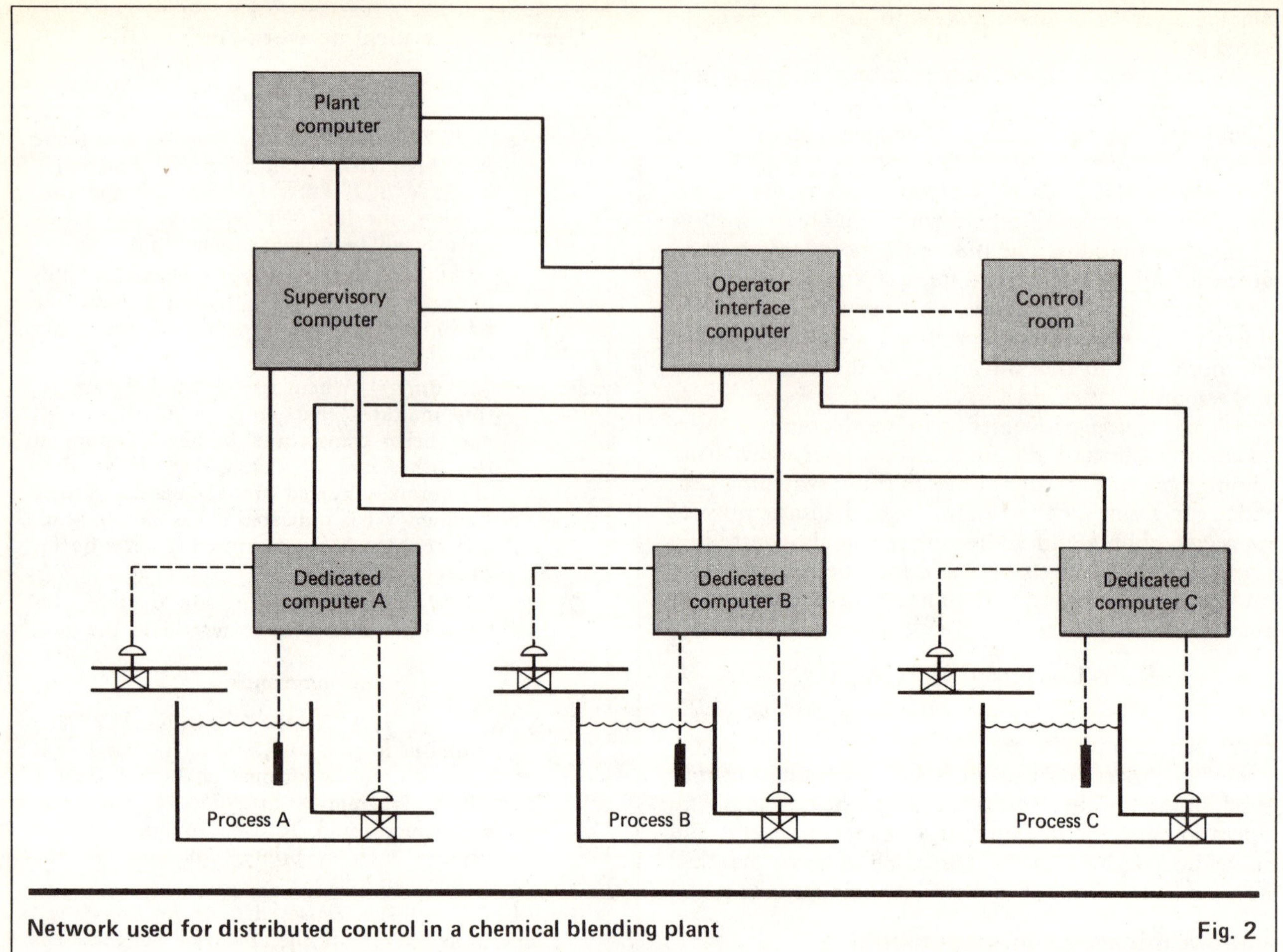

Network used for distributed control in a chemical blending plant **Fig. 2**

makes individual systems elements easier to implement, debug, support and maintain. Further benefits accrue from the high speeds achieved with parallel processing, and the reliability gained from avoiding dependence on a single computer.

At the management level, distributed processing might permit an organization to maintain a company-wide inventory and production-control system, with the data base segmented according to physical location of materials and processing equipment. A computer at each plant might receive inputs from various operator input terminals or process sensors, and perform the file manipulations necessary to define processing performance, capabilities to fill orders, and the need for inventory replenishment.

Communications among the plant computers would not only provide a total picture of corporate production but would also facilitate transferring raw materials and finished product, smoothing loads and demands, and compensating for unexpected bottlenecks and shutdowns.

At the production level, distributed processing generally implies dedicating small computers to specific tasks, and interconnecting machines directly or through the network. Various configurations are possible for such distributed control, depending on plant requirements or preferences. However, all distributed systems tend to be modular, since each controller has well-defined and relatively restricted functions. Flexibility is high because a complex system can be implemented in stages, with computers integrated into the network as required.

The evolution can be downward, beginning with a supervisory or management system, and expanding through addition of more highly dedicated control devices. Upward evolution is also feasible, by adding networking capabilities to a set of independent dedicated computers. After a network is operating, lateral growth is possible by increasing the amount of computation at diverse points.

Distributed control has been attracting particular attention, because of intelligent instruments and actuators coming on the market. Sensors, valves, motors and other components with internal microprocessor logic can be made to act as addressable network nodes, with the ability to execute instructions and transmit or receive data. Buffer registers and other features would typically permit such devices to be polled or to request data without interrupting normal operations.

An illustration would be a control valve with a program that defined installed characteristics based on an internal feedback loop, so that any relationship could be specified between command signal and actual flow capacity. The microcomputer within the valve might also use a feedback calculation to implement direct flow control; the operating signal could then command a volume or mass flowrate, and the valve stem would be positioned to achieve this value regardless of changing stream conditions. Status reports would be returned to the control room only periodically, except when an alarm condition warranted an immediate interrupt.

An example of distributed control is shown in Fig. 2, where a computer is dedicated to the monitoring and control of each vessel in a formulation plant. In operation, a computer might detect that the level in its tank is too low, and perform tests to deduce that the feedstream valve has failed closed. The program in this computer determines that the tank should be placed on hold by closing the outflow valve, and then generating the necessary commands. This status condition is transmitted to the supervisory computer—where a scheduling and feedstock allocation program is initiated—to calculate how the remaining vessels should be operated to optimize performance under the given circumstances.

The supervisory computer sends a new setpoint to the control computer on one of the remaining tanks, and downloads an entire new program to the other. The supervisory computer also notifies the plant computer of the problem, and this machine schedules the necessary maintenance. While these commands are being exchanged, the operator interface computer displays data obtained from the other nodes, accepts and distributes any manual commands, and generates all logs and reports needed by supervisory and management personnel.

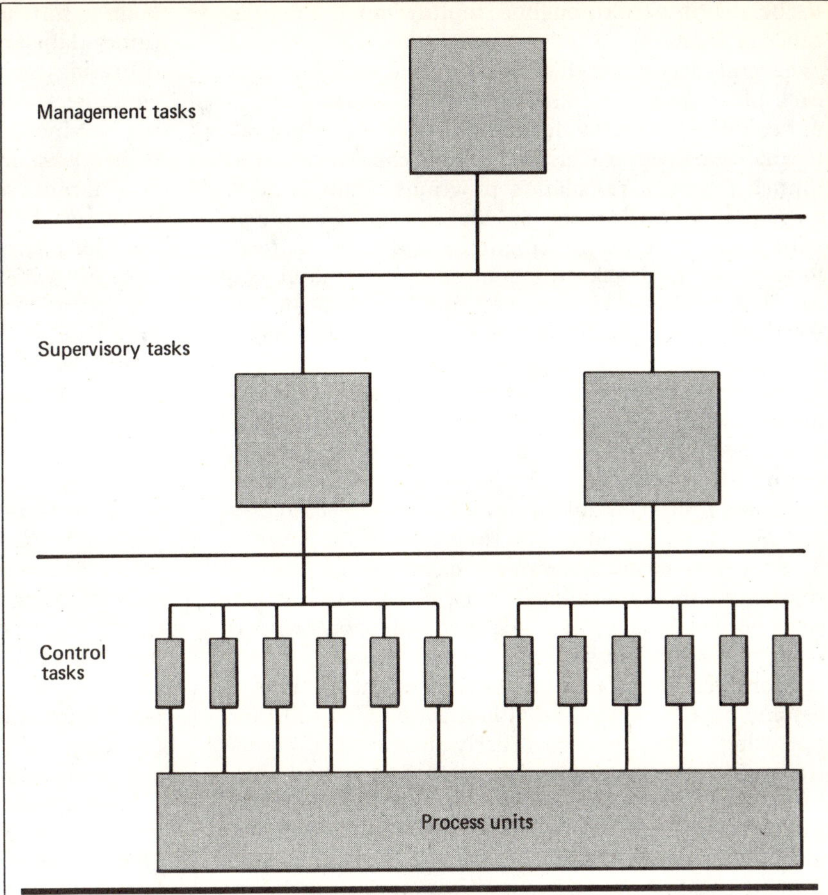

In hierarchical control, functions assigned to computers on various tiers depend on levels of scope and authority within a plant Fig. 3

Hierarchical control

Hierarchical control—a form of distributed computation—is being widely discussed in the process industries. This involves partitioning control functions in tiers, with responsibilities corresponding to levels of scope or authority in the organization (Fig. 3). In addition to the general benefits of distributed control, a hierarchical structure places specific hardware and software resources and data directly in the hands of the users for whom they are most critical.

The lowest levels in the hierarchy comprise intelligent devices or computers for direct interaction with specific process units; for monitoring and control tasks, such as limit-scanning or feedback loop-closure. These machines are accessible to operators, dispatchers and other production personnel close to the plant floor, and would typically have the displays and pushbuttons or knobs needed for manual verification and adjustments.

On immediate tiers in the hierarchy are nodes for such tasks as determining optimum setpoints for lower-level controllers or driving-plant-operating stations. These computers would analyze operating data and management instructions from points all over the plant. Typical computations might include energy or mass balances to establish strategies for maximizing throughput, economically allocating steam or electric utilities throughout the plant, and scheduling startup or shutdown of units. Intermediate processors would be accessible to supervisors or production managers. They would typically include alphanumeric cathode-ray-tube (CRT) keyboard terminals for (1) formatted displays or interactive access to data or entry of new production criteria, (2) low-speed printers for logs and alarm messages, and (3) disk or tape storage for backup or auxiliary programs to be downloaded into the dedicated machines.

The highest tiers have computers to perform such plantwide tasks as system development, production planning, coordinating different plant areas, and monitoring overall efficiencies. Access for entering commands or for receiving data is restricted on these machines to engineering or management personnel. In most cases, resources such as high-speed printers, large-mass memories, and multiprogramming executives tend to be concentrated at the highest tiers of a hierarchical network.

An example can be provided by a network needed for a large petroleum refining operation, having a number of distillation plant sites and a mix of products to be distributed through a multiterminal long-distance pipeline.

Several tiers of intelligent nodes might be used in each plant for control and supervision. Sensors and final-control elements with analog-to-digital and digital-to-analog converters, error-checking capabilities, and limited feedback calculation programs might be used on individual feedstreams or column trays. Computers with local input-output might be dedicated to feed-forward or feedbackward control of individual columns. A higher-level computer could be dedicated to each train, to establish composition setpoints for the individual column controllers.

Machines associated with each train could communicate with a supervisory computer within the plant. The objective at this higher level might be to perform optimization calculations, so that the total throughput of desired products could be maximized, subject to constraints on energy usage or byproduct composition. The supervisory machines might have the capability to reprogram the control computers, if revised strategies were needed because of product specification or operating-objective changes.

Moreover, the higher-tier devices might be able to assume control of any individual column, to continue operating or to direct an orderly shutdown, if conditions indicated that a dedicated computer had malfunctioned. Depending on the size and scope of the plant, this level of control might have one computer, although separate machines could also be used for such functions as logging and report-generation, scheduling distribution of incoming feeds or outgoing product, and monitoring utility demand.

The computer network could profitably extend along the pipeline. The network would not only ensure equitable distribution of product with minimal pumping costs, but could also detect leaks or equipment malfunctions. The computer could then order maintenance, select alternate routings for the product, or close off portions of the particular system in the possible event of emergencies.

Individual digital controllers at unattended booster stations might sequence compressor startup and shutdown, provide suction-discharge pressure to override control, and perform monitoring and logging functions. If all stations along a pipeline were similar, some control programs could be executed by microcomputers with standardized read-only memories, to avoid possibilities of inadvertent erasure. Different dedicated computers might be used at the pipeline terminal points to operate pumps and valves, to perform calculations verifying measurement integrity, and to compensate for possible drift or calibration while in custody transfer.

A supervisory computer could also coordinate all booster stations and terminals from a dispatch station. This supervisory computer would also interrogate or be addressed by production or upper-management computer machines, for information concerning availability of product.

The hierarchy might also extend to central engineering, for applications encompassing control-strategy evaluation; simulation by a research and development team to software development programmers; and laboratory-data analyses by production chemists to study operating trends of chemical engineers. Most of these computers would be time-shared, with terminals having varying degrees of intelligence and amounts of peripheral resources in the work areas.

A number of network nodes might also be found at corporate headquarters. Small computers might be used for payroll and accounting or market analysis. A large central-data-processing system would also be possible, consisting of a single machine or a distributed data-base management system. The central system would be able to supervise the entire network, and would probably have front-end minicomputers to relieve the routine message-handling load.

Synergy

The earliest process-control systems were manual, and included instruments and actuators mounted directly on the production units. Operators spent a lot of time running between the gages and the valves, or shouting across the floor to coworkers performing related functions. Automated process control made it possible to centralize operations, by gathering measurement information and distributing control commands to many points. The cornerstone of the process was communication.

Now, computer networking is on the verge of causing another major step in process control. Networks offer all of the advantages of centralized-data manipulation, but return the control functions to the remote points where they can be most conveniently performed and managed, and simultaneously avoid the risks of depending on a single computer or rack of analog or digital instruments. Added to this step is the expansion of the communications scope, which includes coordination of widely separated plants and incorporates management information as well.

Macroscopic process management has been envisioned since the first digital monitoring and control systems were planned in the late 1950s. But the large computers that were necessary to implement the applications were prohibitive in hardware cost and software complexity. Minicomputers with real-time software made digital control a reality, but scope was necessarily restricted. Now, network communications provides a way of overlaying management upon control—with neither the costs nor the risks of the centralized approach.

The author

Leonard A. Bizarro is Marketing Manager, Process Industries, with Digital Equipment Corp. (146 Main St., Maynard, MA 01754). He is also in charge of marketing, sales and technical support of computer systems and software to the chemical process industries. He has presented technical papers at meetings, and has taught management at Mt. Wachusett College. Before, he was product planning manager, market development manager and technical support specialist. He holds a B.S. degree in marketing from Fairleigh Dickinson University, a B.S. in computer science from Boston University, and an M.B.A. in management from Babson College.

Combined analog and digital control

Digital equipment has unquestionably broadened the capabilities of process control systems. But even sophisticated, multilevel systems look to analog loops for the basic PID (proportional-integral-derivative) control function, and for backup to digital routines.

J. J. Murray, Beckman Instruments, Inc.

☐ To choose the proper mix of analog and digital instrumentation for a process plant, the engineer must design for the needs of all who will use it. For the plant operator, maintenance engineer and process engineer, the hybrid analog-digital system should be convenient, accessible, responsive and flexible. Likewise, for the plant manager, the system should be all of these, but in addition it should enhance his plant's profitability.

The operator

The control system is the operator's only interface with the process. Instrument displays should be positioned so that he can make a quick, accurate assessment of the status of the entire control loop. For convenience, displays of the process variable, setpoint, and valve position should be dedicated, and located near the operator's open/close switches. Also close by should be switches transferring the system from automatic to manual, or from remote to local.

The simplest connection between operator and process is provided by a dedicated control panel having a conventional graphical display format. A graphical display enables the operator to tell at a glance what the deviation of the process variable from setpoint is. A shaded band on the display indicates simply and directly the acceptable bounds of the deviation.

On the other hand, a digital display format can show numerical values in a more accurate and readable fashion. Such a format, however, does not usually permit the operator to compare simultaneously the process variable, the setpoint, and the acceptable deviation. Thus, the digital display format is not generally suited for service in feedback control loops.

Moreover, since digital panel meters and CRT's (cathode ray tubes) are usually time-shared, they do not allow the operator immediate and unencumbered access to all final control elements. Both require a delaying callup procedure, and in an emergency this procedure leaves room for error and possible disaster.

Still, computer-based data-gathering with digital readout can be an important tool for the operator. Rapid calculation and prompt display of plant inventory, efficiency, and material and energy balances strengthens the operator's control over the plant. Data-logging on magnetic tape is an attractive alternative to conventional recording methods. With CRT readout, it enables the operator to refer back to previous plant behavior quickly and easily.

An overall, supervisory computer control system can be helpful in starting up or shutting down a plant in a hurry. Although backup systems are always mandatory, supervisory computer control can relieve the operator from watching a particularly difficult control loop.

If the plant is modeled accurately, supervisory computer control can defend against major upsets. However, many emergencies require operator intervention. In such cases, the operator's easiest access remains the individual, dedicated analog controller and display.

The maintenance engineer

The maintenance engineer's prime objective is zero downtime. In selecting control instrumentation, he must allow enough backup and redundancy so that a processing unit does not go down and stay down because of instrument failure.

In a hierarchical type of control system, there are several levels of control, arranged so that when higher levels are shed, lower levels are still available. A typical hierarchical system might have the following levels:

1. Manual control.
2. Automatic control.
3. Hard-wired logic control.
4. DDC (Direct digital control)/supervisory computer control.
5. Computer-based data acquisition with digital display and control.

Originally published June 21, 1976.

For example, if the DDC/supervisory computer is taken offline for a program de-bug or for routine maintenance, systems 1, 2, 3 and 5 will still permit the plant to run. If only systems 1 and 2 are working, the individual analog controllers will provide automatic control. And if an analog controller is suspected to be faulty, the operator can remove it from the loop and replace it with another controller or a backup manual station without disrupting the process.

It is important to note that individual analog controllers form the base of the control system hierarchy. This arrangement has two advantages. First, the risk of catastrophic control-system failure with ensuing plant outage is sharply reduced. Second, this lower level of control can be maintained by the same staff that tends to the rest of the plant's analog equipment. The computer-based portion of the control system should be maintained by a smaller, specialized maintenance staff.

During the checkout preceding plant startup, systems 1 and 2 allow the maintenance staff to access individual control elements and monitor transmitter outputs. Thus, the startup crew will check out lower-level field devices before proceeding to higher levels in the control system hierarchy. This results in smoother startups and faster turnarounds.

The process engineer

Computer-based control is helpful to the process engineer because it frees him to experiment with more-elaborate control schemes. Multivariable loops, feedforward computations, deadtime compensation, and process optimization can all be realized with a general-purpose computer. If the more-elaborate scheme proves unsatisfactory, the engineer can always return to the simpler scheme of a backup analog system. In experimenting, he will gain greater insight into the process. He must uncover the relationships between process variables, control variables, and system disturbances before he can model the process mathematically.

In compiling information for the model, a computerized data-gathering and display system is invaluable to the process engineer. Information can be stored conveniently and retrieved later on. Plant efficiencies, inventories, and material and energy balances can be calculated, stored and displayed in engineering units. With the proper interface between the analog controllers and the data-gathering and display system, the computer can automatically log controller status, process variable and setpoint, and make status commands.

For simpler instrument loops, the majority of which need the PID control function, analog equipment is still the most popular. This is because computer control uses time-sharing and point-sampling. These techniques introduce deadtime and phase lag into the PID function, and diminish the responsiveness and stability of the control system.

When the analog signal is converted to a digital signal at the computer input (A/D), 14 bits are needed to approach the resolution of the simple analog PID controller; this much resolution is required to handle adequately the derivative, as well as the proportional and integral, part of the control function. At best, the digital PID function offers only a close approximation to the analog PID function. Since dedicated analog controllers are usually supplied in a hierarchical system, it makes sense to use them for PID control and to reserve the digital computer for instances where only elaborate schemes, such as multivariable control, can offer better controllability.

Because it is difficult to construct an accurate mathematical model of a process, feedback control is still commonly used to keep a plant running during minor upsets. For such contingencies, conventional analog controllers provide the best PID controllability. They can also handle some more-sophisticated control responses of the override, gap-action, batch, feedforward and nonlinear types. In a hierarchical system, performing ordinary tasks with analog controllers frees the computer's memory for elaborate tasks like startup, shutdown and optimization, and for special chores like PI control with extremely long reset times.

The plant manager

For the plant manager, a hierarchical system has a number of advantages. First, it enhances the likelihood of an early, trouble-free startup. Added revenue generated by the plant because of an early startup can sometimes equal the cost of the first two levels of control.

Second, as higher levels are added, more members of the organization routinely participate in operating, maintaining and managing the plant control system. This greater organizational depth reduces the chances of having to shut down or curtail production because only one or two programmers know how the control system works. It also allows engineers, as well as operators, to access process controls, while encouraging them to develop strategies for improving throughput.

Finally, in a hierarchical system there are sufficient control levels to prevent shutdown because of either hardware or software failure. Such a system represents only a fraction of total plant cost, yet over its lifetime it can avert millions of dollars of product loss.

The author

Jack J. Murray is Senior Applications Engineer in the Process Instruments Div. of Beckman Instruments, Inc., 2500 Harbor Blvd., Fullerton, CA 92634. Previously he was a project instrument engineer at Monsanto Co.'s Central Engineering Dept. in St. Louis, Mo. A graduate of the University of Notre Dame, he holds a B.S. and a B.A. in electrical engineering. He also holds an M.B.A. from St. Louis University. At present, he is doing graduate work in advanced control systems at California State University in Fullerton, Calif.

Computer control in the design of a grass roots plant

Process control computers have been routinely added to chemical plants and refineries, but seldom are they specified and scoped coincident with the plant design. The purpose of this article is to describe how the techniques of computer control relate to new construction.

J. Patrick Kennedy, Sybron/Taylor Instrument Co.

☐ It costs about $7,000 per loop to make an existing control valve available to a freshly installed computer, and functionally equivalent to a new system. On a new plant, it costs only a few hundred dollars. Since control projects are being justified on existing plants, it is clear that computerization of every new plant should be carefully evaluated.

One must answer the questions of economics and technique, so that the specification of the computer can be done at the initial stage of the project, along with the other design work. In a properly designed project, the computer should be available for the plant startup, and the operation of the plant should be unchanged from the beginning.

Outline of the article

The five sections of this article are as follows:

- *Scoping and specifying the function of the computer*—The way to succeed in developing a project is through good analysis of the computer's function. Elegant solutions to nonexistent problems abound and have no payout. Standard techniques have been developed and used successfully to aid in deciding what problems are to be solved.
- *Equipment design considerations*—There is a considerable range of equipment in the plant control system, including computers, sensors, alarms, switches, charts, controllers and others; and the system philosophy will have a dramatic effect on its design. Operation under abnormal conditions, such as power loss or major failure, must be engineered with the system. The ability of a computer system to provide a payout is integrally tied to its reliability and capability. This section deals with the translation of a functional description into equipment, in a cost-effective manner.
- *Process computer software*—The reliability of a system involves its ability both to function and to operate mechanically. If it fails to operate (a hardware failure), it is a matter of time and money to fix a piece of malfunctioning equipment; but in the case of system software reliability problems (it "bombs" unexpectedly for no apparent reason), this is more severe. If the problem is in the structure of the software, the proper method of "fixing" it is to throw away what there is, and start from scratch with the correct structure. This is a problem that the user cannot totally control, but should address. It also illustrates the need for a complete functional description before any work is done.
- *Advanced control techniques*—There are several advanced control techniques that have general application in plants. Although the correct implementation of these controls is important, the selection of the right algorithms for the problem is a task for the resident control technologist. The methods that the author has found to work in repeated applications are described in this section.
- *Analyzers for control*—Most analyzers are a source of critical information but are unreliable when compared to a transmitted variable. The process engineer alone can judge how important the information from an analyzer is, and whether or not it is worthwhile to incorporate into the system. This section describes the necessary techniques and safeguards for using analyzers.

These sections describe the major areas of potential problems and benefits in the computerization of a new plant.

Specifying the functions of the computer

Over fifty years ago, Will Strunk wrote a book on how to write good English prose called "The Elements of Style" [*1*]. It was notable because it differentiated

Originally published October 15, 1979.

Considerations and warnings in computerization

Feature	Remarks
Operator consoles	CRT consoles tend to be a significant load on the system. As a rule of thumb, the system should be sized so that the CRT's can: be refreshed every 5 s; respond immediately (less than 1 s) to operator changes; and start building on the screen within 2 s without excessive load on the computer.
Control package	The control functions are linear with the number of manipulated variables. The system should be able to process roughly twice the number of loops as control valves every 2 s without excessive load on the computer.
Rotating memory	Most control routines will be either stored in memory or on rotating memory. It is preferable to use fixed head disks/drums for these routines because of speed and reliability. The amount of available fixed-head rotating memory will form one of the limits in a process control system.
Memory	All the programs that have to be run will need to be loaded into memory and thus, in a multi-tasking system, the amount of memory sets a limit on the number of tasks that can be executed concurrently. The choice between solid state memory and core memory is still strongly contested by the individual advocates. The trade-off is lower cost and speed (solid state memory) versus failure mode (core is generally hard and solid state can be intermittent). It is the opinion of the author that if there is enough memory, it does not make a lot of difference.
Multiplexers	The speed of communication between the computer and the field devices is set by the multiplexer hardware. Field multiplexing has the advantage of reducing wiring cost, but it increases the probability of a simultaneous failure of a group of points. Usually it also increases the CPU load and slows down the access time. These devices have a place, but extreme care should be exercised before incorporating multiplexers intended for data logging into a control system. Furthermore, because of the required data flow, field multiplexing will likely prohibit sequencing types of programs.
Instrumentation	There are ways to use pneumatic, conventional electronic and cathode-ray-tube-based electronic controls, but whatever is used should be integrated and staged as a complete system.
Power	Just the instrumentation, or both the computer and the instrumentation, could be backed up. Generally, systems are evolving to projects that have enough payout above automatic control (e.g., polyvinyl chloride production) so that backup power for the complete system is justified, even when pneumatic instrumentation is used for backup automatic operation.
Software	The software system is easily the most dangerous part of the project. Since, to date, most computer systems are after thoughts, deliveries on the order of 6–9 mo are commonly required. This is not enough time to properly design, scope, build, check out, stage, burn in and ship a system. The software is usually the part that suffers because of this, and it is done in the field, or after installation. The best run projects specify all software and hardware at an early stage and some even take delivery at a facility adjacent to the new site before the construction is complete. Realistically, allow 18 mo for a project that is expected to run a normal course without undue expediting or field work. And require that all software is complete and checked, operators are trained and documentation is complete before installation in the control room.
Standard versus special	The computer system has altered some of the trade-offs that exist in this area. It is quite reasonable to have as much application software and hardware as needed to perform the job, and have the system ready for acceptance as defined above. It is not reasonable to perform much designed-from-scratch, serial #000001 work on a large project. What is truly special, and what is application-based, is a difficult question to answer, because it differs between suppliers. Careful definition and scoping by qualified personnel is the only protection.

between English prose and good English prose. Recently a series of books have been published [2] that concern themselves with good computer programs; not simply programs that work, but ones that are structured properly. The book is in the form of a series of rules for writing these programs. Since the scoping and specifying of the computer *function* is actually the scoping and specifying of a computer *program*, these rules are directly applicable.

The rule that is most important is that *a program should be completely defined before the first line of code is written.* Definition includes all specifics—variable definitions, sizes of arrays, calling program, submodules and so on—all the information that a person who is not familiar with the project would need to code the program. This is a difficult rule to follow, but, if accomplished, will result in a properly structured system and eliminate the randomly growing programs that marked the last generation of process control computers. It is primarily because of structure that it has been possible to practically eliminate problems of software reliability after startup.

In a grass roots facility, one has the opportunity to combine the definition of the process control with the definition of the plant operation. It is still possible to fit a computerized operation around existing or slightly modified equipment, but the scope of the project will be limited by the expense and manpower involved.

Control objectives analysis

To look at the proposed function of a process control system, there exists one of the few generalized techniques in the application of computers to processes. A method utilizing game techniques called *control objectives analysis* has been developed and used repeatedly with outstanding success [3]. The object of the analysis is to define the function of the system. The result of a control objectives analysis session is a master specification that defines the objectives of the project. This is used to create a system functional description that identifies all the needed modules, the data base and the needed communication between modules.

A control objective is a concise, precise, true-all-of-the-time statement about the objective of the control system in the plant. The first step is to organize a single meeting in which all the necessary disciplines and backgrounds are gathered at one time. Logistically, this is often a difficult task, but keep in mind that the best package is one that responds to the needs of the plant, its operation, its construction requirements and its management.

This can only happen if all inputs are available during the original scoping. Table I shows the required disciplines for this meeting, keeping in mind that the expertise may be resident in one or more individuals, and certain individuals may have expertise in more than one area. It is critical to the success of the session to have all areas covered so that the scoping meeting can proceed smoothly and need not be stopped to gather inputs from those not present. Proper design of the project is impossible without good communication. Persons attending this meeting must be willing and capable of making decisions in their specialty. The session leader should be experienced in applications of computer systems.

Control-objectives-analysis session attendees — Table I

Control-objectives-analysis session attendees
Control systems engineer (session leader)
Process engineer
Computer systems engineer
Instrumentation engineer
Software applications engineer
Design engineer
Contracting engineer
Operations
Project manager
Plant management

The purpose of the session is to determine the control objectives that will form the basis of the system description. The number of control objectives needed is equal to the number of manipulated variables—called the number of control valves for convenience. The control objectives must be agreed to by all present and *there is no discussion of how an objective is to be met.* It is assumed that, given the task of performing some type of precise control on a plant, the project crew will be able to find the solution. This closely parallels software definition; no code is written until all functions are specified. In this way, the scope of the project can be defined by those interested in the outcome and economics of the plant, leaving the detailed engineering to those qualified to perform the work. This distributes the load and avoids the problem of having only one or two individuals who know the overview. Most importantly, it gives those skilled in the operation, design, management and construction an opportunity to state their objectives and to be fully aware of the task at hand.

Once the list of control objectives is developed and agreed to by all, it is broken down into three main categories:

1. Control objectives to maintain energy and material balances.
2. Control objectives that are a result of maximizing or minimizing an objective.
3. Control objectives that are needed to meet the requirements of management.

These objectives are then used by the control systems engineer to develop the detailed functional description of the system.

Control objectives analysis has been used successfully and establishes a means of standardizing the scoping session of a computerization project. It bridges the gap in communication due to disciplinary differences or job functions, allowing all interested parties to make recommendations and inputs at the front end of the project. It has the beneficial effect of pinpointing decisions in an early stage. If necessary, the entire structure can be modified to accommodate a solution, instead of on-site shoring-up measures.

Ultimately all stages of a computer project will be done with standardized methods, so that once the appropriate goals are selected, the project staff will be able to apply good engineering practice toward solving those problems, and effect a solution. The key to procedures like control objectives analysis is that it routinely accommodates highly nonlinear objectives like ". . . and the plant must be able to change its throughput by 25% without oscillations in the steady state and material balances." These are areas of high payout.

Once the scope of the project is firm, it is necessary for the project engineers to turn the written description of the capabilities of the computer system into a physical, working machine.

In the remainder of the article, this task is discussed.

Equipment design considerations

Integration of the process computer with the ancillary control room equipment is an incentive for incorporating the design of the control system into the design of the plant. In an existing plant, the practical difficulties of tying the equipment together in the field limit the scope of the project, and are of sufficient magnitude to require a separate project and all of the overhead expenses that go with it. From both a cost and function standpoint, the system designed into the plant will be the most cost-effective.

In a recent meeting, a user of a totally computerized PVC plant [4] stated that the entire effort to computerize the facility was around 10% of the plant cost. The plant was capable of both automatic and computerized operation, yet when they ran manually each month for operator training, the throughput of the plant was 50% of normal. These numbers have been verified by similar plants and, in one case, automatic operation (instead of computerized operation) resulted in only 30% of normal production. These plants were all designed from their initial definition specifically for computer control.

There are five areas that must be considered when designing a plant for computer control; these are:

Instrumentation
Operator interface
Function of the system
Abnormal/backup modes of operation
Field equipment.

Each area is discussed below.

Instrumentation

The instrumentation forms the communication link between the process, the computer, the operator and the plant management. Regardless of the type of instrumentation, the concept of "computer compatibility" is largely a myth. Surely, the incorrect choice will limit the feasibility of computerization, but even if it only takes a single wire to connect the instrumentation, numerous other practical problems must be overcome to correctly install computer control into a plant. These include: computer interface, software definition, power system, floor space, functional description of the control, operator philosophy and ancillary equipment. The opportunity to do this work off-site before installation is lost.

In a modern installation, the function of the instrumentation is largely as backup equipment. The computer has complete access to valves, so control algorithms can be done either in the computer system or by the controllers. Since there are algorithms that are impractical to perform with hardwired instrumentation, the control is done in the computer system to avoid splitting the operator interface. If the computer fails, the control of the plant reverts to the instrumentation. This philosophy represents a design decision that must be made at an early stage in the project so that all the needs of the operator can be engineered into the system. Otherwise, the system will evolve to certain functions being done at the instrumentation and others done at the computer console. In practice, this will greatly increase the probability that the functions designed to run in the computer system will be minimal or will be totally defeated by operators.

Fig. 1 shows a control system for a continuous reformer, Klaus unit and SCOT unit for a refinery. In this system, the primary operator interface is the color CRT, which is driven by the computer. The monochromatic CRTs are driven by hardware in the instrumentation and are completely independent of the computer system. During normal operation they are used for trending and an overview of the plant. If there is a computer failure, the system automatically fails over to the instrumentation.

The plant can operate totally on either the computer or the instrumentation consoles. This is the result of the design of the system, since it could have been designed as described above, with some functions on the instrumentation and some on the computer. Instead, the system has fully redundant consoles; the operation is somewhat different, but this would have been the case regardless.

The system was staged complete to the extent that operators could be trained during the burn-in period. The system was shipped to the site as a functional unit with no interfacing in the field required. This resulted in a system that assists in startup rather than one requiring extensive startup itself. The primary payouts from this approach are in the installation costs, equipment requirements, floor space and functionality.

Parenthetically, it is generally assumed that computerization requires electronic instrumentation. This is not true. The question of whether to use pneumatic or electronic instrumentation is independent of the decision to computerize. The trade-offs are the same as they have always been. Initially, it was thought that the advantage of the computer lies in the improved accuracy brought about by the electronic instrumentation. Actually, it is much more important what is done with the measurements *after* they are in the computer.

The question of electronic versus pneumatic must be settled by transmission distances, power backup requirements, reliability and cost. One company insists on using pneumatic instrumentation for a simple reason; their valves are all pneumatic and the addition of electronic components would lower the reliability of the entire system. Furthermore, no functions need be given up when modern pneumatic equipment is used as backup instrumentation, except that the operator interface will be conventional rather than CRT-based. It can still be packaged into a single console as described above.

Operator interface

The operator interface in modern control systems is vastly improved with the advent of the CRT. Anyone who has tried to explain the concept of *range and zero span* versus *engineering units conversion* to operators, or has seen a unit brought down because of the wrong kind of paper in the chart recorders, knows the importance of unambiguous information flow to the operator. The key to effective interface is proper definition of the informational requirements. The operator needs rapidly refreshed displays that respond instantly to alterations and contain as much information as possible.

It was thought at one time that the correct procedure was to only partially fill the CRT screen with information to keep from confusing the operator. It is now apparent, however, that operators *much* prefer to find a piece of data on a CRT page than flip from page to page. The reason for this is that the human brain gains a "set" of the display, i.e., much of the information is scanned automatically by the unconscious and when a display is changed the mind takes some time to regain this position.

Part of the functional description of the system is the definition of these CRT displays. The display system of the instrumentation is generally fixed by the equipment suppliers, but the operator console for the computer is still flexible. The same definition requirements that one would observe with a fixed hardware system should be observed when designing these displays because, although they can be changed in the field, the practical limits of doing so are severe. It is best to assume that no programming expertise will exist in the field and that these displays will never be changed. Assuming that it will be possible to alter the system in the field is a

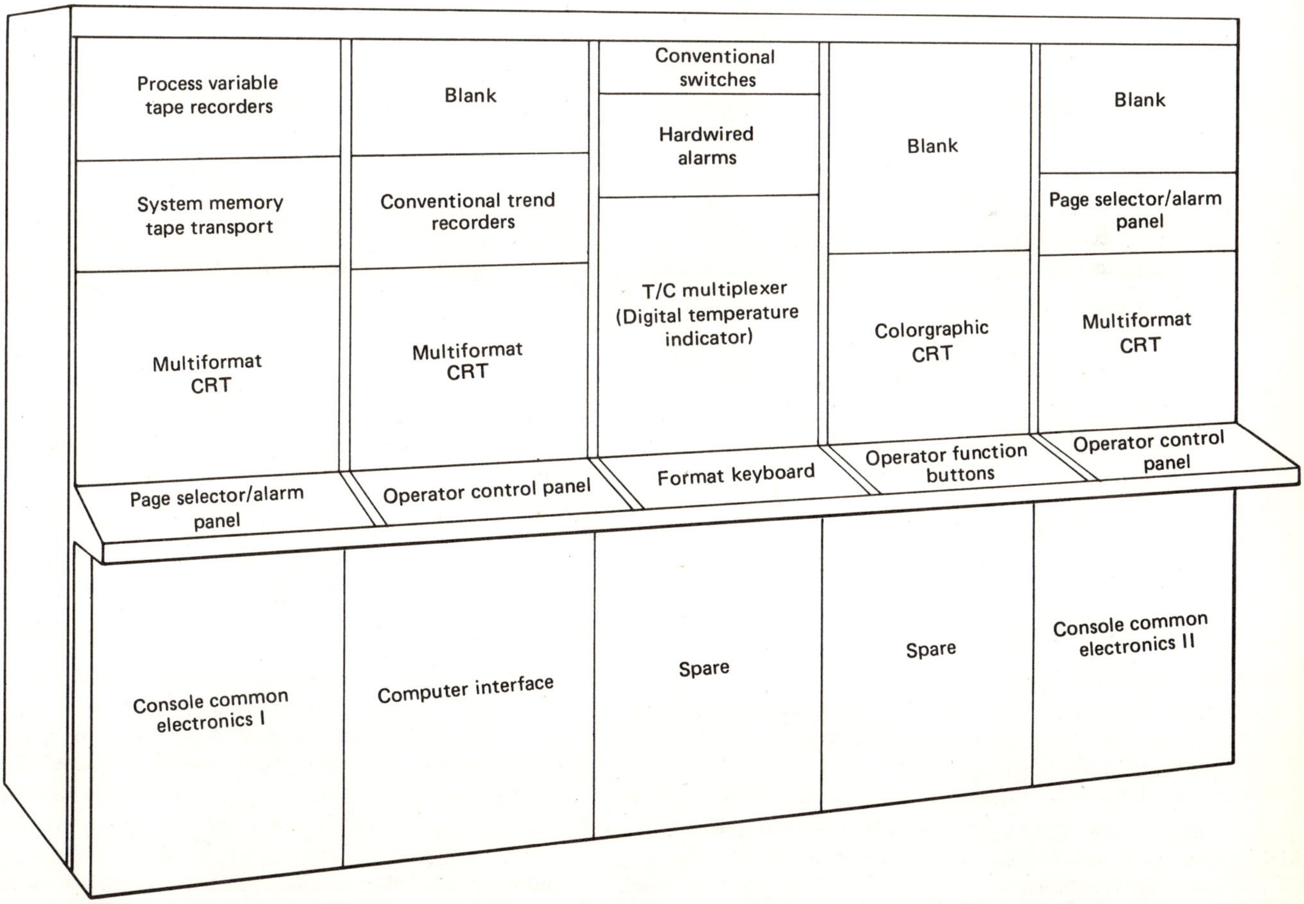

Computer control console for a refinery system **Fig. 1**

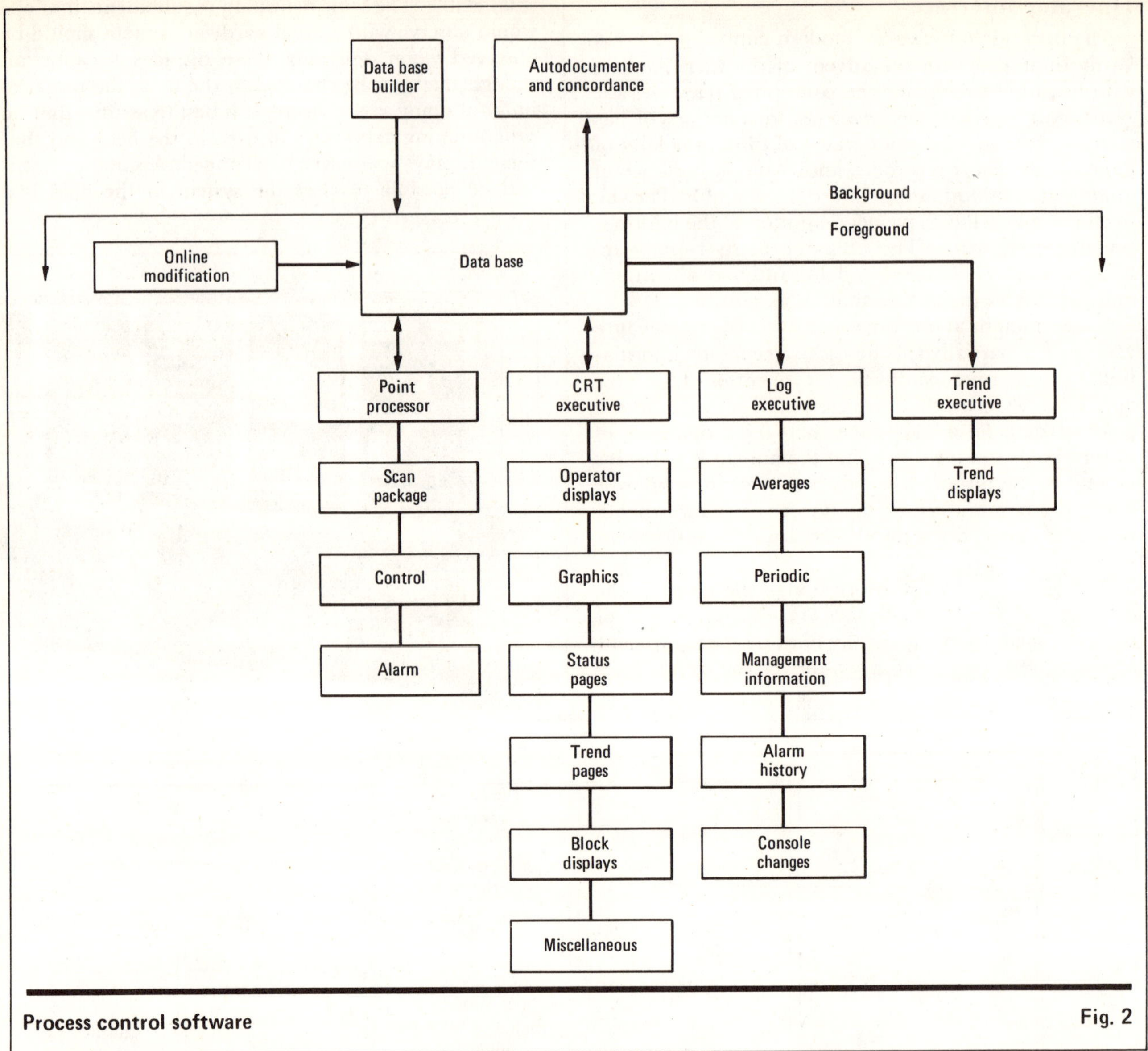

Process control software

Fig. 2

self-fulfilling prophecy and will decrease the profitability of the project by necessitating the cost of a full time programmer.

Function of the system

It is the responsibility of the equipment designers to review the functional specification of the system and provide the equipment capability to accomplish the task. If a system can truly grow without decreasing the probability for success, then it is the nature of a computer that the profitability increases with scope. One must conserve the resources of the system for the projects that have payback.

Guidelines are difficult to evolve because of the vast differences in the equipment called "computers." Some considerations and warnings in computerization are listed in the box.

As with all other branches of design, it is important to provide a safety factor to cover unforeseen problems. In the matching of equipment with the function, there are essentially two finite barriers, computer load and program storage (fixed head disk/drum, and memory). Each should be treated as a resource and there should be allowances for field expansion and modification.

Abnormal/backup modes of operation

The centralization of the control function allows the flexibility to consider several different modes of operation; these are designed to meet the functional requirements of the system. Either dual processor/redundant I/O logging-systems, or single processor control-systems that have no backup, may be unreasonable from a functional point of view. Failure is not simply related to numbers such as MTBF (mean time between failure) and MTTR (mean time to repair). A single failure that happens once every 10 years and causes plant damage or endangers personnel may be unacceptable, while failures that happen daily but are handled in such a way as to not affect operations, may be acceptable. Thus the key is to clearly define the failure modes of the equipment and design the alternate operations into the equipment.

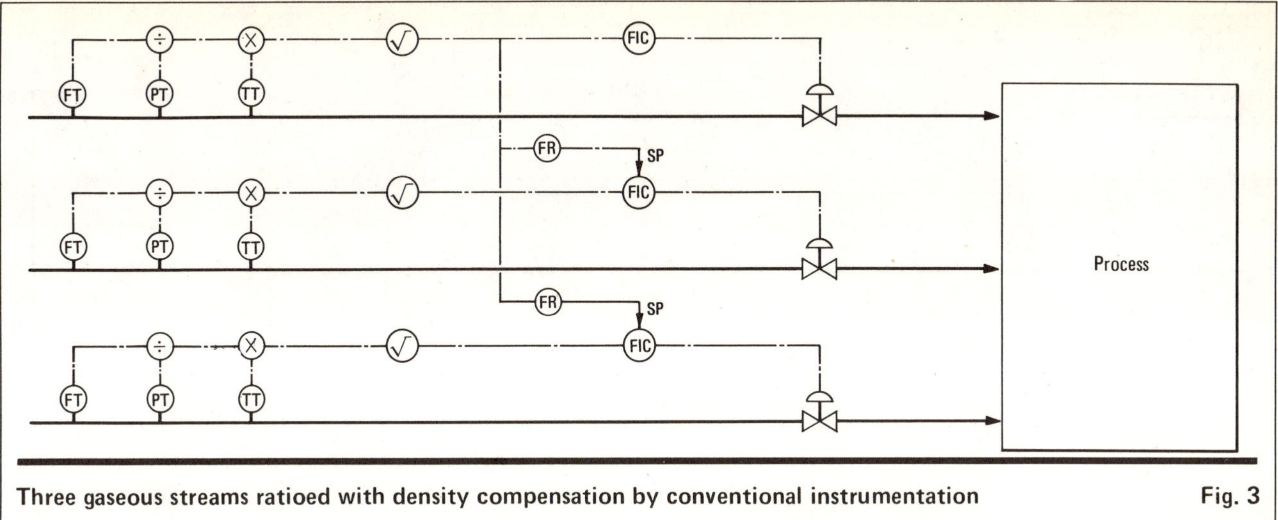

Three gaseous streams ratioed with density compensation by conventional instrumentation **Fig. 3**

The most common severe failure is the failure of a processor; as a result, redundant processor configurations can be supplied as standard. Although it is unwise to make a practice of trying to fit a solution to a problem, the dual processor is a significant advancement and should be considered. In fact, little sequencing is possible without the dual processor because, if the sequencing is lost, a plant is usually down, or is running at a greatly reduced throughput.

Field equipment

The field equipment must be suitable for proper operation by a computer system, particularly the valves and on-off devices. An operator has visual feedback when he operates a device: a flow begins to increase or a temperature falls. The computer needs to have feedback to make sure that what it is doing has not been impeded by equipment failure or manual intervention. Thus the field equipment that is to be used, or might be used, should be purchased with feedback signals. In this way the computer can perform timeouts and other maneuvers to assure itself of the integrity of its actions.

The field equipment selection, like all other aspects of the design, must be based on knowledge of the functional specification and scope of the job at an early stage. It can then be ordered with compatible function and electronics. At the ordering stage, compatible voltage levels and the electrical character of the interface can be selected. When this is not done, it is not uncommon to spend days with voltage dividers and signal isolators to integrate the equipment in the field.

Process control software

The process control software converts a computer system that is physically wired to sensors, controllers and valves, into equipment that helps an operator run the plant. There is quite a lot of routine work that can be done by using standard packages from vendors. There are advantages to using standard software, wherever practical, if care is taken not to try to make the requirements of the plant fit the software. The packages available should be considered only as tools; there still must be a detailed, written functional description of the total system. The purpose of this section is to give an overview of what the standard software is, and describe the techniques for molding this into the final product.

Standard process-control packages are one part of the programs needed to implement the goals described in the functional description. The packages perform a large amount of routine work that is done on each project, plus they are completely self-contained—they have all the pieces and the operating system.

Most process control applications use a data-base driven system. This means that all data associated with the process are maintained in the computer system for utilization by all of the programs, in a structure as shown in Fig. 2. The data base specifies the structure of the system as well as the parameters like the range, zero, scan time and so on. All other programs either modify or read the data base, and the integrity of the system depends on the integrity of the data base. To maintain this integrity, there are no lines of communication between independent branches of the system. This is a top-down structured programming approach, with each task written as an independent modular task.

As stated earlier, the top-down approach allows the system to become extensive without being complex. A system with 10 measurements has the same complexity as a system with 1,000, but, of course, the latter is much more extensive. The primary difference between these examples is the size of the data base.

The advantage of using a standard software system is that the means of displaying and altering the information in the data base is predefined, documented and incorporated as standard programs. There are displays that show the loops, the process graphic, the configuration of the loop and the data within the block. A complete description of these can be found in the system user's manuals [*5*, *6*, *7*, *8*].

A more detailed description of the organization of the data within a process computer can be found in two previous articles that discuss this in detail [*9*, *10*]. This organization of the data is essential to the success of the control project although it is not something easy to see or touch.

Once a data base is built, the system is functional and

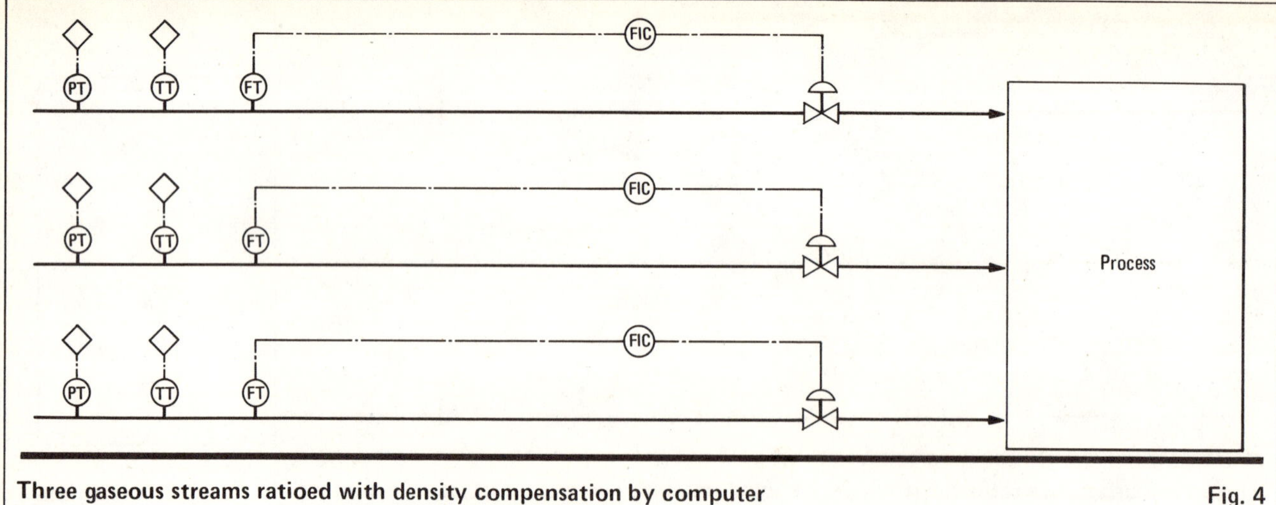

Three gaseous streams ratioed with density compensation by computer Fig. 4

the problem is to make it perform the goals described in the functional description. This area is called *application programming*, and it is a phase that must be rigidly controlled because it is equivalent to the wiring in an instrumentation panel. A few wrong connections can raise havoc in the field. It is yet another reason to stage the system, complete with all ancillary equipment, at some site other than the plant.

Advanced control techniques

Incorporation of process computers into plant design has radically changed the field of control. Complex manipulation of data is trivial to implement and has a unique kind of reliability. If a computer works, everything works, and when it fails, it causes an orderly transfer to a backup mode of control. Compare this failure mode with the advanced control systems that utilize discrete components.

For example, consider the system in Fig. 3, a system to correct three gas flows for density variations due to temperature and pressure, and then ratio them to the primary flow. There are fourteen separate components, any one of which can fail and send the system off-control—a failure that must be diagnosed and corrected by an operator. Compare this with a system based on a computer with instrumentation backup, Fig. 4. The computer has access to any value and can perform the same functions as before; but if the computer fails, the system automatically reverts to the three flow controls. This is a redundant operation that gives the reliability of single loop control, but has all of the advantages of the more complex control. This system is much simpler, and the installed cost is considerably less.

Over-complexity of advanced control creates problems for operators. For example, it is known that oxygen trim control can reduce the energy consumption of furnaces and that chromatograph control can improve the operation of distillation columns. There are, however, very few of these systems actually online and it is primarily for this reason: the handling of abnormal behavior is very difficult without the logical ability of the computer. Reliability is far more important to a plant operator than economics. In fact, there are many locations where these controls were incorporated at considerable expense, and were then turned off.

With the response to partial failure programmed into the logic of the digital computer, the number of advanced controls has been increasing considerably. Some of these methods have had wide enough application to be almost "standard-advanced" controls. In this section, the author will list some of these methods, and appropriate references to instructions for their use. By using the computer to carefully and continuously check the integrity of the measurements as well as perform the control, the reliability of these methods has improved to the point that operators feel confident in using them. Analyzer-based controls fall into this class, and are covered in the next section because of their importance to the operation of the plant.

The techniques covered in this section are:

Control of calculated variables
Logical integral level control
Multivariable constraint control.

These three were selected because they have been successfully used in repeat applications and are continuing to perform well. This implies, but does not prove, that they can be relatively insensitive to disturbance patterns and tuning parameters. There are, of course, other algorithms that perform well under these circumstances and one of the signs of maturity for process control with computers is the recognition of standard algorithms.

Control of calculated variables

Many times a computed variable is more significant to an operator than the sensed temperatures and pressures in a plant. Table II lists some common computed variables in the CPI and makes comments on their use. Care must be taken in the use of computed variables to avoid problems with the dynamics of control. Most computations used for these control programs are steady state energy or material balances and, depending on the character of the disturbances, the control system can exhibit unusual behavior when trying to control variables that are disturbed in the same frequency range as the dynamics of the process. Nevertheless, this

Nomenclature

CPI	chemical process industries
CPU	computer
CRT	cathode ray tube display
FIC	flow indicator controller
FR	flow ratio station
FT	flow transmitter
GLC	gas liquid chromatograph
I	integral control
I/O	input/output
LIC	level indicator controller
LT	level transmitter
P	proportional control
PT	pressure transmitter
PVC	polyvinyl chloride
S.P.	set/point
TT	temperature transmitter

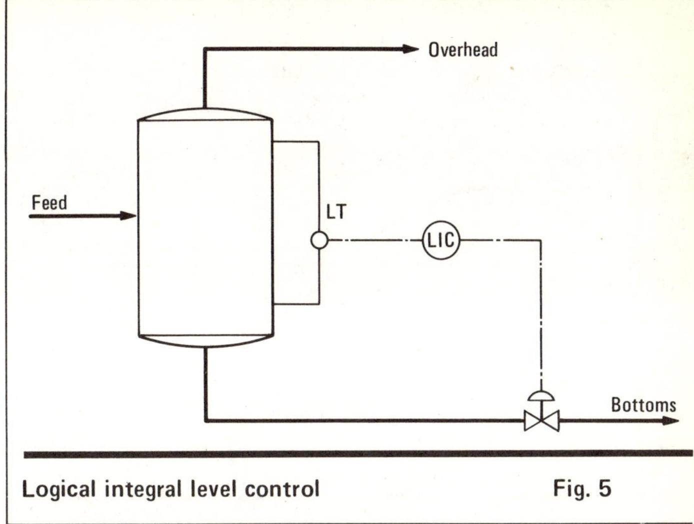

Logical integral level control Fig. 5

is probably the most widespread and useful technique of computer control.

Information can never be generated, only changed in form. For example, if a distillation column cannot be controlled by a temperature control tray because the separation is not sensitive to a boiling point due to the range of distributed components, eliminating the variation of the control tray temperature with pressure does not gain anything. Conversely, if the column can now be controlled by a temperature control tray, use of the computed composition in the computer program will improve the control of the operation. Since the problem of transmitter and computational equipment reliability has been addressed, it then becomes clear that these algorithms must improve control with no chance of degradation.

Logical integral level control

In a continuous plant like a refinery, the loop shown in Fig. 5 is repeated in numerous places. If the level controller is finely tuned, the flow swings with each disturbance. This can be a major problem in a plant, and it will be aggravated with increased energy costs and heat integration.

Since these streams feed downstream units and probably go through heat recovery equipment, this causes simultaneous disturbances in the heat and material balances. On the other hand, if the loop is tuned loosely so that the flow moves slowly, the surge capacity must be increased unreasonably. Sometimes there is no opportunity to increase this surge volume.

When conventional controls are used, the loop is usually tuned to some compromise so that the level is

Computed variables in the CPI — Table II

Variable	Sensed variables	Remarks
Internal reflux	Column top temperature Reflux flow Reflux temperature	Can exhibit "inverse response" if tuned too tightly.
Heat input	Inlet temperature Outlet temperature Exchanger medium flow	Useful for reboilers and circulating refluxes.
Product composition	Control tray temperature Column pressure	Used to eliminate the effect of column pressure on temperature control tray.
Reaction conversion	Integrated heat removal Coolant inlet temperature Coolant outlet temperature	Reactions must be exothermic or endothermic. This is very accurate for polymerization reactions.
Furnace efficiency	Stack temperature Fuel flow Process fluid flow Inlet temperature Outlet temperature	Most responsive measure of furnace tube leak of hydrocarbons.
Mass flow	Volumetric flow Flowing temperature Flowing pressure Specific gravity	Especially useful in streams where density changes with conditions or time.

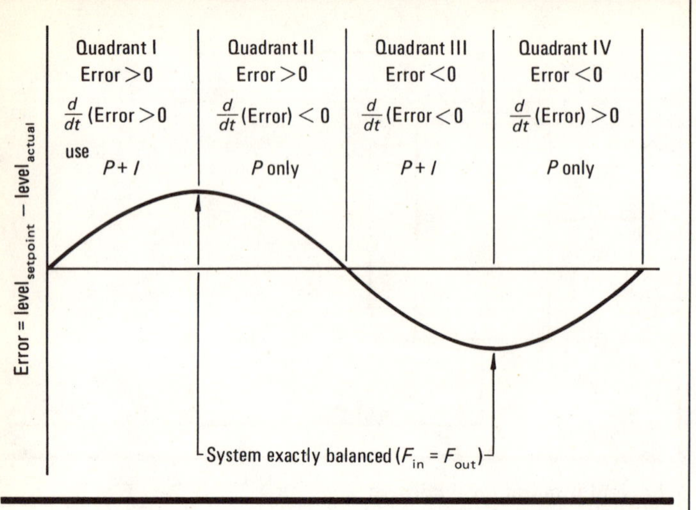

Logical integral level control **Fig. 6**

not often lost and the flow is not so ragged that the downstream units cannot cope. Logical integral control was developed specifically for this application, and is repeated scores of time in a typical refinery.

Consider Fig. 6. In the first quadrant, both the error and the derivative of the error are positive. In other words, the level is away from setpoint and moving away—indicating an imbalanced situation. The same is true during the third quadrant, where the level is below setpoint and decreasing. In the second and fourth quadrant, the level, although off setpoint, is moving back. At the precise point where the level is not moving at all, the inlet and outlet flows are equal and the system is under control.

The logical control algorithm uses two different integral modes, one quite large by conventional standards and the other null. The large value is used in the first and third quadrants, where a responsive action is needed; the null value is used in the second and fourth quadrants, where control has been restored. The problem with large integral actions (windup) is totally eliminated because it is zeroed every time there is a sign reversal. However, if the level moves consistently away from setpoint, it moves the outlet smartly to the new level. This control has worked well in numerous actual installations.

Multivariable constraint control

The last control that will be discussed is a technique to control a plant to a steady state model that utilizes both constraint control and optimizing control in the same algorithm. The technique has been fully described elsewhere [*11*, *12*]. The intent of this section is to show the applications in which it is useful.

Consider the example of three heat exchangers in a crude preheat network as shown in Fig. 7.

The function of a control system would be to:

(1) maintain constant flow.

(2) observe constraints on the temperatures of the products coming from the crude unit.

(3) maximize the amount of heat recovery.

It is necessary to develop a control algorithm that continuously looks at all of the constraints and moves the three flows so as to maintain these. In the event that the constraints are all met and there is a degree of freedom left, the system should move the flows to the most economical point.

The technique does this and has been implemented in several applications (some with as many as 128 constraints), and has been found to be quite stable, and insensitive to the tuning parameters. It has been useful in cases where multivariable control is mixed with constraints, and functionally replaces on-line linear programming methods.

Analyzers for control

Process analyzers have a reputation of being good for process control when they work, but of requiring high maintenance. The reputation is partly deserved, since the analyzer, by the nature of its complexity, can never

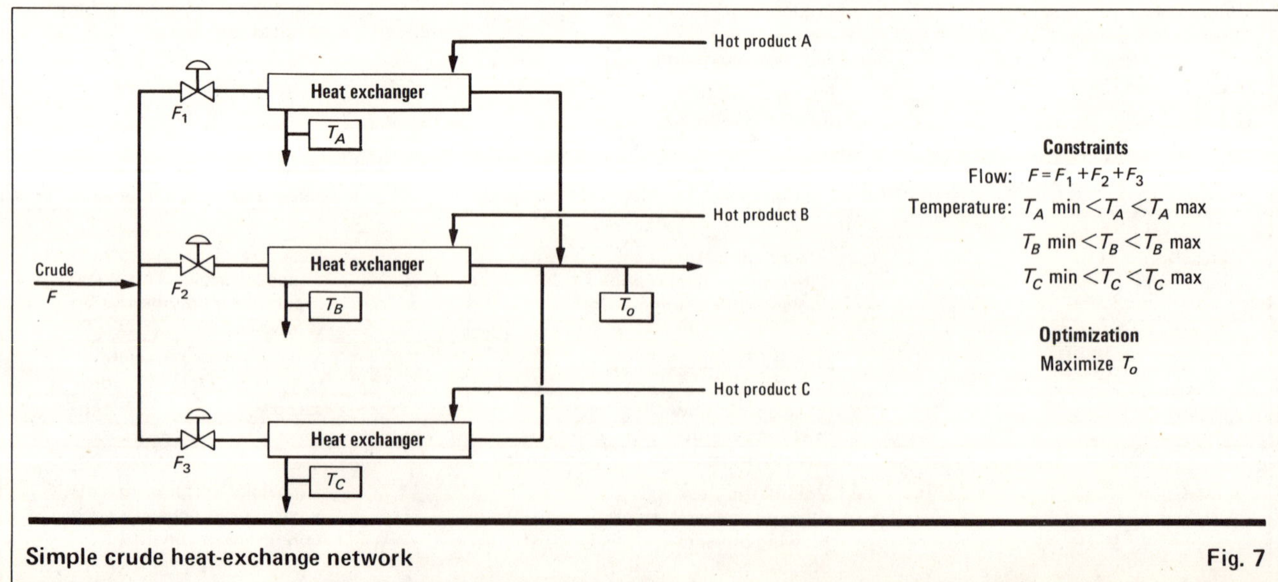

Simple crude heat-exchange network **Fig. 7**

be as reliable as a single transmitter. The correct implementation of analyzer-based control with a process computer can do much to alleviate these problems in the field. Since stability is more important to an operator than economic operation, the key to the application is to fully engineer the operation with, and without, the analyzer, to afford minimal disruption when the device is down. Where analyzers are implemented with conventional controls and there is a measurable transient when the device fails, it will only be used by a committed few who appreciate the value of the information.

There are two aspects of implementing analyzer-based loops that can be augmented by the computer system:

(1) control of a calculated variable that is reset by the analyzer, and

(2) a device handler for the analyzer that keeps track of the integrity of the measurement.

It is best to use both of these methods with any analyzer-based control loop. In this section, the proper use of these techniques will be described by example. The examples of an octane analyzer for a catalytic reformer and of a GLC (gas liquid chromatograph) for a distillation column will be used for the former, and a multistream oxygen analyzer will be used for the latter.

Calculated variables and analyzers

■ In the operation of a catalytic reformer, the octane of the outlet stream is the primary control variable. This is sensitive to the average temperatures of the reactors, feed rate and feed quality. Although the temperatures and feed rate can easily be measured, the feed quality cannot. Hence, some refiners opt for an *octane analyzer* on the product stream. Control is difficult because the point of measurement is downstream from the unit, plus some of the analyzers measure at discrete intervals. This problem was solved by controlling the unit to a computed octane, based on the reactor temperatures and feedrate for the minute-to-minute control.

This is fast enough for precise control of the octane, plus incorporation of numerous constraint programs. When a measurement is available from the analyzer, it is used to reset this calculation rather than set the unit operation directly. An additional advantage is that the amount of correction is directly related to the thermal stability of the catalyst—a very important end-of-run parameter. The control works extremely well. Even controlling an artificial variable, like the calculated octane, to a constant value, has a stabilizing effect on the unit. In addition, the occasional resetting by the analyzer keeps operation close to optimum.

An aspect of this control is that the operator has a way to control the unit, even when the analyzer is out of service—using the calculated value. Even though it may not be as good as the analyzer, it is still better than no octane control at all. The transition from the analyzer being online to offline is not measurable. In fact, the computation is so stable that some installations are running with no online analyzer; instead, laboratory measurements are used to reset the computation.

■ The second example is *the use of a GLC to control the separation in a distillation column*. In some cases, the output from a GLC has been used directly for composition control. While the analyzer is up, it is possible for this control to perform adequately, but the control quality is set by the sampling speed. This requires single stream analyzers for most online applications.

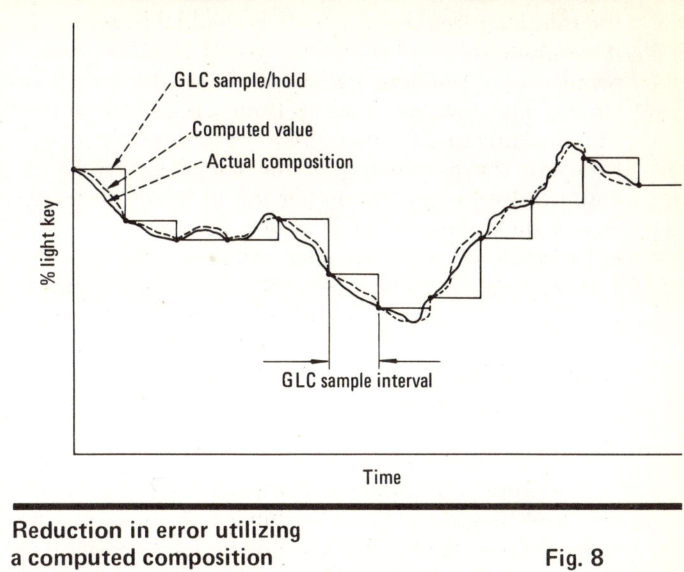

Reduction in error utilizing a computed composition Fig. 8

The other problem with this approach is that the operator must be able to relate temperatures and pressures with composition when the analyzer is down—a very difficult thing to do. In applications where the composition of the key component is sensitive to a control tray temperature, it is possible to compute the composition of the key component based on the last value, the change in pressure and any change in the tray temperature. The specifics of this type of implementation are described in a separate paper [*13*]. In general, the computation will reduce the amount of disturbance caused by the sampled nature of the signal.

Consider the plan in Fig. 8. The GLC output is shown as a series of straight lines representing the output of the sample/hold amplifier between samples; the solid line is the true composition; the dotted line is the computed composition. The disturbance injected into the control system is related to the size of the discrete jumps in the signal at the sample points. As long as the temperature of the control tray changes in the correct direction, the system will be less upset.

In conclusion, the reduction in disturbances caused at the Nyquist frequency by the sampling system is advantageous, but the real advantage is the tremendous improvement in the backup operation while the GLC is out of service.

Analyzer service routines

When an analyzer is brought into the system, the output is not simply measured and put into the system. There can be many checks done on the integrity of measurement, plus it can be put into a useable form.

As an example, consider the *multistream oxygen analyzer* often used for fuel/air ratio control. The sample system is actually stepped by the computer. This allows the computer to determine the time needed on each stream, and how many times per shift to calibrate and change the sampling frequency if there is trouble. In addition, it monitors the "out-of-service" switch and can detect problems by trending the output from the auto-zero circuit. The analyzer is set up from a page that allows the inputting of parameters like the barometric correction and the monitoring of the lamp supply. Every feature added to a program like this increases the integrity of the measurement at almost no cost.

Listed in Table III are some common analyzers and their applications. These are analyzers used primarily in the CPI. Each industry has its own group of analyzers, but their implementation with a process computer closely parallels the above descriptions for analyzers in the refining industry.

Final remarks

The proper application of computers to a process unit requires the same care and definition as for any other piece of large equipment. The functional description of the computer is the key. The problem is deciding what to do to a plant, when dealing with a new technology like computer control. The function of this paper is to collect and present the author's experience, noting the type of functions suited to the computer and the design decisions they require. The complexity of new systems is increasing dramatically, but the technology of computer software seems to be keeping pace.

The algorithms and techniques discussed here are worth large amounts of money in a plant. Projects with a 1- to 2-month payout are not common, but they are also not unheard of. To give all the credit for this to the computer is myopic—the large payouts come from the ability of those familiar with the process to reengineer the unit without the normal restrictions of conventional instrumentation. The controls can be made to conform more closely to the needs of the plant, rather than vice versa.

The most encouraging sign of maturity in this field is the emergence of standard methods. Too long the success or failure of a computer project has rested on the shoulders of a single individual working in an area difficult for normal management to evaluate—an area that has probably too high a risk for investment. These standard techniques will eventually mean that management will be able to expect and to get average success with average engineers doing an average job. At that time the age of computerized plants will be upon us.

Partial list of on-line analyzer control applications — **Table III**

Analyzer	Applications
Gas liquid chromatograph	Distillation column control Sulfur plant control Any trace material control such as chlorine in an ethylene dichloride plant Any light hydrocarbons such as the refrigerant in a liquefied-natural-gas plant
Continuous mass spectrometer	Processes with several streams flowing at a high rate (continuously); with many components, such as ammonia plant control
Hydrogen analyzer	Hydrogen production Reformer recycle hydrogen
Specific gravity	Inferential measurement for changing compositions when the individual gravities vary Hydrogen recycle gas in a reformer Fuel gas for Btu variations
Boiling point analyzer	Applications that are too heavy for gas liquid chromatographs because of end-points above the cracking temperature Crude units Cokers Vacuum units Blenders
Oxygen analyzer	Furnace control
Octane analyzer	Gasoline blenders Reformers
Vapor pressure analyzer	Stabilizer columns Blenders

References

1. Strunk, W. S., and White, E. B., "The Elements of Style," McMillan Pub. Co., New York, New York, (1972).
2. Legrad, H. S., and Chmura, L. J., "Fortran with Style," Computer Programming Series, Hayden Book Co., Rochelle Park, New Jersey, (1978).
3. Hadley, K., Control objectives analysis, Proc. of the NPRA Computer Conference, National Petroleum Refiners Assn., New Orleans, Louisiana, Oct./Nov., 1977.
4. Tsai, T. H., and Lane, J. W., Experience in batch chemical process control, Proc. of the workshop on batch process control, AIChE, Washington, D.C., May, 1976.
5. Sybron/Taylor Instrument Co., COMGEN users manual, Pub. No. IB-22G010, Issue 1.
6. Sybron/Taylor Instrument Co., TABL users manual, Pub. No. IB-22G020, Issue 2.
7. Sybron/Taylor Instrument Co., GRAFICS users manual, Pub. No. IB-22G007, Preliminary Issue 2.
8. Sybron/Taylor Instrument Co., TREND users manual, Pub. No. IB-22G023, Issue 2.
9. Kennedy, J. P., Case histories in computer control: PVC polymerization reactors, Proc. of the winter annual meeting: Case studies in computer control, American Soc. of Mechanical Engineers, San Francisco, California, Dec., 1978.
10. Daniel, R. E., A batch language system for sequential control of multi-unit chemical processes, Proc. of the international instrumentation-automation conference and exhibit, Instrument Soc. of America, New York, New York, Oct., 1974.
11. Kennedy, J. P., A simple steady-state multivariable control algorithm with examples of gasoline blending and an ammonia reform furnace, Summer simulation conference proc., San Francisco, California, July, 1975.
12. Kennedy, J. P., Energy conservation by advanced control in a catalytic reformer, Proc. of the 86th national meeting, AIChE, Houston, Texas, April, 1979.
13. Kennedy, J. P., Computer system for a gasoline plant utilizing generalized methods for advanced control, Proc. of session on computer control of plant operations, 35th National Meeting, AIChE, Philadelphia, Pennsylvania, June, 1978.

The author

J. Patrick Kennedy is a refinery application specialist for Taylor Instrument Co., P.O. Box 2297, 1661 Timothy Drive, San Leandro, CA 94577. Tel: (415)568-5400. He obtained his B.Ch.E. in 1964 and his Ph.D. Ch.E. in 1970 from the University of Kansas, and has held positions in Cities Service Research and Shell Development Co. He is a registered Control Systems Engineer and a member of the Instrument Society of America, the International Association for Hydrogen Energy, and the AIChE.

Computer control of fractionation plants

Does it pay to install a computer? If so, how should installation, startup and personnel training be handled? Here is a useful guide toward providing answers to these and other questions.

***Daniel W. Kemp** and **Donald G. Ellis,** Cities Service Oil Co.*

☐ Economic savings and reliability are the two most powerful arguments in favor of computer control of fractionators. A project engineer contemplating such an application must first justify, at least on paper, the purchase of a computer and accessory hardware. He must also be aware of certain pitfalls concerning installation and operation of the system, and operator training.

In this article, the authors attempt to cover these aspects and others related to computer installation, such as location of sensors, valve-operating devices, sample systems, and flexibility to run without a computer.

Calculating a payout

The first step in the justification procedure is to compare how well the existing control system is operating in comparison with a theoretical ideal. This involves collecting basic operating data over a normal or average period of operation. In a typical fractionating column—e.g., a deethanizer—the data should include the following variables:

- Heavy key component in the overhead product
- Light key component in the bottom discharge
- Temperature, pressure, composition and rate of flow of the feed
- Energy required to run the column—usually, heat to the reboiler.

Deviations from the norm are determined by integrating the composition recordings and measurements of heat to the reboiler, and presenting them on a time-weighted basis. High and low deviations are selected at the points within the period that consumed 90% of the total time. These data define the actual operating circumstances in terms of purity vs. energy required.

Meanwhile, tray-to-tray calculations using feed analyses previously determined over a period of time are plotted as shown in Fig. 2. This graph establishes the relationship between the purity of the overhead product and the energy required at a specific bottom purity—in this case 6% ethane in the propane. In the same fashion, an entire family of bottom-purity curves can be developed—at a constant tray-efficiency, as in Fig. 3, or at varying tray-efficiencies (Fig. 4).

By comparing the actual operating data (Fig. 5) with the theoretical calculation (Fig. 6), a range of deviations can be obtained. This represents conditions previous to the application of advanced controls, and serves as a basis for estimating project payout.

Measuring the value of a computer

At this point, how can the project engineer determine whether investing in a computer is justifiable? Fig. 7, which shows the deviation from set point in relation to maximum product impurity, should provide a good indication. During normal operation, the set point must be set at considerably below the maximum impurity level, which is also that of minimum utility requirements. But if controllability could be improved

Originally published December 8, 1975.

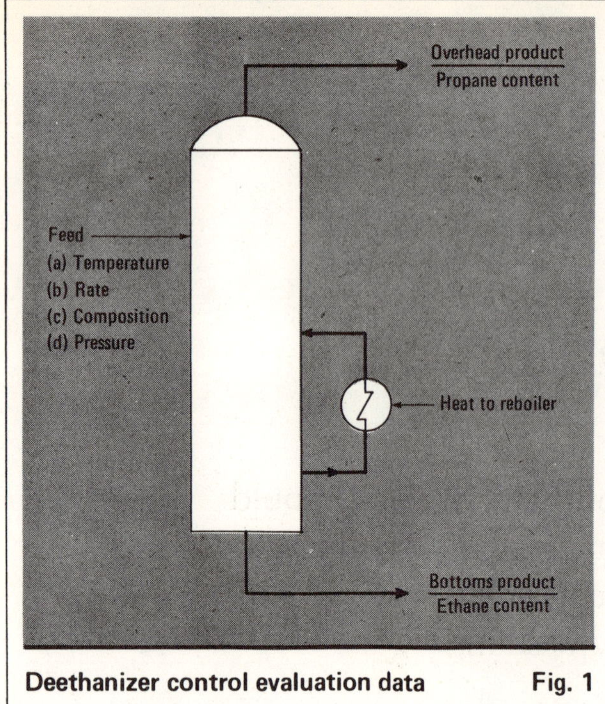

Deethanizer control evaluation data Fig. 1

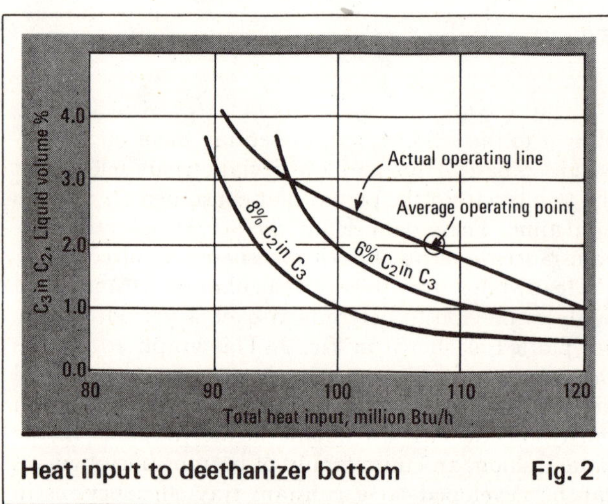

Heat input to deethanizer bottom Fig. 2

to reduce the deviation to much lower limits—as Fig. 7 further illustrates—the set point could be moved much closer to the maximum impurity level.

The improvement is better seen in Fig. 6. The new set point here relates to a much lower energy demand, which leads to lower operating costs and a higher average product-quality. Improving the control at both ends of the column maximizes the light key component in the bottom discharge and the heavy key component in the overhead—all at a much lower energy requirement. Tray efficiency also increases with improved column controllability.

Fig. 8 illustrates the total reduction in utilities that can be achieved by installing advanced controls on a fractionator, as seen from the viewpoint of decreased control-point deviations and increased tray efficiency.

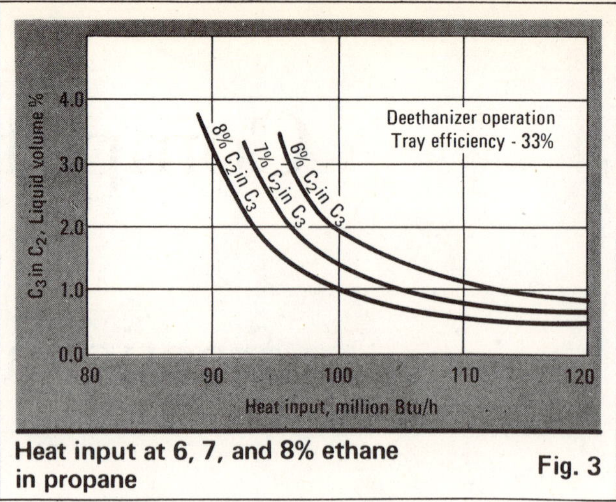

Heat input at 6, 7, and 8% ethane in propane Fig. 3

Reflections before buying

After justification has resulted in management approval, and the computer vendor has been selected, the project engineer is ready to proceed with the purchase. The following factors must first be determined:

- Process inputs to the computer
- Computer outputs for process control
- Power requirements
- Needed reliability

The vendor usually provides an answer to all of the above except for reliability, which is the customer's responsibility.

At the time of purchase, the project engineer should exercise care in choosing hardware for computer interfacing. The input sensors provide data upon which a computer performs its calculation, while the output transducers handle the control valves on a fractionation column. High quality in these items is of vital importance. And the engineer must make sure that the purchased equipment is compatible with the computer.

The computer vendor should always be provided with specifications of any equipment that his customer is buying, and he must approve the purchase.

Measurements provided to the computer should be taken at the best possible locations around the fractionator. Accuracy is important in determining the type of sensor needed. For example, a temperature change of 0.1°F in the overhead line from a deisobutanizer reflects a considerably different composition-shift than does the same change in the overhead line of a deethanizer. At any rate, responsibility for the complete sensor system, including analyzer, programmer, interfacing, sampling system and sampling point should be clearly defined, and preferably handled by one party.

In addition, the individual control points must be evaluated in terms of their effect on the new control package. Additional hardware, if needed, must be specified.

Electrical requirements for a computer installation in a fractionation plant may include main and alternate power supplies for the computer equipment; power for transducers, sensors, chromatographs and related units; and utility outlets. Plants that are required to recover automatically from short-term variations in

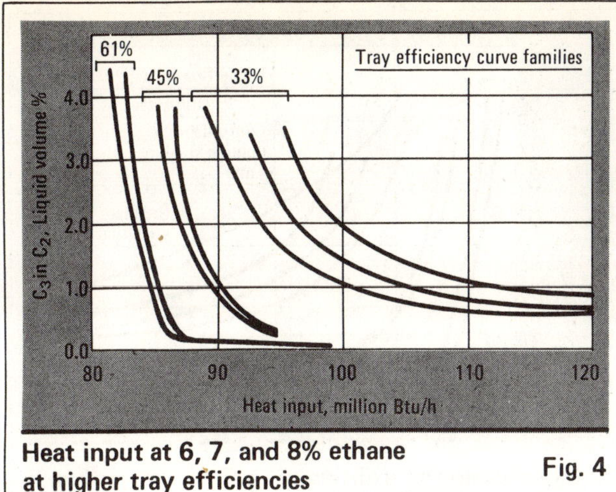

Heat input at 6, 7, and 8% ethane at higher tray efficiencies **Fig. 4**

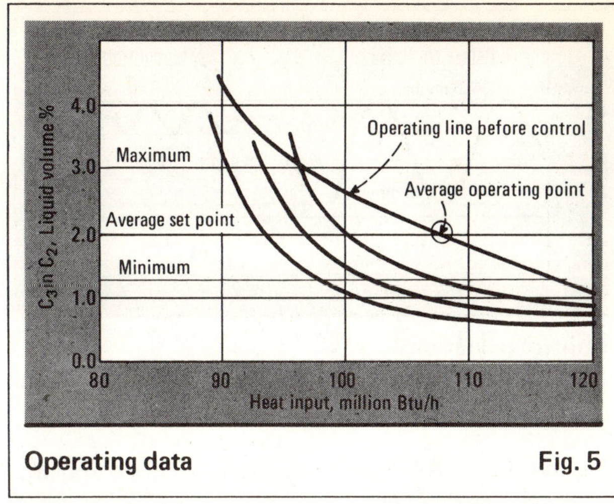

Operating data **Fig. 5**

power supply may need a safeguard mechanism, such as battery backup systems.

Installation and startup

It is essential to have special drawings made for the installation phase, because a great deal of the work involves electrical wiring. Mistakes can be time-consuming and expensive to correct. The engineer should therefore provide good drawings, amend them as the job proceeds, and ensure that it is the "as-built" versions that find their way into the plant files.

Actual installation comprises the computer and associated equipment, input/output hardware, power supplies, and signal and power wiring.

In addition to providing a proper environment for the entire system, the installation should proceed with an eye to operation and maintenance. The wiring, for instance, should be sufficiently flexible. Because installation is usually underway simultaneously with process-control-model development, the sensor inputs are defined at an early stage—usually according to past experience and similar applications. If the computer monitors all of the process variables, there is no need for input wiring spares. But if the installation calls for a limited number of sensors, it may be prudent to install accessories such as spare-signal and thermocouple-extension wires.

Output transducers should present no problem, since they are normally limited by the number of control valves.

Upon completion, the installation job should be checked by the project engineer and the computer vendor. Power supplies must be tested before the equipment is switched on. Input and output connections to the computer should be verified, and tie-ins to the proper process points should be checked.

Depending on the equipment and personnel available, input and output hardware can be calibrated either before or after installation. However, following the verification step, input devices should be run full range, and the measurement on the computer checked to ensure proper operation. Output devices should be similarly checked before startup. This may help detect early failures.

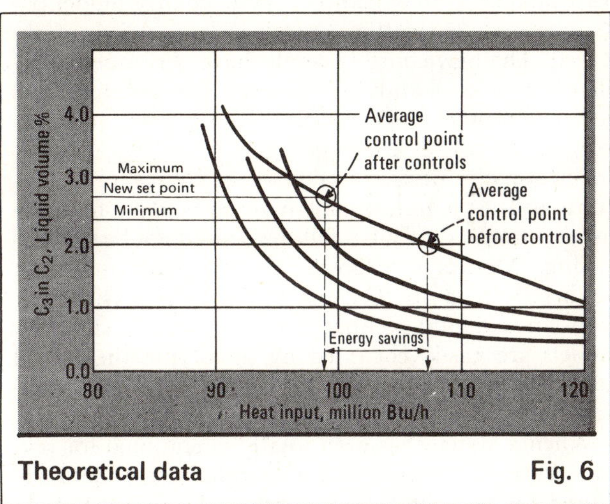

Theoretical data **Fig. 6**

Time required for startup depends on the number of columns to be controlled. In a plant in which the new system replaces an existing pneumatic-control board, the basic sequence is as follows:

- Shift one control loop to the computer
- Tune the local loop on the computer
- Proceed to the next control loop and tune

Whether or not each control loop is left on the computer at this stage depends on the normal mode of operation. For instance, it may be desirable to return a cascade loop to the board after local tuning.

In general, loop tuning can take from ten minutes to several hours, depending on the type of loop and what is happening in the process. Personnel involved in this operation should not leave the console area until performance is satisfactory and the loop is on process control.

Testing the control models

Process-control models should be tested upon completion of control-loop tuning. If possible, independent functions should be tested one at a time. Transfer of controls to the computer should begin as early as possible during the working day to achieve maximum run-

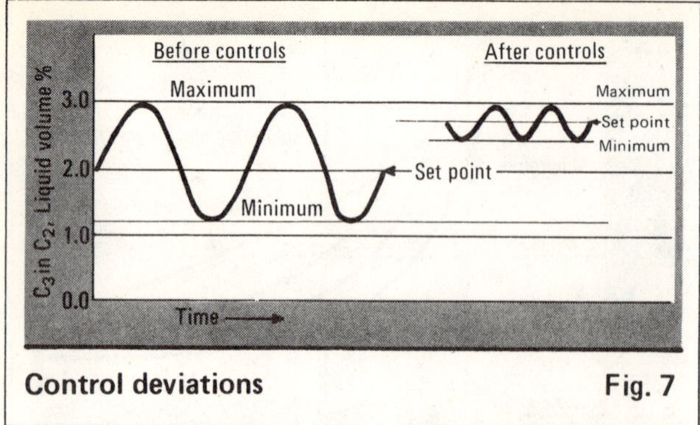

Control deviations Fig. 7

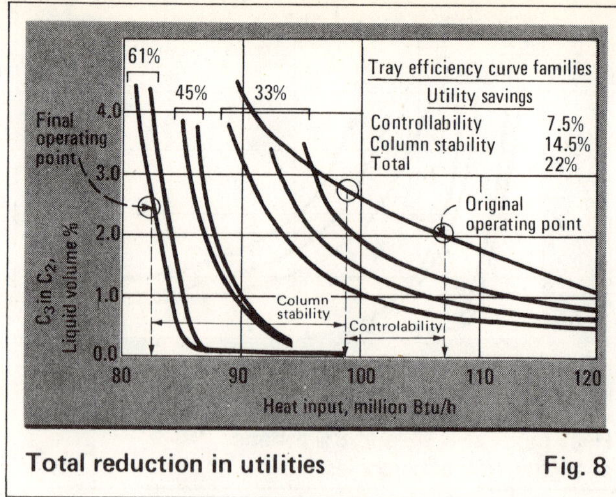

Total reduction in utilities Fig. 8

ning time on a given control-function while the startup personnel is at the plant.

Controls must be closely monitored to verify proper performance. It may be necessary to change model-parameters or override the controls if deficiencies are observed. The new controls should have performed satisfactorily in a "hands-off" manner for several hours before personnel is allowed to leave the computer-console area for any extended period.

Performance of each section of the control model requires at least a 48-h trouble-free run before a final decision is made. Of course, failures should not be too surprising. Models developed for one plant may not perform well in another due to differing physical systems. In a digital computer, failures due to deficient models are easily corrected by modifying the model and feeding a new program. However, a major modification of this kind may take several days.

When a system has been totally operational for several days, the project engineer collects the data required for performance evaluation. These are broken down and overlaid on the curves developed during the justification phase.

Meanwhile, the vendor prepares final documentation for the project. This consists of an update on the "as-built" operating models, and the final configuration of the equipment. The data, along with the manuals and information that the customer needs to make changes is then forwarded to the customer.

Thoughts on personnel training

Training of operating personnel should normally be accomplished in two phases. The first occurs after the process models are ready, but before startup begins. Operators should learn what the project goals are, how the models work, and how to use the system for control and information retrieval.

The second phase coincides with startup. Under the supervision of vendor representatives and the project engineer, the personnel should become familiar with the operation and use of the new control system.

As a rule, operators are skeptical of advanced control systems. They have to see actual improvements in process control before they accept the system. One of the worst mistakes the project engineer or vendor can make before the final stages of a project is to constantly tell operators how well the new system will work. Focus instead on what the system is supposed to accomplish. Openly discuss the possibility of control-logic errors, and disruptions that may occur while shifting over to the new unit. Once the new controls are running well, the fact will speak for itself.

The completion of an advanced control project in a fractionation plant should open a Pandora's box of questions from the operating group and the engineering staff. Here is a list of the most common ones.

1. How can the data-logging capability of the computer be used to upgrade the plant process capabilities?
2. Should the customer develop his own in-house capability for program modifications, or use the vendor's?
3. How many modifications can be added to the installation before system degradation occurs?
4. For future applications, what type of system should be used?
5. How can the rest of the engineering staff be educated in the areas of modern logic packages, sensors and existing advanced applications?

The authors

Daniel W. Kemp is superintendent of operations engineering in the NGL Div. of Cities Service Oil Co. A 1957 graduate in chemical engineering from Purdue University, he is a registered engineer, and has spent his entire career with Cities Service in various engineering and operation positions.

Donald G. Ellis, Staff Engineer, is responsible for evaluation and application of advanced controls in different areas of NGL recovery and fractionation at Cities Service. He obtained a B.S. from the Naval Academy in 1964.

Section XIII
MICROPROCESSORS IN PROCESS CONTROL

Using microprocessors for process control

Microprocessors are here, but applying them to process control is far more complicated than merely installing them online. Before using one for a fairly simple job—regulating three cooling fans to minimize electrical consumption—we had to resolve many difficulties, including substantial reprogramming.

R. L. Wareham, *Monsanto Polymers and Petrochemicals Co.*

☐ The microprocessor was introduced in 1971 as a four-bit, general-purpose circuit to perform as a miniature computer. Newer circuits followed with greater speed and processing capabilities, but the test project discussed here used an early unit, the Intel 4004. Very little of this small machine was needed to execute the cooling system's control function.

Additional electronics augmented the microprocessor to make a useful controller. Unless the final controller design is to be duplicated hundreds of times, the best approach is to purchase preassembled microcomputers. Several companies fabricate microcomputers that incorporate the most popular microprocessors. These assemblies range from an entire microcomputer on a single plug-in card, to larger systems having peripheral options and functional enclosures.

Energy-saving test project

An energy-saving project was chosen for the microprocessor test installation. The microprocessor was not given initial preference for the project, but when its estimated cost came out equal to that for the standard approach, the chance to evaluate this new technology became highly appealing.

Three two-speed fans on a process-water cooling tower were regulated by the microcomputer so that the minimum electrical energy was used to maintain water temperature. The fan motors were controlled at their slowest permissible operating speed, with their operating sequence arranged to ensure equal operating time for each motor. In winter months, each fan was shut off for two hours on a rotating schedule to allow ice buildup in the tower to melt.

The hardware

A single-card microcomputer assembly was used in the test project. With additional memory chips, output drivers, power supplies and card racks, the total cost was less than $1,000. This assembly represents the minimum practical size for a small control application.

Included on the basic microcomputer card are 80 words of random-access memory (RAM), of which only 20 were used; and 256 words of programmable read-only memory (PROM), with expansion space to 1,024 words. The control function program needed only 340 words. Sixteen inputs and sixteen outputs are available on the card but only eight inputs and six outputs were actually used.

Input information concerning contact closures is wired directly into the computer's terminals (Fig. 1). Water-temperature limits are monitored by pressure switches linked to an available pneumatic indicator. Motor status is monitored by relays across each of the motor starter-coils.

Output-port drive capability is limited to about 10

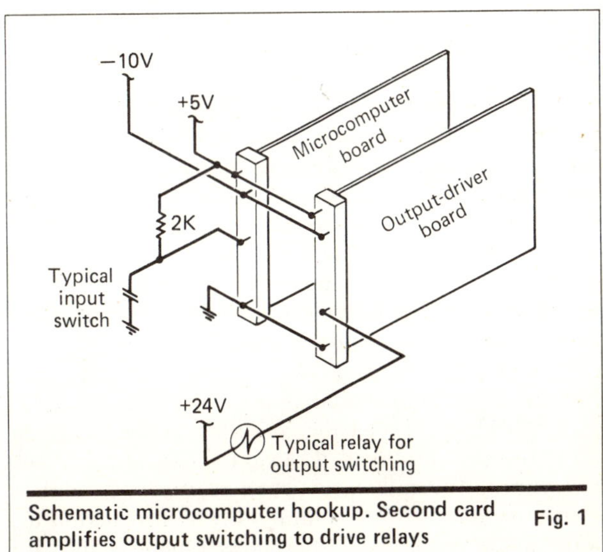

Schematic microcomputer hookup. Second card amplifies output switching to drive relays — Fig. 1

Originally published June 21, 1976.

Allocation of effort for the microprocessor test project

Function	Man-hours
Meetings	20
Problem study	20
Flowsheets	12
Programming	40
PROM preparation	12
Checkout	8
Reprogram	24
PROM revision	4
Final checkout	16
Total	156

mA at 5 V, so another circuit card with 16 power drives was added to the microcomputer to boost the drive capability to 300 mA at 28 V d.c. Twenty-four-V general-purpose relays are operated from the outputs. The relay contact activates the motor starters.

Programming approach

System development costs were kept low by programming without the aid of special development tools. The program was written in the standard Intel assembly-language code. It was transposed manually to hexadecimal coding. An in-house minicomputer was available to format this code and punch it onto paper tape, which was used by the memory chip vendor to enter the program into memory.

Programming the four-bit microcomputer is not difficult, but it does require an understanding of both the language and structure of the microprocessor. Flowcharting the control function is the first step in programming but, as will be illustrated later, this step cannot be considered final when first written. As an example, let us assume that we have a single, two-speed motor driving a cooling-tower fan, and that we have sensed the water temperature to be too high. If we want to raise the fan's speed, the question that must be answered first is, "At what speed is the fan currently running?"

Linked to the first four input terminals are on/off switches that indicate high water-temperature, (terminal 1), low water-temperature (terminal 2), fan on slow speed (terminal 3) and fan on high speed (terminal 4). To answer our question, we must check the status of the switches at terminals 3 and 4. Since this is a four-bit machine, the microcomputer will read four input terminals at once. The four terminals constitute one input port; in this case, port number zero.

In Intel assembly language, the sequence of instructions to answer our question is as follows:

1. FIM 2, 00—To let the computer know which input port we want to read, we load the port number (00) into one of the microprocessor's registers (2) with a fetch immediate (FIM) command. The instruction actually loads two four-bit characters. One goes into the even-numbered register (2); the other into the next register (3). Both characters are zero in this case.

2. SRC 3—We now transfer the number in registers 2 and 3 to the address register. The instruction must refer to an odd-numbered register (3). The transfer will also move the character in the preceding register (2).

3. RDR—We now read-in data from the four terminals of the input port—identified as "00" in the address register. The data is then placed in the accumulator. A closed switch on an input terminal is read in as "1", an open switch as "0." The input word is 0101, with the right-most bit representing input terminal 1. This tells us that the high-temperature switch is closed, and that the slow-speed indicator-switch is closed.

4. RAL—The accumulator has a fifth bit storage-location, the carry. A "rotate left" instruction (RAL) moves the left-most bit of data into carry. The accumulator now reads 0-1010. The carry contains a single bit (0) that indicates that the fan is not on high speed.

5. JCN C1, 38—The computer will examine the carry bit for a "1" or "0" value. If the carry were a "1", the computer would jump on the condition (JCN) that carry equals one (C1) to program step 38. At step 38, we would have written specific instructions for this status. Since the value is "0", the computer continues to the next instruction.

6. RAL—The data are again shifted left. The accumulator word now reads 1-0100.

7. JCN C1, 53—The computer examines the carry bit again. This time, because the value is "1", indicating that the motor is running on slow speed, the computer will jump to program step 53 for the next instruction. At that step, we will have started the instructions for switching the motor to high speed.

As is the case with most computer programs, the first version contained errors. It became necessary to revise the program and rewrite the memory chips—these are erasable with ultraviolet light. The final version worked as intended and went into operation in Sept. 1975.

Engineering effort

Expertise in using the microprocessor combines electronics and programming knowhow. Any engineer with a basic understanding of these subjects can learn to use the microprocessor in three to five days of formal training. Courses are available from vendors, independent lecturers, and some universities for about $300. The most comprehensive training is limited to study of a single manufacturer's microprocessor design. Therefore, it is desirable to select one type early in the project.

Engineering man-hours for this project totaled 156 (Table I). Of this total, 20 man-hours were set aside for selling the new concept to plant engineering personnel. Another 26 h of flowsheeting and programming time were spent learning to use the microcomputer. An additional 30 man-hours of programming time might have been saved had a development system already been available. Thus, a second project of similar size (for which a development system could be purchased) would require only 80 man-hours.

The author

Robert L. Wareham is Specialist in Process Technology for Monsanto Polymers & Petrochemicals Co., 190 Grochmal Ave., Indian Orchard, MA 01051. He provides technical assistance on designing process-control systems for the company's U.S. and foreign plant facilities. In the past, he has worked as instrument and electrical engineer at the Monsanto resin production plant in Springfield, MA. A graduate of the University of Massachusetts, he holds a B.S. in mechanical engineering.

The benefits of microprocessor control

Microprocessor control of distillation columns can save energy, automate control, and yield the benefits of using advanced control algorithms.

M. R. Skrokov, Stone & Webster Engineering Corp.

☐ A number of advantages can be derived from the use of microprocessors on unit operations in chemical process plants. One is that of energy savings. A second is the automation of certain unit operations.

In the control of an olefins plant described later, automation could be applied to cracked-gas dehydration regeneration, propylene-dryer regeneration, etc.

The many desirable features of microprocessor-based control systems include some respectable engineering and design manhour savings during the plant-design phase, as well as savings in later plant operating costs.

During the last few years, owing to increasing energy costs, our company has been performing energy-savings studies for olefins-plant designs. The Control Systems section was assigned the study of specific towers in the distillation column train of a standard plant. Reboiler and condenser energy savings were calculated on a very conservative 5% basis for the following strategies:

- Optimized control of environmental variables.
- Optimized control by material balance.
- Optimized control by heat balance.

Kemp and Ellis [*2*] and others [*1,3*] indicate a 10–22% savings of energy in similar plant designs.

The resulting savings in energy indicated a payout of less than one year. Chargeable system costs assumed the use of microprocessor-based hardware for each of the following towers:

- Demethanizer.
- Deethanizer.
- Ethylene fractionator.
- Depropanizer.

Conclusions reached in the study recommended that all new ethylene-plant proposals include a control block entitled "Column Optimizer" on the flowsheets for these columns, and that the instrumentation-hardware estimating costs reflect the addition of these microprocessor blocks (with some reduction of conventional analog-instrumentation costs).

Energy savings in distillation-column control

The term "microprocessors" is becoming almost a household word. Many, possibly too many, instrument manufacturers claim to have microprocessor-based hardware that is readily applicable to any control problem. But upon evaluation of a spectrum of such systems, we found that currently available hardware fell largely at the ends of a band of systems that might be labeled as: "hardware only" at one end, and "large standardized multiprocessor systems" at the other.

There is a need for relatively inexpensive, dedicated, ROM (read-only memory), microprocessor systems, designed for specific unit operations in process-industry plants, where each system might be as small as the equivalent of approximately 20 analog loops. For example, distributed microprocessor elements dedicated to individual column control, and permanently programmed for feedforward-feedback, heat-and-material-balance strategies might revolutionize both the work of the chemical plant operator and the control systems designer.

Control theoreticians have been complaining for years that practicing control-system engineers have not been fully utilizing the theory that has been available. One answer to this might be the unavailability of practical hardware for easy application of advanced techniques to individual unit operations. Control engineers in operating companies (and those in contracting organizations, in particular), are usually tied to rigid time and manpower schedules. Even simpler programming of the process-control-language type must be carefully planned for, budgeted and controlled.

On downstream units (second-level control), we experienced some difficulty in obtaining quotations on systems that met our specifications. These specifications outlined hardware and software requirements as a system. Control functions, such as ramps, holds, sequences, feedforward-feedback were specified. Many vendors, however, refused to assume systems responsibility even for the "programmers" called for in the downstream-regeneration-type control schemes around the cracked-gas dehydrators and the propylene dryers.

A similar difficulty was experienced on distillation column control. Here, the problem centered on the lack of a system small enough to monitor and control the equivalent of less than twenty control-loops. All systems offered seemed to be designed for five or more towers, to be controlled as one unit operation.

One of the purposes of this article is to influence con-

This article is based on a paper given at the ASME Petroleum Mechanical Engineering Conference, Sept. 19-23, 1976.

Originally published October 11, 1976.

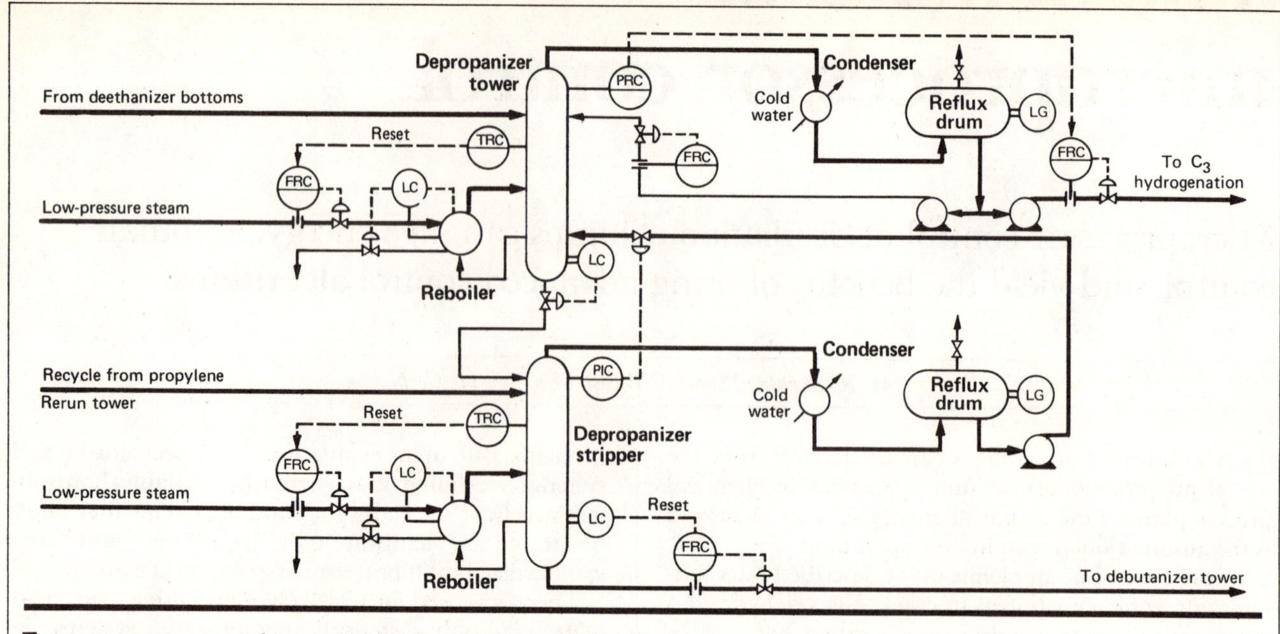

Tower pressure-control system using flooded overheads (reflux drum and overhead condenser) **Fig. 1**

trol-systems manufacturers to produce microprocessor-based systems hardware readily adaptable to almost any distillation column. We propose that the manufacturers produce standardized digital shelf-size instruments dedicated to the control of single columns and yet generalized enough in strategy so that a feedforward-feedback algorithm can be preprogramed onto an ROM, LSI chip and housed in approximately a 6x6-in. panel-mounted instrument.

Control of environmental variables

Control of environmental variables on the four key columns mentioned earlier was based on a first, relatively simple, application of good pressure and temperature control to these towers, with careful consideration to condenser and reflux-drum capacitance.

For example, when controlling tower pressure, two strategies that are sometimes debated between process and control-systems engineers are pressure control by hot-vapor bypass versus control by use of flooded overheads (Fig. 1). In certain cases, the use of hot-vapor-bypass control on towers with totally condensed overheads has caused the tower pressure to oscillate badly enough to upset pressure control. Continuously venting gas from the system (via backpressure control) helps the situation, as does redesign of vapor-tube inlets to minimize liquid surface disturbance—but neither solves the basic problem. Tower pressure regulation of condensed liquid leaving the overheads is a better method in certain cases.

Relative energy levels for key ethylene plant columns **Table I**

Column	Reboiler, % of total (from Btu/h data)	Condenser, % of total (from Btu/h data)
Demethanizer	14%	3%
Debutanizer	7%	8%
Ethylene fractionator	33%	51%
Deethanizer	22%	11%
Depropanizer-rectifier	13%	12%
Depropanizer-stripper	11%	15%
	100%	100%

The flooded-overhead system works by totally flooding the reflux drum and partially flooding the overhead condenser. If the pressure increases, the pressure control will extract more liquid and uncover tube surface—thereby increasing the condensing rate and reducing tower pressure. If the pressure decreases, less liquid is extracted (the overhead-exchanger tube surface is covered), reducing condensation and resulting in an increase in pressure.

From an energy-savings standpoint, the tube surfaces should (optimally) be kept covered. Shinskey [3] proposes the case for a "vapor pressure curve" function-generator to improve efficiency of separation in certain columns. Such a device would manipulate the pressure setpoint as a function of condensate temperature by matching the vapor-pressure curve of the liquid distillate. However, this requires that a constant-boiling liquid feed be present, or that a reliable composition-analyzer be tied to the system to detect feed-composition variations and to provide these as an input to the control system. This is particularly true in the case of the cracked-product feeds present in ethylene plants. This method does, however, add another analog element to the instrument hardware total.

The savings projected by us reflect the use of a column optimizer system.

Optimizing control by material balance

There should be no need to make an excessively pure product merely to ensure that product quality does not

Glossary

AT	Analysis (composition) transmitter
CRT	Cathode ray tube
DDC	Direct digital control
FRC	Flow recorder controller
FT	Flow transmitter
LC	Level control
LCD	Liquid-crystal display
LED	Light-emitting diode
LG	Level gage
LIC	Level indicator controller
LSI	Large-scale integrated circuit
MOS	Metal-oxide semiconductor
PI	Pressure indicator
P&I	Piping and instrumentation
PIC	Pressure indicator controller
PID	Proportional, integral, derivative control
PRC	Pressure recorder controller
PROM	Programmable read-only memory
RAM	Random-access memory
ROM	Read-only memory
TI	Temperature indicator
TRC	Temperature recorder controller

dip below the required purity as it cycles around the setpoint. By using advanced control, product quality can be held close to the setpoint, saving energy previously required to recover from cycling.

By using composition-analysis of the feed, overhead and bottoms, together with feedforward control, boilup can be controlled, thus minimizing the energy needed to produce a required product quality. A feedforward signal is used on the feed flow, multiplied by the feed composition, and delayed by a time-constant, to adjust the top- or bottom-product rate and the bottoms heat rate. A multiplier is used to predict the required change in heat rate. If this fixed multiplier overshoots the goal, a feedback signal is used to change the multiplier to a new value.

Optimizing by heat-balance control

By calculating the heat content (enthalpy) of the feed stream (i.e., the composition of each component multiplied by its enthalpy), a more-advanced computer system can calculate the optimum internal reflux. The feedforward system then changes the internal reflux toward this optimized level. The feedback system is used to hold the internal reflux at the point that produces the required product composition.

A number of operating companies have used these techniques for energy savings [*1,2,4,5*]. Field-reports from these plants generally indicate a 10–15% (and up to 22%) reduction in energy consumed. For our study, we claimed a more conservative (5%) level of savings. The following calculations were based on this savings level:

Table I shows relative energy levels in a typical 1-billion-lb/yr ethylene plant, for six key columns.

The availability of savings in energy levels shown in the table was based upon the following analysis:

Demethanizer—The heat saved by the reboiler was a substitute for cooling water; therefore, no credit or penalty was taken for the reboiler. For the demethanizer condenser, the pumping-energy savings amounted to 640 hp per million Btu.

Debutanizer—Any heat saved in the condenser resulted in a penalty in the reboiler. Therefore, there was no net savings for this column and it was deleted from the final summation.

Ethylene column—The reboiler savings amounted to 80 hp/million Btu, and condenser savings were 190 hp/million Btu.

Deethanizer—The reboiler used free heat from quench water, therefore, one-half duty was taken as low-pressure-steam savings. The condenser saved 150 hp/million Btu (Fig. 2 and 3).

Depropanizer—Our standard plant uses a single column and, therefore, the rectifier was removed from the study. We also took no credit for the condenser. Credit for bottoms in low stream pressure was taken.

Assuming a savings level of 5% of loading gives approximately the tabulation shown in Table II.

Total energy saved amounted to 12.34 million Btu/h. At an energy cost of $2/million Btu, the minimum saving was $24.68/h, or approximately $177,000/yr.

Estimated costs for hardware and engineering for this system were based on the following assumption: Each of the columns shown in Table II would require approximately two process chromagraphs, and the entire system of four columns could be operated by one mini- or microcomputer.

Initially, therefore, the study lumped all columns

Specific energy savings — **Table II**

	10^6Btu/h (10^6kJ/h)	×	Hp/10^6Btu (W/10^6kJ)	×	lb steam (kg)	=	lb-hp steam (kg-W)	×	Btu/lb-hp (kJ/kg-W)	=	10^6btu/h (10^6kJ/h)
Demethanizer	0.270		640		6		1,037		1,400		1.45
Ethylene column reboiler	3.037		80		6		1,458		1,400		2.04
Ethylene column condenser	4.295		190		6		4,896		1,400		6.85
Deethanizer reboiler	1.027		420		—		—		—		0.43
			1,000								
Deethanizer condenser	0.925		150		6		824		1,400		1.15
Depropanizer reboiler	0.99		420		—		—		—		0.42
			1,000								12.34

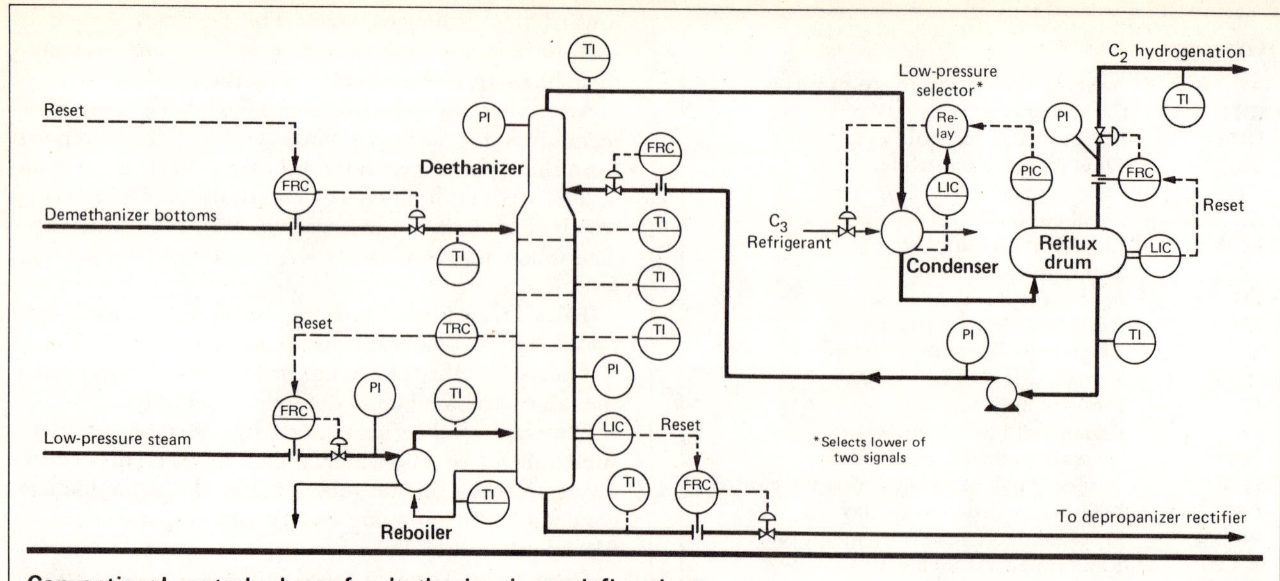

Conventional control scheme for deethanizer in an olefins plant **Fig. 2**

into one computer system, with the resulting figures estimated as:

6 chromatograph systems	= $72,000	(including sampling systems)
1 minicomputer	= 65,000	(complete with applications software)
	$137,000	

Additional engineering was estimated at:

120 h per chromatograph – (for specification)	720 h
250 h per minicomputer	250 h
	970 h (for specification)

Total estimated system cost: $157,000, designed and specified.

At the cost of $157,000 to the client, and a minimum estimated annual return of $177,000, the payout period is less than one year.

One minicomputer handling four columns would also replace some conventional, analog instrument loops. Key loops would, however, be duplicated on a fail-safe redundancy basis.

Control strategies

Which equations should be programmed into the "Column optimizer," and what are the objective functions of this optimizer?

These questions must be answered with the following two objectives in mind: minimum overpurification of the product, combined with a minimum use of process energy.

Woolverton and Murrill [*8*] use a mathematical "relaxation" tray-by-tray analysis, which is ideal for large-digital-computer solution, whereas the simpler heat-and-material-balance strategies can be simulated on either a small analog computer or a digital microcomputer. (The relaxation mathematics require a large machine, such as the IBM 7040, for solution.) We centered on the heat-and-material-balance approach, which is easily handled by a microcomputer and more in line with our objective functions.

Before proceeding further, let us define some terminology.

What is a microprocessor?

A microprocessor is a semiconductor device based on the use of LSI [large-scale, integrated] circuit technology, which allows the placement of literally thousands of bits of command (typically up to 10K) on one MOS [metal oxide semiconductor] chip.

The term "microprocessor," as we are using it, involves the use of a digital circuitry in dedicated, miniaturized systems to control sections of a continuous chemical process such as one (or a group of) distillation column(s), or a cracked-gas-dehydrator reactivation system in an ethylene complex, or related systems in ammonia plants or petroleum refineries.

Why use LSI technology

An affirmative answer to this question is based primarily on the order-of-magnitude decrease in hardware price that has occurred recently for LSI technology. Drastic decreases in the prices of miniature calculators and of LED and LCD wrist watches and calculator displays are an example [*7*].

Most mini- (and micro-) computer systems are RAM- (random access memory) based and, therefore, are inherently designed for reprogramming capability. This increases hardware costs as well as system complexity—system costs may run from $30,000 on up.

We believe that either a PROM (programmable read-only memory) or a dedicated ROM (similar to the LSI circuitry of the hand-held calculator and digital wrist watches, but "masked"—manufactured as standard—for petrochemical-industry usage) is required.

A ROM-chip-based black-box system dedicated to feedforward-feedback distillation-column control, but housing only the equivalent inputs and outputs for approximately 20 analog loops, or less, would be readily

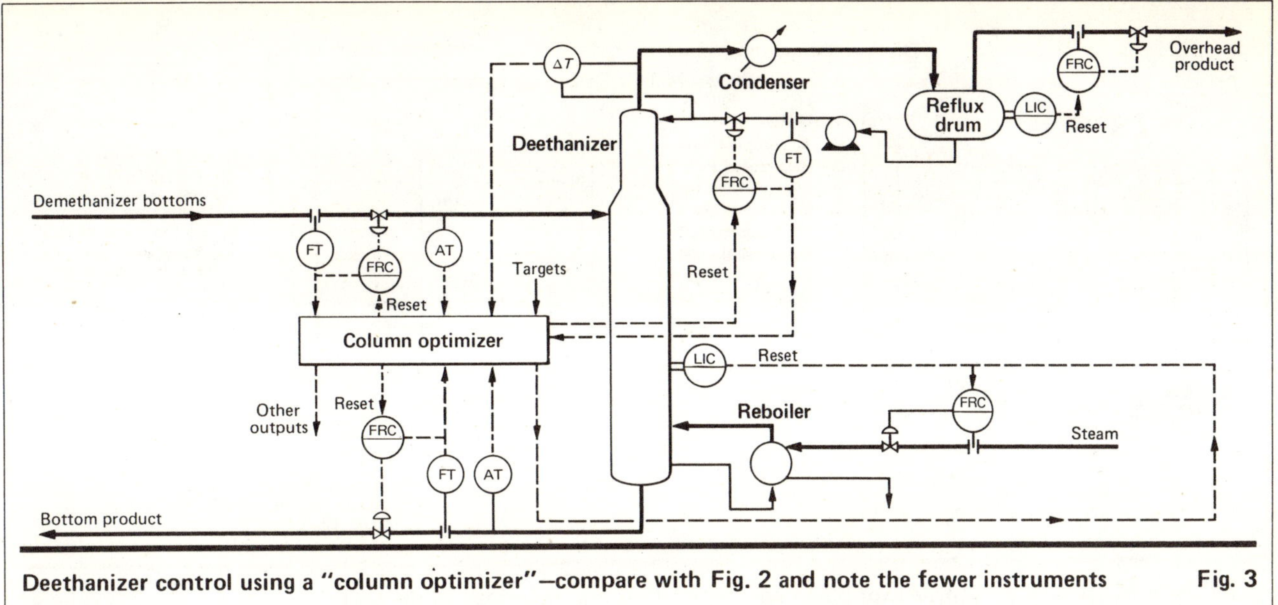

Deethanizer control using a "column optimizer"—compare with Fig. 2 and note the fewer instruments **Fig. 3**

marketable if sold for approximately $3,000 - $5,000 per instrument.

At that price level, many micro-devices could be specified and requisitioned by control-systems engineers as easily as conventional PID analog instruments are today, and treated as such on flowsheets.

In analyzing currently available marketplace hardware, we found that a number of major instrument manufacturers are using microprocessor hardware as the basis for a 1970–1980 equivalent to the 1950–1960 generation of maxi- and minicomputer systems (the Honeywell TDC 2000 is an example). "Data highway" communication links—used to tie sensing, control and readout systems together—are employed in these larger systems. CRT displays have also been redesigned for better communication between the operator and the process, through the control board.

All of these features are improvements over the computer systems of the 1960s. However, the architecture seems to be designed to be sold as a unit and seems aimed at total plant computerization through one system. This sales approach does not readily allow for wide application of DDC, feedforward-feedback combinations, or simple optimization control-strategies on a single unit-operations level, such as one column or one reactor. These larger systems have a further disadvantage in that they usually require a separate project-management effort, extensive software, etc. On a smaller unit-operations level, with microprocessors dedicated to 20 loops or less, the treatment of these controls during the design phase might approach that of the single PID controller; i.e., they could be specified and handled as standardized catalog items.

The other end of the market spectrum for microprocessors centers on loose chips and components that an electronic-instrument designer might specify in order to piece together a subsystem. He can purchase microprocessor chips from several manufacturers (such as Intel, National Semiconductor, etc.) and augment these with memory input/output devices as required, in order to solve the particular systems problem, but this is a highly customed-designed attack for our needs.

A middle level of "unit operation controllers" is also available from various manufacturers such as Bristol (UCS 3000), Robertshaw (UOC), and Rosemount ("Diogenes") among others. These vary from 20- to 80-loop equivalents of RAM systems, programmed by various simplified techniques that allow for easy establishment of control strategies such as feedforward, deadtime compensation, simple optimization, PID, sequencing, and column dynamics. This group is ideal for application to certain downstream units in ethylene and other petrochemical plants—for systems such as cracked-gas dehydration regeneration, propylene-dryer regeneration, etc., which require sequence functions but are not small (distributed) enough to be treated as a single balloon on a flowsheet for individual distillation-column control.

Proposal for a new "Column optimizer" chip

We propose an instrument essentially similar in front-face appearance to an analog control station, which would mount in a conventional control-board shelf. It would house all the lead-lag (dynamic elements) and PID functions necessary to complete the heat-and-material-balance functions around one distillation column in a feedforward-feedback mode, as described earlier (Fig. 4). The control circuitry for such an instrument might be based on one LSI microcomputer chip, possibly as described by Redding and Snyder [*12*], which they call a "Hybrid/LSI Package," and which is essentially a single-component microcomputer. This instrument should sell for between $3,000 and $5,000 per unit and would greatly enhance the application of DDC, feedforward control and simple optimization in distributed systems.

This instrument would, of course, have neither a CRT display nor a data highway, but might be easily hooked into such elements because it would be digital.

In a paper [*6*] given last October at a Stone & Web-

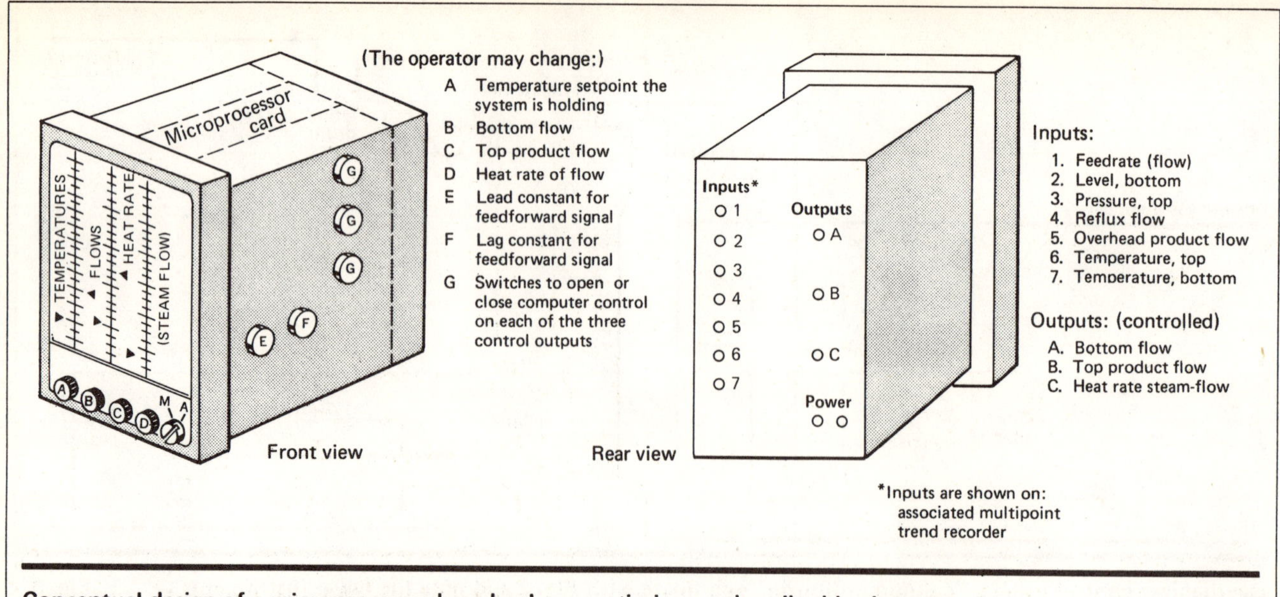

Conceptual design of a microprocessor-based column optimizer as described in the text **Fig. 4**

ster "Olefins Symposium" aimed at ethylene plant design, we listed other possible benefits of microprocessor elements. Some of these were projected savings in control-board space and in engineering-design manhours, plus improved operator/control-board communication.

Control-systems engineering/design savings

We envision a drastic reduction in control-board space within the next five to ten years as a result of development of dedicated microprocessor hardware.

A typical 1-billion-lb/yr ethylene plant can have up to 200 lineal ft of control board, including offsites, for a grass-roots complex. (We are currently designing such a plant with over 150 lineal ft, including consoles; with computer control, this could easily increase to 200 ft of control-board space.) This might be reduceable to approximately 50 to 80 ft, or less.

There are approximately 600 loops with control valves, plus approximately 400 interacting loops in such a complex. Most of the valves will probably never be eliminated, but many of the controllers could be combined through the use of dedicated microprocessors. Assuming a reduction of only 10%, a savings of approximately 60 to 100 controllers is possible—if cascade, and other such loops, are counted. Further, the associated savings in control-systems engineering-and-design manhours required for specification of hardware and for drafting of installation details and loop sketches, etc., could amount to approximately 2,500 h as indicated in Table III.

Control-systems manpower—estimated savings Table III

	Manhours
Control-loop specification	200
Reduction in control diagram effort (engineering)	100
Decrease in flowsheet drafting	50
Decrease in control-board design and specification (engineering and design)	1,200-1,800
Decrease in P&I flowsheet interface	50-100
Decrease in logic-diagram effort	50-100
Decrease in spare-parts effort	50-100
Reduction in instrument index work (design and engineering)	50-100
Reduction in cable routing (design and engineering)	50-100
	2,000
Plus or minus 10%	200
Approximately	1,800–2,200

Purchasing, Project Engineering, and other sections would also have less work to handle as a result of the implementation of this hardware. The manpower savings (in addition to the energy savings) are obvious even neglecting Purchasing, Project Engineering and the other downstream work efforts. This savings would cover the system cost ($50,000 – $60,000) of one minicomputer to handle the entire plant, or ten microprocessors distributed in the plant. The ten microprocessors would be chargeable with approximately one-half this figure ($30,000).

Some of the disadvantages of the large computer systems, such as extensive project-management and applications-programming effort, along with continuing maintenance of massive hardware, should be minimized upon the advent of dedicated microprocessors.

Operator communication

It is well known to anyone with extensive plant-operations or startup experience that plant operators tend to run their units well below optimum level (i.e., well below plant loading point.) Without sophisticated control, they cannot do otherwise. Ambient conditions (rain, etc.), feedstock variability, and heating-systems Btu-content variations are all difficult to anticipate.

Many a large computer system, operating within a similar variable-mix, has failed to achieve predicted savings because of hardware complexity and the lack of an adequate interface with operating personnel.

Dedicated unit-operations control devices, if sized to look like conventional analog instruments, would give the operator the desired control in more-easily-digestible pieces. If he were convinced that regardless of weather, or feedstock variability, his distillation column would perform closer to true target, he would avoid moving individual instrument setpoints below this optimum operating target.

Analog computer systems are available for this purpose and have been used successfully but are not as inexpensive or unobtrusive as microprocessor instruments would be.

One such system, similar in principle but not hardware, for distillation-column optimization, has been developed (in-house) by an operating company. Johncock and Wright [*13*] indicate that they constructed a low-cost modular system as a black box (the "Fractronic") at Amoco Oil Co., Texas City, Tex., in 1971. This system, however, is analog in nature (and is not located on the main control panel) but it does operate on simple material-and-energy-balance principles. Why not use the power of LSI digital technology for such systems? Further, CRT displays would help the operator to see any interactions between the key columns in a train at a glance, if this were necessary to a given plant design. However, CRT hardware would not have to be purchased as part of such a system unless a real advantage for such extra hardware did exist.

Standardized microprocessor blocks would give both the control-system designer and the plant operator more confidence in advanced automation, without the overwhelming features of a huge system.

Conclusions

The use of direct digital control prior to recent developments in electronic microprocessor "chip" technology required the application of a full-blown minicomputer (32K words, at an approximate cost of $60,000, with software) to handle even a small plant with, say, five distillation columns and associated plant calculations.

Analog instrumentation was, and still is in many cases, required for the standby mode. In other words, the computer has been generally used as a setpoint driver. With the advent of inexpensive, card-replaceable microprocessor chip hardware, this may (and should) change so that DDC will become more widely used without extensive analog backup.

The microprocessor computer should then become just another item of instrument hardware—one for which a standardized specification is available, and one that can be shown as just another special balloon on the Piping and Instrumentation (or Process Control) Flowsheet.

We have indicated that a fairly large system can be economically justified for four columns in an ethylene plant on energy savings alone, and that some design-manhour savings could accrue from a routine application of dedicated microprocessor controllers throughout the plant.

The next step, we feel, is for instrument manufacturers to begin the development of the smaller, dedicated microprocessors, in addition to the currently available large multiprocessing systems. Not only will energy savings and design manpower reduction be the result but, by utilizing modern control strategy, operating plants will produce product closer to target.

If the larger instrument manufacturers do not produce the "column optimizer" type of hardware, the gap may be quickly filled by smaller systems houses, using available LSI technology and working with LSI manufacturers to produce a specially "masked" chip for petrochemical purposes. Some smaller systems houses are already beginning to fill such a gap.

The implementation of LSI technology for production of "column optimizer" type hardware by the larger instrument manufacturers might help to eliminate many of the problems associated with computer systems and advanced control and, in the process, possibly revolutionize control-system application-engineering and petrochemical-plant operation.

References

1. Seeman, R. C., and Nisenfeld, A. E., Digital Distillation—Column Control Offers Flexibility, *Oil Gas J.*, Oct. 20, 1975, pp. 57–60.
2. Kemp, D. W., and Ellis, D. G., Computer Controls Fractionation Plants, *Oil Gas J.*, Aug. 11, 1975, pp. 60–64.
3. Shinskey, F. G., Energy—Conserving Control Systems for Distillation, *New Jersey ISA Annual Symposium, Apr. 6, 1976.*
4. Computers Boost Ethylene Yields, *Chem. Week*, Apr. 7, 1976, p. 43.
5. Zinschlog, H.P., others, Potential Impact of LSI on Process Control System, *Advances in Instrumentation, ISA-'75 Annual Conference*, Oct. 6–9, 1975, Vol. 30, Part I, Paper No. 508, pp. 1-5.
6. Skrokov,M. R., Process Control Instrumentation and Computer Systems, Stone and Webster Engineering Corp. *Olefins Symposium*, Great Gorge, N. J., Oct. 2, 1975.
7. McWhorter, E. W., The Small Electronic Calculator, *Sci. American*, Mar. 1976. pp. 88–98.
8. Woolverton, P. F., and Murrill, P. W., An Evaluation of Four Ideas in Feedforward Control, *Inst. Technol.* Jan. 1967, pp. 35–40.
9. Luyben, W. L. Feedforward Control of Distillation Columns, *Chem. Eng. Progr.*, Vol. 61, No. 8, p. 74 (1968).
10. Shinskey, F. G., "Material Balance Control Methods for Distillation Columns," *The Foxboro Co.*
11. Nisenfeld, A. E., Reflux or Distillate: Which to Control?, *Chem. Eng.*, Oct. 6, 1969, pp. 169–171.
12. Redding, F. I., and Snyder, F. G., Hybrid/LSI Package Yields a Single-Component Microcomputer, *Computer Design*, Apr. 1976, pp. 114–118.
13. Johncock, A. W., and Wright, R. M., Low Cost Modular System Controls Distillation of Texas City Refinery, *Oil Gas J.*, Mar. 8, 1976, pp. 56–60.

The author

M. Robert Skrokov is Chief Control Systems Engineer for Stone & Webster Engineering Corp., One Penn Plaza, New York, NY 10001. His work, internationally, centers on petroleum and petrochemical plant design, as well as various organic-chemical plant designs. He holds a B.M.E. from the College of the City of New York and an M.S. in mechanical engineering from Newark College of Engineering. He is a member of the American Soc. of Mechanical Engineers, and the Instrument Soc. of America, and is a registered Professional Engineer in New York and New Jersey.

Microprocessors enhance computer control of plants

Evaluation of computer systems using direct digital control and microprocessor-based technology shows how operators interface and react with each system when running continuous-process plants.

David L. Williams, *Process Control Div., Honeywell Inc.*

☐ Rapid advances in microprocessor technology have had a significant impact on the design of control systems. As a result, the question often arises concerning the proper approach to take in designing such systems.

Overall design of any control system must have:

1. The necessary control-system functions.
2. An adequately designed operator interface.
3. Strong emphasis on system reliability and maintainability.
4. Cost effectiveness.

Let us review how these requirements are met for process control in conventional direct-digital-control (DDC) systems and in distributed systems incorporating microprocessor technology.

Online computer control

There is normally an online computer-based system (Fig. 1) in large chemical, petrochemical and refining plants to control continuous processes.

In the online mode, the computer has all-inclusive responsibilities. In the offline mode, panel-mounted analog backup instrumentation is used. This combination of analog and digital disciplines has formed a system architecture that has provided a better compromise of functions, operator interfaces, reliability, maintainability and cost than previously available.

Under computer control, the operator uses video displays at a control console, and has ready access to the process and the control system via the computer. A digital volt indicator (DVI) and a trend recorder provide additional display. Only when the computer is offline does the operator revert to panel-mounted instrumentation. At the panel, conventional analog devices such as controllers, recorders, indicators, pushbuttons and annunciators are at the operator's disposal.

All field wiring is brought to termination racks (see Fig. 1). The process signals are then wired to the various panel-mounted instruments, digital volt indicator and computer input/output (I/O) hardware. In an application involving 1,400 process variables and 400 control-valve outputs, hundreds of wires are necessary. Much of this

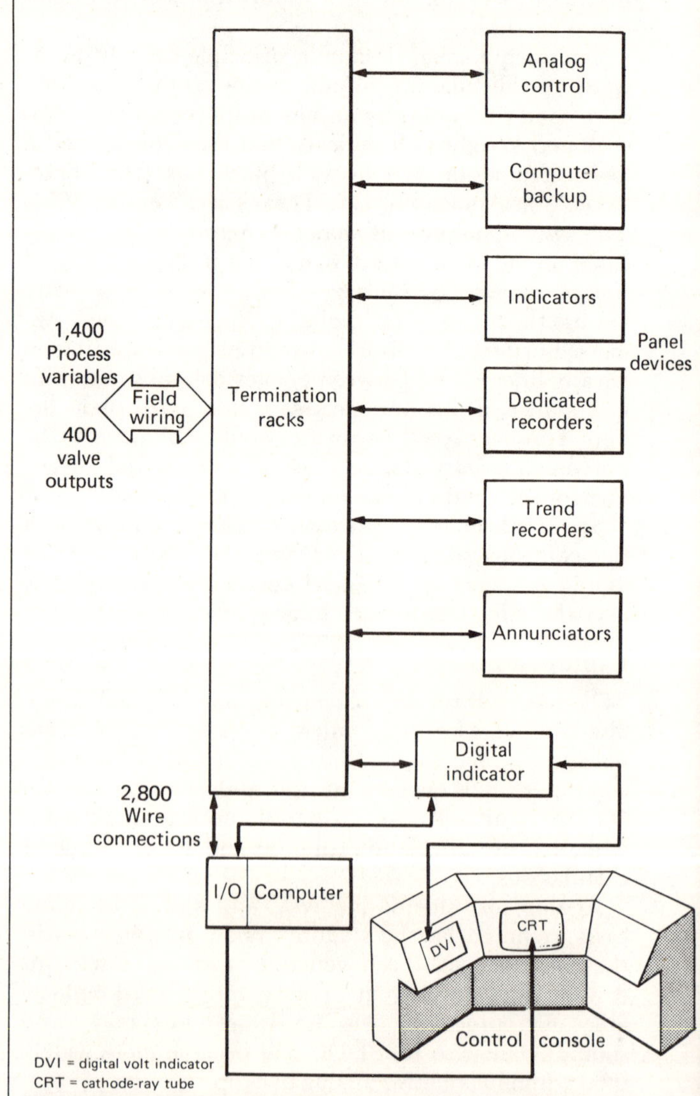

Online direct-digital-computer control system — Fig. 1

Originally published July 18, 1977.

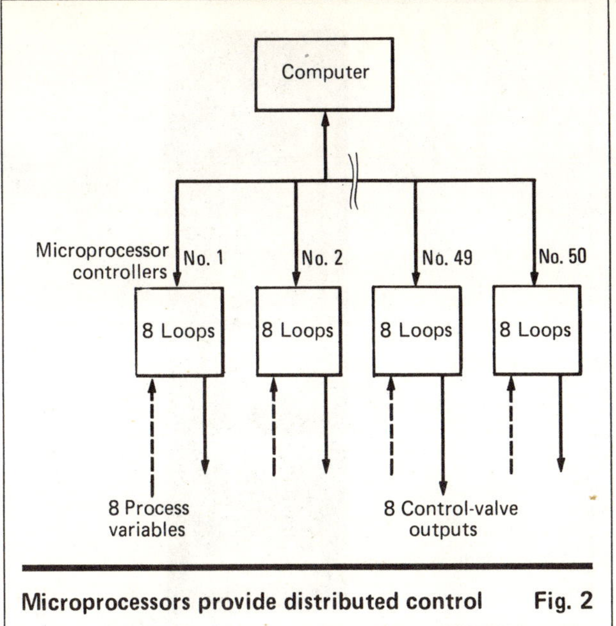

Microprocessors provide distributed control Fig. 2

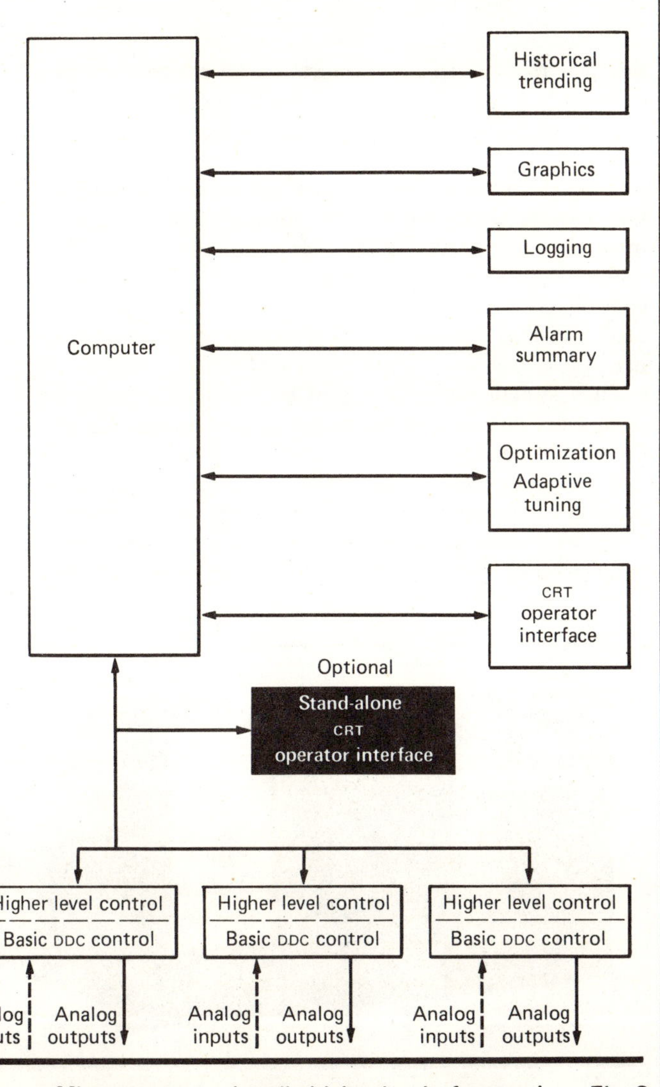

Microprocessors handle higher level of control Fig. 3

wiring represents the need for interfacing the computer I/O with the analog controllers. A total of 2,800 wires are required to bring the signals from the 1,400 process variables to the computer I/O.

Thus, the conventional system satisfies the four system-design requirements. It provides the necessary control functions, has an adequate operator-interface with a combination of a centralized console and an instrument panel, provides an acceptable level of system reliability and maintainability through panel-mounted analog instruments, and has been the most cost-effective way to design a total control system.

Microprocessors control the basic loops

Benefits are gained by removing the routine computer-based DDC and implementing it with microprocessor controllers. These time-shared devices effectively become a smart front end to the computer. Operating independently, they sample and compare the process variables with desired control setpoints to produce error signals. Then, by using one of many configurable control strategies, the microprocessors use the error signals to determine various control-valve outputs.

The intent is not to eliminate the computer but rather to dedicate it to data-manipulation and calculation tasks. Let the computer do the process-optimization and adaptive-tuning functions. The computer is essential for obtaining tight process control and minimizing operating costs.

Microprocessor controllers are much faster in executing DDC loops. A typical computer system executes DDC control at the rate of 100 to 200 different loops per second. The microprocessor can easily do eight different control loops three times per second.

Let us consider a larger system involving 400 simple control loops (Fig. 2). This would require 50 eight-loop microprocessor controllers. Hence, 400 different control loops would be executed three times per second.

Advantages and limitations

The dependence on a central computer for basic loop control is ended. Up to 50% more processing time is provided by freeing the computer from routine DDC control. This has been shown for certain large, continuous processing applications. Extra processing time is now available, so the computer can tackle more-profitable tasks. It is likely that a smaller computer system can be used, which can be reflected as savings in investment.

The computer can provide supporting information to the operator, such as report generation, historical trending, and CRT (cathode-ray tube) graphic displays. This information is desirable but not essential to operations. Distributed control spreads out task responsibility so that failure of one device can not take down the entire system. This suggests good system reliability.

Those functions critical to continuous online control should be removed from the computer (Fig. 3). It is not enough to off-load simple control loops. Higher levels of control can be designed into the microprocessors. Control strategies that would never be implemented in this analog approach because of their complexity are easily handled by microprocessors, and include multilevel cascade and ratio, nonlinear, feedforward, override, constraint, and calculations such as mass flow.

The fundamental difficulty in using analog controllers with digital computers is that analog controllers are inflexible. Once specified, each analog device has a particular purpose that is difficult to modify. A PI (proportional and integral) controller having deviation alarms, pressure-transmitter input, and a 0 to 100-psi range is not easily converted to a PID (proportional, integral and derivative) controller with process-variable alarms, flow-transmitter input and a 0 to 250-gpm range. Likewise, a simple controller cannot readily be converted into a cascade or ratio controller.

Adding the computer interface to an analog controller is difficult if not initially specified. Even with the interface, communication between the computer and the analog controller is limited. Setpoint, output and mode are all that are shared. Adaptive tuning is impossible. Hence, upon computer failure, the analog instruments use preset tuning constants. These may not be the optimum as determined by the computer.

As more knowledge becomes available regarding the process and the desired control, it may be advantageous to modify the control strategy. Such changes are difficult with analog instrumentation. On the other hand, a reconfigurable system can provide the best startup, running, and shutdown control-strategies.

A microprocessor-based control system readily permits the computer and/or operator to make all of these desirable changes. Analog instruments do not provide the overall flexibility and reconfigurability that can be implemented with a microprocessor. When used with or without a computer, microprocessor-based controllers are an improvement over analog ones.

Operator and control-system interface

Unique operator interfaces are possible with a microprocessor-based control system. On the other hand, any radical departure from conventional interfaces is likely to receive a cold reception from the operators. Operator interface devices must look, act and respond like the conventional analog instrumentation (Fig. 4). A display device, dedicated to each loop, must provide process-variable, setpoint and output indication, as well as mode, output and setpoint adjustments, plus alarms.

Other operator interface devices can be provided to use the advantages of the microprocessor. For example, a typical analog-display meter is a ±1% device, readable to ±10° F when using a 0 to 1,000° F range. This is adequate for a profile but not for precise readings. A microprocessor-driven display panel can easily be a ±0.1% device, readable to ±1° F for the same temperature range of 0 to 1,000°F.

The display panel (Fig. 5) can be designed to have the same operator interface capabilities as analog displays. A key-lock provides system security. In the lock position, the operator can observe all loop-detail information but can change only the setpoint, mode and output. An engineer using the key can change: alarm settings, output limits, tuning constants, process-variable filtering values, transmitter ranges, and even the configuration.

Reliability of the control system

More levels of degraded performance are possible with microprocessor controllers than with analog instrumentation. Microprocessor architecture can be designed so that, with almost any hardware failure, indication of the process variables and manual adjustment of outputs are provided.

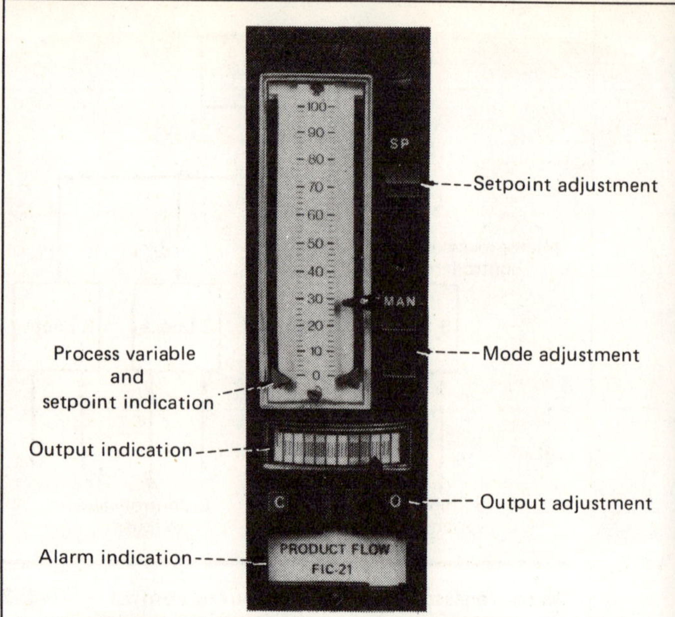

Operator interface must act like analog device **Fig. 4**

The performance levels of analog-control instrumentation include: automatic control, manual control, standby manual, and system failure. Microprocessor controllers include the following performance levels: automatic control, CRT-display manual, operator-panel manual, analog-display manual, standby manual, and system failure.

Such system integrity is possible with microprocessor controllers and represents many "graceful degradation" design concepts. Maximizing the system integrity requires that an independent annunciator system be wired in parallel to the microprocessor controllers, as is currently done with analog-control systems. Alarm points come from process-variable signals, field-mounted pressure and

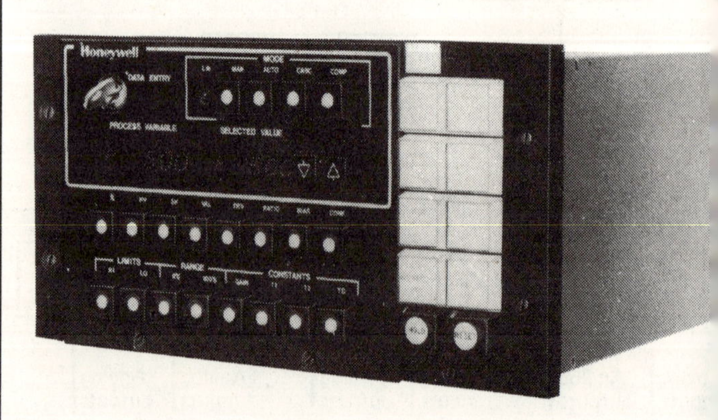

Microprocessor-driven panel for entry and alarm **Fig. 5**

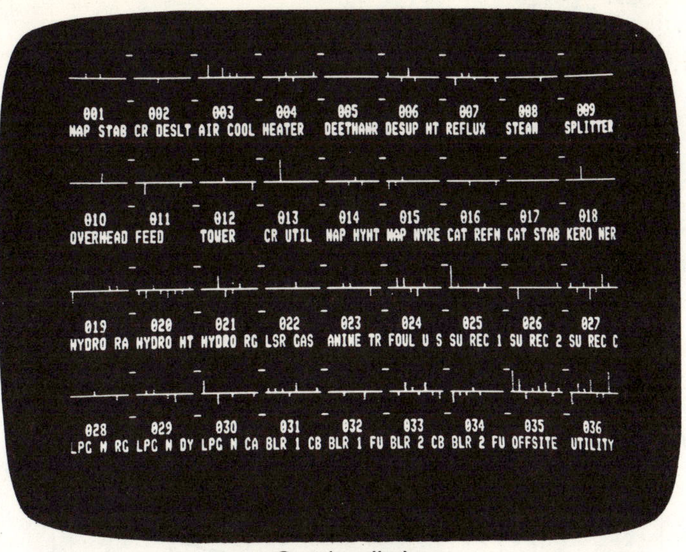

a. Overview display

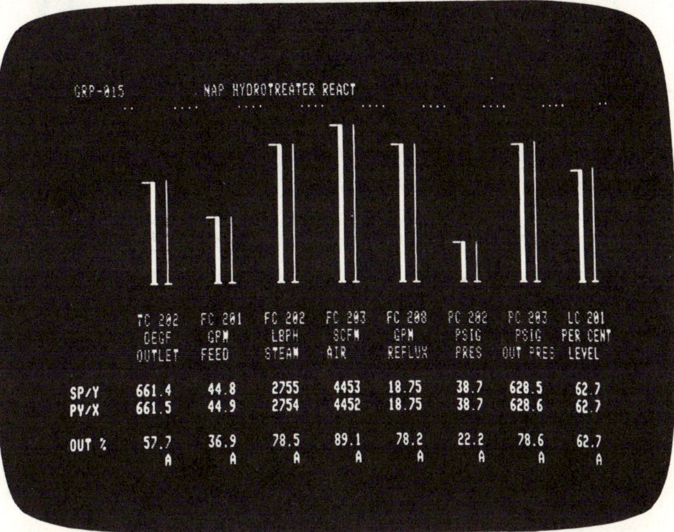

b. Group display

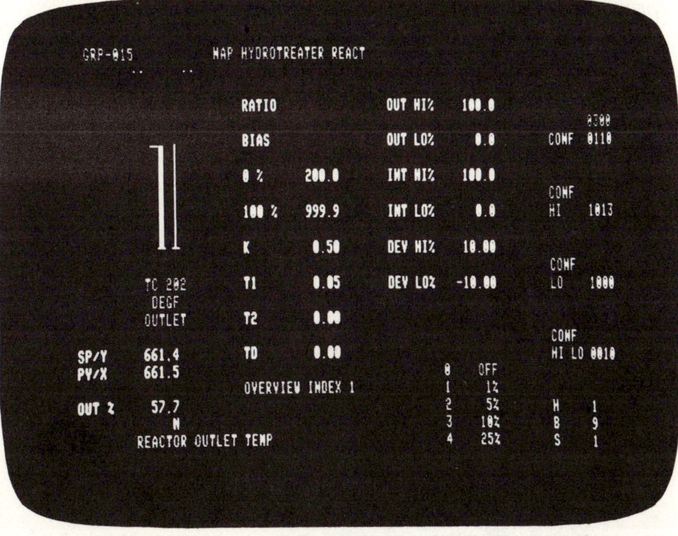

c. Detail display

Operator calls overview, group and detail levels on computer-based CRT display to control process Fig. 6

temperature switches, process-equipment contact closures, and the computer.

Having a maintainable system is almost as important as having a reliable one. Any hardware malfunction must be identified and fixed easily. The maintenance objective is to identify problem circuit cards, replace them, and get back online. With distributed control, each microprocessor-based device performs a set of self-diagnostics. Any malfunctions are summarized for the operator as to the device, its location, and particular problem. Malfunction identification can come from card-mounted LED (light-emitting diode) displays, internal microprocessor-program diagnostics, computer and stand-alone CRT displays. The spare-parts inventory for such a system is minimal because microprocessor controllers in the system are physically identical.

Comparative costs

Results of a study of control costs in a process-plant environment provide insight. The relative costs for an average loop were compared to plant size, measured in loops. (An average loop includes all indicating and recorder points, as well as feedback-control loops.)

There is a tendency for the cost per loop to decrease with plant size because in many types of hardware, a greater number of connections will be made in a single operation. In the study, two types of control installations were compared: conventional analog and a microprocessor design. Each type performs identical functions in the plant, and both use the same kind of analog panel-meter displays.

In smaller plants, the conventional-controller installation has an edge in cost when hardware alone is considered. In larger installations, microprocessor hardware becomes less costly. When the basic installed cost is considered, including wiring and mounting racks, the cost edge is diminished (wiring does not include field wiring or a data highway). When the total cost of installation—including engineering, documentation, startup costs and initial repairs—is considered, the microprocessor system has an economic advantage regardless of system size. Adding the cost of operating, maintaining and upgrading the control instruments over an assumed 10-year life, the microprocessor system shows a 10 to 20% advantage over conventional analog instruments.

It is possible to reduce the amount of control-room wiring involved with conventional systems. The example shown in Fig. 1 (having 1,400 process variables and 400 output signals) required 2,800 wires to the computer's I/O. Since the computer and microprocessors are digital devices, it is possible to connect them via a single redundant coaxial-cable-based highway system. This eliminates almost all wire connections to the computer, and also eliminates some computer I/O hardware.

Both of these improvements result in a cost saving. Once a process variable is connected to the terminal panel of the microprocessor controller, it becomes available to the analog displays via plug-in cables and to the computer via a data highway from the microprocessor memory.

An improved conventional system is possible by using microprocessor controllers, analog displays, operator-entry panel and communication data highway. The four initial-design requirements are better satisfied.

Operator interaction with control panel

Under computer control, the CRT control console is used. When the computer is down, panel-mounted analog displays take over. A greater level of centralization is possible if the CRT control console is independent of the computer (see stand-alone CRT in Fig. 3).

What capabilities should be included in the stand-alone CRT? First, let us consider the situation of an operator facing a conventional control panel from the center of the control room. The operator can see which annunciators are in alarm but cannot read the alarm description. He can observe the relative status of all control loops by noting the pointer position of the process variable relative to the setpoint. To see specific alarm descriptions or the values of the process variable and the exact setpoints, the operator must walk closer to the panel.

From a distance, the operator can see a large amount of general information but very little detail. At four feet from the panel, he can clearly see one entire panel section. This one section represents a process unit, and shows all control modes, process variables, setpoints, outputs, and the status of alarm points. The operator's field of view is narrowed to perhaps 10 to 20 controllers. As more-detailed information is seen, less general information is provided.

Ultimately, if the operator pulls an instrument from its case, all of the detailed information pertaining to that loop is seen. Now, the field of vision is narrowed to a single loop. Overview, group and detail levels of the man/machine interface displays are required. The CRT displays should emulate the familiar control-panel analog displays, and should be as familiar to the operator as possible.

The overview display provides the same indication as the conventional panel board at a distance (Fig. 6a). It consists of deviation displays showing process variables and their setpoints. Such a display can accommodate approximately 300 deviations. Each deviation should be independently scaled, using various sensitivity constants. Thus, for a loop where close control might be required, a maximum deviation of 1% would cause full-scale indication. Other, less-critical loops may tolerate a deviation of 30 to 50%. This approach uses the principle of "operation by exception" to identify rapidly those plant areas that require operator attention.

Once an area has been identified, a process-unit group display can be obtained (Fig. 6b). Both analog representations and digital values for eight loops are shown, as well as their tag numbers and descriptions. At this working level, the operator can change setpoints, modes and outputs.

Group displays must be easily configurable. Individual loops can be used as often as required, depending on what makes the most sense to the operator. Possibly, a temperature variable that is an integral part of a process unit may affect a downstream unit. It may be desirable to display this loop in more than one group. Any control loop can be called up individually as a detail display (Fig. 6c), in order to show pertinent information about that loop. Tuning constants, alarm settings, configuration information, and output limits are samples of such detail.

One standard CRT display can provide an enhanced operator interface in an improved conventional system. Three CRTs can provide a totally integrated system to allow up to 24 control loops to be simultaneously displayed. Since each CRT is identical to the others, any combination of overview, group and detail displays can be shown (Fig. 6). Since each CRT unit is independent, any two can malfunction—the operator still has complete control of the process from the remaining CRT.

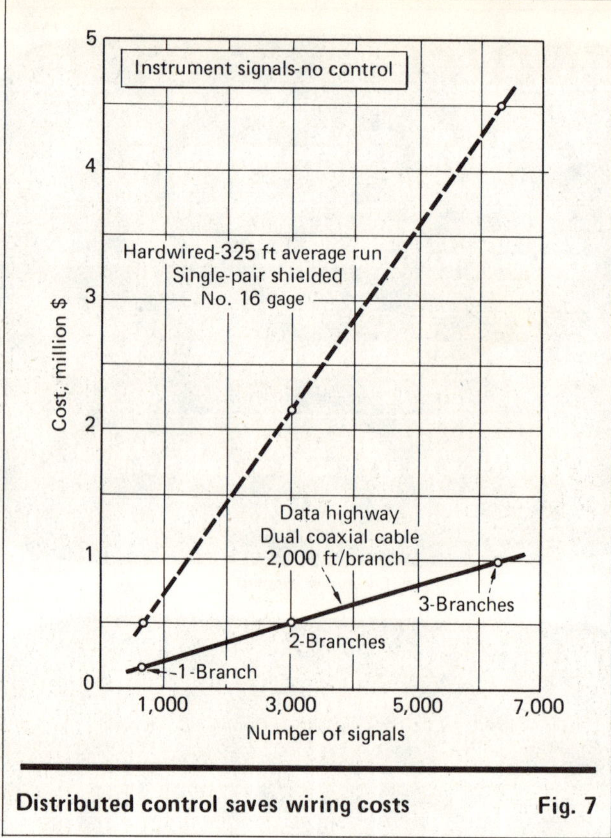

Distributed control saves wiring costs **Fig. 7**

Costs of data acquisition

Eliminating the need for analog displays means that microprocessor controllers can be located remotely in local process areas, and connected to the central control room via a data highway. These remote locations can be 5,000 to 10,000 ft from the central control room. Such distances are possible by using a serially based data-highway link. All field signals are terminated at local control rooms, and transmitted to the control center via the data highway. This results in substantial savings in wire installation (Fig. 7).

The author

David L. Williams is a regional marketing specialist for Honeywell Process Control Div., 1100 Virginia Drive, Fort Washington, PA 19034. He has been with Honeywell since 1973. Previously, he spent six years with Leeds & Northrup Co. as an application engineer for industrial systems. He has a B.S. in chemical engineering from Rutgers University and an M.B.A. in marketing from Temple University.

Distributed digital control—control technology breakthrough

The rapid development and widespread availability of single-chip microprocessors are bringing about an entirely new process control philosophy

Terrence K. McMahon, Market Research Consultant

☐ Distributed digital control (or, DiDC to distinguish it from the widely-used term: DDC, direct digital control) has taken root in industrial control practice, and is rapidly becoming "state of the art" in new plant construction. Moreover, DiDC represents a major breakthrough in process control technology—possibly the most important since the introduction of analog feedback control.

DiDC is a control philosophy rather than a control methodology. While DiDC uses DDC for process-variable regulation, DDC has been and continues to be implemented on a centralized basis, as well as within the context of DiDC systems.

DDC is, of course, the technique wherein a stored-program digital controller (frequently a general purpose digital computer) directly manipulates the ultimate control mechanism (such as a control valve) based on pertinent process-variable measurements and a stored algorithm. DDC can replicate the action of the traditional three-mode analog controller and/or step significantly beyond the capabilities of analog logic to:

- Event-based discontinuous control actions.
- Process-model-based feedforward control actions.

Many other schemes are possible. The DDC technique (its characteristics, advantages, economics, and so on) has been fully examined in the literature.

No concise, definitive consensus has been reached on the exact definition of DiDC. Nevertheless, most observers would agree that it encompasses:

- DDC process-variable regulation.
- Multiple (interconnected) stored-program digital controllers, each handling a limited number of control (variable) loops.
- Electronic process-operator display stations that are typically color cathode-ray-tube (CRT) units.
- Digital data communication between the various DiDC system components.
- Large-scale use of component-redundancy to ensure a high level of overall reliability.

Each of these, in turn, encompasses a multiplicity of more detailed characteristics.

Developments in the process industries

The earliest attempts at controlling industrial processes were manual, and these survive to the present. Analog (feedback) control represented the first attempt at automatic control in the chemical process industries. This technology developed over an extended period, and resulted in the availability of commercial control equipment, embodying feedback technology. Analog control is inherently distributed—one controller per controlled function. Data communication between controllers, however, is difficult, and the process-operator displays are relatively imprecise.

Despite these limitations, analog control enjoyed a long period of supremacy. During this period, successive improvements were realized in going from pneumatic-analog to electronic-analog to miniature-electronic and split-architecture controllers. Control-algorithm complexity advanced from proportional-only to PID (proportional-integral-derivative/3-mode) control to dead-time compensation, cascaded-control (one controller controlling the setpoint of another), and many other technical refinements. Analog controller reliability and cost per control loop were the two factors that turned-back the first digital assault on the analog bastion.

Digital computers entered the process control scene following the pioneering application at Texaco's Port Arthur (Tex.) refinery in 1958, when computer control (then the current term) was viewed as the "ultimate in process control." Actually, computers accomplished many things in industrial processes, but control was not among them—if, by control, actual variable-regulation is understood. These supervisory control systems used analog controllers as their frontline regulators.

Development of the DDC technique (involving the computer in primary control regulation) began as early as the 1960s when Monsanto installed the first DDC-system on a test basis at Chocolate Bayou. Despite a number of highly successful installations, DDC never really challenged analog for the primary controller market because system cost, in most instances, became unjustifiable in order to achieve acceptable levels of reliability.

Process control computers became minicomputers starting in the mid-1960s, and the basic economics of the analog vs. digital tradeoff became more favorable to digital. But, the tradeoff was not favorable enough to generate widespread enthusiasm for DDC. By the early 1970s, single-board minicomputers, using medium-scale

Originally published October 15, 1979.

Distributed digital-control systems **Table I**

Manufacturer	Product	Loops/Controller
ACCO/Bristol	UCS-3000/08A	40
Beckman Instruments	MV 8000	16
EMC Controls	Emcon-D	32
Fischer & Porter	DCI-4000	16
Foxboro	Spectrum/Microspec	40
Honeywell	TDC-2000	8
Powell Industries	S-200/Micon	8
Reliance Electric	UDAC	16
Robertshaw Controls	DCS 1000	1
Sybron/Taylor	MOD III/SMC	1
Texas Instruments	PM 550	8
Westinghouse	Q-Line	25
Xomox	System 1100/DCE	1

circuit integration (MSI), were available at costs that were beginning to look attractive in terms of analog controller replacements. At the same time, the digital microprocessor, using large-scale circuit integration (LSI), burst on the scene. These microprocessors dramatically altered the basic economics of process control systems by making digital logic with its inherent advantages in reliability, precision and flexibility, cost-competitive with analog logic in the context of the relatively complex type of regulating the control actions required in the fluid process industries.

Emergence of DiDC

The announcement by Intel in 1971 of the 4004 (4-bit, P-MOS, single-chip microprocessor) went virtually unnoticed by process control suppliers and users. Intel's subsequent introduction of an 8-bit version (8008) attracted only slightly more interest. However, the announcement of the Intel 8080, in early 1974, offered an order-of-magnitude increase in processing capability over the 8008. Suddenly, the process control community was scrambling to understand this new technology. Intel, of course, was not alone in developing microprocessors, and National Semiconductor, Motorola, Texas Instruments, and many others were also manufacturing these products.

Honeywell had started to investigate the concept of single-loop digital control (SLDC) as early as 1969. The appearance of the microprocessors provided the essential ingredient to make this concept viable. The introduction of Honeywell's TDC-2000, embodying a shared (8-loop) controller package, was of special significance because it represented a commitment by a process-control equipment manufacturer to the DiDC philosophy involving a whole line of specific products. Several other manufacturers, including Bristol and Process Systems Inc., had announced microprocessor-based digital controllers prior to the introduction of TDC-2000. These products, however, were discrete, stand-alone units, not DiDC systems.

Field testing of the TDC-2000 started in 1975 at three user locations. At Imperial Oil's Sarnia, Ontario, refinery, the test involved 16 control loops (pressure, temperature, flow and level), of which 8 were cascade loops in the refinery's lube-oil plant. This test was highly successful and Imperial's parent company, Exxon, has subsequently installed such systems in chemical operations in at least three U.S. locations. By the end of 1978, Honeywell had delivered several hundred TDC systems involving more than 20,000 control loops. Major applications include petrochemical plants, petroleum refineries and pulp mills.

Product trends for DiDC

Table I identifies 13 manufacturers of DiDC products. This list is not claimed to be exclusive. In a recent survey (published in *Instrumentation Technology*), 26 manufacturers of digital controllers were identified, and several of the companies in Table I were not included. Of course, not all digital controllers qualify as DiDC products. In fact, Texas Instruments' PM 550 is not now a distributed-digital product. It was included here to indicate one company's involvement in such products. Manufacturers of process-control equipment known to be developing DiDC products include: Foxboro, Fisher Controls, General Signal/Leeds & Northrup, Babcock & Wilcox/Bailey Controls, and others.

The Honeywell TDC-2000 is the standard for many large users. While this system has been generally well-received, some significant objections have been raised to its design—indicating that there is room in this marketplace for a variety of approaches and suppliers.

Based on a 3- to 4-yr life cycle, the TDC-2000 may be considered a mature product. This is not untypical in high-growth, technology-based products such as DiDC. The next generation TDC-product is believed to be based on a 16-loop shared-controller, and its physical/software design will certainly benefit from the cumulative experience with TDC-2000.

Among the established process-control manufacturers, both Fischer & Porter and Beckman Instruments have adopted DiDC designs that are based on 16-loop shared controllers. The larger controller-nesting allows very complex, cascaded loops and control-loop interactions to be accommodated more effectively. This approach recognizes a continuing decline in user nervousness over the shared-controller concept due to the extensive use of redundancy. It also indicates user awareness that total redundancy is often an unnecessary extravagance, based on emotional perceptions. A careful analysis can substantially narrow redundancy requirements. Beckman's overall MV 8000 design permits the inclusion of its 8800-line (analog control/digital instrumentation), as well as the MVCU (multivariable control unit having 16-loops) shared-digital controllers.

EMC Controls' Emcon-D was probably the strongest competitor to TDC-systems. Both ACCO/Bristol and Powell Industries (Process Systems Inc.) have upgraded their initial dedicated-DDC equipment to DiDC systems. Xomox's System 1100 involves a daisy-chain arrangement of up to 32 digital-control elements (DCE). The DCE mounts directly on the control-valve actuator. Robertshaw's DCS DiDC system is based on its DCM-1000 single-loop digital controller.

Sybron/Taylor's DiDC activities involve combining its SMC (single-loop microprocessor controller) with the MOD III process monitoring and communications (shared display) system. As originally announced,

MOD III interfaced to standard Quick-Scan analog controllers for process-variable regulation. Foxboro's recently announced Spectrum family, including the Microspec digital controller, the Videospec shared-display system, the Foxnet (digital) communications link, and other components, represents another comprehensive attempt to serve the DiDC market.

DiDC products are still rapidly evolving. As of mid-1979, two principal designs are apparent. One design variant is the shared (digital) controller, both 8-loop and 16-loop packages, used by Honeywell, Fischer & Porter, Beckman Instruments, ACCO/Bristol, Powell Industries, EMC Controls (32-loop system), and Foxboro (40-loop system). The other variant is the dedicated (single-loop digital) controller, used by Sybron/Taylor and Robertshaw. Beckman is also straddling the fence by allowing single-loop analog controllers to be incorporated into its MV 8000 DiDC system.

DiDC in practice

At its annual computer conference,* the National Petroleum Refiners Assn. (NPRA) held two sessions on digital instrumentation (DI). DI was defined as a functional replacement for the three-mode analog controller, which operates in the discrete digital realm and which encompasses both DiDC and centralized DDC. Most of the discussion centered on DiDC. The principal areas of interest, relative to DI, were: (1) systems design, (2) operator interface and acceptance, (3) training and organization, and (4) reliability and maintenance.

The future for DiDC

DiDC technology will have and is having a major impact on the direct control-related aspects of operations in the chemical process industries (CPI). More importantly in a larger time frame, it could affect the fundamental design of CPI processes and even chemical engineering education.

The impact of DiDC on the control-related aspects of CPI processes is straightforward. Centralized controller-racks, drawers full of screwdrivers for control-loop tuning, and miles of twisted-pair cabling and associated shielding are gone, replaced by remote controller-nests and digital coaxial cabling.

The control room itself can be scaled-down significantly. With shared-display, vast arrays of indicating/controller panels, strip-chart recorders, gages, dials, process flow diagrams, alarms, and so forth, can be condensed into an integrated console. The console will incorporate CRT displays, display-panels and hard-copy printers.

It is axiomatic that new process-designs do not spring forth from advances in instrumentation technology alone, but rather, they grow from the sum of a complex economic/technological matrix. Economic factors include perceived product value, raw-material availability and/or price, capital availability and/or price, and others. On the technology side, however, instrumentation must rank as one of the most dynamic and consequently influential factors.

A repeating theme in process control has always been that available control sophistication inevitably exceeded that required by the quality and quantity of the basic process measurements. Undoubtedly, this has been true with respect to physical process-variables (such as temperature, pressure, level and flow).

*"1978 Computer Conference: Question and Answer Session on Process Control," Anaheim, Calif., Oct. 29-Nov. 1, 1978, National Petroleum Refiners Assn., Washington, D.C.

Analytical process-variable (stream-composition) measurements, however, have been used sparingly in direct process control. The reasons for this have been sampling uncertainties; instrument reliability, response and precision; as well as the relative complexity of the relationship between the composition measurement and the manipulable parameters of the process.

Newer process analytical technologies are eroding the first two of these obstacles (e.g., *in-situ* analyzers, fast-response multicomponent analyzers). The application of DiDC techniques in combination with these advanced process-analyzer technologies could be an important new process-design tool in the future.

Chemical engineering education has been design intensive since its origin. Instrumentation has clearly been the stepchild of this process and equipment design-orientation for a number of reasons whose enumeration here is beyond the scope of this article. Suffice to say, feedback control, analysis and design went largely by default to electrical and mechanical engineering departments. Thus, chemical engineering has been in many ways divorced from the dynamic characteristics and operational control of the very processes it sired.

Supervisory digital-computer control has involved chemical engineering on the periphery of process control because an understanding of processes is vital to the development of the mathematical relationships on which these systems are based. DiDC extends this opportunity for the exercise of fundamental chemical-engineering analysis to the ultimate control action.

However, it is up to chemical engineering education to bring the practicing chemical engineer a better understanding of process dynamics. The web of mystique woven around this subject by electrical engineers and mathematicians with their Bode diagrams, Nyquist plots, Lyapanov functions, Bellmanian optimality criteria, and the like, can then be exposed to the piercing light of pragmatic (chemical) engineering analysis.

References

1. Selecting Digital Control Systems, *Instrum. Technol.,* May 1979, pp. 28–37.
2. Savas, E. S., "Computer Control of Industrial Processes," McGraw-Hill, New York, 1965.
3. Transcript of Question and Answer Session from 1978 Computer Conference, National Petroleum Refiners Assn., Washington, D.C.

The author

Terrence K. McMahon is a consultant with McMahon Technology Associates, 60 East 42 St., New York, NY 10017. He has been an independent consultant since 1968. Previously, he was employed for a period of seven years by Realtime Systems, Inc. and IBM Corp. His application experience includes industrial automation and control, laboratory automation, and market research studies. He has a B.S. in chemical engineering from Massachusetts Institute of Technology and the M.E. and D.Eng. in chemical engineering from Yale University, and is a member of AIChE, Instrument Soc. of America, and Sigma Xi.

Section XIV
CONTROLLING BATCH PROCESSES

Automatic Batch Control

More highly automated batch process control can raise throughput, increase product quality and cut manpower. Control devices range from simple cam mechanisms for cycle timing to digital computers.

F. G. ENGSTROM and D. A. HAMBLETON, The Foxboro Company

In batch processing, ground rules are needed to establish the automation level and economic justification. The system designer must have a clear understanding of the problem and evaluate all hardware technology from mechanical sequencers to digital computers. Definition of the problem begins by describing the cycle and identifying all the variables for different products.

One technique for design development is an open discussion on current operating problems. This helps to construct a "manual" for process equipment, operation, products, scheduling and inventory procedures. Contributors should include operations personnel, process engineers, instrument personnel, operators, scheduling department, accounting and plant management.

Most batch processes are cyclic, passing through a series of phases: filling a vessel, processing its contents and transferring products to a second vessel. The control system for these operations includes feedback control loops and the necessary sequential logic to vary events step-by-step as a function of product recipes. Many control systems organize variable operations into an automatic duplicating sequence.

Process Applications

The most common examples of batch processing are found in the chemical process industries for production of paints, plastics, adhesives, pharmaceuticals and fine chemicals. Processes vary from a single reactor making multiple products to multiple units dedicated to a single product such as polyvinyl chloride. A typical batch reactor with a cascade temperature control system is represented in Fig. 1.

Other functions performed by a sequential control system include recipe selection, charging, discharging and cleaning. There are numerous examples of catalysts that must be regenerated in situ. These cycles normally proceed through a carefully fixed sequence, possibly with analog overrides to control regeneration temperatures and gas composition. The constant supervision provided by a programmed control system can result in significant time savings and protection of the catalyst.

Atmospheric and pressure type dye becks are used in the textile industry for dying cottons, synthetics and wools. Programmed control systems have been very successful in automating this operation to produce more consistent color control. Another common application is on dryers that remove moisture from gases prior to processing. Programmed cycles can be operated on a fixed cycle or activated by the dry gas quality to avoid off-spec gas.

Programmed Control Advantages

Increased throughput from a batch operation is the primary justification for programmed control. Batch operations require many operator decisions that may occur simultaneously. Automating these decisions saves time and frees the operator for more critical duties. With program controllers, it's possible to rapidly load new programs into the system, and control processing variables to approach maximum production rates.

Improved quality control is another advantage. Repeatability not possible with manual operation results in consistent quality from cycle to cycle, especially for the charging operation. The improvement is evident in fewer spoiled batches and more efficient downstream processing.

Each higher level of programmed control automation increases manpower savings in plant operations. This is

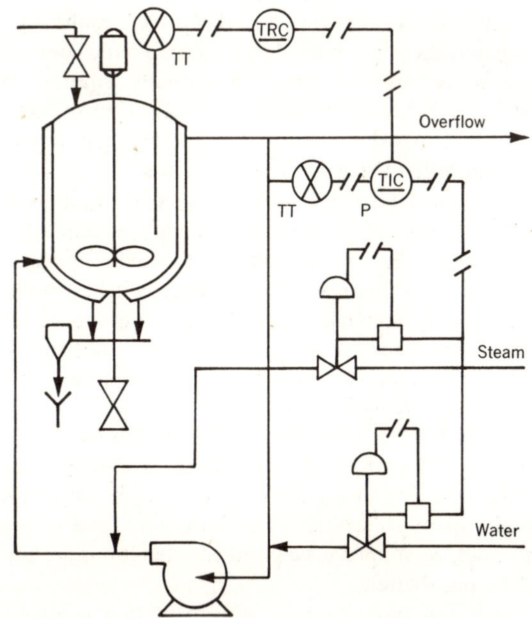

CASCADE temperature control on batch reactor—Fig. 1

Originally published September 11, 1972.

CONTROLS for multiple operations in one location—Fig. 2

particularly true with a large number of reactors and recipes. Fig. 2 illustrates how control of multiple operations can be concentrated in one location with improved operator productivity.

Levels of Control

From the very simple cam-set instrument to the digital computer, there are many levels of automatic control in batch processing. Two types are normally required. The first is time scheduled control (the analog control portion of the system) where the set point is adjusted at a rate to a controlled value for a timed period.

The second type of control is the sequential control logic and its memory capacity relating the cyclic operations in two states, "on" or "off". Individual contacts such as pushbuttons, limit switches, pressure switches, etc., are monitored as input conditions. Logic outputs act as command instructions to open and close valves, start and stop pumps, agitators, etc., and to operate the visual display units indicating cycle status.

In the cam set controller, the control point is automatically moved by a cam follower to a schedule precut on the cam. The automatic sequential logic is limited to simple functions such as cycle start, hold, end of cycle and reset, where process equipment operations are manual by pushbutton.

Motorized set-point rate adjustment to a control set point is made by a synchronous motor driven index. The normal speed of the motor drives the index at the fastest rate, while slower speeds are obtained by interrupting power to the motor with a percentage timer. The output of a deviation contact stops the index at the predetermined set point. Extension of this controlled point for a holding period is accomplished by a preset timer.

Rates of rise and fall may be operated by the same method, but only in the direction the motor is driven. All functions of time and rate are dialed and preset prior to starting the cycle. Limited automatic sequential operations are run from the end point functions such as the end of rise operation or timed period. Typical programs for motorized control profiles and hold times are shown in Fig. 3.

Multiproduct applications frequently require broad variability to meet the needs of input recipe changes. Some examples are parameter changes such as variable set point, hold time and ingredient quantities; procedure changes such as modification in the order of steps; and interlock changes, the arrangement of contact input pattern as used to implement the sequential steps of the program.

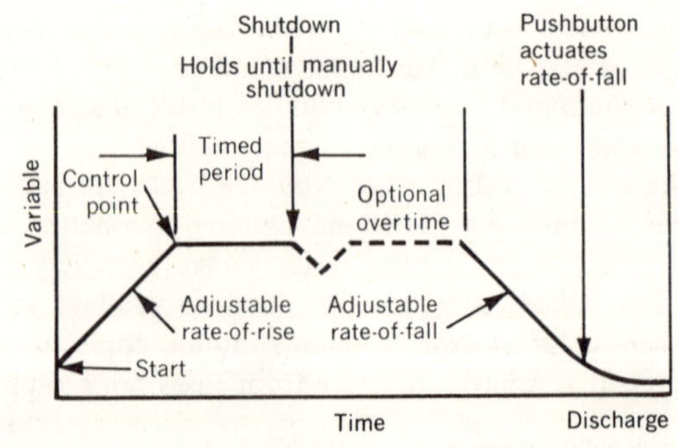

PROGRAMS for motorized control profiles—Fig. 3

Most Systems Are Mechanical

To date the majority of programmed control systems are mechanical. That is, their memory logic relies on switch actions. Relays, timers and mechanical drums are multicontact forms of a switch. Relays can be arranged in an electrical circuit for inhibit or checking actions and to provide interlock functions. Timers and drum programmers provide, from a single input action, multiple prearranged contact outputs.

Other contact forms provide variable memory such as a card or tape reader. Functionally, there is similarity between drums, cards, card-drum combinations or tapes because these devices are all types of switch matrixes. The difference in each's capability lies in the programming method and its variability. Selection of program input devices for a system must consider the simplicity of programming method, memory capacity and ease of maintenance. The scope and size of mechanical systems is often fixed by the dedication of wiring and the programming of the cycle.

Multiproduct control systems require variable forms of memory and program storage. Methods include limited change drum sequences where mechanical positioning of pins on a drum energizes electrical logic circuits to sequence the operations, based on electrical conductivity or photoelectric methods. In the card or tape reader device, procedures and end point parameters can be coded by punching holes with a simple hand punch. On the tape the related sequence of stepped operations may be reorganized. The card drum or tape arrangements do not allow for modification of specific interlock conditions. These may be made with wiring changes or by including a matric board with diode pins inserted for desired operations.

Mechanical systems are limited not only by their wiring and fixed capacity, but in their inability to easily share memory. Two processes identical in operation and end product must be independently started and stopped. Also, because there are many types of program memories, two applications in the same plant may require totally different program control systems. One possible solution lies in the technology of modular design and variable memory capacity to do switch logic and time based sequencing.

Programmed Logic Systems

The programmable logic system is a shared sequential programmer. It uses a magnetic core memory to store control programs, which are made up of lines of relay type logic, numerical counting, timing, comparing and storage functions. Sequential logic control programs are designed with the same format as relays. The common symbol language eliminates special training.

The electrical elementary diagram has been the traditional format to define electrical sequence control. Relay sequence, however, can be easily shown in a circuit resembling a ladder. Each line or rung between the two vertical guides controls power. Contact elements within the rungs control on/off outputs. If the outputs represent the action of a line, the logic electrical sequence can be arranged in an order of events occurring. The ladder, thus, is a graphic presentation of relay logic.

Fig. 4 shows a programmable sequential controller system for operation of two batch reactors. The system will charge measured quantities of reactants, start pumps and agitators, open and close valves. Recipes are inserted using a card reader system.

Plants with a multiplicity of recipes and reactors, whether single or multiple product, probably can justify a digital computer for batch control. A number of computer systems are currently in operation providing continuous process supervision, from scheduling to process control, with increased throughputs and greatly improved product quality and consistency.

Control Subsystems

Many batch processes require precise amounts of reactants added to the vessel. One way to accomplish this is a premeasured system in which materials are weighed or measured manually and placed into hoppers. When the operator presses a button, the program control is signaled that the system is ready to go. If there is no signal the cycle will not proceed to the next step. This procedure is frequently used in dye houses.

Addition of solids or liquids can be completely automated by setting registers on totalizers operating from turbine meters, positive displacement meters, weigh belts or differential pressure transmitters with integrators. The totalizers can be set from the program, providing the operator with utmost flexibility. He can change recipes merely by changing the program input with card, tape or thumbwheel.

SEQUENTIAL controller for two batch reactors—Fig. 4

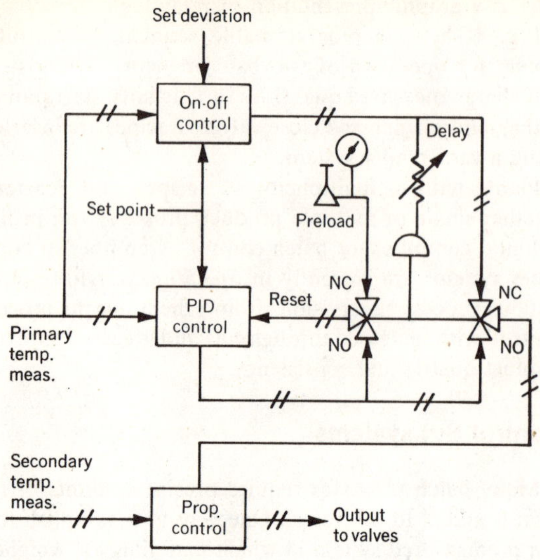

DUAL MODE control system—Fig. 5

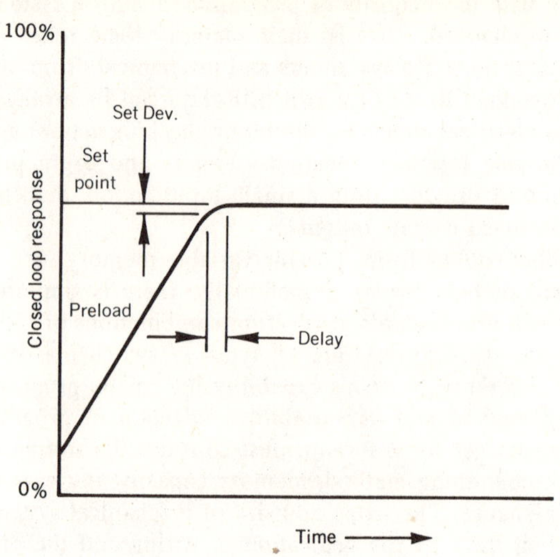

RESPONSES in dual mode control system—Fig. 6

Program control systems can implement most types of analog subsystems for controlling heating and cooling. The basic function of the control system is to provide the set point for temperature control, either as a fixed analog value or programmed in a dynamic manner.

For maximum production in a process making repeat batches, the answer is a special form of cascade called a "dual mode system." This is essentially a two function primary controller in a cascade loop. It has an on/off operation during the heating period, followed by three mode control at the batching temperature. Fig. 5 shows details of the dual-mode system and Fig. 6 illustrates the type of control response provided by the system.

Some plastics manufacturers control quality by totalizing the heat of reaction over a period of time. Totalizing is handled by analog equipment in the case of hardwired, solid state or tape type systems, or digitally with the programmable controller and digital computer.

Some batch reactions also require pressure controls. Depending on the type of system, it is possible to control pressure by either generating a fixed set point or by generating a set point on some type of programmed profile through the rise, hold and fall periods, similar to temperature control.

Display techniques are determined by operator needs. Input variables that the operator may need to change include charge volume, ramp rates, hold times and set points. A manual thumbwheel is the simplest type of input, allowing the operator to set a particular value and enter with a designated function button. The next order of complexity is a card reader, which permits the operator to simultaneously and automatically enter those recipe values necessary to change product volume and specifications.

A tape input program can be very flexible in changing program sequence and set points. With programmable type controllers, it is also possible to have input provided by a computer operating in the supervisory mode and setting recipes, set points and programs.

Outputs to the display system consist of several different devices: numeric indications of totals and set points; lights showing action and status of various portions of the program; and analog recorders indicating the status of set points and measurements continuously throughout the process cycle.

Bibliography

1. Engstrom, Frank G., How to Control Batch Reactors, *Adhesive Age*, May, 1970.
2. Poisson, Norman A., "Programmable Controllers—A New Concept for Batch Process Control," Instrument Society of America, 1970.
3. Shinskey, F. G., and Weinstein, J. L., "A Dual Mode Control System for a Batch Exothermic Reactor," Instrument Society of America, 1965.
4. Amrehn, H., Computers Direct Batching In 80-Reactor PVC Plant, *Instrumentation Technology*, 1967.

Meet the Authors

Frank G. Engstrom is manager of chemical industry sales for The Foxboro Co., Foxboro, MA 02035. Before joining Foxboro in 1965 he was with the Standard Oil Co. of Ohio. Mr. Engstrom was graduated from Michigan State University in 1943 with a B.S. degree in chemical engineering. He is a registered professional engineer in Ohio and Massachusetts, and a member of AIChE and the Instrument Society of America.

Dana A. Hambleton is manager of product systems at Foxboro. He joined the company as a sales engineer in 1959 after graduation from the University of Maine with a B.S. degree in mechanical engineering. Mr. Hambleton is a member of TAPPI and has authored several papers on instrumentation for the pulp and paper industry.

Computer control of batch processes

Batch processes inherently consist of frequent startups and shutdowns, and little steady-state running. Consequently, they are ideally suited to control by computers that can handle the complexity of the operations.

Marcel T. Brodmann, *Economation AG (Switzerland), and* ***Cecil L. Smith,*** *Economation Inc.*

☐ The trend in the chemical process industries is toward continuous processes and away from batch processes. For example, the traditional manufacturing process for TNT was batch, although new TNT plants are continuous [*1*]. Even so, some segments of the industry, such as pharmaceutical manufacture, primarily run batch-process plants.

Batch processes have always been considered backward. They are difficult to operate, offering many opportunities for error. (Murphy's Law* has been experimentally verified in such plants on numerous occasions.) Errors usually result in loss of the batch, although they may even result in fire or explosion.

However, process-control computers are doing much to change the image of batch processes; indeed, they may eventually lead to a fundamental reassessment of the role of batch versus continuous processes.

Recent articles have described computer applications in PVC plants [*2*], and a lube-oil-additives plant [*3*], and other systems are being installed. However, each of these articles treats only a single type of batch process. Unfortunately, such processes differ widely from one to another, and the roles of the computer differ. This article considers several categories of batch processes and discusses the role of the computer in each.

Typical configurations in batch plants

Large variations exist among the equipment configurations in batch plants. However, certain arrangements are relatively common, and we shall classify them into four categories—do not be surprised if a plant does not fit neatly into any one category.

1. *Single reactor, single product*—This is undoubtedly the simplest of batch plants, but probably the least common. Nevertheless, computer control offers several benefits, including improved accuracy in the charging of reagents, reduction in lost batches caused by operator errors, minimization of batch cycle-time, etc.

2. *Multiple reactors, single product*—The benefits are similar to those in the previous case, plus the advantage of solving problems resulting from interactions between the two or more reactors.

3. *Single reactor, multiproduct*—Using the same reactor to produce several products greatly complicates the operator's job. To produce a given product, he must follow a recipe that specifies the amount of each reagent to be charged, the setpoints (for temperature, pressure, etc.), the time required for each phase, etc. A momentary lapse of attention could easily result in the reactor being charged to produce Product A, but using the temperature programmed for Product B instead. The result could be a low-quality product, an unsalable product, a dangerous condition, or some other undesirable outcome.

4. *Multiple reactors, multiproduct*—This is, of course, the most complicated case of all, and it usually offers the greatest incentive for improved control; however, it also usually requires the most effort in implementing a computer-control system.

Within the above four categories, the plant may be configured in one of the following: single line; parallel lines; parallel lines with shared equipment.

Since each line is usually operating at a different stage in the batch cycle, parallel lines always present a difficult problem to the operator. The situation is further complicated in a multiproduct plant since, there, each line may be producing a different product. When equipment such as the reagent charging system is shared among the various lines, scheduling of the batches must also be considered in order to reduce idle time in each of the lines.

Incentives for improved control

Although some potential benefits were discussed in the previous section, we wish to repeat these and expand our list to include some others:

More-consistent process performance—A factor leading to better product reproducibility is closer control of reagent charging. This often involves replacing existing charging systems with more-accurate ones that use turbine meters for flow measurement, load cells on weigh-hoppers from which solids are fed, and similar improvements. In some reactors, brief periods occur when the temperature- or pressure-control systems allow these variables to go outside of the normal ranges. This often leads to reactions that produce byproducts that

* Murphy's Law—Anything that can go wrong, sooner or later will go wrong.

Originally published September 13, 1976.

degrade the main product's color, clarity, or other property. As a result, the product may only be salable at a reduced price, if at all. Improvement in the basic control system can eliminate these situations.

Better equipment utilization—One way to achieve this is to shorten the batch cycle time. This may be done by:

- Minimizing the time required to bring the reactor up to temperature.
- Charging reagents simultaneously.
- Controlling temperatures closer to the high limit (which results in a faster rate of reaction).

Another way to obtain better equipment utilization is to avoid idle time on a reactor. For example:

1. To reduce the risk of operator errors during the charging phase, many plant managers stipulate that charging must not stretch over two shifts. Thus, if charging normally requires 1½ h, and the batch cycle arrives at the charging phase 1 h before shift-change time, the reactor will sit idle for 1 h until the next shift begins. Allowing the computer to supervise the charging sequence eliminates the need for such delay.

2. In many plants with multiple reactors, only one charging system is provided. Thus, if Reactor A is ready to be charged before the charging of Reactor B is completed, Reactor A sits idle until Reactor B is charged. Using a computer to schedule the reactors can reduce such idle time.

Reduced raw-materials and energy consumption—Among the factors affecting raw-materials consumption in a batch process are:

Reactant ratios: In stoichiometric reactions, component ratios should be maintained as accurately as possible. Any excess in one reactant is mostly lost as waste material, e.g., when neutralizing a reaction mixture.

Reaction parameters: Controlling certain parameters, such as temperature, pressure, pH, etc., as close to the desired value as possible often reduces the quantity of byproducts formed in the reaction. In batch processes, improvements can also be obtained by attaining the desired reaction conditions as quickly as possible.

Losses in physical separations: Batch distillation and crystallization are often used to isolate a desired product, remove byproducts, or reclaim solvents from the reaction mixture.

In the above situations, a computer can help to obtain improved control performance over the full range of conditions during the batch cycle, resulting in substantial savings in raw-material consumption.

Energy savings often occur merely as the result of tightening the batch cycle or from reduced amounts of recycled or low-quality material. However, batch processes also offer specific opportunities for reducing energy consumption. An example is the refluxing operation often used to remove a reaction component or to maintain the reaction mixture at a desired temperature. With a computer, flexible control strategies can be implemented—e.g., using the heat removed in the reflux condenser as the feedback signal to a control loop driving the heat-jacket steam valve.

Safety—If large errors are made in the amount or nature of the reagents charged, a potentially hazardous situation exists. Using a computer to supervise the charging of reagents results in fewer errors than are experienced when the operator does it. When such errors could lead to hazardous mixes of chemicals, the computer-supervised system is safer. By checking levels in tanks, and making other secondary measurements, the computer can also perform "reasonableness tests" on process instrumentation such as flowmeters, and thus can detect at least those gross errors that usually must occur for hazardous situations to exist.

There are other ways in which a computer can improve safety. In exothermic reactions, the heat removed from the system can be calculated from temperature- and flow-measurements in the reactor-cooling and reflux-condenser systems. Changes in this heat flux can be detected much faster than corresponding changes in reactor temperature. Thus, the heat flux can be monitored during critical reaction phases and used to trigger some emergency action when safe values are exceeded. The ability of the computer to store and execute emergency programs tailored to specific hazardous situations is of great value.

Process analysis

How does one determine whether his process can be improved by computer control? The answer is a process analysis study whose objective is to resolve the following four questions:

1. What benefits can be achieved?
2. What must be done to obtain these benefits?
3. What are the hardware requirements?
4. What are the software requirements?

To get a meaningful answer to these questions, it is necessary to consider economic as well as technical factors. Among the economic factors, the market situation for a product and its raw materials is of overwhelming importance. Stout [6] distinguishes the following three basic cases, in order of decreasing profit potential:

1. Limited by capacity.
2. Limited by raw materials.
3. Limited by the market.

Obviously, the market situation determines not only the size but also the type of benefits one should look for in the process analysis. For instance, better use of equipment is of little economic value when no raw materials are available, or when the market is already saturated.

Another useful kind of economic information is the cost breakdown for a particular product. Typically, raw materials are responsible for the bulk of production costs, while personnel costs (though higher in batch-process than in continuous plants) trail far behind. The following percentages from actual studies, including a nonautomated batch plant, indicate a possible range:

Product	A	B	C
Raw materials, %	82	96	74
Energy, %	11	3	4
Personnel, %	7	1	22

In the majority of the cases, it seems to be good process-analysis strategy to look for increased production and reduced raw-materials and energy consumption.

Let us illustrate a process analysis with the following example:

In Fig. 1, the sequence of events in the startup of the reactor is:

1. Charge the prescribed amounts of all reagents except *A*.

2. Bring the reactor up to the temperature setpoint, say 95°C.

3. Add Reagent *A* while holding the temperature at 95°C. During this addition, cooling is required (since an exothermic reaction occurs as *A* is added). To keep things simple, we shall ignore the rest of the cycle.

Suppose our plant is in the fortunate situation where sales exceed the capacity of the current plant. We have two options. First, we can expand our plant by adding one or more reactors. Alternatively, we can increase production from our existing plant.

One way to increase production is to decrease the batch cycle time. In the conventional system, we feed Reactant *A* at a constant rate until we have added the desired amount. We might first ask whether we can increase the feedrate of Reactant *A*, and thereby reduce the time required to add it. To find out, we record the signal to the cooling-water valve during one of the normal runs. If the value approaches the fully open condition at any time during the run, then we cannot feed *A* at a higher constant rate. If the valve does not approach fully open, then we can increase *A*'s feedrate, and estimate the maximum rate from the additional capacity left in the cooling system.

If spare capacity is available during the initial addition of *A* but not later, we might try to feed *A* initially at a higher rate, and reduce it later, either by programmed step-decreases in the feedrate or by a continuous decrease. This, however, will require hardware more sophisticated than that now available.

An alternative would be to adjust the feedrate continuously so as to maintain the cooling-water valve at, say, 90% open. If an air-to-close valve is used, this would occur when the output from the TRC is 10%. Using conventional equipment, we could monitor the output from the TRC, make this the feedback variable to a PI controller, and let this controller adjust the setpoint of the flow controller for Reactant *A*. This scheme is illustrated in Fig. 2.

Since we are not now feeding *A* at a constant rate, we need a flow totalizer and a comparator to determine when the desired amount has been added. The output of this comparator can be used to close the flow-control valve. If this valve is an air-to-open valve, then opening a solenoid vent-valve on the air signal to the flow-control valve would cause this valve to close. This system is illustrated in Fig. 2.

Glossary of abbreviations

CRT	Cathode-ray tube
DDC	Direct digital control
FC	Flow controller
PC	Pressure controller
PID	Proportional, integral, derivative control
TI	Temperature indicator
TRC	Temperature recorder-controller

Besides being more complex than the control system in Fig. 1, the system in Fig. 2 suffers several other disadvantages. First, if the amount of *A* to be added varies with the recipe, then the operator must change the setting on the comparator that specifies how much *A* is to be added. If he makes an error, the batch will probably be lost.

Second, if the system works properly, the cooling-water valve would be operated at 90% open all of the time, leaving some unused cooling capacity.

A very simple way to use all of the capacity in the cooling system is illustrated in Fig. 3. The cooling-water valve is now set wide open, and the temperature is controlled by the rate of addition of *A*.

When the specified amount of *A* has been added, the solenoid valve on the air supply is opened, which (as in Fig. 2) causes the flow-control valve for Reagent *A* to close. Unfortunately, this leaves us without any temperature control in the reactor (which is unacceptable). Somehow, at this time we must switch over to temperature control, using the cooling-water flowrate as the manipulated variable. Although such a switch could be attained with conventional hardware, it is far simpler when a digital system is used. The sequence-control logic for this phase of the reaction sequence is readily implemented in a computer system.

As for how much the reaction cycle time can be reduced, we can again estimate this from the spare capacity in the cooling system. Knowing the current total-batch cycle time, we can then compute how much time is saved during an entire year, and then how much additional product can be produced during the time saved. Knowing the value of the product and the value of the raw materials, we can calculate how much has been saved over one year.

Batch control requires versatility

One of the reasons that batch control has lagged behind continuous control is that control of batch processes is inherently more complex. Referring to the relatively simple system in Fig. 3, the following capabilities are required:

Continuous control—Since the computer system will be responsible for controlling the temperature in the reactor, a feedback control equation must be implemented in the computer. This problem is much the same as that for continuous processes.

Sequence control—The computer must be able to supervise the sequence of events that occur in the batch process. The sequence of events (starting after the initial charge of reagents) is as follows:

- Start agitator (if not already on).
- Set steam valve to full open.
- Wait until temperature gets to 5° of desired value.
- Close temperature cascade to feed flow-control loop, resetting flow totalizer to zero.
- Close steam valve.
- Wait until temperature gets within 2° of desired value.
- Set cooling-water valve to full open.
- Wait until required amount of Reagent *A* has been fed.
- Open temperature cascade to feed flow-control loop.

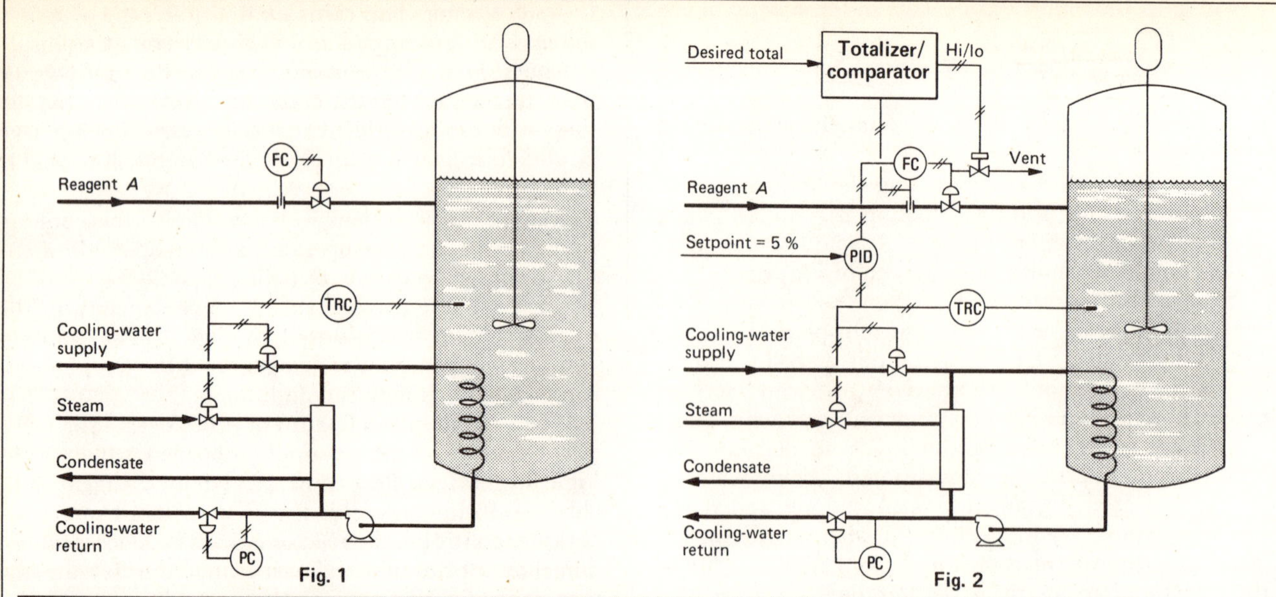

Cooling-limited reactor operating under three modes of control

- Close feed flow-valve.
- Close temperature cascade to steam and cooling-water valves.

This logic can be implemented in a high-level language such as FORTRAN or, alternatively, it can be more easily implemented using the special software for batch-process control that is available from several control-system vendors. Using the latter approach, only one statement would usually be required for each of the above steps.

Problems in batch control

Non-steady-state effects—In most continuous processes, the units run at or near the same operating conditions essentially all of the time, the main exceptions being utility plants and recovery units. However, no batch units operate under constant conditions. Throughout the cycle, the conditions within the process are changing, and thus the characteristics of the process also change. This is due to such factors as changes in viscosity, catalyst activity, heat transfer, etc.

Most conventionally controlled batch-reactors suffer from the change in control-system performance as the batch progresses. Thus, the conventional controller must be tuned so as to be stable during the most difficult conditions. Under other conditions, the conventional instrument does not control the process in the best manner possible. Three approaches can be taken to remedy this:

- Use different controller tuning-constants in different phases of the cycle. A few plants do this by having the operator modify these parameters as the batch progresses. A computer could do this automatically.
- Redesign the control system so that it is less sensitive to changes in process characteristics. Any such system is likely to use techniques other than the conventional PID control equation, and thus would be most readily implemented on a computer.
- Use different modes of control during each phase of the batch cycle. For example, using conventional controls, the warmup phase is likely to be initiated by making a step change in the setpoint to the PID temperature controller. With the computer, it is just as easy to use a different control mode during warmup, e.g. fully open the steam valve until the reactor reaches, say, 5° of the final setpoint.

Unfortunately, none of the above three approaches is especially easy. However, a good degree of control is a necessary prerequisite to ensure smooth process operation.

Higher-level effects—Many batch plants present other problems, which we might refer to as higher-level effects. Some batch plants (an example being a polyester plant) involve two batch reactors in series. Within certain limits, the extent of the reaction can be shifted from one reactor to the other. The problem now becomes that of finding the proper balance so as to attain the most efficient operation of the plant as a whole.

Another problem in the category of high-level effects is that caused by the interaction between equipment. Some plants are limited by the capacity of the central cooling-unit or central generating equipment. In such cases, the batch cycles must be carefully scheduled so as not to overload the system; or alternatively, we may encounter different times for the heatup phase, depending upon the current total-heat-demand in the plant.

Scheduling problems also occur at the interface between batch production and continuous processing and, above all, in multiproduct plants. Switching plant-equipment from one product to another invariably results in idle time and loss of material. In order to get the most out of a plant, intelligent decisions must be made regarding production sequence, length of production runs, and allocation of equipment to particular products. In complex plants producing a large variety of products, a computer can help in reaching the best

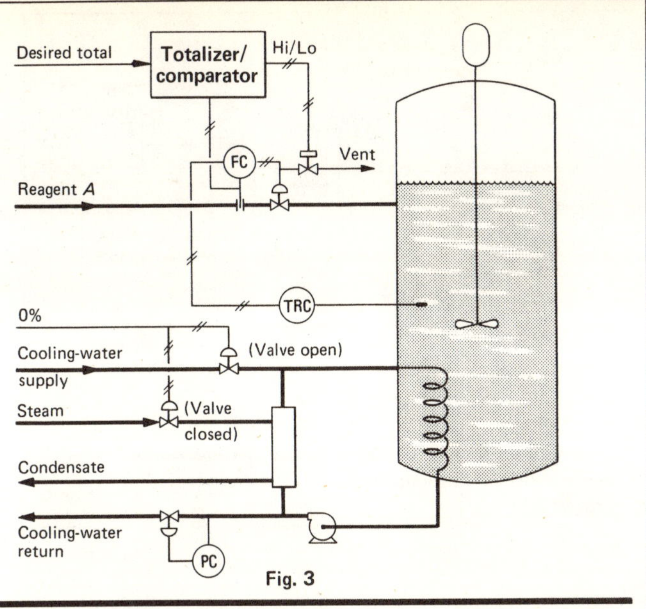

Fig. 1-3

decisions. However, computer-aided scheduling should not be considered a primary control problem and is best done on a separate computer.

Backup for computer systems

Although the reliability of computer systems is steadily increasing, we must still prepare for the time when the computer is not performing. Unfortunately, the situation is not as simple in batch processes as it is in continuous ones.

In a continuous process, when the computer-control system fails, it is usually desirable to continue to operate essentially at the current conditions. Therefore, the process interface hardware is normally designed to simply maintain the existing output levels until another value is obtained from the computer. Such hardware is quite simple to design and therefore economical.

When the computer-control system in a batch plant fails, it is usually *not* desirable simply to operate at the conditions existing when the computer failed. For example, if the computer is charging one of the reagents, it would usually be more desirable to shut off the feed. Design of such systems is usually somewhat difficult. In fact, such equipment has not generally been available because the suppliers basically have designed their systems for continuous processes and then adapted them for batch processes.

In plants with multiple reactors, the backup problem is even more complicated. With a small number of reactors, we might expect the operator to manually oversee the steps in each reactor when the computer fails. For this to be possible, however, the operator must somehow be able to ascertain the status of each of the reactors at the instant the computer fails.

The simplest approach would be to log each change in status in each reactor on a logging typewriter. This has the disadvantage that the operator must look back through these messages to find the last change in status for each reactor. This takes time, and offers a potential for error in that the operator may overlook an entry.

An alternative approach is to provide a separate logging typewriter for each reactor. Now the operator need only examine the last message on each typewriter to determine the status of the corresponding reactor. Especially for plants with several reactors, the cost of this approach would be a drawback.

Another alternative would be to use a video display unit or cathode ray tube (CRT). On the CRT would be displayed the current status of each reactor. If this CRT were supplied with its own memory and an independent source of power, then the image would be preserved when the computer failed. Now the operator can quickly ascertain the status in each reactor. We could also display information such as temperature setpoint, time into the current step in the cycle, the total amount of each reagent charged, etc. If these entries were updated on a frequent basis (such as once every five seconds) the operator would know much more about the conditions within each reactor at the time the computer failed.

When there are many reactors in the plant, it may be impossible for the operator to run the plant manually, because he simply does not have enough time to oversee every step in each reactor. In such plants, the computer may have been installed because conventional controls were not up to the task. Therefore, conventional controls are not entirely suitable as backup, and are likely to be costly; also, a smooth switch from computer to conventional controls would be difficult to implement.

In such cases, the most logical backup may be a second computer—resulting in what is known as a dual-processor system. In addition to a second computer, we must also provide in the interface for dual or redundant components such as two analog multiplexers. The engineering costs in such systems increase; but, because the same control programs are used in each computer, the costs do not double. The suppliers of such systems also provide the software support necessary to achieve as smooth a switchover as possible.

Computer roles in batch processes

Before using a computer in a batch process, it is important to define the functions it will perform. This is normally done as part of the process analysis described earlier. The hardware and software required to implement these functions will largely depend on the role assigned to the computer.

In its simplest configuration, a batch cycle consists of the following sequence:

- Vessel preparation (cleaning, pressure testing, etc.).
- Reagent charging.
- Reaction phase.
- Cooldown and dump.

Although the computer may be useful in all of these phases, we shall only consider the reaction phase.

Fig. 4 shows a simple, jacketed reactor. The control loop of primary interest is for temperature control of the reactor. A cascade control system is used, with the reactor temperature-controller providing the setpoint

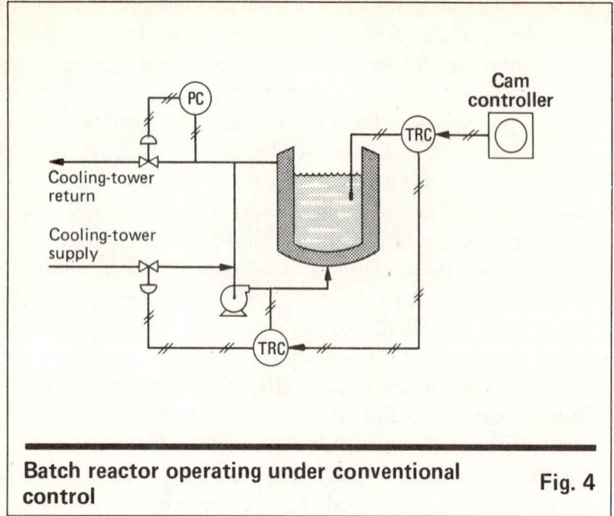

Batch reactor operating under conventional control Fig. 4

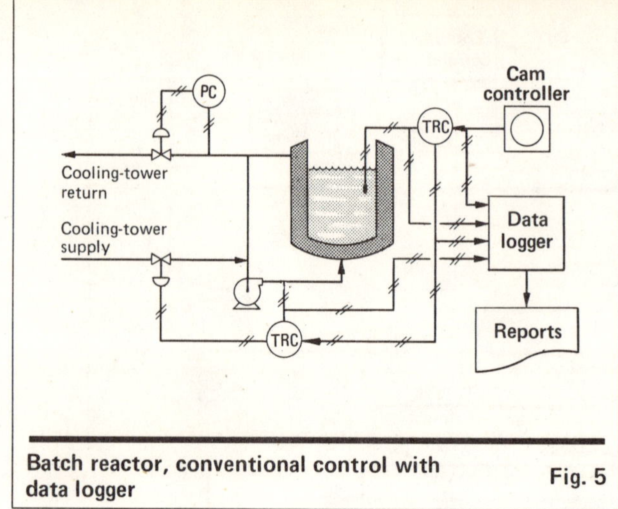

Batch reactor, conventional control with data logger Fig. 5

for the jacket water-temperature controller. The output from the latter controller is split-ranged—with the steam valve being open when the controller output is above midrange, and the cooling-water valve open when the controller output is below midrange. The setpoint for the reactor temperature-controller is provided by a cam programmer.

We shall use this system in examining the question: "How can I use a computer to generate a larger economic return?" We want to examine four potential system configurations: (1) data logger, (2) supervisory control, (3) direct digital control, and (4) hierarchical computer-control systems.

Data logging

Fig. 5 illustrates the application of a data logger to our sample process. Wherry and Parsons [*4*] have stated the case for (or perhaps we should say against) the data logger very eloquently. The usual sales pitch generally goes something like this: "From the data generated by simply logging all pertinent variables, you will learn so much about your process that the optimum operating strategy will be obvious." Unfortunately, life is not so simple. Most of the data will be redundant taken at or near the normal process operating conditions.

Although a data logger can be a valuable asset during efforts such as developing a process model, it must take a back seat. To develop a model entails the careful design and execution of process tests; such conditions never just "happen." Furthermore, during these tests, it is usually necessary to augment the output of the data logger by collecting samples for subsequent analysis in the laboratory.

Although the data logging concept is simple, implementation is deceptively expensive in terms of computer resources. Large files of data on an auxiliary storage unit are normally necessary, and the programs to format the reports are large (although writing the programs has been simplified by standard software packages offered by most computer-system suppliers).

The problem in justifying the system stems from the difficulty (or impossibility) of placing a value on the data. The basic question is, "Who will do what with these data?" In some industries, this question is easily answered, because a governmental agency mandates that the data be obtained and kept. For example, the U.S. Nuclear Regulatory Commission (formerly the Atomic Energy Commission) requires operators of nuclear power plants to keep certain data, and the U.S. Food and Drug Administration places similar requirements on producers of pharmaceuticals. Under such circumstances, management often concludes that a data logger is desirable simply to be sure that the proper data are collected.

In the final analysis, management must place a value on the data. Tangible returns from improved process operation simply do not exist. Therefore, management must weigh the value of the data against the cost of the system and answer the question, "Is it worth it?" We neither say nor imply that this is easy to answer.

Supervisory control

As classically defined, supervisory computer control applies to the situation where the computer provides the setpoint for an analog controller. In Fig. 6, the supervisory control computer has replaced the cam programmer in the conventional system illustrated in Fig. 4. The computer also monitors the two process temperatures for purposes of alarming.

What advantages does this have over the conventional system in Fig. 4? That depends upon the application.

First, the conventional cam programmers have their limitations. For example, it is not normally possible to achieve a step increase in the temperature. Instead, the friction on the lever that rides the cam determines the maximum rate that the setpoint can be increased. If the rate is too rapid, the cam "hangs."

In batch plants that produce a variety of products, it is usually desirable for the computer to also supervise the charging sequence. In the conventional plant, the operator has to charge the proper materials and then place the appropriate cam on the cam controller. On numerous occasions, experimental data have been obtained indicating that Murphy's Law is valid in this

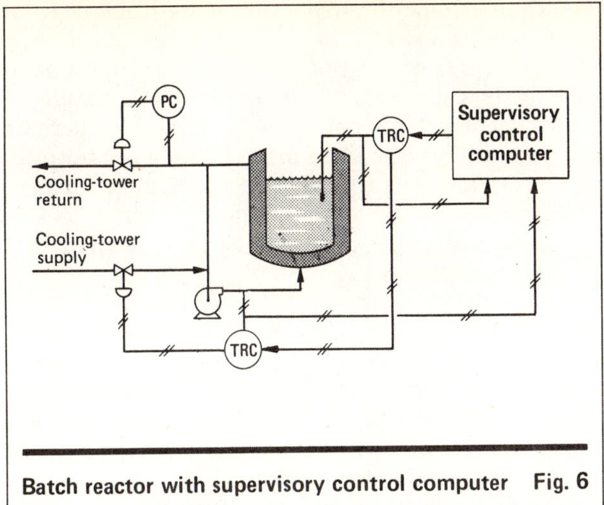

Batch reactor with supervisory control computer Fig. 6

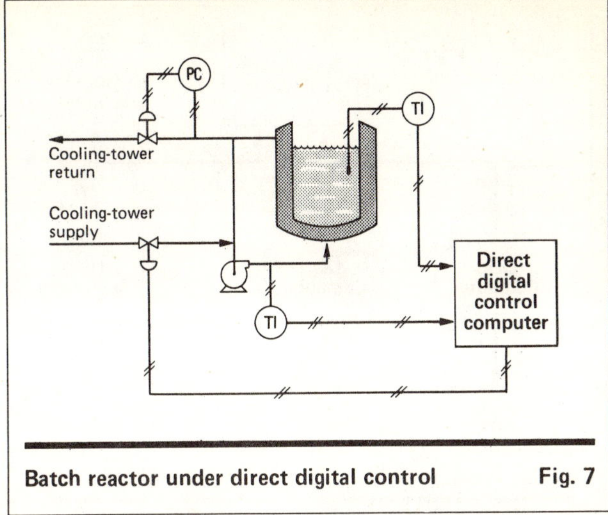

Batch reactor under direct digital control Fig. 7

mode of operation, i.e., the reactor will be charged to produce Product A and will be controlled with a cam to produce Product B. With a supervisory control computer, the operator tells the computer to make the product, and the computer will charge the proper reagents and use the proper temperature program.

Direct digital control

As classically defined, direct digital control applies to the case where the computer output signals are sent directly to the final actuator, usually a valve. Fig. 7 illustrates the direct digital control system for our reactor. The computer monitors the reactor temperature and the jacket-water inlet-temperature, and directly manipulates the control valves.

How is this arrangement superior to the conventional system in Fig. 4 or the supervisory control system in Fig. 6? In a production-limited plant, the production can be increased by shortening the batch cycle-time. One way to do this is to minimize the warmup time and cooldown time. We might use the following simple logic:

Warmup—Set steam valve fully open and leave open until reactor comes to within, say, 5° of the desired temperature, and then switch in the digital equivalent of the cascade system in Fig. 4 to bring the temperature to its desired value.

Run—Use the cascade system in Fig.4.

Cooldown—Set the cooling-water valve fully open and leave open until batch is cooled to the desired temperature, and then close cooling valve (computer output is set at midrange).

While the above sequence is easily implemented with a digital computer control system, it is not easy using conventional hardware.

Direct digital control has received its black eye largely from the experiences gained in the petroleum industry, an example being that reported by Parsons et al. [*5*] for a continuous ethylene plant. However, such plants are ideally suited for conventional analog hardware. The plant always runs at or near the same operating conditions, the control strategy rarely needs to be modified, and little or no sequencing must be done. These conditions simply do not exist in batch plants, and thus the shortcomings of conventional control systems are more evident; the use of DDC is perhaps the most viable way to overcome these shortcomings.

This is not to imply, however, that a total or 100% DDC system is necessary or desirable. For example, in Fig. 7, the back-pressure regulator on the jacket is still implemented in analog hardware. The wise designer only implements on the computer those loops for which the computer offers a potential advantage.

In effect, the above paragraph suggests that the computer be used only where it is attractive to do so.

■ Though investment cost for the digital control system rarely exceeds that of an analog system, but the digital system can seldom be justified by hardware savings alone. If any significant amount of analog backup hardware is provided, these savings will be nil.

■ Personnel costs do not generally decrease with the installation of the computer.

The above two statements essentially mean that a digital system can only be justified by improved process performance. That is, we must look for a more-capable control system, not a cheaper one.

Using a digital computer to implement the same functions as performed by an analog or conventional control system does not inherently lead to improved performance, and consequently not to improved economic returns. Referring to the system in Fig. 1, there is no incentive from a direct control standpoint to use DDC for the temperature control loop as configured. However, if the scheme in Fig. 3 is used, the computer becomes necessary. A switch from the arrangement in Fig. 3 to the arrangement in Fig. 1 must be made when all of Reagent *A* is added. Such switches are conveniently implemented only with a computer.

Hierarchical computer-control-systems

In a previous section, we discussed the difficulties in providing backup for batch reactor systems. The difficulty basically stems from the operator having to absorb a large amount of information in a short period of time. (If only one reactor were involved, the situation would be much simpler.)

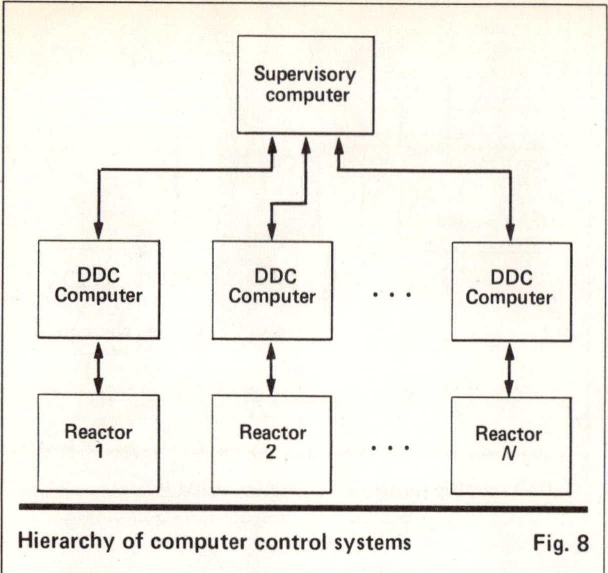

Hierarchy of computer control systems Fig. 8

In a multireactor plant, a possible approach would be to divide the responsibilities among several computers, as illustrated in Fig. 8. One of the computers provides the supervisory functions such as scheduling the various reactors, considering interactive effects between reactors, and the like. In addition, each of the reactors is provided with its own DDC computer.

The system could be arranged so that when the supervisory control computer decides to produce a particular product in a given reactor, it transmits the necessary information in the recipe to the DDC computer, which can then proceed to produce the product without further direction.

In this environment, a failure of the supervisory control computer would not interrupt any batch in progress. However, as no batch would begin without some direction to the DDC computers, we must provide a way for the operator to initiate a batch. Perhaps he could enter the necessary data into the DDC computer, or merely instruct the DDC computer to repeat the previous batch.

If one of the DDC computers fails, it would be feasible for the operator to assume control of that particular reactor. If speedy repair were necessary, it might be desirable to keep a spare DDC system that could be quickly installed in place of the system that failed. Another alternative would be for the supervisory computer to assume the responsibility until the DDC system were repaired.

At the present, systems such as in Fig. 8 must be specially designed and installed, thereby increasing the costs. Hence, this is not normally a feasible alternative at present, but should be in the not-too-distant future. For example, the capabilities of the Remote Intelligent Multiplexer available from General Automation Corp. could be easily extended to provide the functions of the DDC computers in Fig. 8.

Economics

One factor not considered in this article is cost. This was omitted for two reasons. First, it is very difficult to discuss costs except when a specific application has been defined. Second, the capabilities of computers are rising steadily and their costs are decreasing.

To give a range, a minimal system for batch process control would cost from $50,000 to $100,000 for the computer and process interface. A medium-size system that would be required in most multi-line batch plants of any significant size would cost between $100,000 and $250,000. A dual processor system would cost in the vicinity of $500,000.

To this must be added the engineering and installation costs. These will easily equal the hardware costs but may run much higher. Especially in plants with pneumatic instrumentation, the installation costs will be high, although pneumatic multiplexers can often be used to reduce these costs to an acceptable level. However, the benefits derived from using a computer are usually larger in those plants having pneumatic instrumentation.

References

1. Malone, K. B., others, Dynamic Simulation of the Continuous TNT Process, presented at the Summer Computer Simulation Conference, Montreal, Canada, July 17-19, 1973.
2. Kennedy, J. P., Tighter Process Design via Computer Controls, *Chem. Eng.*, Mar. 17, 1975, pp. 54-60.
3. Spellman, R. A., and Quinn, J. B., Computer Control of Batch Reactors, Proceedings of the Second Joint Spring Conference, Instrument Soc. of America, Montreal, Apr. 23-24, 1975.
4. Wherry, T. C., and Parsons, J. R., Guidelines for Profitable Computer Control, *Hydrocarbon Proc.*, Vol. 46, No. 4 (Apr. 1967), pp. 179-182.
5. Parsons, J. R., others, Performance of a Direct Digital Control System, presented at the 25th Annual ISA Conference, Philadelphia, Oct. 26-29, 1970.
6. Stout, T. M., Economic Justification for Computer Control Systems, presented at 1972 IFAC (Intel. Fed. on Automatic Control) 5th World Congress. (Proceedings available from ISA, 400 Stanwix St., Pittsburgh, PA, 15222).

The authors

Marcel T. Brodmann is President of the firm Economation AG, Postfach 246, CH-7002 Chur, Switzerland, a company specializing in industrial computer applications. A chemical engineering graduate of the Swiss Federal Institute of Technology, he gained his first industrial experience with Emser Werke AG. Later, he worked for Exxon in various managerial positions related to computing and process computer control. Dr. Brodmann is a member of AIChE, International Management Assn., and the Schweizerische Ingenieur-und Architektenverein.

Cecil L. Smith is Professor of Chemical Engineering and of Computer Science, Louisiana State University, Baton Rouge, LA 70863. He also works with Economation AG and its U.S. affiliate, Economation Inc., on computer control projects. His interests lie mainly in the areas of digital control and simulation of industrial processes. He holds B.S., M.S. and Ph.D. degrees in chemical engineering from Louisiana State and is the recipient of the 1972 Donald P. Eckman Award for outstanding contributions to the field of automatic control. He is a member of AIChE, Instrument Soc. of America, and Assn. for Computing Machinery.

Batch-Control Improvement Through Computers

CONTROL CONSOLE of representative digital-control system (lower right-hand corner) contrasted with conventional-control instrument panel in rear.

According to many users, computer-based systems are the only economically and technically feasible means of automating numerous batch and blending processes.

JOHN BALL, CHARLES BREZ and JERRY CASSIDAY, Fisher Controls Co.

Plant and process engineers are designing more-complex production equipment to meet the demands for higher-quality products at lower costs, and to satisfy increasingly strict requirements for materials and energy usage. This calls for more-sophisticated systems to sequence the operation of valves, motors, heaters and other control elements, as well as to regulate process variables such as pressure, temperature and chemical composition.

In addition to performing the above operations, computer-based systems minimize variation in batch-to-batch product quality, permit the scheduling of multireactor plants with shared equipment, allow the implementation of frequent product changes, and perform special control calculations.*

Digital Control

Particularly important for batch operations are control systems that incorporate digital computers that store program sequences, perform signal handling and command generation functions. Some reasons for the importance of these systems are:

- Flexibility for batch sequences, modification of control strategies, and formulation changes. In conventional systems, control programs are hard-wired, or fixed in the design of electrical circuits and mechanical elements, and are therefore not readily altered to suit changing application requirements.
- Increased throughput and product consistency due to extensive computational capabilities. Input data are manipulated—according to advanced-control strategies—to produce operating commands. Examples are the computation of mass or energy balances online, or the use of Smith predictor algorithms to compensate for dead times in adding ingredients†.
- High-speed scanning. This, along with computational capabilities, can promote efficient manpower utilization by performing routine and time-consuming functions. These features also produce operator displays, which present more-complete information about the process in a form easier to understand. For instance, a single cathode-ray tube (CRT) terminal—used as a computer output device—can display present and past values of any selected process variables (see picture above).
- Automated logging and reporting. These features provide good documentation of processing conditions with minimum manual effort.

An example of the need for advanced control systems is the polymerization of styrene or vinyl chloride. Since consumption of polystyrene and polyvinyl chloride has been growing at about 10% a year, and the costs of the raw materials have been rising, these plastics provide a strong incentive for high-productivity equipment.

Polymerization Steps

Charging—Because polymerization begins when the reactor is charged, close control over the quantities of reactants added is needed to determine final-product characteristics. Automated operation of valves based on signals from flow totalizers can ensure precise metering of ingredients in minimum time.

However, the control problem is often complicated by the need to accommodate a product mix with several different ratios of reactants, or to operate various combinations of valves on headers to transfer material into se-

*Garibian, S.K., and Koeninger, E.C., Computers and Batch Systemization, *Chem. Eng. Progr.*, March 1972.

†The application of a dead-time compensator using predictor control techniques requires that a process model be used as part of the control system.

Originally published August 5, 1974.

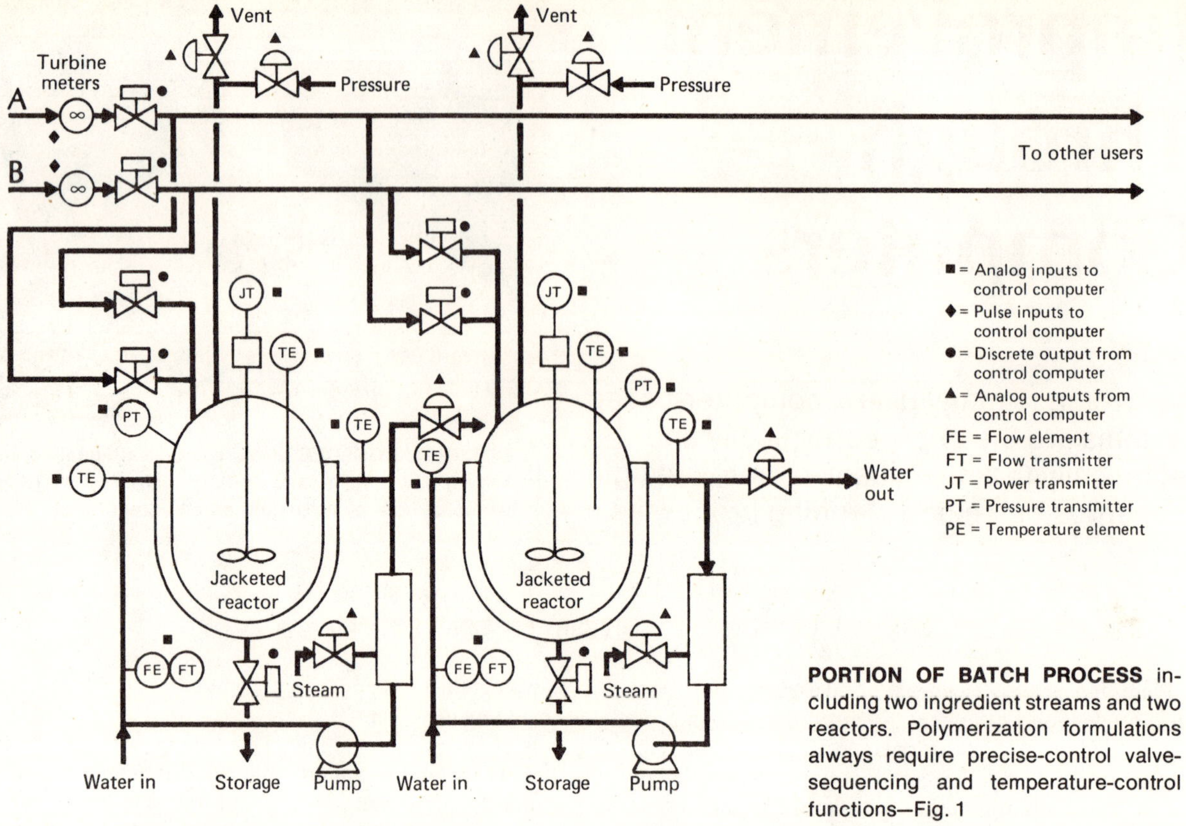

PORTION OF BATCH PROCESS including two ingredient streams and two reactors. Polymerization formulations always require precise-control valve-sequencing and temperature-control functions—Fig. 1

lected lines or vessels. An example of such complexity is shown in Fig. 1, by a portion of a typical process that has two ingredient-streams and two reactors.

Reaction—Usually, heat is added to the ingredient mixture to start reaction, by a flow of steam or hot water through the vessel jacket. But as the reaction proceeds, the process becomes exothermic, and coolant must be forced through the jacket.

Different control algorithms are normally required for the heating and cooling processes. Implementing the change in control mode is critical, not only to maintain the desired temperature but also to prevent material solidification or even explosion. This could happen if the reaction became unstable due to heat generation or insufficient energy removal.

Rapid heating can shorten the processing cycle and increase throughput without affecting the product. However, in manual or semiautomated systems, a slow approach to the final temperature is generally recommended, to avoid overshooting (and consequent product loss), equipment disabling, or safety hazards. More-advanced control techniques can provide heat at the maximum available rate, using measured data to calculate when cooling must begin (Fig. 2). In one instance, use of a computer system in this manner reduced the heating phase of a polystyrene process by two hours, and increased reactor production by 15%.

Curing and Transfer—After the reaction is completed, the polymer is cured and then cooled and transferred to a blowdown tank. Here, again, close temperature control is needed during the curing. To ensure uniformity, advanced-control systems can often be justified for processes in large vessels, or for materials having low heat-transfer coefficients.

It may also be desirable to control the cooling process to optimize the rate of heat removal, consistent with expenditure of energy. This may be accomplished by means of temperature or heat-flux sensors in the vessel walls, and by regulating the coolant flow in the jacket.

Economics can be particularly significant if several reactors share a raw-material-charging system, because balanced scheduling can appreciably lower overall demand. Finally, if the product can be stored in a number of tanks, some form of control is needed to operate the proper combination of valves.

Popular Myths Versus Facts

Early computers were expensive, prone to malfunction—especially in plant environments—and difficult to program. Control specialists were forced to purchase computers as components, assemble and integrate the necessary additional hardware and software, and provide process interfaces before the problem of control could even be approached. Such origins have fostered a number of popular myths about the state of the art in digital-computer control:

Costs—One widespread misconception is that the capital investment in computer systems is high and can only be justified in terms of subsequent returns. In many in-

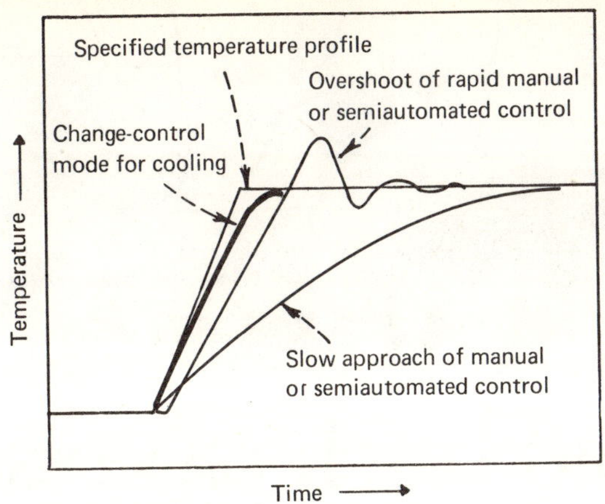

ADVANCED-CONTROL TECHNIQUES make it possible to reach final temperature rapidly, with no overshoot—Fig. 2

stances, initial expenditures for computer hardware and software are less than those for equivalent hard-wired equipment. As an example, Coca-Cola reports that the low bid on an installed computer system to control a four-unit, beverage-base mixing plant was $120,000, while the minimum estimate for a functionally-similar relay system was $200,000.*

The smallest practical computer-control systems presently on the market tend to cost about $50,000, installed and ready to program for specific tasks. Although this price is likely to exceed that of a drum or relay programmer (with associated analog controllers) for a small batch process, as the amount or the complexity of equipment under control grows, the costs of hardware-oriented systems rise proportionally. By contrast, those of digital controllers increase only slightly.

Such cost differences can be particularly noticeable when a system is expanded to upgrade an operation or increase the amount of equipment under control. As a result, computer systems may often be justified on the basis of capital costs alone, particularly if the processes are large or complex, or if the operational scope is expected to grow during the expected life of the control equipment.

Reliability—The use of solid-state electronic equipment, and the growing awareness by manufacturers of environmental conditions in processing plants, has led to computers with highly reliable hardware. Software failures can be difficult to attack in a systematic manner, but many operating and application packages are available that have been essentially debugged through extensive use. Electromechanical peripherals—such as input-output devices—are somewhat more likely to fail in service, but such faults can be isolated in a manner that prevents overall control-system disruption.

Programming—Most digital controllers are preprogrammed with software executive systems, which facilitate operation in process environments. For example, these systems make it possible for control and process specialists to specify formulation conditions or to designate operating equipment, without using computer-oriented instructions. The techniques available to users range from diagrammatic methods—which are similar to the drawing of relay ladders or matrix-board connection paths—to language-based concepts that involve control statements written in terminology familiar to practically all process engineers.

The system software makes programming simpler for computers than for hard-wired controllers, because the work is conceptually easy and can be done from keyboard terminals without wiring or assembly of mechanical elements. A further advantage is that after a program is written it can be used repeatedly with little further development or implementation costs.

Blue Sky or Silver Lining—A decade ago, digital control was heralded as a momentous wave of the future in the chemical processing industries. Since, however, other process revolutions had come and gone, and neither the technical nor the economic feasibility of the computer had yet been demonstrated, the computer was either blue-sky thinking or a cloud with a silver price, but with no lining to match.

The situation has now changed, largely due to the minicomputer and the practicality of dedicating general-purpose machines to specific classes of applications. A number of manufacturers offer standardized control systems—including integrated computing and interface hardware and software—which make the advantages of digital processing available to virtually any user. The systems are optimized for control tasks and are highly flexible in the options that can be exercised over reactor and process operation.

The computer in an integrated system is less visible to the manager, engineer and operator than are the solenoids and transistors of conventional controllers. Instead, the entire package appears to be a window between the user and the process, which facilitates giving instructions and obtaining information.

*Reed, V.W., Beverage Plant Batching by Minicomputer, *Instrumentation Technology*, Sept. 1972.

Meet the Authors

J. BALL

C. BREZ

J. CASSIDAY

John Ball is a sales engineer, digital systems, at Fisher Controls Co., Marshalltown, IA 50158, where he has had eight years of process-control experience. He has a Certified Engineering Technology degree from Ryerson Polytechnical Institute, Toronto, Can., and is a member of the Instrument Soc. of America.

Charles Brez is a group manager of systems application at Fisher Controls Co., Marshalltown, where he has had eight years of experience in the fields of applied computer process-control, batch and continuous control, chemical processes, gas-distribution control, and missile-control systems. He holds M.S.E. and B.S.E.E. degrees from St. Louis University, and is a member of the Instrument Soc. of America and of the Institute of Electrical and Electronic Engineers.

Jerry Cassiday is a digital-system project engineer at Fisher Controls Co., Marshalltown, where he has had 10 years of experience applying analog and digital instrumentation to process-control systems. He attended the University of Vincennes (Vincennes, Ind.) and is a member of the Instrument Soc. of America.

Operator-oriented design in computer-controlled batch processes

Although computers are increasingly being used to control batch processes, automation is never so complete as for continuous processes. Hence the operator is quite important to the process, and the system must be designed for easy communication between him and the computer.

Charles Brez, Ralph Draves, and Fred Gore, Fisher Controls Co.

☐ Batch processing continues to be widely used for producing chemicals, foods and other products. The importance of the technique results largely from the flexibility of equipment, which enables the handling of different formulations or recipes, and from the ability of modern control instrumentation to follow predetermined reaction paths.

Considerable work is being devoted to improving the profitability of batch process plants. For example, larger vessels are being built to increase output volume and lower incremental capital costs. Likewise, simulations are being used, to aid in devising equipment that runs safely at high throughput rates. New instrumentation is also being developed, to implement automated control and sequencing, which improves yield and quality while safeguarding the operation.

Operator-process interfaces

Despite these equipment-oriented improvements, batch plant performance ultimately depends on manual interactions with the equipment. Many critical operations do not lend themselves technically or economically to automation. Examples are: regulation of variables that are not easily measured online, such as percent conversion; implementation of management decisions, such as changes in recipes, to accommodate market demands that are not readily programmed; and protection against dangers or losses during situations that have not been anticipated. To plan for effective performance, computer interfaces must be provided that (1) facilitate exchanges of data and commands between the operator and the process under normal conditions and (2) permit continuing operation or safe shutdown and troubleshooting under abnormal circumstances.

Computer-controlled plants

Computers are now being commonly used to automate the sequencing and control of batch-process reactions. The arithmetic, logic and memory capabilities of these high-speed signal-handling computer systems are exploited to provide constant surveillance, rapid response, and sophisticated control actions. A well-conceived process-computer system relieves much of the burden on plant personnel by implementing routine procedures. These include simple but tedious tasks such as logging variables or checking valves, and extend to complex functions such as ramping temperatures or estimating when to change from a heating to a cooling mode. In this manner, the computer minimizes the amount of information that must be transferred between operator and process, reducing chances for human delay and error.

However, a digital control system can go further. Some exchanges of supervisory information and commands are always necessary, because of external events and decisions, and the capabilities of the computer can be exploited to enhance the operator interface. This

Originally published December 20, 1976.

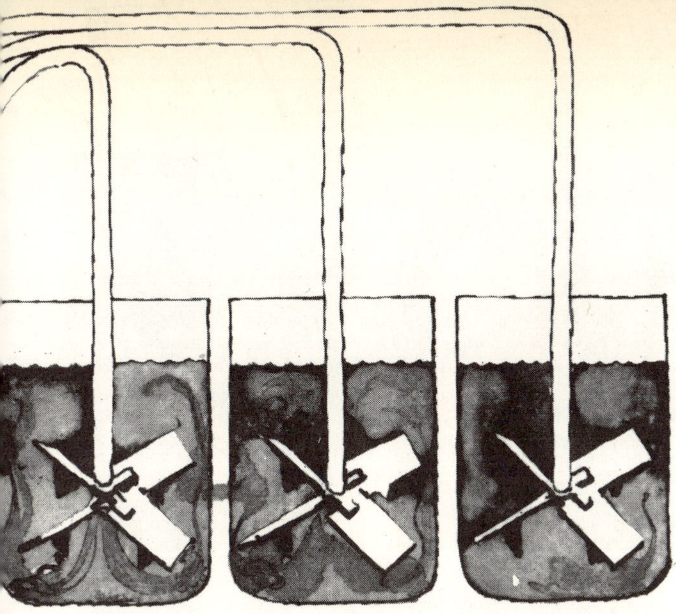

may involve such features as automatic checks on the reasonableness of all operator inputs, to prevent accidental entry of potentially dangerous changes. Moreover, the computer may provide means of implementing human-engineered capabilities by generating and accepting information in the form of English-language phrases, tables of numbers, graphics, and the like.

Digital-control-system process interfaces

A digital control system has the elements of Fig. 1. (These elements include the hardware and software necessary to perform the indicated functions.)

Since batch processes are dynamic, control system outputs must be continuously changing to yield satisfactory performance. This affects the need for backup in case the computer becomes unavailable. The input-output system may accordingly include separate analog controllers and manual pushbuttons or other discrete devices as shown in Fig. 2. These operate in a supervisory sense, with analog setpoints and on/off states determined by the digital system, and can function independently if no valid commands are received from the computer. Computer/manual or computer/automatic/manual stations may also be placed between the digital control system and the process. These provide the switching, as well as the manual input or automatic control functions, necessary to continue operation if a computer failure is suspected.

Regardless of the analog or discrete equipment used in a digitally-oriented control system, maximum capabilities for information exchange can be obtained if the computer drives the primary interface between the operator and the process. All communications, even those involving such actions as setpoint changes on analog controllers, are therefore implemented in a unified manner and are monitored by the computer for logging and other purposes.

Interface hardware

Specification of interface configuration is influenced by the amount and type of information to be transferred, the need for documentation, and the ultimate disposal of logs and reports.

A great variety of hardware is available for use on computer-oriented systems. Some equipment involves standardized computer terminals while other equipment has been developed specifically for control room use; likewise, complete modules and individual components may be obtained. In any case, designers and users have wide latitude in meeting individual requirements and preferences.

Command entry devices range from dedicated pushbuttons and general-purpose keyboards to adjustment dials and numerical-thumbwheel switches. These components may be used in configurations that provide purely visual feedback in such forms as pointers, backlights, or alphanumeric displays; the same entry elements may also be used with printout devices to supply hard-copy documentation of actions.

Indicators range from alarm lights and numerical panel displays, to cathode-ray-tube (CRT) screens or printed documents. Most computer-driven interfaces use a combination of such elements, as shown in Fig. 3. Often, the operator has a choice of the devices to be used for certain functions. For example, visual indications may be selected during routine operation, while hard copy might be required for product changeover. In either case, a record of all conditions will probably be stored automatically.

Interface software

The software used to implement interface functions is also important, since the formats chosen for data entry influence the manner in which the selected consoles will be used. In general, the cost of interface software is inverse to the effort demanded of the operator; design tradeoffs accordingly involve balancing both the expense of customizing software to the application and the expense of the extra load placed on the computer,

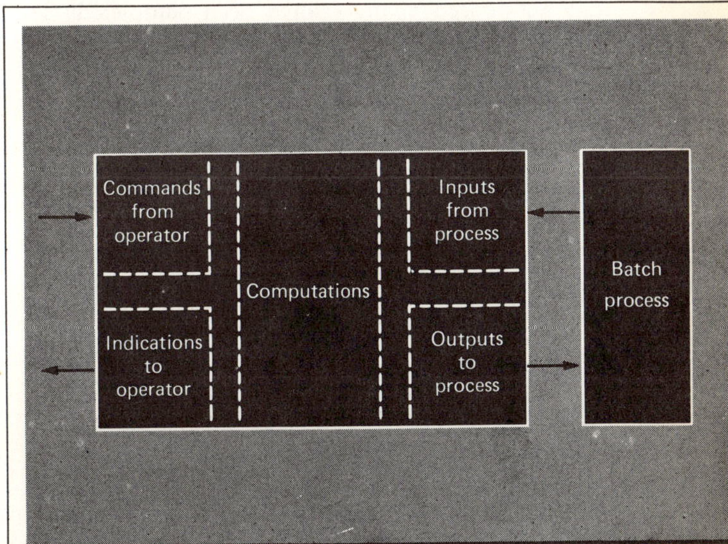

Functional elements of a digital controller **Fig. 1**

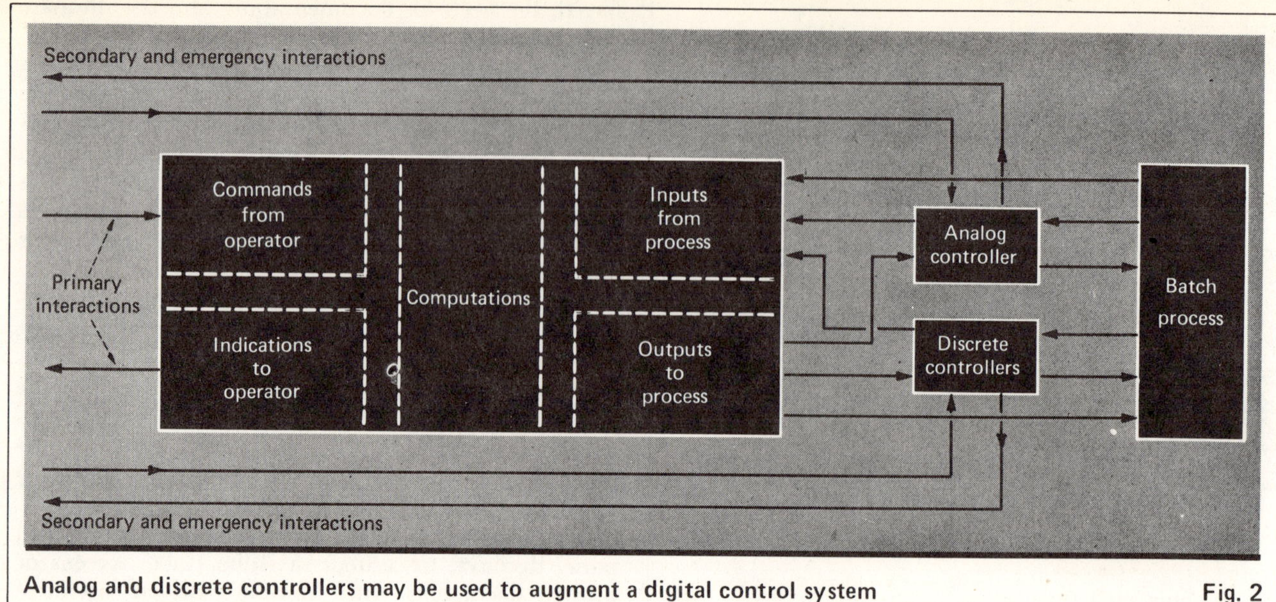

Analog and discrete controllers may be used to augment a digital control system **Fig. 2**

against both the amount and complexity of information to be exchanged and the consequence of errors.

Parameter modification can be employed for data entry in systems having standard computer-terminal hardware and software. The operator uses a typewriter-like keyboard, to enter identifying codes and values of any data to be changed from the previous batch. A CRT display or character printer may be used to provide the required visual feedback.

A typical parameter identification transaction is shown in Table I. In this transaction, the system software handles the input of parameters in a standard format, without special program statements. The computer accepts the entered values directly and updates the stored parameter list. The process begins after the operator uses a pushbutton or enters a code, to indicate that all desired changes have been completed. The chief shortcomings are: the operator has to remember the symbols and the meanings of various parameters; time requirements can become excessive if many values are to be modified; and erroneous entries can cause serious damage to control unless the software is written to prevent operator changes of critical parameters.

Representative parameter modification interaction **Table I**

Operator's console keyboard entry		Comments
NØ:3	(CR)*	Reactor No. 3 to be charged
P/3,1/: 101.2	(CR)	Code No. of batch used for product identification
P/3,2/: 667	(CR)	Product type to be made
P/3,3/: 230	(CR)	Batch size to be made
P/3,4/: 13	(CR)	Product to be discharged to Tank No. 13
N1:1	(CR)	Setting this flag causes program to check entries for reasonableness, make necessary charge calculation, and proceed with batch

*Carriage return

Interpretive data-entry involves a special console, with software specifically written or customized to recognize and accept predetermined classes of entries. The process operator's station normally contains pushbuttons indicating functions, a keyboard to enter codes and numerical data, and a thumbwheel or other device to increase or decrease values. The feedback may again be either visual or printed.

An example of an interpretive data-entry transaction is shown in Table II. This technique is easier for the operator than parameter modification, since pushbuttons are labeled with process-related functions so that few special terms must be learned. Further, the fixed data-format ensures that only specified memory locations can be accessed from the console, so errors are unlikely to destroy or interfere seriously with the control program. However, the specialized hardware and software tend to raise cost, and can restrict flexibility for system modification unless expansion capability is provided as part of the design.

Interactive data entry places the entire responsibility for information exchange on the system software. The operator need only start the entry routine, then respond to questions appearing on his display. Printers can be used; however, the amount of conversation is often voluminous, so visual displays are preferred to raise speed and reduce complexity.

An example of interactive transaction is shown in Table III. Questions are stated in terms familiar to the operator, so concentration is entirely on the desired process data. The user therefore need know little about control-system details, beyond the means of initiating the data entry program, operating the keyboard, and interrupting the computer for high-priority commands. No special consoles are needed, so some hardware economy is possible and the system can be modified relatively easily. However, special programming and con-

Representative interpretive format interaction Table II

Operator's console pushbutton and keyboard entry	Comments
REACTOR	Operator pushes REACTOR button and observes that it lights up. Computer is then ready to receive the reactor number
3	Numerical keyboard entry of Reactor No. 3
ENTER	Operator pushes ENTER button to signify completion of keyboard entry
CODE	Operator pushes CODE button and observes that it lights up
101.2	Numerical keyboard entry of the desired code number for identification
ENTER	Operator pushes ENTER button to signify completion of keyboard entry
PRODUCT	Operator pushes PRODUCT button and observes that it lights up
667	Numerical keyboard entry of product type
ENTER	Operator pushes ENTER button to signify completion of keyboard entry
BATCH SIZE	Operator pushes BATCH SIZE button and observes that it lights up
230	Numerical keyboard entry of batch size (in this case, liters)
ENTER	Operator pushes ENTER button to signify completion of keyboard entry
STORAGE TANK	Operator pushes STORAGE TANK button and observes that it lights up
13	Numerical keyboard entry of tank to be used for storage
ENTER	Operator pushes ENTER button to signify completion of keyboard entry
ENTRY COMPLETE	Operator pushes ENTRY COMPLETE button to indicate that all data have been provided

Note: In each case, the computer examines entry for high and low limits and illuminates an illegal-entry alarm if a value is out of range.

siderable software support are necessary, so interactive data entry is usually the most expensive technique to implement.

Interface functions

A number of distinct classes of functions are usually assigned to an operator in a batch processing plant. The details of these dictate the choice of interface required. Generally, all expected functions should be considered as a unit, and a single console selected whenever practical. This saves money and helps ensure

Representative interactive format interaction Table III

Operator's console CRT or Teletype output	Operator's console CRT or teletype input	
	BATCH	(CR)*
ENTER REACTOR NUMBER	3	(CR)
ENTER IDENTIFICATION CODE	101.2	(CR)
ENTER PRODUCT TYPE	667	(CR)
ENTER BATCH SIZE	230	(CR)
ENTER STORAGE TANK	13	(CR)
IS ENTRY COMPLETE?	YES	(CR)

*Carriage return

In each case, the computer examines entries for high and low limits and repeats request if bounds are exceeded or if entry is otherwise improper.

operator familiarity with the equipment. Here are some of the classes of functions:

Initialization and supervision

In normal processing, the operator must frequently initialize batching and provide overall supervision by entering data into the system. The least expensive interfaces are suitable when formulations and processing conditions are repetitive, since little modification is normally required after startup. More-sophisticated systems are called for when the types and amounts of ingredients vary, and when details must be given concerning processing steps.

Operator instructions

Another important interface function in a batch plant is for the computer to instruct the operator to perform manual tasks, such as drawing samples or visually inspecting reactors, as part of the batch sequence. Displays are needed to present the instructions, and pushbuttons or keyboards are necessary to allow the operator to either acknowledge that tasks are completed or to enter the results of a directed action. General-purpose displays may be used for most functions,

Process/operator interface having several types of entry and display devices on a single console Fig. 3

but one or more dedicated readouts may be provided for variables that warrant particular attention.

Alarm acknowledgement

The interface must draw operator attention to alarm conditions. This is most often done visually with flashing CRT displays or colored lights, or aurally with horns or bells. The interface design must make it easy for the operator to identify and diagnose the cause of an alarm, and to take appropriate response actions. For this purpose, the part of the process or piece of equipment causing the alarm should be indicated, and appropriate command-entry devices should be mounted near the corresponding indicators. Alarm conditions and acknowledgements should also be automatically logged as part of the interface function.

Process emergencies

Extensive fail-safe and diagnostic measures are incorporated in most digital control systems, for example using limit checks that indicate abnormal conditions, or calculations to predict effects such as fouling of heat-transfer surfaces. However, most true emergency conditions are unanticipated, so the interface should include the means for providing information to the operator and for carrying out actions based on his response. The main goal must be to ensure safety, and a possible secondary objective is to permit economic evaluation of operation or shutdown. These tasks require a display that calls attention to problems, an entry station that permits the operator to branch to emergency sequences, and a means of documenting the events associated with the incident.

Process failures typically include valve or pump malfunctions. These are difficult to identify because batch reactions are transient. Moreover, such failures inevitably cause process interruptions, and may therefore be dangerous or expensive. The computer can conceivably be programmed to carry out the emergency sequence and provide appropriate alarms. However, more often, the responsibility will fall on the operator to interrogate the system, deduce the possible problems, and take corrective action.

Control system failures

Failures can also occur in the control system; these demand operator intervention. Computers occasionally encounter problems, even though equipment is now considered extremely reliable and tolerant to external disturbances. Unfortunately, even minor difficulties can seriously affect overall plant operation, since the same computer provides many different functions. If batch interruptions are not acceptable and computer control is essential, a redundant central processor can be specified. This can be switched, by a watchdog timer, to take over all control and scheduling if the control program is not cycling properly, or switched by the operator if some malfunction is suspected. Since the central processor represents only a fraction of total system cost, this approach does not necessarily imply a heavy economic penalty.

However, in most batch applications, an extra computer is not needed, since the operator can temporarily assume control and scheduling functions, or a few analog and discrete controllers can operate critical loops, in the event of a central processor failure. While process performance may be downgraded under such circumstances, the approach may be economically preferable to complete automated backup.

Failures in the process input and output subsystems are more common than those in the central processor, because of their vulnerability to external disturbances. Problems can be avoided if hardware and software are designed to minimize interactions between channels, for example by operating the various output files independently. However, bypass or backup can be provided for these conditions in cases where analog controllers are included in the system. The interface must then provide a means of switching control to the analog elements, and of exchanging information with them.

Electromechanical peripherals are somewhat prone to malfunction. Most such devices are used for the operator interface only, so failures are inconvenient but need not affect a batch in process. In addition, precautions can be taken that enhance system integrity, at relatively low cost. For example, the interface can be designed so that general-purpose keyboard logging-devices back up specialized operator-data-entry consoles.

Analog controller failures can also occur. If these devices are being used in a supervisory mode, problems can often be brought to the attention of the operator at the interface, by using limit sensors on process conditions or by monitoring actual output signals. Backup failures are difficult to detect, because operation is only occasional. It is therefore advisable that backup functions be tested regularly—automatically or by the operator—and special indicators can be included in the interface for this purpose.

The authors

C. J. Brez

R. H. Draves

F. E. Gore

Charles J. Brez is manager of electronics product marketing at Fisher Controls Co., 1800 Fisher Bldg., Marshalltown, IA 50158, where he is primarily responsible for planning new analog and digital hardware, software and application technology. He holds a B.S.E.E. and an M.S.E. from St. Louis University and is a member of the Institute of Electrical and Electronics Engineers and the Instrument Soc. of America.

Ralph H. Draves is a control engineer at Fisher Controls Co., specifying and implementing digital control systems for process industry applications. A chemical engineering graduate of Iowa State University, he has had ten years of experience at Northwestern States Portland Cement Co. in the computer control of raw mix and cement kilns, as well as quality-control experience at Fort Dodge Laboratories and Bestwall Gypsum Co.

Frederic E. Gore is a technical consultant at Fisher Controls Co. and provides recommendations for analog and digital instrumentation for chemical reaction systems and unit operations. Prior to joining Fisher in 1970, he filled several engineering positions for Monsanto Co. in the chemicals, plastics and industrial-electronics areas. He holds a B.S.Ch.E. from the University of Cincinnati, an M.S.Ch.E. from the University of Wisconsin and an M.S. in Engineering (Servo) from St. Louis University.

BATCH-POLYMERIZATION process often requires computer control to handle complex series of operations—Fig. 1

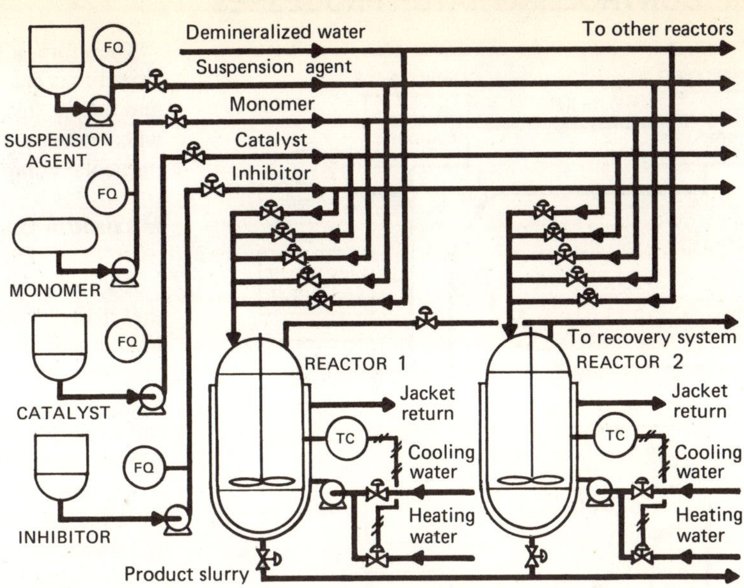

Backup For Batches

A reliable backup system is essential for computer-controlled batch operations. Design can vary from a simple arrangement that holds the process at a safe point, to a complete secondary circuit that allows uninterrupted operation.

RALPH MAYFIELD, Brown & Root, Inc.

Batch-polymerization processes today involve reaction vessels in the 40,000-gal range, often with several operating simultaneously. There may be as many as 40 remotely actuated on/off valves for each reactor, and 300 operations per batch during normal conditions. Because of the complexity and size of new batch systems, computer control is widely accepted.

In addition to the primary computer system, some type of backup control must be provided. Design philosophies vary from the use of a standby sequencer that can maintain production at a reduced rate, to computer-controlled shutdown of the process. Selecting the best system for a new polymer plant is difficult because of the many safety and economic factors involved.

Although the backup system should be designed to avoid loss of production, safety is the prime consideration. Among the dangerous conditions that can occur are loss of agitation, cooling water or temperature control. Also, overcharging the catalyst can seriously speed up the reaction. All of these conditions can result from computer failure. Without some type of backup system, emergency situations could require venting the reaction mass, which is the last possible resort.

From the production standpoint, computer systems usually have long delivery times and can require extensive debugging. On the other hand, management often insists that a new facility be put in operation as soon as it is mechanically complete, even if the computer system is not functioning. Also, if special development batches are to be run, perhaps on a one-shot basis, the computer could require extensive reprogramming. In these situations, total manual backup may be justified.

Types of Computer Systems

Fig. 1 illustrates a typical batch-polymerization facility. The basic operation consists of charging the reactors with demineralized water, suspension agent, monomer and catalyst. Heat is normally needed to bring the mass to reaction temperatures, and then the exothermic mass is held at constant temperature by agitation and external cooling. Finally, an inhibitor is added to terminate the reaction. Unreacted monomer is vented to a recovery unit, and the product slurry is removed for drying and storage. Total cycle time can be up to 8 or 10 h.

An eight-reactor plant with computer control could easily require 1,200 digital or discrete inputs, 1,000 digital or discrete outputs, 350 analog inputs, 45 analog outputs, and 15 pulse inputs. The system might be asked to monitor and transfer electrical power loads, determine abnormal reaction rates, control emergency actions, sequence large electrical equipment back online after

This article is based on a presentation at the 29th Annual Symposium on Instrumentation for the Process Industries, held at Texas A&M University on Jan. 16.

Originally published June 10, 1974.

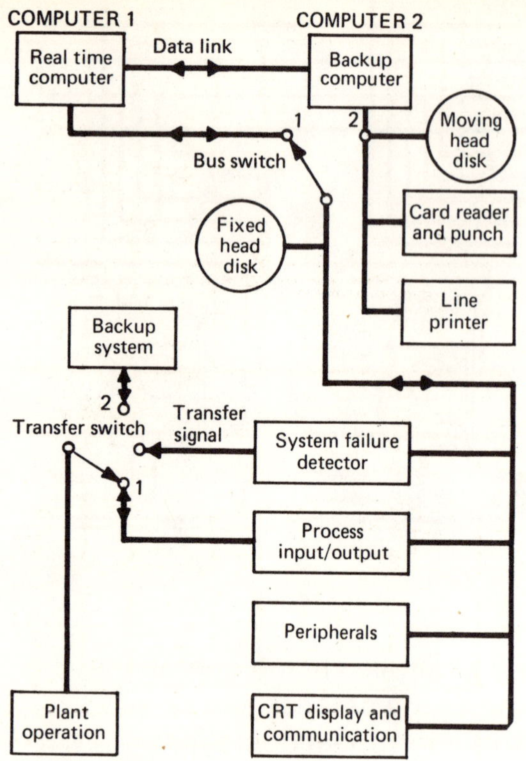

DUAL COMPUTER system for process control—Fig. 2

power interruption, calculate heat balances, store recipe information, and maintain overall process sequence. To handle these jobs, many companies favor a minicomputer system with relatively small core memory and high-speed access to magnetic-disk memory.

Once the primary control system is selected, the next step is to decide whether it is economically feasible to have a backup computer. Studies show that if a single computer system has a mean time before failure of 4 to 8 mo, two computers with a data link have a mean time before failure of 19 to 23 mo. The second computer and associated hardware cost in the neighborhood of $50,000, a rather small price to pay for the additional reliability, considering total plant cost.

A dual computer system is shown in Fig. 2. This consists of two central processing units connected by a data link, a fixed-head memory disk, and a backup moving-head memory disk. The backup computer monitors the status of the primary system and also can be utilized offline for various functions. If there is a malfunction in the first computer, the bus switch transfers the process to the backup system.

It can be difficult to determine computer failure in this type of system. Although the computer can be designed to make routine parity checks that give some indication of the system's integrity, it is possible to maintain parity and still output erroneous information. Another problem is that after a transfer to the backup system the second computer may be unable to continue. In this case, the last outputs of the primary computer might be maintained and go undetected for some time.

Therefore, it is obviously necessary to design a failure circuit that can determine the primary system's integrity and output this information to take the computer offline when necessary. Design of this circuit depends on the particular computer system involved.

Maximum Backup

A system with maximum backup control is designed so that if the primary computer fails the process continues uninterrupted, with a minimum of operator intervention. One way to achieve this is by using a solid-state logic sequencer or programmable controller. The computer only initiates a sequence; the sequencer actually performs the process operations. These operations continue until another initiate command is required. This could come from the operator. The sequences are set to hold at noncritical steps in the process.

The charging operation can be designed so that the computer only calculates quantities, and outputs this information to batch counters, which store the data until the sequencer initiates the charge. With loss of the computer, the charging continues normally. If the computer is not available to calculate charge quantities, the operator can refer to a recipe table and manually set the values.

Although this type of system has advantages from a backup standpoint, it is probably somewhat slower than direct computer charging. The reason is that the sequencer is not as flexible as the computer when making charges to different reactors. In addition, the batch counters are rather complex and require setpoint updates in either binary form or pulse train. This involves more computer outputs or more computer time.

The reaction-cycle controls can be designed so that the computer only calculates temperature setpoint and total heat flux. The setpoint can be fed to a supervisory-type controller, and the heat flux can be used to initiate the end-of-reaction sequence and start product recovery.

With computer failure, the above system would perform normally. Setpoint information and approximate reaction time can be made available to the operator in a recipe table so that he can set these values manually. While this system affords maximum backup, a direct-digital temperature control loop offers advantages, since the tuning constants can be adjusted according to ingredient concentrations, type of product, and ambient conditions. This should allow for faster heating, thus reducing total batch time.

The emergency inhibitor system can be arranged so that the sequencer monitors all necessary process variables and starts the addition of inhibitor when required. The unit might have a dedicated, uninterruptible power supply and use nitrogen for cylinder-valve operation.

Additional backup for the entire control system might include an uninterruptible power supply for both the computer and plant controls. The normal operator interface would be via a cathode-ray-tube (CRT) display, but since plant operations must continue upon loss of computer, a redundant display would be required. This probably would be a control panel or console for each reactor, which would continuously show the status of all

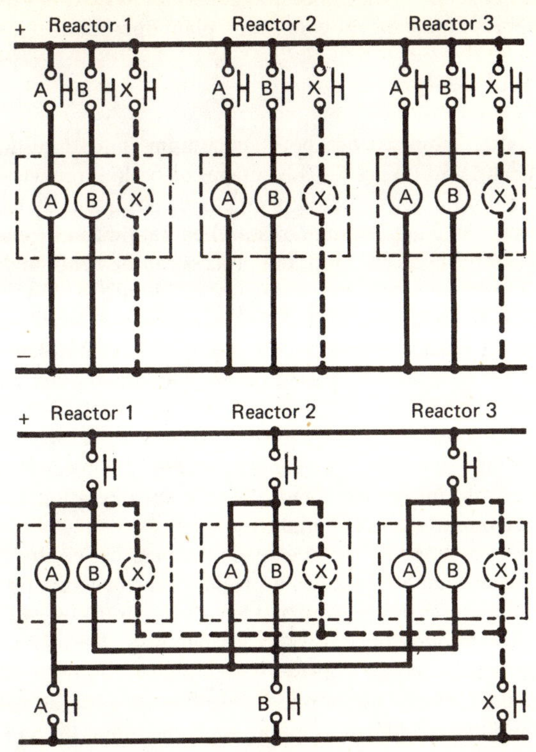

MATRIX-WIRING system (bottom) gives substantial savings in wiring compared with conventional design—Fig. 3

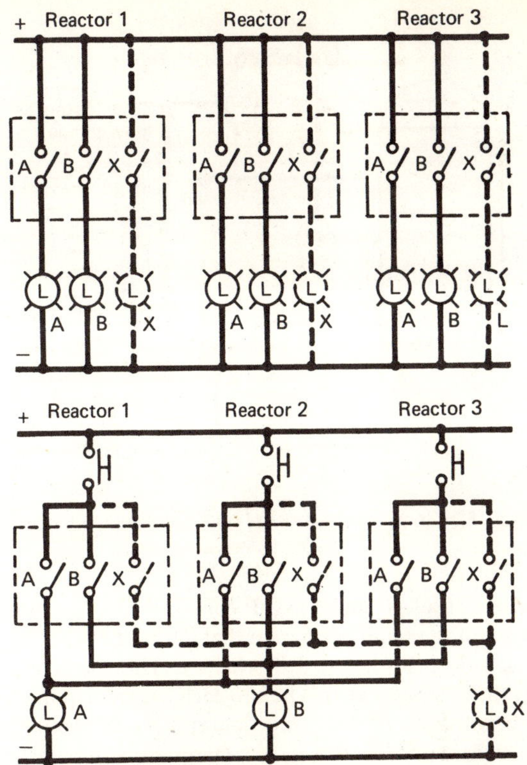

MATRIX-SYSTEM (bottom) simplifies input wiring—Fig. 4

valves and motors, reactor temperatures, reactor pressure, and agitator speed. These panels would also display the sequence step of each reactor and the time the reactor had been in this step. These counters would be updated by the sequencer so that, upon computer failure, the operator could determine the exact status of each unit.

The designer should consider that a control panel with the above display—possibly as many as 150 indicator lights per reactor—would be extremely complex. Even with the best possible arrangement, it could be quite confusing to the operator. Overall, the system outlined here would offer maximum backup but would be complicated and expensive.

Minimum Backup Design

A minimum backup system can be designed in which, upon computer failure, the process is taken to a safe condition and then shut down until the computer is again available. In other words, the system would discontinue all operations, except for maintaining safe temperature control of the reactors.

Such a system might be designed so that all digital outputs from the computer would be momentary. By using a cross-point or matrix-wiring system, substantial savings could be made—in field wiring and in the number of digital inputs/outputs and manual backup switches.

Fig. 3 shows an example of a conventional system compared with a matrix design. The devices labeled A, B and X could be solenoid valves in the field, and the switches could be for manual backup or digital outputs from the computer. The figure illustrates that adding one device (X) on each reactor requires an additional switch for each reactor in the conventional system, but only one switch overall in the matrix-wiring design. The matrix system calls for bistable devices such as a dual-coil solenoid or set-reset relay.

The same type of wiring applied to input circuits is shown in Fig. 4. While the conventional circuit uses nine indicator lights for limit-switch status, the matrix system requires only three lights and three pushbuttons. The lights can be for backup display or digital inputs, and the pushbuttons for manual display-command or computer digital-outputs.

Because of the high speed at which a computer system operates, the matrix design can offer performance equivalent to a conventional system. The disadvantage in manual operation is that continuous status is not available—the operator can only display one reactor at a time and only control one discrete device.

This type of system will maintain outputs at their last position, which is not desirable in some cases. In the event of computer failure, most of the on/off valves can be held in their last position, but all charge valves must be closed. This can be accomplished by switching from a computer power bus to a manual one, and utilizing monostable output devices such as a standard solenoid valve or relay.

The charging operation can be designed so that before initiating a sequence the computer prints out the quantities required for the next reactor in the sequence. A

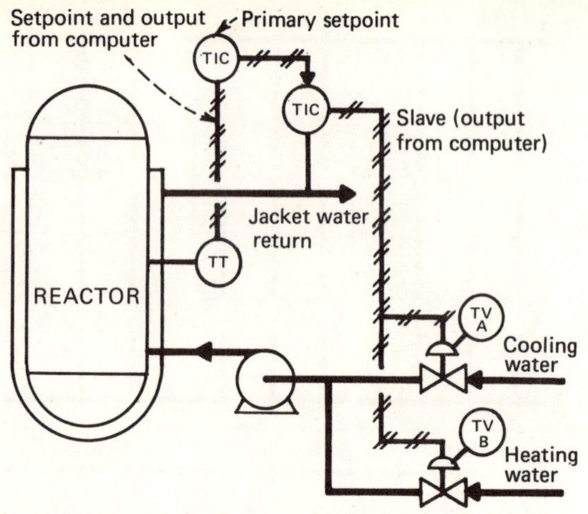

CASCADE temperature control on a batch reactor—Fig. 5

series of electromechanical counters can be used to monitor the same pulse-train that the computer is receiving from field-mounted flowmeters. These can be reset by the computer to zero before each charge. Upon loss of the computer system, the operator would have exact knowledge of charge status. Once the computer were again available, he could enter this information in the system via the CRT keyboard. While this design is quite simple, it results in total loss of production until the computer is back on line.

Direct-Digital Control

The reaction-cycle control system can be set up to incorporate a direct-digital control system, which can be adaptively tuned to meet various process requirements. This can give faster heatup with a reduction in batch time. Some type of backup must be furnished.

A simple arrangement is to have only manual backup. Upon computer failure, the control station opens the cooling-water valves to all reactors, and the operator determines the necessary valve settings.

The emergency inhibitor system can be designed so that the computer monitors all necessary process variables and initiates inhibitor addition. This type of system is simple, but relies on operator intervention for backup and could be dangerous since the emergency inhibitor must be added immediately after agitation stops.

Additional backup for the system described above would probably be very similar to that used in a maximum backup situation. One way to reduce cost is to install a gas- or diesel-driven water pump that has sufficient head to circulate water through the reactor jackets.

Because plant operations would be curtailed upon loss of the computer, display could be kept to a minimum. It probably would consist of one console from which the operator could obtain the status of valves, motors, and reactor temperature on command. Since all communication with the process would be with a CRT, a system with this configuration could be dealt with most easily by the operator during normal situations. The problem with this concept is that once the computer system is down, there is no means for continuing plant operations.

Final Design

After considering both maximum and minimum backup and taking the advantages of both approaches, a safe production-oriented plant can be designed. By modifying the charge system, manual operation can be easily accomplished. The charging process can be continued by using a batch counter along with a totalizer. If this is done, the operator must charge one reactor at a time, and once a reactor is selected for charging, the complete sequence must be completed before going to another reactor.

This design is much slower than computer operation and may possibly require more manpower. However, it is relatively inexpensive and allows plant operation when the computer is not available.

Fig. 5 shows a typical cascade temperature-control for a batch reactor. This has a primary reactor-temperature controller that is cascaded to a slave controller, which monitors outlet-jacket water temperature. A backup for this system could consist of a primary controller that would have two set-and-hold amplifiers. One receives the computer-calculated setpoint and the other the calculated output, which is the setpoint for the slave controller.

The slave controller would be a direct-digital-control station, with a remote set to maintain output signal to the heating and cooling control valves. Should computer failure occur during the reaction, the control system has the necessary information to continue the batch until it is complete.

If different tuning constants are required for heatup and reaction conditions, the backup system can be designed to meet these requirements. However, the switching circuits should be operational when the computer is online. If a malfunction takes place, the computer can alarm the operator through an alarm printer. Although a system of this type is expensive, it offers the advantages of direct-digital control and total backup.

A backup inhibitor system can be arranged so that a circuit is armed as the catalyst valve is opened. If the agitator fails before the circuit is disarmed when the recovery valve opens, inhibitor is automatically added. This design is relatively inexpensive but still provides backup for agitation loss, which probably is the most critical variable.

Meet the Author

Ralph P. Mayfield is a senior instrument engineer at Brown & Root, Inc., P. O. Box Three, Houston, TX 77001. Previously he held engineering positions at Geigy Chemical and W. R. Grace. Mr. Mayfield has a degree in mechanical engineering from the University of Mississippi and is a registered professional engineer in Texas.

Section XV
CONTROL ROOM DESIGN

How to design control panels

There is an art to getting all the instruments in a control room laid out so that the operators will note problems and be able to handle them easily. Besides this, the instruments must be readily maintainable.

John A. Masek, *Consultant, Philadelphia, Pa.*

Instrument control panels are an important part of the design of any chemical process plant. The panels should be set up for the convenience of the operators who will control the plant.

Some companies planning a new plant may leave the control panel layout to the engineering company that will design and build the plant. Others, however, prefer to have their own engineers design the front of the control panel to assure that it will meet the operators' needs.

The plant may be controlled by a computer, operated from a console with appropriate pushbuttons and readout devices. In such a case, many companies will insist on a complete backup of analog instruments on a well-designed control panel, which can be relied upon to continue to operate the plant in the event of a computer failure.

The company must assure that the panel incorporate all of the best operating design features acquired through years of operating experience. This may include the touch-dialing of several hundreds of thermocouple temperature points that can be monitored on a digital temperature-indicating system mounted on the front of the control panel. Many operators prefer a panel location to one on the desk or the control console because they are likely to be on their feet, scanning the panel, and it is distracting to have to go back to the console for such readings. Design of the control center must be such that operations can be handled by a minimum number of operators on each of the shifts.

Control panels

Control panels are essentially designed for mounting the instrument cases. Operators are normally expected

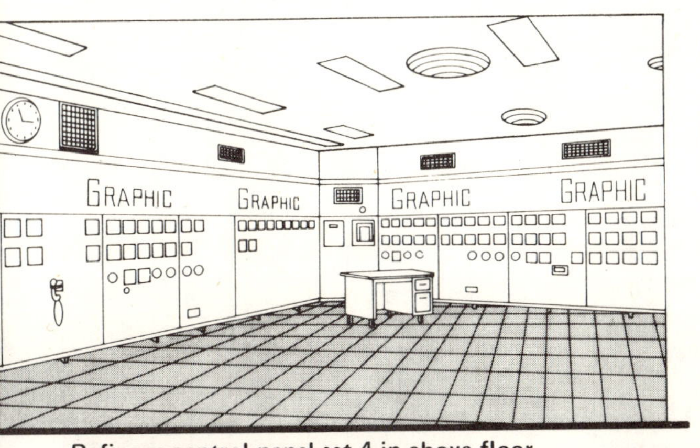

Refinery control panel set 4-in above floor for air circulation and cleaning ease **Fig. 1**

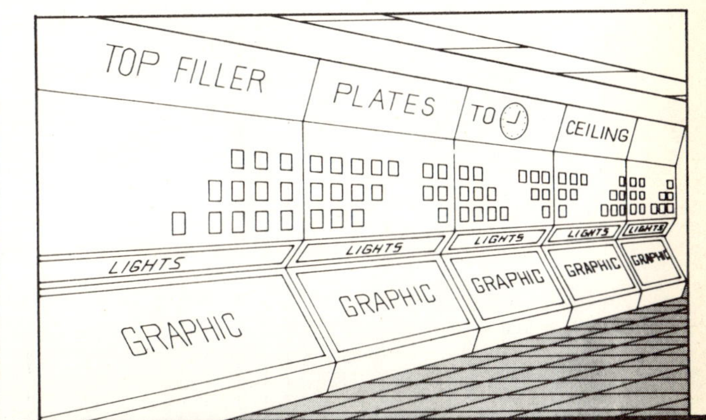

Five assembled sections make up this panel with bottom process-graphics section **Fig. 2**

Originally published June 21, 1976.

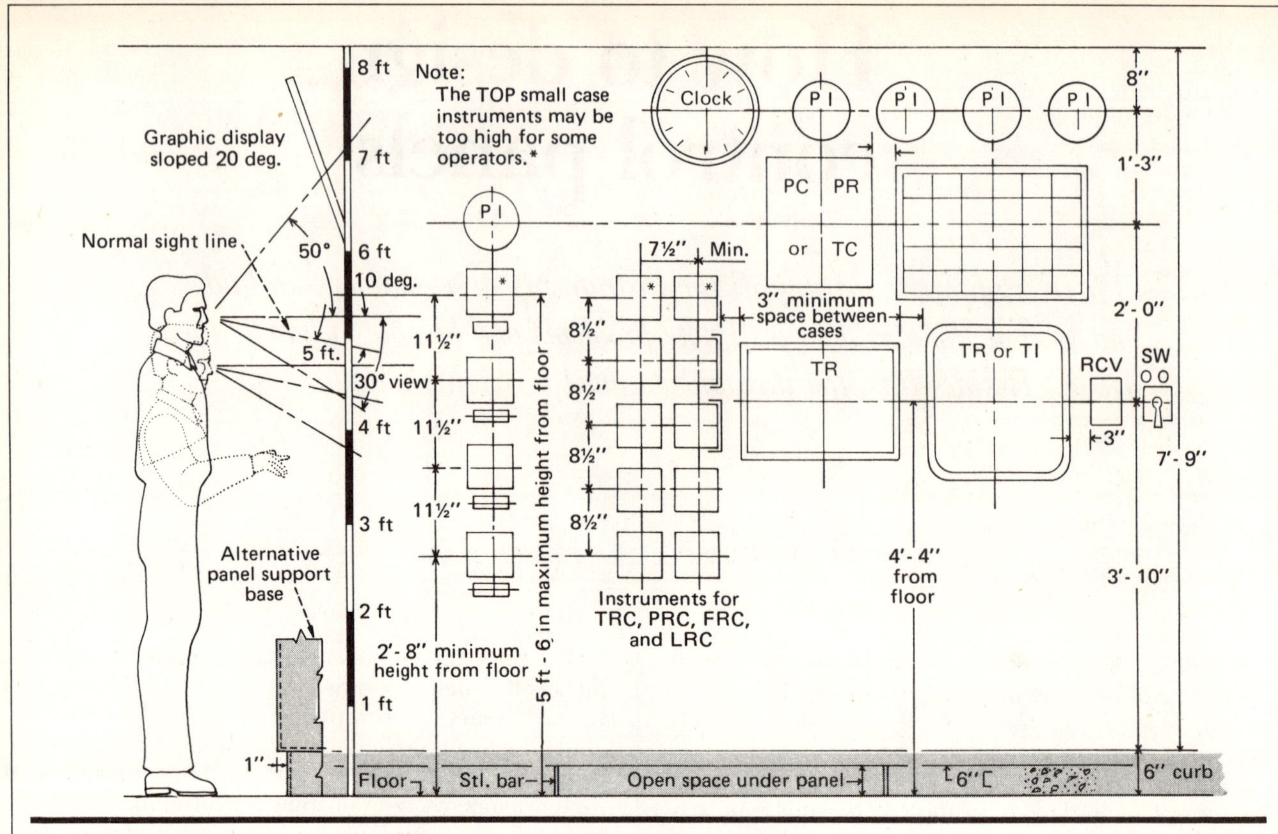

Flat-faced control panel showing typical instrument-locating dimensions — Fig. 3

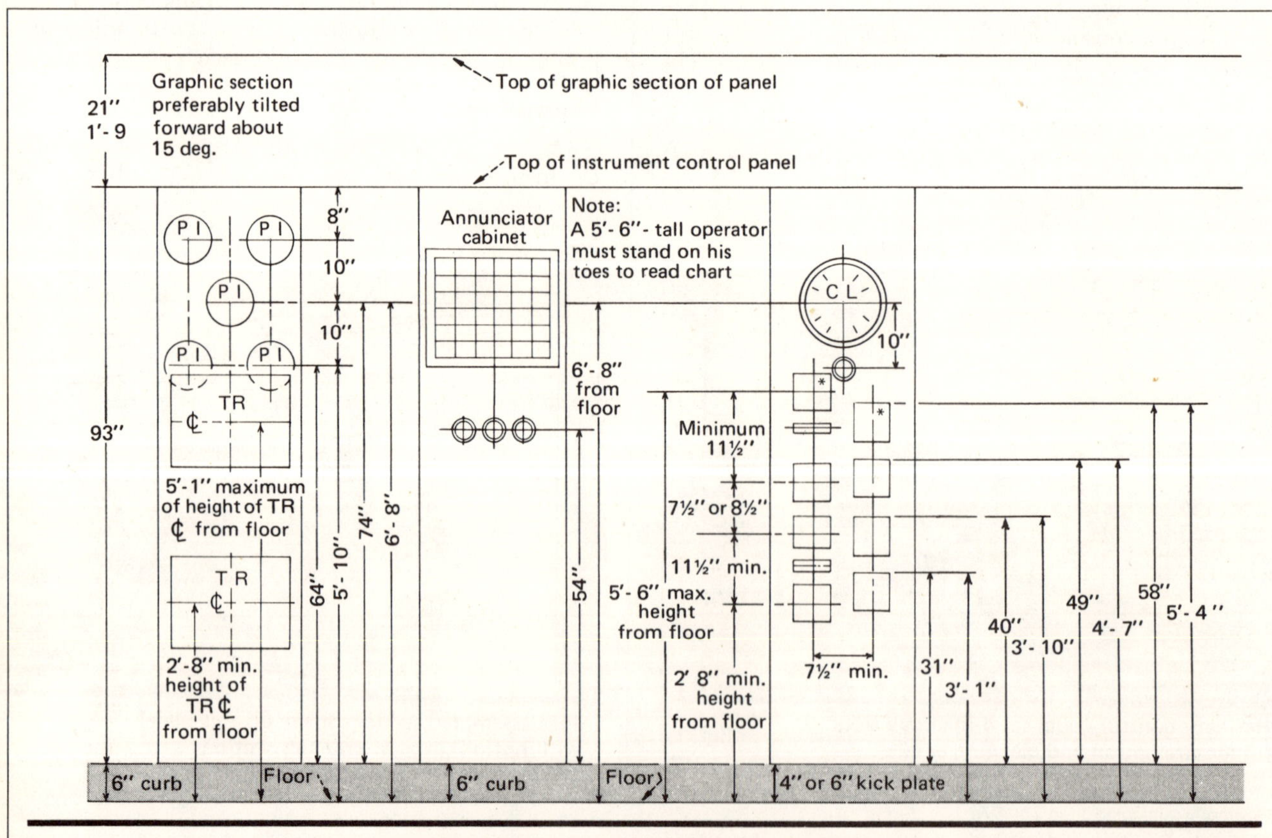

Additional typical dimensions for use in designing control panels for standing operators — Fig. 4

A. Steel filler plate furnished with control panel (installed between top of panel and control-room ceiling).
B. Sloped panel section for semigraphic flowsheet. Alarm lights may be incorporated within the diagram or in separate annunciator cabinets, with rows of windows, that are installed between sections of the flowsheet.
C. The top row of most panels usually contains the pneumatic-receiver type of pressure, level and other indicating dial-gages.
D. The three main rows of instruments include the indicating and recording controllers. These are installed in groups of ten units that are shelf mounted and installed in large panel-cutout openings. Temperature recorders are installed in the upper row.
E. On large panel cutouts, reinforced-steel bars are welded at the bottom back of the panel. The back end of the shelf is supported on a steel cross-brace.
F. Auxiliary switches for pumps and valves, as well as a telephone receptacle with connections to the field operators.
G. Open ventilating grille at bottom of spare openings.
H. Toe recess base, 4 x 4-in, at bottom of panel.
I. Fluorescent lights for servicing operations. The fixtures, with 4-ft-long tubes, are fastened inside the fabricated steel channels.
J. A minimum of 4 to 5 ft between panel and rear wall should be provided. Where CRTs are installed, additional space must be provided for passage of a 5-ft-wide service handtruck.
K. Panel base is installed on vibration-proof material and bolted to floor.

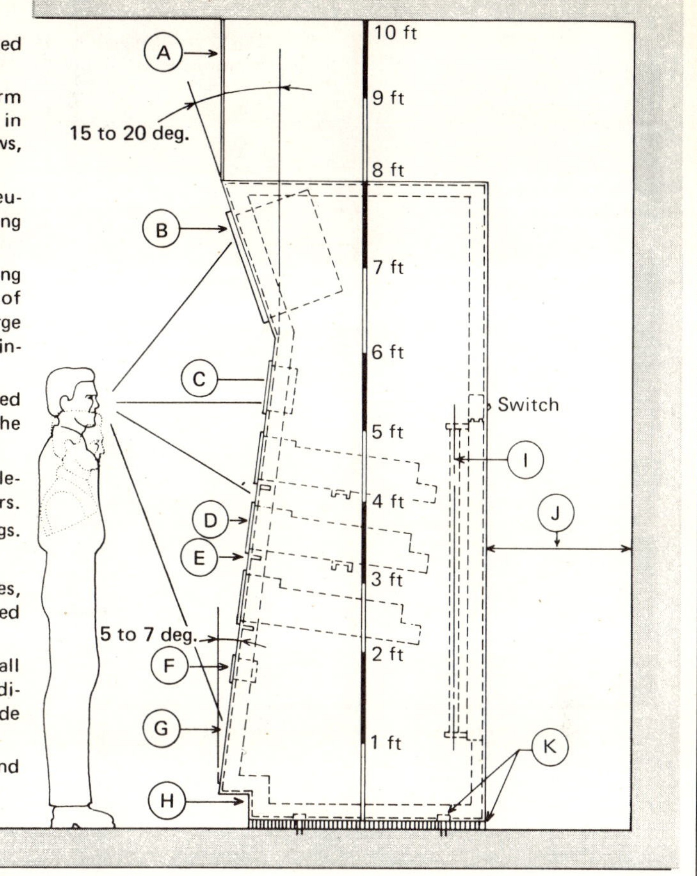

A modern instrument control-panel incorporating a sloped front **Fig. 5**

to manipulate setpoint knobs, submanual transfer switches and pushbutton switches, and these must be located at convenient elevations for easy operation.

Also, these must be laid out on the front of the panel in a systematic manner that will enable the operators to easily scan the process flow. Semigraphic diagrams of the process may be shown on the upper part of the panel, as in Fig. 1, or on the lower part, as in Fig. 2. These diagrams are references that will help in startup of new operations by men who may be unfamiliar with the process. They are also valuable for breaking-in trainees.

The diagrammatic drawings

The planning of a new process plant will always start with the development of the process flowsheet, which will set forth the basic concepts of the process. The next important drawings will be the piping diagram and the instrumentation diagram, which is similar to the piping diagram, but with the main emphasis on the instruments and control points located on process piping and equipment.

Instrument locations are indicated with ISA (Instrument Soc. of America) standard symbols [*1*] that are lettered within 7/16-in-dia. circles that have a horizontal line through the center, indicating a panel-mounted instrument. In the upper half of the circles are the instrument abbreviations such as TRC (temperature recorder-controller) or PC (pressure controller). The lower half of the circles are used for the instrument index numbers.

Most process and instrumentation flow diagrams start the flow from the left and continue in proper sequence toward the right of the drawing. It is customary to lay out instrument panel boards in the same way, unless there is a need to group the instruments on a station-to-station basis.

If an engineering construction firm is to do the design, this will normally be based on conferences with the process company's engineers. They will discuss whether the panel will be a conventional straight-front free-standing one, such as in Fig. 3, and 4, or one having a break-front (Fig. 5, 6). There may be need for a bench-board-type panel (for mounting the controllers and manually operated switches) (Fig. 7), that may also have a flow semigraphic display at its top. Or, the engineers may require the parabolic-front type of panel (Fig. 8), which is receiving attention because it is designed with human-engineering data [*2*].

The practical heights above the floor for installing instrument cases must always be considered, and some are noted in Fig. 3 and 4. There is a temptation to place instruments too high—this has been a source of complaint with panels shipped to Asiatic countries,

A hydrocracking-refinery process instrument panel handling over 120 operating loops

Fig. 6

where shorter operators are the rule. With some types of trend recorders installed at a 5½-ft centerline level, a 5-ft 6-in operator will be forced to stand on his toes to read the recordings.

The semigraphic flow diagrams allow the operator to respond without hesitation in times of emergency. Such panels are generally installed at the top of the board and are preferably sloped for better visibility. It is good practice to locate the instruments so that they will be approximately vertically above or below where the instrument point appears on the diagram. With some graphic displays, built-in annunciator alarm lights are incorporated within the diagram (Fig. 9), and can have connections to the alarm annunciator cabinet at the top of the panel as a double verification.

Alternatives to the semigraphic panels include the projector-type display where process flow references and data are stored on 2 × 2-in slides, and are selected and projected at will. Another is the cathode ray tube (CRT) that can present a four-color display of variables or data either graphically or digitally. If a CRT system is installed in the panel, a minimum 5-ft clearance must be provided behind the panel for passage of a service handtruck.

Steel plates for panels

Most panels can be fabricated from ⅛-in-thick hot or cold-rolled steel plates. The plates are stocked by the steel warehouse in 8-ft lengths, and these can be formed into panels by bending a 1 or 2-in length at the ends at a 90-deg angle. This leaves either a 7-ft 8-in or 7-ft 10-in front length on which instrument cutouts can be made. Sections of this length are easily handled and shipped by truck, and bolted together in the control room. Very long panels may present problems—they have fallen off trucks and have had to be remade. Also, they cannot be put into ships' holds and must be handled as deck cargo, which may result in many of the instruments having to be replaced after a long voyage.

Steps in control-panel design

After a flowsheet and instrumentation diagram are completed, the specific instruments themselves must be selected and specified. This prepares the way for developing the front of the control panel. The instrumentation engineer will start by making freehand sketches of the panel layout. The panel designers will then draft the layout, usually on a 1 in. equals 1 ft scale. Preferred instrument heights are shown on Fig. 3 and 4 (usable on most straight-front panels). To accommodate all the instruments, it may be necessary to go to a U- or L-shaped panel, depending on the control loops and control-room floor area.

Some instrument makers will provide rubber stamps with scale drawings of their instrument cases. Such stamps can be used to reduce drafting time. The instrument symbol and index number, as well as operating service description, are lettered in after the instrument outline is stamped on the tracing.

On the panel drawing, the vertical instrument rows are frequently numbered at the top (1, 2, 3, etc.), while the horizontal rows are identified with capital letters (A, B, C, etc.). The top (A) row is generally used for 6- or 8-in indicating dial pressure-gages, draft gages and alarm lights, or annunciator windows in a cabinet. The B row will contain space for manually monitored temperature indicators, alarm-annunciator reset buttons, and recorders for temperature, flow, pressure and level. The C row will be used for multipoint pyrometer temperature recorders, manually operated switches and phones. The D row is for large-case recording flow-controllers or manually operated switches—if there is a wet meter, it must be positioned here. An instrument containing mercury must never be installed over an electrical instrument.

At the bottom of the panel (Fig. 1), a 4-in-high kickplate is recessed 4 in, or this space can be left open (except for steel supporting bars) for ease of floor cleaning.

The tabulations of instruments should contain the horizontal and vertical row designations, which will facilitate locating the device on the panel drawings.

In making cutouts the minimum space between instruments should preferably never be less than 3 in. Also, the visible parts of the instruments should never cover more than half of the total panel surface. This will assure having sufficient space in back of the panel for wiring, tubing runs, relays and pressure switches.

Types of instruments

The instruments mounted on the face of the panel fall into three general types, and the average panel will carry some of each. Some will be large, fullscale cases—for example, the multipoint temperature-recording pyrometer or the recorder-controller flowmeter with a circular chart (Fig. 3, 4). Then there will be miniature-case indicating and recording controllers having 6 × 6-

in cases as in Fig. 3 and 4. The more recent trend is toward the high-density types of indicating controllers that are 2 or 3-in wide and 6-in high, and recorders that are 6-in wide and 6-in high, and 24-in deep.

The high-density instruments are assembled in groups of 6 to 10 in rectangular panel-cutouts that hold metal-box shelves with grooved tray-tracks for sliding the instruments in or out from the front of the panel (Fig. 5, 6, 8).

Some operators claim that the high concentration of instruments is difficult to use because this requires more-careful attention when observing readings or changing setpoints. This crowding can be overcome by using blank spacers between instruments (Fig. 9). This blanked-off space may prove useful if a modification of the process calls for additional instruments.

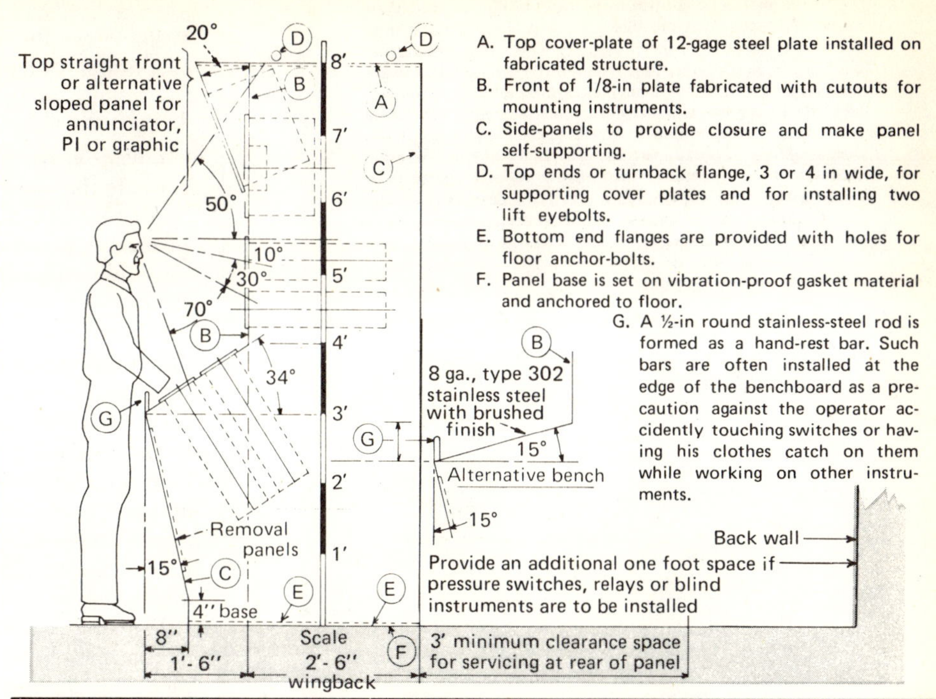

Typical dimensions for control panels using a bench-board arrangement **Fig. 7**

Arranging panels

Making the general arrangement of the panel can be time-consuming because the drawings often have to be remade several times as problems are solved and improvements incorporated. So, instead of drawings, a photo technique is sometimes used. Small photos of the panel instruments are made (to scale) and are mounted on cardboard, backed with an adhesive substance. These photographs are experimentally laid out on a drawing of the panel until a suitable arrangement is agreed upon and approved. The final layout is photographed and this picture is submitted to the client for its approval, and then to the panel vendor (along with schematics and instrument lists) for bid. After receiving the order, the panel fabricator makes his steel-fabricating design drawings, basing them on the photograph.

Full-scale panel mockups

Some companies feel that they cannot approve a complicated control panel simply on the basis of looking at drawings or photographs. They insist on full-scale mockups. These are usually made of steel-surfaced plywood, shaped like the front of the panel. Then full-scale photographs of each instrument are either obtained from the manufacturers or made. These are mounted on cardboard and backed with magnetic tape. The combination of magnetic tape and steel-surfaced plywood enables the photographs to be hung as they appear on the general arrangement drawing for the panel.

These mockups can be studied by the engineers as well as maintenance personnel and control-room operators. Instruments can be rearranged or respaced easily since the photographs are magnetically self-supporting.

The service manager will have an opportunity to see that there will be sufficient clearance around instruments for finger movement and for wrench and screwdriver accommodation so that cases can be removed easily for repair when necessary.

In addition to the cutouts for the instruments that are finally decided upon, it is well to provide an additional 20% of blanked-off cutouts to provide for future expansion or instrument changes.

Startup considerations

A mockup is especially useful in simulating startup operations. Often, four or five men will be working to-

Parabolic-front panel puts operators' vision in focal plane of each instrument **Fig. 8**

This semigraphic control panel handles 650 process-plant loops. Indicating dial pressure-gages and alarm annunciators are at the tops of the panel sections. The semigraphic panel has built-in alarm lights that relate to the diagram. Below are two rows of indicating, analog, electronic solid-state controllers. The 1 x 6-in control instruments are located in functional groupings, corresponding to the process operations of the diagram above. The 12-instrument capacity of the mounting cabinets allows for fewer instruments to be installed with blank covers over the unused spaces. This permits panel design flexibility—future changes can be accommodated without alteration to the main panel. The bottom row of cabinets contains the process recorders. The console shows only the more-important process controllers.

In this panel, alarm lights are incorporated in the semigraphic process flowsheet

Fig. 9

gether in front of a panel, analyzing operating conditions. Each will watch his assigned instruments, and many of the controls will have to be manipulated manually. Many of the adjustments will have to be made with the instrument covers open.

With some types of instrument cases, an open cover will overlap the adjacent instrument and can lead to difficulties during the tense periods of startup, particularly if the density of instruments also makes for crowding of operators. (Sometimes the instruments are so densely arranged on a console that it resembles a modern airplane cockpit.)

The design of the panel should be determined by operator needs during abnormal conditions—startups and upsets—so that the operators will have quick and easy access to switches and protective controls.

Control-room size

The size of the control room is dictated by the size of the control panel and will vary from plant to plant. If preliminary panel layouts have been completed, this is not a great problem; but, sometimes, the architect will have to plan a control room even before an instrument engineer has been assigned to the project. In such a case, the control room will have to be sized on the basis of those used in previously executed projects. Sometimes, this is satisfactory, but there have been cases where the estimated dimensions have been so small that instruments have had to be crammed on to the panel—even the wall space behind the panel has had to be pressed into service.

Engineering blueprints

Customarily, the designers and builders of a plant provide the plant's operating management with a complete reference set of all the prints used in building the project. The drawings that will be referred to most frequently will be the process piping and instrumentation flow diagrams and the instrument control-panel drawings.

If these drawings are furnished as blueprints, they will never withstand the wear and tear in handling during years of plant operations. Consequently, many contracts call for these drawings to be furnished as either cloth or plastic reproduced tracings. These will last as long as the plant, and will always be available for study when future modifications are planned. Such modifications can be incorporated into the tracings, and prints made for use in the field.

When only the panel fabricator's prints are furnished as the reference for the general arrangement of instrument panels, the drawings will contain steel construction details that are not of much interest to the process and instrumentation engineers. What they need to know is the general arrangement of instruments on the panel, with their index numbers and general operating service notations.

In such cases, the engineering department should remake the drawings, preferably on a larger scale that will permit lettering the complete operating service description within the instrument case outline. Only the overall instrument panel dimensions need be shown, as well as the centerlines of the instrument rows. When modifications must be made to the instrument layout, they are noted on the tracing, and blueprints are made for the personnel who will make the actual changes.

References

1. "Instrumentation Symbols and Identification," ISA Standard S5.1. Revised 1973. Instrument Soc. of America, 400 Stanwix St., Pittsburgh, PA 15222.
2. "Human Engineering Guide to Equipment Design," Revised ed., 1972. Supt. of Documents, U. S. Govt. Printing Office, Washington, DC 20402.
3. "Panel Design Concepts," PUB 116B, The Foxboro Co., Foxboro, MA 02035.
4. "Panel Consideration, Design and Systems Philosphy, Quick-Scan 1400," #PDS-3M002, and "Taylor Quick Scan 1300," #PDS-3E001, Taylor Instrument Process Control, Rochester, NY 14601.
5. Sellwood, F. E., "Instruments Panel and Console Design," Pub. TP 5060, Kent Instruments Ltd., Luton, Bedfordshire, England.

The author

John A. Masek, The Drake—Box 202, 1512 Spruce St., Philadelphia, PA 19102, has had a long career as a piping and instrumentation designer for oil refineries and chemical and power plants. He is presently retired from Gulf Oil Corp.'s Philadelphia Refinery where he accumulated 25 years of service. After leaving Gulf, he was associated with United Engineers and Constructors, Gilbert Associates, Kuljian Corp and W. F. Koelle & Sons.

He studied at Northwestern University and Armour Institute, and is a charter member (retired) of the Instrument Soc. of America.

CRT Consoles: New Look in Control Rooms

CONSOLE includes graphic, alphanumeric CRTs—Fig. 1

More capable, flexible and reliable than panels of instruments, interactive cathode-ray-tube operator consoles can return a higher rate on investment. Savings are also possible from less training, and higher product quality and yield.

NOEL R. STRADER II, Houston Engineering Research Corp.

Cathode-ray-tube (CRT) systems currently available can be categorized as digital TV or stroke-write—both of which are shown in Fig. 1. These categories can further be identified as full graphic, limited graphic, or alphanumeric.

Operator-Console CRT Types

Information is "printed" on the **digital TV** screen when the electron beam that is moving in a fixed pattern over the face of the screen is turned on to illuminate only points where information is desired. The beam is moved in a fixed pattern of horizontal lines that are similar to the standard 525-line raster used in home TV sets. The resolution of individual dots in the horizontal direction depends on video amplifier bandwidth, CRT deflection speeds, and screen phosphors. Commercial systems are usually limited to 580 display-bits/line.

With the *full-graphic digital TV,* the beam has access to each dot individually (not in a fixed pattern). Consequently, process-control schematics and system block-diagrams are represented on the CRT screen by arrangements of dots. Digitally generating such a display from randomly located vectors requires that memory be provided for each horizontal display-point on the screen. The complexity of such a system makes it unacceptably costly.

The *limited-graphic digital TV* is similar to the alphanumeric (discussed next) except that it provides additional special symbols. Typically, horizontal and vertical lines, as well as special blocks and other symbols, are formed by a 7x9 or 5x7 dot matrix. This adds to complexity and cost but does provide minimal graphic capability.

This article is based on a paper presented by the author at the Jan. 1973 meeting of the 28th Annual Symposium on Instrumentation sponsored by the Dept. of Chemical Engineering of Texas A&M University.

Originally published April 30, 1973.

The *alphanumeric digital TV,* because it makes possible access to groups of dots (7x9 or 5x7 dot-matrix characters), can display from 2,000 to 3,500 characters. Each dot matrix can represent any alphanumeric character or special symbol. By limiting the number of characters to approximately 64 different combinations, the memory required for the continuous display of alphanumeric information is minimized. The alphanumeric digital TV affords an economic way of displaying alarm and other messages to the control-room operator; however, its ability to display process schematics or system block-diagrams is extremely limited.

With the **stroke-write CRT,** the beam moves only as necessary to display information. It is turned on to make one presentation, then turned off, moved to the next location, and turned on to make another.

The screen of the *full-graphic stroke-write CRT* is organized into a specific number (usually 1024 by 1024) of addressable vertical and horizontal points. The beam can begin at any point and continue to any other, and go over the same location more than once. This CRT provides the most complete capability of any for presenting schematics, flow diagrams, bar charts, trend graphs, alphanumerics and other information.

The screen of the *limited-graphic stroke-write CRT* is also organized into a specific number (usually 1024 x 1024) of addressable vertical and horizontal points. Unlike the full graphic, this CRT does not provide for moving from any point to another with a single vector. It, instead, permits the beam to "draw" at several different vector angles (for example, 24 angles) from any point with multiples of short-line segments. More than one line segment may be needed to connect a pair of points. If

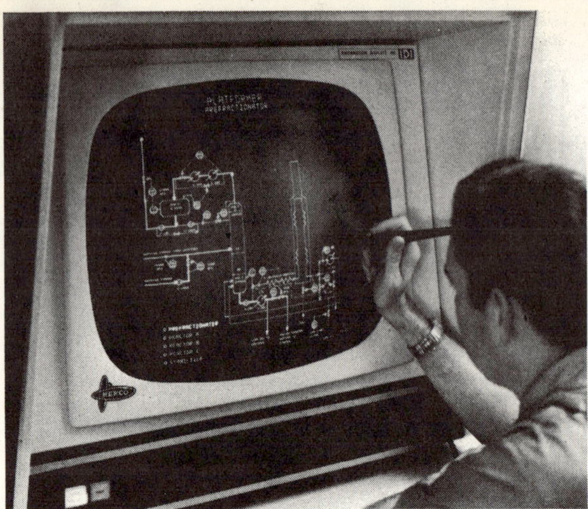

OPERATOR calls prefractionator with light pen—Fig. 2

PARAMETERS of stabilizer appear at upper right—Fig. 3

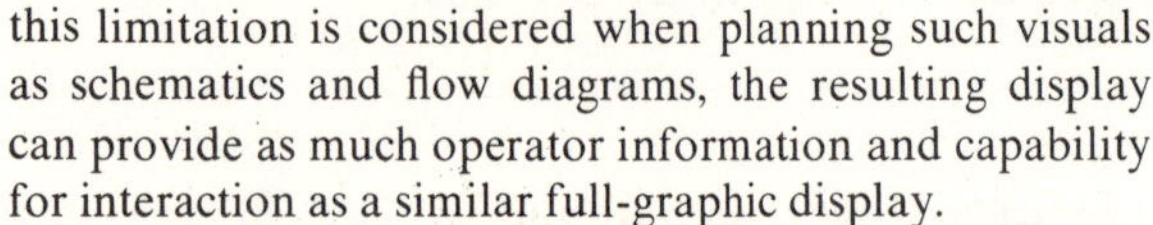

this limitation is considered when planning such visuals as schematics and flow diagrams, the resulting display can provide as much operator information and capability for interaction as a similar full-graphic display.

The third type of *stroke-write CRT,* the *alphanumeric* is not cost effective because alphanumeric capability is generally provided on both full- and limited-graphic CRTs and because the alphanumeric digital TV usually costs much less than the alphanumeric stroke write.

The *alphanumeric digital TV* is probably the most widely used CRT, and is the least expensive of the types discussed.

Alphanumeric units cannot generally provide the detail required for good interactive control; therefore, the full-graphic and limited-graphic stroke-write CRTs are most often used in interactive CRT operator consoles. The limited-graphic is now the most effective stroke-write CRT because its excellent operator-interaction capability is available at significantly less cost than that for a full-graphic system.

Process-Plant Applications

An interactive CRT operator console in a petroleum refinery results in a simple and economical means of performing complex operator functions. The CRT displays block diagrams and process-flow schematics of the plant. Standard ISA instrument balloons and identification appear on the display, allowing alarming and monitoring.

Fig. 2 and 3 show typical graphic CRT displays for the prefractionator and stabilizer of a Platformer. The system "menu" at the lower lefthand corner gives a list of the sections of the Platformer that can be displayed. The operator can display any section of the Platformer by pointing to its label with the light pen. After the display is in view, the operator can call up the parameters for any instrument by touching the instrument balloon with the pen. The parameters will appear at the upper right.

When a setpoint control instrument is selected, the system's response will be to display a submenu with RAISE and LOWER functions. (Alternatively, a 10-key keyboard can be displayed. The current value of the setpoint will be displayed for either option.) The operator can change the setpoint by pointing to RAISE or LOWER. The label of the action selected will be intensified and an EXECUTE label will appear. The operator will then point the light pen to EXECUTE to vary the setpoint value in the direction chosen. When the setpoint reaches the value selected, the operator lifts the light pen to maintain it.

With the 10-key keyboard, the operator can select successive digits (which appear as numerals in the display) with the light pen. When the desired setpoint is selected, the operator points to EXECUTE to enter the new value.

A system alarm commands immediate operator attention by sounding an annunciator and blinking the name of the Platformer section causing it. If the name Stabilizer were blinking, the operator would respond by silencing the alarm and selecting the stabilizer section. The alarming element will blink when the section display appears. If, for example, both the instrument balloon for the level-indicating controller and the label HLA were blinking, a high-level alarm for a stabilizer column would be indicated. The operator would examine the stabilizer, and all the information displayed for it, and then determine the cause of the alarm and the action necessary to correct the condition. If insufficient flow from the column caused the alarm, the operator would probably ask a field operator to check the bottoms pump and the flow-control valve operated by the level-indicating controller.

The versatility of the graphic CRT in process control permits elimination of all the indicators, recorders, annunciators and graphic panels from the control room. Manual backup for the CRT console is provided by analog controllers installed as blind instruments in relay racks in the equipment room. Redundant failover techniques assure overall system uptime of 100% at reasonable costs.

Interactive CRT vs. Conventional Control

The conventional refinery control-room—with its array of graphic panels, controllers, recorders, indicators, and annunciator boxes—is usually no more than a consolidation of control panels and instruments for each of the units of a multi-unit process plant. It, nevertheless, represents the "modern" approach for many processors to safe, stable and profitable control of operating units.

Large numbers of panel-mounted instruments obviously require substantial direct-capital investment in instrumentation, as well as in building space and utilities. Operating and maintenance costs to support such a facility are also significant.

An alternate method of control-room consolidation (Fig. 4) depends on interactive graphic CRTs to perform monitoring and control operations. Peripherals, such as printers, alphanumeric CRTs, teletypewriters, and push-button keyboards, provide for alarm messages, system logs and operator interaction. Interfacing to field instruments and controls can be made directly from the system controller or concentrated in remote stations to minimize wiring, instrumentation, and system overhead, while retaining centralized monitoring and control.

Fig. 5 shows an interactive CRT operator console containing two interactive full-graphic, stroke-write CRTs and an interactive alphanumeric digital TV. The full-graphic units are equipped with light pens for operator interaction with the process, and the alphanumeric CRT with an alphanumeric keyboard. Such consoles, as an integral part of a centralized control system, have demonstrated the ability to perform the functions of conventional control-room equipment more flexibly, reliably and profitably.[2,7]

The design basis for a consolidated control center must specify a system that meets the following operator interface criteria:

- Graphic displays must enable the operator to visualize the process, and to associate flow, pressure, temperature and other variables with the instruments that monitor and control them.
- The operator must have immediate access to these instruments at any time for monitoring and control purposes.
- All critical monitoring, display and control functions must be backed up.

Economics Favor CRT Consoles

Economic incentives for installing graphic CRTs are illustrated by an example that compares returns on investment for consolidating the control of seven process units by conventional methods and by interactive CRT operator consoles.

The following is assumed:

- Economic justification for consolidating the control of seven existing process units into a central control-room is being considered.
- The process units, although scattered, can be conveniently grouped into three main areas.
- Only operator-initiated actions and controls are to be considered in the economics.
- Credit is not included for the lower cost of training operators for interactive CRT control.
- Credits are not claimed for increased operating efficiency, improved product quality or reduced offgrade products with interactive CRT controls.

The cost of consolidating control conventionally is estimated as:

Instrumentation	$1,100,000
Wiring to central center	$185,000
Building renovation (4,000 sq.ft.)	$200,000
Total cost	$1,485,000

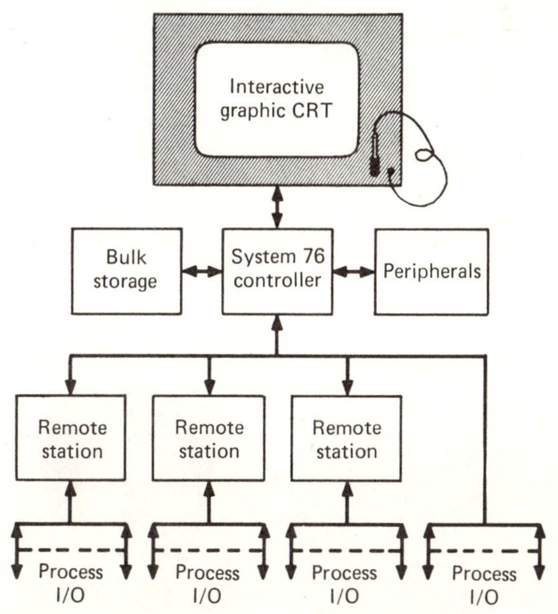

SYSTEM is typical for monitoring and control—Fig. 4

CONSOLE contains two graphic stroke-writes—Fig. 5

These figures are based on an actual customer cost study. Substantial additional instrumentation would have to be added to accomplish the consolidation, the study revealed.

The plant consists of process units concentrated in three main areas. A system configuration for consolidating the seven process units, based on using a remote station for each of the three main areas, is shown in Fig. 6. Placing remote stations at these locations minimizes wiring and instrumentation costs. The cost for a redundant interactive CRT installation for this plant with three remote stations is estimated at:

Three programmable remote stations (including wiring)	$270,000
Three CRT operator consoles with redundant controller and bulk storage	$303,000
Building renovation (1,000 sq.ft.)	$50,000
Total cost	$623,000

Returns on investment between conventional and CRT-console consolidations:

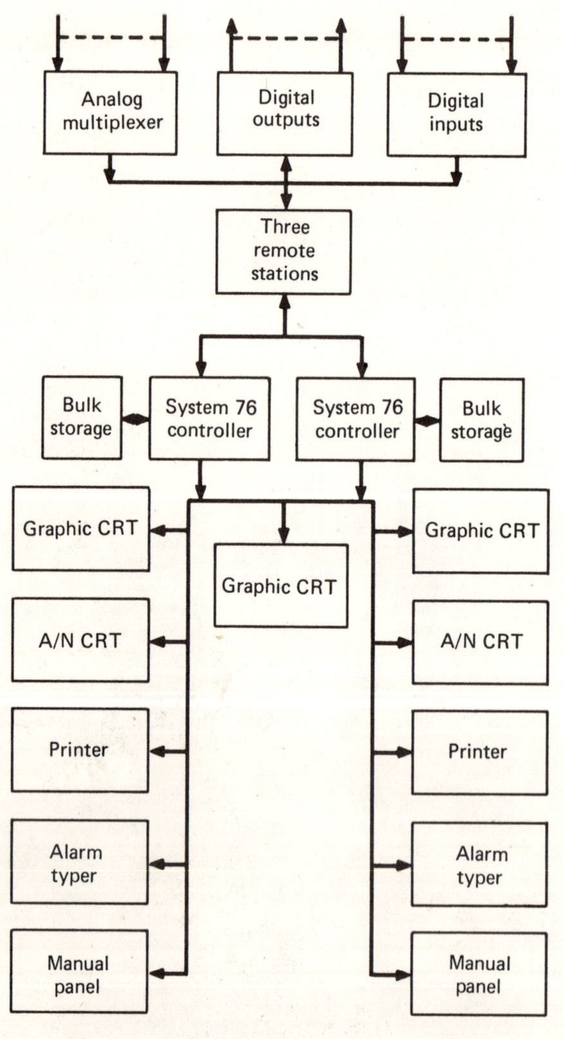

CONFIGURATION combines seven process units—Fig. 6

Conventional Consolidation

Total investment, dollars	1,485,000
Savings:	
Shift-position reduction from 14 to 9	200,000
Less depreciation, 5%	−75,000
Less maintenance, 5%	−75,000
Savings before taxes	50,000
Less taxes, 50%	−25,000
Net savings	25,000
Return on investment	2%

CRT Operator Console Consolidation

Total investment, dollars	623,000
Savings:	
Shift-position reduction from 14 to 3	440,000
Less depreciation, 5%	−32,000
Less maintenance, 5%	−32,000
Savings before taxes	376,000
Less taxes, 50%	−188,000
Net savings:	188,000
Return on investment	30%

The returns are based on reduced operator, maintenance, depreciation and tax costs. For this case, it is clear that conventional consolidation cannot be justified, because of the high cost of additional instrumentation. Three shift positions are typical for the equipment configuration chosen in the example. Even if five shift positions were retained, the return would still be 24%.

The proposed configuration provides for 100% uptime. The three operator interface CRTs and dual controller and storage afford full redundancy for the consolidated control center. Manual backup at reduced capability is possible with manual panels at each remote station.

References

1. Crowder, R. S., CRT Interfaces for a Continuous Plant, *Instr. Tech.*, Jan. 1971, pp. 58-62.
2. Hale, D., Energy Control Center, *Pipeline & Gas J.*, Apr., May 1972.
3. Kelleway, J. L. and McMorris, A. H., Economics Favor Replacing Refinery Control Rooms with Interactive CRT Operator Consoles, Proceed. ISA Spring Distillation Symp., 1972.
4. Knipp, R. S., An Advanced System for Process Instrumentation Handling, Proceed. Texas A&M Instrumentation Symp., Jan. 1972, pp. 45-53.
5. Lauher, V. A., Weinrich, C. W. and Underwood, R. K., CRT Control Center for a Multi-Loop Process, *Instr. Tech.*, Sept. 1970, pp. 33-38.
6. McMorris, A. H., Kelleway, J. L., Tapadia, B. and Dohmann, E. L. Are process Control Rooms Obsolete?, *Control Engr.*, July 1971, pp. 42-47.
7. Standard Oil Co. (N. J.) (now Exxon Co.), 1969 Annual Report.

Meet the Author

Noel R. Strader II is Asst. Manager of Software Systems Engineering for Houston Engineering Research Corp. (P.O. Box 3246, Houston, TX 77001). He has served as project engineer involved with CRT interactive control systems and participated in design studies for their installation and operation. He received B.S., M.S. and Ph.D. degrees in electrical engineering from the University of Houston. His major areas of research have included digital systems and hybrid-control-systems analysis.

Section XVI
INSTRUMENTATION AND SAFETY

An Evaluation of Intrinsically Safe Instrumentation

The concept of intrinsic safety requires limiting the electrical energy that may be released from process control instruments and associated wiring so as to prevent ignition, under normal or abnormal conditions of operation, of combustible gases, vapors, dusts and fibers in hazardous locations.

H. C. DELONEY, Texaco Inc.

Most engineers realize the potential that intrinsic safety (I.S.) has for reducing hazards to property and personnel. Somehow, acceptance and application of intrinsically safe equipment have been on a smaller scale than had been forecast. Many users still seem to prefer installations other than intrinsically safe ones because they feel these best suit the situation involved.

Our purpose here is not to deal with the design or testing of intrinsically safe equipment, but to look at:

- Past acceptance of intrinsic safety in the U.S.
- Present availability of intrinsically safe equipment.
- Expected trend toward intrinsically safe installations by industry to meet new safety regulations.
- Hopes for greater standardization in design and approval of I.S. systems to get better acceptance by users.

The word "intrinsic" is here defined as meaning "inwardly" or "inherent." The Instrument Soc. of America in RP12.2[3] defines intrinsic safety (I.S.) as follows:

"I.S. equipment and wiring is that which is incapable of releasing sufficient electrical or thermal energy under normal or abnormal conditions to cause ignition of a specific hazardous atmospheric mixture in its easily ignited concentration."

These provisions are also recognized by NFPA-493.[4]

Since the concept of I.S. depends on the limitation of energy, it cannot be applied to power circuits such as used for lighting, motors, etc., but is useful for low energy levels used in electronic instruments.

Hazardous Areas in Plants

Section 500 of the National Electrical Code[2] defines hazardous areas in the following manner:

Class I areas involve gases and/or vapors.
Class II areas involve dusts.
Class III areas involve fibers or flyings.

Class I consists of four different groups:

Group A: Acetylene. Most hazardous.
Group B: Hydrogen or similar gases and vapors.
Group C: Ethyl ether vapors, cyclopropane, ethylene.
Group D: Hexane, gasoline, naphtha, propane, etc.

Division 1 areas may contain ignitable concentrations of gas, dust or fibers under normal conditions. Division 2 areas may contain ignitable concentrations of gas under abnormal conditions, or where dust or fibers may be present, stored or handled.

During the design of a process unit, each section is classified by a combination of the above definitions. For example, Class I, Group D, Div. 1 would be a hazardous area containing a gas such as pentane, under normal conditions.

Any electrical equipment installed must be suitable for use in the specific area.

Reduction of Hazards

Explosion-proof installations allow safe operations of electrical equipment in potentially hazardous areas. They accomplish this by being designed so as to confine any explosion (ignited by the equipment) within a strong, heavy, relatively tight enclosure, and thereby prevent spread of the explosion outside the enclosure.

Although explosion-proofing permits the use of high-level voltages and currents and is still very useful for this service, it has many disadvantages for low-energy circuits that are common in electronic instruments. Some of the drawbacks: (1) enclosures are heavy, bulky, expensive; (2) power must be turned off before removing covers in hazardous areas, and this limits the ability to test, check and maintain the instrument without removal to a safe area; (3) maintenance is a big problem.

Fig. 1 shows a typical explosion-proof installation required to meet the National Electrical Code.

Air purging is another method sometimes used for reducing or limiting hazards with electronic instruments. The National Electrical Code[2] in Article 500, Par. 1, recognizes this method, provided it meets the applicable portion of NFPA 496,[4] which is adopted from ISA S-12.4.[3] This method is more often used for a particular instrument than a complete system.

This article is based on a paper originally presented at the 27th Annual Symposium on Instrumentation for the Process Industries, Jan. 19-21, 1972 at Texas A&M University, College Station, TX 77843.

Originally published May 29, 1972.

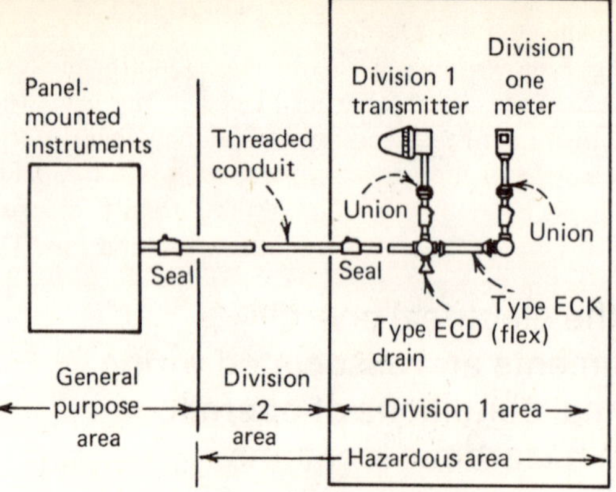

EXPLOSION-PROOF installation meets N.E.C.—Fig. 1

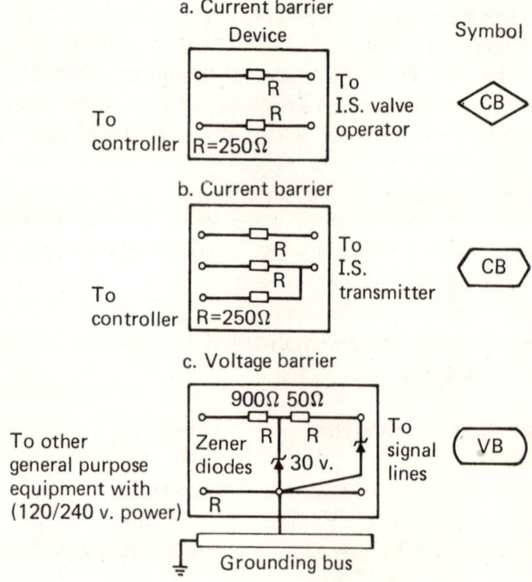

BARRIERS limit current or voltage—Fig. 2

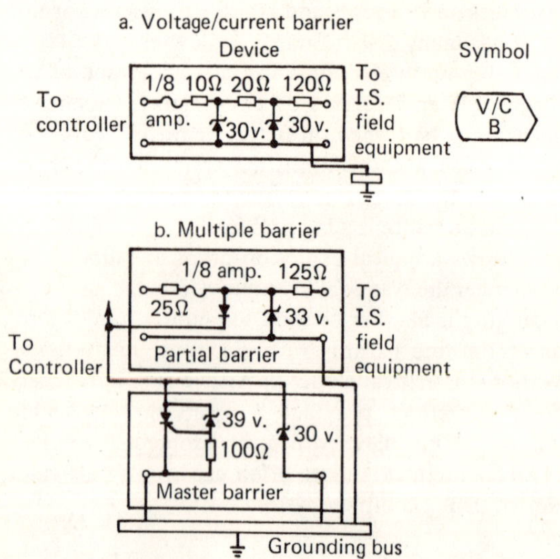

COMBINATION voltage and current barriers—Fig. 3

Basic Intrinsic-Safety Designs

Based on the definition outlined in ISA-RP1.2 (previously cited), the following descriptions generally represent the designs of U.S. instrument manufacturers.

1. Normally, the control room is located in a nonhazardous area. Even though the instruments are I.S. type in the control room and limit energy in the wiring to a hazardous area, field instruments such as transmitters, valve actuators, etc., must be designed and installed so that the system will not store enough energy (even when there are faults in wiring or components) to cause ignition of a specific gas.

2. One method used to meet I.S. requires that each instrument in the loop be certified. Any useful combination of certified equipment may be interconnected to form a control loop.

In this approach to I.S., energy limiting is accomplished by a transformer so constructed that there is no possibility of a line-voltage-to-secondary failure. For example, the transformer passes the NFPA burnout test or is otherwise acceptable to the testing agency.

Transformer leads require specific thicknesses of insulation to prevent insulation failure from shorting leads together. Resistors in circuits leading to the field are used along with the transformer to limit energy to the hazardous area.

Separation is provided between power and signal terminals to ensure no dangerous cross-connections in the instrument. Panel wiring is arranged to separate nonintrinsically safe and intrinsically safe wiring so as to eliminate the possibility of transmitting dangerous energy levels to the field.

3. Another method used to limit electrical energy to hazardous areas is known as the barrier method. By using barriers, only the equipment connected to Div. 1 sides of the barrier has to be I.S. The equipment in Div. 2 or nonhazardous areas may be the conventional type.

In order for the barriers to meet the energy-limiting requirements associated with combinations of electronic hardware, several types are required. Barriers are designed and built to prevent accidental bypassing of them. The entire system is sometimes encapsulated, or installed in such a way as to limit accidental or deliberate shorting after the system is assembled and tested.

Fig. 2a shows a current barrier that is used in some cases to limit current output from an I.S. controller to an I.S. valve operator. It contains a 250-ohm wire-wound resistor in each lead. These resistors are wire-wound to ensure that resistance value cannot become less, nor fail in a dangerous direction, nor short-out during a fault and allow excess current flow.

Fig. 2b shows another type of current barrier that is sometimes used between an I.S. transmitter and an I.S. receiver. It contains three 250-ohm resistors, one of which is a dropping resistor for the receiver.

The voltage barrier in Fig. 2c contains a fuse, two Zener diodes, and two resistors. This device is capable of limiting voltage under any condition to 30 v. It is used between general-purpose auxiliary equipment and I.S. panel-mounted instruments.

An encapsulated-type combination voltage-and-current (Redding) barrier is shown in Fig. 3a. It is commonly used between a conventional controller located in a Div. 2 area, or a nonhazardous area, and an I.S. field instrument. This barrier contains a fuse, two Zener diodes and three resistors. It is capable of protecting against a line voltage of 250 v. rms. (354-v. peak) at the input terminals, and thereby limits the electrical energy to the hazardous area to safe limits.

The barrier shown in Fig. 3a is a positive type and is used in systems where the negative side of the transmitter is grounded. A negative-type barrier is similar except that the Zener diodes are inverted where the positive side of the transmitter is grounded. This type of barrier passes a signal current of 4 to 20 ma. at 24 v. d.c., and does not degrade the transmitted signal appreciably (less than 0.1%). This minor inaccuracy is caused by leakage of the Zener diodes.

The multiple barrier shown in Fig. 3b has the advantage over the single type (Fig. 3a) of reduced unit cost. It does this by having a single high-wattage Zener diode shared with several barriers, each containing two 30-v., 10-w. Zener diodes along with a ⅛-amp. fuse and current-limiting resistors. This allows redundant protection at a lower cost for each barrier.

The barrier shown in Fig. 4 is a repeater barrier that contains an amplifier capable of accurately reproducing a signal from an I.S. transmitter to a general-purpose instrument. A second amplifier is available for use with the controller's output to the I.S. valve operator. This sytem features 10,000-ohm resistors in each input lead, which are large enough to prevent excessive energy from the conventional controller from reaching the hazardous area. The input current from the field is converted to a voltage signal that is fed into the high-impedance input. This signal is amplified and converted back to a current signal for retransmission to control-room devices. The barrier also provides an I.S. power supply that furnishes power to field equipment.

Types of Intrinsic-Safety Circuits

General circuit design varies from manufacturer to manufacturer. The following sketches depict the various ways that barriers, and other design methods, are used in I.S. applications. In the previous illustrations, each type

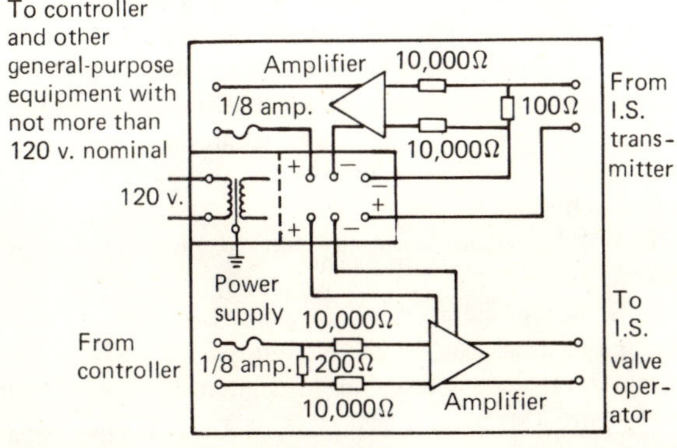

REPEATER barrier amplifies signals—Fig. 4

of barrier had a specific symbol. These will be used in the following circuit diagrams.

Fig. 5 shows how one major manufacturer makes use of the voltage-current barrier placed in the leads to the field instruments. The equipment in the general-purpose area can be of the conventional type, while the field instruments have to be I.S. Barriers are usually located behind the panel, bolted to a solidly grounded bus. These barriers limit energy to safe levels even with line voltages of 250 v. rms.

Another system is shown in Fig. 6. Here, the power-supply transformer is designed as I.S. by having a grounded shield so as to limit the voltage to a safe level. Then, the field leads need only current barriers to limit energy to I.S. field instruments. The voltage barrier is required when connections are made to other general-purpose instruments with 120-v. power. Barriers are usually located behind the panel, bolted to a grounded bus.

An arrangement used by at least two manufacturers is shown in Fig. 7, whereby the instrument housing contains the current barriers to the field leads. These barriers are separated in the instrument case by a physical barrier to isolate field leads from power leads. This version also uses an I.S. power supply, but still requires an external voltage barrier in leads to other general-purpose equipment.

The arrangement of Fig. 8 is used by one manufacturer and is similar to Fig. 6. The main difference is that a voltage-current barrier is used instead of a voltage barrier where other general-purpose equipment is involved. This system has an I.S. power supply, and uses current barriers to field leads.

Fig. 9 depicts a system using a repeater barrier in connection with a conventional instrument in the general-purpose area. Field instruments and associated wiring are intrinsically safe.

The system shown in Fig. 10 represents a complete loop approach that is used by one manufacturer. In this system, each and every device is certified and may be used in any workable combination. The energy-limiting features are integrally designed into the system to meet the requirements of I.S., according to the several standards in use today. No substitutes for energy-limiting components are allowed.

In Fig. 11, the diagram shows an alarm system that has an I.S. power supply, and 10,000-ohm resistors in each lead to the field. The capacity of the wiring to the field contacts has to be considered and kept within safe limits.

Certification of I.S. Systems

The various methods of design used in the above circuits show that there are many possible safe ways to build I.S. instruments or systems. The same also applies to getting certification of equipment by the recognized approval agencies.

Two agencies that test and certify I.S. systems in the U.S. are Factory Mutual Research Corp. (F.M.) and Underwriters Laboratories (U.L.). The requirements of these two agencies are different in many ways, but both are considered equal by insurance companies, code authorities and users.

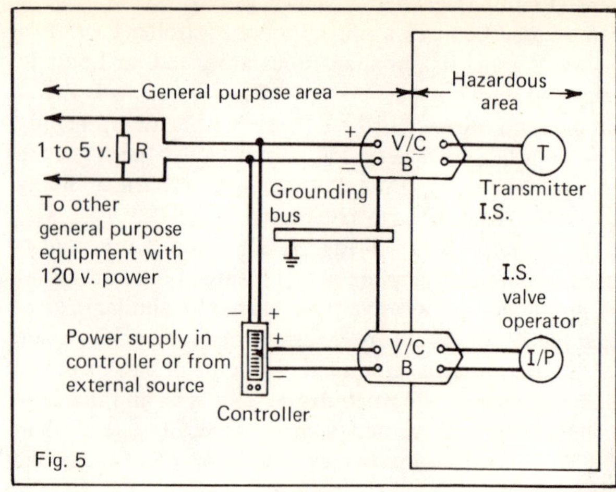

Fig. 5

INTRINSIC SAFETY SYSTEMS

Combined current and voltage barrier–Fig. 5
Separate current and voltage barriers–Fig. 6
Current barrier in controller housing–Fig. 7
Separate current and voltage/current barriers–Fig. 8
Repeater barrier–Fig. 9

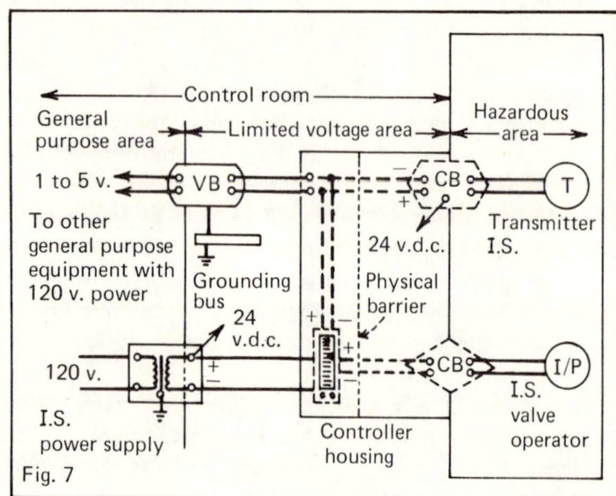

Fig. 7

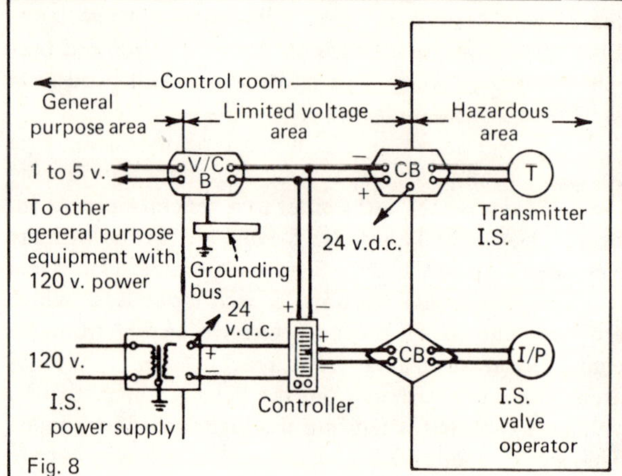

Fig. 8

Canadian Standards Assn. in Canada, and Safety in Mines Research Establishment in Great Britain are testing agencies for I.S. equipment but are not on the approved list of the *Federal Register*.[1]

Intrinsic Safety in the Past

There has been a vast improvement in electronic equipment design in the past few years. By using solid-state components, integrated circuits, printed circuits and other innovations available today, dependable low-energy-level systems have been developed in the 24-v. d.c. range that is becoming the standard voltage for the majority of U.S. manufacturers.

Although pneumatic instruments still represent a large share of the total market, many users over the past ten years have tried different makes of electronic instruments.

Most users have learned that electronic instruments are dependable and have advantages over the pneumatic type in many installations. Even though I.S. systems have been available (at least, by one company since 1965), acceptance by users has been rather slow. In 1970, four instrument manufacturers offered I.S. instruments. Today, there are at least six companies with listed systems, with other systems being tested.

The writer polled a number of instrument engineers on their company's usage and thinking about I.S. systems. Of those representing the oil companies, nine had not installed any I.S. systems; most had considered but could not justify the I.S. system for various reasons. Three companies reportedly had installed several hundred loops. The number of electronic instruments reported installed as I.S. was only a small percent of the total. Some of the reasons given for not using I.S. instruments are:

1. Local, state or company rules require conduit and seals.

2. Some engineers think that I.S. instruments would cost more than the methods being used. Where low-energy-level instruments are installed in general-purpose, nonhazardous areas, wire is run in open trays. However, explosion-proof transmitters are used in the field.

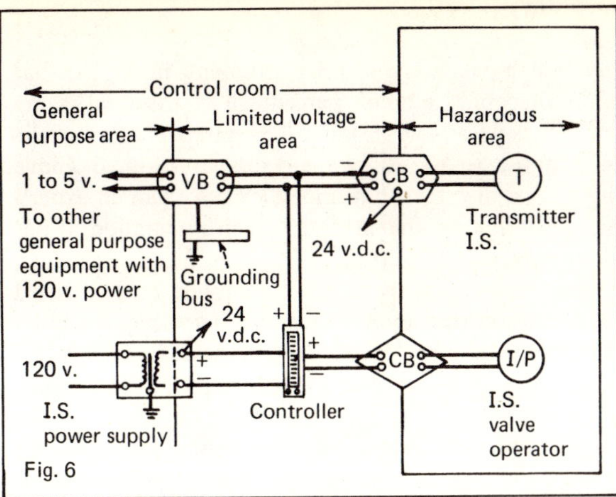

Fig. 6

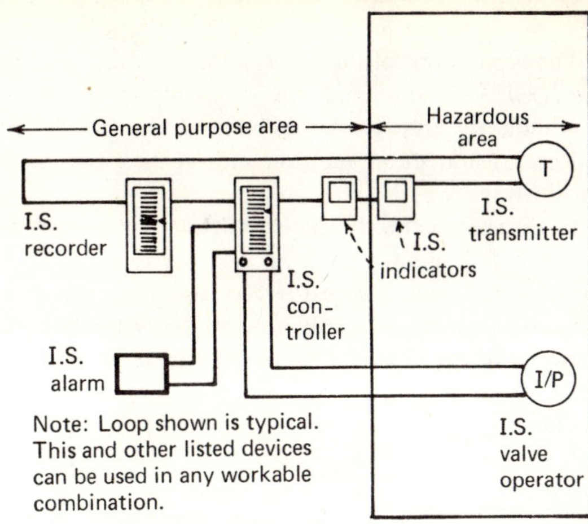

CONTROL LOOP in which all instruments are listed as intrinsically safe—Fig. 10

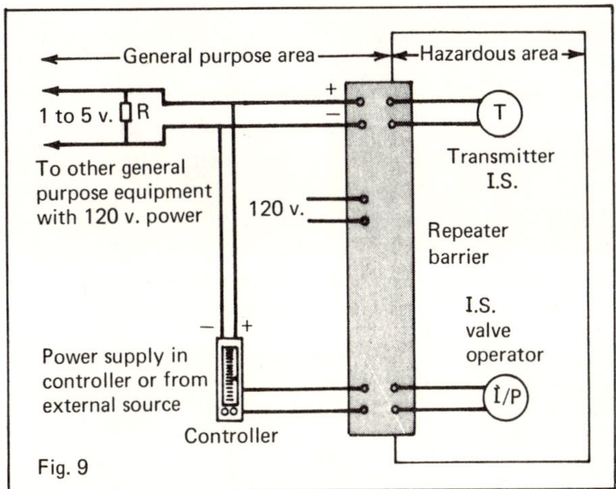

Fig. 9

3. Others think that there is too much confusion on how to meet the complete requirements of I.S., since no one has done anything about alarms, analyzers, multipoint temperature recorders, etc., which represent about 40% of the leads to the field.

4. Most agree that their company officials and insurance companies are not pushing the I.S. concept since no overall standard has evolved.

5. Final design of instrument systems is being left up to the engineers to do as they see fit.

6. No fires or explosions have been caused by fault of conventional electrical instruments, according to the report.

One contractor had built a large plant with over a thousand loops. Conventional panel-board instruments were designed as I.S. However, due to the present state of the art, alarms, multipoint thermocouple instruments, analyzers, etc., were not made I.S. because of extra problems that would have been involved. Another contractor said that with the experience gained on the first job, the second job would go in easier and with less expense, but that it definitely required considerable extra engineering and drafting time to go I.S. on the first job.

According to the latest information available to the author, there are at least six companies in the U.S. with a full line of electronic instruments that have been listed as I.S. (three by U.L. and three by F.M.; still other lines are being tested).

Also, several annunciators are available that have been listed as I.S. Although at least three manufacturers of multipoint, thermocouple-type recorders are working on a design to meet I.S. requirements, apparently none is available. In order to meet the requirements of the National Electrical Code for using thermocouple wiring in hazardous areas where electronic recorders are involved, it appears that explosion-proof installations are required. However, analyzers, computers, multipoint recorders, etc., can be used by adapting some of the barrier systems previously described.

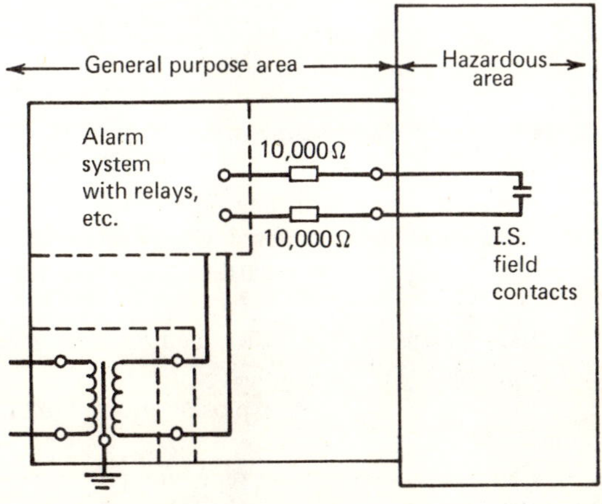

TYPICAL alarm system for intrinsic safety—Fig. 11

Occupational Safety and Health Act

The regulations of the Occupational Safety and Health Act (OSHA) are now in force, and compliance with them is mandatory. In the case of electrical equipment, installations must meet the requirements of the National Electrical Code (N.E.C.).

In discussing the details of meeting the requirements of the N.E.C. with contractors and users, the author has found some areas where there is disagreement on what the code requires. Not only must we have a better understanding of the N.E.C., but the interpretation of it has to be uniform among contractors, users and authorities if the OSHA regulations are to serve their purpose.

If electronic instruments are to be installed in hazardous areas, meeting the full requirements of OSHA regulations,[1] then the user has no reasonable choice but to investigate what it takes to install I.S. instruments. Although explosion-proof installations still meet N.E.C. requirements, it is generally agreed by contractors, users and manufacturers that I.S. systems can be installed at a lower cost than explosion-proof systems.

Future of Intrinsic Safety

• In considering all of the above aspects of designing instrument systems, it seems that instrument engineers who have the responsibility of installing electronic instruments in hazardous locations should become thoroughly acquainted with the I.S. concept. They should investigate the approved electronic systems that are available and decide which one meets their requirements. Obviously, management must be convinced of the advantages of I.S. systems.

• Manufacturers have the responsibility of continued steps toward standardization if I.S. is to become more practical and less expensive to use in hazardous areas.

• If testing agencies would lean more toward a standard method of testing, and promoting design philosophies, then the user and manufacturer would not be so far apart in understanding and applying I.S.

• Code authorities at federal, state and local levels should as far as possible interpret and recognize the same type of installations for meeting N.E.C. requirements. For various reasons, there are still different requirements in certain cities and counties in the U.S.

Advantages of Intrinsic Safety

In order to leave a lasting impression of why I.S. should be carefully considered when electronic instruments are being used in hazardous areas, the following advantages are claimed:

1. Less expensive than other methods that meet the full requirements of the N.E.C.
2. Eliminates need for explosion-proof boxes, conduits, seals, etc.
3. Can be serviced without removing power or getting safety clearance.
4. Less maintenance required to keep in working order than is the case with explosion-proof boxes, conduits, seals, etc.

Summary

Although past usage of I.S. instruments in the U.S. has represented only a small percentage of the total, some companies who have installed I.S. systems almost exclusively agree to all of phe advantages claimed above. These users also feel that since they have gained experience of designing and using I.S. instrumentation, it has become a standard method.

With the availability of more complete lines of listed I.S. electronic instruments, and the increased flexibility offered over earlier designs, it is becoming easier to install all the necessary components required in complex instrument loops.

With continued improvement in design, better understanding of I.S. philosophy by concerned parties, and the added pressure brought upon industry to meet OSHA regulations, I.S. instruments will probably show a steady rise in usage.

Acknowledgements

The following companies have furnished technical bulletins, answered questions from a letter poll and contributed information that was helpful in preparing this article: The Bristol Co., Barber Coleman Co., Fisher Controls Co., Fischer and Porter Co., The Foxboro Co., General Electric Co., Honeywell, Inc., Motorola Instrument and Control Div., Taylor Instrument Cos.

References

1. Standards of the Occupational Safety and Health Act, Vol. 36, No. 105, Part 2, May 29, 1971; Vol. 37, No. 32, Feb. 16, 1972, *Federal Register*, Washington.
2. "National Electrical Code, 1971," National Fire Protection Assn., Boston.
3. ISA-RP12.2 and ISA-S12.4, Instrument Soc. of America, Pittsburgh.
4. NFPA: No. 493 and NFPA: No. 496, National Fire Protection Assn., Boston.
5. Hickes, W. F., "New Developments in Intrinsic Safety," The Foxboro Co., Foxboro, Mass.
6. Leeds, R. E., "Intrinsic Safety and Its Maintenance Aspect," Taylor Instrument Cos., Rochester, N.Y.
7. "Instrumentation—Engineering Report No. 201A," Motorola Instrument and Control Div., Phoenix, Ariz.

Meet the Author

H. C. Deloney is chief instrument engineer for Texaco Inc., P. O. Box 52332, Houston, TX 77052. He joined Texaco in 1937 at its Port Arthur refinery where he held instrument jobs from helper to supervisor of instrument engineering. He was promoted to his present position in 1969. He has a B.S. in electrical engineering from Louisiana Tech. He is a member of the Instrument Soc. of America, and is presently chairman in charge of rewriting the American Petroleum Institute's RP 550.

Alarm and Shutdown Devices Protect Process Equipment

EDWARD J. RASMUSSEN, Fluor Engineers and Constructors, Inc.

Many operations in process plants require some means of detecting an out-of-limit condition in the process variables. Instruments for monitoring these variables are termed alarm and shutdown devices.

Because of the great variety of such devices, we will confine our discussion to those monitoring pressure, flow, level, temperature and vibration. In doing so, we will describe the operating conditions for the devices and indicate suitable applications for them.

Occasionally, the installed system will not perform as anticipated or cannot be set at process values suiting the abnormal condition. Usually, this happens because an alarm-sensing device such as a pressure switch or flow switch has been improperly specified. However, other factors can also contribute to a poorly operating alarm or shutdown system. These include problems with annunciator systems, shutdown-actuator devices, and the power supply for the shutdown or alarm system.

Alarm-detection devices are generally referred to as switches. This does not mean that they are always electrical. Most are also available as pneumatic switches, and therefore can be applied to pneumatic-shutdown circuits.

Pressure Switches

Pressure switches are the most common of all alarm-detection devices for process control use. They can detect out-of-limit pressures, and are often connected into pneumatic-instrument systems for alarm detection of pneumatically transmitted, measured variables.

The operating characteristics of a pressure switch are of greatest concern to the chemical engineer, since they determine an instrument's function relative to the process. Pressure switches are manufactured in two basic types in relation to setpoint. They are either indicating or blind (i.e., nonindicating). Most often, the indicating type of pressure switch has a scale (graduated in psi, or other units) that permits the setpoint to be easily adjusted to a specific value; but such scales are usually disappointing in terms of accuracy. However, a few pressure switches do have an indicating scale of satisfactory accuracy.

Originally published May 12, 1975.

If the indicating pressure switch is required (i.e., a switch that the operator can easily change to a predetermined setting without calibration equipment), it is essential that the scale of the switch have proper resolution and that the adjustment be easily accessible to the operator. For most applications, a pressure switch with a blind setpoint is adequate.

Generally, settings of blind pressure switches or pressure switches with poor indicating scales are not intended to be adjusted by a process operator. Often, these settings are made with a screwdriver and provided with lock nuts or other fixtures to prevent accidental change of the setpoint. The primary method of adjusting pressure switches is to calibrate them with a standard reference such as a precision pressure gage, manometer, or deadweight tester. This is done by an instrument technician, and the setting recorded.

Differential action is an often-critical characteristic when applying pressure switches. This is the gap of operation between actuation in one direction as the pressure rises and actuation in the opposite direction as the pressure falls. The engineer can select either a fixed or an adjustable differential-action-type switch, depending on requirements. Some differential action is generally desirable because without it the switch contacts would alternately actuate and reset if the measured pressure were operating around the switch's setpoint.

Generally, the switch differential is so selected that the pressure switch will not reset until the process pressure has changed significantly to the safe side of the operating point. When adjustable differential is provided, this function can be changed to provide a variable differential gap, so that the gap may be set to suit the process.

For any specific adjustable or fixed differential-action switch, there is a minimum differential gap that cannot be reduced. This gap is necessary to provide the force to actuate the switch. The minimum differential can become as much as 25% of the instrument span in some devices. It is one of the difficulties often leading to misapplication of pressure switches.

If the differential characteristics are so great that the switch cannot reset itself to clear the alarm after the pro-

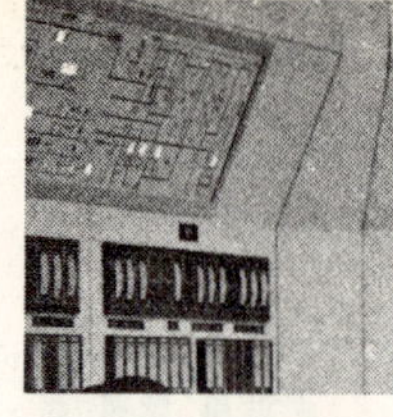

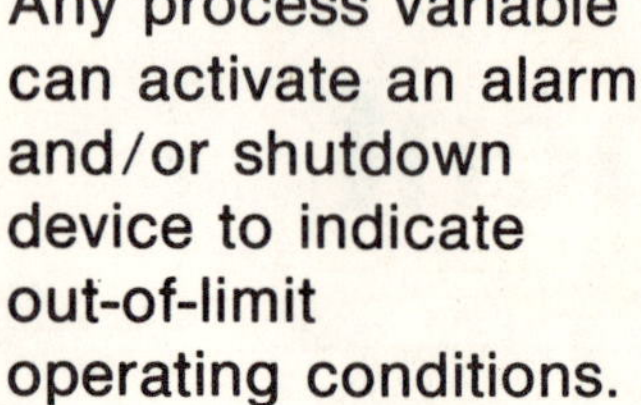

Any process variable can activate an alarm and/or shutdown device to indicate out-of-limit operating conditions.

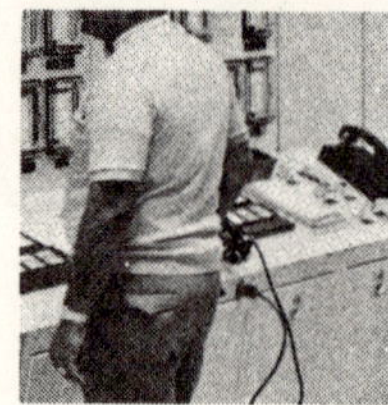

cess operating pressure returns to normal, the switch will not serve its purpose satisfactorily. This problem can arise from two sources: (1) specifying the wrong type of switch, and (2) selecting a switch having an excessively wide operating range.

The differential gap is proportional to the operating range. For example, a 0 to 50-psi switch will have a narrower differential gap than a 0 to 100-psi switch. We must use the manufacturer's data to determine whether a selected switch has suitable differential action.

Another characteristic that must often be included is the overrange characteristic or proof pressure. Manufacturers specify the maximum pressure that the pressure switch can handle without distorting its calibration or primary element. If this pressure is exceeded, the switch will be damaged.

If the application is such that the highest possible operating pressure (possibly a setting of a relief valve or a shut-in head of a pump) is significantly higher than the pressure setpoint, we have to make a tradeoff between the proof-pressure capability of the switch and the desirable minimum differential action. The relationship between these varies widely with manufacturer and design of the switch.

Fig. 1 shows some representative pressure switches and some of their characteristics. These are available in a variety of materials of construction to suit corrosion requirements and process temperatures. Some are available with an optional, manual, reset lever. This is useful in a shutdown system where it is not desirable to have the switch return to its original state automatically after actuation. Instead, the operator must reset the switch after the process pressure returns to normal.

Any pressure switch must be properly applied so that it will sense a sizable change in pressure from the normal operating condition to the alarm condition. Often, pressure switches are used in pump-discharge systems to provide an alarm on pump failure and/or to start a standby pump. However, in some arrangements such as a pump circulating system with a low-head pump, the pressure when the pump shuts off or fails is only slightly different from the pressure when the pump is operating. In this instance, a pressure switch would be a poor choice. Since flow is the variable most affected by pump failure, a flow switch would be suitable. When applying any type of alarm device, we must choose a parameter for measurement that will show a significant change for the condition that we want to alarm.

Flow Switches

Flow switches fall into one of two categories: a variable-area device or a velocity-sensing device. A variable-area flow switch can be a rotameter with a switch attachment, or a simple, nonindicating variable-area device having a very limited range of adjustment. The simple variable-area devices and some velocity-sensing devices (such as a vane or paddle that protrudes into a flowing stream) are generally the lower-cost flow switches. These are not designed for accurate settings but are quite adequate for many applications. For example, they are commonly used in fire-protection systems and are also suitable in lubricating-oil systems to determine whether or not flow exists.

For flow switches that must be set accurately at a predetermined flow, the standard rotameter equipped with a switch, or the orifice-type flowmeter, would generally be selected. The rotameter has a 10 to 1 flow range and is one of the best devices for flow measurement or flow-switch application. However, rotameters are not made in sizes suitable for large flows. Here, the orifice-type flowmeter is the general choice.

The orifice meter requires a restriction of known characteristics in the pipeline. The differential pressure created across the restriction by the velocity of the flowing fluid is measured on an instrument. The differential-pressure-measuring flow switch can provide either a direct indication and mechanical operation of the switch or can be an electronic or pneumatic flow-transmitter. A transmitter converts the differential-pressure measurement into a standard instrument signal, and then an alarm switch can be adjusted to operate when that signal reaches predetermined value.

There are special problems in using the differential-head flowmeter as a flow switch. The indication on the meter is proportional to the square of the flow. This is the cause of problems at the low-flow end of the measurement, and also is the cause of the limited rangeability of this type of meter. For flows less than about 25% of the meter's maximum design flow, there will not be adequate sensing for a dependable alarm function. Due to the square-law characteristic, 25% of the flowrate is only 5% of the differential-head meter scale. Operation this close to the end of an instrument scale is unsuitable for most alarm and shutdown devices. In addition, resolution of the instrument becomes very poor at the low end of the scale, due to the same square-law characteristic.

Therefore, in low-flow applications, the range of the instrument must be chosen so that the differential pressure across the orifice plate at the alarm-point flowrate is no less than 10% of the instrument range. In many cases, this will cause the switch to be over-ranged at normal flowrates. Two flowmeters can then be connected to the same orifice taps. One might have a meter range from 0 to 100 in of water, and under normal operating conditions would read midscale. The other instrument, dedicated to a low-flow switch function, could be a meter with a 0 to 25-in range. At normal flowrates, it would operate above the upper limit of its range. However, as flowrates fell to the low-alarm value, this instrument would provide an accurate measurement of the critical low-flow situation.

Flow switches are available blind or with an indicating scale. They also can be provided with adjustable or fixed differential-switching actions. The same principle is involved as with a pressure switch. However, the problem of over-ranging the instrument generally does not exist. Most flow switches can be over-ranged without permanent damage or even loss of calibration. They are available for almost any static-pressure requirement. And, static-pressure characteristics of the flow switch have no effect on instrument sensitivity or differential action. This is just a matter of having adequate strength in the meter body, and suitable seals. Some examples of flow switches are shown in Fig. 2.

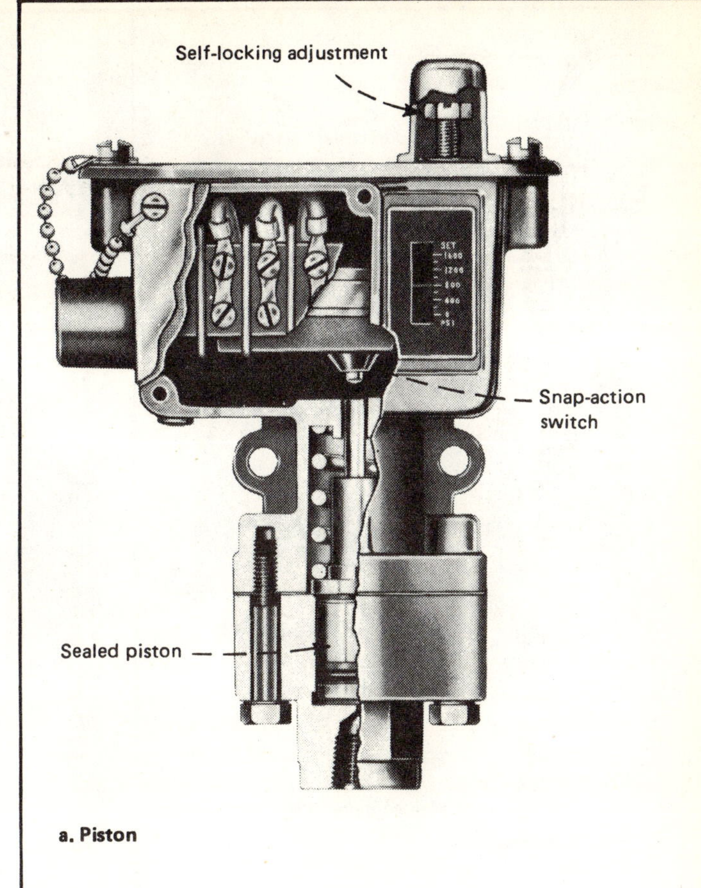

a. Piston

Level Switches

The most common level switch is the float type, in which a buoyant element in the liquid to be measured is coupled to a switch-operating mechanism. Float switches can be the internal type, with the float operating inside the process vessel; or the external type, with the float in its own pressure-containing chamber that is connected to the process vessel. For a high-level alarm, the float switch would be mounted on the vessel somewhere above the normal operating level. Hence, as the level rises in the process vessel, the float will rise and on reaching the limiting value would cause the switch to operate.

Conductivity-probe and capacitance-probe level detectors have the advantage of no moving parts in the process liquid. They are often the most economical selection for applications involving highly corrosive materials or very high pressures, in food processing, and for sewage sumps. Both conductivity and capacitance switches use a probe entering the vessel wall through an insulated fitting, and both have a sensitive detector for determining whether the liquid level is above or below the end of the probe. The conductivity unit has a bare metallic probe or probes, and detects liquid level by sensing the conductivity of the process liquid either between the probe and the vessel wall or between two probes. Obviously, this system is useless for nonconductive liquids. The capacitance unit generally has an insulated metallic probe for conductive liquids, and may have a bare probe for nonconductive liquids. Level is sensed by the change in capacitive reactance between the probe and the vessel wall.

Level switches can also detect the level of solids such as grains, sand, rock and coal in storage tanks or hoppers. A very common switch for this service uses a small paddle wheel mounted within the tank close to its wall, and driven by a small motor from the outside. When the paddle wheel is covered by solids in the storage vessel, it cannot turn. This condition is detected by a torque switch on the motor mounting, and indicates that the level is above the height of the paddle-wheel location. Nuclear gages are also quite effective in solids measurement. Another device is a flexible diaphragm mounted flush with the inside wall of the tank, and fixed with a switch mechanism to detect the pressure of solids against the diaphragm.

Nuclear level detectors use a beam from a radioactive source directed across the vessel to a radiation detector. When the level is high enough to interfere with the path

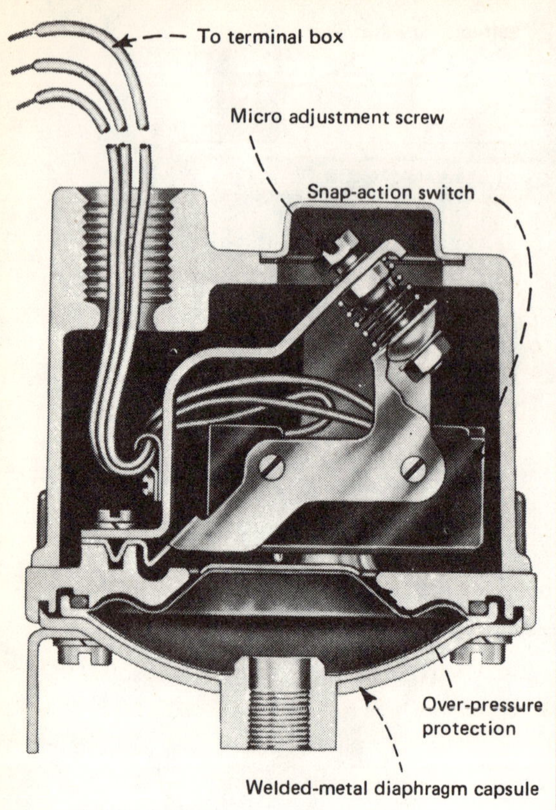

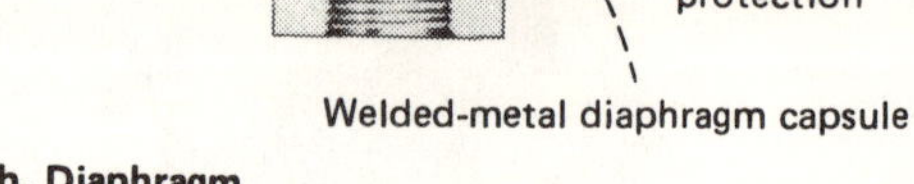

b. Diaphragm

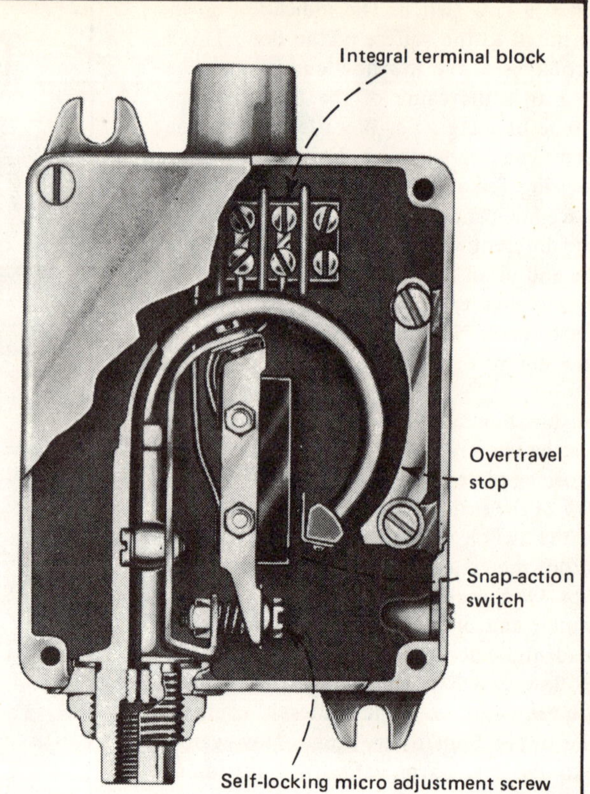

c. Bourdon Tube

d. Bourdon Tube ▶

PRESSURE switches have various means for sensing the operating pressure in process equipment—Fig. 1

between source and detector, some radioactive energy is absorbed by the material. The reduced signal to the detector operates the alarm relay. These instruments can be used for both liquids and solids. They require no contact with the measured material and no holes in the vessel wall. They are very expensive compared with other level switches.

Some other level-indicating devices rely on ultrasonics, photo cells, light refraction, and weight.

The key to instrument selection is twofold. The selected unit should be economical and still provide dependable, maintenance-free service. The tradeoff between initial cost and dependability sometimes must be considered, because alternate selections may vary widely in cost. In these cases, we must judge the consequences of the switch not operating in the function that it was designed for. If failure of the switch to operate could result in a very expensive or catastrophic problem, then the most dependable switch regardless of expense should be selected.

Let us also consider that sophisticated measuring techniques (often in conjunction with solid-state electronics) may not be the most reliable choice over the simple mechanism. Excessive heat or a little moisture in the probe or in the amplifier/relay unit can disable the best electronic equipment. Hence, let us keep this in mind when evaluating any device for alarm and shutdown systems. Let us also apply simple rugged mechanisms in preference to delicate and complex ones whenever the simple one is suitable. See Fig. 3 for examples of level switches.

Temperature Switches

Temperature switches generally operate on one of four principles: thermal expansion, differential thermal expansion, vapor pressure, or thermocouple.

The thermal-expansion device can be a mercury-filled bulb somewhat like an ordinary glass laboratory thermometer.

In the industrial temperature switch, this would normally be a mercury-filled or gas-filled system consisting of a copper or stainless-steel bulb connected to a pressure switch through a small capillary tube. Expansion of the mercury or gas with an increase in temperature would in-

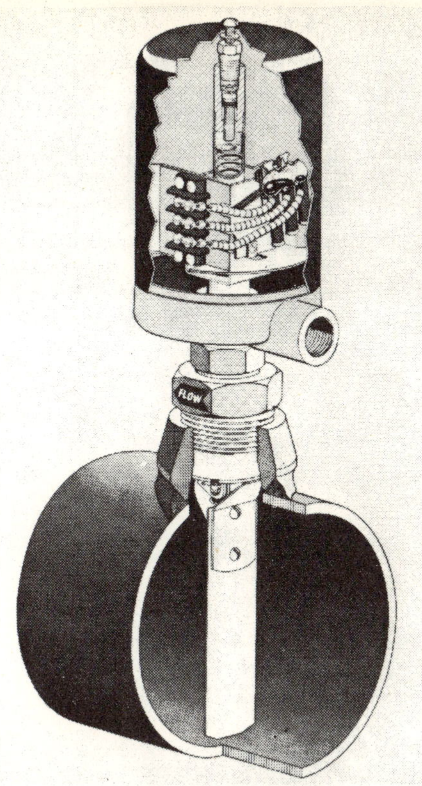

a. Vane Actuated

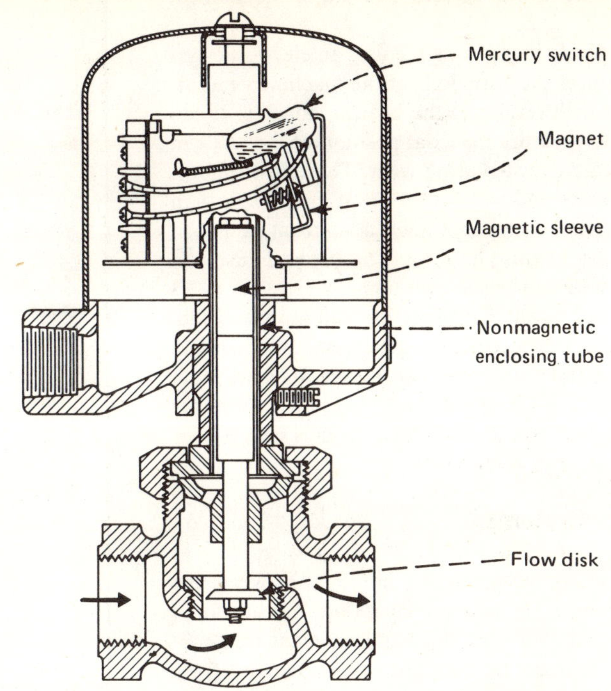

b. Flow Actuated

FLOW switches detect the moving stream either by using a velocity-sensing device or a variable-area device to provide the alarm function—Fig. 2

crease the filled-system pressure, which then actuates the switch.

The vapor-pressure switch is a filled-system having the same configuration as the mercury thermal-expansion type. In this case, the bulb is filled with a liquid operating in equilibrium with its liquid and vapor phases. As the temperature increases, the vapor pressure of the liquid increases and causes the pressure-sensing element to operate the switch at the predetermined temperature.

Thermal-expansion and vapor-pressure temperature switches cover a wide range of temperatures and provide ample force to operate a dependable switch mechanism. The capillary tube, which may be up to 100 ft long in some models, permits the switch to be installed at an easily accessible location.

Bimetallic temperature switches are made in many configurations. They operate on differential thermal expansion between two metal parts that are thermally coupled to the process stream. The motion of the differential expansion is transferred to the switch through an adjustable mechanism. Bimetallic temperature switches are generally small, low cost, and useful over a wide range of temperature. However, they must be mounted directly on the process line or equipment. If vibration in the line is present, the situation should be discussed with the manufacturer before a bimetallic switch is selected. Line mounting can also cause problems in providing accessibility for maintenance.

Thermocouples can be used as a temperature-measurement element because they generate a minute voltage that is a function of the temperature. These sensors are good for a wide range of temperatures and are very rugged and inexpensive. As a primary element, they are one of the best temperature-sensing devices for general use having reasonable accuracy. The measurement voltage for a thermocouple is the difference between the voltage at the point of measurement and the reference-junction voltage at some other point in the circuit. This voltage must be measured on a very sensitive meter relay or it can be measured by using a stabilized electronic amplifier—with a comparator system to operate a relay when the temperature crosses the alarm value.

Most of this electronic equipment is not suited for outdoor installations. The thermocouple is installed in a thermowell on the process equipment. Usually, the thermocouple extension wires go to the electronic alarm detector in the control room. The thermocouple switch requires an external power source, usually 120-v a.c., which sets it apart from the others, since they are all self-operating. Fig. 4 shows examples of a filled-system temperature switch and a bimetallic type.

Vibration Switches

Vibration switches are used for alarm and/or shutdown of mechanical equipment due to out-of-balance or bearing problems. The simplest detect a large unbalance in a machine and are installed directly on the machine frame. These types use the inertia of a mass within the switch to activate it when the switch housing is subjected to vibration or acceleration beyond the established limit. Vibration switches are simple, self-contained mechanisms that require no external power source for operation. They are not very sensitive or precise and are, therefore, limited to the detection of major mechanical failure.

Another type of vibration switch uses an electronic system that measures the variations of the machine's radial-shaft position with respect to the bearing housing. It can also be used to measure the axial position of the shaft in order to indicate thrust-bearing wear. These instruments are quite accurate and will detect vibrations of less than 0.1 mil. Vibration is measured by a noncontact probe mounted in the bearing housing. The probe is used in conjunction with an electronic unit, which provides probe drive, and signal demodulation and amplification. The electronic-indicator/alarm-detection unit usually has two alarm functions. One can be set for a warning alarm, the other for a shutdown. They are commonly used on large, high-speed, rotating machines such as turbines and centrifugal compressors.

Annunciator Systems

The annunciator system is the interface between the alarm instrumentation and the operator. Annunciators are available with many standard and nonstandard function sequences to meet the purchaser's requirements.*

Most annunciators provide both an audible and a visual signal to get the operator's attention and inform him of the alert condition. The operator is generally required to activate a pushbutton switch to silence the audible signal and to change the initial flashing visual signal to a steady light. Fig. 5 shows a typical general-purpose annunciator system.

Many useful options are available. Alarm indicator lights can be built into a semigraphic display of the process-flow scheme, so that the operator can easily associ-

*A table of common sequences is available in "Specifications and Guides for the Use of General Purpose Annunciators," ISA-RP 18.1, Instrument Soc. of America, Pittsburgh, PA 15222, 1965.

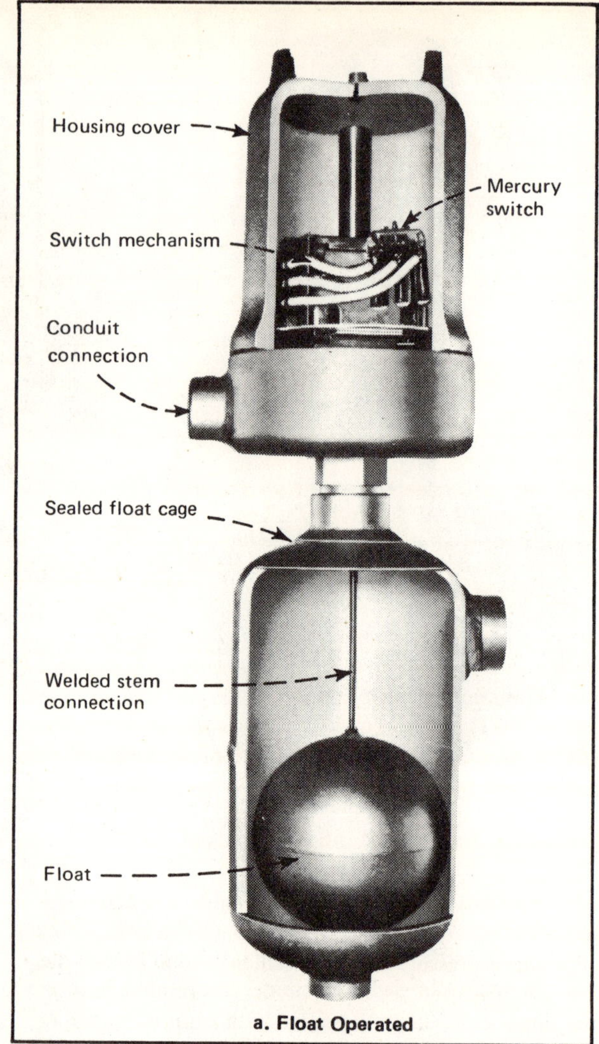

a. Float Operated

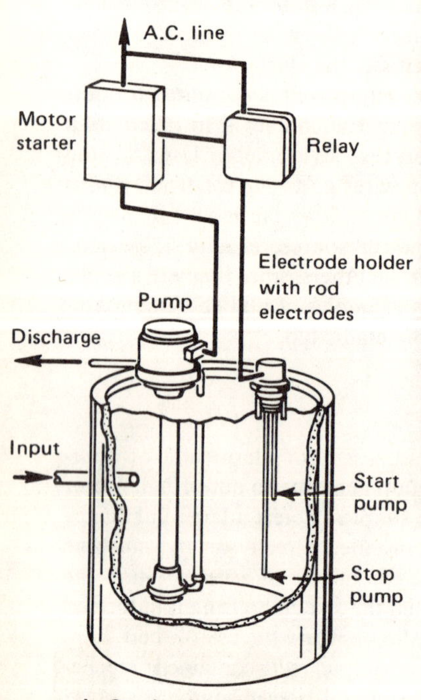

b. Conductivity Probe

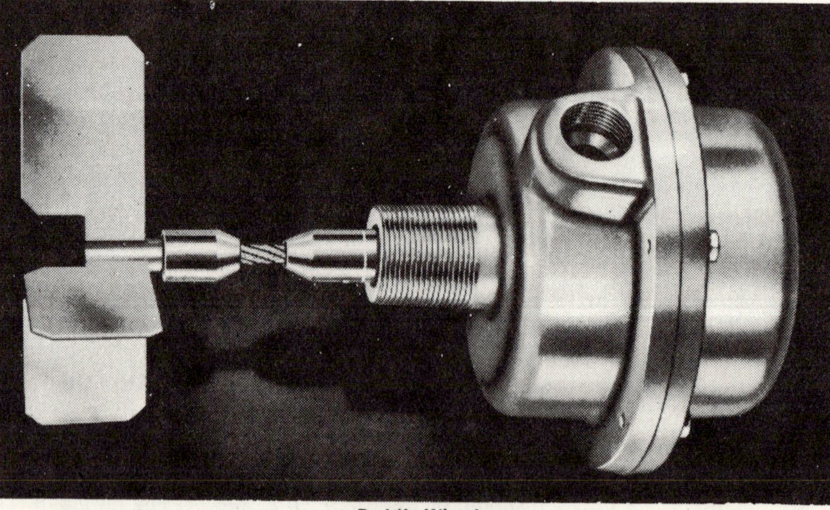

c. Paddle Wheel

LEVEL-sensing instruments for liquids and solids rely on physical or electrical properties to detect high or low levels of liquids or granular solids that are processed in plant operating equipment or stored in suitable vessels—Fig. 3

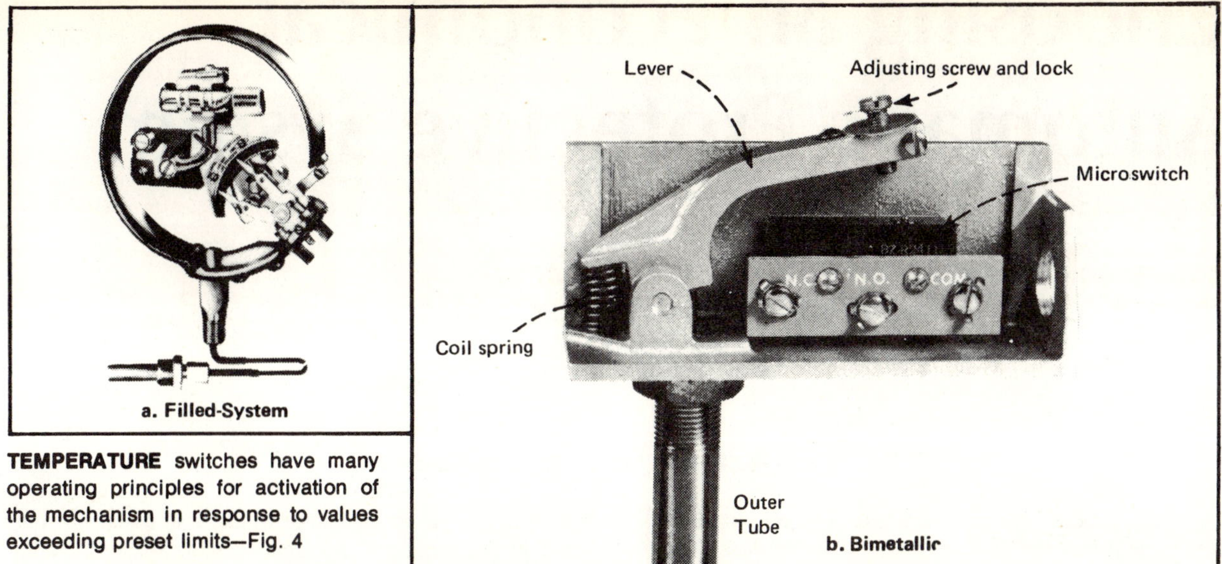

TEMPERATURE switches have many operating principles for activation of the mechanism in response to values exceeding preset limits—Fig. 4

ate the alert indication with the abnormal process condition. The alert message can be typed on an automatic printer, or displayed on a CRT (cathode ray tube) console by means of a computer. Voice-alert messages can be provided by a special recorder system.

Power Sources and Failure Modes

An electrical or pneumatic power source usually operates the alarm and shutdown systems. Often, both electrical and pneumatic equipment are used. For critical systems, the engineer must determine that failures in alarm and shutdown devices and in power sources will not create an unacceptable safety hazard. If a practical fail-safe design cannot be developed, it will be necessary to use redundant systems and/or to design the system for high reliability.

The effect of various types of power failures should be considered. For example, total failure of plant power would shut down all equipment except that provided with an emergency power source. This creates a different set of problems than a power failure affecting only a critical shutdown system. A defective circuit breaker or other component could cause a failure in a critical shutdown system even though the power came from a redundant, uninterruptible power source.

ANNUNCIATORS for general purpose use may have many standard and nonstandard function sequences—Fig. 5

Vendor Participation

Considering the variety of devices discussed here, and that most manufacturers produce several designs (each with various options), it is wise to consult the manufacturers' engineers regarding applications and limitations of their equipment. Cost is also part of any engineering evaluation, and manufacturers will provide cost and cost/quality tradeoff data.

Acknowledgements

The following firms have supplied information and illustrative material for this article: Barksdale Control Div., Delaval Turbine Inc. (Fig. 1a, 1b, 1c); The Mercoid Corp. (Fig. 1d, 4c); Magnetrol Div., Schaub Engineering (Fig. 2a, 2b, 3a); Robertshaw Controls Co.; Monitor Manufacturing Inc. (Fig. 3c); Burling Instrument Co. (Fig. 4b); and Panalarm Div., The Riley Co, (Fig. 5).

Meet the Author

Edward J. Rasmussen is a senior instrument engineer for Fluor Engineers and Constructors, Inc., 2500 S. Atlantic Blvd., Los Angeles, CA 90040. He joined Fluor in 1956, and has been lead instrument engineer on major refinery projects. He has also supervised field-construction instrument personnel on additional refinery and petrochemical projects.

Choosing an Economical Automatic Protective System

How reliable should a process' protective system be? The author shows how to compare the costs of redundancy with the financial consequences of a system failure.

H. J. de HEER, DSM, Geleen, The Netherlands

Critical process variables are usually monitored by automatic protective devices. These have the task of returning the process to a safe condition in the event one or more of the critical variables exceeds a preset limit. However, because all devices are fallible, one must consider that the protective system can fail, specifically in two essentially different ways:

- Functionally–failing to act when it should.
- Operationally–acting when it should not.

One must appreciate that the protective device has a passive role. It spends most of its life waiting for a limit-transgression on which to act. Therefore, of the two types of failures, the functional defect is more obnoxious because it does not reveal itself until confronted with a limit-transgression–the very situation the device should control.

When limit-transgression and functional failure coincide, the results range from an annoying spill to a major calamity. But one thing is sure: Such a coincidence will always cost money, usually a considerable amount since management would not have installed a rather costly protective system unless the consequences of limit-transgression were substantial.

Operational failure of the protective device can be dealt with more easily for two reasons. The failure reveals itself by shutting down part of the process, and there is no element of danger to personnel and equipment. Of course, there probably will be a production loss, which can be expensive.

Economic Aspects

Most of the literature on failures of automatic protective systems for industrial processes is written against a background of saving human life and limb. Such an approach naturally emphasizes functional failures because they are inherently dangerous, and treats operational defects as annoying side-effects, which they are.

Measures to minimize functional failures, which increase the complexity of the protective system, tend to raise the rate of operational problems. There are ways to get around this dilemma, but only at a cost.

Originally published March 17, 1975.

In many cases, critical limit-transgressions do not involve physical danger, but only damage to equipment, loss of production, or government fines. Typical situations include pileups in conveyor systems, lines blocked by polymerized material, column flooding, accidental stream pollution, and equipment fouling or corrosion caused by improper feedwater.

In the following discussion, we will assume that unchecked limit-transgressions will only damage property, and that operational failures will cause financial losses because of reduced production.

Predicting Potential Losses

In a properly controlled and maintained plant, functional and operational failures of protective equipment can be treated as rare, random events. Although no one can predict the exact moment of their occurrence, one may assess the potential losses by proceeding as insurance companies do–calculating averages from observed events and processing these data with the help of equations derived from statistical theory.

An appropriate theory involves assessing the coincidence probability of events in time where the order of occurrence is relevant. Not finding an adequate or prac-

Failure Rates of Protective Systems – Table I

System	Rate of Harmful Coincidences	Rate of Operational Failures
Single channel	$\lambda\mu\tau$	m
1-from-2	$\lambda(\mu\tau)^2$	$2m$
2-from-2	$2\lambda\mu\tau$	$2m^2\tau$
2-from-3	$3\lambda(\mu\tau)^2$	$6m^2\tau$

λ = Rate of limit-transgressions of process variable.
μ = Functional failure rate of a single channel.
m = Operational failure rate of a single channel.
τ = Mean duration of failures in channels.

tical theory in the literature, we developed one of our own [1,2].

The theory maintains that coincidences of limit-transgressions and functional failures on the one hand, and the occurrence of operational failures on the other hand, can be expressed with sufficient approximation in mean (yearly) rates. This leads to relatively simple expressions. Those pertaining to the present discussion, expressed in the fundamental parameters of a single protective channel, are given in Table I.

As can be seen in the table, every improvement (or deterioration) of a more-elaborate arrangement can be judged against single-channel performance. This also simplifies the mutual comparison of multiple-channel performance.

Fundamental Parameters

Although it is not feasible to reestablish the theory in this article, it is necessary to explain the fundamental parameters. With respect to harmful coincidences caused by functional failures, there are:

- Mean rate of occurrence of limit-transgressions, λ.
- Mean rate of occurrence of functional failures, μ.
- Mean duration of functional failures, τ.

The first two parameters must be assessed from actual operating data by dividing the number of occurrences by the observation time period. In the absence of such records, others may provide data on similar situations, which may help in estimating the parameters.

The third number, the mean duration of failures, requires some explaining. Protective systems in the chemical process industries (CPI) usually are inspected regularly. We assume that all failures are detected and remedied. Any other approach, such as allowing for overlooked failures, is less in keeping with experience and leads to endless harangues about what percentage should be taken into account.

Since functional failures are not "self-announcing," those discovered may have developed just after the last inspection, just before the current one, or at any time in between, with equal probability. So, if τ_i is the inspection interval, on the average:

$$\tau = \tfrac{1}{2}\tau_i$$

That is, the mean duration of functional failures is statistically equal to half the inspection interval. This is valid because in the CPI the time required for inspection and repair is a small fraction of the inspection interval.

Single Protective Channel

As shown in Table I, the mean rate of harmful coincidences with a single protective channel is:

$$\lambda\mu\tau$$

This plausible relation results from careful reasoning and simplifications that, for the sake of brevity, will not be repeated here. However, there is no doubt that the criteria for the validity of the expressions in Table I are met in the CPI. More important, the simplified relations always err on the safe side. They give values slightly higher than those derived from the complete but unwieldy equations.

The mean rate of operational failures for a single channel is given by one parameter, m. This is obvious since the failures are self-revealing, and do not depend on either limit-transgressions or inspection intervals, They occur, period.

Multiple-Channel Systems

General theory treats multiple-channel systems as so-called M-from-N systems. This assumes that systems are built from N identical single channels, guarding the same process variable, arranged in such a way that at least M out of N channels have to be activated to give a protective response.

To have M greater than unity has obvious advantages for operational reliability because a failure in one of the channels does not immediately disrupt or incapacitate the system. When $M = 1$, every operational failure in any one of the N channels causes a system failure, resulting in a shutdown. The larger N is, the higher the probability of such shutdowns. On the other hand, increasing N always increases the functional capability of the system, since it takes at least $N-M+1$ failing channels to leave the process unprotected.

Only four arrangements are practicable in the CPI:

- 1-from-1, the single protective channel ($M=1$, $N=1$).
- 1-from-2, in which two channels guard a single process variable in such a way that if one channel fails functionally, the other takes over ($M=1$, $N=2$).
- 2-from-2, in which the two channels are arranged in such a way that both must fail operationally to cause a shutdown ($M=2$, $N=2$).
- 2-from-3, in which three channels are arranged in such a manner that at least two must fail to cause a sys-

Average Yearly Cost of Four Protective Systems — Table II

C(1-from-1) = $c + \lambda\mu\tau D + mL$
C(1-from-2) = $2c + \lambda(\mu\tau)^2 D + 2mL$
C(2-from-2) = $2c + 2\lambda\mu\tau D + 2m^2\tau L$
C(2-from-2) = $3c + 3\lambda(\mu\tau)^2 + 6m^2\tau L$

c = Yearly cost of a single protective channel.
D = Amount of damage from a harmful coincidence.
L = Amount of loss from an operational failure.

Approximate Cost in Units of c Dollars — Table III

C/c(1-from-1) = $1 + \lambda\mu\tau D/c + mL/c$
C/c(1-from-2) = $2 + 2mL/c$
C/c(2-from-2) = $2 + 2\lambda\mu\tau D/c$
C/c(2-from-3) = 3

tem failure, either functionally or operationally ($M=2$, $N=3$).

Evaluation of Table I

To appreciate the relative values of the expressions in Table I, it is necessary to know the order of magnitude of the four parameters. The mean yearly rate of limit-transgressions, λ, will be about unity—perhaps more frequent for small plants and less for large ones.

The mean yearly rate of functional failures of a single protective channel, μ, will range from 1 for a very complicated apparatus, to 0.02 for the kind of equipment that fails once in 50 years on the average. The mean yearly rate of operational failures of a single channel, m, will cover about the same range as μ. The mean downtime, τ, will range from approximately 0.001 (daily care) to 0.5 (yearly overhaul). Since monthly inspection is common, a figure of about 0.04 will be encountered often.

One should note that in actual cases the numerical value of $\mu\tau$ and $m\tau$ will be much smaller than unity. For example, channels consisting of simple, robust pressure switches, followed by an equally robust relay and shutdown valve, have shown failure rates of $\mu = 0.02$. The reciprocal, the mean time between failures (MTBF), is 50 years. When checking such a channel every month, the mean downtime, τ, will be about 0.04. Consequently, $\mu\tau = 0.02 \times 0.04 = 0.0008$.

The table shows that a 1-from-2 system reduces the mean rate of harmful coincidences by a factor of $\mu\tau$ in comparison with a single channel. With the previous example of 0.0008, the reduction amounts to more than a thousand times! In practice, one will seldom encounter values greater than 0.01, so the reduction is always considerable. The table also shows that for this system the operational failure rate increases by a factor of 2. This is plausible since either channel will activate the system.

The 2-from-2 system acts the other way around. With respect to a single channel, it reduces functional capability by a factor of 2, since one channel cannot activate the system on its own. Operational reliability, however, is considerably increased because the failure rate is reduced by a factor of $2m\tau$.

Finally, the 2-from-3 system cuts the coincidence rate by $3\mu\tau$ and the operational failure rate by $6m\tau$, with respect to a single channel. These reductions are less than $\mu\tau$ and $2m\tau$, respectively, but contrary to the 2-channel systems, this arrangement reduces both rates. It does so, however, at the cost of an additional channel.

Total Cost of Protection

Three cost items should be assigned to protective systems. The first is a down-to-earth one. If the cost of a single channel totals c dollars per year (including interest, amortization of maintenance), the yearly cost of N identical channels will be Nc. The two other items are potential costs calculated from statistical probability.

If one estimates the financial loss caused by a single harmful coincidence at D dollars, the average yearly cost will be the product of this amount multiplied by the average yearly rate of harmful coincidences. Likewise, if

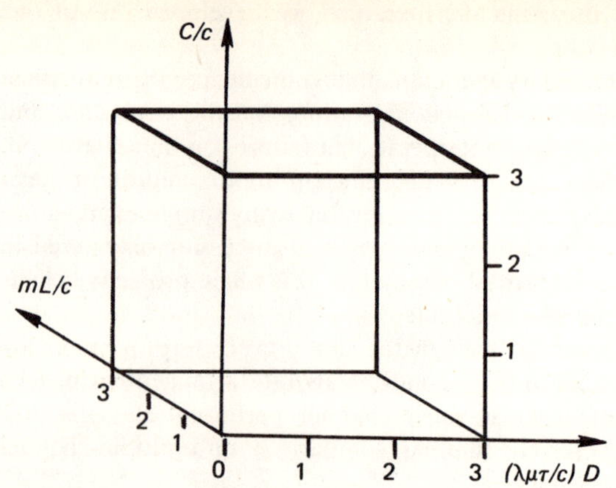

PLANE at level 3 represents relative cost of a 2-from-3 system, independent of both D and L—Fig. 1

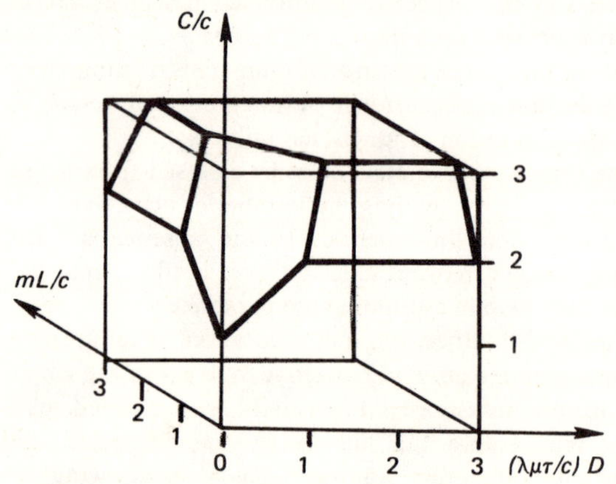

ROOFLIKE structure formed by cost planes of all four systems represents the unavoidable cost—Fig. 2

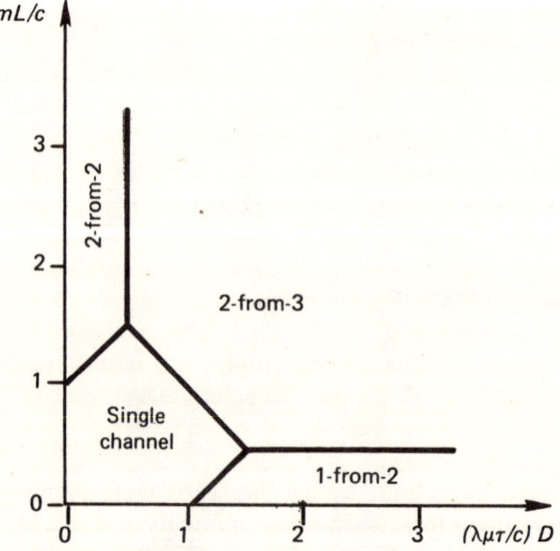

COMPARING the four systems by projecting the intersections of their representative planes—Fig. 3

one places the loss from a single shutdown caused by an operational failure at L dollars, the average yearly cost is calculated by multiplying L by the average yearly rate of operational failures. The yearly average cost, C, of the four relevant systems can be expressed as in Table II.

Most critical process variables are safeguarded by a single protective channel. The following discussion will show that this is the correct approach. Applying additional channels obviously increases investment and maintenance costs. This must be offset by a decrease in the potential for damage and lost production. Although Table II fully presents the individual contribution of each system, the relationships are not transparent.

Remember that the numberical values of $\mu\tau$ and $m\tau$ are always very small. Thus, omitting terms that include these values as a multiplier seems attractive. And with the uncertainty in assessing all parameters except c, this simplification is also sound practice. Moreover, knowing the cost of a single protective channel, one can use c as a base to compare all other costs with.

Carrying out these proposals gives the equations in Table III. These show that a 1-from-2 system practically removes the potential cost of damage by limit-transgressions, but adds to the losses of shutdowns caused by operational failures. With a 2-from-2 system, the reverse is true. Finally, a 2-from-3 system practically nullifies all but the cost of its own three channels.

Using the Equations

A graphical presentation of the equations in Table III can simplify a quantitative survey. From the table, it is obvious that the maximum permissible cost, expressed in c cost units, equals 3. When one of the other systems exceeds this value, the 2-from-3 arrangement should be chosen.

Fig. 1 is a three-dimensional presentation having C/c on the vertical axis. One horizontal axis is divided into units $\lambda\mu\tau D/c$, and the other into mL/c. The main independent variables clearly are D and L, while C is the main dependent variable. This agrees with the equations in Table III.

The top plane of the figure, at level 3, represents the cost of a 2-from-3 system. According to the simplified relations in Table III, this cost is independent of both D and L. Therefore, the representative cost plane runs parallel to the base of the figure.

Fig. 2 contains the representative planes of all four systems. Their intersections form the edges of a rooflike structure, the surfaces of which represent the minimum unavoidable cost. Therefore, the total cost is only of interest for accounting purposes, not for comparing the four systems. To make comparisons, it is helpful to project the intersections of the planes on the base plane (Fig. 3). The artificial limitations caused by drawing the representative planes within a cube are omitted in this figure.

Evaluating Fig. 3

To get a bearing on the magnitudes of D and L that would cause us to abandon the single channel for one of the other systems, we can focus on the point having the coordinates (1,1) on the borderline between the area covered by a single channel and the area of a 2-from-3 system. For the basic parameters, let us assume that $\lambda = 2$, $\mu = 0.02$, $m = 0.02$ and $\tau = 0.04$. Let us further assume that c (yearly cost of a single channel) is \$1,000. Then, a unit on the vertical axis means:

$$(m/c)L = (0.02/1{,}000)L = 2 \times 10^{-5}L = 1$$
$$\text{or, } L = \$50{,}000$$

For the horizontal scale, one unit means:

$$(\lambda\mu\tau/c)D = (0.0016/1{,}000)D = 16 \times 10^{-7}D = 1$$
$$\text{or, } D = \$625{,}000$$

Assuming that we have picked typical values for the parameters, it is obvious that the "corner" covered by a single channel represents the majority of applications to be found in the CPI. This is true because a loss of \$50,000 in one shutdown would only occur in a huge plant, and a harmful coincidence that led to a loss of \$625,000 also would be likely to happen only in a very large production unit.

From the standpoint of cost, it appears that the 1-from-2 and 2-from-2 systems are more-or-less special cases. In other words, choosing between a single channel and a 2-from-3 system seems more likely than between one of the two-channel systems. Thus, the common practice of protecting most processes with just one channel looks like a sensible approach. (With a very large plant or where there are unusual potential hazards, it might pay to check the situation against Fig. 3.)

In the foregoing discussion, our reasoning led to conclude that the relative cost C/c was unavoidable and, therefore, could be eliminated from the factors that determine the optimal choice. However, this is not entirely true, as must be apparent from Fig. 2 and Table III. If D and L are so small that:

$$(\lambda\mu\tau/c)D + (m\mathrm{M}/c)\,L < 1$$

then the estimated potential losses are smaller than the cost of a single protective channel. In this case, one could decide to do without automatic protective devices.

References

1. de Heer, H. J., Calculating How Much Safety is Enough, *Chem. Eng.*, Feb. 19, 1973, p. 121.
2. de Heer, H. J., A Basic Theory on the Probability of Failure of Safeguarding Systems, Proceedings of the First International Symposium on Loss Prevention and Safety Promotion in the Process Industries, Elsevier Scientific Publishing Co., Amsterdam and New New York., 1974.

Meet the Author

H. J. de Heer is head of the Instrumentation Chemicals Div., DSM, P.O. Box 18, Geleen, The Netherlands, where he has worked since 1947. Prior to his present position, he worked on development of coke-oven and fuel-gas generators, and as head of instrumentation for DSM's Fertilizer Section. A graduate in technical physics (1946) from the Delft Technical University, Mr. de Heer worked on development of electron microscopes (1942-44) and was assistant manager of the Institute for Electron Microscopy (1944-47).

Designing safe interlock systems

Commonly used hardware, logic diagrams, and examples of typical systems are all covered in this overview of safety interlocks.

J. Victor Becker, Scientific Design Co.

☐ Even though they may not have to draw up safety interlock systems, most engineers have to review such designs to see that they will actually do what they are intended to do. To carry out a review of this kind, the engineer needs to know something of the hardware involved, and be able to read interlock logic diagrams. Both of these areas are covered in this article. Also, there are a number of examples of typical interlock systems.

First, however, let us look at the relationship of the plant operator with interlock systems.

The plant operator

The operator's function in any safety interlock system must be limited only to the preparation of equipment required to clear a system for startup, and to push-button actions such as reset, start and stop. He must never be allowed discretionary powers to change alarm, interlock and response-time setpoints or sequences; all interlock systems must be made "operator proof." This entails both securing the physical installation of the systems and their components from unauthorized access and providing for the worst cases of operator dereliction or interference as defined by fault-tree analysis. A review of these analyses, at this phase of engineering, must also consider the operator's physical relationship to the developed plant layout, his work routes and mandatory actions.

From these data some plant redesign might be indicated—such as relocation of access routes, or relocation of components in order to direct his work routines through the best plant-observation patterns.

Hardware—general considerations

Power supplies to electronic instruments are usually 24 V d.c., with the same voltage for the interlock circuits. Long-distance equipment-control power, from the control room to solenoids, motor starters, etc. in the plant, may be 24 or 110 V a.c. in order to minimize the size of the carrier wiring.

Electrical contacts should always be referred to and shown in their failed- or shelf-position and are to be designed to fail safe, that is by always moving the initiating contacts into a safe condition on component failure. A mechanically operated low-flow-alarm switch for example will have normally open (NO) contacts that will close when flow is above the setpoint (i.e., within the allowable process condition) and open when the flow is lower than this parameter. In other words, the normal condition of the process holds the contacts closed and an abnormal condition opens them. Also, power failure in the switch circuit produces the same alarm condition as an open switch. A mechanically operated high-flow-alarm switch, however, will have normally closed (NC) contacts. They *open* when the flow becomes too high. Thus, high flow and other similar high normally-closed alarm contacts are much less safe than NO switches because the contacts are *not* held-in by normal process conditions but by the mechanical construction of the switch. This type of device may be employed only if a failure to alarm and shut down has acceptable consequences, such as the overflow of a water tank.

Where intrinsically fail-safe conditions are essential, the switching devices must be electronic. The contacts of such devices fail open for both high and low abnormal process conditions by deenergizing an internal relay. Where the employment of such fail-open contacts is mandatory, the contacts of the initiating device must be so shown and the correct hardware provided to accomplish this failure mode.

Switch actuation considerations

A typical control "loop" may have three or more elements that can actuate a switch, namely, a transmitter, a controller-indicator-recorder and a valve-position indicator. Thus, a switch responding to a process variable that controls an interlock circuit can be initiated by several different methods.

When a switch is actuated by a process transmitter output, it usually takes its signal from the input to the receiver located in the process control panel, and thus such switches are most often located behind the panel. Tilting, mercury pressure switches are most commonly used if the transmitter is pneumatic. Electronic switches are used if the transmitter produces an electrical analog signal.

Where a critical "on-off" electrical signal is required, the interlock switch must be actuated by an independent and separate process-contacting device with its own transmitter and switch output. In other words, it must be a totally independent backup to a process control when (not if) that loop fails.

For background on this topic, see "Fundamentals of interlock systems," in the Oct. 8, 1979, issue of *Chem. Eng.*, p. 101.

Originally published October 15, 1979.

It is permissible that an alarm switch be actuated by a signal or contact in the receiving instrument (indicator, recorder, or controller). For example, in certain electronic temperature circuits the construction is such that the interlock switch is best located in the recorder or controller, but it must be pointed out that it is the least reliable switching method because any fault in the sensing element transmitting mechanism or receiving instrument can impose a false signal on the switch. This method should be used for advisory alarms only, and never for interlocks.

Valve-position indicators are useful in alarming, limiting the effect of overcontrol and determining current valve positions, but they do not measure the process. They are, therefore, secondary-effect mechanisms and should only be used as status devices.

Pushbuttons are provided in most critical interlock circuits to allow the operator to trip (shut down) a circuit if necessary, even though the process variables are within their safe limits. Pushbuttons are also used for reset and system startup.

These services employ two basic types of pushbuttons, "momentary" (make or break) and "maintained" (make or break). A momentary-make pushbutton is spring-loaded so that the switch returns to its normal position when released. An electrical holding latch (actuated by the pushbutton switch) keeps the circuit energized after button release (a second, momentary-break, button is required to deenergize the latch). Release of this button allows the first button to reactivate the holding latch at any time thereafter. "Maintained contact" pushbutton switches are mechanically interconnected pairs of buttons (e.g., start, stop). Pushing one (closing) causes the other button's switch to open and vice versa. Once pressed, such a switch stays in its last position until the other button is pushed.

These pushbuttons are used where interlocking from alarms or other circuits is not required, such as in an emergency shutdown circuit. In these applications, they have the advantage (over momentary-contact pushbuttons) of eliminating the electrical holding latch, as well as by showing the circuit condition from the button positions. They are sometimes used in manual reset and interlock circuits to provide additional protection against the inadvertent reset of the circuit.

Rotary or selector switches have a wide range of applications, from relatively simple devices to complicated multicontact ones, with the position of the handle telling the operator which circuits are energized. Selector switches usually maintain their last position, but they can be equipped with mechanical or electrical actuators that will cause them to return to their first position in the event of a tripout. They are not used extensively in chemical process plants (except for the key-operated form described below) because most circuits are of the stop-start or the stop-reset interlock variety for which momentary pushbuttons are more suitable and safer.

Key-operated switches are similar to automobile ignition switches and although they can have multiple positions, are usually restricted to two. They are used for bypassing components of interlock circuits for maintenance—for instance, for the temporary isolation of a process analyzer for calibration.

Key-operated switches are used instead of maintained-contact pushbuttons or selector switches for such bypass services, because the need for a key limits unauthorized switch operation. In general practice, the lock is made so that the key cannot be removed unless the switch is in the "normal" position and panel lights are provided to emphasize the switch's condition.

Relays

The relays used in wired interlock systems are generally electromechanical devices consisting of an armature coil, an electromagnet plunger and one or more switching contacts. These contacts open or close discrete and separate circuits when the coil is energized and deenergized by a signal voltage. A relay is, essentially, a solenoid with switch contacts. Where possible, the type that allows visual checking is preferred because it permits relatively easy trouble-shooting.

Solid state relays (SCRs—silicon controlled rectifiers), which replace electro-mechanical relays in digital processors, act in the same manner as NO relay contacts; that is, in the conducting stage they represent closed contacts and in the nonconducting stage, or failed condition, open contacts. Like mechanical relays, SCRs can be activated by either an outside voltage, current or contact closure. Where a.c.-power switching is required, triac components may be used; however, they should be used with caution because they inherently fail closed.

Solenoid valves

Where an interlock circuit is set up to rapidly stop or start the flow of some process fluid or utility, the means for doing this is almost always a solenoid valve. These devices combine an electromechanical coil and solenoid-plunger core with a valve, consisting of a disk or plug, which is positioned by the solenoid plunger to arrest or permit flow. Power to these valves is usually 24- or 110-V a.c. The electrical enclosures can be either general-purpose or explosion-proof. The three basic types of solenoid valves available are:

Two-way solenoid valves—These fail either normally open or closed, and are sometimes used to shut off or permit the flow of a utility stream. They are limited in size and are seldom used directly in process streams because of their pressure-, temperature- and modulation-limitations. This type of valve can be furnished with an electrical or a manual reset, which must be actuated before the valve can return to its operating position. This feature is useful in the case of a furnace fuel shut-off, for example, because the operator can be required to go out to the furnace to reset the latch before permitting fuel to flow to the burner again. This requirement is a good safety practice because the operator should inspect the furnace to see that everything is satisfactory before reset; the proper placement of such a device directs the operator's mandatory actions.

Three-way solenoid valves—These are used as intermediate devices between most interlock circuits that involve the stopping or starting of a process fluid. Here the solenoid valve acts as an air pilot to another valve, usually a control valve. It has three pipe connections, an air supply and two pathways, one path being open

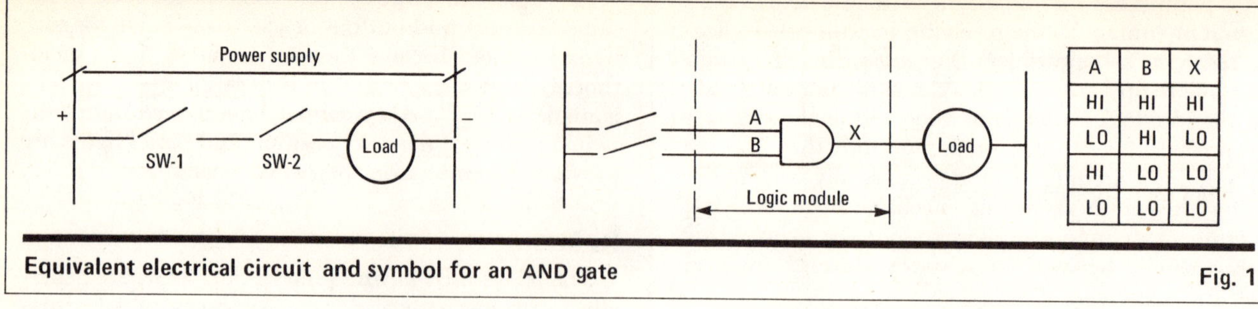

Equivalent electrical circuit and symbol for an AND gate

Fig. 1

while the other is closed. Such valves are available in three types: (1) normally open, (2) normally closed, and (3) universal. The normally-open type is used to vent the control valve diaphragm motor when the solenoid is energized; the normally-closed type to vent the diaphragm when the solenoid is deenergized. Normally-open or -closed types are fixed-action—i.e., their construction does not permit repiping and the internals cannot be reversed. The universal type can be arranged to operate either way, and may be needed when pressure is required at the vent port, but are not commonly used because of the possibility of error in assembly.

The three-way valve sizes used for diaphragm-motor actuation are usually ¼ and ½ in. although sizes are available up to 2 in. For very quick control-valve venting actuation, an air-operated 3-way quick-exhaust valve, actuated in turn by a solenoid pilot, may be employed. This arrangement vents or pressurizes the diaphragm motor considerably faster than a common solenoid, and valve stroking times of 200 milliseconds or less may be achieved even with large valves.

Four-way solenoid valves—These are generally used to operate double-acting piston operators on valves. They have four pipe connections: one supply connection, two cylinder connections and one exhaust. In the first valve position, pressure is applied to one end of an operator cylinder, while the other end exhausts. In the other valve position the pressure- and exhaust-cylinder connections are reversed. Sizes are available from ¼ to 1 in. Double-acting cylinders, so actuated, are primarily used in process plant instrumentation for remotely-operated open or shut block valves that do not have fail-safe requirements.

Interlock logic-diagrams

These diagrams are the construction drawings and engineering end-documents that symbolically indicate all of a system's components, such as switches, relays, lights, solenoids, etc., together with their interlock logic sequencing and relationship. The interlock logic diagram does not show the actual physical locations of the control devices; it does, however, combine in schematic form all the necessary information required for someone to logically follow the operation of the various devices in the circuit and system. The symbols are adapted from ISA, military and ANSI standards and may be used to design and represent any control logic statement regardless of the hardware employed.

The logic modules, or gates, represent the components (SCRs, relays) of a logic statement on which external inputs (contacts, switches) act to give a desired output; they can be arranged in any order that will adequately describe a sequence of events. They are of a binary nature (that is having two states; on or off, conducting or nonconducting), and provide (through simple Boolean algebra) a structured and rigorous means for presenting interlock statements. A basic review of this methodology follows.

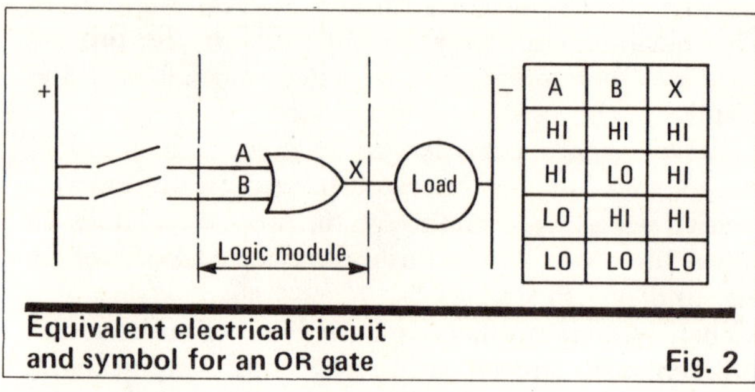

Equivalent electrical circuit and symbol for an OR gate

Fig. 2

HI refers to a positive signal level (>0 V), LO to a false or negative level (0 V).

A basic logic review

AND gates—The first basic logic module is the AND gate, which may have two or more inputs but only one output. The output of a basic AND gate is HI (true) only when all of the inputs are at the HI (true) state. A LO (false) signal on any of the input terminals produces a LO (false) signal at the output. Thus, the AND gate performs the logical operation of producing a true (HI) output signal only when all the input signals are simultaneously true. Fig. 1 illustrates a circuit in which no power is delivered to the load unless both switches in series are closed, as well as the equivalent AND logic module, with its associated truth table.

OR gates—The second basic module is the OR gate, which may have two or more inputs but only one output. The output of a basic OR gate is HI when one or more of the inputs are at the HI state. Thus, the OR gate performs the logical operation of producing a true output for the time duration that one or more of its inputs are true, and a false (LO) output for the time duration in which none of its inputs are true. Fig. 2 illustrates a circuit in which voltage is delivered to the load when either or both switches, in parallel, are closed.

Signal inversion—The inverter is a logic device with one input and one output that is used to invert the sense of the input signal; that is, the output of an inverter provides a signal that is the complement of the input. Thus, the inverter amplifier performs the logic operation of producing a LO output for the time duration

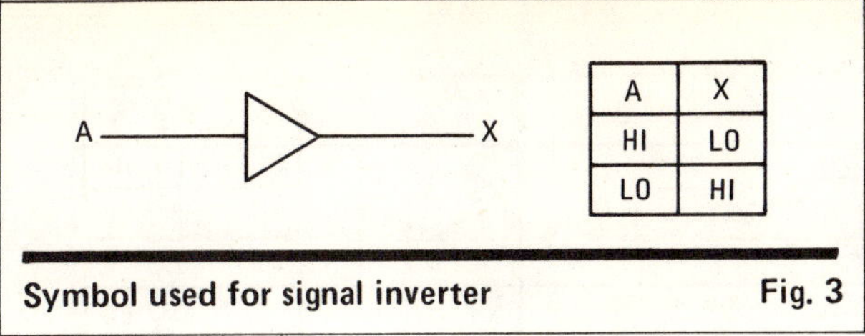

A	X
HI	LO
LO	HI

Symbol used for signal inverter Fig. 3

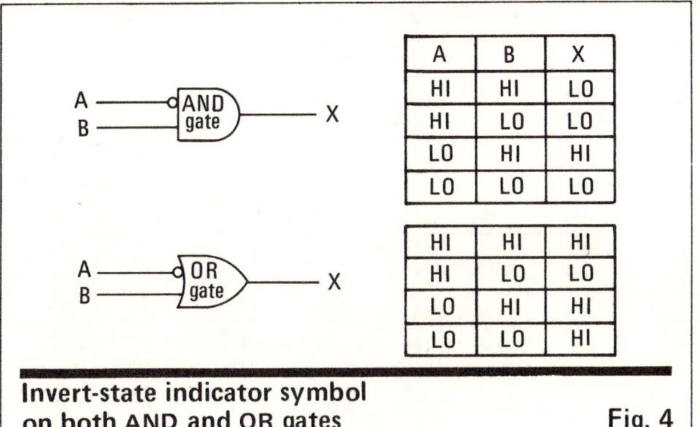

A	B	X
HI	HI	LO
HI	LO	LO
LO	HI	HI
LO	LO	LO

A	B	X
HI	HI	HI
HI	LO	LO
LO	HI	HI
LO	LO	HI

Invert-state indicator symbol on both AND and OR gates Fig. 4

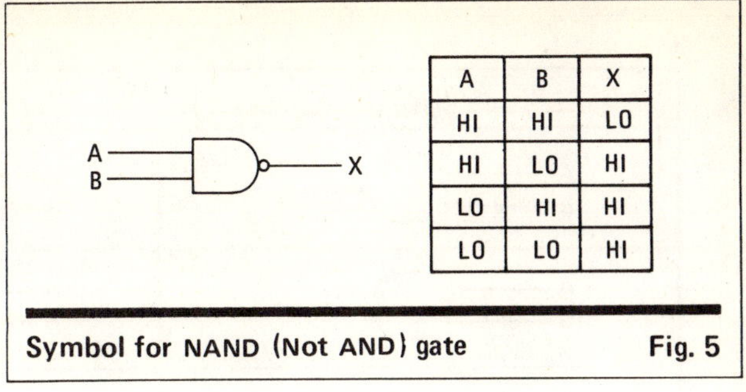

A	B	X
HI	HI	LO
HI	LO	HI
LO	HI	HI
LO	LO	HI

Symbol for NAND (Not AND) gate Fig. 5

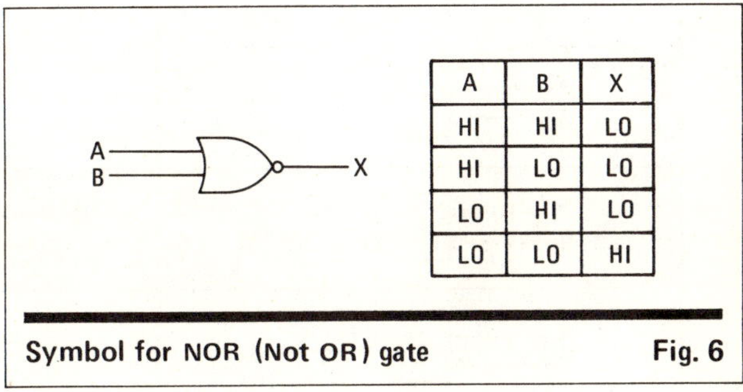

A	B	X
HI	HI	LO
HI	LO	LO
LO	HI	LO
LO	LO	HI

Symbol for NOR (Not OR) gate Fig. 6

that the input is HI and vice versa. The inverter is presented in Fig. 3.

Invert-state indicator—Logic diagrams make extensive use of the invert-state symbol, which expresses the same logic statement as the inverter above, but which is shown as a small circle at an input or output of a gate. An AND gate with LO-state indicator at the *A* input is shown in Fig. 4. The output of this gate is only HI when the *A* input is LO and the *B* input is HI. The second example is an OR gate shown with a LO-state indicator at the *A* input. The output of this gate is HI if either the *A* input is LO or the *B* input is HI.

NAND gate—A NAND (Not AND) gate is the functional equivalent of an AND gate followed by an inverter, and performs the logic function of producing a LO at the output only when all the input signals are HI. Thus a LO applied to any input results in a HI output (Fig. 5).

NOR gate—A NOR (Not OR) gate is the functional equivalent of an OR gate followed by an inverter, and fulfills the logic function of producing a HI at the output only when all the input signals are simultaneously LO; thus a HI applied to any input results in a LO output (Fig. 6).

The truth tables (Fig. 7) show all the logic statement variations possible from the two basic AND/OR modules and their input and output signal inversions.

EXCLUSIVE OR gate—An EXCLUSIVE OR gate produces a true output when the input states are not identical, with the output going LO if the inputs are identical. Thus, the EXCLUSIVE OR gate (Fig. 8a) fulfills the logical function of producing a HI if one and only one input is LO. The second configuration (Fig. 8b) of the EXCLUSIVE OR with an inverted output produces the complement (COINCIDENCE).

Time delays—When a switching time delay is required after a signal change, the symbol of Fig. 9 is employed. In the ON delay example given below an input change to HI will produce a change in output to HI after the time specified has elapsed. When the input signal goes LO the output goes LO immediately. An OFF delay is indicated when the output signal state must go HI immediately on a HI input; and after the delay time stated, change to LO. By adding the inverter symbol to the input or output the signals become the complements of each other.

Preparing interlock logic diagrams*—The general arrangement of these drawings is with all the system inputs shown on the left-hand side, with all interlock logic shown in the center and with all the outputs arranged on the right. Inputs are the contact condition information from pressure-, flow-, level-switches, etc., and any other on/off-type switching devices affecting a given sequence. The logic section of the diagram represents, in symbolic form, the decision-making functions required from the status of a given set of inputs to produce a predicted set of outputs. As previously mentioned, this section may be composed of electromechanical or pneumatic relay components arranged to provide series and/or parallel logic wiring, or to express a solid-state processor program. Such processors, which permit a large variety of logic conditions at lower cost, higher speed and greater reliability than hard-wired relay installations, are rapidly replacing even small hard-wired or piped relay switch functions. The outputs are actuating signals required to stop and start motor controls, operate solenoid pilots on control valves, sequence process starts, sound alarms and initiate partial or total shutdowns.

Examples of interlock circuitry

Selfcancelling interlock

In Fig. 10, a very low level in Tank U-330, as detected by level instrument LALL-330, requires immediate

*See Oct. 8 article, Fig. 4.

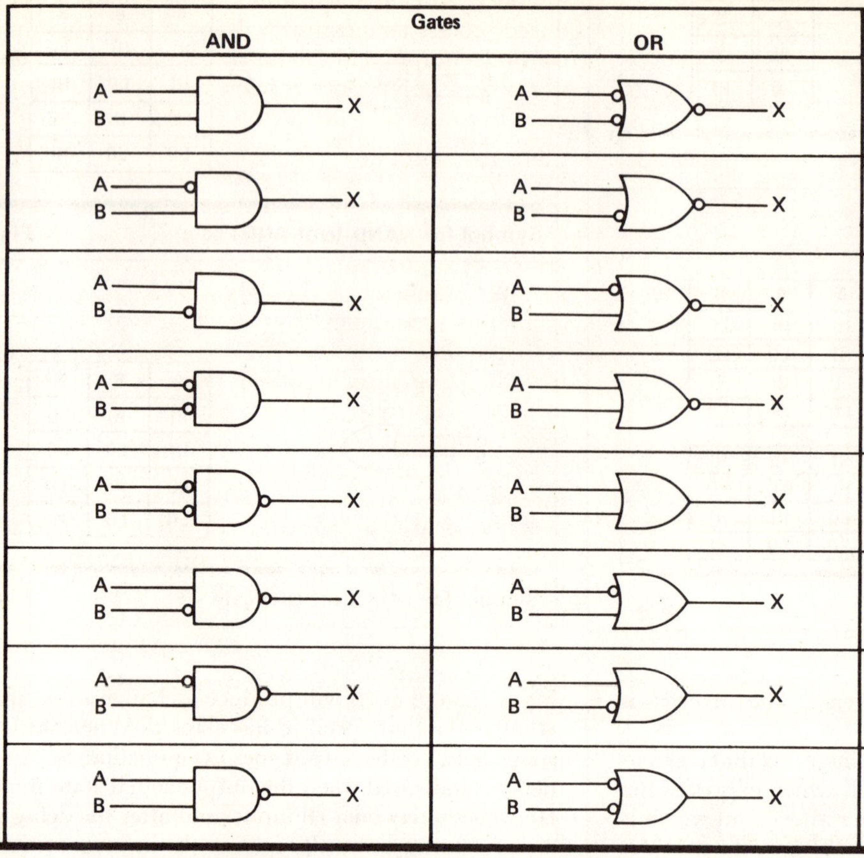

Gates AND	Gates OR	A	B	X
		HI HI LO LO	HI LO HI LO	HI LO LO LO
		HI HI LO LO	HI LO HI LO	LO LO HI LO
		HI HI LO LO	HI LO HI LO	LO HI LO LO
		HI HI LO LO	HI LO HI LO	LO LO LO HI
		HI HI LO LO	HI LO HI LO	HI HI HI LO
		HI HI LO LO	HI LO HI LO	HI LO HI HI
		HI HI LO LO	HI LO HI LO	HI HI LO HI
		HI HI LO LO	HI LO HI LO	LO HI HI HI

Basic logic diagrams and their truth tables **Fig. 7**

action to correct the situation. LAL-330, a low level alarm (not shown) will have already warned the operator that an abnormal situation is occurring, and it is assumed here that no corrective action has been taken. So the low-low level sensor opens contacts LALL-330 in the circuit to the annunciator and LY-330 solenoid valve. This in turn vents air from the diaphragm of the makeup valve, letting it open to admit more liquid to U-330.

The interlock is self-cancelling since, as soon as this minimum level is regained, LALL-330 contacts will reclose and reenergize the solenoid pilot valve and, in turn, close the makeup valve. However, in accordance with standard practice, the annunciator will continue to alarm until the operator acknowledges it.

The example given is primarily for illustrative purposes. Self-cancelling interlocks are *not permitted if the corrective action can create a hazard,* e.g., in the example given, unabated makeup could cause tank overflow. When they are used, they should be identified as such by engineering flowsheet notes; they should *not* be provided unless specifically acceptable.

The self-cancelling system described has also a significant disadvantage in that the interlock will continuously alternate between "activated" and "cancelled" if the process condition which causes the interlock action is cycling around the critical value. This can, with unfavorable conditions, lead to equipment damage. To prevent this, most interlock circuitry is set up to require positive action by the plant operator in order to cancel the interlock once conditions return to normal (although in some cases a reset differential gap may be furnished by another level switch set at a different value). This gap may be the minimum required to prevent rapid cycling, to the full height of the vessel, such as required in a pumpdown and refill circuit.

In Fig. 11, HI pressure will open PAH-430 and remove the signal from one side of the first AND gate. As both the pressure switch and the stop button HS-430A have to be HI (closed in series) the gate output will go to LO. This signal removal from the second AND gate's input will, in turn, deenergize SV-430.

The first AND gate will be automatically reenergized by the restoration of PAH-430 to a HI condition. However, the second AND gate has to have a momentary, or reset, closure at HS-430B to provide it with the other AND input and thus make a HI signal appear on the gate output. This is then returned to the reset input OR gate, which latches itself to the second AND gate and thus maintains the circuit. The system may be manually dumped by the process operator's pushing of HS-430A.

Sometimes after an interlock has occurred, a controller must be manually set so that its output corresponds to the interlock position of the valve. This ensures that the valve does not move when the interlock is reset.

The example considers the case of a steam-flow controller, controlling live steam injection into a column by

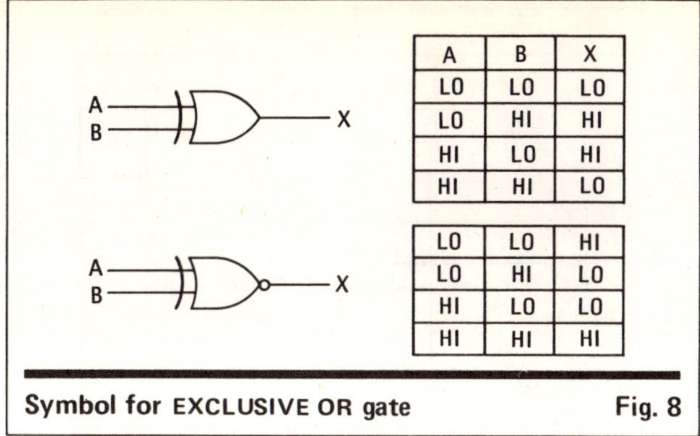

A	B	X
LO	LO	LO
LO	HI	HI
HI	LO	HI
HI	HI	LO

A	B	X
LO	LO	HI
LO	HI	LO
HI	LO	LO
HI	HI	HI

Symbol for EXCLUSIVE OR gate **Fig. 8**

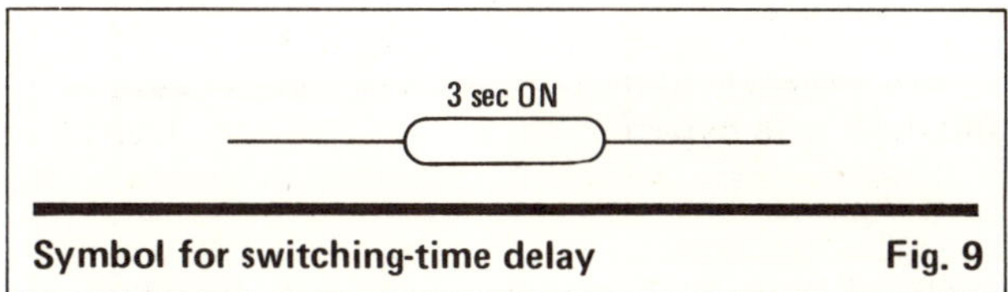

Symbol for switching-time delay **Fig. 9**

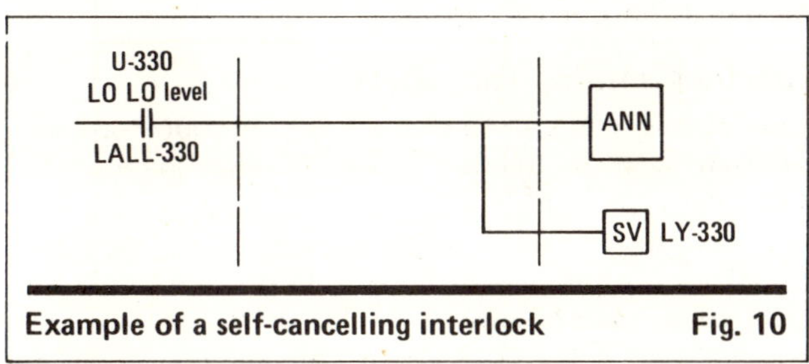

Example of a self-cancelling interlock **Fig. 10**

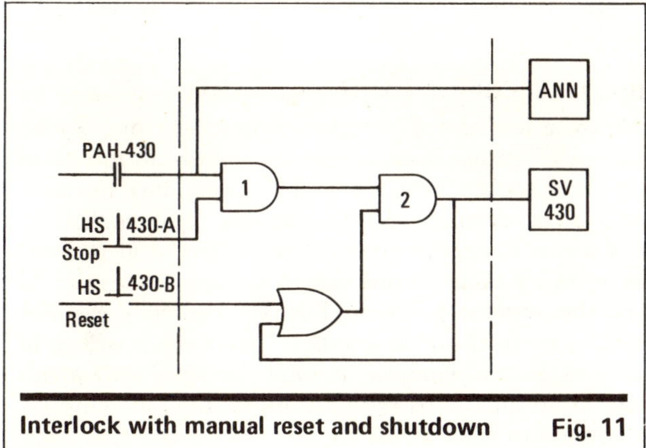

Interlock with manual reset and shutdown **Fig. 11**

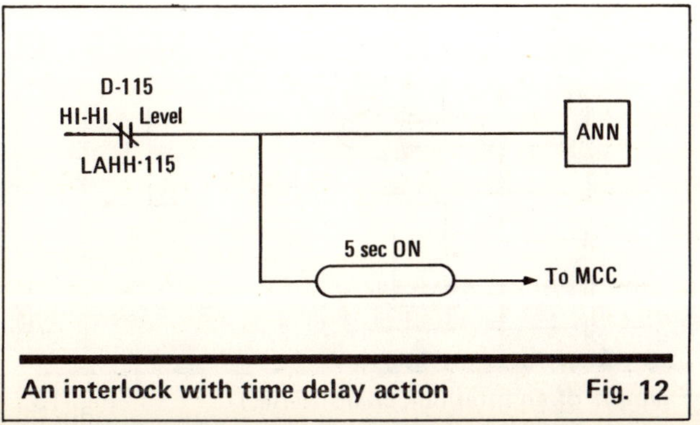

An interlock with time delay action **Fig. 12**

means of a valve that is closed by a high-temperature interlock. After the activation of the interlock, the reduced process temperature will inherently wind up the controller output and call for the control valve to open; thus when the interlock is reset, the valve will fly wide open from the presence of this "wound-up" signal. To avoid this, "reset lockout" has to be employed. If this precaution is overlooked, when the interlock is cancelled, the valve will immediately take the position called for by the controller. This may result in a hazardous or undesirable situation, e.g., live steam surge into the column which could damage its trays or lead to additional process upsets.

In this case a "lockout" feature should be incorporated in the circuitry to prevent the reset of the interlock until the controller output satisfies a predetermined restart condition.

In this example, HI temperature opens TAH-520 thereby deenergizing one input to the first AND module. When this gate's output goes LO, the second and third AND gates also go LO, thereby deenergizing solenoid valve SV-520. Note that the annunciator connection from TAH-520 is direct, so that reclosing this switch will return the annunciator to normal separately from any condition in the interlock circuit.

An additional set of contacts (PAH-520) has been added to the circuit. These contacts are actuated by the steam-flow-controller output-sensing device—in this case a pressure switch in the pneumatic signal line to the control valve. If the controller has not been set to the desired output, pressing the reset button HS-520B will *not* complete the electrical circuit and thus the system can be reenergized only after the temperature alarm TAH-520 closes and the control signal alarm PSH-520 is satisfied (low output). That means, in practice, that the operator has to reset the steam-flow controller to zero flow before the interlock condition can be cancelled.

Once normal operation is resumed, the controller output must be free to vary over its whole range without triggering the interlock, hence, the output of #2 gate is returned to itself via an OR gate, which latches it in the HI condition and maintains signal continuity in the circuit until either HS-520A is pushed or TAH-520 opens.

The example shows a startup bypass of an interlock for a low-compressor-speed shutdown. When the compressor is stopped, or below interlock speed, contacts SAL-110 are open and Gate 1 has a LO output. This signal is inverted to HI at Gate 2, which, when energized and latched by pushbutton HS-110C, also energizes Gate 4 and a bypass advisory panel light. After the reset (start-HS-110A) button is pushed and the compressor speed exceeds the trip point, SAL-110 contacts will close, energize Gate 1, maintain a HI signal to Gate 4, while the now LO (inverted) signal to Gate 2 deenergizes Gate 2, the panel light and the startup latch (Gate 3).

Interlock with time-delay action

It is sometimes necessary to allow a time delay between various interlock actions. In Fig. 12, high level in D-115 opens LAHH-115 contacts and the signal to the

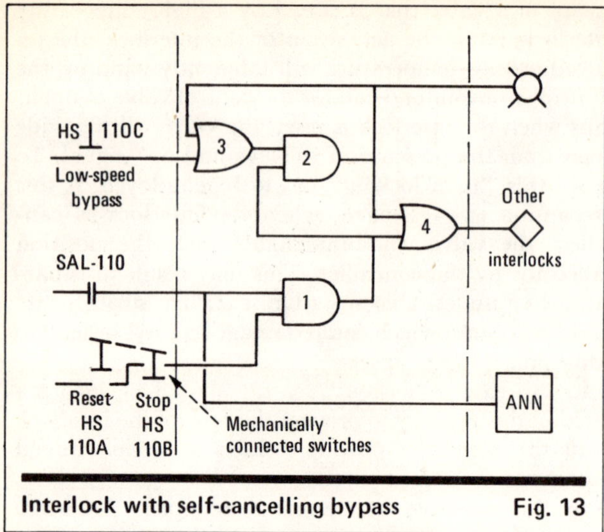

Interlock with self-cancelling bypass Fig. 13

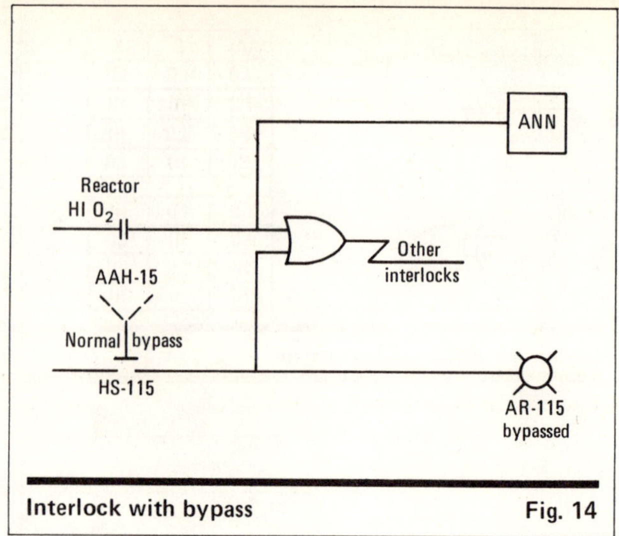

Interlock with bypass Fig. 14

MCC goes LO. A time delay is included in the circuit and specified as "5 seconds ON," which means that the HI signal to the MCC will not be restored until 5 seconds after LAHH-115 contacts reclose. Conversely, if power is required to continue for 5 seconds after LAHH-115 opens, a deenergized delay notation is employed, viz. "5 seconds OFF."

Interlock with self-cancelling bypass

Occasionally, an interlock may have to be bypassed for only a short period, e.g., in order to start up a system, but must be operable at all other times. To avoid the possibility of a manually operated bypass being forgotten or used incorrectly, a self-cancelling bypass should be provided (Fig. 13).

Interlock with bypass

It may be necessary to carry out maintenance on an instrument associated with an interlock to ensure that it operates reliably. Where such maintenance is required frequently, e.g., for analyzers, it is often advisable to include a bypass in the interlock circuitry to avoid unnecessary interruptions in plant operation, as shown in Fig. 14.

In this example, HI oxygen concentration will open AAH-115 which switches the OR gate to LO output, which causes actions not shown here. When key-operated HS-115 is switched to bypass, it keeps the gate turned on thereby preserving the electrical continuity to the process while the analyzer is being serviced. A panel advisory lamp is provided to show when the bypass is being used. (Note: HS-115 maintains the HI output of the interlock, and *not* the alarm contacts AAH-115. This allows the operators to verify that the alarm is satisfied—low O_2 concentration—before cancelling the bypass.)

Interlock chains (series)

In most systems, several interrelated process conditions must be satisfied simultaneously for safe operation. Deviation of any variable outside its normal range must trigger the interlock, i.e., cause emergency shutdown. In the system shown in Fig. 15, the contacts associated with each process alarm are arranged in series (AND, AND, AND, etc.), hence, if any one goes LO, the system output will go LO.

Interlock chains (parallel)

In some systems deviation of one or more process parameters can be tolerated, provided that some other parameter is still within allowable limits, whereby quite complex interrelationships can be catered for (Fig. 16).

In this example, it will be seen that the alternative signal paths may be any of the following: 1 AND 2, 3 AND 4, 1 AND 4, 2 AND 3, OR 5. Thus three, or four, or five process variables must be outside tolerable limits to trigger the interlock by deenergizing all the OR gate inputs.

Interlock chains (voting systems)

A variation of the parallel system described above can be used in situations where it is important to ensure reliable operation of an interlock depending on only one process variable, while minimizing the possibility of an instrument failure giving a false signal and triggering an unnecessary shutdown.

In the system shown in Fig. 17, two of the three flow-alarms must signal low-flow before the interlock is initiated. In this example, the flow switches act on the gates to produce a HI signal if any of the following HI combinations exist across the flow switches—1 AND 2, OR, 2 AND 3, OR, 1 AND 3.

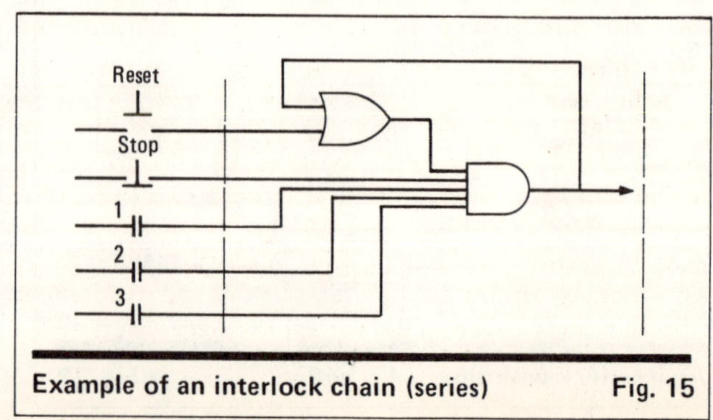

Example of an interlock chain (series) Fig. 15

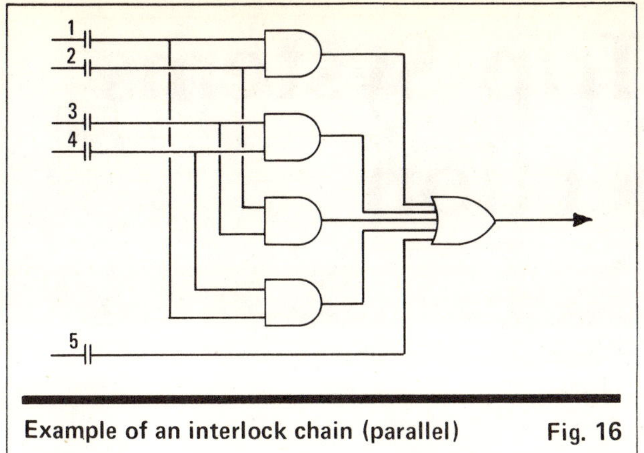

Example of an interlock chain (parallel) Fig. 16

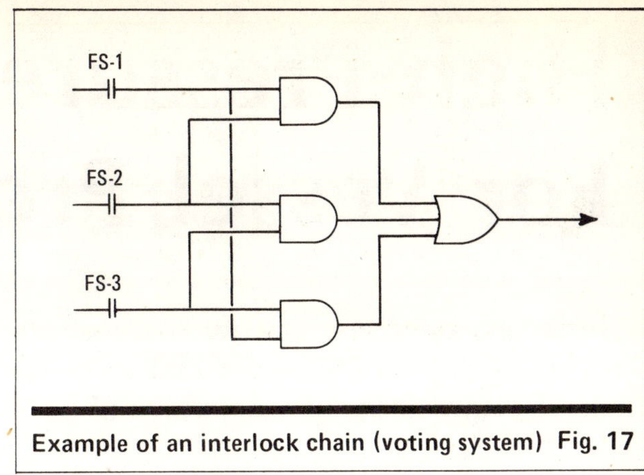

Example of an interlock chain (voting system) Fig. 17

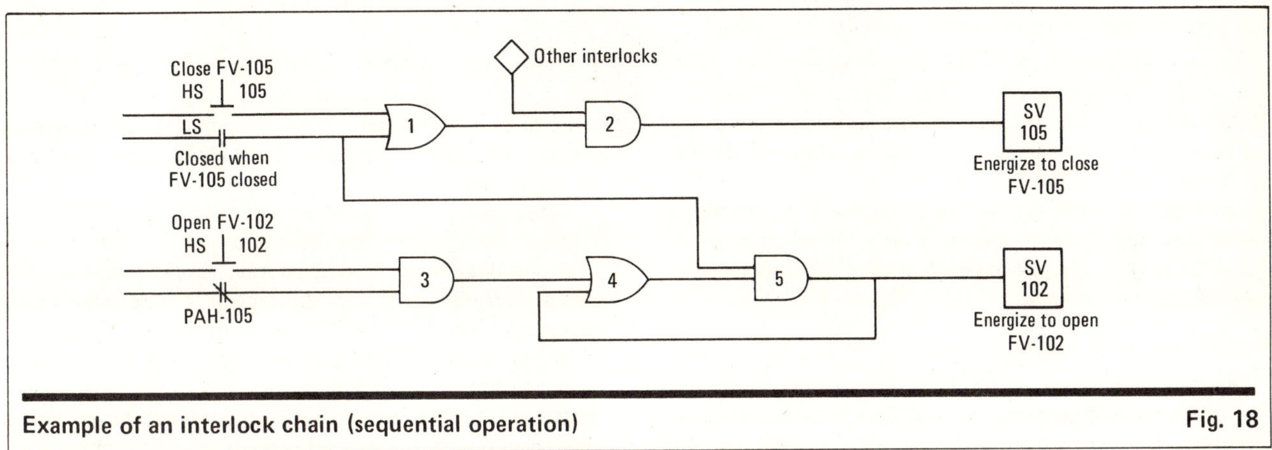

Example of an interlock chain (sequential operation) Fig. 18

Interlock chains (sequential operation)

Interlocks are not necessarily used exclusively to initiate shutdowns; they can also be used during start-up, to ensure that certain critical operations are performed or to ensure that the startup steps are carried out in a predetermined sequence. In the system shown in Fig. 18, it is required that "fail-open" valve FV-105 be fully closed before "fail-close" FCV-102 is opened.

SV-105 will be energized (to close) by pushing HS-105, provided other interlocks are satisfied and position switch LS-105 (on FV-105) closes when the valve is closed, thus keeping SV-105 energized on HS-105 pushbutton release. HS-102 energizes and latches power to SV-102, provided that PAH-104 and LS-105 are closed and HI, and the system remains in this condition until tripped, which action returns the valves to their fail-safe positions. It should be noted that even if the operator pushes both buttons at the same time, the correct sequence ensues.

The foregoing examples are only a few of the many process situations that may have to be protected by safety interlocks but, generally speaking, illustrate most of the unit logic statements that may be expressed in a process-plant safety-interlock logic diagram. Such logic configurations as counters, and multiple-choice decision-making routes should be avoided because these complications will reduce the system's reliability. The finished drawing should be rechecked against the requirements of the control documents—the sequential logic diagrams—and clearly marked with a prohibitional note against changes without authority from the Safety Committee.

Conclusion

This article and "Fundamentals of Interlock Design," in the Oct. 8 issue, have reviewed the basic requirements for engineering safe interlock systems. Because of the enormous variety of hazardous situations that can exist in the chemical process industries, these reviews could not cover all specifics. However, the required technical approaches to any interlock system design may be summed up as follows:

1. Establishment and rigid control of the project safety engineering documents and formats. 2. Examination and safety analysis of the process before general engineering commences. 3. Determination of intrinsically fail-safe requirements and the hardware involved. 4. Incorporation of fuse requirements into discrete backup systems separate from plant process control. 5. Provision of parallel, series, or series-parallel redundant components, systems or utilities, as required. 6. Examination and complementary agreement from both positive and negative logic viewpoints. 7. Incorporation of the finalized sequential logic diagrams into the *mandatory* practices of the Plant Operations Manual. 8. Reexamination of these practices after plant startup and shakedown, with possible modification to provide additional reliability.

High-Pressure-Trip Systems For Vessel Protection

Pressure switches, solenoid valves, electrical relays, etc., can isolate process pressure sources just as reliably—at considerable savings—as do the conventional relief-valve and flare systems.

HERBERT G. LAWLEY and TREVOR A. KLETZ, Imperial Chemical Industries Ltd.

Great interest has been shown in replacing relief valves by protective systems that isolate sources of pressure. The reason is that, as plants get larger, relief valves and associated flare systems become larger and considerably more expensive.

Here, we shall discuss the reliability that protective systems that isolate pressure sources should have, and will show—by an example—the detailed design of one such system.

For many years, universal practice has been to fit relief devices on all pressure vessels, as required by pressure-vessel codes. However, as plants and equipment get bigger, the cost of providing such devices rises. Relief valves are not so expensive in themselves but, if flammable materials are handled, they may require flare systems. These are expensive, they sterilize a lot of ground, and produce complaints about noise and light. And if toxic materials are being handled, the relief-valve discharge may have to pass through a scrubbing system.

This is no reflection on the quality of relief valves; they are some of the most reliable and well-engineered equipment we have. The problem is disposing of the material that comes out of the tail pipes of the relief valves, because we must reduce the production of unwanted materials.

How To Avoid Relief Valves

One method of avoiding relief devices is the installation of stronger vessels. For instance, if a distillation column is heated by steam, the size of the relief valve is ordinarily determined by assuming that there is a loss of cooling water to the condenser, or a loss of reflux while the heat input continues. Then, a large relief is usually called for. But if the column can be made strong enough to withstand the vapor pressure at the temperature of the steam, there is no need for a relief valve to protect against such pressure; only a smaller fire-relief valve is then needed.

In other cases, relief valves can be replaced by instrumented protective systems (trips) that detect a rise in pressure and shut off the source of pressure. On a distillation column, for example, the rise in pressure can be used to isolate the heat input to the base. Then, only a small relief valve is needed to cope with other sources of overpressure.

When runaway reactions—such as oxidation—are liable to occur, it may be difficult to install relief valves that are large enough and operate quickly enough to avoid overpressuring the equipment. A trip can then isolate the supply of one of the reactants.

Many engineers, however, still reject such solutions because they feel instruments are unreliable. The truth is that relief valves are not 100% reliable either, and that it is easy to design a protective system with a reliability as good as, or even better than, that of a relief valve. Table I shows some typical "hazard rate" figures (the rate at which the vessel is overpressured), assuming in Example 1 that, if there were no relief valve or protective system, the vessel would be overpressured once every year. Example 2 shows typical figures, assuming a hazard rate—with no relief valve or protective system—of once every five years.

Reliability Comparison of Protective Systems
Table I

Type of System	Example 1, Hazard Rate	Example 2, Hazard Rate
No relief valve or trip	Once/yr (assumed)	Once/5-yr (assumed)
Relief valve with annual testing	Once/250-yr	Once/1,000-yr
Simple trip with:		
annual testing	Once/4-yr	Once/15-yr
monthly testing	Once/36-yr	Once/180-yr
weekly testing	Once/150-yr	Once/750-yr
Duplicated trip with weekly testing	Once/18,000-yr	Once/90,000-yr

Originally published May 12, 1975.

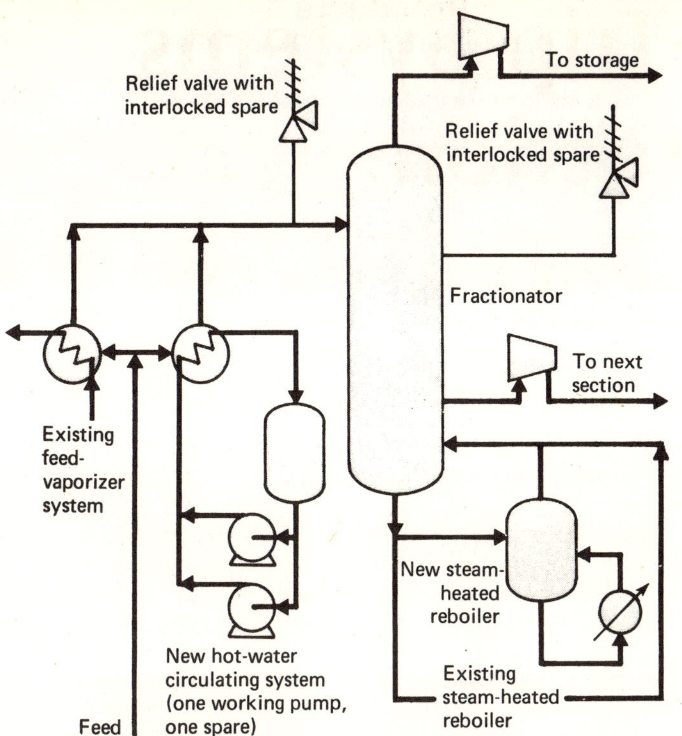

UPRATING SCHEME required a new hot-water vaporizer and a new steam-heated reboiler—Fig. 1

A duplicate trip is actually safer than a relief valve, although complete duplication is not always necessary. Regular testing, however, is essential. It is recommended that any trip taking the place of a relief valve have a reliability 10 times greater than the figure quoted in Table I for a relief valve. The reason for this is the uncertainty of the figures, and the difference in the mode of failure of relief valves and protective systems. A relief valve that fails to open at the set pressure may well open at a considerably higher pressure. A trip that fails to operate at the set point, however, is more likely to have failed completely.

Designing High-Pressure-Trip Systems

Following is an example of how a trip system was designed (when a plant was uprated) to take the place of a larger relief-valve installation.

Two new sources of heat were required on an existing plant—i.e., a feed vaporizer heated by pump-circulated hot water, and a new steam-heated reboiler (Fig. 1). Calculations indicated that the two existing relief valves on this section of the plant would not afford adequate protection in the uprated system against a downstream block-in.

Because of limitations in the relief-header system, additional protection could not be achieved by resizing the existing relief valves. One expensive possibility was the installation of a third relief valve with its own header, and possibly its own knockout tank. The alternative means was the installation of a high- (rising) pressure-trip system—in parallel with the existing relief valves—to trip the proposed new, hot-water circulating pump (and its spare), and to shut off the steam to the new reboiler in the event of an abnormal rise in system pressure (Fig. 2).

When a conventional process-relief method is to be replaced in its entirety by a high-pressure-trip system, fire relief must always be taken into consideration. Unlike process relief, fire relief can usually be vented directly to the atmosphere, without an expensive header and blowdown system.

An optimum trip system is the one that yields the desired level of protection at minimum cost. In other words, the most favorable compromise between installed cost, cost of proof testing and routine maintenance, and consequential losses resulting from spurious operation. Thus, to arrive at the optimum design, it is usually necessary to consider various alternative systems capable of meeting the primary-reliability criterion, as well as to carry out a full-cost evaluation on each.

For the purpose of our example, assume that capital

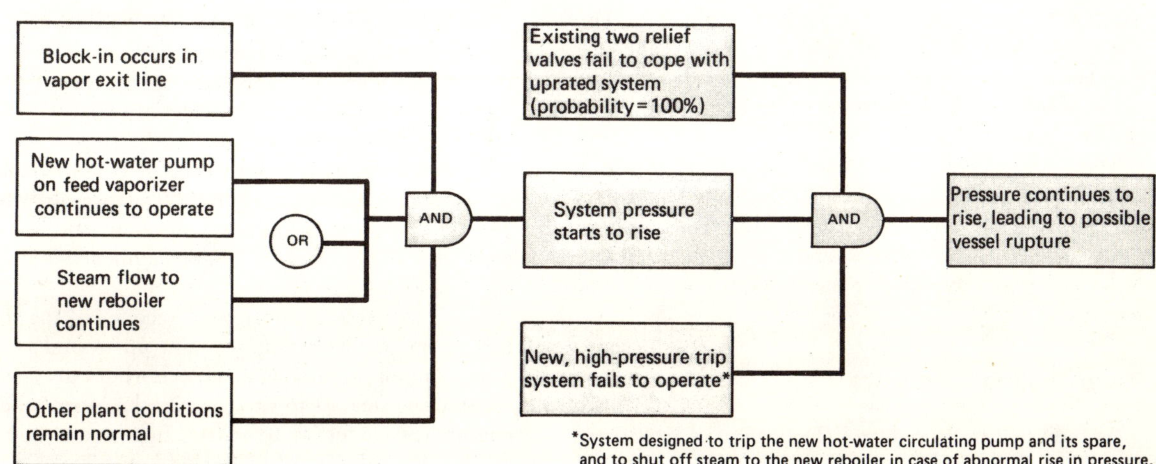

FAULT-CONDITION DIAGRAM for possible vessel rupture (due to overpressure) in the uprated-plant system—Fig. 2

cost is of overriding importance when compared to the cost of proof testing or spurious operation, provided that the frequency of spurious trips is no more than once per year. Thus, the design requirement is a trip system of minimum complexity that will provide the desired level of protection at a spurious trip rate not exceeding about once per year, subject to any constraints on proof testing as deemed important by management.

On a completely new plant, process-relief protection against vessel rupture in our area of interest would be two conventional relief valves (with interlocked spares), which would be required to lift together on a two-out-of-two basis to be effective. For the uprated plant, this protection—rather than three relief valves operating on a three-out-of-three basis—should be the yardstick for the design of a high-pressure-trip system. The reason for this is that a two-out-of-two relief-valve system is more reliable.

Assume the following fail-danger fault-rates for each relief valve with respect to overpressure (based on our experience and on data from the U.K. Atomic Energy Authority Data Bank): 0.001 faults/yr leading to stuck shut; 0.004 faults/yr leading to lifting heavy at, say, greater than 130% of set pressure; 0.005 faults/yr total for failure to relieve safely. If we further assume a random fault incidence and proof testing of once every two years (which is the present custom), the fractional dead time (FDT) for a two-out-of-two relief-valve protective system—as calculated in Procedure 1—would be:

$$\text{FDT} = 2(0.5 \times 0.005 \times 2.0) = 0.01$$

Since, as stated earlier, a trip-protective system should have a reliability ten times higher than that of a relief-valve one, this would be equivalent (assuming a total fail-danger rate of only 0.0005 faults/yr per relief valve) to a target FDT for alternative high-pressure-trip protection of 0.001 maximum.

Procedure 1

Deriving Fractional Dead Time

It can be shown that the mean proportion of time for which a system will be in the failed state—i.e., the fractional dead time (FDT)—is given by:

$$\text{FDT} = (1/T)\int_0^T p(t)dt$$

where T is the proof-test interval; and $p(t)$ is the cumulative probability of the system having failed by time t:*

For a single device with constant failure rate f:

$$p(t) = 1 - e^{-ft}$$

which, for small values of ft, approximates to:

$$p(t) = ft$$

Hence, the FDT for a single device is:

$$\text{FDT} = (1/T)\int_0^T ft\,dt = 0.5fT, \text{ or } 0.5p(T)$$

For n redundant devices with equal f's:

$$p_n(t) = [p(t)]^n$$

which approximates, as above, to:

$$p_n(t) = (ft)^n$$

Therefore, the fractional dead time for n redundant devices with identical proof-test intervals is:

$$\text{FDT} = (1/T)\int_0^T f^n t^n\,dt = \left(\frac{1}{n+1}\right)f^n T^n, \text{ or } \left(\frac{1}{n+1}\right)p_n(T)$$

Thus, for two such redundant devices ($n = 2$):

$$\text{FDT} = (1/3)f^2T^2, \text{ or } (4/3)[(1/2)fT]^2 = (4/3)(\text{FDT of a single device})^2$$

*Note that the expression $p(t)$ is not the product of $p \times t$; it is *one term* indicating the probability p of the system failing in time t.

Basic Components of a Trip System

In the event of a block-in situation in the uprated plant, continued operation of the new hot-water circulation pump *or* continued flow of steam to the new reboiler would lead to a rate of overpressure exceeding the combined capacities of the existing two relief valves. The trip system must therefore be capable of tripping the new circulation pump (and its spare), *and* of shutting off steam to the new reboiler in the event of an abnormal rise in system pressure (Fig. 2).

Therefore, the minimum basic requirements are a high- (rising) pressure sensor on the overhead vapor line, designed to operate a relay-contactor in the power supply to the new circulating pumps, and a solenoid valve and associated trip valve on the steam supply to the new reboiler.

Proof-Testing Frequency

As indicated in Procedure 1 (Derivation of Fractional Dead Time), the reliability of a system depends not only on the fail-danger fault-rates of the components but also on the frequency of proof testing. Any design must therefore take into account testing constraints thought to be desirable by management. In our example, these were:

1—Movement of the steam trip valve(s) had to be closely restricted (e.g., by chocking), if proof testing were carried out with the plant online; the reason for this was the pronounced adverse effect of variations in steam rate on plant operation.

2—Proof testing for pump shutoff could be tolerated online—with the operator in attendance to minimize process upsets—at a frequency of about once every two weeks maximum. Such a schedule happened to fit in with the limited availability of craftsmen.

This proof testing covered all sections of the trip system, from pressure-impulse point up to and including pump shutoff, as well as movement (but only slight closure) of the steam trip valve(s). On the other hand, fully comprehensive testing to include tight steam shutoff was permissible only during plant shutdowns, which occurred about once every 16 weeks.

For the situation under consideration, the optimum

Procedure 2
Calculating Minimum FDT for One-Out-of-One High-Pressure Trip System

The FDT for the Fig. 3 system (bottom of column), when proof tested at regular intervals of T years, is:

FDT for initiation $= 0.095T + 0.001 + 0.0001/T$
FDT for pump shutoff $= 0.005T$
FDT for steam shutoff $= 0.110T$
FDT for total system $= 0.210T + 0.001 + 0.0001/T$

For such a relationship, the total FDT passes through a minimum with a change in T. The value of T for the minimum FDT is obtained by differentiating with respect to T, and equating the differential to zero:

$$d(\text{total FDT})/dT = 0.21 - (0.0001/T^2) = 0$$

for which $T = \sqrt{0.0001/0.21} = (1/45.8)$ yr (or 8 days).

Thus, the minimum total FDT would be achieved by proof testing every 8 days, and would amount to:

$$(0.21/45.8) + 0.001 + (0.0001 \times 45.8) = 0.0102$$

Even with special additional precautions to reduce the dead time due to inadvertent isolation of the pressure-sensing point (0.001 assumed above), and with no limitations with respect to proof-testing frequency, the system would be incapable of meeting the target reliability (FDT $= 0.0102 - 0.001 = 0.0092$ absolute minimum, compared with the target FDT of 0.001 maximum, as shown below). By the same procedure, the minimum achievable FDT for the initiation section of the system ($0.095T + 0.001 + 0.0001/T = 0.0031 + 0.001 + 0.0031 = 0.0072$) would become 0.0062 minimum, with the elimination of the dead time due to inadvertent isolation of the impulse point.

---

trip system would be the simplest configuration meeting the target reliability (FTD $= 1.0 \times 10^{-3}$ maximum) at a spurious trip rate not exceeding once per year, subject to the constraints on proof testing mentioned above. Possible systems were therefore considered in order of increasing complexity.

One-Out-of-One System

The simplest trip configuration (one-out-of-one throughout) comprises: (1) a single pressure switch on the overhead vapor line with contacts open to give trip signal on rising pressure; (2) a single relay and contact on the power supply to the circulating pumps (to deenergize to trip); and (3) a single relay and contact to the solenoid valve (trip valve) on the steam supply to the new reboiler (to deenergize to trip), as shown in Fig. 3.

From the reliability data in Table II, the FDT due to equipment faults—when proof tested at regular intervals of T years—would amount to:

FDT for trip initiation $= 0.5 \times 0.19\ T = 0.095\ T$
FDT for pump shutoff $= 0.5 \times 0.01\ T = 0.005\ T$
FDT for steam shutoff $= 0.5 \times 0.22\ T = 0.110\ T$
Total FDT due to equipment failure $= 0.210\ T$

Allowance must also be made for other factors contributing to the FDT for trip initiation. One factor is inadvertently leaving closed the isolation valve that is in the in-impulse line to the pressure switch. This causes the initiation section to remain dead until the next proof test. If only one such error occurs every 1,000 tests (due to special check procedure, valve key, etc.), this gives an additional FDT for trip initiation of $1/1{,}000 = 0.001$, which is independent of the proof-test frequency.

Another factor is the disarmed time* when proof testing online. If this requires an average of about one hour per test (i.e., 1.0×10^{-4} yr/test), then at $1/T$ proof tests per year, this yields an "initiator disarmed FDT" of $0.0001/T$ for testing, plus associated online maintenance.

*The term "disarmed time" is applied to instances when the initiator is being tested, because it is then isolated from the process and is not capable of detecting changes in pressure.

Assumed Equipment Fault-Rates – Table II

Type of Fault	Total Rate, Faults/Yr	Fail-Danger Faults/Yr	Fail-Safe Faults/Yr
Trip Initiator			
Impulse-line blockage	0.03	0.03	Nil
Impulse-line leakage	0.06	0.06	Nil
Pressure switch (contacts open to give trip signal on rising pressure)	0.13	0.10	0.03
Cable fractured or severed	0.03	Nil	0.03
Loss of electrical supply	0.05	Nil	0.05
Total for single initiator	0.30	0.19	0.11
Steam Shutoff System			
Relay coil (deenergize to trip)	0.05	Nil	0.05
Relay contact	0.02	0.01	0.01
Relay terminals and wires	0.01	Nil	0.01
SOV* (deenergize to trip)	0.30	0.10	0.20
Loss of power supply to SOV circuit	0.05	Nil	0.05
Trip valve (closes on air failure)	0.25	0.10	0.15
Blocked or crushed air line	0.01	0.01	Nil
Fractured or holed air line	0.01	Nil	0.01
Loss of air supply	0.05	Nil	0.05
Total for single shutoff system	0.75	0.22	0.53
Pump Shutoff System			
Relay coil, contact, terminals and wires (as above)	0.08	0.01	0.07
Total for single shutoff system	0.08	0.01	0.07

*Solenoid valve

The total FDT for a simple one-out-of-one system would thus amount to $0.21T + 0.001 + 0.0001/T$, which is greater than the target FDT for all values of T. This would also be true if the dead time due to inadvertent isolation of the pressure switch at a proof test were reduced to a negligible level by special additional precautions (e.g., installation of a nonreturn valve in the branch used for proof testing, to hold the test pressure until the route to the process were opened, as indicated in Procedure 2). Duplication of those sections making a significant contribution to the total system FDT is therefore essential. These constitute the initiator section (minimum achievable FDT at optimum proof-testing frequency = 0.0062–0.0072, as shown in Procedure 2), and the steam shutoff section (with an FDT or $0.11T$ for a one-out-of-one steam shutoff, fully comprehensive testing would have to be at about three-day intervals for the FDT for this section alone to fall below target).

To summarize, a simple one-out-of-one trip system would be totally incapable of meeting the target reliability, mainly because of the high FDT associated with the initiation and steam shutoff sections. Having demonstrated the need for duplication of the latter sections, it seemed logical to next consider a system completely duplicated throughout. This would have been relatively inexpensive because duplication of the pump shutoff section would require only one extra relay and contact. Subsequent analysis (see below) demonstrated that the best completely duplicated one-out-of-two trip system would just be capable of meeting the target reliability within the constraints on proof testing discussed above.

Duplicated One-Out-of-Two Trip Systems

For a trip system comprising two pressure-switch initiators, two pump-shutoff relays, two solenoid valves, and two steam-trip valves, several one-out-of-two configurations are possible, depending on the extent of cross-connection both within and between the component sections. The extent of cross-connection determines the proof-testing frequency needed to achieve a desired over-

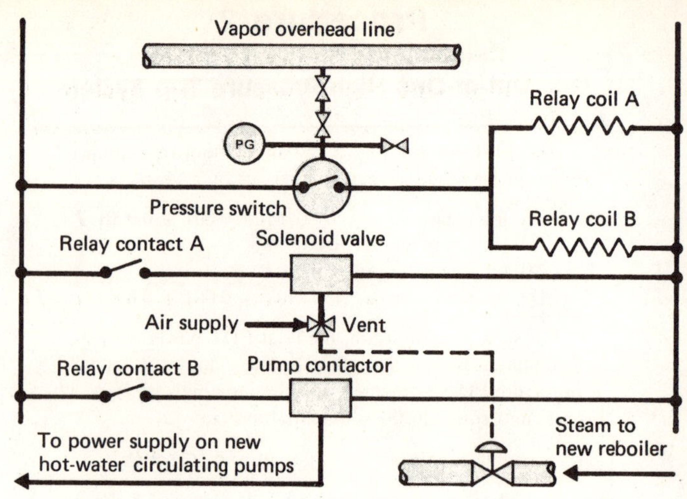

ONE-OUT-OF-ONE high-pressure trip system—Fig. 3

all reliability, and also the manner in which the proof testing must be carried out.

Because testing for full steam shutoff would only be permissible during a plant shutdown, the first step is to determine whether or not the most reliable of the various possible one-out-of-two configurations (i.e., the one with the highest degree of cross-connection) would be capable of meeting the target fractional dead time (0.001 maximum), when proof tested only during a shutdown. This simplifies the testing procedure and also avoids problems that online testing may present. Failing this, the next step is to determine whether this system or an alternative one-out-of-two configuration would be able to meet the target, when proof tested in an acceptable fashion with the plant online (i.e., online testing to cover all sections except steam-trip-valve closure and tight shutoff, supplemented by fully comprehensive testing for tight steam shutoff during a shutdown, as described above).

The most reliable cross-connected one-out-of-two system would be that in which either of the two independent

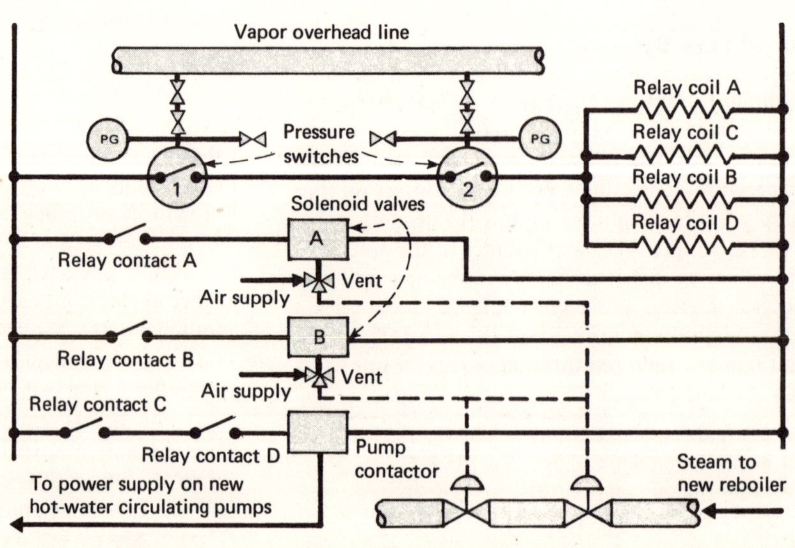

ONE - OUT - OF - TWO cross-connected, high-pressure trip system. Most reliable is the one in which either of the two independent initiators can trip the power supply to the pumps, and can shut off either of the two steam-trip valves—Fig. 4

Procedure 3

Calculating FDT for One-Out-of-Two Cross-Connected, High-Pressure-Trip System

Taking the data in Table II, and assuming the so-called "simultaneous" full-proof testing at regular intervals of T years (i.e., testing via the second initiator is always done after the first initiator testing is complete), the FDTs for the Fig. 4 system—as in Procedure 1—are:

(*a*) *FDT for Trip Initiation*—The FDT for a single initiator, including inadvertent isolation at a proof test (as derived in p. 84) is:

$$(0.5 \times 0.19T) + 0.001 = 0.095T + 0.001$$

For two such initiators operating on a one-out-of-two basis, the FDT is:

$$(4/3)(0.095T + 0.001)^2 = 0.012033T^2 + 0.000253T + 0.0000013$$

(*b*) *FDT for Pump Shutoff*—For a single pump-shutoff system, the FDT is:

$$0.5 \times 0.01T = 0.005T$$

Therefore, the FDT for two such systems operating on a one-out-of-two basis is:

$$(4/3)(0.005T)^2 = 0.000033T^2$$

(*c*) *FDT for Steam Shutoff*—The FDT for venting of a common, trip-valve air-manifold by a single solenoid system, assuming (as a close approximation) that an air-line blockage would always occur on the solenoid-valve side of the crossover line, would be:

$$0.5(0.01, \text{ for relay contact } + 0.10, \text{ for solenoid valve } + 0.01, \text{ for air-line blockage})T = 0.06T$$

Thus, the FDT for two such solenoid systems operating on a one-out-of-two basis is:

$$(4/3)(0.06T)^2 = 0.0048T^2$$

And the FDT for operation of a single trip-valve, when the air manifold is vented, is $0.5 \times 0.10T = 0.05T$. Therefore, the FDT for the operation of two such trip valves on a one-out-of-two basis is:

$$(4/3)(0.05T)^2 = 0.003333T^2$$

Thus, the total FDT for the steam shutoff system is:

$$0.0048T^2 + 0.003333T^2 = 0.008133T^2$$

(*d*) *FDT for Total System*—The total FDT for the Fig. 4 system—due to equipment faults plus inadvertent isolation of an initiator at a proof test, but excluding disarmed time (see footnote on p. 84 for definition) when proof testing (which would be zero if testing were only carried out during a plant shutdown)—is given by adding (a), (b) and (c):

$$\begin{aligned}\text{Total FDT} &= 0.012033T^2 + 0.000253T + 0.0000013 \\ &\quad + 0.000033T^2 + 0.008133T^2 \\ &= 0.020199T^2 + 0.000253T + 0.0000013\end{aligned}$$

Therefore, to meet the target FDT of 0.001 maximum (target reliability, p. 84), the required frequency for fully comprehensive proof testing, when carried out in the so-called "simultaneous" fashion, is given by:

$$0.020199T^2 + 0.000253T + 0.0000013 \leq 0.001$$

for which $T = 0.216$ yr maximum (or about 11 weeks). Comprehensive proof testing during a plant shutdown (estimated average interval between shutdowns is about 16 weeks) would therefore be insufficient to meet the target reliability. The FDT for the Fig. 4 system, with shutdown testing only at this frequency, is:

$$(0.020199)(16/52)^2 + (0.000253)(16/52) + 0.0000013 = 0.0020$$

---

high-pressure-trip initiators was capable of tripping the power supply to the circulating pumps, and of shutting either of the two steam trip valves. The arrangement to meet this requirement is shown in Fig. 4.

Assuming the reliability data in Table II are correct, it can be shown that the Fig. 4 system would be incapable of meeting the target FDT when proof tested only during a plant shutdown (Procedure 3). Routine online testing, supplemented by comprehensive testing for tight steam shutoff during a plant shutdown, would therefore be required for this system or for any of the other one-out-of-two configurations.

For the Fig. 4 system or any other possible one-out-of-two systems—in which a pressure switch is capable of actuating either steam trip valve—online proof testing would need to restrict the movement of both trip valves simultaneously to avoid a steam shutoff (e.g., by chocking the valves, or by fitting pressure gages on the solenoid-valve vent ports). During the test period, there would be no means of tripping steam in the event of process overpressure. This immediately places an upper limit on the allowable period for simultaneous valve restriction, equal to the target FDT of $1.0 \times 10^{-3} \times 8{,}760 = 8.76$ h/yr.

At about 1 h/test per initiator, this would allow only about four online proof tests per channel per year, which would be inadequate. Thus, it follows that a trip system in which each pressure switch operates only a single steam trip valve is essential. It is then possible—by keeping one initiator active while the other is proof tested—to have one-out-of-one protection through the period of the test. The alternative possibility of always having someone in attendance to shut off steam in the event of a process overpressure during testing was considered unreliable (person not always available, etc.).

The simplest, completely duplicated one-out-of-two system in which each pressure switch is capable of actuating only one of the two steam trip valves is shown in Fig. 5. Calculations based on the data in Table II show that, with a combination of online proof testing of both channels at fortnightly intervals (to cover all sections except steam-trip-valve closure and tight shutoff), plus fully comprehensive testing every 16 weeks, the Fig. 5 system would just meet the target FDT of 1.0×10^{-3} maximum (see Procedure 4). This falls within the con-

Procedure 4

Calculating FDT for One-Out-of-Two Separate-Channel, High-Pressure-Trip System

As indicated on p. 86 for the Fig. 5 system, a combination of online proof testing and proof testing during a plant shutdown would be necessary for the system to meet the target FDT. The following would have to be done: (1) online testing to cover all sections, from pressure-impulse point up to and including pump shutoff, as well as slight movement (but not closure) of the steam trip-valves (with the trip valves chocked); (2) testing during a plant shutdown (preferably just before the plant is brought back online), to cover all sections of the system comprehensively, including tight steam shutoff.

Because of the different proof-testing procedures and their frequencies, it is necessary at the outset to distinguish between steam-trip-valve fail-danger faults that prevent valve movement (which would be revealed by the online test procedure), and faults preventing full valve travel and tight shutoff (which would only be revealed during a shutdown test). Thus, for the purpose of this analysis, it is assumed that the total fail-danger fault-rate for a trip valve (0.10 faults/yr, Table II) is made up of 0.07 total faults/yr, which would prevent valve movement (valve seized, spring failure, actuator vent blocked, etc.), and of 0.03 faults/yr, which would prevent full valve travel and tight shutoff (valve plug detached or damaged, valve seat badly scored or fouled, etc.).

Assuming the other reliability data in Table II to be correct (with trip-valve failure apportioned as above), and "simultaneous" proof testing of both channels at regular intervals of T years (defined in Procedure 3), then the FDT—calculated as in Procedure 1—is as follows:

(*a*) Total FDT for a single channel—from pressure-impulse point right through to pump shutoff and to steam shutoff, including inadvertent isolation of the pressure switch at a proof test:

0.5(0.19, for initiator faults + 0.22, for steam shutoff + 0.01, for pump shutoff)T + 0.001, for inadvertent isolation = $0.21T + 0.001$

Hence, the total FDT for two such channels operating on a one-out-of-two basis is:

$$(4/3)(0.21T + 0.001)^2 = 0.0588T^2 + 0.00056T + 0.0000013$$

(*b*) The total FDT for a single channel, from pressure-impulse point right through to pump shutoff and to steam-trip-valve movement (but not closure), including inadvertent isolation of the pressure switch at a proof test is:

$$0.5[0.19 + (0.22 - 0.03) + 0.01]T + 0.001 = 0.195T + 0.001$$

Therefore, the total FDT for two such channels operating on a one-out-of-two basis is:

$$(4/3)(0.195T + 0.001)^2 = 0.0507T^2 + 0.00052T + 0.0000013$$

(*c*) The total FDT effectively associated with tight steam shutoff for the one-out-of-two system is:

$$0.0588T^2 + 0.00056T + (1.3 \times 10^{-6}) - [0.0507T^2 + 0.00052T + (1.3 \times 10^{-6})] = 0.0081T^2 + 0.00004T$$

Note that this last FDT is quite different from that which would be derived by fail-danger faults associated with tight steam shutoff, which would yield an FDT for the one-out-of-two configuration of only:

$$(4/3)(0.5 \times 0.03T)^2 = 0.0003T^2$$

It is therefore essential to consider the system as a whole, as in calculations (a) through (c) above.

Taking the maximum-allowable online test frequency of once every two weeks, and an average interval between plant shutdowns of 16 weeks, the FDT for the one-out-of-two system, due to all faults except equipment faults associated with tight steam shutoff is:

$$(0.0507)(2/52)^2 + (0.00052)(2/52) + 0.0000013 = 0.000096$$

The FDT for the one-out-of-two system due to equipment faults associated with tight steam shutoff is:

$$(0.0081)(16/52)^2 + (0.00004)(16/52) = 0.000779$$

Hence, the total FDT for the one-out-of-two system due to equipment faults, plus inadvertent isolation of an initiator at a proof test would be:

$$0.000096 + 0.000779 = 0.000875$$

To complete the calculation, it is now necessary to determine the additional FDT due to partial disarming at a proof test (one initiator kept active while the other is tested). Assuming the normal practice that one channel would always be kept in the active state while the other one were being proof tested, then, at $52/2 = 26$ online tests/yr/channel (ignoring plant outage), and 1 h average/test, there would be only one-out-of-one trip protection for $26 \times 2 = 52$ h/yr. In other words, this would be the equivalent of $(52)(100/8{,}760) = 0.6\%$ of the online time, which corresponds to an FDT—see calculation (b) above—for this period of:

($0.195T + 0.001$, for all sections except tight steam shutoff) + ($0.5 \times 0.03T$, for tight steam shutoff)

Assuming the above frequencies of once/2 weeks for online testing, and once/16 weeks for shutdown testing, the FDT—while disarmed for testing—is:

$$(0.195)(2/52) + 0.001 + [(0.5)(0.03)(16/52)] = 0.0131$$

Hence, the overall FDT for the Fig. 5 system, with allowance for partial disarming at a proof test, is given by the summation of: (1) the FDT when not testing (one-out-of-two protection for 99.4% of the time) =

$$0.994 \times 0.000875 = 0.00087$$

and (2), the FDT when proof testing (one-out-of-one protection for 0.6% of the time) =

$$0.006 \times 0.0131 = 0.00008$$

which gives a total FDT of:

$$0.00087 + 0.00008 = 0.95 \times 10^{-3}$$

Thus, with "simultaneous" online proof testing of both channels at the maximum allowable fortnightly frequency as per test procedure (1), above—supplemented by comprehensive proof testing at a plant shutdown—the Fig. 5 system would just about meet the target FDT of 1.0×10^{-3} maximum, if the average interval between shutdowns does not exceed the expected 16 weeks.

Regular comprehensive testing at every shutdown to maintain an average frequency of once every 16 online weeks would be particularly important, because of the high sensitivity of the overall FDT to faults associated with tight-steam shutoff, when this section were tested less frequently (see above calculations). It is also clear from the results that the frequency for online testing could not be significantly reduced below the fortnightly level.

For simplicity, the use of staggered online testing (e.g., Channel 1 only at time zero; Channel 2 only at time $T/2$; Channel 1 only at time T; Channel 2 only at time $3T/2$; etc.) has not been considered here.

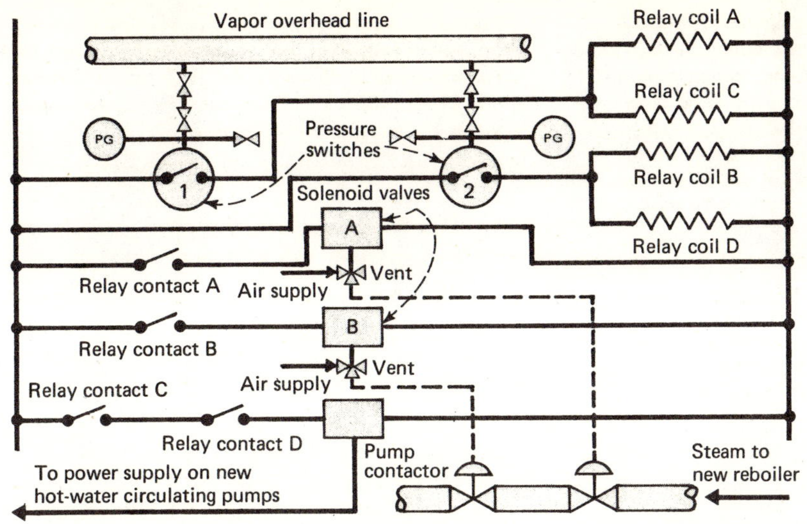

SEPARATE - CHANNEL, duplicated one-out-of-two trip system, in which each pressure switch actuates only one of the two steam valves—Fig. 5

straints on testing, and would therefore be acceptable.

However, it would be particularly important to maintain an average frequency of once every 16 online weeks for the full test, because of the high sensitivity of overall FDT to fail-danger faults associated with tight steam shutoff, when the steam section is tested less frequently. Comprehensive testing should therefore be carried out at every shutdown to maintain this average.

Spurious-Trip Rates

Fail-safe faults in either channel of the Fig. 5 system would lead to a spurious trip, the estimated frequency of which—based on the reliability data of Table II—would be:

Initiator Section—(pressure switches, 2×0.03) + (cable runs, 2×0.03) + (loss of electrical supply, 0.05) = 0.17 faults/yr. Therefore, the probability of failure in a proof-test interval of T years is $0.17T$, which, at $1/T$ proof-test intervals/yr, yields a spurious-trip rate of $0.17T \times 1/T = 0.17$ trips/yr (pump and steam simultaneously).

Pump Shutoff Section—(relay coils, 2×0.05) + (contacts, 2×0.01) + (terminals plus wires, 2×0.01) = 0.14 spurious trips/yr (pump only).

Steam Shutoff Section—(relay coils, 2×0.05) + (contacts, 2×0.01) + (terminals plus wires, 2×0.01) + (solenoid valves, 2×0.20) + (trip valves, 2×0.15) + (air-line fractures, 2×0.01) + (loss of power to solenoid valves, 0.05) + (loss of instrument-air supply, 0.05) = 0.96 spurious trips/yr (steam only).

The total spurious-trip rate (pump and/or steam) would thus amount to $0.17 + 0.14 + 0.96 = 1.27$ trips/yr, which would be acceptable.

Conclusions

The one-out-of-two, separate-channel, high- (rising) pressure trip system shown in Fig. 5 would be capable of safeguarding against vessel rupture, with a reliability some 10 times greater than conventional relief protection (in this case, two relief valves, which would be required to lift together to be fully effective). This would conform with the policy outlined earlier.

Such level of reliability could be achieved within the constraints on proof testing thought desirable by management, by a combination of: (a) online proof testing at regular fortnightly intervals to cover all sections of the trip system (except steam-trip-valve closure), and tight shutoff (with one channel always kept active while the other is tested with its trip valve chocked); and (b) fully comprehensive proof testing at a plant shutdown to cover all sections, including steam shutoff. The spurious trip rate for the Fig. 5 system would be about once per year.

Meet the Authors

◀ **Herbert G. Lawley** is Safety Adviser with Imperial Chemical Industries Ltd. (P.O. Box 90, Wilton Middlesbrough, Cleveland TS6 8JE, England). Previously, he was process investigation leader, and before that he did process-development research and exploratory research. He has B.Sc. and Ph.D. degrees in chemistry from the University of Wales, where he received first-class honors. During World War II, he served in North Germany and Palestine.

Trevor A. Kletz is Safety Adviser to the Petrochemicals Div. of Imperial Chemical Industries Ltd. (P.O. Box 90, Wilton Middlesbrough, Cleveland TS6 8JE, England), where his special interest lies in the application of quantitative methods to safety problems. He joined ICI in 1944 and engaged in research work. Prior to his present position, he also worked in production management. Educated at the University of Liverpool, he received a B.Sc. in chemistry with first-class honors. ▶

Section XVII
INSTRUMENTATION FOR SPECIFIC APPLICATIONS

How to select a system for drying instrument air
Instruments for environmental monitoring
Designing for pH control
Selection and care of pH electrodes
Precise combustion control saves fuel and power
Selecting temperature controls for heaters
Ground-level detector tames flare stack flames

How to select a system for drying instrument air

Water, or other contaminants, in instrument air can cause all sorts of problems in a process plant. Here are guidelines for correcting an inadequate drying system or engineering a new system.

DECKER G. McALLISTER, JR., Consultant

Instrument air in a process plant seldom receives the attention it deserves—until a plant is shut down by instrument failure. Poor-quality instrument air can increase maintenance costs, produce inconsistent product quality, lost product, or create an unstable plant operation. The added costs due to such situations can far outweigh the time and costs necessary for engineering a new, adequate air drying system, or correcting an inadequate, old one.

Gas Drying

There are innumerable methods and variations available for drying gases. (See Table, p. 39) But, in general, the significant methods are: chemical (reaction, hydration); adsorption (silica gel, activated alumina, molecular sieves); and mechanical (compression, refrigeration).

Instrument-air drying differs from process-gas drying in two significant ways: The volumes are smaller and the moisture-content requirements are usually not as critical as for processing.

Instrument Air Systems

The function of an instrument air system is to properly condition air, and deliver it to the valve operator or other instruments (Fig. 1).

A number of important variables are common to most air-drying systems:

Pressure: Air should be available, at the typical instrument, at 20 psig. Since most installations have an airset, the pressure delivered to the airset should be 25 psig. minimum. Therefore, the pressure should be at least 40 psig. going into the distribution header to allow for line losses. However, many systems operate at 80–100 psig. These higher pressures have several advantages:

- The distribution system acts as a receiver.
- Higher pressure is available for actuators.
- Smaller dryers and filters may be used.
- More moisture is removed by the compressors.

Particulate Matter: Compressor intake filters, aftercoolers, dryer prefilters, and afterfilters remove particles

Typical Process-Gas Drying Methods

Drying Method	Product or Process	Stream	Dessicant/Method
Absorption	Sulfuric acid manufacture (Direct sulfur combustion)	Inlet air	93–98% sulfuric acid
	Sulfuric acid manufacture Wet process	SO_2, roaster gas	93–98% sulfuric acid
Adsorption	Liquified air, O_2, N_2, etc.	Various streams	Activated alumina Silica gel Molecular sieves
	Petroleum refining Petrochemical processing Chemical processing	Various feed and product streams	Molecular sieves extensively used
Refrigeration	Liquified air, O_2, N_2, etc. (High pressure process)	Compressed air after aftercooling	Freon
	Liquified air, O_2, N_2, etc. (Low pressure process)	Compressed air after aftercooling	Freezeout with reversing heat exchangers
Compression	Liquified air, O_2, N_2, etc. (High pressure & low pressure)	Inlet air	Aftercooling used with all compressors

Originally published February 26, 1973.

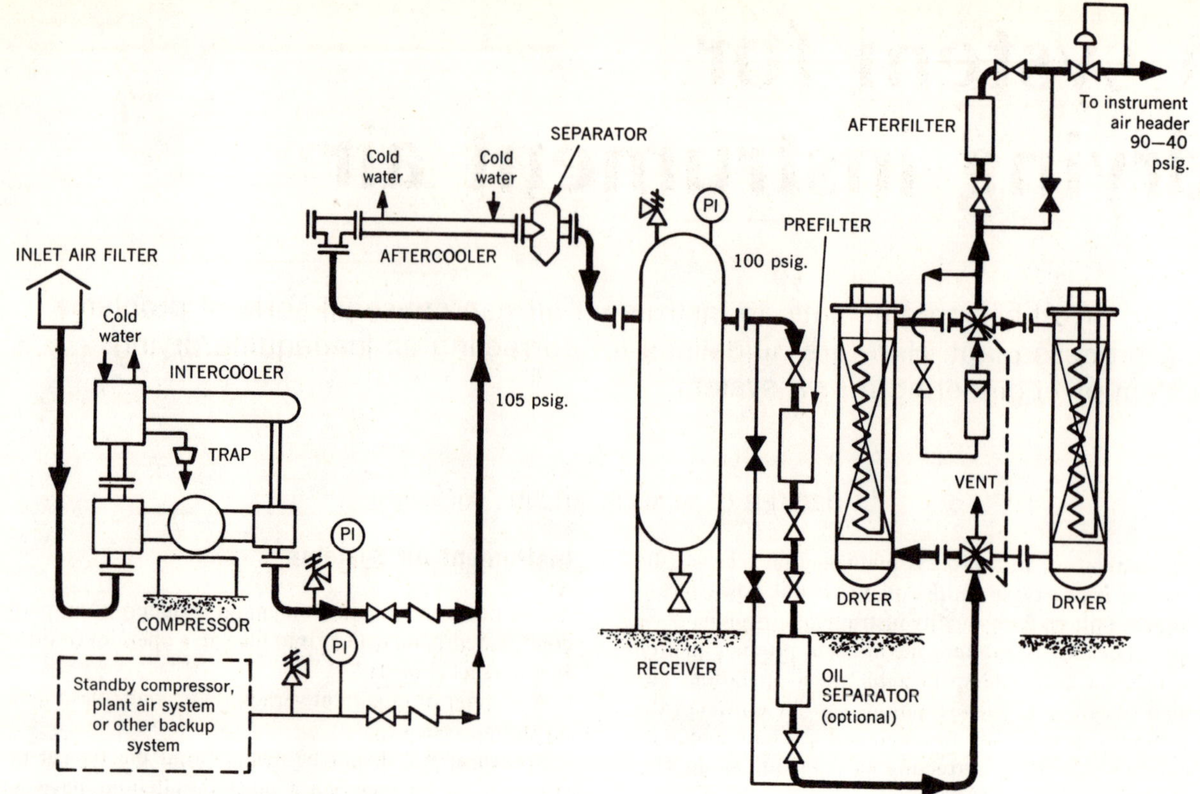

INSTRUMENT AIR supply system—Fig. 1

in the intake air. Additional particles are generated within the system by breakdown of the compressor piston rings and the dessicant bed (adsorption dryers). Filters, included in airsets located at the point of use, remove particles generated within the distribution piping.

Water Content: As long as water is present in the vapor phase, it isn't a problem. The problems occur when water condenses. Therefore, it is best to maintain the water content equivalent to a dew point 10 F. below the lowest ambient temperature expected at any point in the complete system. This dew point is not the atmospheric dew point (ADP), but the dew point for the pressure (pressure dew point, PDP), at that point in the system. Conversion data are shown in Fig. 2.

Oil Removal: Oil creates the most difficult and frequent problems in any air system. This is particularly true for instrument air. Special separators and activated-carbon filters can remove both entrained and vaporized oil from instrument air. The oil is removed before air drying, because it fouls and reduces the drying capacity of both deliquescent and adsorption dryers.

Generally, nonlubricated compressors are used. However, oil can get in through the air intake and through a backup system (lubricated utility-plant air compressors) on nonlubricated compressor systems.

Corrosive Gases: Check the location of the compressor air intake. In the past, before the emphasis on environment controls, it was possible to suck reactive gases into the system. This should be considered when checking out a system that requires excessive maintenance.

Compressors: It is usually best to have a completely independent instrument air system. And most designs include nonlubricated reciprocating compressors. But lower-cost lubricated compressors can be used when proper precautions are taken. Rotary Nash water-ring compressors are used for low-pressure systems and do not require aftercoolers. Centrifugal compressors are very reliable and popular for plant air systems. They are oil-free, and should be considered for large instrument air systems of 500 scfm, or larger.

Backup Systems: Many different types and configurations of backup and control systems are available. Usually, pressure controllers and other sensors detect power failure, equipment failure, and low air pressure. The controllers will cut in either standby compressors, air from the utility air system, nitrogen, etc. Instrument air receivers should also be sized adequately to allow safe plant shutdown when all systems go down.

Instrument Air Dryers

Different drying methods, of course, can be combined in a particular system. Before describing the various drying methods in some detail, it should be emphasized that compression and cooling is a drying technique often taken for granted. The cooled, compressed gas is then dried further if necessary. Also, the drying techniques available can be combined in a system or used singly.

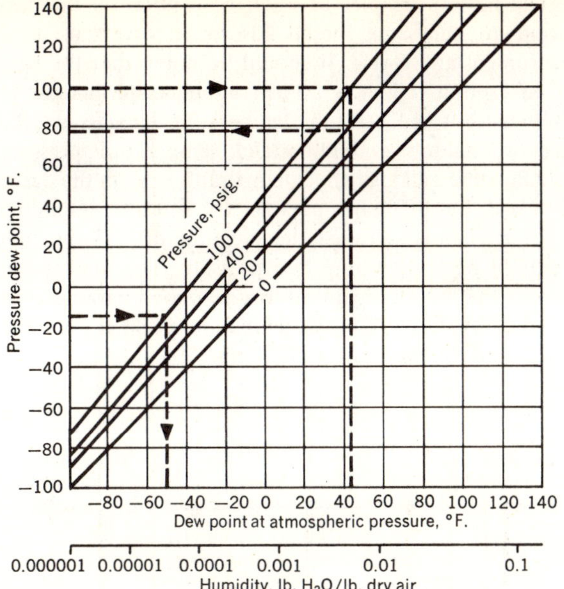

CONVERSION CHART for figuring dew points—Fig. 2

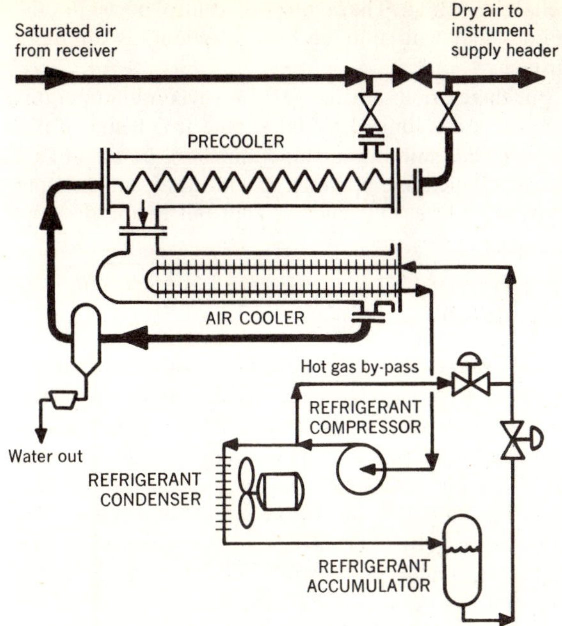

REFRIGERATION type air dryer—Fig. 3

Compression and Cooling

Water will condense when air is compressed and aftercooled to the point where the partial pressure of water vapor exceeds the vapor pressure. Under most conditions, the water content in the air will be reduced by at least 50%.

Air leaving the aftercooler is usually saturated, and it is at its pressure dew point or PDP. If this air were reduced to atmospheric pressure and cooled to the temperature at which the water vapor begins to condense, that temperature would be the atmospheric dew point or ADP.

For example (refer to Fig. 2), air at 75 F. and 50% relative humidity is compressed to 100 psig., and aftercooled to 100 F. The corresponding pressure dew point is 100 F. and the atmospheric dew point is 42 F. If this same air is reduced to 40 psig., the corresponding PDP is 78 F. Since this temperature is above the ambient temperature, some water will condense. This analysis is based on the assumptions that (1) the air is actually cooled to 100 F. and (2) that the aftercooler separator is 100% efficient. Unfortunately, it is not uncommon to find inefficient aftercoolers and separators. But it is more economical to remove water with efficient aftercoolers and separators than with dryers. It pays to keep them clean and check their efficiencies.

Some plants have operated satisfactorily for years without additional dryers. It just depends on plant conditions and a good engineering analysis of the system.

Refrigeration

Refrigeration dryers (Fig. 3) are really nothing more than an aftercooler capable of cooling air to 35 to 40 F. They do not cool below 35 F. because of water freezeup. Conservatively, they can always be counted on to cool air to 40 F.

With a PDP of 40 F. at 100 psig., air will have a PDP of 22 F. at 40 psig., which is quite adequate for most applications.

The small size units do not necessarily come with economizers (or precoolers). Economics dictate use of these exchangers (warm wet inlet air to cold dry air) in the larger sizes.

It is mandatory to use efficient oil separators with refrigeration dryers. Oil can form emulsions and plug water-removal traps. Furthermore, oil causes problems in the instruments.

Absorption: Solid and Liquid

There are a variety of chemical absorbents for drying gases, but the two major types are liquid and solid dessicants. The liquid absorbents include sulfuric acid, salt solutions and glycols. The solid absorbents include phosphorus pentoxide, magnesium perchlorate and anhydrous calcium sulfate. A special class of solid absorbents are the deliquescent absorbents such as calcium chloride.

All absorbents dry air by means of hydration, chemical reaction, absorption and a combination of these mechanisms. But many absorbents are ruled out for use on instrument air-drying systems because of toxicity, corrosiveness, cost, and cost of regeneration systems.

Deliquescent absorbents (calcium chloride) should be considered for moderate amounts of drying. The attainable dew point reduction decreases as the dessicant is used up. The dessicant absorbs water, dissolves, and is discharged from the dryer as a salt solution. Normally, additional dessicant is added to the bed every 2–3 months. More frequent charging may be desired based on inlet conditions and dew-point reduction.

The advantages of the deliquescent dryers (Fig. 4) are low installed cost and low operating cost in the smaller sizes. And they are used in combination with other dryers.

Reliability is high: They require no control power or other utilities and will continue to dry as long as air passes through.

But there are disadvantages. The maximum operating temperature is limited to 100 F., and it is better to plan on 90 F. The minimum temperature will be about 25 F. or lower depending upon the specific dessicant and operating procedure. Oil must be kept out of the dessicant bed.

Adsorption

Adsorption dryers use adsorbents such as silica gel, activated alumina, clay, or molecular sieves. Adsorbent selection depends on the design of the dryer, type of regeneration system, and the desired moisture content of the dried air. Molecular sieves can generally achieve lower moisture contents (below −80 F. dew points) than either silica gel or activated alumina (down to −80 F. PDP). Molecular sieves can also dry gases at higher temperatures than the other adsorbents.

The adsorption dryer is very popular for instrument-air service. It is generally applied in a dual-column configuration (Figs. 1, 5, 6). However, this configuration is not always necessary. Small batch operations, or even large cyclical operations, will work very successfully with a single column. It can be regenerated while the process is down.

Prefilters, oil separators and afterfilters are necessary with adsorbers. The afterfilters catch dessicant particles usually generated during regeneration. Some dryer designs minimize effect of pressure shocks on the desiccant.

The adsorbent dryer has a fixed bed of dessicant that can hold just so much water. Therefore, it will usually be designed for the worst situation of low inlet pressure and high inlet temperature. The inlet air is usually assumed to be saturated. Exit moisture content (dew point) will depend upon a number of factors including regeneration temperatures, purge flowrates and operating pressures.

The basic principles of water adsorption on the dessicant are the same for all adsorption dryers used on instrument-air service. It should be noted that the heat of adsorption will show up as a temperature increase in the dessicant bed. And the temperature increase will be proportional to the inlet humidity. Temperature, pressure and flowrate will also affect temperature rise in the dessicant bed; but the major parameter is still inlet humidity. If the inlet air is saturated, the inlet humidity is a function of both the temperature and pressure. The main point here is that bed-temperature rise is a good parameter to monitor. A decrease in the temperature rise at the same inlet conditions indicates problems.

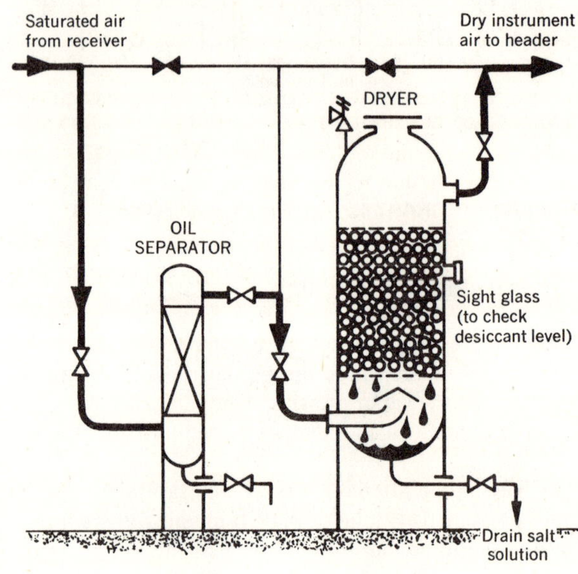

CHEMICAL DRYER is low cost—Fig. 4

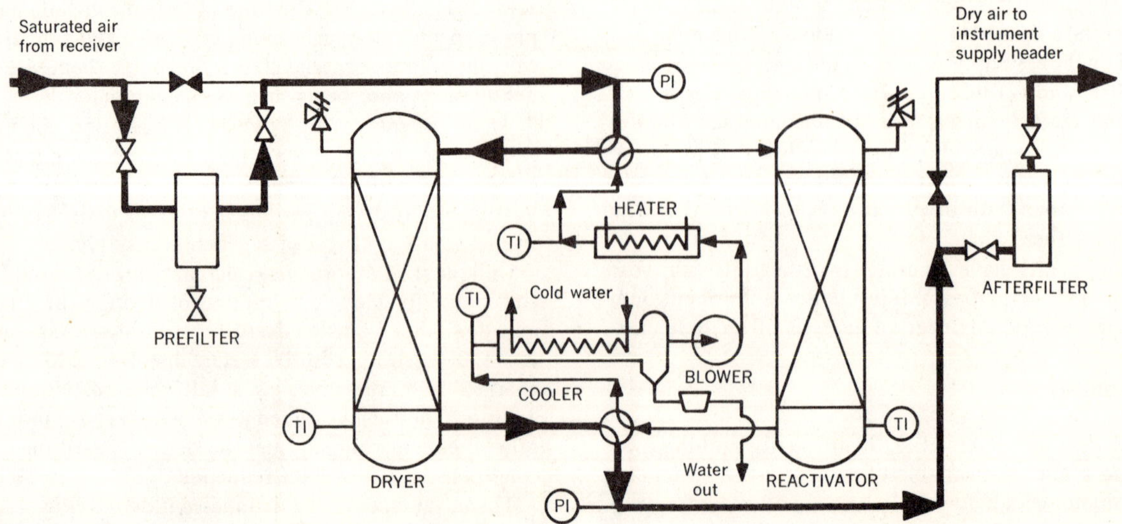

REACTIVATING SYSTEM for adsorption dryers—Fig. 5

Regenerating the Dryer

Basic differences between different adsorption dryers show up mainly in the method of dessicant regeneration. There are basically three techniques:

- Heat conduction
- Heat convection
- Heatless

Heat Conduction: Electrically-heated conduction dryers (Fig. 1) appear to be the most popular. Electrical heaters, placed in the dessicant bed, provide heat for regeneration. A small flow of dried air (expanded to atmospheric pressure) is purged through the bed during regeneration. This purged air serves two purposes. It sweeps out the water vapor during regeneration, and cools the bed after water removal. The purged air usually amounts to 2–3% of the dry air flow from the dryer.

All adsorption dryers operate on cycles of from 4 to 10 minutes (heatless) to either 8 or 16 hour cycles (heat regenerated). The longer the cycle, the larger and more costly the equipment. The longer time required for heated dryers is due to the time required to heat up to regenerating temperature and then cool down.

Electrically heated adsorption dryers are relatively trouble free with the exception of heating-element hot spots and burnouts. This doesn't occur with all makes, but it can happen. Therefore, be sure to check dryer designs for ease of element replacement. Actually, most of the maintenance will be in changing the filters and freeing gummed-up and sticking switching-valve operators.

Heat Convection: The heated convection-regeneration-type dryer (Fig. 5) uses an external heater, blower and condenser to remove water from the dessicant and the system. No air (or process gas) is vented to the atmosphere and consequently there is no air purge. Regeneration takes place at operating pressure, and this is a significant difference from other types of adsorption dryers. This also means that this method is limited to the dew point (PDP) that can be achieved, by the temperature of the cooling-water available, operating pressure and the limitation on regeneration temperature.

Since this dryer has significantly higher capital costs than other types in sizes under 1,000 scfm., it isn't often used for instrument air. It might be considered for larger systems, but the trend to electronics in instrumentation indicates smaller, not larger, instrument air systems.

Heatless: Heatless adsorption dryers (Fig. 6) are also called "pressure-swing" dryers. Operation is similar to the heated conduction dryer, but with two major differences. First, they operate on a 5 to 10 minute cycle rather than the 8 or 16 hour cycles. Second and most important, they operate without heating the dessicant bed. This requires a larger air purge of 10 to 20% or more of the dried instrument air, compared to 2–3% for the heated type. Thus, a larger compressor is necessary and the increased air purge is a major economic factor (Fig. 7).

During regeneration, the beds are switched by a cycle timer that actuates the inlet four-way valve (Fig. 6). When the beds are switched, check valves at the outlet automatically divert a stream of metered dry air to the regenerating bed. The regenerating bed vents directly to the atmosphere.

When the metered dry air enters the regenerating dryer countercurrently, it expands from operating pressure to atmospheric pressure.

It can be shown that there is ample driving force available to remove water from the dessicant bed. This driving force is then a function of the operating pressure and increases with pressure.

The purge rate is essentially determined by the inlet conditions during the drying cycle. Dessicant equilibrium loading is a function of dryer temperature (inlet air temperature plus temperature rise from heat of adsorption) and the inlet partial pressure of water vapor. Since the inlet air is usually saturated, the partial pressure of the inlet air is the vapor pressure of water at the inlet tem-

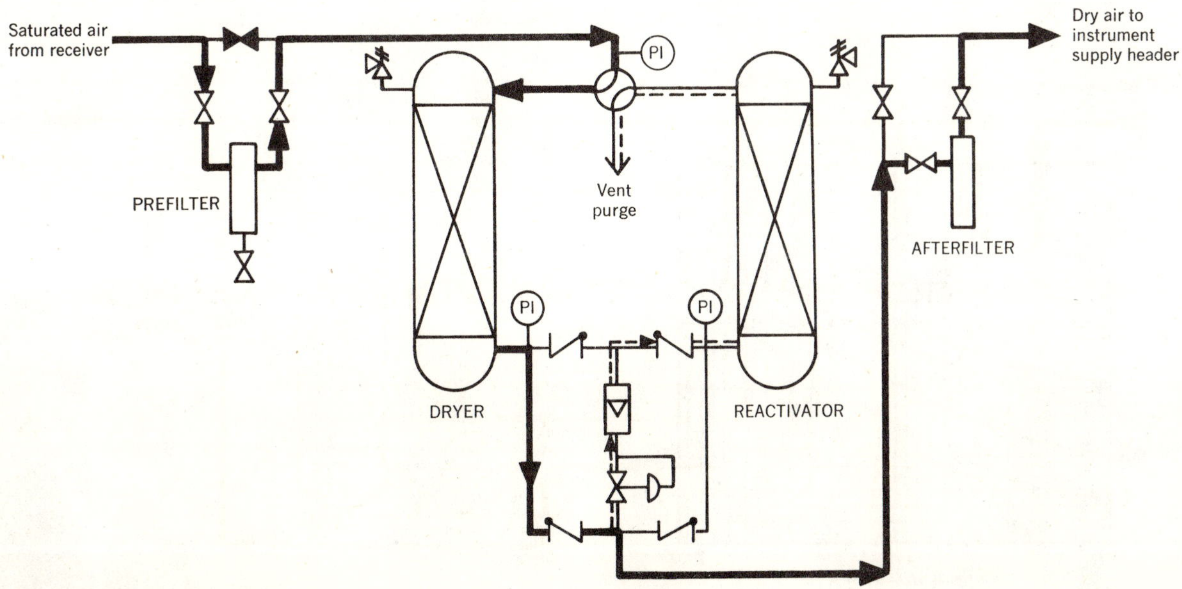

HEATLESS adsorption dryer—Fig. 6

perature. The minimum possible air purge will then be saturated at atmospheric pressure and bed temperature. The water content corresponds to the ADP at bed temperature and the vapor pressure of water at that temperature.

Heatless dryers require filter changes and maintenance on the switching-valve operator just as with the heated dryer. Check valves have "hung up" in some installations, but usually a light tap will free them.

As far as reliability goes, all of the adsorption dryers require electricity either for controls or for heat regeneration. Obviously, the larger heat activated beds that can go 4 to 8 hours between regenerations, give better backup than a bed that needs to be regenerated every 5 minutes. However, this is something that can be worked out with the supplier when a dryer is purchased.

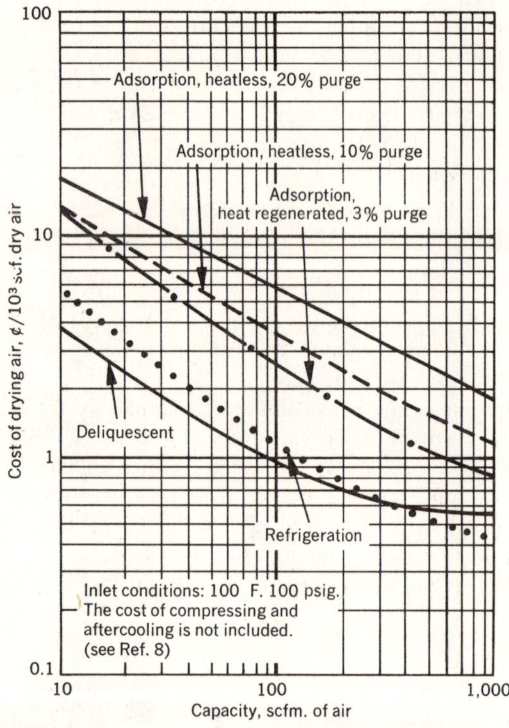

COST of drying instrument air—Fig. 7

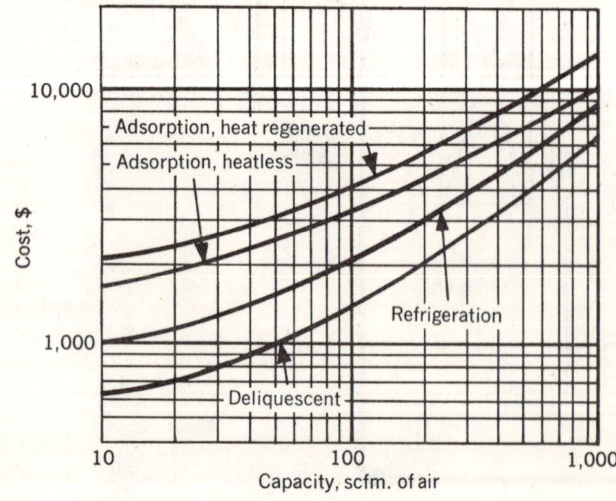

INSTALLED COST of instrument air dryers—Fig. 8

Other Methods

Some plants do not dry air. Of course, the environment must always be above freezing or this will not work.

These plants use aftercoolers and galvanized piping that is sloped, with drip legs and traps. The takeoff for the instruments comes off the top of the piping. In addition, these installations typically use a larger, more efficient, filter-separator such as the Bailey Filter Sets along with the regular airset. Some plants are also using inexpensive filter holders with an element consisting of an ordinary roll of toilet paper.

Costs

Operating and installed costs for various dryers are shown in Figs. 7 and 8. These costs, based on compressing the air to 100 psig. and aftercooling to 100 F., give a general comparison between dryers. These data do not reflect relative costs at lower inlet humidities, lower operating pressures, and different exit pressure dew points.

Acknowledgements

The cooperation of the following companies and their representatives is greatly appreciated: Bailey Meter Co., Champlin Oil Co., Compressor Systems, Inc., Deltech Engineering, Inc., Dollinger Corp., Fischer & Porter Co., Filter Engineering Co., General Air Systems Div./Zurn Ind., W. A. Hammond Drierite Co., Ingersoll-Rand Co., C. M. Kemp Mfg. Co., Kel Air, Inc., Lectrodryer Div./McGraw-Edison, Linde Div./Union Carbide Corp., Monsanto Co., Pall Trinity Micro Corp., Union Oil of California, Van-Air, Inc.

Bibliography

1. "ASHRAE Guide & Data Book—Fundamentals," American Society of Heating, Refrigerating and Air Conditioning Engineers, New York.
2. Perry, J. H., "Chemical Engineers' Handbook," 3rd Ed., pp. 877–884, McGraw-Hill, N.Y., 1950.
3. "Compressed Air and Gas Handbook," Compressed Air and Gas Institute.
4. "Manual on Installation of Refinery Instruments and Control Systems: Part I—Process Instrumentation and Control," API RP550, American Petroleum Institute.
5. "Instrument Air Supply Systems," Bulletin 91-53P-01, Fischer & Porter.
6. Considine, D. M., "Handbook of Applied Instrumentation," McGraw-Hill, New York, 1964.
7. "Desiccation of Air For Air Control Instruments," Bulletin 226, Lectrodryer Division/McGraw-Edison.
8. McAllister, D. G., Jr., "Air," *Chem. Eng.*, Dec. 14, 1970, pp. 138–141.

Meet the Author

Decker G. McAllister, Jr. is a consulting engineer (Suite 406, 122 West Fifth Street, Long Beach, CA. 90812) with over 18 years of extensive experience in plant engineering, economic evaluations, equipment design and procurement, construction, and startup supervision. He holds a B.S. (Ch.E.) from Massachusetts Institute of Technology and belongs to AIChE and American Chemical Society. Mr. McAllister is also a director of the Consulting Chemists Assoc. and a registered engineer in the State of Calif.

Instruments for environmental monitoring

Plants are required to monitor many pollutants in both air and water. Here is a rundown of the instruments available for the job.

D. M. Ottmers, Jr., D. C. Jones, L. H. Keith and R. C. Hall, Radian Corp.

☐ Government regulations require that industry monitor for various gaseous pollutants, particulates, lead, and the so-called "priority pollutants." Hence, monitoring of air and water qualities has become increasingly important during the past decade. Thus, it is important to know:

- The instrumental methods that are currently accepted for pollutant monitoring.
- The instrumentation available for this purpose.

This article discusses instrumentation for monitoring both air and water quality. Emphasis is placed upon how these instruments function, their approximate costs, and what systems may be used or developed in the near future.

Air-quality monitoring

Air-quality monitoring consists primarily of measuring various gaseous pollutants, particulates, and lead. Instrumentation for monitoring these atmospheric pollutants are discussed in the sections that follow. Quality assurance considerations and research in air-quality monitoring are also discussed briefly.

Gaseous pollutants

Federal regulations have established allowable limits for sulfur dioxide, nitrogen dioxide, carbon monoxide, and ozone [*1,2*]. To aid in attaining compliance with the oxidant standard, a standard has also been established for non-methane hydrocarbons [*3*].

Monitors for these gases usually are continuous, i.e., the indicated pollutant concentration may vary continuously with time. These monitors are automated, and are capable of unattended operation for days or weeks.

Also, manual samplers are available that determine an average pollutant concentration. They do this by chemically scrubbing the pollutant from a known quantity of air collected at a constant rate during the sampling period, with subsequent analysis of the captured material. However, the servicing and analysis to support these samplers is manpower intensive, and they are generally not well suited to the observation of frequent short-term averages. With improvements in the reliability and accuracy of the continuous monitors, manual samplers have gradually lost favor with most users. Consequently, this article will be limited to a discussion of continuous analyses.

Originally published October 15, 1979.

The basic elements of a pollutant monitoring system are shown in Fig. 1. Air is taken in by the sample manifold, which is made of an inert material such as glass or Teflon. The sample manifold brings a fast-moving stream of air to the immediate vicinity of the monitor in order to minimize the residence time of the air sample in that portion of the system upstream of the measurement cell. Even though the sample-handling system is inert, reactions may occur between gaseous species in the air, e.g., reaction of ozone and nitric oxide to produce nitrogen dioxide. Thus, the residence time in the air transport system should be minimized.

Monitors may include filters to remove particulates that might otherwise cloud optical windows, plug orifices, and the like. Some monitors may use a scrubber to remove a gas in the air that would interfere with measurement of the one of interest.

Support gases are required for some measurement methods. These are consumed during the measurement and thus may increase operational costs and logistics problems.

Measurement cells can be based on many different methods for the pollutants of interest. Basically, the pollutant is introduced into an environment where it will undergo a chemical or physical reaction whose output can be converted to an electrical signal. Examples include:

Flame photometry—The gas of interest is pyrolyzed by a flame, with degradation products reacting to produce a particular wavelength of light. This light is detected by a photomultiplier tube and converted to an electrical signal.

Infrared—The gas of interest absorbs infrared light, which is beamed through the sample cell. This absorption results in heating, which is measured as a pressure imbalance, and converted to an electrical signal.

Chemiluminescence—The gas of interest is mixed with a gas with which it undergoes a spontaneous reaction. The reaction gives off light of a particular wavelength, which is measured via a photomultiplier tube and converted to an electrical signal.

Electrochemical—The gas of interest is dissolved in a cell with electrodes where it will react with other dissolved species. This reaction, or auxiliary reactions, generates an electric current at the electrodes that is proportional to the concentration of the gas of interest.

Fluorescence—Ultraviolet light is beamed through the

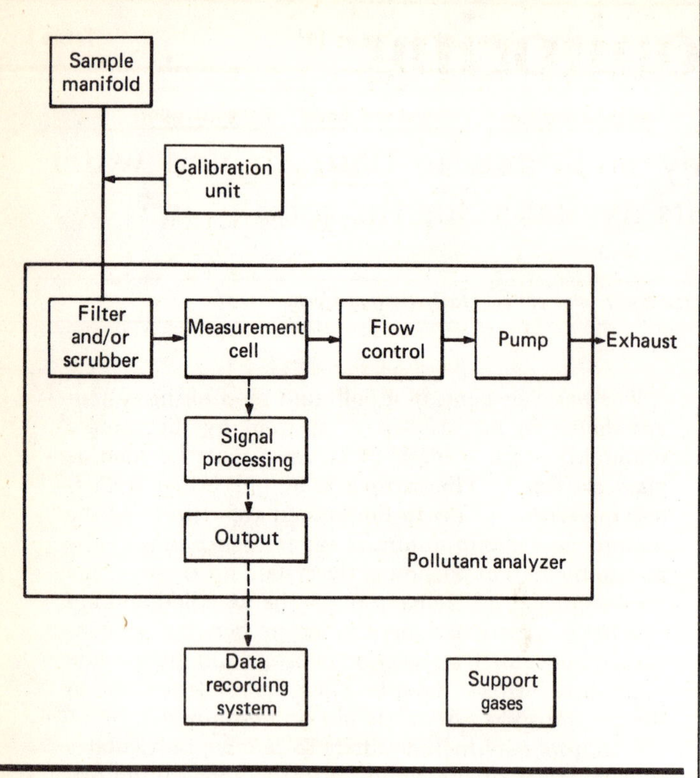

Elements of pollutant monitoring system **Fig. 1**

sample cell, either continuously or in pulses. Absorption of this light induces emission of fluorescent light by the species of interest. This is detected by a photo-multiplier tube and converted to an electrical signal.

Second-derivative spectroscopy—Ultraviolet light is absorbed by the gas of interest in the sample cell. Measurements are based on the shape characteristics of the absorption spectrum rather than on the change in intensity. The output signal is proportional to the second derivative of intensity with respect to wavelength.

Ultraviolet spectroscopy—Ultraviolet light passes through a sample and is measured by a photoelectric detector. The decrease in intensity—which is exponentially proportional to the concentration of the absorbing pollutant—is then converted to an electrical signal.

Generally, only one type of measurement cell is used in any particular monitor. As shown in Fig. 1, other items in the system include a flow-control system and a sample pump. Flow control may be achieved via a capillary orifice, needle valve, and so on, with suitable pressure control.

The calibration unit must supply zero air, and gases of known concentration to check or adjust instrument response. This unit may be manual or automated.

Various types of outputs may be supplied with air-quality instrumentation. Panel meters, digital displays, recorder jacks, and terminal strips are all used in various brands.

The data recording system provides a permanent record of the monitoring results. Strip-chart recorders provide a continuous record but are manpower intensive in terms of data reduction. Data loggers typically sample each data channel at some frequency. Their output may be merely a voltage printed on a paper tape; more-sophisticated systems may provide a machine-readable output. Small computerized data-acquisition systems based on microprocessors can provide on-site reduction, with data storage on magnetic tape or disk. These units also may assume control functions such as automatic calibration of the instruments, calculation of zero and span drift, sequencing of high-volume samplers, and so forth.

Table I provides a listing of monitors that meet EPA's reference or equivalency specifications for the four gaseous criteria pollutants. The type of detection system used is specified for each monitor, along with remarks on support gases, scrubbers, etc.

Non-methane hydrocarbons are typically analyzed via a gas chromatograph with a flame ionization detector. No EPA-approved instruments are available for this measurement.

Particulates

Regulations also have been established for total suspended particulates [*5*]. Only one measurement technique is approved, the high-volume sampler. This technique uses a filter to trap particulates during a 24-h sampling period. The filter is contained in an all-weather housing, built according to a specific design. The air flowrate is measured by a chart recorder (continuously) or a rotameter (manually). This flowrate must be calibrated periodically. The filters are weighed (at a controlled humidity) before and after exposure to determine the particulate loading. The air-flow volume is converted to standard conditions (1 atm pressure and 25 °C) and used to determine the particulate concentration in the sampled air.

High-volume samplers are typically run every sixth day, every third day, or daily. The sampling period is midnight to midnight.

Lead

A federal standard for ambient lead has recently been promulgated [*6*]. The measurement procedure involves chemical analysis of filters collected in the manner (described above) for particulates. The standard is based on a three-month average of lead on filters collected no less frequently than once every six days.

Quality-assurance considerations

Air-quality monitoring should be supported by an adequate quality-assurance program. Included are functions such as calibration method and frequency, zero/span checks, traceability of standards, analyzer audits, reports, and documentation. Quality-assurance guidelines are available from EPA [*7-9*]. However, EPA regional offices, or other permitting offices, may have considerable latitude in specifying particular procedures.

Research in air-quality monitoring

Research in air-quality monitoring is currently very active. For gaseous pollutants there is a major activity in remote sensing, i.e., measurement of the pollutant concentration in the air at some point far removed from the monitor. These systems are typically based on lasers or correlation spectroscopy. For particulates, much research has been done on size fractionating devices, which allow respirable and inhalable particulates to be quantified. New regulations will probably be required before these new techniques enjoy widespread application.

The analysis of specific organic compounds in ambient air is receiving much attention. This is currently achieved by collecting air samples and transporting them to a laboratory. However, the use of special gas-chromatograph–mass-spectrometer systems for real-time analysis shows considerable promise.

Water-quality monitoring

The chemical parameters for monitoring water pollutants are very extensive. They include traditional measurements such as pH, color, biochemical oxygen demand (BOD), chemical oxygen demand (COD), total solids (TS), nitrogen, oxygen, sulfate, sulfite, sulfide, chloride, ammonia, hardness, etc. The instruments for measuring and monitoring these parameters are common and have been the subject of many past articles, books, and reference manuals [*10*]. They will not be further discussed here. Instead, this discussion will be limited to a new list of chemical parameters that both industry and the EPA are required by law to begin monitoring in the near future.

These new chemical parameters are the so-called "priority pollutants," and are listed in Table II. They are distinguished from chemical parameters previously monitored for, by being specific organic compounds and isomers. (The elements are still measured as total copper, zinc, etc.)

The priority pollutants resulted from a 1976 court settlement after the EPA was sued by several environmental groups for failing to implement portions of the Federal Water Pollution Control Act. This history of the development of the list of 129 priority pollutants, and present methodology used to analyze for them is summarized in a recent review [*11*].

Regulations for promulgation of new-source performance standards and pretreatment standards started to be issued in June 1979, and are to be completed for 21 major industrial categories by June 1981. These regulations will require application of best available technology (BAT) economically achievable by June 1984. The EPA is spending about $90 million to gather analytical, engineering, and economic data on which to base these regulations. It is estimated that the economic impact to American industry to meet the treatment and monitoring requirements of these regulations will be between $40 and $60 billion.

Sample treatment

Before the instrumentation needed to monitor for the priority pollutants is discussed, a brief description of sample treatment is necessary to put the complexity of the monitoring requirements in perspective. There are eight different procedures required to prepare a single sample for analysis.

The organic compounds are divided into four fractions. The very-volatile organic compounds are col-

EPA-approved pollutant monitors [4] **Table I**

Method and cost	Brand and model	Range(s), ppm	Notes
	Sulfur dioxide		
Flame photometry $4,000-6,000	Meloy SA185-2A	0.5, 1.0	1,2
	Meloy SA285E	0.05, 0.1, 0.5, 1.0	
	Monitor Labs 8450	0.5	
	Bendix 8303	0.5, 1.0	
Fluorescence $4,000-6,000	Thermo Electron 43	0.5	3
	Beckman 953	0.5, 1.0	
	Meloy		
	Monitor Labs		
Electrochemical $7,000-8,000	Phillips PW 9755	0.5	3,2
	Phillips PW 9700	0.5	
Derivative spectroscopy $8,000	Lear-Siegler SM 1000	0.5	3
Other	Asarco 500	0.5	
	Nitrogen dioxide		
Chemiluminescence $5,000-6,000	Monitor Labs 8440E	0.5	3
	Bendix 8101-C	0.5	
	CSI 1600	0.5	
	Meloy NA530R	0.1, 0.25, 0.5, 1.0	
	Beckman 952A		
	Thermo Electron 14 B/E	0.5	
	Thermo Electron 14 D/E	0.5	
	Ozone		
Chemiluminescence $4,000-7,000	Meloy 0A325-2R	0.5	4
	Meloy 0A350-2R	0.5	
	Bendix 8002	0.5	
	McMillan 1100-1	0.5	
	McMillian 1100-2	0.5	
	McMillian 1100-3	0.5	
	Monitor Labs 8410E	0.5	
	Beckman 950A	0.5	
	CSI 2000	0.5	
	Phillips FP 9771	0.5	
Ultraviolet spectroscopy $3,000-4,000	Dasibi 1003-AH	0.5, 0.1	3
	Dasibi 1003-PC	0.5, 1.0	
Infrared $6,000-7,000	Bendix 8501-5CA	50	3
	Beckman 866	50	
	MSA 202S	50	
	Horiba AQM-10	50	
	Horiba AQM-11	50	
	Horiba AQM-12	50	

Notes:
1. Support gas (hydrogen) is needed.
2. Scrubber is required.
3. No support gases needed.
4. Support gas (ethylene) is needed (except Phillips 9771).

lected in a headspace-free 40-milliliter (mL) vial. They are stripped from solution by nitrogen or helium which is bubbled through a 5-mL aliquot. These compounds are trapped in a stainless-steel tube packed with Tenax and silica gel, or a similar solid adsorbent. When the purging is complete (12 min) the trap is rapidly heated to 180°C; the adsorbed compounds are flashed off the trap and retrapped on the head of a gas chromatographic column after which they are analyzed by one of the instrumental methods discussed later.

The remaining organic sample is collected in a one-gallon jug. A 1-L aliquot is extracted with a hexane/methylene-chloride mixture to remove the pesticides and polychlorinated biphenyls (PCBs). The extract is concentrated to 10 mL using a Kuderna-Danish concentrator on a steam bath. Florisil chromatography of this concentrate provides three fractions that contain all the pesticides and PCBs. Each of these fractions is concentrated again to 10 mL prior to instrumental analysis.

A 2-L aliquot of the water is made alkaline to pH 10 and extracted with methylene chloride. Then it is made strongly acidic with hydrochloric acid and reextracted with methylene chloride. The first (alkaline) extract contains both the basic and neutral priority pollutants and the second (acidic) extract contains the phenols. Each of these extracts is concentrated to about 10 mL using a Kuderna-Danish concentrator and then each is concentrated to exactly 1.0 mL using a micro-Snyder 3-ball condenser column. These concentrated extracts contain the majority of the priority pollutants.

Total cyanides are determined by reflexing 500 mL of an acidic water solution. Cyanide, as HCN, is vaporized and trapped in a sodium hydroxide solution. Buffer is added, followed by chloramine-T solution and pyridine-barbituric acid reagent. A red-blue dye is formed and the concentration is determined by using a spectrophotometer.

Total phenols are determined by distilling 500 mL of the water to remove most interferences (the phenols steam distill). Solutions of 4-aminoantipyridine and potassium ferricyanide are added to form a yellow dye. This, in turn, is extracted with chloroform and the concentration is determined by use of a spectrophotometer.

The metals are digested three different ways before they are analyzed by atomic absorption. In addition, three different techniques of atomic-absorption analysis are used: flame excitation, graphite furnace, and cold-vapor analysis. One group of metals (cadmium, chromium, copper, nickel, lead, and zinc) are reanalyzed using the graphite furnace if they are not first detected using flame excitation.

Finally, the asbestos samples are filtered through Nuclepore filters and the retained particles are carbon coated under vacuum. The organic filter is dissolved with chloroform, leaving the fibers embedded in a carbon film. A portion of the film is magnified 20,000 times with a transmission electron microscope, and the asbestos fibers are identified by selective-area electron diffraction. A representative area of the electron microscope grid is counted, and the concentration of asbestos, in millions of fibers per liter, can be calculated from the size of the water sample.

Analytical measurements

The instruments currently used to provide measurements of the priority pollutants are:

- Spectrophotometers, for total cyanides and total phenols.
- Atomic-absorption spectrophotometers, for the elements.
- Transmission electron microscopes, for asbestos.
- Gas chromatographs (GC) with electron capture detectors, for pesticides.
- Computerized gas-chromatograph–mass-spectrometers (GC-MS), for the purgeables, base/neutrals and phenols.

All of these instruments—except the spectrometer and gas chromatograph—are relatively expensive. The EPA is moving to substitute less expensive or more cost-effective instruments for monitoring purposes whenever possible. For instance, gas chromatographs with specific detectors and high-pressure liquid chromatographs will, in some cases, be able to substitute for gas-chromatograph–mass-spectrometers (GC-MS). These instruments are cheaper than the GC-MS and, if only a few of the organic priority pollutants need to be monitored, can offer significant savings. However, if many of the organic priority pollutants must be monitored, the cost of all these instruments plus the more labor-intensive sample preparation requirements may more than offset the investment in a GC-MS instrument.

Another substitution forthcoming will be inductively-coupled argon-plasma emission-spectroscopy (ICAP). Although an ICAP is many times more expensive than an atomic absorption spectrophotometer, it can analyze many of the elements at once and the labor savings, if thousands of samples are to be analyzed, will offset the extra investment in the instrument.

The instrumental methods presently in use, and those under consideration as alternative methods, are summarized in Table III.

Priority pollutant analysis

Instrumentation for measuring the priority pollutants is discussed in this section. Here, the discussions involve: the equipment types and their approximate costs, associated data systems, principles of operation, and sample detection.

Equipment types and approximate costs

Gas chromatography (GC) and gas chromatography coupled with mass spectroscopy (GC-MS) are the primary instrument systems used for analysis of the organic priority pollutants. High performance liquid chromatography (HPLC) is also used, but at present is recommended for only a few of the priority pollutants. These systems are extremely flexible and can be equipped with a variety of options to alter their characteristics and meet different analytical needs. The major features of these systems are compared in Table IV.

The prices of gas and liquid chromatographs are approximately the same. Basic instruments can be purchased for approximately $6,000. These instruments, however, include few options and cannot be used for the analysis of all the pollutants. More-flexible units

Priority pollutants as listed by EPA **Table II**

Purgeable organics

Acrolein
Acrylonitrile
Benzene
Toluene
Ethylbenzene
Carbon tetrachloride
Chlorobenzene
1,2-Dichloroethane
1,1,1-Trichloroethane
1,1-Dichloroethane
1,1-Dichloroethylene
1,1,2-Trichloroethane
1,1,2,2-Tetrachloroethane
Chloroethane
2-Chloroethyl vinyl ether
Chloroform
1,2-Dichloropropane
1,3-Dichloropropene
Methylene chloride
Methyl chloride
Methyl bromide
Bromoform
Dichlorobromomethane
Trichlorofluoromethane
Dichlorodifluoromethane
Chlorodibromomethane
Tetrachloroethylene
Trichloroethylene
Vinyl chloride
1,2-*trans*-Dichloroethylene
bis(Chloromethyl) ether

Base/neutral-extractable organics

1,2-Dichlorobenzene
1,3-Dichlorobenzene
1,4-Dichlorobenzene
Hexachloroethane
Hexachlorobutadiene
Hexachlorobenzene
1,2,4-Trichlorobenzene
bis(2-Chloroethoxy) methane
Naphthalene
2-Chloronaphthalene
Isophorone
Nitrobenzene
2,4-Dinitrotoluene
2,6-Dinitrotoluene
4-Bromophenyl phenyl ether
bis(2-Ethylhexyl) phthalate
Di-*n*-octyl phthalate
Dimethyl phthalate
Diethyl phthalate
Di-*n*-butyl phthalate
Acenaphthylene
Acenaphthene
Butyl benzyl phthalate
Fluorene
Fluoranthene
Chrysene
Pyrene
Phenanthrene
Anthracene
Benzo(a)anthracene
Benzo(b)fluoranthene
Benzo(k)fluoranthene
Benzo(a)pyrene
Indeno(1,2,3-c,d)pyrene
Dibenzo(a,h)anthracene
Benzo(g,h,i)perylene
4-Chlorophenyl phenyl ether
3,3′-Dichlorobenzidine
Benzidine
bis(2-Chloroethyl) ether
1,2-Diphenylhydrazine
Hexachlorocyclopentadiene
N-Nitrosodiphenylamine
N-Nitrosodimethylamine
N-Nitrosodi-*n*-propylamine
bis(2-Chloroisopropyl) ether

Acid-extractable organics

Phenol
2-Nitrophenol
4-Nitrophenol
2,4-Dinitrophenol
4,6-Dinitro-*o*-cresol
Pentachlorophenol
p-Chloro-*m*-cresol
2-Chlorophenol
2,4-Dichlorophenol
2,4,6-Trichlorophenol
2,4-Dimethylphenol

Pesticides/PCBs

α-Endosulfan
β-Endosulfan
Endosulfan sulfate
α-BHC
β-BHC
δ-BHC
γ-BHC
Aldrin
Dieldrin
4,4′-DDE
4,4′-DDD
4,4′-DDT
Endrin
Endrin aldehyde
Heptachlor
Heptachlor epoxide
Chlordane
Toxaphene
Aroclor 1016
Aroclor 1221
Aroclor 1232
Aroclor 1242
Aroclor 1248
Aroclor 1254
Aroclor 1260
2,3,7,8-Tetrachlorodibenzo-*p*-dioxin (TCDD)

Metals

Antimony
Arsenic
Beryllium
Cadmium
Chromium
Copper
Lead
Mercury
Nickel
Selenium
Silver
Thallium
Zinc

Miscellaneous

Asbestos (fibrous)
Total cyanides
Total phenols

Instrument for analyzing priority pollutants **Table III**

Priority pollutants	Present instrument	Alternative instrument
Phthalate esters	GC-MS	GC with EC detector
Haloethers	GC-MS	GC with Hall detector
Chlorinated hydrocarbons	GC-MS	GC with EC detector
Nitrobenzenes	GC-MS	GC with EC detector
Nitrosamines	GC-MS	GC with alkali FI detector
Benzidines	GC-MS	HPLC with electrochemical detector
Phenols	GC-MS	GC with EC and FI detector
Polynuclear aromatics	GC-MS	HPLC with fluorescence detector
Pesticides	GC, GC-MS	– – –
Halogenated purgeables	GC-MS	GC with Hall or EC detector
Nonhalogenated purgeables	GC-MS	GC with FI detector
Nonhalogenated hydrocarbons	GC-MS	GC with FI detector
Elements	AA	ICAP
Asbestos	Electron microscope	– – –
Cyanide	Spectrophotometer	– – –
Total phenols	Spectrophotometer	– – –

GC-MS, gas-chromatograph-mass-spectrometer
GC, gas chromatograph
AA, atomic absorption spectrophotometer
EC, electron capture
FI, flame-ionization
HPLC, high-performance liquid chromatography
ICAP, inductively coupled argon-plasma emission-spectroscopy

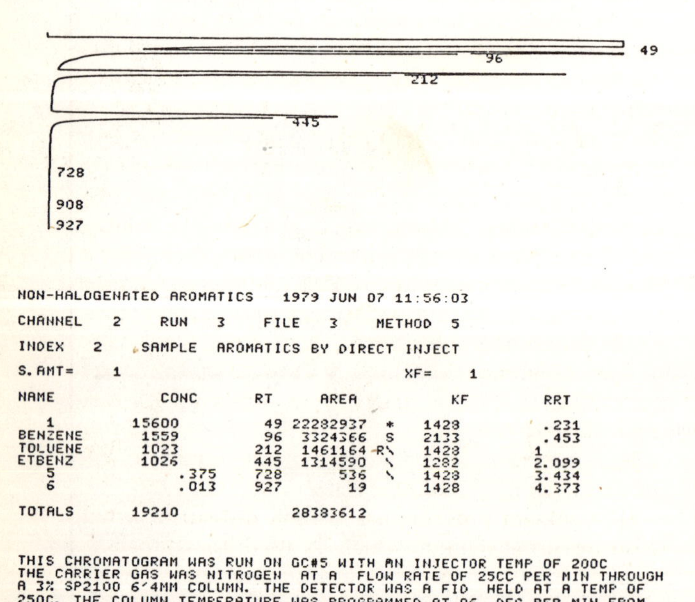

NON-HALOGENATED AROMATICS 1979 JUN 07 11:56:03

CHANNEL 2 RUN 3 FILE 3 METHOD 5

INDEX 2 SAMPLE AROMATICS BY DIRECT INJECT

S. AMT= 1 XF= 1

NAME	CONC	RT	AREA		KF	RRT
1	15600	49	22282937	*	1428	.231
BENZENE	1559	96	3324366	S	2133	.453
TOLUENE	1023	212	1461164	R\	1428	1
ETBENZ	1026	445	1314590	\	1282	2.099
5	.375	728	536	\	1428	3.434
6	.013	927	19		1428	4.373
TOTALS	19210		28383612			

THIS CHROMATOGRAM WAS RUN ON GC#5 WITH AN INJECTOR TEMP OF 200C THE CARRIER GAS WAS NITROGEN AT A FLOW RATE OF 25CC PER MIN THROUGH A 3% SP2100 6′4MM COLUMN. THE DETECTOR WAS A FID HELD AT A TEMP OF 250C. THE COLUMN TEMPERATURE WAS PROGRAMMED AT 06 DEG PER MIN FROM 35 C TO 60 C WITH AN INITIAL HOLD OF 08 MIN AND A FINAL HOLD OF 15 MIN THE ATTENUATION USED WAS X100.

Printer/plotter output **Fig. 2**

capable of multiple pollutant analysis generally cost from \$10,000 to \$15,000. The addition of a data system for automated data processing and reporting of results is an important consideration and adds another \$5,000 to \$10,000. If a data system is not purchased, a strip-chart recorder is required, which costs \$750 to \$1,500.

Gas-chromatograph–mass-spectrometers are considerably more expensive than the other chromatograph systems and cost approximately \$80,000 to \$120,000 for an instrument suitable for analyzing all the organic priority pollutants. The higher cost is primarily due to the complexity of the mass spectrometer. Not only is the mass spectrometer complicated in terms of components such as vacuum pumps, data system and the mass analyzer, but it is also quite complicated to operate and requires a skilled operator. In addition to their higher cost, GC-MS systems are usually floor-mounted and require a space two to four times that occupied by a gas or liquid chromatograph.

Although all of the present priority pollutants can be analyzed by either GC or GC-MS techniques, the choice between the two is not particularly easy. Such factors as the number and type of compounds to be analyzed, nature of the samples, type of personnel available for operating the equipment, and degree of confirmation required must be considered. After these factors have been considered, the chromatographic capability required for the analyses must be determined. In general, a minimum of two gas chromatographs equipped with a total of four different detectors is required to analyze all compounds (see Table II). The cost effectiveness of GC versus GS-MS techniques have been discussed in detail elsewhere [*12,13*] and will not be presented in this article.

Data systems

The data system is an essential component of all modern gas-chromatograph–mass-spectrograph instruments. Analytical capability and ease of use are directly related to the data system. Specifications related to specific ion-monitoring, background-subtraction capability, spectra searching and compound identification techniques, size of reference library, ability to use central spectra libraries via telephone communication, and computer instrument tuning ability should be thoroughly investigated.

In contrast to the gas-chromatograph–mass-spectrometer, a data system is not absolutely essential to gas chromatographs or high-performance liquid chromatographs. Nevertheless, a data system should be seriously considered as an integral part of any chromatograph system. Besides extending the analytical capability of the chromatograph, these systems can generate detailed final reports of analysis with accompanying chromatograms and be used to totally automate the analysis. In so doing they greatly reduce the amount of operator expertise required and enable most analyses to be routinely performed.

A variety of data systems are available for chromatography. Instruments range from simple digital integrators to complex computing integrators with printer/plotter outputs. The more-complex systems may be able to handle data from more than one input

Forms of chromatography available for pollutant analyses **Table V**

Chromatography	Mobile phase	Stationary phase	Classification	Major use
Gas	Gas	Solid	Gas-solid	Separation of gases and hydrocarbons
	Gas	Liquid	Gas-liquid	Various organic separations except gases
Liquid	Liquid	Solid	Normal phase (absorption)	Nonpolar to moderately polar compounds
	Liquid	Liquid	Reverse phase (liquid-liquid)	Most nonionic compounds
	Liquid	Ion exchange resin	Ion exchange	Ionic compounds
	Liquid	Polymer (gel)	Steric exclusion (gel permeation)	Macromolecules

gas-liquid and gas-solid chromatography; whereas liquid chromatography is subdivided into normal phase, reverse phase, ion exchange, and steric exclusion.

The basic process responsible for the separation can also be used to categorize the chromatography. For instance, partition chromatography is the separation process due to equilibria established between the mobile phase and a liquid, and adsorption chromatography involves equilibria between the mobile phase and a solid.

The separation process occurs during transport of the sample through the chromatographic column. Two major types of columns have been developed for gas chromatography, packed and capillary. Packed columns, as their name suggests, contain a packing that either acts as an adsorbent or supports a stationary liquid phase. These columns are commonly coiled tubes (2 to 4 mm I.D.) made of metal or glass. Capillary columns contain the liquid phase coated on the inside wall of the tube. They are considerably smaller in diameter (0.25 to 0.5 mm) than packed columns, and vary in length from approximately 10 to 150 m. In contrast to gas chromatographs, modern columns for high-performance liquid chromatographs are almost exclusively made of metal, packed with very small particles (5 to 10 μm) and only 15 to 25 cm in length.

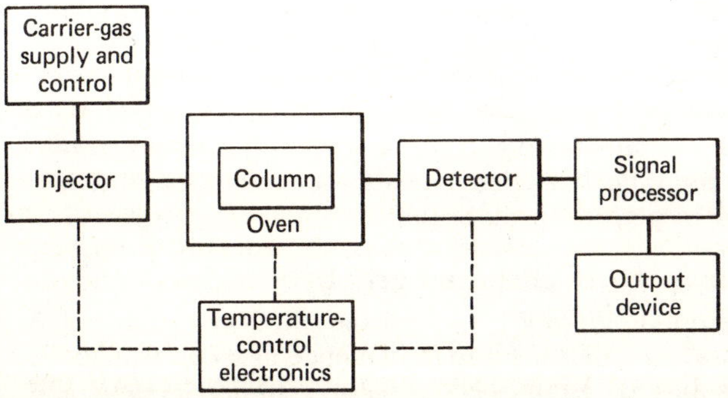

Block diagram of a basic gas-chromatograph system **Fig. 3**

As shown in the block diagram of Fig. 3, the components of a basic gas-chromatograph system consist of a carrier-gas supply with pressure and flow controls, injector (injection port), column, column oven, detector, signal processor, temperature-control electronics, and output device. The carrier gas serves as the mobile phase that transports the sample through the column and into the detector. The injector is normally a heated septum inlet and forms a gas-tight seal between the column and the ambient environment.

The column and detector form the "heart" of the chromatograph system. The column performs the analytical separation, and the detector transforms the separated constituents into electrical signals. These signals are usually very small currents that require amplification by the signal processor before they are capable of driving an output device. The temperature-control electronics maintain the injector and detector at fixed temperatures, and enable the column oven to be operated at a fixed temperature (isothermal operation) or to be temperature programmed at a fixed rate from an initial to a final temperature. The great advantage of temperature programming is that it allows compounds of significantly different volatilities to be analyzed in a single chromatographic run.

The components of a liquid chromatograph system are similar to these in gas chromatography with three main differences:

1) A pump is used for driving the liquid mobile-phase through the column.

2) A solvent programmer is used instead of a temperature programmer, which is used to change the mobile phase composition during an analysis.

3) A column oven is not essential.

Two main techniques are used for introducing priority pollutant samples into a chromatograph. These are syringe injection and the use of a sample valve. Syringe injection is used to introduce small aliquots (1 to 10 μL)

Instruments available for priority pollutant analysis **Table IV**

Instrument	Use	Price range*	Major U.S.† manufacturers	Comment
Gas chromatograph	All organic priority pollutants	$6,000-25,000	Hewlett-Packard Perkin-Elmer Tracor Varian	Instrument specificity depends upon the detector used
Liquid chromatograph	Benzidines and polynuclear aromatics	$6,000-28,000	Altex Du Pont Hewlett-Packard Micromeritics Perkin-Elmer Spectra-Physics Tracor Varian Waters	Not compatible with spraying techniques of analysis. Compounds must have a high ultraviolet absorbence or be fluorescent
Gas chromatograph-mass spectrometer	All organic priority pollutants	$50,000-150,000	Finnigan Hewlett-Packard	Requires a skilled operator

*Price depends on the options chosen.

†Listings are in alphabetical order and do not include manufacturers of speciality instruments.

(multichanneled). Level of data-processing sophistication is usually not related to the number of data channels that can be handled, however. Simple digital integrators usually do no more than print peak areas, retention times, and area percents or concentrations. The more sophisticated printer/plotter systems are capable of peak identifications, multilevel calibrations based on one or more internal or external standards, flexible integration and baseline-correction algorithms, and detailed reporting with custom headings. Most of these systems can be user-programmed for total control of integration and reporting. In addition, many instruments can be equipped with floppy-disk or magnetic-tape systems for retaining programs and raw chromatographic data. This allows the data to be reprocessed using different integration algorithms and reporting formats.

A typical printer/plotter output of a computing integrator is shown in Fig. 2. The first information presented is the chromatogram with retention times of the components printed at each peak. Printing the retention times on the chromatogram greatly facilitates locating integration results for a given peak. After the chromatogram, a heading is printed that may contain information regarding the type of analysis (pesticides, etc.), date and time of analysis, details of the instrument (channel, run, file, method), and the sample analyzed. Other information such as the analyst, correction factors, etc., can also be included. Results of the analysis can be presented in a variety of formats. Information presented usually includes peak names, concentrations, retention times, integration areas, method of baseline correction, and compound response factors. The report can be terminated with a statement that identifies the chromatograph used and all operating parameters of the chromatographic analysis.

Peaks are identified in the report by either name or number. They are identified by name according to information supplied by the user prior to the analysis. This information consists of the name to be used, the expected retention time, and the window of retention time for which a given name is to be applied to a peak. If a peak does not fall within one of these retention time windows, it is assigned a numerical value.

It should be understood that identification of a peak by name does not necessarily mean the peak is actually the compound identified, because retention times are only characteristic of a compound's identity. Usually several other compounds can be found that will have the same retention time as the compound of interest under a specific set of chromatographic conditions, and theoretically hundreds of compounds may have that same retention time simply because there are so many organic compounds (~3,000,000) known.

Information concerning component concentrations, retention times, peak areas, and method of baseline correction can be presented in various ways depending upon preference and the particular integrator used. Concentrations can be in parts per million, parts per billion, weight percent etc. Retention times of peaks may be output in units of time (seconds, minutes) or converted to other parameters such as relative retention times or Kovat's Retention Indices. Peak areas are normally output as total integrator counts. A symbol or code usually follows the peak area that tells how the peak was integrated and the method of baseline correction used. Peak heights may also be shown in addition to or in place of peak areas.

Principles of chromatography

Chromatography [*14,15*] is the physical method of separating a mixture into its components by an equilibrium process established between a stationary and a mobile phase. The mobile phase is a gas in gas chromatography and a liquid in liquid chromatography. As presented in Table V, gas chromatography consists of

Characteristics of chromatograph detectors commonly used for priority pollutant analysis **Table VI**

Detector	Type	Classification[1]	Use	Sensitivity[2]	Specificity	Complexity[3]
Flame ionization	GC	Universal	Hydrocarbons	Nanogram	None	2
Photoionization	GC	Universal/Selective	Aromatics	High picogram	Low	1
Electron capture	GC	Selective	Pesticides, PCBs[1] chlorinated hydrocarbons, phthalate esters	Picogram	Variable	2
Nitrogen-phosphorus	GC	Specific	Nitrogen-containing compounds	High picogram	High	3
Hall detector	GC	Specific	Halogen-containing compounds	High picogram	High	3
Mass spectrometer	GC	Specific	All volatile compounds	Nanogram	High	4
Absorbance	HPLC	Selective	Polynuclear aromatics	Nanogram	Variable[4]	1
Fluorescence	HPLC	Selective	Polynuclear aromatics	High picogram	Variable	2
Electrochemical	HPLC	Selective	Benzidines	High picogram	Variable	3

[1] Universal, response is approximately the same for all compounds; selective response is greater for certain compounds; specific, response is specific for compounds that contain a given element.

[2] Quantity of material that can normally be detected in a typical analysis.

[3] Complexity: 1, requires few adjustments and little maintenance; 2, requires several adjustments; 3, requires several adjustments and routine maintenance; 4, requires numerous adjustments and routine maintenance by a skilled operator.

[4] Specificity depends upon the nature of the compounds present.

of liquid samples, primarily into gas chromatographs. Sample valves are used to introduce samples into both gas and liquid chromatographs. In gas chromatography the valve forms part of a gas sparging apparatus and directs the thermally desorbed volatile pollutants onto the head of a gas chromatography column. On the other hand, in liquid chromatography a valve is used to inject a liquid sample, normally 10 to 100 microliters (μL) in volume.

Sample detection

A wide variety of detectors exist for chromatography. Characteristics of the nine detectors commonly used for analysis of priority pollutants are summarized in Table VI. As shown in this table, the detectors used for analyzing the various categories of pollutants (see Table I) range from non-specific to highly specific. They also range from those needing little maintenance or knowledge for use to those requiring a skilled operator. Although the mass spectrometer should not be considered as a standard chromatograph detector, it is included since it is often used as such in the analysis of pollutants. A brief description of each of these nine detectors is presented below.

The *flame ionization detector* (FID) is a universal detector that responds to all organic compounds, but it is primarily used for the detection of hydrocarbons. Sensitivity is in the low nanogram-per-peak range (1×10^{-11} g/s). The detector consists of an air/hydrogen flame burning in a polarized environment (Fig. 4). During normal combustion the flame produces very few ions. However, when an organic compound enters the flame, ions are created that are collected by a collector electrode to produce an ion current. The ion current, which is directly proportional to the mass of the compound entering the flame, is then amplified by an electrometer. The detector response is linear over a very wide range (approximately 10^4). The FID requires hydrogen and air burner gases at flowrates of approximately 30 and 300 cm^3/min, respectively. Detector maintenance requires occasional optimization of burner gases and cleaning of the flame jet and collector electrode.

The *photoionization detector* (PID) is similar to the FID in response characteristics. It is primarily used for the detection of organic compounds. Response is approximately ten times greater for aromatic compounds than it is for other compounds and for this reason is recommended for the detection of aromatic pollutants. The PID like the FID is a simple device that is easy to operate and maintain. The detector consists of a high energy UV light source, polarizing electrode and collector electrodes which are contained in a heated housing.

The detector signal is generated from an ion current that is formed by the high energy UV ionization of the compound. The detector is slightly more linear and sensitive than the FID, and detector operation requires no additional gases.

The *electron-capture detector* (EC) is a selective detector that responds to compounds that have a high electron affinity. The detector contains a radioactive foil (usually Ni-63) that ionizes nitrogen or argon/methane carrier gas to produce an electron cloud. This electron cloud is sampled with a collector electrode to produce a standing current. Electron negative compounds, such as the chlorinated pesticides and PCBs, that enter the detector cavity, deplete the electron population that can

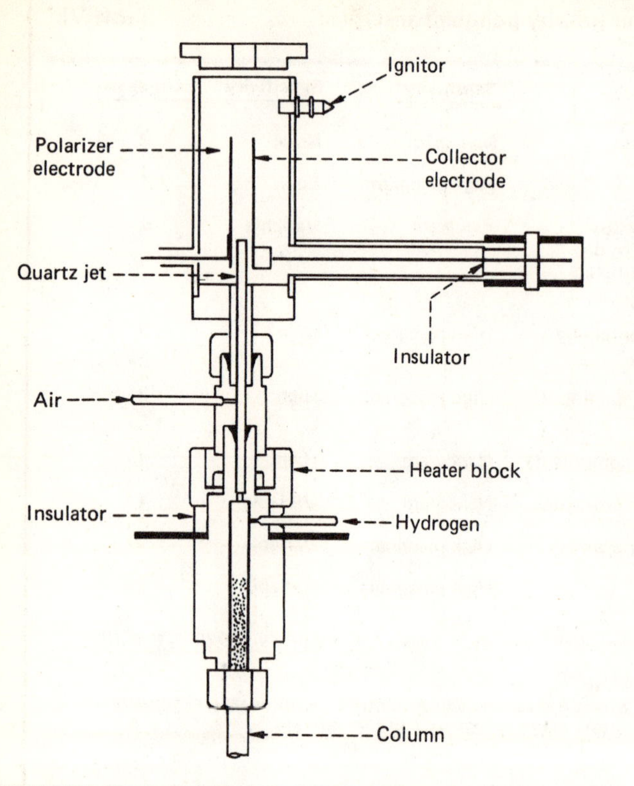

Cross-section of the flame-ionization detector **Fig. 4**

be sampled, thereby resulting in a detector signal. Modern electron-capture detectors are operated in a constant-current mode in order to extend their linear range to approximately 10^4. This is done by varying the sampling rate as a function of electron demand (compound concentration in the detector). The detector is extremely sensitive and will respond to sample quantities of sensitive materials as low as 10^{-12} or 10^{-13} grams. Due to its high sensitivity, the detector is also sensitive to contamination from GC column bleed and to dirty samples, but with reasonable caution is easy to use and requires little maintenance.

The *nitrogen-phosphorus detector* (N-P) is a FID that has been modified to contain an alkali source between the flame jet and the collector electrode. In recent designs the source contains a nonvolatile alkali such as rubidium silicate. The source is electrically heated to a dull red and held at a negative electrical potential with respect to the collector electrode. Hydrogen flowrates of only 1 to 5 cm^3/min are used, because such a low flow will not support a flame as classically perceived. Instead, a plasma is created around the source. The alkali acts as a catalyst and selectively ionizes certain nitrogen and phosphorus species that are created in the plasma. In this manner, an ion current similar to that in a FID is established. Selectivity to nitrogen compounds relative to hydrocarbons is usually 10^4. The N-P detector is very sensitive and can be used to detect subnanogram quantities of most nitrogen compounds. Response is approximately ten times greater for phosphorus than it is for nitrogen. Frequent adjustment of the source temperature and hydrogen flowrate may be required to maintain the same detector-response characteristics. In addition, the alkali source may need replacement every one to six months.

The *Hall electrolytic conductivity detector* (HECD) is a specific detector that can be used for the detection of halogens, sulfur, and nitrogen containing compounds. Detector components include a reactor, electrolytic conductivity cell, electrolyte system, and electronics for measuring electrolytic conductivity and heating the reactor. The relationship of these components is shown in Fig. 5. The detector operates by catalytically converting organic halogen, sulfur, or nitrogen to HX, SO_2, and NH_3, respectively. These molecules are then extracted into an electrolyte stream and the resulting change in electrolytic conductivity output is measured as the detector signal. Selectivity is achieved by either removing unwanted reaction products with a scrubber, converting them to substances that do not change the electrolyte conductivity (i.e., CH_4), or suppressing the conductivity of certain ions by using nonaqueous electrolytes. The detector can be operated in two main modes, reductive and oxidative. Halogen and nitrogen compounds are detected in the reductive mode, using hydrogen as the reaction gas. Sulfur compounds are detected in the oxidative mode with air as the reaction gas. The detector exhibits subnanogram sensitivity and selectivities of 10^5 or greater (relative to hydrocarbons). Linear range of operation is from 10^3 to 10^5, depending upon the element detected. Maintenance involves periodic replacement of the electrolyte, reaction tube, scrubber (not used in detecting halogen compounds), and the ion exchange resin.

The *quadrapole mass spectrometer* is the only system that is commonly employed as a chromatographic detector for priority pollutant analysis. This system can be considered as a mass-selective gas-chromatograph detector. Compounds entering the ion source (Fig. 6) are either directly ionized by electrons emitted from a heated filament (electron-impact mode), or indirectly by interaction of a reaction gas that has been previously ionized by electron bombardment (chemical-ionization mode). When the compound is ionized, it fragments into characteristic ions that are separated by the quadrapole filter. The separation process is very rapid and usually takes only a few milliseconds. This process results in the formation of an ion spectrum that can be used to confirm the identity of a compound. The data system can be used to record and compare spectra to those contained in a library (memory), or it can selectively monitor specific ions, resulting in a mass-specific chromatogram. Selection of characteristic ions or ratios of certain ions to be monitored results in high specificity for most compounds.

The ion information, separation, and detection must be performed under high vacuum with the aid of sophisticated electronics. Consequently, most mass spectrometers are fairly complex to operate and maintain. Sensitivity is normally in the nanogram range for most of the priority pollutants. Specificity is very high, but depends upon the ions chosen for monitoring the individual compounds.

The *ultraviolet absorption* (UV) detector is probably the

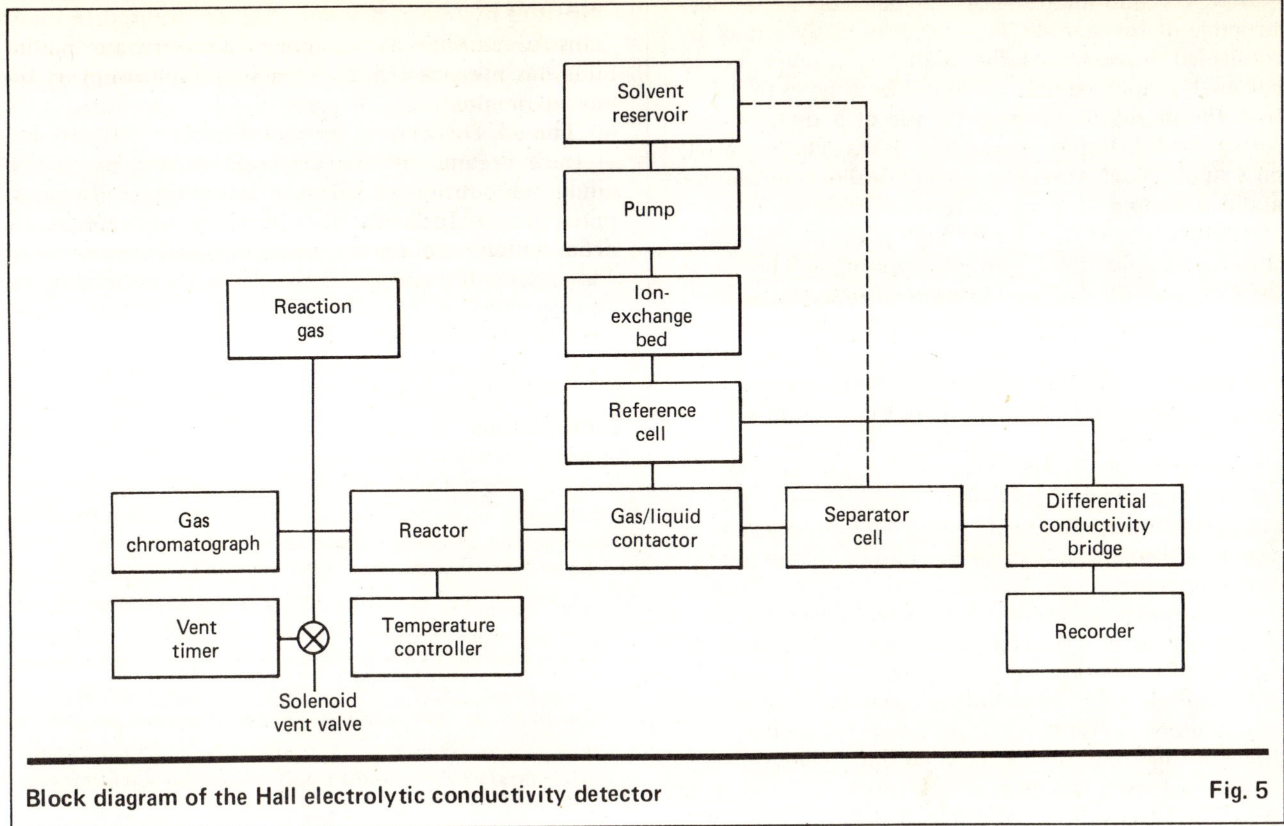

Block diagram of the Hall electrolytic conductivity detector **Fig. 5**

most commonly used detector for high performance liquid chromatography (HPLC). UV detectors fall into two classes; namely, fixed wavelength and variable wavelength. Both types of UV detectors operate on the well known principles of photometry, which involve the absorption of light by certain types of organic and inorganic compounds. The amount of light absorbed by the sample compounds is directly proportional to the concentration of the compound in solution. The solvent used as the mobile phase in the chromatographic process must of course be transparent to the ultraviolet light.

Fixed-wavelength detectors operate at a single wavelength of light, which is provided by the emission of special lamps containing appropriate elements. For example, the most commonly used lamp in this type of detector is the mercury-vapor lamp, which has primary emission at 254 nm wavelength. In order to obtain other wavelengths, the lamp must be changed to another type.

Variable wavelength detectors incorporate a broad-spectrum lamp that emits a continuous band of wavelengths. The desired wavelength is then obtained by use of a grating or prism monochromator which, in effect, transmits only the selected wavelength to the sample cell.

Both types of UV detectors are moderately selective in the types of compounds they will detect. Fixed-wavelength detectors are best suited to the detection of aromatic compounds, and can be very sensitive for certain strong UV-absorbing compounds such as polynuclear aromatic hydrocarbons (PNAs). Variable wavelength detectors offer additional selectivity, and sometimes added sensitivity, for many other types of compounds.

Fluorescence detectors depend on the fact that certain types of compounds, when irradiated with light of a certain wavelength, emit light of a longer wavelength, that is, they fluoresce. Fluorescence detectors are somewhat similar to UV absorption detectors in that they require fixed or variable light sources and some type of monochromator. In fluorescence detectors, however, the amount of light *emitted* by the sample is measured rather than the amount of light absorbed.

Since relatively few types of compounds fluoresce, the degree of selectivity provided by fluorescence detectors is generally much higher than that of UV detectors. In addition, for certain types of compounds, fluorescence-detection-sensitivity is much better than by UV absorption.

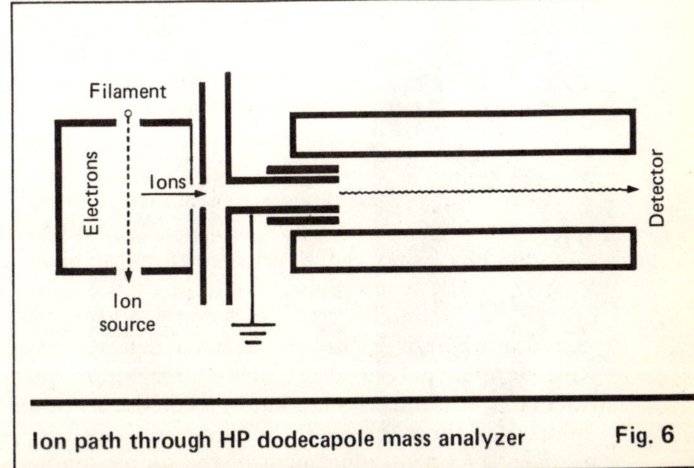

Ion path through HP dodecapole mass analyzer **Fig. 6**

Both UV and fluorescence-type detectors are nondestructive of the sample. For this reason, they may be connected in series to the high-performance liquid chromatography column to obtain both types of detection almost simultaneously. By use of a dual-pen recorder, both UV and fluorescence peaks can be plotted on a single chart, providing a great deal of information about a sample.

Another type of detector for high-performance liquid chromatography that is currently gaining acceptance is the *electrochemical detector*. This type of detector is based on the fact that many compounds undergo chemical reactions that cause them to conduct a current when exposed to an electrical potential. The amount of current, or the voltage produced in the sample cell, is measured and then related to the concentration of sample. Detectors of this type may be designed to measure only a type of electrochemical reaction such as, for example, conductance or they may be multipurpose to permit measurement of oxidation and reduction as well as conductance.

A multipurpose type of detector will provide a controlled source of electrical voltage or current and be capable of measuring the current or voltage produced in the sample cell. The sample cell generally consists of two or more electrodes that are immersed in the solution flowing from the chromatographic column. The presence of sample solute in the carrier solvent then causes the electrodes to produce an electrical potential or current, which is measured.

Electrochemical detectors can be very sensitive and specific for certain types of compounds. Detection limits in the picogram range are sometimes obtainable. The use of these detectors, however, puts a number of restraints on the type of mobile phase that may be used for the chromatographic separation since the mobile phase plays an integral part in the electrochemical reactions taking place in the cell

Summary

Instrumentation for monitoring air and water pollutants has been described. As noted, monitoring of atmospheric pollutants is more firmly established and automated. However, more-sophisticated measurement of trace organics in the atmosphere may be in the offing. Monitoring of water pollutants is in a transitional phase. Much effort is currently being expended to define suitable, yet less-expensive measurement systems. This article has described the currently used systems and presented a general description of the type of instrumentation available.

References

1. *Federal Register,* Vol. 36, Nov. 25, 1971.
2. *Federal Register,* Vol. 44, Feb. 8, 1979, for NAAQS, Ref. and Cal. Method—(Chemiluminescent was also reference method in *Fed. Reg.,* 11/25/71)
3. *Federal Register,* Vol. 36, Nov. 25, 1971.
4. *Federal Register,* Vol. 40, Apr. 25, 1975.
5. *Federal Register,* Vol. 36, Nov. 25, 1971.
6. *Federal Register,* Vol. 43, Oct. 5, 1978.
7. "Quality Assurance Handbook for Air Pollution Measurement Systems, Vol. I, Principles," EPA Environmental Monitoring and Support Laboratory, Research Triangle Park, NC 27711.
8. "Quality Assurance Handbook for Air Pollution Measurement Systems, Vol. II, Ambient Air Specific Methods," EPA-600/4-77-027a, May 1977, EPA Environmental Monitoring and Support Laboratory, Research Triangle Park, NC 27711.
9. "Guidelines for Development of a Quality Assurance Program: Reference Method for the Determination of Suspended Particulates in the Atmosphere (High Volume Method)," EPA-R4-73-0286, Office of Research and Monitoring, U.S. EPA, Washington, D.C., June 1973.
10. American Public Health Assn., American Water Works Assn., and Water Pollution Control Federation, "Standard Methods for the Examination of Water and Wastewater," American Public Health Assn., Washington, D.C. 20036, 1976, pp. 1193.
11. Keith, L. H., and Tellieard, W. A., "Priority Pollutants I—A Perspective View; *Environ. Sci. Technol.* Vol. 13, No. 4, pp. 416–423 (1979).
12. Brenner, N., others, "Gas Chromatography," Academic Press, New York, 1962.
13. Budde, W. L., and Eichelberger, N. W., *Anal. Chem.,* Vol. 51, p. 567A (1979).
14. Finnigan, R. E., others, *Environ. Sci. Technol.* Vol. 13, p. 534 (1979).
15. Snyder, L. R., and Kirkland, J. J., "Introduction to Modern Liquid Chromatography," John Wiley & Sons, New York, 1974.

The authors

Delbert M. Ottmers, Jr. is Assistant Vice-President, Engineering and Chemistry, Radian Corp., 8500 Shoal Creek Blvd., P.O. Box 9948, Austin, TX 78776, telephone (512) 454-4797. He has been involved, at Radian, in process design and development activities, with primary emphasis on air/water pollution problems. He holds B.S., M.S. and Ph.D. degrees in chemical engineering from the University of Texas. He is a member of AIChE, Tau Beta Pi, Phi Lambda Upsilon, Omega Chi Epsilon, Sigma Xi and Phi Kappa Phi.

David C. Jones is Assistant Vice-President, Technical Staff, Radian Corp. His major interest at Radian has been ambient air quality monitoring. He is an authority on instrument selection, calibration procedures, and quality assurance studies for air monitoring stations. He holds a B.S. in chemistry and a Ph.D. in physical chemistry from the University of Texas, and is a member of Alpha Chi Sigma and the National Assn. of Corrosion Engineers.

Lawrence H. Keith is Manager, Analytical Chemistry Div., Radian Corp., where he is responsible for maintaining the technical quality of the analytical staff and assuring that the laboratories keep pace with advances in instrumentation and technology. He holds a B.S. in chemistry from Stetson University, an M.S. in organic chemistry from Clemson University and a Ph.D. in natural products chemistry from the University of Georgia. He is a member of Sigma Xi, Gamma Sigma Epsilon, Kappa Kappa Psi, and ACS (Chairman, Executive Committee, Div. of Environmental Chemistry, and Past Chairman, Central Texas Section).

Randall C. Hall is Department Head/Senior Staff Scientist, Radian Corp., where he serves as head of the Organic Chemistry Dept. and as acting group leader of the Analytical Separation group within that department. He holds a B.S. in chemistry and a Ph.D. in analytical-organic and physical-organic chemistry from Texas A&M University, and is a member of ACS and Sigma Xi.

Designing for pH control

Process design is an essential aspect of controlling the pH of plant streams. Here are some guidelines whereby the process engineer can aid the control engineer in achieving a suitable system at minimum expense.

D. L. Hoyle, The Foxboro Co.

☐ The automatic control of pH has become established throughout the chemical process industries, in plants ranging from ore beneficiation to food processing. It is used to control reagent additions to flotation cells, lime additions to ball mills, the incoming water to paper mills, the feedstocks for synthetic fiber and crystallization processes, and the process water for sugar refining. Environmental protection regulations have brought increased emphasis to automatic pH control of waste streams from all sorts of plants.

Despite such widespread application, pH control often appears to be difficult and costly when applied to the narrowing quality limits required of modern processes and governmental regulatory agencies. The near-universal basis of these difficulties is revealed in Fig. 1, where pH is plotted against the reagent demand necessary to bring the pH to neutrality (pH 7).

With current technology, the average reagent delivery system exhibits at least 1% hysteresis—i.e., the smallest incremental change in reagent flow that can be metered is 1% of the total flowrate. Thus, when the influent has a pH of 2.0 (see Fig. 1), this 1% hysteresis represents (0.01)(10,000) or 100 units of reagent. The effect of these 100 units on a neutral product is equivalent to a pH of 10 (on the basic side) or a pH of 4 (on the acid side). If the desired control limits are a pH of 6–8, some way must be found to reduce the error.

Usually, this involves the complete system: stirred tanks, type of mixing, the baffles and agitators, vessel inlet and outlet locations, tank sizes and configurations, number of vessels, the control system, the reagent, and (possibly) lagoons.

Stirred tanks and process gain

At least one well-stirred tank is generally required for adequate performance in difficult pH-control applica-

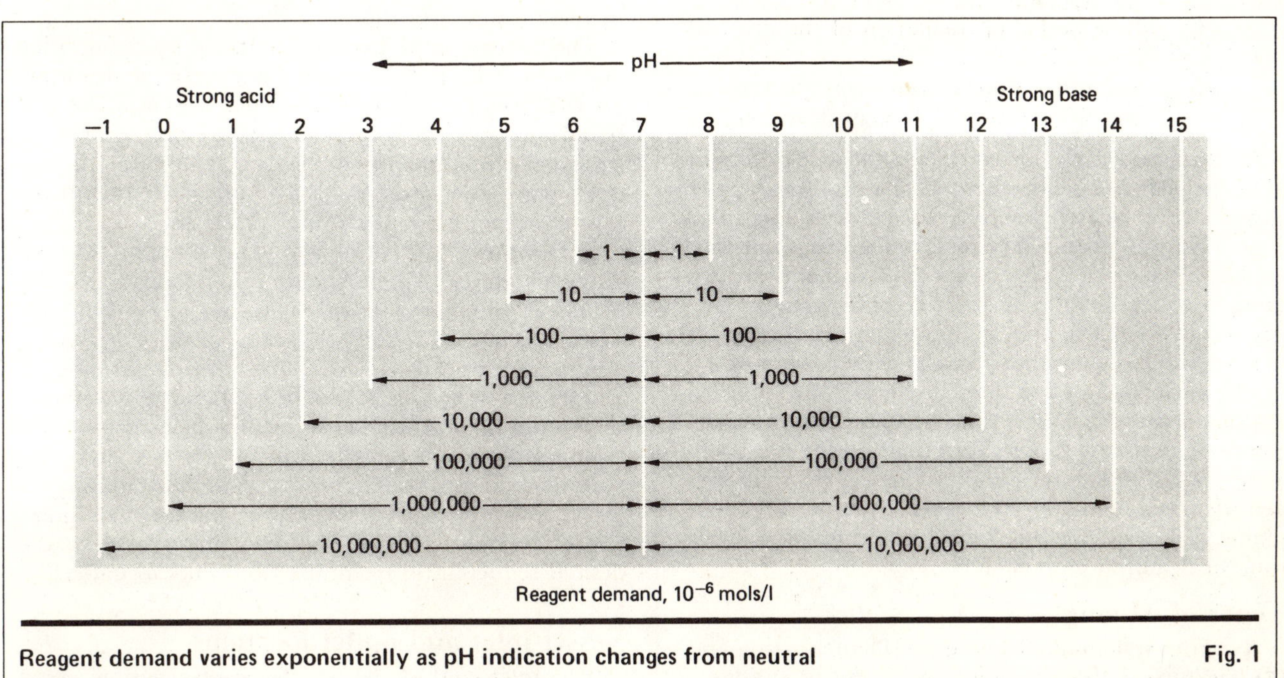

Reagent demand varies exponentially as pH indication changes from neutral **Fig. 1**

Originally published November 8, 1976.

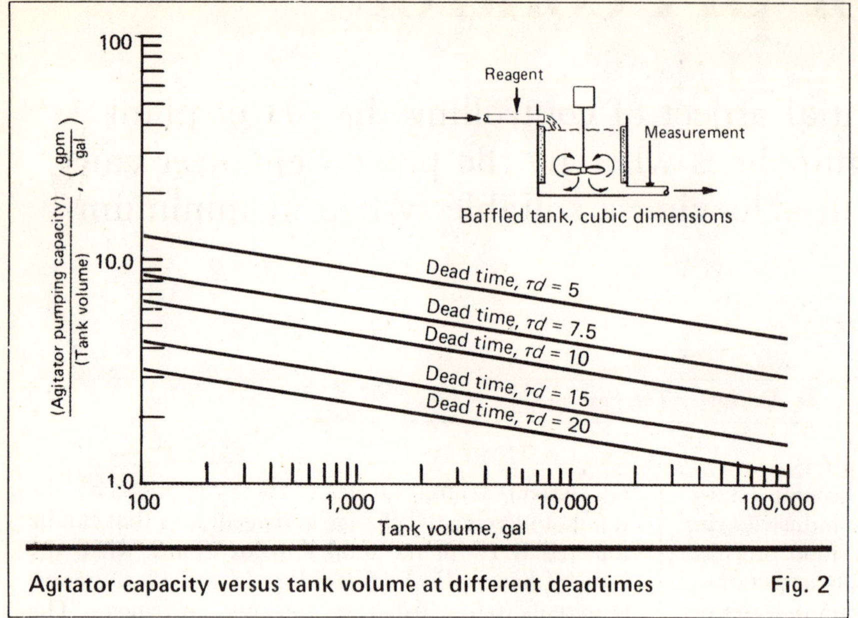

Agitator capacity versus tank volume at different deadtimes Fig. 2

tions. The effect of a stirred tank on periodic disturbances in the entering flow can be calculated in terms of its deadtime, retention time, and dynamic gain.

Dead time is the interval between the addition of a reagent and the first observable change in the effluent measurement; it is affected by agitation and tank geometry as well as retention time. Retention time is the volume of the vessel divided by the flow through the vessel. Dynamic gain is: (% change in output)/(% change in input). These terms are related by [*1*]:

$$G_d = t_0/(2\pi\tau_1)$$

where: G_d = the dynamic gain, %/%
τ_0 = time period of oscillation of the disturbance
τ_1 = first-order time constant of the tank ≅ (retention time) — (system deadtime)

As illustrated in the boxed calculation (see p. 123), a dynamic gain of 0.037 can easily be obtained for a stirred tank with only 3 min holdup, or retention, time. This is equivalent to a reduction in the overall process gain by a factor of (1.0/0.037) = 27. Furthermore, two tanks used in series reduce the process gain by the product of their individual gains, so that two such tanks in series would reduce the process gain by a factor of 27^2 or 729.

Consequently, the use of stirred tanks permits process engineers to reduce process gain to a controllable level, as well as to smooth out high-frequency errors in reagent delivery caused by measurement noise. There remains, however, the problem of obtaining a suitable ratio of deadtime to retention time.

Types of Mixing

A control system is influenced by (1) intermixing and (2) backmixing. Intermixing is necessary to eliminate areas of unreacted compounds or untreated effluent, when reagent is added to a stream. One or two right-angle bends in the piping, or a free fall of one or two feet to the surface of a stirred tank, are usually sufficient to provide good intermixing. Poor intermixing causes an erratic, or noisy, signal to be observed in the effluent pH measurement.

Backmixing is more important to accurate pH control than is intermixing. In general, the degree of backmixing can be defined in terms of the pumping capacity of an agitator with respect to the flow and volume of the neutralization vessel. In practice, however, this definition has limited usefulness because of variables such as agitator construction and blade pitch, baffling of the neutralization vessel, and placement of inlet and outlet measuring electrodes.

Experience has shown that the best way to define backmixing for control purposes is via the ratio of the system deadtime to retention time of the vessel. The treated stream must be held in a vessel long enough for the reagent to react and be backmixed. A ratio of deadtime to retention time of 0.05 approaches an optimum value.

Baffles and agitators

Suitable baffling and agitator positioning should prevent a whirlpool effect in mixed vessels, since the power is supplied to turn over the contents of the vessel rather than to whirl them about. Thus, a propeller or axial-flow impeller should be used to direct mixing flow toward the bottom of the vessel, whereas a flat-bladed or radial-flow impeller should be avoided, since it generally tends to divide the vessel into two sections and increase system deadtime.

The relationships between agitator pumping time and tank volume for various parameters of deadtime are shown in Fig. 2. The data, which are empirical and based on tanks of 200, 1,000, 10,000 and 18,000-gal capacity, apply only to baffled tanks of cubic dimensions, with the inlet at the surface and the outlet at the bottom of the opposite side of the tank, as in Fig. 2.

In these tests, the ratio of impeller diameter to tank diameter varied from 0.25 to 0.4. Square-pitch impellers at an average peripheral speed of 1,500 ft/min were used in the tanks up to 1,000-gal capacity. Axial-flow turbine impellers at an average peripheral speed of 700 ft/min were used in the larger tanks. Fig. 3 shows the minimum net shaft horsepower to be expended as a function of tank volume. It is necessary to specify this parameter as well as agitator pumping capacity, since a large, slow-moving blade will move the specified quantity of fluid but not be fast enough to achieve the desired reduction in deadtime from inlet to outlet.

Vessel inlet and outlet locations

The inlet and outlet in the neutralization vessel should be located at opposite sides, one high and one low with respect to the bottom of the vessel. It is often most convenient to introduce the influent stream at the surface of the tank, with the outlet at the bottom.

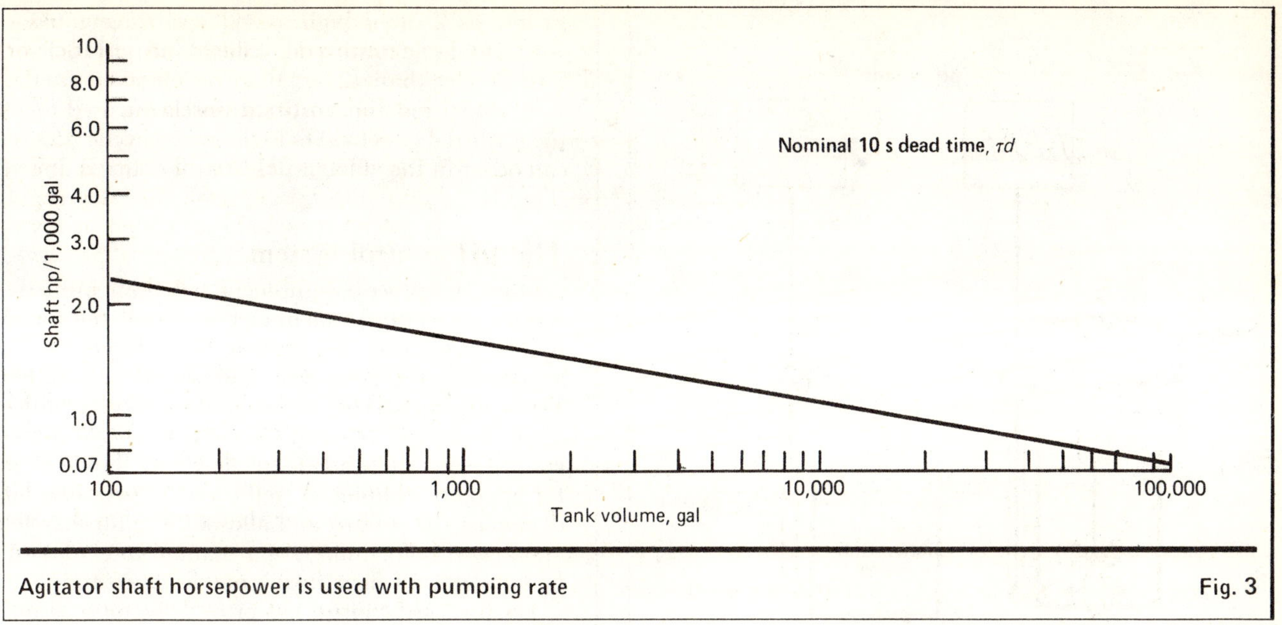

Agitator shaft horsepower is used with pumping rate Fig. 3

Estimating the dynamic gain of a stirred vessel

Define the total system deadtime, t_{dt}, as:

$$\tau_{dt} = \tau d_1 + \tau d_2$$

where: τd_1 = deadtime in tank, inlet to outlet
τd_2 = remaining loop deadtime (sampling, electrode, valve response, etc.)

For good control, let:

$$\tau_{dt} = 0.05\ (V/F)$$

where: V = vessel volume
F = flow through vessel

The first-order time constant, τ_1, for an agitated vessel can be expressed as its retention time minus its deadtime, or [*1*]:

$$\tau_1 = (V/F) - \tau d_1$$

If the retention time (V/F) is assumed to be 3 min, $0.01(V/F) = 1.8$ s is a reasonable value for the loop deadtime, τd_2, if good engineering practice is followed in installation design. However, from the above expression for total system deadtime:

$$\tau d_1 = \tau_{dt} - \tau d_2 = 0.05\ (V/F) - 0.01\ (V/F) = 0.04\ (V/F)$$

Similarly, the first-order time constant can be expressed as:

$$\tau_1 = (V/F) - 0.04\ (V/F) = 0.96\ (V/F)$$

Thus, both the total system deadtime, τ_{dt}, and the first-order time constant, τ_1, are expressed in terms of retention time. We can further relate these two terms as:

$$(V/F) = \tau_{dt}/0.05 = \tau_1/0.96$$
$$\tau_1 = 19.2\ \tau_{dt}$$

Furthermore, the period of oscillation, τ_0, of a typical process composition (under closed-loop control with an optimally tuned 3-mode controller) can be approximated as a function of the system deadtime, as for example [*1*]:

$$\tau_o = 4.5\ \tau_{dt}$$

We thus have values necessary for substituting into the formula for dynamic gain, as:

$$G_d = \tau_o/(2\pi\tau_1) = 4.5\ \tau_{dt}/(2\pi\ 19.2\tau_{dt}) = 0.0373$$

Variations in the location of inlet and outlet can change the deadtime considerably. Reversing the flow through the tank, for example, so that the inlet is at the bottom and the outlet on the surface, causes the deadtime to increase by a factor of 2.5–3.0. In effect, the path from inlet to outlet has been doubled by this change (see Fig. 4). The additional deadtime is attrib-

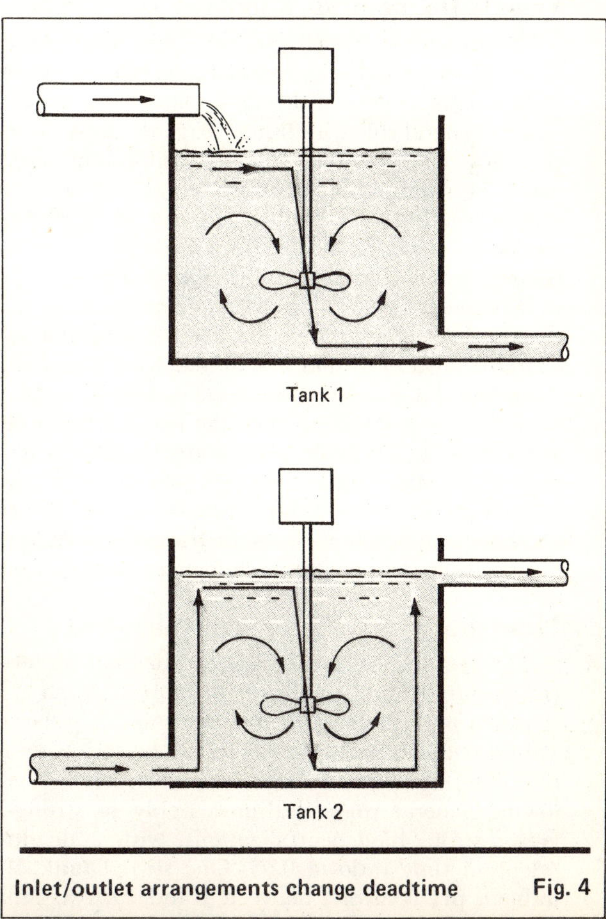

Inlet/outlet arrangements change deadtime Fig. 4

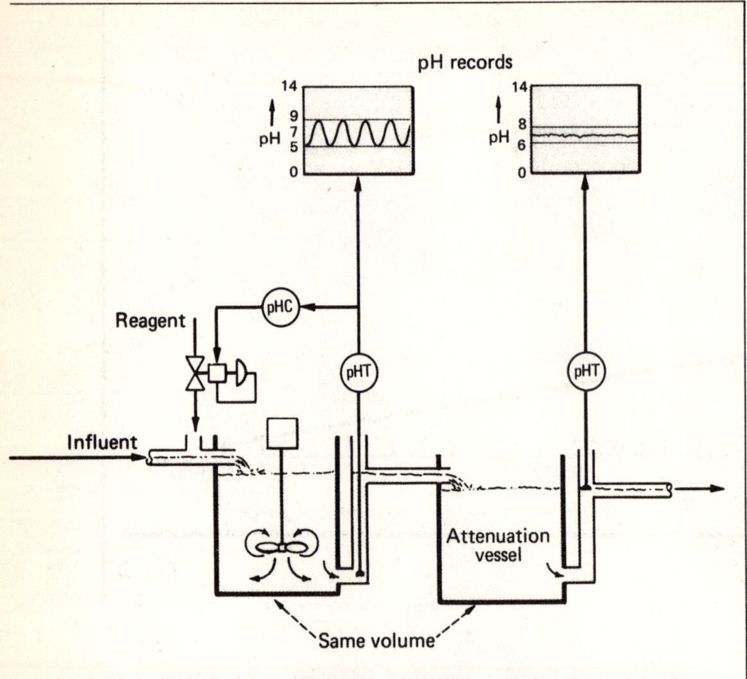

Feedforward control with a two-tank system Fig. 5

utable to the swirl caused by the agitator, this swirl being minimized but not eliminated by baffling.

Vessels for neutralization

Neutralization vessels should have approximately cubic dimensions. If a cylindrical tank is used, the depth of liquid should equal the tank diameter.

As a general rule with liquid reagents, a 3-min retention time will give the smallest feasible tank size, because the combination of much less than 3 min retention time plus a requirement for a (deadtime)/(retention time) ratio of 0.05 will cause considerable splashing and trap large quantities of air.

Otherwise, the size of the vessel depends on the rate of reaction of the neutralization being performed, as for example, at least 5 min is required with high-calcium lime; and if a dolomitic limestone is used, 25 or 30 min would be required because of the low solubility of the reagent. Also, insoluble precipitates, such as iron hydroxide, require larger tanks; the precipitate traps unreacted reagent and/or influent material and causes an extended reaction time, since the trapped material must diffuse from the precipitate before reaction can occur.

How many vessels for neutralization

The required number of neutralization vessels depends entirely on the difficulty of control, which in turn depends on the maximum and/or minimum pH or pIon values coupled with the degree of buffering or reaction complexity present at the control point. The following general rules of thumb apply to strong-acid versus strong-base neutralizations, with a deadtime/retention time ratio of 0.05: One stirred tank, if the influent pH is always between 4 and 10; two tanks—one stirred and one unstirred—if the influent pH can be as low as 2 or as high as 12; and three tanks—two stirred and one unstirred—if the influent pH is less than 2 or greater than 12.

When stirred and unstirred vessels are used together, the unstirred vessel serves to damp out cyclic upsets that can occur in the effluent pH from the stirred first vessel (Fig. 5).

The pH control system

Once the process equipment has been properly designed, the control system can be considered for maximum effectiveness. This system should be kept as simple as possible for both cost and maintenance reasons. While the capital cost of the neutralization facilities can often be greatly reduced through the use of advanced control techniques, adequate allowance should be made for operator training. A well-trained operator who understands the process and allows the control system to operate without unnecessary manual intervention can be very helpful in reducing operating costs.

Feedforward control has proved the most useful advanced technique in the waste-treatment field. As the name feedforward implies, measurements of the process variables (i.e., composition and flow) are made upstream of the neutralization facilities; reagent requirements are calculated from this information by a mathematical model of the process; and reagents are added directly to the influent stream.

The need for feedforward control is determined by the rate of change of process variables, so that these must be measured or accurately estimated. A black box labeled feedforward computer cannot by itself solve all pH control problems, because of accuracy limitations. Proper process design and a feedback loop are still very necessary parts of the overall control system. Feedforward control systems using nonlinear and adaptive controllers have been described in detail [*2*].

The importance of a good definition of the control problem cannot be overemphasized. Titration curves should be prepared representing the extremes in buffering changes, and records of flow and pH variations should be collected over a period of several days. If the process is new and such data cannot be assembled, the best possible estimates must be made. Without such data, the system design engineer can only design the process facilities for the worst possible case. This often results in overdesign and added expense for a processing facility that yields no financial gain.

Reagent selection

The best reagents will incur the least overall cost, including not only the cost of the reagent itself but also the cost of maintaining the reagent system, and the facilities to use the reagent. The combination of all such factors may make a slightly more expensive reagent less expensive overall. For example, liquid 98% sulfuric acid and 20% sodium hydroxide require only filtration before use, react almost instantaneously, and therefore can be used with a minimum-size reaction vessel, reducing initial cost. Lime slurry, by contrast, clogs valves, causes density control problems, and requires at least 5-min retention time in the reaction vessel. Consequently, lime slurry is generally economical only for large installa-

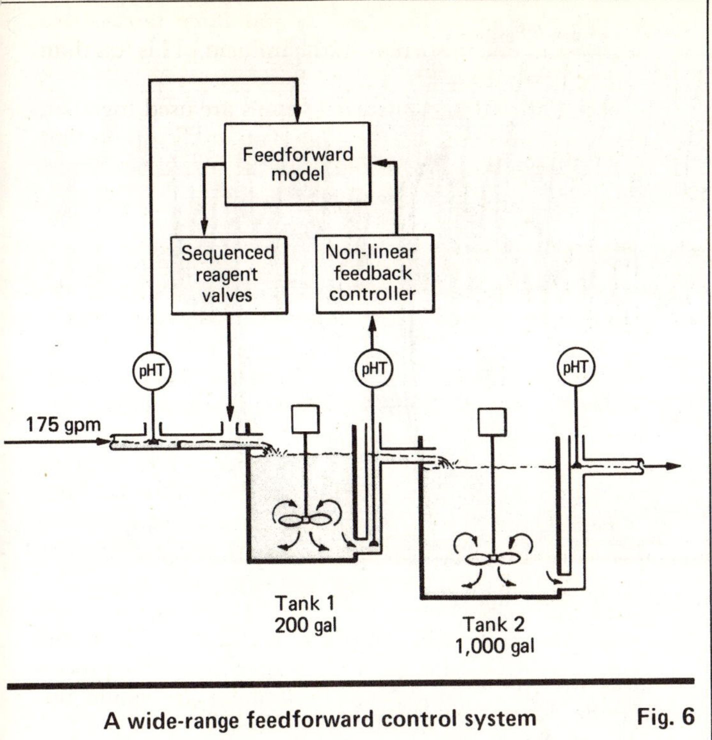

A wide-range feedforward control system Fig. 6

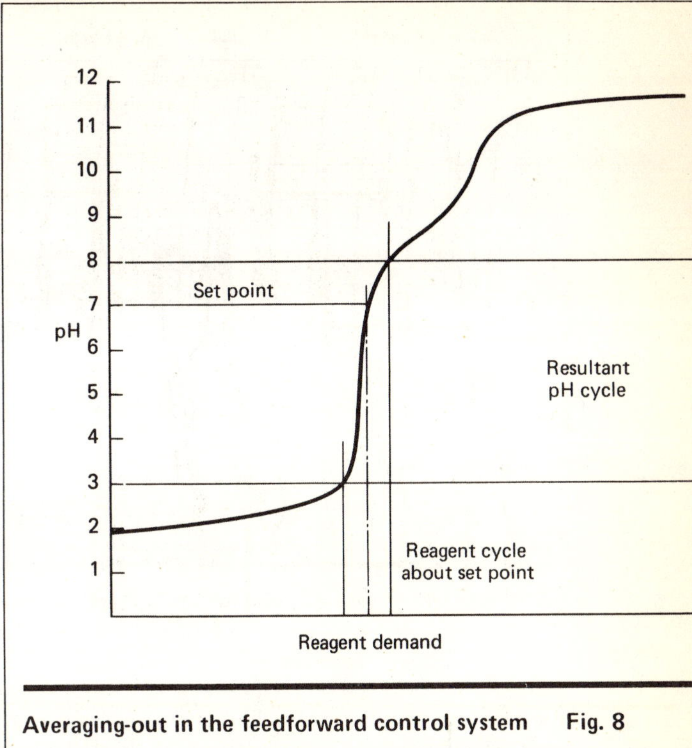

Averaging-out in the feedforward control system Fig. 8

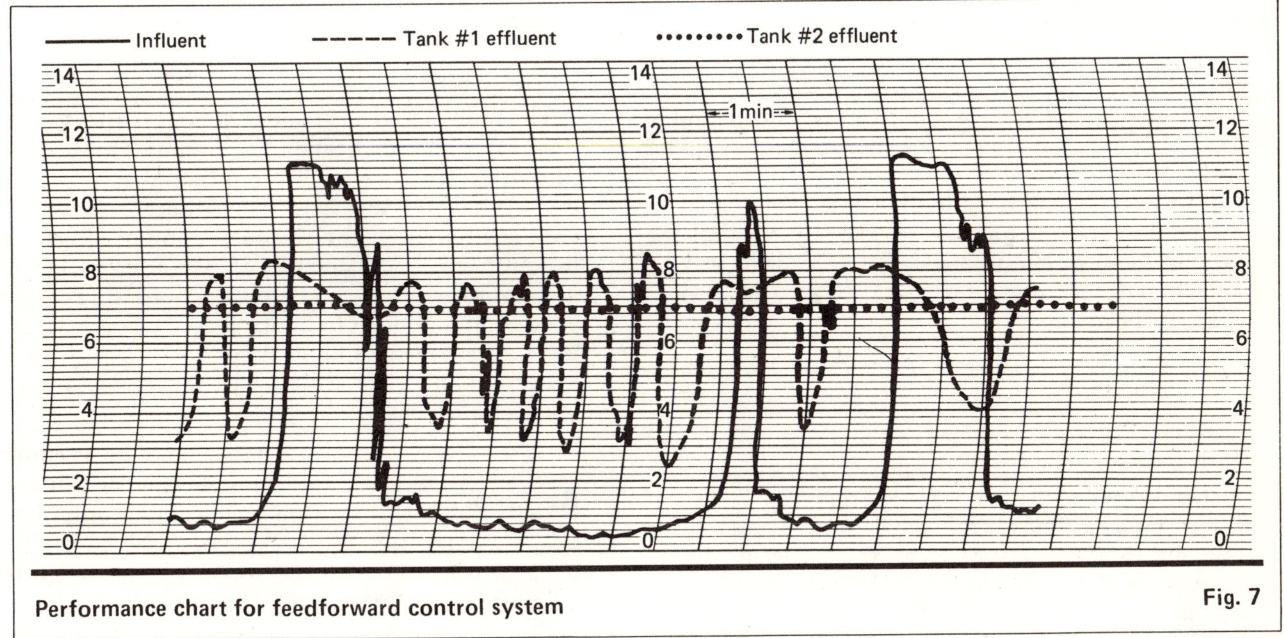

Performance chart for feedforward control system Fig. 7

tions where the reagent cost is much larger than the maintenance expense.

What good is a lagoon?

A lagoon or holding vessel upstream of a stirred neutralization vessel can prove very useful by smoothing out upsets in influent pH and flow. It thus allows the use of a simple feedback control system rather than more costly feedforward control. The choice between feedforward control or a lagoon can usually be based on a simple cost analysis, except where space considerations or waste decomposition precludes using a lagoon.

Generally, small flowrates will make the lagoon approach more attractive, since control-system costs are almost independent of the quantity of waste to be treated, whereas lagoon costs increase with flowrate. The lagoon can also be used to store material bypassed around the neutralization process in the case of failure—a very important consideration if off-specification effluent can cause a plant shutdown.

One thing a lagoon cannot do is replace a mixing vessel as part of a control system. Any attempt to control the pH of a lagoon by closed-loop feedback can only result in an effluent pH that oscillates between the

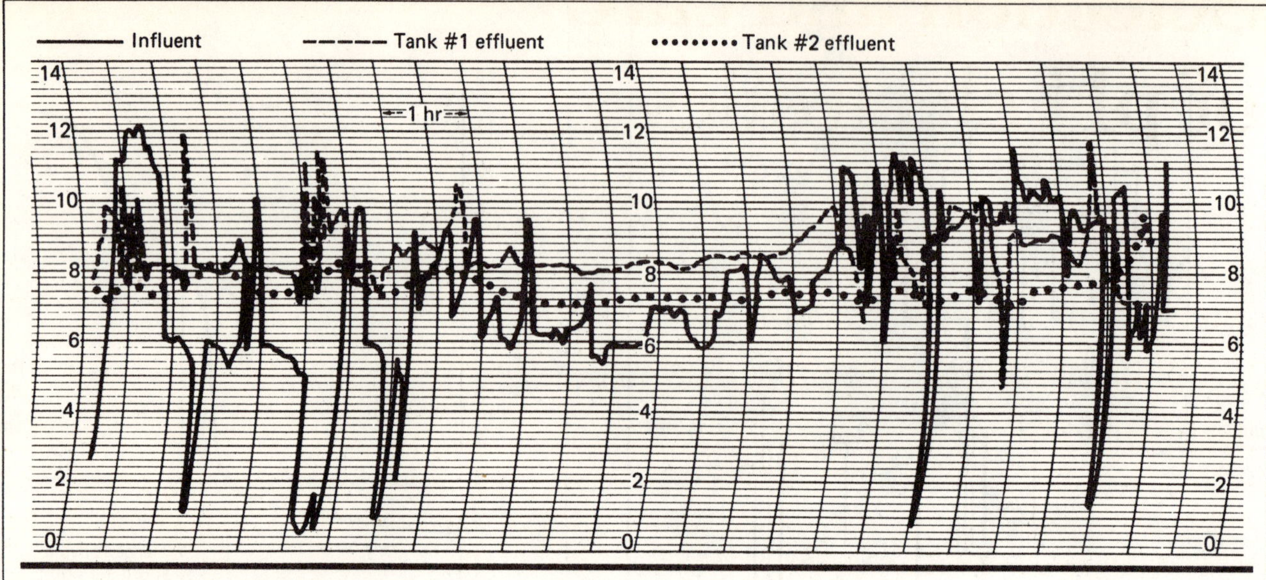

A 12-hr performance of a wide-range feedforward control system with broad limits **Fig. 9**

influent pH and a complementary pH value on the opposite side of neutrality. The period of such oscillations will depend on the deadtime of the lagoon, but typically will be on the order of hours.

Typical systems

Fig. 6 shows a simplified wide-range feedforward control system, and Fig. 7 the performance of this system. Note that the effluent Tank 1 has a time period of oscillation of about 0.5 min for changes between pH 3 and 8.

If we apply the formula for dynamic gain to analyze the performance of Tank 2, we can assume that the ratio of its deadtime to reaction time is 0.05, and we note that its retention time is 1,000/175 = 5.7 min (Fig. 6). Therefore:

$$G_d = \tau_0/2\pi\tau_1 = 0.5/2\pi(5.7 - 0.05) = 0.5/11.4\pi = 0.014$$

From this, we would expect that the effluent of Tank 2 would show a percentage change that was 0.014 that of the effluent from Tank 1. From Fig. 1, the pH 3–8 variation is equivalent to 1,001 units of reagent, so that the expected change from Tank 2 would be 14 units of reagent, equivalent to a pH ranging between 5 and 8.

Actually, this pH cycles between 0.2 and 0.3 around a set point of 7, as shown in Fig. 7. This corresponds to a reagent-unit variation less than 1.0. Fig. 8 shows why. Equal reagent cycles caused by closed-loop control in Tank 1 around the setpoint of pH 7 are transformed by the titration curve into pH variations between 3 and 8. Tank 2 averages out the reagent cycles to produce a final pH cycling about pH 7.

Fig. 9 shows the 12-hour performance of another wide-range feedforward pH control system. The process arrangement was similar to that shown in Fig. 4, except that the attenuation vessel was of the same volume as the neutralization vessel. The effluent limits required by this application were very broad: 5 to 10 pH.

Note that the fast cycles are completely attenuated, while the slower upsets can be detected in the effluent pH record. Increasing the size of the attenuation vessel would have reduced the effluent upsets even more, but was hardly necessary considering the broad effluent pH limits.

Conclusion

The problems of pH control are not limited to waste treatment. This application has been used in this article since it is very current, and substantial experience has been gained in this field. Government requirements have assisted in focusing attention on the effluent control problem. The economies derived from proper handling of pH control can be substantial. This requires a good combination of both design and control engineering to achieve a suitable system, which will lead to better conformance with government regulations at minimum expense.

References

1. Shinskey, F. G., "Process Control Systems," McGraw-Hill Book Co., New York, 1967, pp. 82–86.
2. Shinskey, F. G., "pH and pIon Control in Process & Waste Streams," John Wiley & Sons, Inc., New York, 1973, Chap. 9.

The author

D. L. Hoyle is Senior Systems Design Engineer, The Foxboro Co., Foxboro, MA 02035, where he is presently responsible for design of advanced control systems from concept to startup. Earlier experience at Foxboro included two years of panel design and five years of chromatograph application and development. He joined Foxboro in 1959, following experience with Union Carbide Corp.'s Plastics Div. as a Product Development Engineer. An author of several articles on systems application for continuous processes, he holds a B.S. in chemical engineering from Worcester Polytechnic Institute.

Selection and care of pH electrodes

The choice of electrodes for an effluent-stream pH analyzer or control system depends on chemical and physical interactions between stream and electrodes. Proper maintenance and calibration ensure good electrode service.

Samuel C. Creason, Beckman Instruments, Inc.

☐ Recently enacted federal, state and municipal regulations require pH monitoring and control of effluent streams. The heart of a pH analyzer or control system is a sensor that provides a signal related to the pH of the waste stream. The sensor is basically a pair of electrodes that are in constant contact with the stream, which may abrade, coat or chemically attack them. Proper selection and care of electrodes and auxiliary equipment will provide a system for nearly any industrial waste stream. A typical pH control system is shown in Fig. 1.

The effluent stream and control system

There is no typical effluent stream. pH values vary from stream to stream, as do contents. It is essential to know the contents of your stream.

Streams containing grease or oil foul pH electrodes. The addition of neutralization reagents such as lime in acidic streams or sulfuric acid in basic streams will almost certainly produce suspended solids. These coat the electrodes or, if the stream velocity is high enough, abrade them. If acidic fluorides are present, they will chemically attack the electrodes.

Controlling effluent-stream pH may be relatively simple or very difficult. For pH values below 4 or above 11, more than one stage of control is usually needed to bring the pH within an acceptable range (typically 6 to 9). Frequently, two or three individual control subsystems are used in series, each accomplishing a part of the total pH adjustment. In these, only the electrodes in the first subsystem are exposed to extremes in pH. This type of system is discussed later in connection with streams containing acidic fluorides.

The electrochemical cell

A pH sensor consists of a pair of electrodes and a thermocompensator, which is usually a temperature-sensitive resistive element (see Fig. 2). When the sensor is submerged in the effluent stream, an electrochemical cell is formed. The potential of the cell is directly proportional to the pH and temperature of the stream. Since the temperature dependence is well known and consistent for similar electrode pairs, the effect of temperature change on cell potential can be nullified by the thermocompensator, which accurately adjusts the gain of the pH analyzer. The effect of temperature change on the true pH cannot be compensated for, because this depends on the composition of the stream. Whether the effect is significant in a particular case is best determined by experiment.

Although two electrodes make up the cell, only one is sensitive to pH. The other, the reference electrode, merely completes the electrical circuit through the cell.

The pH-sensitive electrode

The most commonly used pH-sensitive electrode, and the only one recognized by the U.S. Environmental Protection Agency[*1*], is the glass electrode. Its principal advantage is interference-free response over a very wide pH range.

The glass electrode is a thin-walled bulb of glass containing a pH-buffered solution and the elements of a half cell (typically silver/silver chloride). When the electrode is placed in the effluent stream, the bulb acts as a thin glass membrane separating two solutions, and a potential is developed across it. This potential is directly proportional to the difference in pH values between the two solutions.

Most commercially available glass electrodes can perform over a pH range of 0 to 14 with about 0.1 pH error. The error is confined largely to values above pH 11, at which some pH-sensitive glasses are affected by sodium ions. At a pH of 14, this "sodium-error" can be limited to 0.2 to 0.3 by proper choice of electrode. At a pH of 11, the sodium error is 0.02 or less.

All glass electrodes are vulnerable to abrasive agents. In difficult applications, for example copper-ore-processing effluent streams, they will last only a few months. As mentioned, they are totally unsuited for making measurements in acidic fluoride streams, because glass is rapidly etched by hydrofluoric acid. In a fluoride-containing stream with a pH of 4, it may survive no more than a week. But if the electrode is used in

Originally published October 23, 1978.

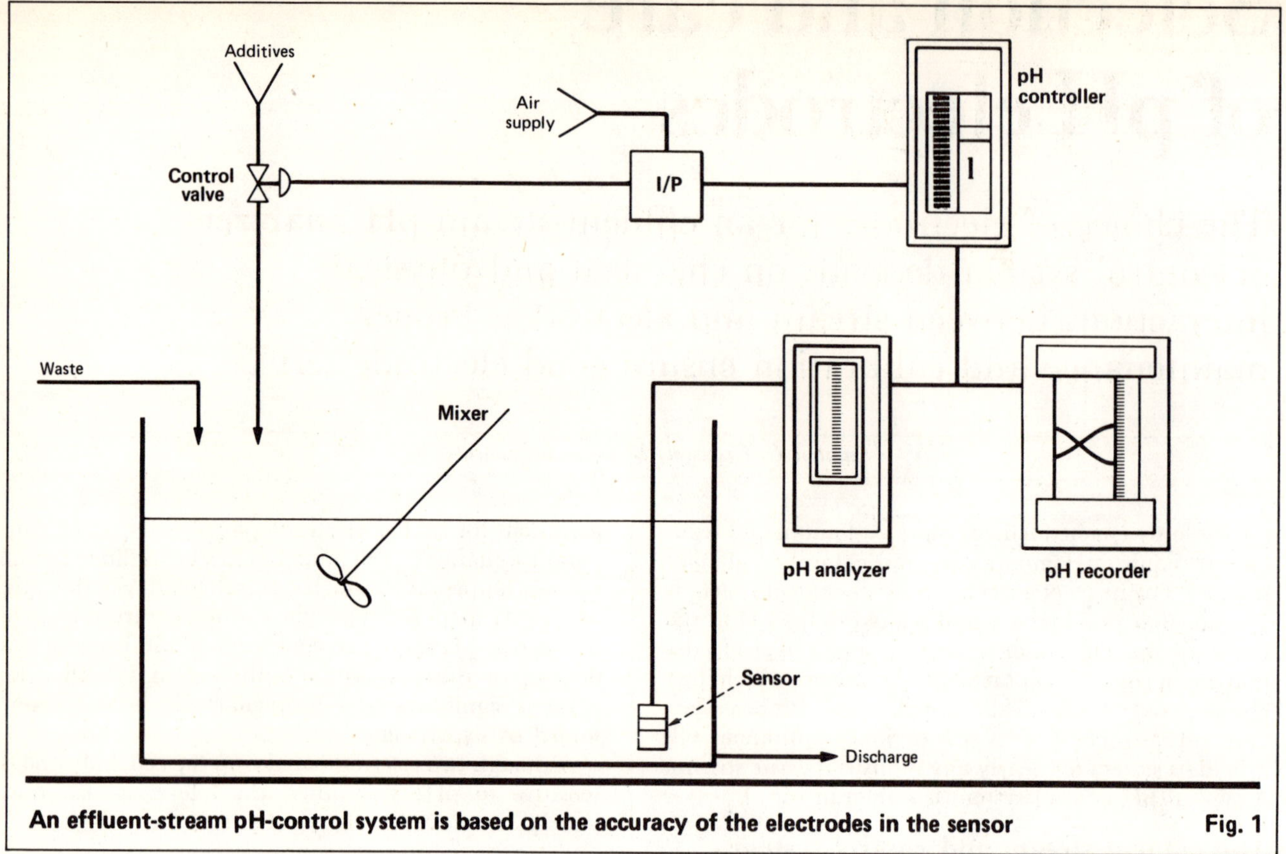

An effluent-stream pH-control system is based on the accuracy of the electrodes in the sensor **Fig. 1**

a pH-control system, the presence of fluorides may be of less concern. In a one-stage control system that properly controls pH above 7, the glass electrode will probably not be exposed to acidic fluorides for extended periods of time. Of course, feedforward control based on glass-electrode pH measurements in an acidic fluoride stream should not be attempted.

Glass electrodes are also susceptible to attack by hot caustic streams (pH greater than 12), but generally these are not effluent streams. Pre-discharge pH control of such a stream would require precautions similar to those for glass electrodes in acidic fluoride streams.

Alternatives to the glass pH electrode are generally the metal/metal-oxide type, which uses the pH dependence of the couple. The metal is more durable than the glass, but is sensitive not only to the hydrogen ions that determine pH, but also to other ions.

A commercially available electrode of this type is the antimony electrode. Because its principal advantage is ruggedness, it has been used successfully in ore slurry applications. The antimony electrode is sensitive to oxidizing and reducing agents such as found in plating waste, and cannot be used at extreme pH values.

The reference electrode

The reference electrode is designed to be as insensitive as possible to all ions. The most commonly used reference electrode is the non-flowing-junction type. It consists of an inert envelope containing a filling solution and the elements of a silver/silver chloride or mercury/mercurous chloride (calomel) half-cell. Electrical contact between the solution and stream is provided by a porous junction area, which is usually a ceramic disk, wooden plug or conductive polymeric matrix that is part of the envelope. Porosity is controlled during manufacture to prevent significant loss of filling solution and gross contamination of the electrode.

Commercially available non-flowing electrodes can withstand temperatures in excess of 100°C (silver/silver chloride internals, but not calomel) and pressures of 150 psi. They are rugged and impervious to chemical attack. Non-flowing-junction electrodes are suitable for perhaps 90% of all industrial effluent-monitoring and control applications.

In some applications a flowing-junction electrode may be useful. In this type, filling solution is continuously (but slowly) forced through a controlled leak in the inert envelope. Thus, contamination of the electrode by the stream is completely avoided. An external reservoir of filling solution replenishes that which is lost. The flowing-junction reference electrode requires frequent maintenance to keep the reservoir full of filling solution. It is somewhat difficult to use in applications involving pressurized streams.

The flowing-junction and non-flowing-junction reference electrodes have one major shortcoming; the junction will be blocked if the stream contains any material that reacts with the filling solution to form a precipitate. Particularly troublesome are silver, lead and mercury, which form insoluble chlorides, and sulfides, which form insoluble silver salts. In some cases, a blocked electrode can be rejuvenated by boiling the junction in concentrated acid.

The double-junction reference electrode is designed

to avoid blocking. It is essentially a complete electrode within an electrode-like outer body, both using non-flowing junctions. A "non-reactive" electrolyte (typically potassium nitrate) is the filling solution in the outer body and only this solution is in contact with the stream. Blocking of the outer junction is extremely unlikely, since neither potassium ions nor nitrate ions form insoluble compounds with the overwhelming majority of materials found in effluent streams. Blocking of the inner junction is not possible.

Maintenance

Maintaining a pH analyzer is basically electrode cleaning. The exact interval between cleanings is best determined by trial and error, based on the accuracy that is required and the loss of response that can be tolerated.

Methods of manual cleaning vary with the manufacturer. Automated devices that clean the electrodes without interrupting pH measurement are available. For instance, a jet of water may provide sufficient cleaning action for lightly coated electrodes. Typically, the jet is controlled by a solenoid valve actuated by a cam timer. The jet is turned on a few minutes each hour. During cleaning it may be necessary to disable the control system, since the water from the jet will probably not be at the same pH as the stream.

Some manufacturers provide mechanically operated brushes or wipers as an integral part of the electrode station. Such devices are useful in many applications, but in streams containing a lot of abrasive material they may scratch a glass electrode.

Ultrasonic cleaners attached directly to the electrode station can lengthen the interval between manual cleanings by a factor of ten or so, depending on the stream. The benefit is most pronounced when the electrode is coated by hard, crystalline material. In the treatment of pond water containing tailings in a copper mining operation, an ultrasonic cleaner reduced the need for manual cleaning from once every four days to once a month. The electrode was being coated by calcium sulfate. In the waste stream from a steel mill, an ultrasonic cleaner reduced manual cleaning from once a day to once every four days. Here the coating contained much palm oil.

There are no comparable methods for dealing with abrasion of electrodes. The only useful approach is reducing the stream velocity near the electrodes. An open, submersible electrode station can be protected by a cylindrical guard cut to extend below the edge of the station. A relatively stagnant zone forms in the stream at the tips of the electrodes. If the electrodes are part of a pH control system, this technique should be used carefully, since a significant amount of dead time may be introduced into the control loop.

Calibration

There are two principal methods for calibrating a pH analyzer. The first is particularly useful if disassembling the electrode station is inconvenient. A grab-sample of the stream is taken from as near the electrodes as possible. The reading of the pH analyzer is noted at the time the sample is taken. The pH of the sample is then determined by means of an accurately-calibrated second analyzer. The appropriate correction (if any) is made to the first analyzer by adjusting its "standardize" control.

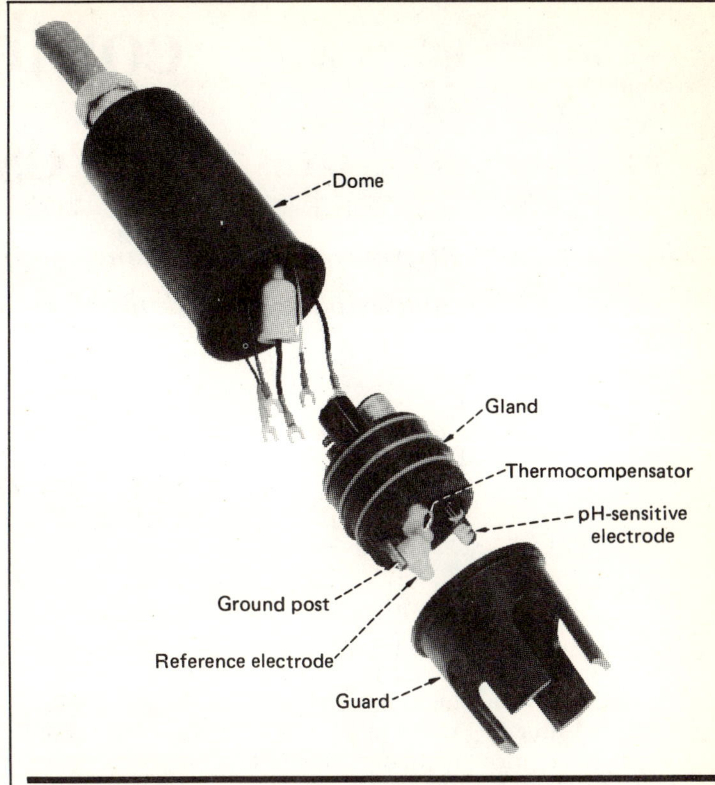

The internals of a pH sensor — not all sensors use a ground post **Fig. 2**

Of course, calibrating a pH analyzer by means of a grab-sample provides only a one-point calibration. As long as the electrodes are in good condition, this will suffice. However, if the condition of the electrodes is at all suspect, the system should be calibrated by means of two or more pH buffer solutions. The interval between calibration by buffers depends on the application. A good rule of thumb is to perform this type of calibration each time the electrodes are manually cleaned.

Reference

1. "Water Programs—Guidelines Establishing Test Procedures for the Analysis of Pollutants," Federal Register, 41 (232), Dec. 1, 1976, p. 52779.

The author

Samuel C. Creason is a principal engineer at Beckman Instruments, Inc., Process Instruments Division, Fullerton, Calif. He is responsible for applications of electrochemical instrumentation manufactured by the division.

Dr. Creason received a Ph.D. in analytical chemistry from Northwestern University, Evanston, Ill., and an M.S. in chemistry from the California State University, Sacramento.

He is a member of ACS, ISA, Amer. Soc. for Testing and Materials, and the Technical Assn. of the Pulp and Paper Industry.

Precise combustion control saves fuel and power

Instrumentation assuring a minimum amount of excess air for combustion of fuels improves the performance and increases the thermal efficiency of boilers, furnaces and kilns.

Lyman F. Gilbert, Environmental Data Corp.

☐ When the cost of fuel was $0.15/10^6 Btu, it did not make sense to install laboratory-type equipment on a fuel-burning system because it would have taken more than 10 yr to recover the investment.

During the fuel embargo of 1974, however, some low-sulfur fuel oils sold for $4.13/10^6 Btu. (The more typical price today for low-sulfur fuel oil is about $2.10/10^6 Btu.) Suddenly, fuel-burning efficiency became important. And as fuels become scarcer and costlier, good fuel-burning efficiency can have a large impact on plant profitability.

The fuel user can now justify scientific instruments to save fuel. For example, recovery of the capital investment for combustion instruments and associated control equipment on utility boilers having steam capacities of over 10^6 lb/h will occur in one year from fuel savings. In plants having steam capacities of less than 10^6 lb/h, recovery of the investment will take about three years. In plants using large single-burner fuel systems such as package boilers and kilns, recovery of the capital investment often takes less than 90 days.

Controlling the combustion process

To achieve good combustion control and the best fuel-burning efficiency requires that two fundamental conditions be established:

- Exact balancing of the air/fuel ratio from burner to burner.
- Precise control of the chemical composition of the flue gases leaving the boiler or process.

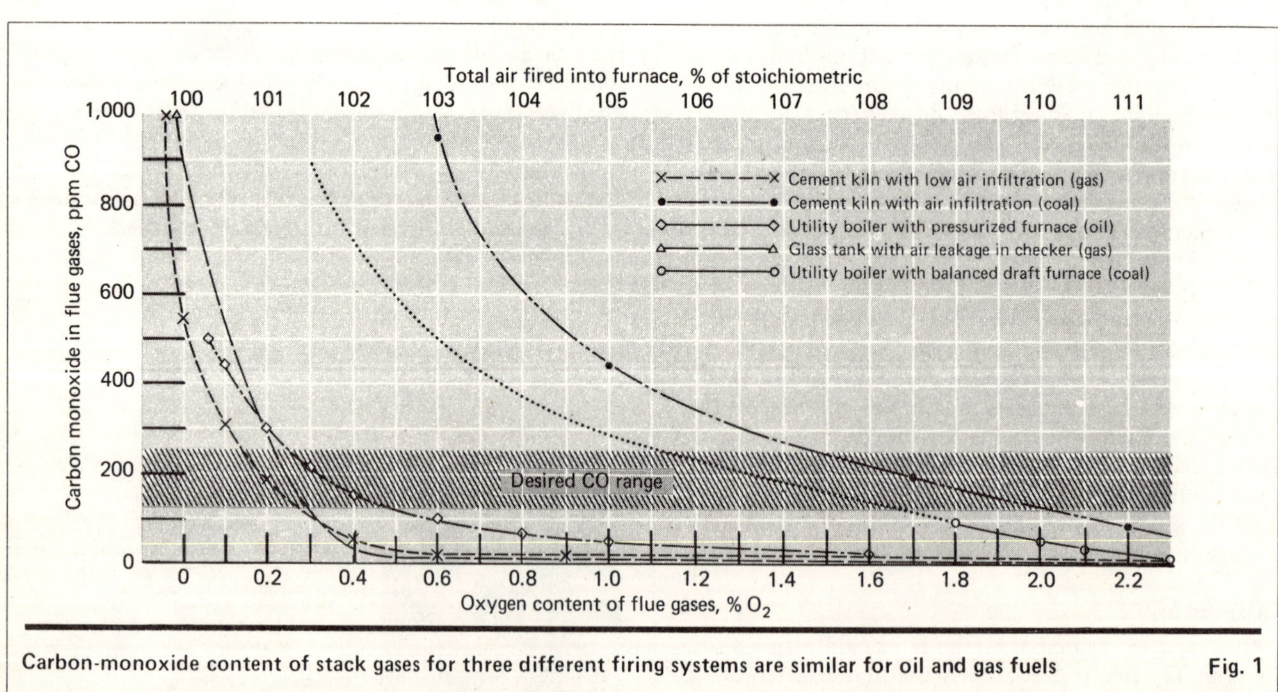

Carbon-monoxide content of stack gases for three different firing systems are similar for oil and gas fuels **Fig. 1**

Originally published June 21, 1976.

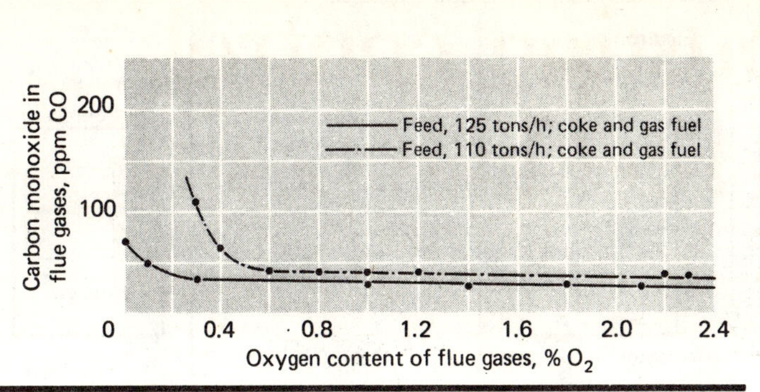

Firing system burns mixtures of gas and coke Fig. 2

These two conditions are completely dependent upon each other. Let us explore them in greater detail.

Burner condition affects combustion

If the burners have unequal air or fuel distribution, smoking will occur on those having insufficient air to burn the fuel—even though the overall furnace has large quantities of excess air.

A tangential furnace provides some degree of turbulent mixing from burner corner to burner corner. However, the final composition of the flue gases is dependent upon a good distribution of air and fuel at each burner nozzle.

Wall-fired boilers with burners at front and rear, or downfired turbofurnaces, have little mixing from burner to burner. Any burner with a slight maldistribution of air and fuel will cause smoking and a disproportionately high level of CO emission despite high excess air in the furnace.

In a boiler with multiple fuel burners, the fuel nozzles may have a random-size distribution of ±5%. This variance in nozzle size will require a minimum excess air of 7 to 10% in order to operate the burners without smoking, assuming there is a uniform distribution of air to each burner. The delivery of air to the burners can also have a random distribution of ±5%. Therefore, the designed excess air is increased to 15% to 20% above stoichiometric in order to provide a cushion between the tolerances for burner-tip sizing and air-flow distribution.

From the foregoing discussion, it is apparent that such effects as dirty burner tips, tip wear from abrasives in the oil, and air-register deterioration all contribute to reduce firing efficiency.

The following procedures provide the operator with a method for determining the relative balance of each burner's air/fuel ratio and its condition:

1. Decrease the furnace air supply until the CO increases to about 300 ppm.
2. Turn off a burner and close its air register. Note the change in CO. Restart the burner.
3. Turn off another burner and close its air register. Note the change in CO. Restart the burner.
4. Repeat Step 3 for all remaining burners, one burner at a time.

Those burners causing the largest change in CO are dirty, have oversized tips, have less air flow, or have poor air/fuel mixing.

Those burners causing the smallest change in CO are partially plugged (dirty), have undersized tips, or a larger than average air flow, broken register linkage, etc.

These procedures can alert the operator to burners that need attention. In systems with remote burner controls, the operator can determine the condition of the burners for an entire boiler in 20 min, or less.

With each burner providing approximately the same level of CO, it is now possible to maintain overall low-excess-air firing conditions.

Flue-gas control

The second condition for good fuel burning is precise control of the chemical composition of the flue gases leaving the boiler or process furnace. This requirement for efficient combustion cannot be effected until balanced burner-firing conditions are fully achieved.

Flue gases from a boiler or furnace contain the end-products of the chemical combustion process, plus the infiltration and dilution gases. Complete combustion of a fuel at stoichiometric conditions should react all of the hydrogen to water, all of the carbon to carbon dioxide, etc. There should be no free oxygen, free hydrogen or carbon.

In a practical burner-furnace system, these reactions are never perfect, i.e., they never go to completion. There are always some traces of incomplete combustion such as C, O_2 and CO. Of these trace products, CO is the most useful in determining the efficiency of the fuel-burning reaction. Let us examine some of the reasons why this is so.

Carbon monoxide in flue gases can be measured in parts-per-million (ppm) concentrations even in the presence of air infiltration or dilution. Oxygen in flue gases is a combination of excess air through the burners plus the infiltration of air through idle burners, observation ports, expansion joints, casing leaks, flue-gas recirculation ducts, soot-blower openings and the like.

Carbon monoxide is only formed in the flame envelope of the burners. Therefore, the traces of CO provide a direct measurement of the fuel-burning-system's performance. Fig. 1 illustrates CO vs. excess air (O_2) for a number of furnace systems. Note the similarity of the oil and gas curves for three different firing systems. Fig. 2 shows a firing system burning a mixture of gas and coke. Note the similarity to the firing systems of Fig. 1.

Most plants have used "O_2 in the flue gas" as a measure of the excess air in the fuel-burning zone. A 1% infiltration of air into the flue-gas stream near the point of O_2 measurement will cause the O_2 analyzer to read 20% high. A 5% infiltration of air near the O_2 analyzer can cause it to read 95% high. The total air infiltration may be only a fraction of a percent of the total flue-gas flow, but if the point of infiltration is very near the O_2 analyzer's sample probe, gross errors in the readout can occur even though the analyzer is in perfect calibration. The basic problem is that measured O_2 can be from excess air in the furnace or from infiltration. The O_2 analyzer cannot differentiate one source of oxygen from another.

Characteristics of analyzers for controlling excess air during combustion **Table I**

	Analyzer	
	CO	O_2
Affected by infiltration of air	No	Yes
Dilution effects	Linear	Exponential
Measurement	Averaged across duct	Single point in duct*
Sampling system required	None	Depends on instrument
Method	Optical spectrometry	Several†
Response time (63% change), s	1	45
Idenfifies "dirty" burners	Yes	No
Indicates flame stoichiometry	Yes	No

*If measured *in situ.*
†Paramagnetic, electrochemical

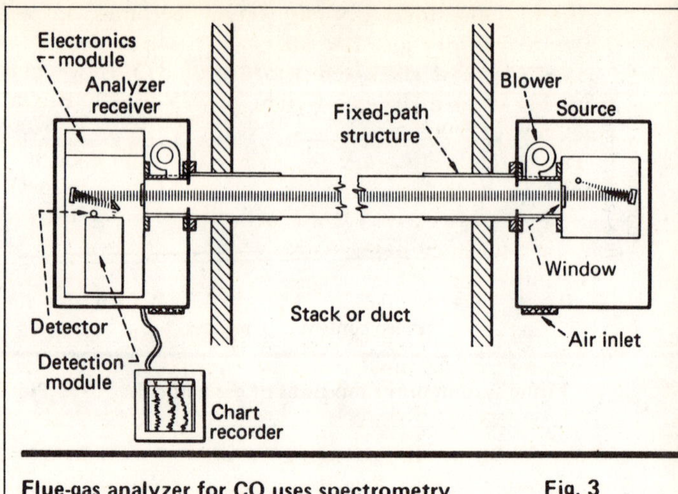

Flue-gas analyzer for CO uses spectrometry **Fig. 3**

Carbon monoxide in a furnace only comes from the flame envelope; it does not leak in from the atmosphere. This gas is linearly diluted by infiltration. Because the CO level is held so low (i.e., 150 ppm), a 5% dilution of the CO by air infiltration causes only a 7.5-ppm reduction in the CO reading. From Fig. 1, we can see that the curve most affected would be the coal-fired cement kiln. A 7.5-ppm error at 150 ppm CO would reduce the available excess air by less than 0.2%.

On a typical boiler, the CO should be controlled at 100 to 200 ppm (0.01 to 0.02% CO). These very low levels of CO are generated when the air/fuel ratio is very close to stoichiometric. At these levels, the excess air will usually be under 5% for gas or oil firing, and under 10% for coal firing. The most favorable CO level is computed for each installation, and is a tradeoff between the heat lost from excess air that is raised to the flue-gas exit temperature and the heat lost from unburned fuel.

Furthermore, carbon monoxide analysis can warn of high combustibles in the furnace, which are caused by malfunctioning burners. This is particularly important at lightoff, when the furnace is cold and the burners are not stable. When the CO exceeds 1,000 ppm, a warning should sound, and at 1,500 ppm a nonsilencing alarm should be set off. These CO levels are so low that they do not present an ignitable concentration (i.e., they are not an explosion hazard), but they can warn of burners that are on the verge of going out. The CO analyzer is not a substitute for flame scanners, but rather complements them. It provides an early warning of impending problems.

A summary of the important advantages of carbon-monoxide measurement over oxygen measurement, as a firing guide, is given in Table I. A schematic arrangement for an *in situ* flue-gas analyzer, based on optical spectrometry, is shown in Fig. 3. This system analyzes the flue gas across the entire stack or duct without contacting the gas and requires no sampling system.

Examples illustrate savings

Example 1—A power-topping boiler in a pulp and paper mill generates 800,000 lb/h steam at 600 psi and 900°F. The boiler includes an economizer and a small air heater. Feedwater temperature is 275°F, and flue-gas exit temperature is 375°F. Boiler efficiency = 84.5%, load factor = 75%, oxygen in the flue gas = 4.5%. The unit is fired with #6 fuel oil costing $2.10/10^6$ Btu. Let us evaluate the savings that will occur if we reduce the excess air from 27.5% (4.5% O_2) to 3% (0.6% O_2).

We begin by calculating the fuel-burning efficiency of the unit under the original conditions (i.e., 4.5% excess O_2). For convenience, we can use the following data (as required) to approximate fuel weights:

Fuel	Weight
Natural gas	45 lb/10^6 Btu
Fuel oil #6	55 lb/10^6 Btu
Coal (10,000 Btu/lb)	100 lb/10^6 Btu

Step 1—To determine the amount of combustion air, we use the data of Fig. 4. Hence, for an oil fuel fired so as to give 4.5% excess oxygen, we obtain from Fig. 4: 960 lb air/10^6 Btu.

Step 2—Next, we find the approximate dry-gas weight by adding the weight of fuel to the weight of air, and find this to be 55 + 960 = 1,015 lb dry gas/10^6 Btu. (Moisture can be neglected because its effect cancels when calculating efficiency improvements.)

Step 3—Fig. 5 enables us to determine the approximate dry-gas loss at a stack temperature of 375°F. Hence, for a dry-gas weight of 1,015 lb/10^6 Btu, the stack loss becomes 7.25%.

Now, we repeat the computations of Steps 1 to 3 in relation to the following: We use an estimated value of 0.6% O_2 (3% excess air) for gas and oil, and 2% O_2 (10% excess air) for coal. It is known that the stack temperatures will be lowered as the excess air is reduced. The amount that the stack temperature is lowered is highly dependent upon the unit's design. For our purposes, it will be conservative to assume a constant stack temperature as the excess air is reduced. In actual practice, an additional fuel savings will occur.

For this example, we use 0.6% excess O_2 and repeat the computations as follows:

Step 1—From Fig. 4, we find for an oil fuel that 770 lb air/10^6 Btu are needed.

Step 2—The total dry-gas weight becomes 55 + 770 = 825 lb dry gas/10^6 Btu.

Step 3—For a stack temperature of 375°F, we get from Fig. 5 at a dry-gas weight of 825 lb/10^6 Btu a stack loss of 5.90%.

The gain in efficiency represents the difference between the stack loss when 4.5% excess O_2 was present and when 0.6% excess O_2 was achieved. For our example, the efficiency gain is 1.35%, i.e.:

$$7.25\% - 5.90\% = 1.35\%$$

In order to find the savings resulting from a decrease in excess air, we must calculate the hourly fuel cost:

$$(2.10/10^6)(800 \times 10^3)(1{,}462.5 - 244)(1/0.845) = \$2{,}422.58/\text{h}$$

where \$2.10/10^6 Btu = fuel cost; (800 × 10^3) lb/h = steam rate; 1,462.5 Btu/lb = heat content of steam; 244 Btu/lb = heat content of feedwater; and 0.845 is boiler efficiency.

Annual operating hours at load are:

$$0.75(24)(365 - 15) = 6{,}120 \text{ h/yr}$$

where 0.75 is load factor, and 15 days outage/yr is assumed.

The fuel savings due to improved efficiency of firing are calculated as:

$$2{,}422.58(6{,}120)(1.35/100) = \$200{,}153.55/\text{yr}$$

where \$2,422.58/h is fuel cost, 6,120 h/yr is operating time, and 1.35% is gain in efficiency.

A further savings is also realized due to the smaller quantity of excess air that is handled by the forced-draft and induced-draft fans. For estimating purposes, we can assume that fan power represents approximately 1% of the maximum continuous rating (MCR) of the boiler. In terms of heat savings, the value for this example becomes:

$$\left(\frac{1}{100}\right)(8 \times 10^5)(1{,}462.5 - 244)\left(\frac{1}{0.845}\right) \times \left[0.75\left(\frac{128 - 103}{103}\right)\right] = 2.10 \times 10^6 \text{ Btu/h}$$

where 8 × 10^5 lb/h = steam rate; 1,462.5 Btu/lb = heat content of steam; 244 Btu/lb = heat content of feedwater; 0.845 is boiler efficiency; 0.75 is fan characteristic, 128% is the total air initially, and 103% is the total air finally.

For a fuel oil costing \$2.10/10^6 Btu, this reduction in fan power on an annual basis amounts to:

$$(2.10/10^6)(2.1 \times 10^6)(6{,}120) = \$26{,}989$$

By reducing the excess air from 27.5% (4.5% O_2) to 3% (0.6% O_2), the dry-gas losses were reduced from 7.25% to 5.90%. Total air flow was reduced by 19.2%. This required 1.35% less fuel to heat the excess air, resulting in a fuel savings of \$200,154/yr.

Fan power was reduced by 19.2%. If the fans are driven by a steam turbine or from inhouse electric power, this provides an additional savings of \$26,989/yr. Hence, the total savings are \$227,143/yr.

To achieve these savings, the plant will have to make a capital investment of approximately \$80,000 plus an annual maintenance expense of about \$10,000 to \$15,000. Therefore, the net savings become:

$$\$227{,}143 - \$15{,}000 = \$212{,}143/\text{yr}$$

Payback time for the initial year of operation is:

$$\frac{\$80{,}000 + \$15{,}000}{\$212{,}143} = 0.45 \text{ yr, or } 5\tfrac{1}{3} \text{ mo}$$

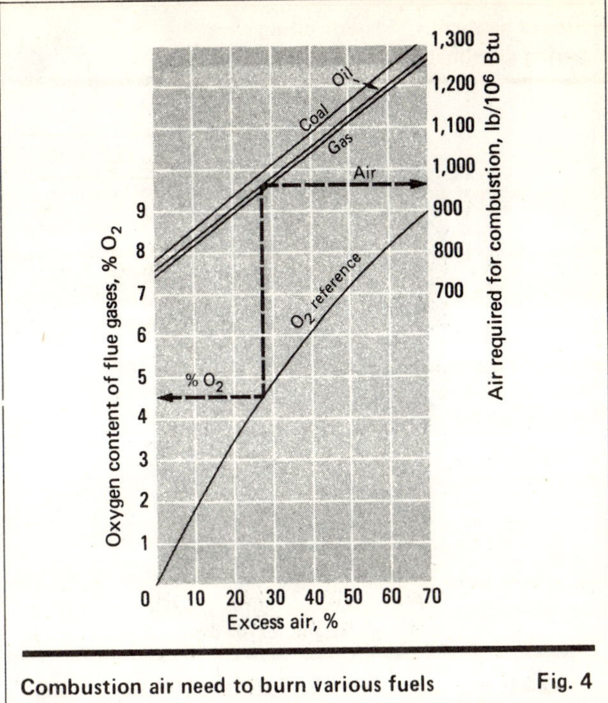

Combustion air need to burn various fuels Fig. 4

Example 2—A rotary cement kiln, handling 100 tons/h of clinker, is to be oil or coal fired. Its heat rate will be 5.66 × 10^6 Btu/ton of clinker, and its load factor is 92%. When oil-fired, the flue gas contains 2.5% O_2; when coal-fired, 3.5% O_2. Flue-gas exit temperature is 1,120°F; outage is 20 d/yr. Fuel cost for #6 oil will be \$2.10/10^6 Btu; for 10,000 Btu/lb coal, \$1.25/10^6 Btu. What savings can be achieved if the oxygen content of the flue gas is reduced to 0.2% for the oil fuel, and to 0.4% for the coal?

We first obtain the information for Steps 1 to 3 in accordance with the procedures in Example 1 for the initial conditions of this problem. The results are:

Step	Oil	Coal
1—Combustion air, lb/10^6 Btu	860	930
2—Dry-gas weight, lb/10^6 Btu	915	1,030
3—Stack loss, %	23.06	25.96

For Step 3, we evaluate the stack loss from the following relation because Fig. 5 does not show exit-gas temperatures greater than 700°F:

$$0.24 \times 10^{-6}(915)(1{,}120 - 70)100 = 23.06\% \text{ for oil}$$

$$0.24 \times 10^{-6}(1{,}030)(1{,}120 - 70)100 = 25.96\% \text{ for coal}$$

where the product of 0.24 × 10^{-6} (specific heat of flue gases, Btu/(lb)(°F) per Btu of fuel fired), the weight of dry gas, and the difference between the exit-gas temperature and ambient temperature, yields the stack loss.

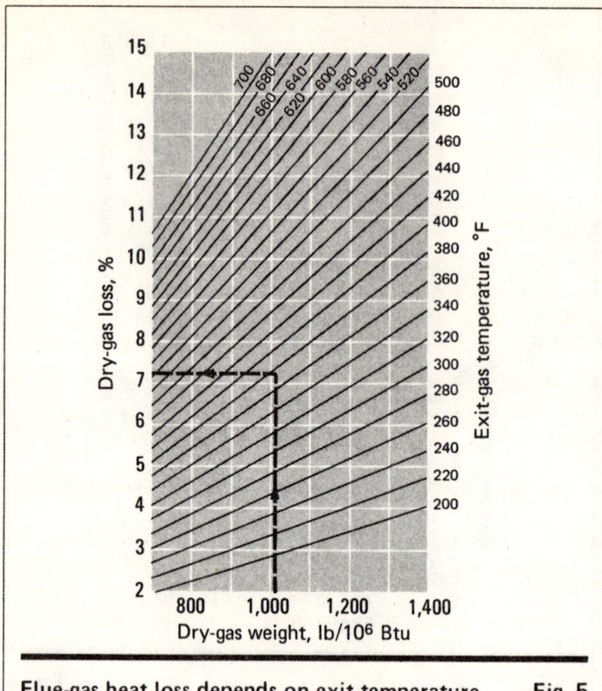

Flue-gas heat loss depends on exit temperature **Fig. 5**

In order to determine our savings, we must now compute the stack losses when the flue gas contains 0.2% O_2 for oil firing and 0.4% O_2 for coal firing. The computations for these conditions yield:

Step	Oil	Coal
1—Combustion air, lb/10^6 Btu	765	795
2—Dry-gas weight, lb/10^6 Btu	820	895
3—Stack loss, %	20.66	22.55

Gain in efficiency represents the difference between stack losses when the initial amount of O_2 in the flue gas is subtracted from the reduced amount of O_2. Hence, the efficiency gain for:

$$\text{Oil: } 23.06\% - 20.66\% = 2.40\%$$

$$\text{Coal: } 25.96\% - 22.55\% = 3.41\%$$

We calculate the hourly fuel cost in order to determine the actual savings when firing with oil or coal as:

$$\text{Oil: } (2.10/10^6)(5.66 \times 10^6)(100) = \$1{,}188.60/\text{h}$$

where $2.10/$10^6$ Btu = fuel cost; (5.66×10^6) Btu/ton = heat rate, and 100 tons/h is clinker capacity.

$$\text{Coal: } (1.25/10^6)(5.66 \times 10^6)(100) = \$707.50$$

where $1.25/$10^6$ Btu = coal cost.

Annual operating hours at load are:

$$(92/100)(24)(365 - 20) = 7{,}617.6 \text{ h/yr}$$

where 92% is load factor, and 20 represents d/yr outage.

The annual fuel savings due to improved efficiency of firing can now be computed for:

$$\text{Oil: } (1{,}188.60)(7{,}617.6)(2.40/100) = \$217{,}302.70/\text{yr}$$

$$\text{Coal: } (707.50)(7{,}617.6)(3.41/100) = \$183{,}780.31/\text{yr}$$

Since the fans will now handle a smaller load of air due to improved firing conditions in the kiln, we realize an additional savings. For this example, we will assume that fan power is 800 hp, and power cost to drive the fans is $228/(hp)(hr). We will also need the following data for the computations:

	Oil-fired		Coal-fired	
	Original	Final	Original	Final
O_2 in flue gas, %	2.5	0.2	3.5	0.4
Excess air, %	12.5	1.0	17.5	2.0
Total air, %	112.5	101.0	117.5	102.0

For these variable-speed fans, the savings (i.e., reduction) in fan power becomes:

$$\text{Oil: } \left(1 - \frac{101\%}{112.5\%}\right)100 = 10.22\%$$

$$\text{Coal: } \left(1 - \frac{102\%}{117.5\%}\right)100 = 13.19\%$$

And the actual cost savings in fan power for each fuel become:

$$\text{Oil: } (228)(800)(10.22/100) = \$18{,}641/\text{yr}$$

$$\text{Coal: } (228)(800)(13.19/100) = \$24{,}058/\text{yr}$$

The direct savings made possible by reducing the amount of excess air fired with each fuel, and the additional savings resulting from the decreased load on the fans, are summarized as:

	Oil	Coal
Fuel reduction, %	2.40	3.41
Fan-power reduction, %	10.22	13.19
Fuel savings, $/yr	217,302.70	183,780.31
Fan-power savings, $/yr	18,641.00	24,058.00
Total savings, $/yr	235,943.70	207,838.31

To achieve the savings in this example, a typical oil-fired kiln will require a capital investment of $45,000 to $50,000, plus an annual maintenance expense of $7,000 to $12,000. The net savings becomes:

$$\$235{,}944 - \$12{,}000 = \$223{,}944/\text{yr}$$

Payback time for the initial year of operation is:

$$\frac{\$50{,}000 + \$12{,}000}{\$223{,}944} = 0.28 \text{ yr, or } 3.3 \text{ mo}$$

Improved performance of rotary kiln

Other benefits arising from a reduction in the required amount of excess air for combustion are:

- Reduced dust loading in the flue gases. Dust loading is velocity dependent. It is estimated that a 15% reduction in flue-gas flow causes a 21% reduction in dust flow (depending on particle size) into the baghouse or precipitator. This alone has paid for the instrumentation and maintenance in one year.
- Longer burning zone. A reduction in the excess air for combustion creates a longer flame path. This greatly stabilizes kiln temperatures. The kiln is less prone to get into upsets, thus permitting higher clinker throughput (i.e., higher load factors).
- Better clinker quality. The reduced excess air provides more-uniform clinker production because of greater flame stability.
- Longer refractory life. Both the reduced excess air and longer flame path spread the heat more uniformly,

and thereby increase refractory life. Fewer hot spots develop because of the lower flame temperatures.

■ Reduced NO_x pollution. Less excess air and lower flame temperatures reduce nitrogen oxides.

Example 3—A gas-fired glass-melting furnace has a capacity of 11 tons/h of melt, operates at a heat rate of 4.7×10^6 Btu/ton of melt, and has a load factor of 95%. Fuel input is 48,000 ft³/h of 1,080 Btu/ft³ gas. The flue gas contains 5.5% O_2 initially at an exit temperature of 1,000°F. Outage corresponds to 4 d/yr. Fuel cost is $1.67/10^6 Btu. What savings can be achieved if the oxygen in the flue gas is maintained at 0.2%?

Following the corresponding procedures of Example 2, we calculate the following relations at the initial conditions for this example:

Step	Gas fuel
1—Combustion air, lb/10⁶ Btu	1,005
2—Dry-gas weight, lb/10⁶ Btu	1,050
3—Stack loss, %	23.18

We repeat these calculations for the reduced-oxygen content of 0.2% in the flue gas. The results are:

Step	Gas fuel
1—Combustion air, lb/10⁶ Btu	750
2—Dry-gas weight, lb/10⁶ Btu	795
3—Stack loss, %	17.55

We compute the hourly fuel cost as:

$$(1.67/10^6)(1{,}080)(48{,}000) = \$86.57/\text{h}$$

where ($1.67/10^6 Btu) = fuel cost; 1,080 Btu/ft³ = heating value of gas; and 48,000 ft³/h = fuel input.

Annual operating hours at load are:

$$(95/100)(24)(365 - 4) = 8{,}230.8 \text{ h}$$

where 95% is load factor, and 4 is d/yr outage.

We find the gain in efficiency from the difference in the stack loss at the original conditions and that at the reduced-oxygen conditions. The efficiency gain is:

$$23.18\% - 17.55\% = 5.63\%$$

The annual fuel savings due to improved efficiency in firing can now be computed as:

$$(86.57)(8{,}230.8)(5.63/100) = \$40{,}116/\text{yr}$$

In this example, we will assume that fan-power needs are 9.6 kWh/ton of clinker, power cost for the fans = $0.03/kWh. We will also use the following data for the computation:

	Original	Final
O_2 in flue gas, %	5.5	0.2
Excess air, %	35.5	1.0
Total air, %	135.5	101.0

Therefore, the reduction in fan power resulting from a decrease in the amount of excess air becomes:

$$\left\{1 - \left[\frac{101}{135.5}(0.75) + 0.25\right]\right\} 100 = 18.9\%$$

where the no-load power requirements of a constant-speed fan are assumed to be 0.25 of full-load power, thus leaving 0.75 of full-load power to actually move air at full load.

The actual savings in fan power for this glass furnace are:

$$0.03(18.9/100)(9.6)(11)(8{,}230.8) = \$4{,}928.21/\text{yr}$$

By reducing the excess air (use Fig. 4) from 35.5% (5.5% O_2) to 1% (0.2% O_2), the dry-gas losses were reduced from 23.18% to 17.55%. This required 5.63% less fuel, and 18.9% less fan power. Fuel savings were $40,116/yr, and fan-power savings were $4,928.21/yr, making a total of $45,044/yr.

To achieve these savings, a typical glass furnace will require a capital investment of $20,000 to $25,000 plus an annual maintenance expense of $3,000 to $5,000. The net savings are, therefore:

$$\$45{,}044 - \$5{,}000 = \$40{,}044/\text{yr}$$

Payback time for the initial year of operation is:

$$\frac{\$25{,}000 + \$5{,}000}{\$40{,}044} = 0.75 \text{ yr} \approx 9 \text{ mo}$$

Most of the benefits resulting from a reduction in the amount of excess air in Example 2 also apply to the glass furnace. These are:

■ Reduced dust carryover. The reduced air flow through the glass furnace will cause an estimated reduction of 49% in the dust flow (depending on particle size) to the baghouse.

■ Reduced plugging of checkers. With reduced flue-gas flow and dust loading, there is much less dust per ton of glass melted to adhere to the checkers. This greatly reduces fan power per ton of glass produced, over the life of the furnace and its associated checkers.

■ Improved combustion-air temperature. The smaller air flow reduces the rate of cooling of the checkers. This provides higher average combustion-air temperatures between reversals.

■ Longer checker life. The reduced air flow recovers more of the stored heat in the checkers for the combustion process. Also, there is less of a temperature change in the checkers between reversals, thereby reducing cracks in the checker walls from thermal shocks.

■ Longer furnace life. The carbon monoxide control of the air flow eliminates differences in air/fuel ratio that occur between firings from alternate sides. This provides more-uniform firing conditions and less thermal stress on the furnace at each firing reversal.

■ Reduced NO_x pollution. The reduced excess air and associated lower flame temperatures decrease the production of nitrogen oxides.

The author

Lyman F. Gilbert is Director of Control Operations for Environmental Data Corp., 608 Fig Ave., Monrovia, CA 91016. For a period of 27 years, he was engaged in research and development of fuel-burning systems and associated controls. Mr. Gilbert has 34 U.S. and foreign patents in controls and firing systems. He graduated from Washington University with a B.S. in education, and majored in mathematics and physics. He is a member of the Institute of Electrical and Electronic Engineers and the Instrument Soc. of America.

REGULATOR with narrow proportional control.

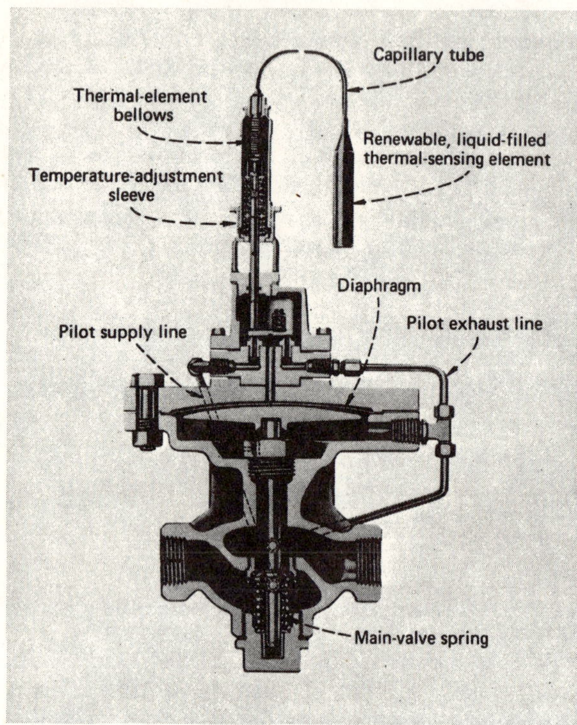

Selecting Temperature Controls For Heaters

Temperature controls should be selected according to the type of heater to be used—storage or instantaneous—as well as the requirements of steam flow, temperature range and promptness of action of the instrument.

WILLIAM L. SCULL, Leslie Co.

Successful application of storage or instantaneous heater temperature-controls depends on the right regulator or controller being specified for the job. Of the three common fluid-control applications—pressure, temperature and level—the greatest number of field problems are found in temperature control. Most of these can be traced directly to the use of the wrong regulator or controller.

Some temperature controls can be used on storage heaters but not on instantaneous ones. Others will work on both storage and instantaneous heaters. A few basics can set the ground rules for considering either new or retrofit temperature controls for storage or instantaneous heaters:

- Storage heaters can be controlled by on/off or proportional-type regulators or controllers. The type selected depends upon the desired recovery rate and the ability of the boiler to supply fast-changing loads.
- Instantaneous heaters require wide proportional band control for stability.
- When fast load changes occur in instantaneous heaters, combined pressure-temperature control should be used to provide proportional response, proportioned steam pressure, and pressure limiting.

Originally published May 26, 1975.

Controls for Storage Heaters

A storage heater consists of a steam coil or steam jacket to heat a tank, vat, or kettle with a large volume of liquid. Temperature control of storage heaters is relatively simple, because the large volume of liquid—compared to the heat-transfer capacity of the coils—will absorb small errors in the amount of steam delivered. There are several important factors, however, which influence the selection of such controls.

For storage heaters, an on/off or high-gain controller provides full steam flow for very small changes in temperature. This control mode produces the fastest temperature recovery, because the valve remains fully open until the temperature approaches the setpoint.

The disadvantage of on/off control is that it produces very rapid changes in steam demand. If the heater represents a large portion of the total steam load, this control can upset the boiler or other components in the system. Fig. 1(a) shows the response of a typical on/off controller as it returns the temperature to the setpoint. The lower curve shows the variations of steam flow.

On/off (high-gain) control can be both self-contained and air-operated. The self-contained types are usually pi-

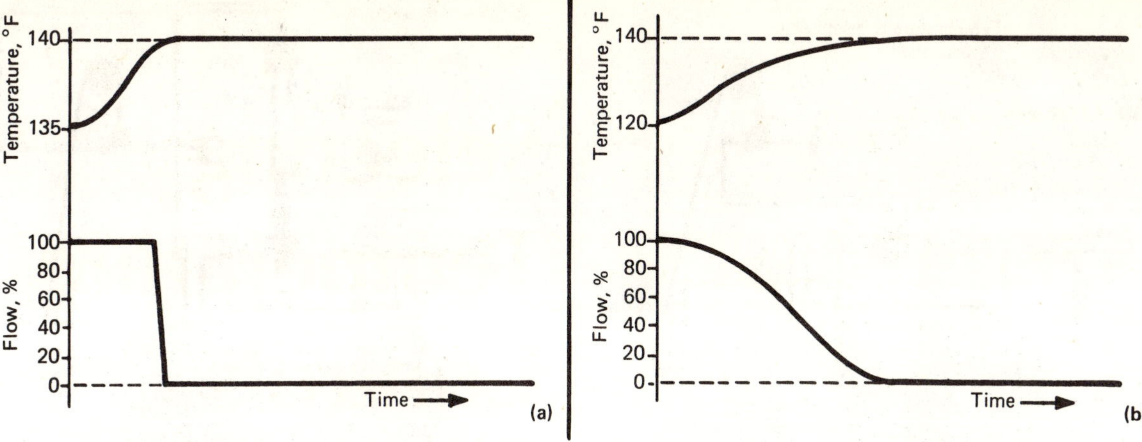

TEMPERATURE RECOVERY and steam flow for a storage heater with: (a) on/off control, and (b) proportional control—Fig. 1

lot operated to provide full valve travel for temperature changes as small as 3 to 7°F (see figure on p. 128). Air-operated systems use a diaphragm control valve with a high-gain (on/off) controller to position the valve. Most air-operated controllers can be adjusted to vary the gain (i.e., the amount of temperature change needed to fully stroke the control valve).

Proportional Controls

The proportional (or throttling) controller for storage-heater applications produces steam flow in proportion to the change in temperature. As shown in Fig. 1(b), this control mode reduces the steam flow as the setpoint is approached, and therefore takes longer to restore the temperature. For storage heaters, this has the advantage of producing slower changes in steam flow than the on/off type, making it easier for the boiler to keep up with the load.

The most common self-contained proportional regulators are the direct-operated type, with either liquid- or vapor-filled thermal elements. Depending on the type of element used, the regulators may require from 10 to 40°F change to fully stroke the main valve. The amount of change needed for a specific regulator is not usually adjustable. Air-operated systems use proportional controllers that are generally supplied with adjustable gains.

Steam-Pressure Limiting

In some cases, it is desirable to limit the maximum steam pressure delivered to the coils. These may have a maximum pressure rating that is lower than the steam-supply pressure, or the pressure may be limited to prevent caking of heat-sensitive materials on the coils or steam jacket.

Pressure limiting can be accomplished in two ways: one of them includes the installation of a pressure regulator; this method, however, requires the expense of a second valve, and also increases the required size of the temperature controller because the available pressure drop is reduced.

The second method of pressure limiting uses a combination pressure-temperature regulator. Here, a self-contained reducing valve is preset to limit the maximum steam pressure delivered. An integral temperature-measuring element automatically reduces the pressure setting of the regulator, as the temperature setpoint is approached. When the temperature returns to the setpoint, the pressure regulator setting is reduced to 0 psig so the regulator closes. This type of control not only regulates pressure and temperature, but does it with a smaller valve, because the entire pressure drop is taken across one valve. Air-operated pressure-temperature control systems—which are also available—use a temperature transmitter to adjust the setpoint of a pressure controller. These systems are discussed further on.

Controls for Instantaneous Heaters

The most common instantaneous heaters are the steam-shell-and-liquid-tube type. Since they are sensitive to small errors in the amount of steam delivered, they need temperature controls matched to the characteristics of the heater.

The storage heaters discussed above have a large reservoir of liquid to absorb errors in the quantity of steam delivered. Instantaneous heaters, on the other hand, contain a very small quantity of liquid to be heated, compared to their heating capabilities. For instance, if the liquid flow through an instantaneous heater is suddenly stopped, steam contained within the shell at that instant will transfer heat to the liquid in the coils, causing the temperature to rise above the setpoint.

If on/off control is applied to an instantaneous heater, cycling of the controlled temperature results. On startup, the regulator will go wide open, causing maximum steam to be passed to the shell, which results in fluid overheating. When the fluid reaches the sensing bulb, the regulator will close fully.

Just as the above happens, the liquid going through the heater falls below the setpoint temperature, so that when this cooler liquid reaches the sensing element, the regulator will again go wide open. This on/off action will continue indefinitely, producing wide fluctuations in the output temperature.

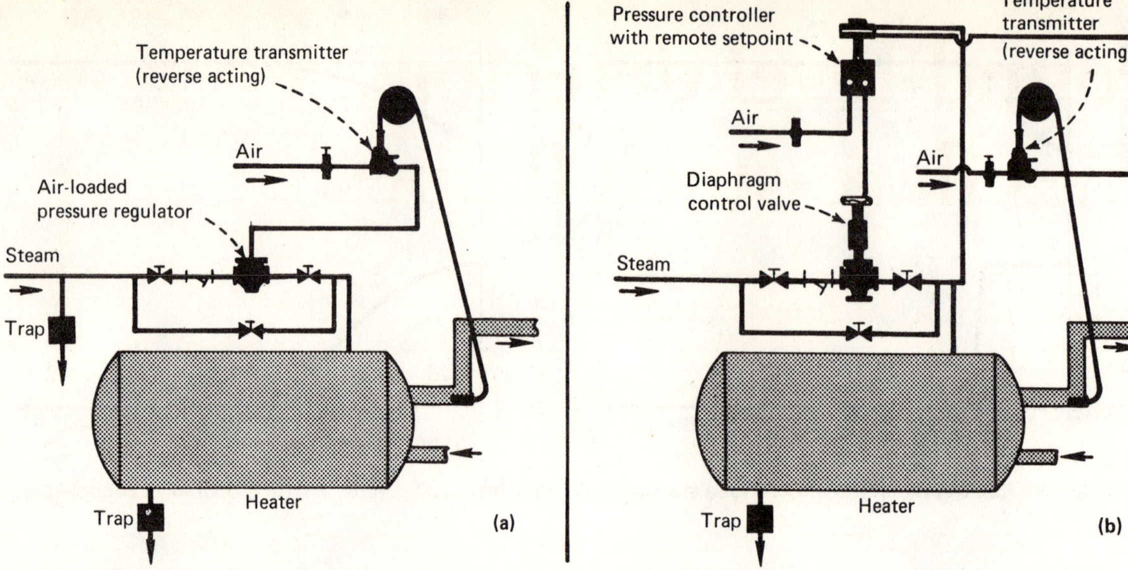

INSTANT HEATER pressure-temperature control: (a) air-loaded pressure regulator, (b) diaphragm control valve—Fig. 2

To properly control an instantaneous heater, the following throttling (proportional) control characteristics are required:

- Reduced sensitivity of the control to prevent on/off or cycling action. This control mode increases the steam flow gradually as the temperature drops, preventing an upset in temperature.
- Control of steam pressure in relation to the load. The reason for this is that the heat-transfer rate, from the steam to the liquid, is directly related to the steam pressure in the shell. With combined pressure-temperature controls, steam pressure is increased as the temperature drops, to match the heat-transfer rate to the load. As the load decreases, the temperature rises, and the pressure setting of the controller is reduced.
- Corrective action with anticipated changes in load. As load changes, the condensation rate of the steam increases or decreases, producing a change in the steam pressure. A pressure-temperature controller senses this change in pressure and immediately repositions itself to correct for the change. This corrective action takes place before any temperature variation is felt at the thermal-sensing element.

Pressure-temperature controllers can be preset to limit the maximum steam pressure delivered to the heater. This feature eliminates the requirements for a separate reducing valve ahead of the temperature control, and can also compensate for oversized heaters. By limiting the maximum steam pressure delivered to the shell of the heater, the tendency to overheat when rapid shutdown occurs is also reduced.

Instantaneous heaters operating with steady or slowly changing loads can usually be controlled with standard, proportional regulators or controllers. When fast load changes occur, all three control characteristics—proportional response, proportioned steam pressure, and pressure limiting—may be required.

Providing Pressure-Temperature Control

There are three basic ways of providing combined pressure-temperature control. In the first, a self-contained pressure-temperature regulator—consisting of an internal pilot reducing valve with a thermal element and lever assembly—adjusts the pressure setpoints as the temperature changes. A new development in this regulator is an adjustable gain feature, which provides the same flexibility of control as air-operated systems.

The second pressure-temperature control method has an air-loaded pressure regulator, which is adjusted by the output of a pneumatic temperature transmitter. The transmitter is provided with adjustable gain (Fig. 2a). This system provides the flexibility of an air-operated control without the expense of a separate pressure controller.

The third method is used where large valves are needed—i.e., above the available size of self-contained pressure-temperature regulators or air-loaded pressure regulators. This method includes a diaphragm control valve operated by a pressure controller. The setpoint of the pressure controller is determined by the output of the temperature transmitter (Fig. 2b).

Meet the Author

William L. Scull is manager of research and development of Leslie Co., 399 Jefferson Rd., Parsippany, NJ 07054. Holder of a B.S. degree in mechanical engineering from Fairleigh Dickinson University, Teaneck, N.J., he is a registered professional engineer in New Jersey, and a member of the American Soc. of Mechanical Engineers and the Instrument Soc. of America.

Ground-level detector tames flare-stack flames

An optical-radiation sensor coupled to a control system for a smoke suppressant (usually steam) promotes smokeless combustion.

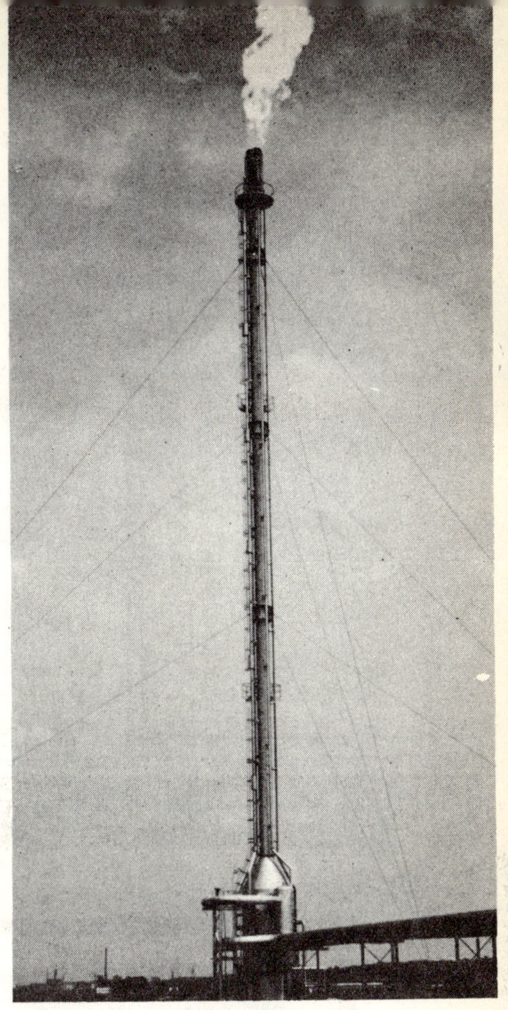

Thomas R. Schmidt, *Shell Development Co.*

☐ The occasional large flame from a flare stack is one of the more noticeable features of a refinery or chemical plant. Often, a flare results from an unexpected shutdown of a section of the plant, and the consequent necessity of rapidly disposing of large amounts of gaseous materials. An increasing effort has been underway for some time to save and use these materials, but occasions will continue to arise when burning them in a flare is the safest expedient.

When materials are burned in a flare, the combustion must be clean and smokeless. This is done by adding a smoke suppressant, usually steam, to the material at the flare tip, to cause increased mixing of the gases with air, and to affect the chemistry of burning so as to promote smokeless combustion.

A control system is required to proportion the suppressant into the flare gas to produce clean burning, but to do this without any unnecessary excess because the suppressant creates an added energy loss and can increase flare noise.

In the past, such systems have centered about some form of flare-gas flow-measuring device. Requirements for a flowmeter in this type of service are formidable. Since the flowmeter is mounted in an emergency dump line, the flow-measuring device must not obstruct the line or reduce its capacity. Further, it must operate over an unusually wide range of flows. And since a single flare may serve several units, there is seldom any opportunity to install, inspect or maintain a flowmeter. Too often, these systems have ended up on manual control—permitting occasional smoke or, more likely, excessive expenditure of steam. An added complication is that different materials require different amounts of suppressant for smokeless combustion. Heavier hydrocarbon gases generally require added suppressant, and sometimes a density-measuring instrument is installed with the flowmeter to provide this correction.

Several years ago, we undertook the development of a new approach to flare control that has proven to be not only substantially simpler than previous systems but more effective [*1*]. The concept is based on measuring the radiant-heat energy from a portion of the flame with a ground-level sensor. The heart of the system is an optical monitor located at a moderate distance from the base of the flare stack and trained on the base region of the flame (Fig. 1). Since this monitor is at ground level, it can be safely installed or checked at any time without taking the flare out of service. The monitor is essentially a rugged telescope equipped with a photocell to measure total radiation in the near-infrared from a restricted

This article is based on a paper originally presented at the 32nd Annual Symposium on Instrumentation for the Process Industries, Jan. 19–21, 1977 at Texas A&M University, College Station, TX 77843.

Originally published April 11, 1977.

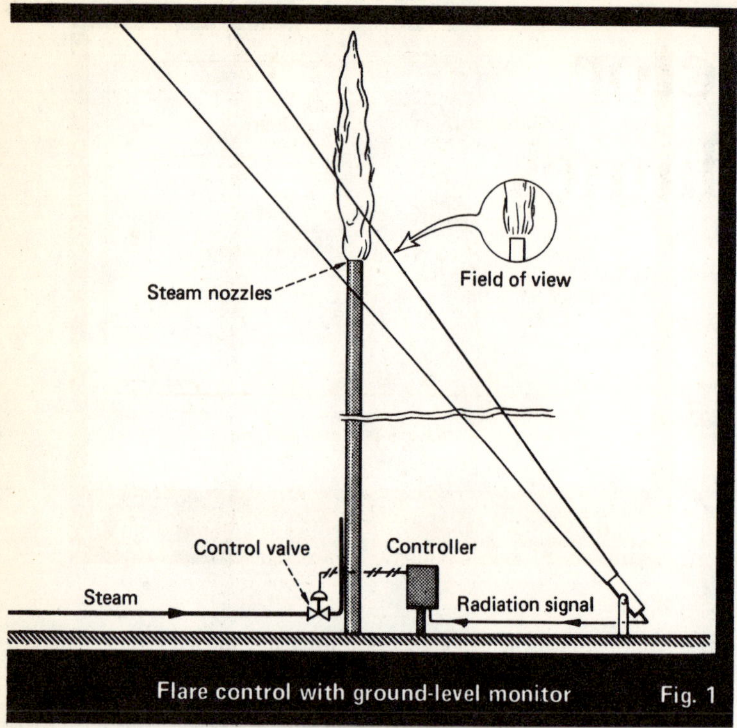

Flare control with ground-level monitor Fig. 1

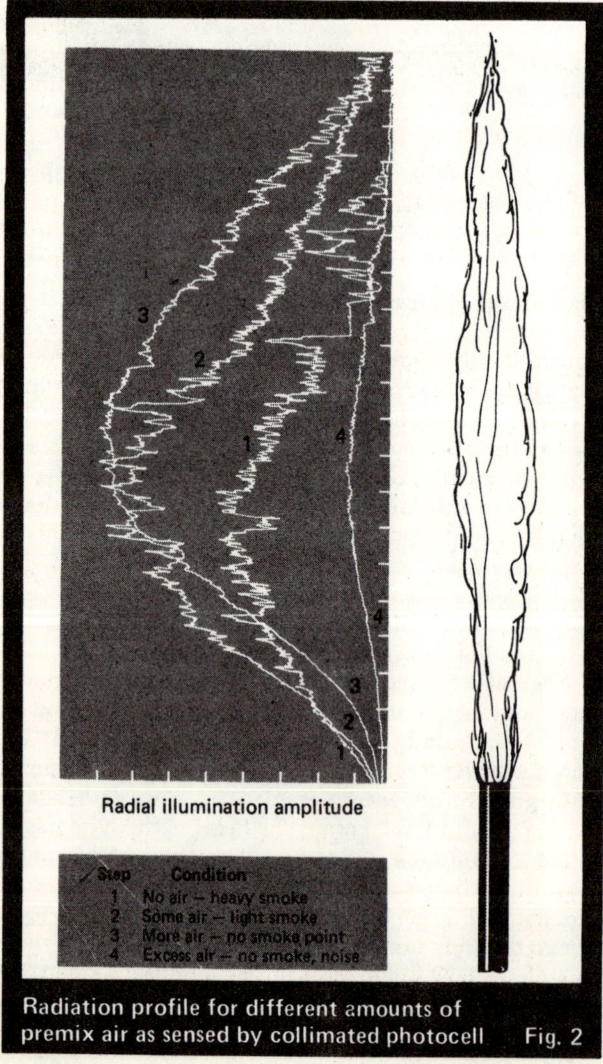

Radiation profile for different amounts of premix air as sensed by collimated photocell Fig. 2

field of view. The signal from this monitor is a combined measure of flow, density, and steam effect, and is used to control suppressant valve-position or flow.

Flare characteristics affect radiation

The base of the flare (that part of the flame immediately above the tip of the stack) is the region of interest for control purposes. This is the flame area most sensitive to the addition of steam. For example, note the apparent gap between the base of the flame and the tip of the stack in the photo on p. 489. This gap and area of reduced illumination is largely caused by the steam injection at the flare tip, resulting in improved combustion, flame liftoff, and dilution.

Fig. 2 shows the results of a test conducted with a small flame in the laboratory during the development of this system. In this test, the radial illumination from different levels of the flame was measured with a highly collimated photocell scanning the length of the flame. With a fixed gas-flow, air injection was increased in four steps, from zero to a high flow, and the radiation profile plotted for each condition.

The important feature to notice is the behavior of the base region of the flame. Radiation from this zone decreases with increased air flow, even though the total radiation from the flame increases over the first three steps. Similar behavior has been observed in large plant flares. This reduction in radiation from the base region of the flame with increased suppressant flow provides the negative feedback for control purposes.

Radiation from this zone and the regions above it will increase with increased gas flow, and to a lesser extent increased gas density. Kent [*2*] has suggested that the fraction of energy released in a flare as heat radiation can be approximated by:

$$f = 0.20\left(\frac{50m + 100}{900}\right)^{1/2}$$

where f = fraction of total energy release appearing as heat radiation, and m = molecular weight of hydrocarbon stream.

Burning heavier molecules in a flare requires increased suppressant for smokeless combustion. Controlling the suppressant flow with a flare-radiation measurement fulfills, at least in part, this requirement. A more exact compensation is provided in one installation in which the flare gas contains a substantial amount of nitrogen. Since nitrogen does not enter into combustion and is invisible to the monitor, no suppressant is provided for this portion of the flow.

The heat-radiation signal developed by this monitor is a composite of the effects of hydrocarbon flow and density in the positive direction, and of smoke-suppressant flow in the negative direction.

Locating the monitor

Since the monitor is an optical device that can be designed for a specific field of view, the choice of locations around a flare is relatively open. Typically, a ground-level location at a safe distance such that the monitor looks up at the flare at about a 45° angle is chosen. Considering the different heights and capacities of flares, ground-level illumination does not vary from

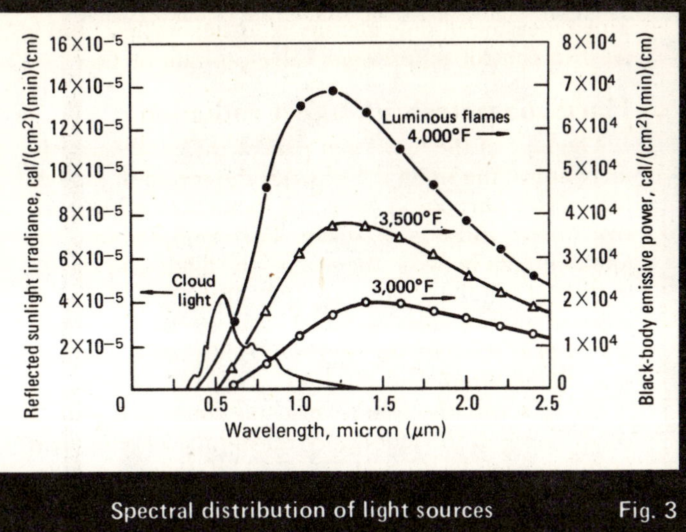

Spectral distribution of light sources Fig. 3

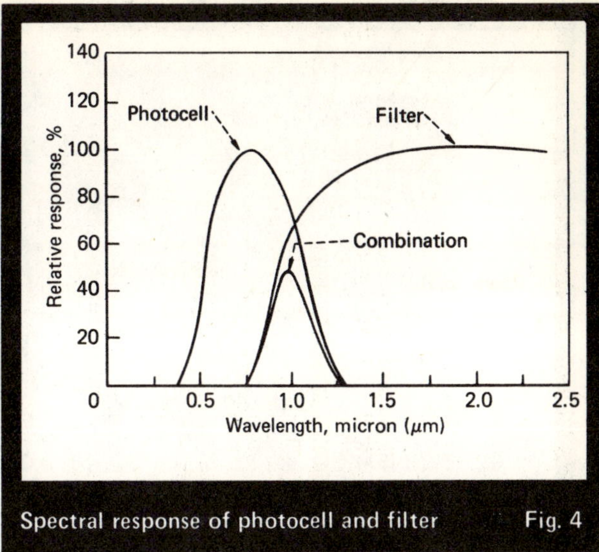

Spectral response of photocell and filter Fig. 4

flare to flare as widely as one might suppose. Flares are normally designed with the height necessary to produce a safe radiant-heat flux in the base region. This also gives each flare roughly the same sensitivity for these monitors. The usual relationship (reciprocal of distance squared) for radiation patterns from a point source holds for typical monitor locations, so excessive distances from the flare stack to the monitor are avoided.

A location must also be picked so that the sun never shines into the monitor. In the Northern Hemisphere, this is to the south of the flare. In a few locations, there may be concern over something like the plume from a cooling tower occasionally obstructing the optical path between the monitor and the flare tip. This has not been a serious problem, but two or more monitors can be located about a flare, and the largest signal picked by electronic circuitry.

Monitor design and characteristics

The optical monitor (photo on p. 489) is a rugged telescope that is sealed and constructed of stainless steel. A photocell, located at the focal plane, develops a signal that is proportional to the total radiation.

The sun is not the only extraneous source of radiation in the sky. Sunlight reflected from white clouds or even the blue sky can provide a background radiation level from the area in and about the flare. The radiation from these sources is a form of background noise through which the control systems must operate. It is not a major problem, but two steps are taken in the design of the monitor to minimize it. First, the field of view of the monitor is narrowed to include only that portion of the flare useful for control, and second, optical filtering is used.

Fig. 3 shows a curve of spectral distribution for cloud light [*3*], and curves for the black-body radiation that is more representative of luminous flares. Radiation from the flares tends toward the yellows, reds and infrareds, while the cloud light is skewed more toward the blues. An infrared filter is used before the photocell, peaking its sensitivity at a wavelength of about one micron (1 μm) in the near infrared. Individual and combined responses for the photocell and filter are shown in Fig. 4. These two steps provide a substantial reduction in background noise (extraneous illumination) in the monitor signal.

The photocell is a silicon solar cell and is self-generating. No external power is required for the monitor, which typically develops a 25-mV output signal at full scale. Output impedance of the monitor is low, and the signal can be treated much like a thermocouple signal.

Design of flare-stack control system

The control system associated with this monitor is conventional, except that some form of low-pass filtering or other lag element is required in the control loop. Because the monitor responds at electronic speed, each

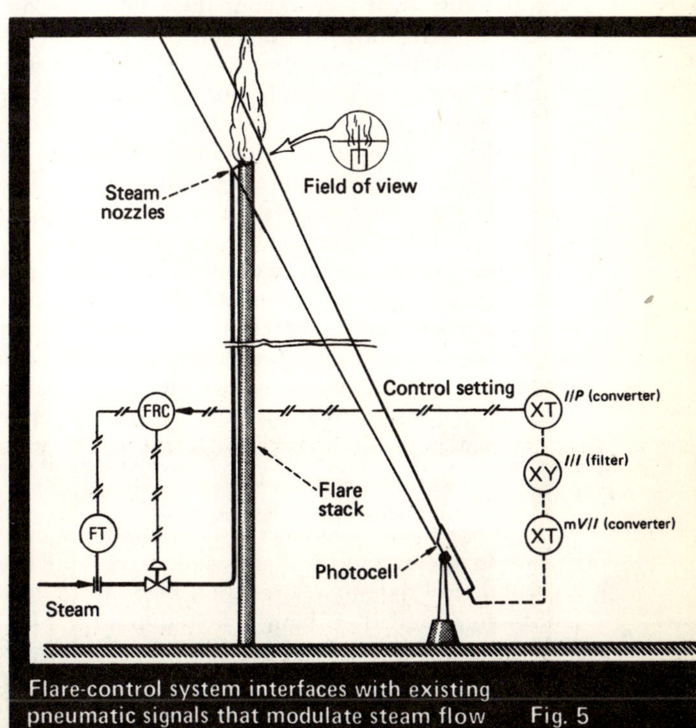

Flare-control system interfaces with existing pneumatic signals that modulate steam flow Fig. 5

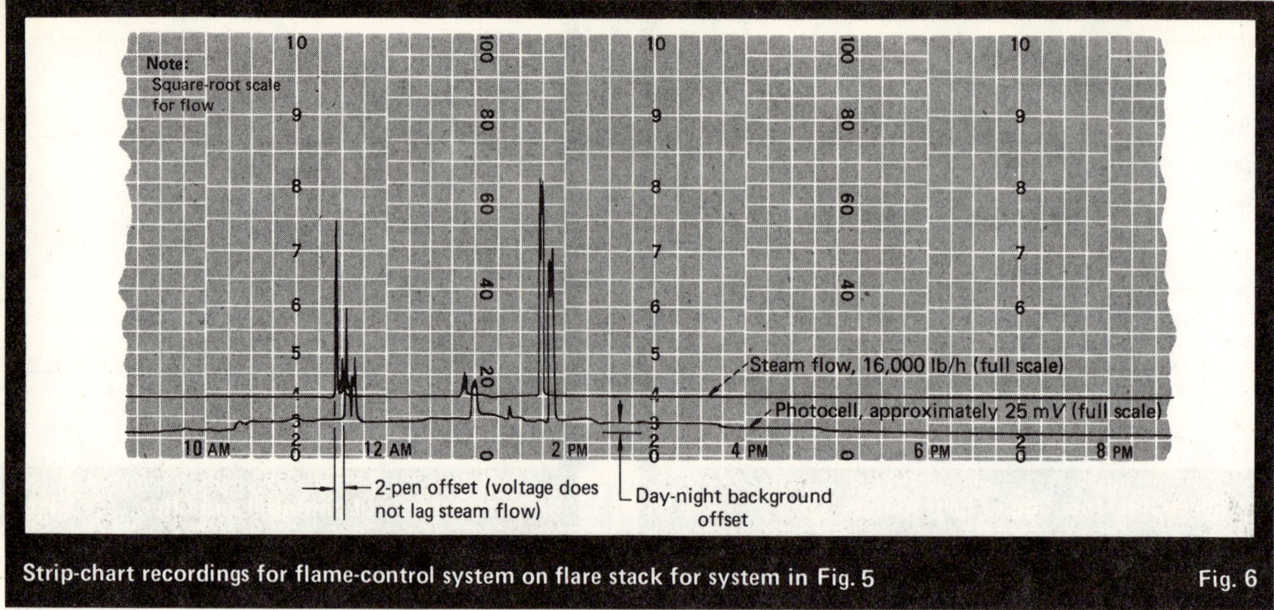

Strip-chart recordings for flame-control system on flare stack for system in Fig. 5 **Fig. 6**

flicker or pulsation of the flare is measured and appears in the electrical output of the unit. Filtering, either electrical or pneumatic (if the signal has been converted to pneumatic), causes the system to respond to a running average, rather than to every variation.

Some gain may be realized due to the steam effect on radiant energy in the field of view sensed by the monitor. Most systems operate with moderate proportional gain in the controller.

In the typical system (Fig. 5), the electrical signal from the monitor has been converted to a pneumatic one in order to accommodate existing equipment. Fig. 6 shows records of three flares and the steam response to them for this system.

A question often asked about the system is: Can it operate in fog? Infrared radiation can penetrate fog better than visible radiation. Systems have been observed to operate successfully during heavy rains and fog. This fog observation, incidentally, was with a ground-level fog, the flare being observed from a high vantage point. In the rare event of both a flare and an extremely heavy fog, the system can still be operated manually to add more than enough suppressant.

The monitor is typically mounted on a post or some other rugged support. Alignment is done by removing the photocell assembly and temporarily substituting an eyepiece with cross-hairs. Once the proper aiming point is achieved, the monitor is bolted into place and the photocell reinstalled. A simple test for system activity can be made by shining a flashlight into the monitor.

Conclusions

We at Shell have a number of the radiation-monitor systems in operation at various locations, ranging from a ground-level flare to a flare 350 ft high. More installations are planned on both new and existing plants. Operating experience has been excellent. Primary advantages of the system are:

1. Relatively simple, low maintenance. No components exposed to flare-gas stream.
2. Can be installed or inspected without shutting down flare system.
3. Ground-level installation.
4. Rapid response to burning conditions (both for flow and, with some compensation, for density).
5. More-exact proportioning of suppressant reduces operating costs and flare noise while providing smokeless burning.

Limitations of the system are:

1. Does not provide a flare-gas flow measurement.
2. Cannot anticipate arrival of flare gas—delays of a few seconds (the transit time of the steam to the flare tip) are experienced before control becomes effective.

Acknowledgement

The author wishes to acknowledge the contributions of G. B. Schinman of the Shell Deer Park Manufacturing Complex, Houston, Tex. in the development of this system, and particularly for the initial field installations and tests.

References

1. Schmidt, T. R. and Shinman, G. B. (to Shell Oil Co.), "Method and Apparatus for Controlling the Addition of a Combustion Assisting Fluid to a Flare," U.S. Patent 3,891,847 (June 24, 1975).
2. Kent, G. R., Practical Design of Flare Stacks, *Hydrocarbon Process. Petrol. Refiner,* Aug. 1964, pp. 121–125.
3. Remote Sensing with Special Reference to Agriculture and Forestry, National Academy of Sciences, Washington, D.C. 1970.

The author

Thomas R. Schmidt is a member of the Process Control and Instrumentation Dept., Shell Development Co., Westhollow Research Center, P. O. Box 1380, Houston, TX 77001. His primary areas of interest are measurement and instrumentation development. He has a B.S. and an M.S. in mechanical engineering from the University of Wisconsin, and is a member of the Instrument Soc. of America.

INDEX